W9-BSO-352

THE ECONOMY OF NATURE

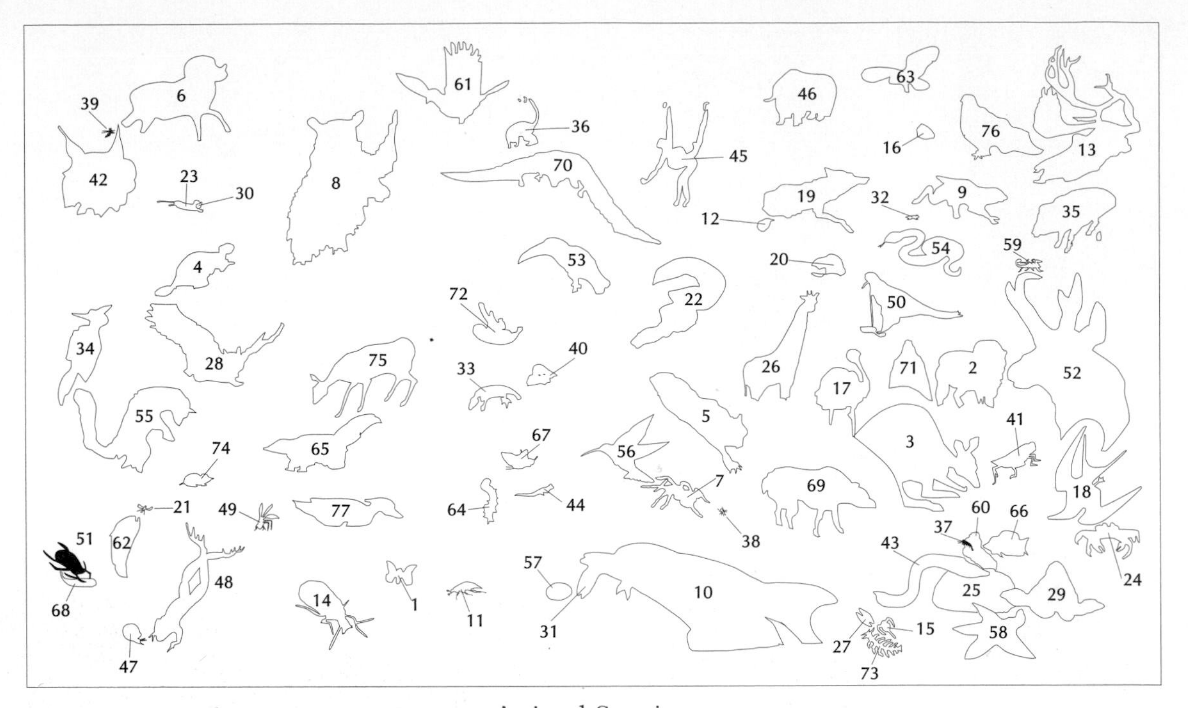

Animal Species

1 Io Moth
2 African Lion
3 Agile Wallabies
4 Beaver
5 Bengal Tiger
6 Bighorn Sheep
7 Bird-eating Spider
8 Black Bear
9 Black-tailed Jackrabbit
10 Bottlenosed Dolphin
11 Burying Beetle
12 Cactus Wren
13 Caribou Skull
14 Caterpiller Hunter
15 Ceratium Dinoflagellate
16 Collard Lemming
17 Common Rhea
18 Common Tern
19 Coyote
20 Desert Kangaroo Rat
21 Dragonfly
22 Emerald Tree Boa
23 Ermine
24 Ghost Crab
25 Giant Clam
26 Giraffe
27 Gomphonema Diatom
28 Great Horned Owl
29 Grouper
30 Harvest Mouse
31 Herring
32 Horned Toad
33 Iguana
34 Ivory-billed Woodpecker
35 Javelina
36 Kinkajou
37 Krill
38 Leaf-cutter Ant
39 Mosquito
40 Lowland Gorilla
41 Lubber Grasshopper
42 Lynx
43 Moray Eel
44 Mudskipper
45 Murique
46 Musk Ox
47 Orb Snail
48 Pickerel Frog
49 Potter Wasp
50 Prairie Falcon
51 Predacious Diving Beetle
52 Pronghorn Antelope
53 Red-breasted Toucan
54 Red Diamondback Rattlesnake
55 Red Fox
56 Ruby and Topaz Hummingbird
57 Sand Dollar
58 Scarlet Starfish
59 Scorpion
60 Sea Fan
61 Siberian Ruby Throat
62 Smallmouth Bass
63 Snowy Owl
64 Sphinx Larva w/ cocoons of Brachonid Wasp
65 Striped Skunk
66 Surgeon Fish
67 Swallowtail Butterfly
68 Tadpole
69 Tapir
70 Tent-making Bat
71 Termite and Mound
72 Three-toed Tree Sloth
73 Tomopteris Worm
74 Water Schrew
75 White-tailed Deer
76 Willow Ptarmigans
77 Wood Duck

This piece of art mirrors the balance of biological life on the planet. Although many species depicted here are threatened or endangered, the future of this planet belongs to them, not to us. Hence, the only manufactured object depicted in the mural is the littered drink can, washed up on an otherwise unfettered shore, a shore that has borne the ceaseless pounding of the sea for eons before humankind's emergence as a distinct species. Because the balance of biological life on Earth is not dependent upon humans (the "top" of the evolutionary ladder), but rather is dependent upon plants, bacteria, and insects, the certainty of Life's future is thus ensured, with or without humans. We obviously now have some choices to make with regard to our continued participation in that balance, a participation that is clearly a privilege and not a right.

The ivory-billed woodpecker, a species believed extinct, is included as a reminder of the intense fragility of Life's legacies, and how the unthinking destruction of wildlife habitats can drastically affect the intricate beauty of creation. The purple vine, which delineates the biomes, does not terminate with a leaf in the upper right quadrant of the picture, but instead reaches toward the Light as it grows into an unknown domain, representing the undetermined possibilities of Life and its undaunted indestructibility. The synthesis of these elements shows that Life has an indeterminate will, a cosmic mandate to expand its own limits, to test the nebulous structures of its subtle quantum designs within an unfathomable system in which success and failure are not among its measuring tools. The glory of this infinite force is ours as human beings only to behold, embrace, and consecrate; nothing else is possible in the wake of its irresistible pilgrimage toward the unknown. As conservators only of our own small heritage, we humans on this Spaceship Earth now have the nascent opportunity to come together and sing harmoniously in celebration of that glorious chance at Life.

THE ECONOMY OF NATURE FOURTH EDITION

A Textbook in Basic Ecology

ROBERT E. RICKLEFS *University of Missouri–St. Louis*

W. H. FREEMAN AND COMPANY
New York

Senior Editor: Deborah Allen
Project Editor: Christine Hastings
Text and Cover Designer: Blake Logan
Text Illustration: Network Graphics
Cover Art: Marquetry mural at the Boston University Biological Sciences
Center by Lara Hunt and Spider Johnson
Illustration Coordinator: Susan Wein
Production Coordinator: Paul Rohloff
Composition: W. H. Freeman Electronic Publishing Center/Sheridan Sellers
Manufacturing: RR Donnelley & Sons Company

Library of Congress Cataloging-in-Publication Data

Ricklefs, Robert E.
The economy of nature : a textbook in basic ecology / Robert E.
Ricklefs. — 4th ed.
p. cm.
Includes index.
ISBN 0-7167-2815-X
1. Ecology. I. Title.
QH541.R54 1996 96-33300
574.5—dc20 CIP

Printed in the United States of America

Third printing 1997

CONTENTS

PART IV Populations 299

PART V Species Interactions 397

PART VI Communities 497

PART VII Ecological Applications 591

PREFACE

Although it has been only four years since the publication of the third edition of *The Economy of Nature*, the fourth edition includes details of many new developments in ecology as well as broader coverage of traditional topics, rearrangement of some material to provide a more logical organization, and changes in the presentation to improve clarity. *The Economy of Nature* remains a basic textbook in ecology for undergraduate students, with explanations of the general principles of ecology and narrative accounts of the lives of organisms and the workings of ecological systems. As in previous editions, I have endeavored to develop topics, including mathematical models of ecological processes, in a clear and logical fashion, and to balance theory with experimental studies and empirical examples of ecological patterns.

More than ever, ecology is a vital and growing discipline, with new concepts, discoveries, and basic information accumulating at a rapid pace. Furthermore, the connections between the basic principles of ecology and the causes of environmental problems are becoming better understood. This provides some hope for solutions to the dilemma of providing for a growing human population within the constraints of limited space and natural resources. This book emphasizes the critical need for basic understanding of ecological principles and the commitment to act on them.

The fourth edition of *The Economy of Nature* has been revised from cover to cover, with substantial changes in many chapters, including updated coverage and editing for clarity. There are some new topics in the fourth edition, such as a treatment of the biome concept in ecology with a detailed comparison of the terrestrial and aquatic biomes of the earth (Chapter 5). You will also find new sections on phenotypic plasticity (reaction norms) (Chapter 11) and phylogenetic reconstruction (cladistic analysis) (Chapter 21), among other subjects. Some examples, such as the obligate mutualistic relationship between the yucca and its moth pollinator, have been expanded considerably to illustrate general principles, in this case that of coevolution.

A major change in the fourth edition is the addition of a two-color art program throughout the book. In addition to making the text more attractive, I hope that the use of a second color will enhance the graphical presentation of data and concepts. The full-color photo essays have also been retained and somewhat revised with a new selection of photographs. I have also increased the use of "Boxes" to set apart from the main text certain statistical or methodological topics.

As in the past, I have been very pleased with the response of teachers and students to the third edition of *The Economy of Nature* and hope that this new edition will serve your needs even better. Many of the changes have been in direct response to comments and suggestions received from those of you who have used the book, and I look forward to hearing from more of you in the future. This book is an important part of my professional life because of the immense amount I learn from working on it and the many personal interactions it has brought. I hope that *The Economy of Nature* will continue to be current, relevant, and engaging for those who study ecology in the future.

Finally, as I wrote in the preface to the third edition, I hope that *The Economy of Nature* will encourage students to appreciate the natural world they live in, and the increasing impact of human activities on the natural world—our world. Progress toward an ecological balance depends on the rational application of knowledge and understanding. We are the technological species; now we must become the ecological species and assume a responsible position in the economy of nature.

St. Louis
August 1996

ACKNOWLEDGMENTS

I have received much help in preparing this edition. A large part of the credit belongs to my editor at W. H. Freeman and Company, Deborah Allen, who was a constant source of encouragement, gentle prodding, and good advice. The production staff at W. H. Freeman, particularly Christine Hastings, were wholly professional, proficient, and personable. I am grateful for the hard work and interest that they have contributed to this edition. Norma Roche undertook the task of copyediting with thoroughness, precision, and insight.

Of particular importance to me were the many useful comments of individuals who critically read parts of the revised manuscript: Martin B. Berg, Robert Curry, Robert Holt, Michael Mazurkiewicz, Wayne F. McDiffett, Laszlo J. Szijj, Stefan Sommer, and David C. Wartinbee. Helpful suggestions on particular topics came from Robert Curry, Tom Getty, Jon E. Keeley, Olle Pellmyr, and Truman P. Young. One of the strongest elements of *The Economy of Nature* are the photographs, many of which were generously provided by colleagues. My thanks for this to D. N. Alstad, A. Basolo, R. Boonstra, O. Brown, D. H. Clayton, H. Cogger, P. Dayton, G. F. Edmunds, J. R. Ehleringer, T. Eisner, Z. Glowacinski, M. A. Guerra, E. Hanauer, C. C. Hansen, W. H. Haseler, D. H. Janzen, H. B. D. Kettlewell, J. A. McGowan, C. H. Mueller, O. Pellmyr, M. V. Parthasarathy, D. Pimentel, J. W. Porter, D. W. Schindler, P. W. Sherman, W. J. Smith, W. P. Sousa, D. G. Sprugel, K. Vepsalainen, L. T. Wasserthal, and R. H. Whittaker. In addition, I obtained many fine photographs from various agencies of the United States government, including the Bureau of Sport Fisheries and Wildlife, Department of Agriculture, Department of the Interior, Fish and Wildlife Service, Forest Service, National Park Service, Smithsonian Tropical Research Institute, and the Soil Conservation Service.

CHAPTER 1

INTRODUCTION

The English word *ecology* is taken from the Greek *oikos,* meaning "house," our immediate environment. In 1870, the German zoologist Ernst Haeckel gave the word a broader meaning: the study of the natural environment and of the relations of organisms to one another and to their surroundings. Haeckel wrote:

> *By ecology, we mean the body of knowledge concerning the economy of nature—the investigation of the total relations of the animal both to its organic and to its inorganic environment; including above all, its friendly and inimical relation with those animals and plants with which it comes directly or indirectly into contact—in a word, ecology is the study of all the complex interrelationships referred to by Darwin as the conditions of the struggle for existence.*

Thus **ecology** is the science by which we study how organisms (animals, plants, and microbes) interact in and with the natural world.

The word *ecology* came into general use only in the late 1800s, when European and American scientists began to call themselves ecologists. The first societies and journals devoted to ecology appeared in the early decades of the twentieth century. Since that time, ecology has undergone immense growth and diversification, and professional ecologists now number in the tens of thousands.

With the rapid growth of the human population and the quickening deterioration of the earth's environment, ecological understanding is now urgently needed to maintain the condition of the environmental support systems upon which humanity depends for food, water, protection against natural catastrophes, and public health. Management of biotic resources in a way that sustains a reasonable quality of human life depends on the wise application of ecological principles to solve or prevent environmental problems and to inform our economic, political, and social thought and practice.

This text, *The Economy of Nature,* presents the basic principles of the scientific discipline of ecology. These principles have been defined through more than a century of observation, experimentation, and theoretical exploration of natural systems. They explain the processes that maintain the structure and function of these systems; they tell us how each part fits into the whole, emphasizing the interrelatedness of all of nature; and they help us to understand why the functioning of natural systems can break down under certain stresses. In this way, ecological principles offer guidelines for the preservation of biodiversity and management of the environment for sustained use.

To begin with, this chapter outlines a general framework for the study of ecology to start you on the road to ecological thinking. We shall first discuss several vantage points from which ecological knowledge and insight can be viewed—for example, as different levels of complexity, varieties of organism, types of habitat, and dimensions in time and space. We shall see how we can regard many different entities as **ecological systems,** by which we mean any organism or assemblage of organisms, including their surroundings, united by some form of regular interaction or interdependence. While the extent and complexity of ecological systems varies from the single microbe to the entire biosphere blanketing the surface of the earth, all obey similar principles of ecological functioning. Some of the most important of these principles are taken up next: these concern physical and chemical attributes of ecological systems, regulation of structure and function in ecological systems, and evolutionary change in ecological systems. Finally, a few examples will illustrate how an understanding of ecology can help us to meet the challenge of maintaining a healthy environment for natural systems and for ourselves in the face of increasing ecological stresses.

Ecology provides a framework for interpreting the flood of information that comes our way about the natural environment. It also gives us the insight we need to envision the consequences of our activities for natural systems. Haeckel's analogy of the economy of nature emphasizes the interconnectedness of everything on the surface of the earth, just as human ventures are interrelated and defined by economic principles. We and our enterprises directly affect natural processes. Humankind itself is an important part of the economy of nature.

Levels of ecological organization

The organism is the most fundamental unit of ecology, the elemental ecological system. No smaller unit in biology, such as the organ, cell, or molecule, has a separate life in the environment (although, in the case of single-celled protoctists and bacteria, cell and organism are synonymous). The structure and functioning of the organism—whether it is a plant, animal, or microbe—are determined by a set of genetic instructions inherited from its parents and by the influence of many factors in its environment. Every organism is bounded by a membrane or other covering across which it exchanges energy and materials with its surroundings. To be successful as an ecological entity, the organism must have a positive balance of energy and materials to support its maintenance, growth, and reproduction. In Part 1 of this book, we will examine factors that influence exchanges between the organism and the physical environment, and we will see how organisms solve problems involving temperature, water loss, salt balance, and other environmental challenges. We shall return to these themes again in Part 3 to consider how organisms cope with life in heterogeneous and varying environments.

In the course of their lives, organisms transform energy and process materials as they metabolize, grow, and reproduce. In doing so, they modify the conditions of the environment and the amounts of resources available for other organisms, and they contribute to energy fluxes and the cycling of elements in the natural world. Assemblages of organisms together with their physical and chemical environments make up an **ecosystem.** Ecosystems are immensely large and complex ecological systems, including up to many thousands of different kinds of organisms living in a great variety of individual surroundings. A warbler overhead searching for caterpillars, and a bacterium helping to decompose the organic soil underfoot are both part of the same ecosystem. Because of this complexity, the **ecosystem approach** to ecology describes organisms and their activities in terms of common "currencies," which are amounts of energy and chemical elements, by which the activities of organisms as different as bacteria and birds can be compared. Thus, the ecosystem approach provides a framework for studying the transformation of energy and the cycling of elements within ecological systems.

We may speak of a forest ecosystem, a prairie ecosystem, and an estuarine ecosystem as distinct units because relatively little exchange of energy or substances occurs *between* these units compared with the innumerable transformations going on *within* each of them. Ultimately, however, all ecosystems are linked together in a single **biosphere** that includes all the environments and organisms at the surface of the earth. The far-flung parts of the biosphere are linked together by the energy and nutrients carried by currents of wind and water and the movements of organisms. Water flowing from a headwater to an estuary connects the terrestrial and aquatic ecosystems of the watershed to those of the marine realm. The migrations of gray whales link the ecosystems of the Bering Sea and the Gulf of California. The importance of movement of materials between ecosystems within the biosphere is underscored by the global consequences of human activities. Industrial and agricultural wastes spread far from their points of origin, inflicting harm on all regions of the earth. Ecosystem processes are the subject of Part 2 of this book.

Many organisms of the same kind together constitute a **population.** Populations differ from organisms in that they are potentially immortal, their numbers being maintained over time by the birth of new individuals that replace those that die. Birth and death within populations depend on how well organisms function in their environments, as we shall see in Part 3. Populations also have collective properties, such as geographic boundaries, densities (number of individuals per unit of area), and variations in size or composition (for example, evolutionary responses to environmental change and periodic cycles of numbers in some cases) that are not exhibited by individual organisms. The population approach to ecology, the subject of Part 4, is concerned with numbers of individuals and their variations through time, including evolutionary changes within populations.

Many populations of different kinds living in the same place constitute an **ecological community.** The populations within a community interact in various ways, as explained in Part 5. Many species are predators that eat other kinds of organisms. Almost all are themselves prey. Some—such as bees and the plants whose flowers they pollinate, and many microbes living together with plants and animals—enter into cooperative arrangements, called **mutualisms,** in which both parties benefit from the interaction. All these interactions influence the numbers of individuals in populations—that is, the **dynamics** of populations. The **community approach** to ecology differs from the ecosystem approach in that the units of measurement are population sizes and the focus is on interactions between populations: predation, competition, mutualism.

Unlike organisms, communities have no rigidly defined boundaries; no skin separates a community from what surrounds it. The total interconnectedness of ecological systems means that interactions between populations spread across the globe as individuals and materials move between habitats and regions. Thus, the community is an abstraction representing a level of organization rather than a discrete unit of structure in ecology, as explained in Part 6.

Community and ecosystem approaches to ecology provide different ways of looking at the natural world. We may speak of a forest ecosystem, or we may speak of the community of animals and plants that live in the forest, using different jargon and referring to different facets of the same ecological system. The study of ecosystems deals with movements of energy and materials within the environment, which result from activities of organisms and from physical and chemical transformations in the soil, atmosphere, and water. The study of populations and communities is concerned with the development of ecological structures and with the regulation of ecological processes by means of population growth and the interaction of populations with their environments and with each other. These processes also reflect the activities of organisms going about their everyday lives.

The kinds of organisms

From the standpoint of ecosystem function, organisms play a great many distinct roles. Characteristic differences in structure and function between plants, animals, fungi, protoctists, and bacteria (prokaryotes) have important ecological implications. All ecological systems depend on transformations of energy to function. For most systems, the ultimate source of that energy is sunlight. On land, plants capture the energy of sunlight and use it to synthesize carbon dioxide and water into carbohydrates and other organic molecules. In aquatic systems, photosynthesis is accomplished mostly by various protoctists (algae) and bacteria. The organic carbon produced by photosynthetic organisms provides food, either directly or indirectly, for the rest of the ecological community. Some animals consume plants; others consume animals that have eaten plants. Many fungi and bacteria feed on the dead remains of plants and animals, which are called **detritus;** others feed on living individuals, in which case we refer to them as parasites or pathogens.

Animals and plants differ in many important ways besides their sources of energy (Figure 1.1). Plants expose large surfaces to capture the energy of sunlight, and they feature thin leaves and stiff, supportive stems. Plants also need enormous quantities of water to replace water lost by evaporation from leaf surfaces; not surprisingly, most plants are firmly rooted in the ground, in constant touch with supplies of water and nutrients. Those that are not, such as orchids and other tropical "air plants" (epiphytes), can survive only in humid environments bathed in cloud mists much of the year.

Animals also need large surfaces to exchange energy and materials with the environment, but because animals do not require light, their exchange surfaces can be enclosed within the body. A modest pair of human lungs has a surface area of about 100 square meters, which is close to the size of a tennis court. The gut also presents a large surface across which nutrients are assimilated into the body. For example, the intestine of a robin is about 30 centimeters long and has an absorptive surface area of over 200 square centimeters, or about half the size of this page. By internalizing exchange

Figure 1.1 This photo of a giraffe browsing on a tree in eastern Africa emphasizes the fundamental differences between plants, which assimilate the energy of light and convert carbon dioxide to organic carbon compounds, and animals, which derive their energy from the production of plants.

surfaces, animals can achieve bulk and streamlined body shapes, and they can develop the skeletal and muscular systems that make mobility possible. In terrestrial environments, the internalized surfaces of animals also are protected from evaporative water loss, so land animals don't need roots to keep them continuously supplied with water.

Although plants and animals are similar in having organized, complex body plans with diverse, interconnected parts, they have contrasting mechanisms of growth and reproduction. Animals grow predominantly by the multiplication of cells in many tissues and organs throughout their bodies. During growth, the size of each part of an animal is precisely regulated in proportion to the rest of the body, and most animals have a characteristic adult size. In contrast, most plants grow in a modular fashion from many independent growth centers, which are called **meristems,** located at the tips of branches and roots (Figure 1.2). This modular organization enables plants to withstand loss of tissue to grazers and yet continue to grow and reproduce. Each meristem can produce a precisely regulated structure, such as a leaf or a flower, but the form of the whole plant may vary depending on which of the shoot tips receive the most light or nutrients from the roots, or escape being eaten.

Figure 1.2 Each of the growing tips of this Joshua tree can produce reproductive structures and, potentially, an entire tree. The death of one or more of the branches does not directly affect the others. Joshua trees are native to the Mojave Desert of southern California. Photograph by J. Boucher, courtesy of the U.S. National Park Service.

Any one growing shoot can also reproduce the entire structure of the plant. Thus a branch cut from a tree may take root when placed in soil and become a separate tree (Figure 1.3). This ability allows most plants to reproduce asexually—to engage in what biologists often refer to as **vegetative reproduction,** or **cloning.** Most animals reproduce by means of specialized sexual cells (eggs and sperm) formed in the sexual organs, or gonads. Despite this general difference between plants and animals, most plants also employ sexual reproduction in their life cycles, and many animals have the capacity to clone themselves, either by modifying the sexual process to eliminate fertilization or by mechanisms resembling vegetative growth. The so-called clonal animals—hydroids and corals are familiar examples—illustrate the difficulty of drawing sharp lines between animal function and plant function (Figure 1.4; Figure 1.6d).

The fungi, which like plants and animals are multicellular organisms, assume unique roles in the ecosystem because of their distinctive growth form. Unlike plants and animals, the fungus grows from a microscopic spore without passing through an embryonic stage. The fungal organism is made up of threadlike structures called **hyphae** (singular, **hypha**). These hyphae may form a loose network, called a **mycelium,** which can invade plant tissues or dead leaves and wood on the soil surface, or grow together into the reproductive structures that we recognize as mushrooms (Figure 1.5). Because fungal hyphae can penetrate deeply, fungi readily decompose dead plant material, recycling many of the nutrients in detritus. Fungi digest their foods externally, secreting acids and enzymes into their immediate surroundings, cutting through dead wood like biochemical blowtorches and dissolving recalcitrant nutrients from soil minerals. Fungi are the primary agents of rot—unpleasant to our senses and sensibilities, perhaps, but very important to ecosystem function.

The protoctists are a highly diverse group of mostly single-celled organisms that includes the algae, slime molds, and protozoa. Algae can form large structures—some seaweeds can be up to 100 meters in length (see, for example, Figure 1.12)—but their cells lack the organization into specialized components—tissues and organs—that one sees in plants and animals. Algae, including diatoms, are the primary photosynthetic organisms in most aquatic systems. Foraminifera and radiolarians feed on tiny particles of organic matter or absorb small dissolved organic molecules. Some of the ciliates are effective predators—on other microorganisms, of course.

Bacteria, or prokaryotes, are the biochemical specialists of the ecosystem. Their structure is a simple, single cell, lacking a proper nucleus and chromosomes to organize their DNA; however, their enormous range of metabolic capabilities enables them to accomplish many unique biochemical transformations essential to the ecosystem. These transformations include the assimilation of molecular nitrogen (the common form found in the atmosphere), which is used to synthesize proteins and nucleic acids, and the use of inorganic compounds such as hydrogen sulfide as sources of energy. Plants, animals, fungi, and most protoctists cannot accomplish these feats. Furthermore, many bacteria live under anaerobic conditions (lacking free oxygen) in mucky soils and sediments, where their metabolic activities regenerate nutrients and make them available for plants. We will have much more to say in later chapters about the special place of microorganisms in the functioning of the ecosystem.

Because each type of organism is specialized to a particular way of life, it is not surprising that there are many examples of different types of organisms joining together to their mutual benefit, each partner in the symbiosis providing something that the other lacks. For example, the specialized organelles so characteristic of the eukaryotic cell—chloroplasts for photosynthesis, mitochondria for various oxidative energy transformations—originated as symbiotic prokaryotes (bacteria) living within the cytoplasm. Other familiar examples include lichens, which comprise a fungus and an alga in one organism; bacteria that ferment plant material in the guts of

Figure 1.3 A living fence, which results when freshly cut fence posts take root and sprout. Because such fences provide shade and resist termite damage and rot, they are commonly used throughout the Tropics. Courtesy of J. Blake.

Figure 1.4 (a) Colony of a marine bryozoan. (b) Close-up of individual zooids. Each zooid was produced by asexual reproduction at the edge of the colony, which is about 15 mm in diameter. Courtesy of D. Harvell.

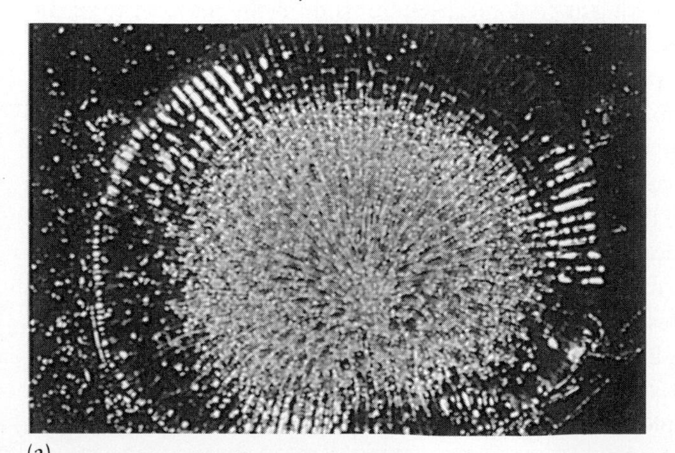

(a)

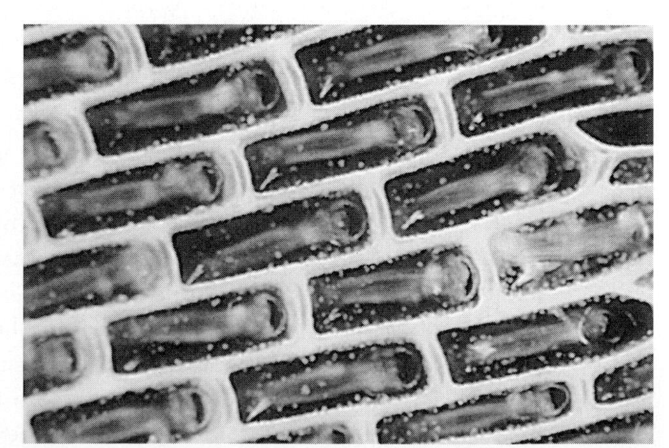

(b)

(a)

(b)

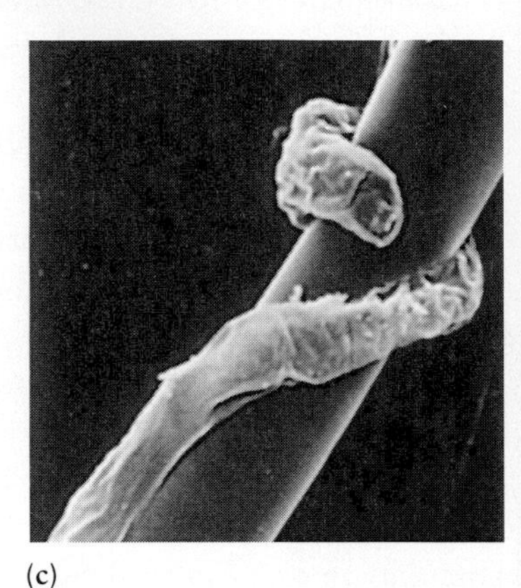

(c)

Figure 1.5 Many fungi, such as the *Amanita* mushroom fungus (a) and the shelf fungus (b), obtain nutrients from dead organic material. Others are pathogenic and attack living tissues of plants and animals, or even other fungi. (c) A hypha of one fungus attacks that of a larger plant pathogen. The hypha of the larger fungus has a diameter of about 5 μm. Courtesy of the U.S. Department of Agriculture.

cows; protozoans that digest wood in the guts of termites; fungi associated with the roots of plants that help them to extract mineral nutrients from the soil in return for a carbohydrate energy source from the plant; photosynthetic algae in the flesh of corals and giant clams; and nitrogen-fixing bacteria in the root nodules of legumes.

Habitats and niches

Another way of looking at ecological systems is from the perspective of the organism itself. A **habitat** is the place, or physical setting, in which an organism lives. A **niche** is a representation of the ranges of conditions that an organism can tolerate and the ways of life that an organism can pursue—that is, its role in the ecological system. A simple analogy might help. A worker's habitat could be an office building, and his or her niche in that building could be described in terms of the physical office space (fifth floor, 2 by 3 meters, a long way from the coffee machine), the office furnishings (desk, typewriter, filing cabinet), office hours, the particular task performed, and his or her boss and coworkers. A katydid's "office building"—its habitat—is the forest, but its niche within the forest includes the range of temperatures it can tolerate, the types of plants it can eat, and its enemies. An important principle of ecology is that each species has a distinctive niche: no two are exactly the same, because each species has distinctive attributes of form and function that determine the conditions it can tolerate, how it feeds, and how it escapes its enemies.

Ecologists characterize habitats by their conspicuous physical features, often including the predominant form of plant life or, sometimes, animal life (Figure 1.6). Thus, we speak of forest habitats, desert habitats, and coral reef habitats. How we define a habitat depends on the point of reference: it is strictly a matter of convenience. The habitat of an earthworm is the soil, whereas that of the bear treading on the soil is the forest. The habitat viewpoint differs from the ecosystem and community viewpoints because of this focus on the organism and its way of life rather than on the functioning of the ecological system within which it lives.

Ecologists have devoted much effort to classifying habitats. For example, one may distinguish terrestrial and aquatic habitats; among aquatic habitats, freshwater and marine; among marine habitats, ocean and estuary; among ocean habitats, benthic (on or within the ocean bottom) or pelagic (in the open sea). Such classifications ultimately break down, however, because habitat types overlap broadly and absolute distinctions between them do not exist. The idea of habitat nonetheless emphasizes the variety of conditions to which organisms are exposed at the earth's surface. Inhabitants of abyssal ocean depths and tropical rain forest canopies experience vastly different conditions of light, pressure, temperature, oxygen concentration, moisture, viscosity, and salts, not to mention food resources and enemies. Thus, although the same principles of ecological interaction and evolution apply to both habitats, the expressions of these processes differ utterly. The variety of habitats holds the key to much of the diversity of living organisms. No one organism can live under all the conditions of the earth's different habitats; each must specialize, with respect to both the range of habitats within which it can live and the size of the niche that it can occupy within a habitat.

Scale in time and space

The natural world varies in time and space. We perceive temporal variations in our environment as the alternation of day and night and the seasonal progression of temperature and precipitation. Superimposed on these cycles are irregular and unpredictable variations. Winter weather is generally cold and wet, but the weather at any particular time cannot be predicted much in advance; it varies perceptibly over intervals of a few hours or days as cold fronts and other atmospheric phenomena pass through. Some irregularities in conditions, such as the alternation of series of especially wet and dry years, occur over longer periods. Some events of great ecological consequence, such as fires and tornadoes, strike a particular place only at very long intervals.

Each type of variation in the environment has a characteristic dimension or scale. Variation between night and day has a dimension of 24 hours; seasonal variation has a dimension of 365 days. Waves pound a rocky shore at intervals of seconds; winter storms bringing rain or snow may follow one another at intervals of days or weeks; hurricanes may strike a particular

Figure 1.6 Four distinctly different habitats: (a) a semiarid scrub-grassland in southern Texas, (b) a red cedar forest in northern Idaho, (c) an open acacia woodland in Kenya, and (d) a coral reef in Panama. The dominant plant form in each of the terrestrial habitats is determined by climatic factors, particularly temperature and precipitation. Note the similarity between the growth forms of the acacia trees and the corals. Like the acacia, the coral, which is a colonial animal, requires light; its tissues contain symbiotic photosynthetic algae. Photographs courtesy of (a) the U.S. Forest Service, (b) the U.S. National Park Service, and (d) J. W. Porter.

coast at intervals of decades. In general, the more extreme the condition, the lower its frequency and the longer the interval between events.

Both the severity and the frequency of events are relative measures, depending on the organism that experiences them. Fire may touch a tree many times within its life span but skip dozens of generations of an insect population. As we shall see, how organisms and populations respond to change in their environments depends on the scale of temporal variation. The dimensions of ecological processes may also be intrinsic properties of the systems themselves. For example, in pine woodlands, the probability of a destructive fire increases with time since the last such event. As litter and other fuels accumulate, they produce a characteristic fire cycle for a particular habitat. Similarly, the rapid spread of contagious disease through a population often depends on the accumulation of young, nonimmunized individuals following the last epidemic.

The environment also differs from place to place. Variations in climate, topography, and soil type cause large-scale heterogeneity (across meters to hundreds of kilometers). At smaller scales, much heterogeneity is generated by the structures of plants, by the activities of animals, and even by the particle structures of soil (Figure 1.7). A particular dimension of spatial variation may be important to one animal and not to another. The difference between the top and the underside of a leaf is important to an aphid, but not to a moose, which happily eats the whole leaf, aphid and all.

Spatial heterogeneity combined with speed determines how frequently a moving individual encounters new environments. That is, spatial variation is perceived as temporal variation by an animal traveling through the environment. For a plant, the dimension of spatial variation determines the variety of conditions that its roots encounter in the soil and the variety of habitats in which its offspring germinate, which depends on how far its pollen and seeds travel.

In recent years, a new branch of ecology, called **landscape ecology,** has begun to deal with the obvious fact that animals and plants live in

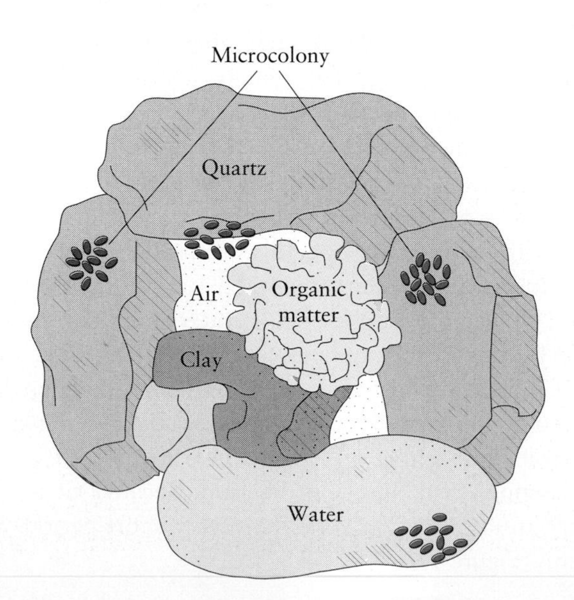

Figure 1.7 Section through a soil particle showing the patchy distribution of bacterial microcolonies on grains of quartz and clay. Note the porous structure of the soil particle.

Figure 1.8 A thunderstorm over the desert near Tucson, Arizona, may dump several centimeters of rain over a small area in less than an hour.

spatially varied environments and that the particular size and arrangement of habitat patches has important repercussions for the activities of individuals, the growth and regulation of populations, and the interactions between species. The lessons of landscape ecology have been driven home by studies of how birds have responded to the fragmentation of their habitats as human land use patterns have changed. The isolation of subpopulations in small habitat fragments increases their vulnerability to extinction caused by random environmental perturbations. Habitat fragmentation also means that each area of a particular habitat is closer to an edge where it abuts a different kind of habitat. For forest birds in North America, this has meant that some nest predators, which normally live in fields and do not venture far into the forest, now prey upon a larger proportion of eggs and young; as a result, populations of some species have declined precipitously.

The temporal and spatial dimensions of ecologically important phenomena are often correlated—that is, duration usually increases with size. For example, tornadoes last only a few minutes and affect small areas compared with the devastation inflicted by hurricanes over periods of days or weeks (Figure 1.8). In the oceans, at one extreme, small eddy currents may last only a few days; at the other extreme, ocean gyres (circulating currents encompassing entire ocean basins) are stable over millennia.

Compared with marine and, especially, atmospheric phenomena, variations in landforms have very long temporal dimensions at a particular spatial dimension. The reason is simple: spatial patterns in the terrestrial realm are determined by underlying topography and geology, which are transformed at a snail's pace by such processes as mountain building, lava flows, erosion, and even continental drift. In contrast, spatial heterogeneity in the open oceans results from physical processes in water, which are obviously more fluid (changing) than those on the land. Because air is even more fluid than water, atmospheric processes have very short periods at a given spatial dimension (Figure 1.9).

A principle related to the space-time correlation states that the frequency of a phenomenon is generally inversely related to its spatial dimension or local severity. Thus tornadoes and hurricanes occur at longer intervals, on average, than do winter storms. The frequency of forest fires or brush fires is inversely related to the area they burn. Such disturbances create patches of habitat in various stages of ecological development, or succession, thereby contributing to the spatial heterogeneity of the environment on many scales of time and space.

Some general principles of ecology

Because ecological systems are so complex and varied, the study of ecology can seem bewildering. The secret to coping with this complexity is to understand that all ecological systems are governed by a manageable number of basic principles. Throughout this book, you will see how these principles are expressed in different ecosystems and at different levels of ecological organization. A brief consideration of three of the most general principles will illustrate for you the underlying unity of ecology.

Ecological systems are physical entities

Life builds upon the physical properties and chemical reactions of matter. The physical laws of thermodynamics, which apply to energy in nature, govern physical and chemical transformations in biological systems, including the diffusion of oxygen across body surfaces, rates of chemical reactions, resistance of vessels to the flow of fluids, and transmission of nerve impulses. Biological systems are powerless to alter these fundamental physical qualities of matter and energy, but, within the broad limits imposed by physical constraints, life can pursue many options, and it has done so with astounding invention.

Figure 1.9 Temporal and spatial scales of variation in (a) atmospheric and (b) marine systems. Notice the general correlation between temporal and spatial dimensions. The large spatial dimensions of droughts and El Niño events reflect their tight link to changes in ocean circulation patterns. With respect to marine phenomena (b), the ellipses enclose the temporal and spatial dimensions of fluctuations in population density. The shorter time scales of population fluctuations as compared with physical variation in the marine realm result from biological causes of population change. After J. H. Steele, *J. Theoret. Biol.* 153:425–436 (1991).

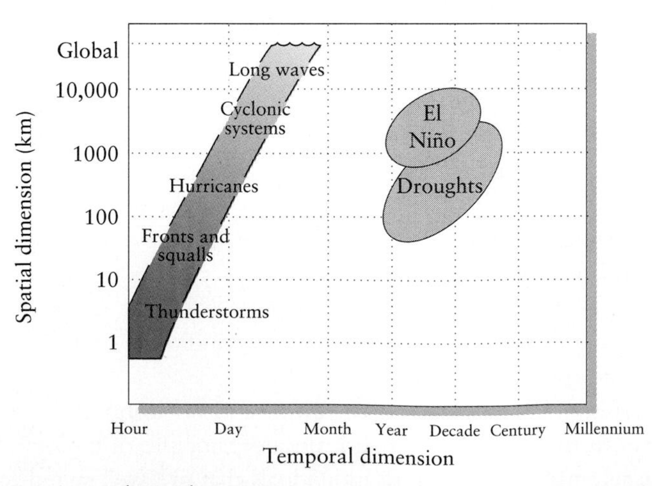

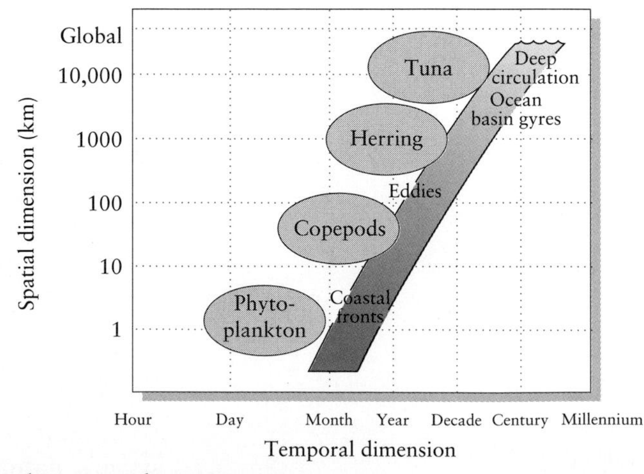

Ecological systems exist in steady states

Whether one's attention is focused on the organism, the population, the ecosystem, or the entire biosphere, each of these ecological entities continually exchanges matter and energy with its surroundings. That ecological systems remain more or less unchanged implies that gains by, and losses from, these systems are more or less balanced. When a system changes, gains and losses are out of balance. A warm-blooded animal continually loses heat to a cold environment; this loss is balanced by heat gained from the metabolism of foodstuffs. When gains do not add up to losses, the body cools. The proteins of our bodies are continually broken down and replaced by synthesis of new proteins; we lose some of the nitrogen of metabolized proteins in urine and replace it through sources of nitrogen in our food. Much of the material in the bodies we all carried around a year ago has been replaced, although we still look pretty much the same.

This idea of maintaining a steady state in the face of continuous flux of materials and energy between an ecological system and its surroundings applies to all levels of ecological organization. For the population, gains and losses are births and deaths. The diversity of a biological community decreases when species become extinct, and it increases when new species invade the habitat of the community. The biosphere itself could not exist without the energy received from the sun, yet this gain is balanced by heat energy radiated at infrared wavelengths back out into space. How the steady states of ecological systems are maintained and regulated is one of the most important questions posed by ecologists, one to which we will return frequently throughout this book.

Ecological systems undergo evolutionary change over time

The history of life on earth has shown that the attributes of organisms change over time by the process of **evolution.** Although the physical and chemical properties of matter and energy are immutable, what living systems do with matter and energy is as variable as all the forms of organisms that have existed in the past, are alive today, or might evolve in the future. The structures and functions of organisms evolve in response to the features of the environment with which they must contend. Such features include both the physical conditions that prevail and the various other kinds of organisms with which each population interacts. For example, animals that have visually hunting predators are often colored in such a way that they blend in with their backgrounds and escape notice (Figure 1.10). Many plants that grow in hot, dry climates have thick, waxy cuticles that reduce the loss of water by evaporation across their leaf surfaces. These features of structure and function that suit an organism to the conditions of its environment are called **adaptations.**

The close correspondence between organism and environment is no accident. It derives from a fundamental and unique principle of biological systems: **natural selection.** Only those individuals that are well suited to their environments survive and produce offspring. The inherited traits that

(a)

(b)

Figure 1.10 The green coloration of a mantid (a) and a tree frog (b) blend in with the backgrounds of these tropical animals and reduce their risk of being seen by visually hunting predators. Note the leaflike projections from the thorax of the mantid and the speckling on the frog, which resembles the fungi and other blemishes on natural leaves. Photographs by C. C. Hansen, courtesy of the Smithsonian Tropical Research Institute.

they pass on to their progeny are preserved. Unsuccessful individuals do not survive, or they produce fewer offspring, and their less suitable traits therefore disappear from the population as a whole. Charles Darwin was the first to recognize that this process allowed populations to respond, over periods of many generations, to changes in their environments. The wonderful thing about natural selection and evolution is that as each species changes, new possibilities for further change are opened up, and the environment of other species with which the changed species interacts is altered. In this way, the complexity of ecological communities and ecosystems builds upon, and is fostered by, existing complexity.

Generation of ecological diversity

A distinctive quality of ecological systems is their **diversity.** The earth is inhabited by millions of different kinds, or species, of organism. Presumably, all of these descended from a much smaller number (probably only one) at some very remote time in the past. The process by which species proliferate, which is called **speciation,** involves independent evolutionary change within isolated populations of a single species. The isolation may be brought about by physical barriers to dispersal, such as a rising mountain range, an expanding ice sheet, or a changing sea level, or by an ecological shift resulting, for example, from the spread of an herbivore to a new host plant. Regardless of the particulars that create isolation, sufficient genetic differences may eventually accumulate that individuals in the isolated populations could no longer interbreed successfully should they renew their contact. This sequence, repeated over and over again, has produced an astounding number of living things.

Isolated subpopulations diverge ecologically as well as genetically when different habitats or other environmental factors steer evolutionary change along different paths. Thus, a part of biological diversity results from the variety of physical environments at the surface of the earth to which organisms adapt. This process is self-accelerating because biological structures create further environmental heterogeneity and establish opportunities for continuing evolutionary diversification. Biological interactions within ecological communities also promote diversification. For example, various enemies (predators, herbivores, and disease organisms) unwittingly favor those individuals among their potential victims that can best avoid them, and thus constantly promote evolutionary change in other species. In this way, a process such as natural selection, which is a common feature of all ecological systems but acts in varied ecological settings, can contribute to the wonderful variety of life on earth.

The study of ecology

To understand natural diversity, ecologists must learn about the common principles of function that all ecological systems share, and also about the manner in which each system uniquely responds to its particular environment. Like other scientists, ecologists apply many methods to learn about nature. Most of these methods reflect four facets of scientific investigation: (1) observation and description, (2) development of hypotheses or explanations, (3) testing of these hypotheses, and (4) application of general knowledge to solve specific problems. Most research programs begin with a set of facts about nature that invite explanation. Usually these facts describe a consistent pattern. For example, measurements of rainfall and plant growth over several years might reveal a correlation between precipitation and plant production. To cite another example, exploration during the nineteenth century established that the number of animal and plant species in tropical regions greatly exceeds that in temperate regions. Recognition of this relationship between biodiversity and latitude grew out of comparisons of the accumulated observations of many scientists until it became confirmed as a general pattern. Like the relationship between rainfall and plant growth, this is a pattern that invites explanation. Because many explanations are plausible, it is then necessary to conduct experiments or other kinds of investigations to determine which best accounts for the facts.

Hypotheses are ideas about how a system works—that is, they are explanations. If correct, a hypothesis may help us to understand the cause of an observed pattern. Suppose we observe this pattern: male frogs sing on warm nights after periods of rain. If a reasonable amount of observation produces no exceptions to the pattern, it may be regarded as a generalization that enables us to predict the behavior of frogs from the weather. Having established the existence of a pattern, we may wish to understand it better. For example, we may wish to know how a frog responds to temperature and rainfall; we may also wish to know why a frog responds the way it does. The "how" part of this particular phenomenon involves details of sensory perception, the interplay between environmental stimuli and hormonal status, and neuromotor effectors—in other words, it involves physiological processes.

The "why" question deals with the costs and benefits of the behavior to the individual; it is more ecological and evolutionary in nature. If we suspect that male frogs attract females by singing, we may entertain the idea that males sing after rains because that is when females look for mates. If frogs chorused at other times, they might attract few mates (low benefit) but still expose themselves to predation or other risks (high cost) in the attempt. We now have generated a number of hypotheses about how the frog system works: (1) singing by the males attracts females and leads to mating; (2) females actively search for males only after rains (we could also propose some explanations for this pattern if it turned out to be true); (3) singing imposes a cost, which compels males to save their singing for times when it will do them the most good.

If we are to convince ourselves that a hypothesis is valid, we must put it to the test. Only rarely can a particular idea be proved beyond a doubt, but our confidence increases the more we explore the implications of a hypothesis and find it to be consistent with the facts. If our second hypothesis about frog singing were true, we would expect to observe more receptive females on nights following rains than on nights following fair weather. This is a **prediction,** which is a statement that follows logically from a hypothesis. If observations of the activity of females confirm this prediction, then the hypothesis is strengthened; if not, the hypothesis is weakened, or perhaps it may be rejected altogether.

The strongest tests of hypotheses are often the outcomes of **experiments,** in which one or a small number of variables are manipulated independently of others to reveal their unique effects. Thus we may design experiments to test predictions of a hypothesis specifically and unambiguously. In the frog example, the number of females hopping about does not provide a direct measure of male mating success, which is the point of the hypothesis. A stronger test of the hypothesis would be to determine whether mating success is lower when a male sings after fair weather than it is when a male sings after rains. Unfortunately, males don't sing after fair weather. Perhaps, by some suitable manipulation, we could trick a male into singing on the "wrong" night. This would be a good experiment if we could make frogs sing without altering other aspects of their behavior. Another experiment that comes to mind would be to record the songs of male frogs on a tape, play them through speakers on different nights, and tally the numbers of females that are attracted to the calls.

Tests of predictions generate new information that often initiates additional rounds of hypothesis formation and testing. For example, if we find that female frogs are more active after rainy weather, we have discovered a new pattern that invites explanation. In this way scientific discovery builds upon itself, generating a rich understanding of the workings of natural systems.

Although the ways of acquiring scientific knowledge appear to be straightforward, many pitfalls exist. For example, the existence of a correlation between variables does not imply a causal relationship; the mechanism of causation must be determined independently by suitable investigation. Different hypotheses may explain a particular observation equally well, and one must make predictions that distinguish among the alternatives. In addition, many predictions cannot be tested by experimental methods. Thus we may be unable to determine the consequences of variation in particular factors for the structure or functioning of a system. These limitations become particularly critical with patterns that have evolved over long periods and with systems that are too large for practical manipulation.

The observation that biodiversity decreases at higher latitudes has suggested as many explanations as there are physical attributes of the environment that also vary with latitude. As one travels north from the equator, average temperature and precipitation decrease, light intensity and biological production decrease, and seasonality increases. Each of these factors may interact with biological systems in ways that could affect the number of species that can coexist in a locality, and dozens of hypotheses based on these attributes have been proposed. Isolating the effects of each of these factors has proved difficult because each tends to vary in parallel with the others.

Faced with these difficulties, ecologists have resorted to several alternative approaches to hypothesis testing. One of these is the **microcosm** experiment, which attempts to replicate the essential features of a system in a simplified laboratory setting. Thus, an aquarium with five animal species may behave like the more complex natural system in a pond or even like ecological systems more generally; if so, experimental manipulations of the microcosm may yield results that can be generalized to the larger system. The hypothesis that diversity decreases as environmental variation increases might be approached in a microcosm experiment by determining whether variations in temperature, light, acidity, or nutrient resources cause species to disappear from the system. Of course, it is a long stretch to generalize from an aquarium to a "real" ecological system, but if variation consistently reduced diversity in a variety of microcosms, the hypothesis would be considerably strengthened.

Another approach is to construct a **mathematical model** of a complex system, in which the investigator represents the system as a set of equations. These equations portray our understanding of how a system works in the sense that they describe the relationship of each of the system's components to other components and to outside influences. A mathematical model is a hypothesis; it provides an explanation of the observed structure and functioning of a system. Models can be tested by comparing the predictions they yield with what is actually observed. Most models make predictions about attributes of the system that have not been measured or about the response of the system to perturbation. Whether these predictions are consistent with observations determines whether the hypothesis on which they are based is supported or rejected.

Why do we do all this? The wonders of the natural world summon our natural curiosity about life and our desire to understand our surroundings. For many of us, our curiosity about nature and the challenges of taking a scientific approach to its study are reason enough. In addition, however, an understanding of nature is becoming more and more urgent as the growing human population stresses the capacity of natural systems to maintain their structure and functioning. Environments that human activities either dominate or have produced—including our urban and suburban living places, our agricultural breadbaskets, our recreation areas, tree farms, and fisheries—are also ecological systems. The welfare of humanity depends on maintaining the health of these systems, whether they are natural or artificial. Virtually all of the earth's surface is, or soon will be, strongly influenced by people, if not fully under their control. Already, humans usurp more than 40% of the biological productivity of the biosphere. We cannot take this responsibility lightly.

Human ecology

To say the least, humans are a prominent part of the biosphere. The human population is approaching 6 billion, and it consumes energy and resources, and produces wastes, far in excess of needs dictated by biological metabolism. This has caused two related problems of global dimension. The first is the impact of human activity on natural systems, including the disruption of ecological processes and the extermination of species. The second is the steady deterioration of humankind's own environment as we push the limits of what ecological systems can sustain. Understanding ecological principles is a necessary step in dealing with these problems. Consider the following examples.

Several years ago, the Nile perch was introduced into Lake Victoria in East Africa. This was done with the well-intentioned purpose of providing additional food for the people living in the area and additional income from export of the surplus catch. However, because basic ecological principles were ignored, the introduction ended up destroying most of the lake's fishery. Until the introduction of the Nile perch, Lake Victoria supported a sustainable catch of a variety of local fishes, mostly of species belonging to the family Cichlidae, which feed primarily on detritus and plants. Nile perch eat other fish: the smaller cichlids in this instance. However, because energy is lost with each step in the food chain, predatory fish cannot be harvested at so high a rate as herbivorous prey species. Furthermore, the perch was alien to Lake Victoria, and the local cichlids had no innate behaviors to help them escape predation. Inevitably, the perch annihilated the cichlid populations, destroying the native fishery and all but eliminating its own food supply. Consequently, the perch's voracious habits among defenseless prey brought about its own demise as an exploitable fish species. To be sure, the native fishery was already precariously close to being overexploited as a result of the burgeoning local human population and the use of advanced, nontraditional fishing technologies. However, the appropriate solution to these problems would have been better management of the cichlids and development of food sources other than fish, not the introduction of an efficient predator upon them.

Other consequences of the introduction of the Nile perch turn the story into a tragicomedy of errors. The flesh of the perch is not favored by the peoples living near the shores of Lake Victoria, who prefer the familiar texture and flavor of the native fishes. Moreover, the oily meat of the perch must be preserved by smoking rather than sun-drying, so local forests have been cut rapidly for firewood. Because the larger perch must be fished with larger and more elaborate nets, local subsistence fishermen were not able to compete with more prosperous outsiders who were equipped for commercial fishing. The lesson is painful, yet simple: humans are an integral part of the ecology of the Lake Victoria area. Traditional local fishing had been sustained for thousands of years, until population pressure and the opportunity to develop an export fishery prompted an ecologically unsound decision that spelled economic and social disaster.

Half a world away, efforts to save the California sea otter illustrate the intricate intermingling of ecology and other human concerns (Figure 1.11). The sea otter was once widely distributed around the North Pacific Rim from Japan to Baja, California; in the 1700s and 1800s, intense hunting for fine otter pelts reduced the population to near extinction. Predictably, the fur industry collapsed as it overexploited its economic base. Subsequent protection enabled the California sea otter population to multiply to several thousand individuals by the 1990s, well above the danger level. The sea otter's success, alas, has irked some California fishers, who claim that the otters—which do not need commercial fishing licenses—drastically reduce stocks of valuable abalone, sea urchins, and spiny lobsters. Matters deteriorated at one point to the marine equivalent of a range war between the fishing industry and conservationists, with the otter caught in the line of fire, often fatally.

Ironically, the otters benefit a different commercial marine enterprise, the harvesting of kelps, which are large seaweeds used in making fertilizer. Kelps grow in shallow waters in stands called kelp forests, which provide refuge and feeding grounds for larval fish (Figure 1.12). Kelps are also grazed by sea urchins, which, when abundant, can denude an area. The sea otter is a principal predator of sea urchins. When the expanding otter population has spread into new areas, urchin populations have been controlled, allowing kelp forests to regrow. Human involvement in the ecosystem of the sea otter requires that management of otter populations and various commercial enterprises be balanced in an economically and socially acceptable manner. This will require a detailed understanding of the complex ecological role of the otter in the system.

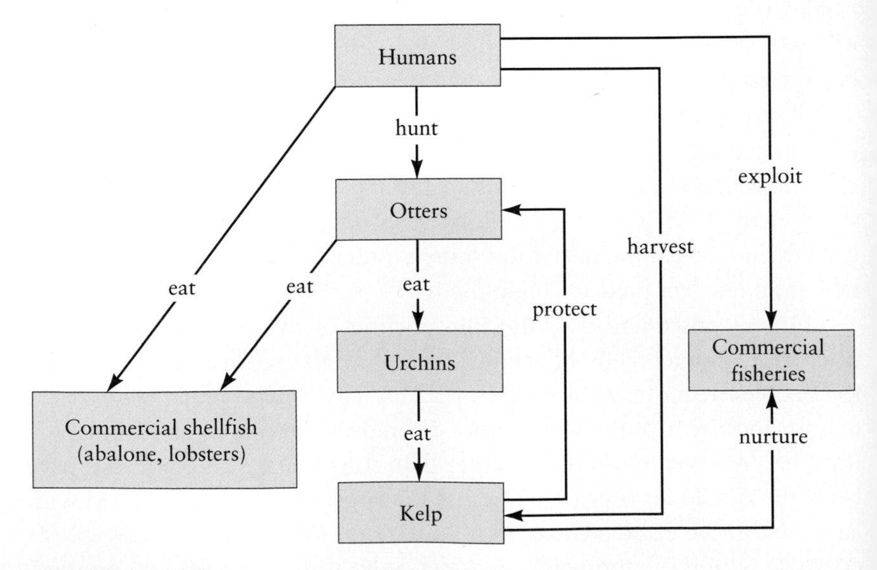

Figure 1.11 A simple diagram of the otter-urchin-kelp system.

Figure 1.12 The kelp forest provides feeding grounds and refuge for many species of fish and invertebrates. The integrity of the habitat depends on the presence of sea otters, which eat sea urchins, which in turn eat the young kelp. When sea urchins are abundant, they may prevent the regeneration of the kelp forest. Photograph by E. Hanauer, courtesy of P. Dayton.

Although the plight of endangered species may arouse us emotionally, ecologists are increasingly coming to realize that the only effective means of preserving and using natural resources is through the conservation of entire ecological systems and the management of broad-scale ecological processes in a sustainable fashion. Individual species, including the ones that humans rely on for food and other products, themselves depend on the maintenance of a healthy ecological support system. We have already seen how predators such as the Nile perch and the sea otter may assume key roles in the functioning of natural systems, for better or worse, depending on the circumstances. On land, pollinators and seed dispersers are necessary to the long-term maintenance of plant formations. Many species of tropical vines in the genus *Gurania,* in the cucumber family, are pollinated exclusively by *Heliconius* butterflies, whose larvae feed only on passion vines, *Passiflora.* If passion vines were to disappear, many species of *Gurania* would fail to set fruit and eventually would also disappear. It is ecologically naive to try to manage forests without paying attention to the populations of insects, birds, mammals, and myriad other organisms on which the persistence of the forest depends. Natural systems are diverse and highly evolved in the sense that each species is adapted to its particular role within the system; as a result, the system's integrity depends to some degree on each player acting out its role.

The natural balance of nature includes continuous change. Most natural systems sustain cycles of disturbance (fire, hurricanes, disease) and rejuvenation. Restoration depends on patches of intact habitat from which species can recolonize disturbed areas. But as habitats become more and more fragmented, as forests in some regions have been by the spread of agriculture and urban development, disturbances can so thoroughly disrupt the remaining habitat as to leave it little chance of complete recovery. Disturbance cycles, whatever their spatial and temporal dimensions, are as much a part of the natural landscape as are key species.

Many advances in technology that appear to improve the quality of human life also make humans more efficient consumers who increasingly strain the earth's ecological support systems. In the end, these advances might reduce rather than enhance the quality of our lives. Viewed on any scale, the deterioration of the earth's environment is alarming. On a global scale, it is sometimes difficult to connect human activities with changes in the biosphere as a whole. On a local scale, particularly in partially enclosed systems, such as the estuaries of major rivers, it is not hard to see the changes or pinpoint their causes. The San Francisco Bay ecosystem, for example, has gradually but profoundly deteriorated under the pressures of population and agriculture to the point at which it would be unrecognizable to nineteenth-century inhabitants of the region. The diking of wetlands to create farmland, filling to create land for urban construction, pollution from sewage and from agricultural and industrial waste, and the introduction of exotic species have transformed a healthy, productive, and economically viable system into an ecological disaster area whose greatly diminished value is based solely on recreation (which is itself diminishing in appeal) and commercial shipping. The bay's once generous biotic resources are practically gone. True, the present human population of the San Francisco Bay area may be incompatible with a pristine estuary. However, the worst deterioration came about through state subsidy of unprofitable agricultural ventures, a lack of ecologically and economically sensible city planning, careless introduction of pest species, use of the bay as a cheap sewer, and uncontrolled harvesting of fish and shellfish. Ecological disaster need not have happened.

The multiplication of technology and population has brought ecological crisis to far larger regions of the earth than San Francisco Bay. One of the most critical is the Sahel region of Africa. The Sahel is a vast band of semiarid habitat lying at the southern edge of the Sahara Desert and extending from Gambia in the west to Ethiopia in the east. Population pressure, and with it intensified grazing and collection of firewood, has caused extensive deforestation of the entire region. As humans and domesticated animals clear vegetation from such vast areas, climate is affected. Rainfall decreases because less water vapor is recirculated to the atmosphere and because the heating and cooling patterns of the land surface are changed. Even when plentiful rains result in good crop yields, the abundance may be snatched away by plagues of grasshoppers and locusts, whose populations grow rapidly under the moister conditions. The Sahel is at present truly devastated; the quality of human life there, not to mention the state of natural ecosystems, is unthinkably poor.

In wetter parts of the world, tropical rain forests are being cleared at an alarming rate for forest products and largely unsuitable agriculture, bringing profit to a few in the short term but predictable disaster in the long term. The destruction of rain forests dismantles an intricate self-sustaining system that cannot easily be re-created. Species, many of them unknown to science and some of them with potential commercial value, are being driven to extinction. Burning of cut vegetation adds greatly to the already worrisome level of carbon dioxide in the atmosphere. Rampant erosion strips the unprotected land of valuable topsoil and chokes rivers with silt. In this instance, the ecological principles that should govern rational use of the rain forest are generally understood. The destruction of tropical rain forests, as well as the deterioration of other habitats and resources, derives from the short-term need to provide for poor and growing human populations and from the failure to add long-term costs to the immediate price of exploitation. We are borrowing heavily on our future without the prospect of ever being able to pay back the debt.

The destruction of rain forests is scarcely alone in its far-reaching effect. Automobile emissions reduce tree and crop production over large areas. Sulfurous gases from burning coal have acidified the rain and snow over large areas of eastern North America and Europe, killing forests and degrading freshwater ecosystems. Chlorofluorocarbons (CFCs) used to pressurize spray cans and as refrigerants have reduced the ozone concentration in the upper atmosphere, increasing the amount of dangerous ultraviolet radiation that reaches the earth's surface.

Industrial growth and deforestation have caused carbon dioxide, methane, and CFCs in the atmosphere to increase to such levels that they may be changing the earth's climate, although this is the subject of considerable scientific debate. These so-called greenhouse gases absorb long-wave radiation from the earth's surface, which may lead to an increase in the temperature of the atmosphere. Some scientists have warned that the temperature of the earth may rise by 2 to 6 degrees Celsius during the next century. Such a change would cause a rise in sea level as the polar ice caps melt and warmer ocean water expands; it could disrupt agriculture and shift ecological habitats across the landscape as temperature and moisture patterns change; and it might cause widespread extinction of forms of life that are unable to adjust to the temperature change or to its indirect effects.

Our capacity to wreak havoc on ourselves and the world in which we live is virtually unlimited. The most appalling manifestation of this is our ability to wage global nuclear war, wherein destruction of ecological systems and extinction of species, possibly including ourselves, would be virtually without parallel in the history of the earth. Nuclear war remains unthinkable, even if we ignore the ecological consequences, and its likelihood has greatly diminished in recent years. But, tragically, the gradual degradation of a once livable planet may in the end be just as punishing. Understanding ecology will not by itself solve our environmental problems in all their political, economic, and social dimensions. However, as we contemplate the need for global management of natural systems, our success will hinge on our understanding of their structure and functioning, an understanding that depends on knowing the principles of ecology.

Summary

1. Ecology is the scientific study of the natural environment and of the relationships of organisms to each other and to their surroundings.

2. Organism, population, community, ecosystem, and biosphere represent levels of organization of ecological structure and functioning. They form a hierarchy of progressively more complex entities.

3. Different kinds of organisms play different roles in the functioning of ecosystems. Plants fix the energy of sunlight; animals consume biological forms of energy. Fungi are able to penetrate soil and dead plant material and so play an important role in breaking down biological materials and regenerating nutrients in the ecosystem. Protoctists of various sorts are single-celled analogues of plants (algae) and animals (protozoa). Bacteria are the biochemical specialists of the ecosystem, able to accomplish such transformations as the biological assimilation of nitrogen and the use of hydrogen sulfide as an energy source, which are essential components of ecosystem function.

4. An individual's habitat is the place in which that individual lives. The habitat concept emphasizes the structure of the environment as it is experienced by each type of organism. An individual's niche is a representation of the ranges of conditions that it can tolerate and the ways of life that it can pursue—that is, its role in the natural system.

5. Ecological processes and structures have characteristic dimensions of time and space, which ecologists refer to as their scale. Landscape ecology addresses the ways in which spatial and temporal heterogeneity affect the outcomes of ecological processes.

6. The variety and complexity of ecological systems is understandable in terms of a small number of basic ecological principles. Among these are the idea that ecological systems are physical entities and that they function within the physical and chemical constraints governing energy transformations. Furthermore, all ecological systems exchange materials and energy with their surroundings. When inputs and outputs are balanced, the system is said to be in a steady state.

7. All ecological systems are subject to evolutionary change, which results from the differential survival and reproduction, within populations, of individuals that exhibit different genetically determined traits. As a result of this natural selection, organisms exhibit adaptations of structure and function that suit them to the conditions of their environments.

8. The great diversity of ecological systems is generated by the proliferation of species in a heterogeneous environment. Spatial variation in conditions promotes differences between species that live in different habitats. Interactions among populations within habitats also promote local diversification of species.

9. Ecologists employ a variety of techniques to study natural systems. The most important of these are observation, the development of hypotheses to explain observations, and the testing of hypotheses by attempting to confirm the predictions they generate. Experiments are an important tool in testing hypotheses. When natural systems do not lend themselves readily to experimentation, ecologists may work with microcosms or mathematical models of systems.

10. Humans play a dominant role in the functioning of the biosphere, and human activities have created an environmental crisis of global proportions. Solving our acute environmental problems will require the intelligent application of general principles of ecology within the framework of political, economic, and social action.

Suggested readings

Barel, C. D. N., et al. 1985. Destruction of fisheries in Africa's lakes. *Nature* 315:19–20. (Introduction of the Nile perch into Lake Victoria.)

Bartholomew, G. A. 1986. The role of natural history in contemporary biology. *BioScience* 36:324–329.

Booth, W. 1988. Reintroducing a political animal. *Science* 241:156–158. (The ecological role of sea otters in kelp communities.)

Estes, J. A., and D. O. Duggins. 1995. Sea otters and kelp forests in Alaska: Generality and variation in a community ecological paradigm. *Ecological Monographs* 65:75–100.

Franklin, J. F., C. S. Bledsoe, and J. T. Callahan. 1990. Contributions of the long-term ecological research program. *BioScience* 40:509–523.

Harley, J. L. 1972. Fungi in ecosystems. *Journal of Animal Ecology* 41:1–16.

Margulis, L., D. Chase, and R. Guerrero. 1986. Microbial communities. *BioScience* 36:160–170.

Margulis, L., and K. V. Schwartz. 1988. *Five Kingdoms: An Illustrated Guide to the Phyla of Life on Earth,* 2d ed. W. H. Freeman, New York.

McIntosh, R. P. 1985. *The Background of Ecology: Concept and Theory.* Cambridge University Press, New York.

Nichols, F. H., J. E. Cloern, S. N. Luoma, and D. H. Peterson. 1986. The modification of an estuary. *Science* 231:567–573. (San Francisco Bay.)

Sai, F. T. 1984. The population factor in Africa's development dilemma. *Science* 226:801–805.

Sinclair, A. R. E., and J. M. Frywell. 1985. The Sahel of Africa: Ecology of a disaster. *Canadian Journal of Zoology* 63:987–994.

Urban, D. L., R. V. O'Neill, and H. H. Shugart, Jr. 1987. Landscape ecology. *BioScience* 37:119–127.

PART I

LIFE AND THE PHYSICAL ENVIRONMENT

CHAPTER 2

THE PHYSICAL ENVIRONMENT

This chapter considers physical factors in the environment that influence ecological systems. We often speak of the living and the nonliving as opposites: biological versus physical and chemical, organic versus inorganic, biotic versus abiotic, animate versus inanimate. But although we can easily distinguish these two great realms of the natural world, they do not exist in isolation from each other. Life depends on the physical world. Living beings also affect the physical world: soils, the atmosphere, lakes and oceans, and many sedimentary rocks owe their properties in part to the activities of plants and animals. We take for granted the earth's oxygen-rich atmosphere, but it owes its character to the oxygen produced by photosynthetic organisms over the past 3.5 billion years. When life first evolved, the concentration of oxygen in the atmosphere was only one-thousandth of its present level.

Although distinct from physical systems, life forms nonetheless function within limits set by physical laws. Like internal combustion engines, organisms transform energy to perform work. An automobile engine burns gasoline chemically, and it transmits power from the cylinder to the tires mechanically. When an organism metabolizes carbohydrates or moves its appendages, it follows related chemical and mechanical principles.

Although biological systems operate on the same principles as physical systems, one important difference sets them apart. In physical systems, energy transformations follow paths of least resistance and tend to even out variations in energy level throughout the system. Thus, heat flows from points of high temperature to points of low temperature; a chemical reaction yields products having less energy than the reactants. In biological systems, the organism transforms energy in such a way that it keeps itself out of equilibrium with the physical environment. Indeed, it often uses its energy to counteract the physical forces of gravity, heat flow, diffusion, and chemical reaction. As a result, the physical conditions inside the organism's body differ dramatically from those of its surroundings. When organisms move, they overcome gravity and work against the resistance of the physical world. Thus the physical world both provides the context for life and constrains its expression.

In a sense, the way in which the organism uses energy is its secret of life. To make this point, let's first consider a simple physical example. A boulder rolling down a steep slope releases energy during its descent, but it performs no useful work in doing so. The source of the energy—in this case, gravity—is external, and as soon as the boulder comes to rest in the valley below, it regains an equilibrium with the forces in its physical environment. In distinct contrast, a bird in flight constantly expends energy to maintain itself aloft against the pull of gravity. The source of the bird's energy—the food it has assimilated—comes from within, and the bird uses that energy to perform useful work: pursuit of prey, escape from predators, or migration.

The ability to act against external physical forces distinguishes the living from the nonliving. A bird in flight supremely expresses this quality, but plants just as surely perform work to counter physical forces when they absorb soil minerals into their roots and synthesize the highly complex carbohydrates and proteins that make up their structure. Also, unlike physical systems, living organisms have a purposeful existence. Their structures, physiology, and behavior are directed toward procuring energy and resources and producing offspring. Certainly, life is constrained by physics and chemistry, just as architecture is constrained by the properties of building materials; however, as in biological systems, the purpose of the design of a building is unrelated to, and transcends, the qualities of bricks and mortar.

In the final analysis, life is a special part of the physical world, but it exists in a state of constant tension with its physical surroundings. Organisms ultimately receive their energy from sunlight and their nutrients from the soil and water, and they also must tolerate extremes of temperature, moisture, salinity, and the other physical factors of their surroundings. The heat and dryness of deserts exclude most species, just as the bitter cold of polar regions discourages all but the most hardy, but we need not search so far as such extreme conditions for evidence of the tension between the physical and biological realms. The form and function of all plants and animals have evolved partly in response to conditions prevailing in the physical world. In this chapter, we shall explore those attributes of the physical environment that are most consequential for life. Because life processes take place in an aqueous environment, and because water makes up the largest part of all organisms, water seems a logical place to start.

Properties of water

Water is abundant over most of the earth's surface, and within the temperature range usually encountered it is liquid. Many of the thermal properties of water are favorable to life. For example, one must add or remove a large amount of heat energy to change the temperature of water, and water conducts heat rapidly. Because of these two properties, which are referred to as the **specific heat** and **thermal conductivity** of water (Box 2.1), the temperatures of organisms and aquatic environments tend to remain relatively constant and homogeneous. Water also resists change of state between solid (ice), liquid, and gaseous (water vapor) phases. Over 500 times as much energy must be added to evaporate a quantity of water (the **heat of vaporization**) as to raise its temperature by 1°C! Freezing requires the removal of 80 times as much heat (the **heat of melting**) as that needed to lower the temperature of the same quantity of water by 1°C. Another curious, but serendipitous, thermal property of water is that, whereas most substances

BOX 2.1 Thermal properties of water

Specific heat is the quantity of heat energy required to raise the temperature of 1 g of water 1°C: 1 calorie (cal) or 4.2 joules (J).

Heat of melting is the quantity of heat energy that must be added to ice to melt 1 g of water at 0°C: 80 cal or 335 J.

Heat of vaporization is the quantity of heat energy that must be added to evaporate 1 g of water: 597 cal or 2,498 J at 0°C, 536 cal or 2,243 J at 100°C.

Thermal conductivity is the flux of heat through a 1 cm^2 cross section at a gradient of 1°C cm^{-1} (units are J cm^{-1} s^{-1} $°C^{-1}$): 0.0055 at 0°C, 0.0060 at 20°C, 0.0063 at 40°C, and 0.022 for ice at 0°C.

Density is the mass per unit of volume:

Water at 30°C	=	0.99565 g cm^{-3}
20°C	=	0.99821
10°C	=	0.99970
4°C	=	0.99997 (maximum density)
0°C	=	0.99984
Ice at 0°C	=	0.917

Figure 2.1 Water becomes less dense as it freezes, and ice therefore floats. But because the density of ice is 0.92 g cm^{-3} (not very far from that of water, at 1 g cm^{-3}), more than 90% of the bulk of this antarctic iceberg lies below the surface.

become more dense at colder temperatures, water becomes less dense as it cools below 4°C. Water also expands and becomes even less dense upon freezing. Consequently, ice floats (Figure 2.1), which not only makes ice skating possible but also prevents the bottoms of lakes and oceans from freezing and enables aquatic plants and animals to find refuge there in winter.

Water has an immense capacity to dissolve substances, making them accessible to living systems and providing a medium within which they can react to form new compounds. The formidable solvent properties of water derive from the strong attraction of water molecules for other compounds. Molecules consist of electrically charged atoms or groups of atoms called **ions.** Common table salt, sodium chloride (NaCl), contains a positively charged sodium atom (Na^+) and a negatively charged chlorine atom (Cl^-). In the absence of water, the sodium and chlorine attract each other and combine to form molecules of sodium chloride. In water, however, the charged sodium and chlorine atoms are so powerfully attracted by water molecules, compared with the strength of the bonds that hold salt molecules together, that salt readily dissociates into its component atoms—another way of saying that the salt dissolves.

Water and soil

Most terrestrial plants obtain the water they need from the soil. The amount of water that soil holds, and its availability to plants, varies with the physical structure of the soil. Soil consists of grains of clay, silt, and sand and particles of organic material (see Figure 1.7). Grains of clay, produced by the weathering of minerals in certain types of bedrock, are the smallest; grains of sand, derived from quartz crystals that remain after minerals more susceptible to weathering dissolve out of rock, are often the largest; silt particles are intermediate in size. Soil scientists define particles smaller than 0.002 mm as clay and those larger than 0.05 mm as sand. Collectively, these particles make up the **soil skeleton.** As the name implies, the soil skeleton is a stable component that influences the physical structure of the soil and its water-holding ability, but does not play a major role in its chemical transformations.

Water is sticky. The capacity of water molecules to cling to one another (the basis for surface tension) and to surfaces they touch (capillary action) causes water to rise in capillary tubes against the pull of gravity. Water also clings tightly to the surfaces of the soil skeleton. The more surface area, the more water a soil can hold. Because the total surface area of particles in a given volume of soil increases as their size decreases, clay soils and silty soils hold more water than coarse sands, through which water drains quickly.

The availability of water in the soil is only partly determined by the amount present. Plant roots easily take up water that clings loosely to soil particles by surface tension, but water very close to the surface of the particles adheres tightly to the soil skeleton by more powerful forces. The strength of these forces is called the **water potential** of the soil. Soil scientists quantify soil water potential, and the strength with which the cells of root hairs can absorb water from the soil, in terms of equivalents of atmospheric pressure. The weight of the air above us is 14.7 pounds per square inch at sea level; in the International System of Units (see Appendix A), this is 101,325 pascals (0.1 megapascals). Capillary attraction holds water in the soil with a force equivalent to a pressure of about 0.1 atmosphere (atm), or 10^4 pascals (Pa). Water that is drawn to soil particles with a force of less than 0.1 atm (water in the interstices between large soil particles, generally more than 0.005 mm from their surfaces) drains out of the soil under the pull of gravity and joins the groundwater in the crevices of the bedrock below. The amount of water held against gravity by forces of attraction greater than 0.1 atm is called the **field capacity** of the soil. Imagine a particle of silt with a diameter of 0.01 mm enlarged to the size of this page (×25,000); the film of water held at field capacity by forces of capillary attraction would be as thick as half the width of the page.

A force equivalent to 0.1 atm can raise a column of water nearly 1 meter. We may surmise, however, that plant roots can exert a much greater pull on water in the soil, because the tallest trees raise water to leaves more than 100 meters above the ground, which is equivalent to a pressure of about 10 atm. In fact, most plants can exert a pull of about 15 atm on soil water. During drought stress, the water potential of the soil steadily increases as plants remove water that is held by forces weaker than 15 atm. When the soil water potential finally exceeds 15 atm, most plants can no longer obtain water, and they wilt, even though some water still remains in the soil. Thus ecologists refer to a water potential of 15 atm as the **wilting coefficient,** or **wilting point,** of the soil.

As soil water decreases, the remainder is held by increasingly stronger forces, on average, because a greater proportion of the water lies close to the surfaces of soil particles. Figure 2.2 shows the relationship between water content and water potential for a typical soil with a more or less even distribution of soil particle sizes from clay through silt to sand (such soils are called loams). When saturated just after a heavy rain, a loam contains about

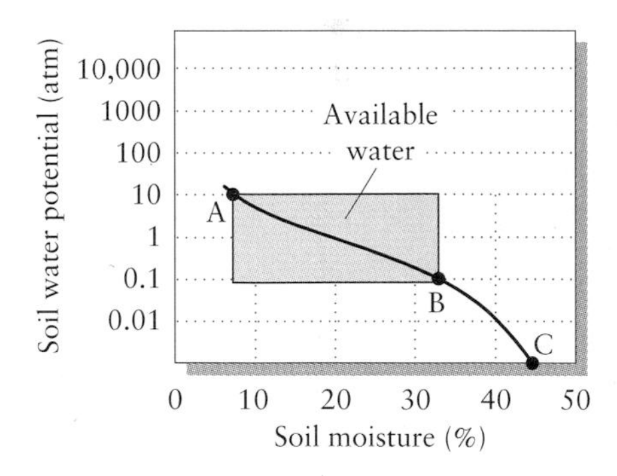

Figure 2.2 Relationship between the water content of a loam and the average force of attraction of the water to soil particles (water potential). The difference between the soil water content at field capacity (B, 0.1 atm) and the wilting coefficient (A, 15 atm) is the water available to plants. Point C is the saturation capacity of the soil. After N. C. Brady, *Nature and Properties of Soils,* 8th ed., Macmillan, New York (1974).

45 grams (g) of water per 100 g of oven-dried soil (45% water). As water drains from such a soil under gravity, the water content decreases to the field capacity of about 32% water. The wilting coefficient of a typical loam is about 7% water. The difference between the field capacity and the wilting coefficient, about 25% in this type of soil, measures the water available to plants. Of course, plants obtain water most readily when the soil moisture is close to the field capacity because only a small percentage is held by strong forces.

Temperature

Life processes, as we know them, occur only within the range of temperatures at which water is liquid: 0°–100°C at the earth's surface. Relatively few plants and animals can survive body temperatures above 45°C. Some photosynthetic cyanobacteria tolerate temperatures as high as 75°C, however, and some archaebacteria can live in hot springs at temperatures up to 110°C (Figure 2.3). Hot water imparts a high kinetic energy to living systems, which tends to open up, or denature, the structure of biological molecules. Thus, existence at high temperatures requires that proteins and biological membranes have strong forces of attraction within and between molecules to resist being shaken apart. The proteins of thermophilic ("heat-loving") bacteria have subtly different proportions of amino acids compared with other, heat-intolerant organisms; as a result, the structures of these proteins remain stable at temperatures up to 95°C or higher.

Figure 2.3 A hot spring in Yellowstone National Park. Even though the temperature of the water approaches the boiling point, some thermophilic bacteria thrive in this environment.

Figure 2.4 In the antarctic fish *Trematomus,* blood and tissues are prevented from freezing by the accumulation of high concentrations of glycoproteins, which lower the freezing point of body fluids to below the minimum temperature of seawater (1.8°C) and prevent ice crystal formation. Photograph courtesy of P. Dayton.

Although temperatures on the earth rarely exceed 50°C, except in hot springs and at the soil surface in hot deserts, temperatures below the freezing point of water occur commonly over large portions of the earth's surface. When living cells freeze, the crystal structure of ice disrupts most life processes and may damage delicate cell structures, eventually causing death. Many kinds of organisms successfully cope with freezing temperatures either by maintaining their body temperatures above the freezing point of water or by activating mechanisms that enable them to resist freezing or tolerate its effects.

The freezing point of water may be depressed by dissolved substances that interfere with the formation of ice. For example, the freezing point of seawater, which contains about 3.5% dissolved salts, is –1.9°C. The blood and body tissues of most vertebrates contain less than half the salt content of seawater and thus may freeze at a higher temperature than the freezing point of the ocean. This creates a problem for fish living in polar seas. Saltier blood would help out; for example, a 10% solution of sodium chloride reduces the freezing point of water by 10.5°C. But protein structure and function are sensitive to high salt concentrations, so this a physiologically impractical solution. Instead, the freezing points of the body fluids of many marine organisms are reduced to below the freezing point of water by high concentrations of glycerol and glycoproteins. A 10% glycerol solution, for example, lowers the freezing point of water by about 2.3°C. The presence of these antifreeze-like compounds in their blood and tissues allows antarctic fish to remain active in seawater that is colder than the normal freezing point of the blood of fish inhabiting temperate or tropical seas (Figure 2.4). Terrestrial invertebrates also use the antifreeze approach, and their body fluids may contain up to 30% glycerol, in extreme cases, as winter approaches.

Supercooling provides a second solution to the problem of freezing. Under certain circumstances, fluids can cool below the freezing point without ice crystals developing. Ice generally forms around some object, called a seed, which can be a small ice crystal or other particle. In the absence of seeds, pure water may cool to more than 20°C below its melting point without freezing. Supercooling has been recorded to –8°C in reptiles and to –18°C in invertebrates. Glycoproteins in the blood of these cold-adapted animals impede ice formation by coating developing crystals, which would otherwise act as seeds.

Finally, some organisms, such as the overwintering stages of many temperate and boreal insects, can tolerate the freezing of most or all of the water in their bodies. Such organisms restrict ice formation to the spaces between cells; as a result, ice does not destroy cell structure. However, because salts are excluded from ice when it freezes, the salts in their bodies are concentrated in the liquid water within cells. Thus, freezing-tolerant organisms must cope with extremely high salt levels in their tissues during the winter.

Temperature has several opposing effects on life processes. Heat increases the kinetic energy of molecules and thereby accelerates chemical reactions; the rate of any biological process commonly increases between two and four times for each 10°C rise in temperature throughout the physiological range (Figure 2.5). This factor of increase is called the $\mathbf{Q_{10}}$ of a process, and it is estimated by the relationship between the rates of physiological processes, plotted on a logarithmic scale, and temperature. Also, enzymes and other proteins become less stable at high temperatures and may not function properly or retain their structure. In addition, the level of heat energy in the cell influences the conformations of proteins, which balance the natural kinetic motions induced by heat and the forces of chemical attraction between different parts of the molecule. Similarly, the physical properties of the fat molecules in cell membranes and those that many animals accumulate as a reserve of food energy also depend on temperature. When cold, fats become stiff (picture in your mind the fat on a piece of meat taken from the refrigerator); when warm, they become fluid.

Enzymes function well only when they assume the proper shape. Too hot, and the molecule may open its structure and tend to unfold; too cold, and it may close up, preventing substrates from binding properly to it. In short, the structures of enzymes and other molecules enable them to function best within the normal range of body temperature of the organism.

Acidity

Acidity refers to the concentration of hydrogen ions (H^+), whether in the environment or in the organism. Acidity is commonly measured on a scale of **pH,** which is the negative common logarithm of the concentration of hydrogen ions. In pure water, a small fraction of water molecules (H_2O) are dissociated at any given time into their hydrogen and hydroxide (OH^-) ions. The pH of pure water, which is defined as neutral pH, is 7, which means that the concentration of hydrogen ions is 10^{-7} (0.0000001) moles

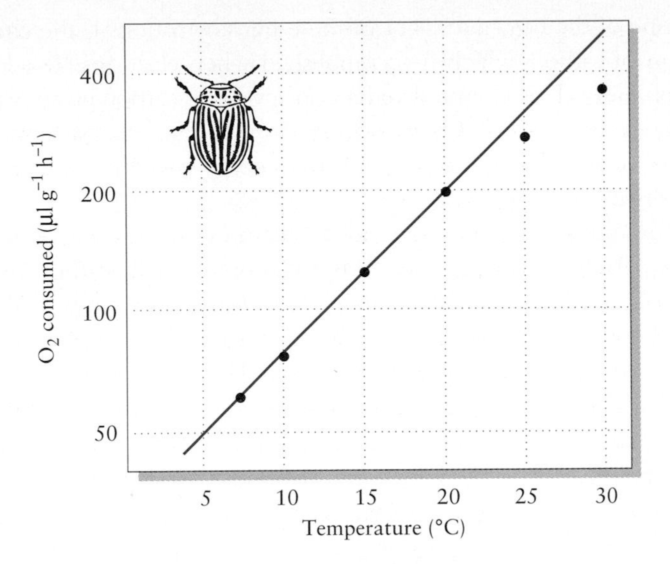

Figure 2.5 The rate of oxygen consumption by the Colorado potato beetle as a function of temperature. At lower temperatures the rise in oxygen consumption is exponential, increasing by a factor of about 2.5 for each 10°C. After K. Marzusch, *Zeits. vergl. Physiol.* 34:75–92 (1952).

per liter. For comparison, a liter of water contains almost 56 moles of water molecules. Strong acids, such as sulfuric acid (H_2SO_4), dissociate much more readily when dissolved in water than does water itself, and they produce high concentrations of hydrogen ions and a low pH. Most natural waters contain weak acids, such as carbonic acid (H_2CO_3) and various organic acids, and tend to have pH values closer to neutral. Many natural waters are somewhat alkaline (pH > 7, resulting from an excess of OH^- over H^+). The normal range of pH is between 6 and 9, although small ponds and streams in regions with acid rainfall, or which receive drainage from coal mining areas, can reach pH values as low as 4.

Hydrogen ions are extremely reactive, and in high concentrations, they affect the activities of most enzymes and have other, generally negative, consequences for life processes. The limiting acidity for photosynthetic cyanobacteria is about pH 4. Other kinds of bacteria, which are called acidophilic bacteria, tolerate acidity down to almost pH 0, but they maintain their internal pH in the range of 6 to 7. In addition to the direct detrimental effects of hydrogen ions on living systems, they also help to dissolve highly toxic heavy metals, such as arsenic, cadmium, and mercury, from rocks and soils, and they enhance the leaching of beneficial cationic nutrients such as calcium (Ca^{2+}).

Carbon, oxygen, and biological energy transformations

Living organisms consist of many elements joined together into organic molecules that make up the structure of the individual organism. Such organic compounds also contain energy needed to maintain the organism in the form of chemical bonds between atoms. These energy-containing bonds are made possible by chemical changes in the atoms that constitute carbohydrates, lipids, proteins, and other biological molecules. In biological systems, one of the most prevalent of these transformations is the chemical **reduction** of carbon, which is accomplished when electrons are added to the carbon atom. This chemical reduction gives the carbon atom a higher chemical energy potential. **Oxidation** of carbon strips electrons from carbon atoms, at the same time releasing the energy potential of carbon for other biochemical work.

Carbon dioxide (CO_2) is an oxidized form of carbon. During **photosynthesis,** plants chemically reduce the carbon atom in carbon dioxide. This altered atom can be used by the plant to form new compounds, such as the carbohydrate glucose ($C_6H_{12}O_6$), that have elevated energy levels. The added energy thus stored comes from light. To release this stored energy for other purposes, both plants and animals undo the results of photosynthesis by oxidizing carbon back to carbon dioxide; this process is known as **respiration.** The oxidation of carbon during respiration releases energy, a portion of which organisms harness for their own purposes; the rest escapes as heat.

Photosynthesis and respiration involve the complementary reduction and oxidation of carbon and oxygen. Oxygen's common oxidized state is molecular oxygen (O_2), which occurs as a gas in the atmosphere and dissolved in water. In a reduced state, oxygen readily forms water molecules (H_2O). Thus, as carbon is reduced during photosynthesis, oxygen is oxidized from its form in water to its molecular form. During respiration, inhaled or absorbed oxygen is reduced to the form contained in the compound we call water, and carbon is oxidized to the form it exhibits in carbon dioxide. Why does the coupling of an oxidation reaction to a reduction reaction result in a net release of energy that the organism can use to perform other work? The reduction of oxygen is thermodynamically more favorable (it requires less energy input) than the reduction of carbon. Therefore, the oxidation of carbon releases more energy than the reduction of oxygen requires.

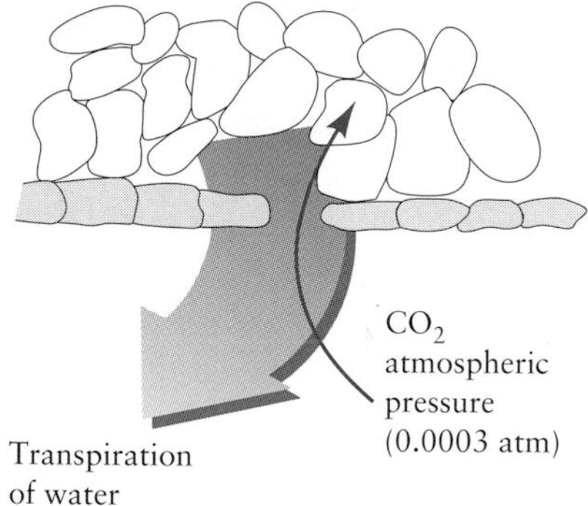

Figure 2.6 Schematic cross section of the lower portion of a leaf, showing the different pressures that carbon dioxide and water exert in the tissue of a leaf and the surrounding air. The shaded cells represent the lower epidermis of the leaf, which is relatively impermeable to water. Gas exchange occurs primarily through pores (stomates) on the undersurface of the leaf. Because the plant uses carbon dioxide in photosynthesis, the concentration of that gas remains at a low level in the leaf. The vapor pressure of water in the atmosphere may be as low as 0 atm (dry air) and at the surface of leaf tissues as high as 0.1 atm (at 45°C).

Plants assimilate more carbon in photosynthesis than they oxidize in respiration (otherwise they would not grow), so they require an external source of carbon. The only practical source of inorganic carbon, carbon dioxide, has an extremely low concentration in the atmosphere (about 0.03%). Therefore, the concentration of CO_2 in the atmosphere, which drives the movement of CO_2 into plant cells, is much, much less than the gradient of concentration that drives water vapor from the plant into the surrounding atmosphere. This makes water conservation a problem for terrestrial plants, especially in arid environments, and it accounts for the fact that plants evaporate 500 g of water from their leaves, more or less, for every gram of carbon assimilated (Figure 2.6).

Carbon poses different problems of availability for aquatic plants. The solubility of carbon dioxide in fresh water is about 0.0003 cubic centimeters per cubic centimeter of water, or 0.03% by volume—about the same as its concentration in the atmosphere. When carbon dioxide dissolves in water, however, most of the molecules form carbonic acid (H_2CO_3), which provides a reservoir of inorganic carbon. Depending on the acidity of the water, carbonic acid molecules dissociate into bicarbonate ions (HCO_3^-) and carbonate ions (CO_3^{2-}). Within the range of acidity of most natural waters (pH values between 6 and 9), bicarbonate is the more common form, and it dissolves readily in water (69 g of sodium bicarbonate [$NaHCO_3$] per liter of water, for example). As a result, seawater normally contains concentrations of bicarbonate ions equivalent to 0.03–0.06 cm^3 of carbon dioxide gas per cubic centimeter of water (3–6%), more than 100 times the concentration of carbon dioxide in air (Figure 2.7).

Unfortunately, the rate of diffusion of carbon dioxide through unstirred water is about 10,000 times less than it is in air, and the bicarbonate ion diffuses even more slowly. Every surface of an aquatic plant, alga, or microbe has a **boundary layer** of unstirred water, which may range from as little as 10 micrometers (μm) for single-celled algae in turbulent waters to 500 μm for a large aquatic plant in stagnant water (Figure 2.8). Thus, in spite of the high concentration of bicarbonate ions in the water surrounding these organisms, photosynthesis may nonetheless be limited by a diffusion barrier of still water at the surface of the plant or photosynthetic microbe. Once inside the cell, bicarbonate ions can be used directly as a source of carbon for photosynthesis, although at only 10–40% of the efficiency of utilizing carbon dioxide. As carbon dioxide itself is depleted in the cell environment, the bicarbonate ion also produces carbon dioxide at the site of photosynthesis. Bicarbonate ions and carbon dioxide exist in a chemical equilibrium, which represents the balance achieved between H^+ and HCO_3^-, on one hand, and CO_2 and H_2O on the other. Symbolically, we can represent this equilibrium as $H^+ + HCO_3^- \rightleftharpoons CO_2 + H_2O$. As CO_2 is used in photosynthesis, some of the bicarbonate ions reassociate with hydrogen ions to replenish the carbon dioxide (Figure 2.9).

Oxygen often limits animals in aquatic habitats because of its low solubility, compounded by the vastly lower rate of diffusion of oxygen in water than in air. Compared with its concentration of 0.21 cm^3 cm^{-3} in the atmosphere, the solubility of oxygen reaches a maximum (at 0°C in fresh water) of 0.01 cm^3 cm^{-3}—only 14 parts per million by weight. Furthermore, below the limit of light penetration in deep bodies of water and in waterlogged sediments and soils, no oxygen can be produced by photosynthesis. Therefore, as animals and microbes use oxygen

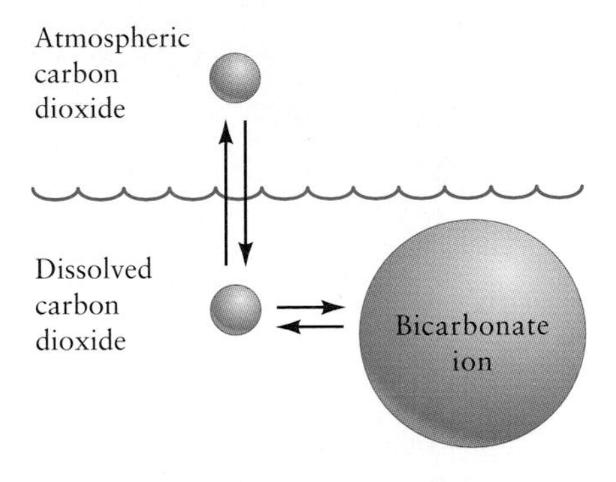

Figure 2.7 Schematic diagram of the relative amounts of carbon dioxide that are available in the atmosphere, that are dissolved in water, and that occur in the form of bicarbonate ions in water at equilibrium. The fact that bicarbonate forms when carbon dioxide dissolves in water, and itself readily dissolves in water, greatly increases the reservoir of carbon dioxide available to aquatic plants and algae.

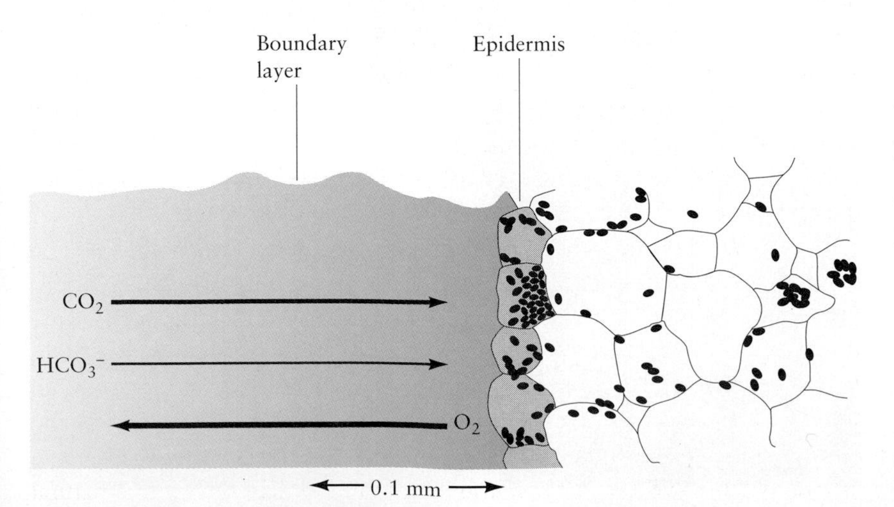

Figure 2.8 Diagrammatic cross section of a leaf of the aquatic plant *Vallisneria spiralis,* showing the thickness of the boundary layer in still water relative to the sizes of cells. The arrows are proportional to the rates of diffusion of carbon dioxide (CO_2), bicarbonate ions (HCO_3^-), and oxygen (O_2) through the boundary layer. Note that the diffusion constants are inversely proportional to the size of the molecule: the larger bicarbonate ions diffuse more slowly through the boundary layer. After H. B. A. Prins and J. T. M. Elzenga, *Aquatic Botany* 34:59–83 (1989).

to metabolize organic materials, such habitats may become severely depleted of dissolved oxygen. Habitats such as deeper layers of water in lakes and mucky sediments of marshes that are devoid of oxygen are referred to as **anaerobic** or **anoxic** habitats. Such conditions in waterlogged soils of swamps pose problems for terrestrial plants, whose roots need oxygen for respiration. In these habitats, many plants have specialized vascular tissues, called **aerenchyma,** that conduct air directly from the atmosphere to the roots. The roots of cypress trees and many mangroves grow vertical extensions that project above the anoxic soil and conduct oxygen directly from the atmosphere to the roots (Figure 2.10).

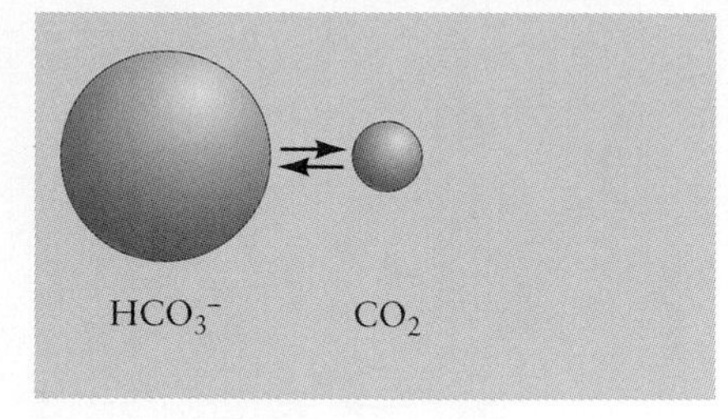

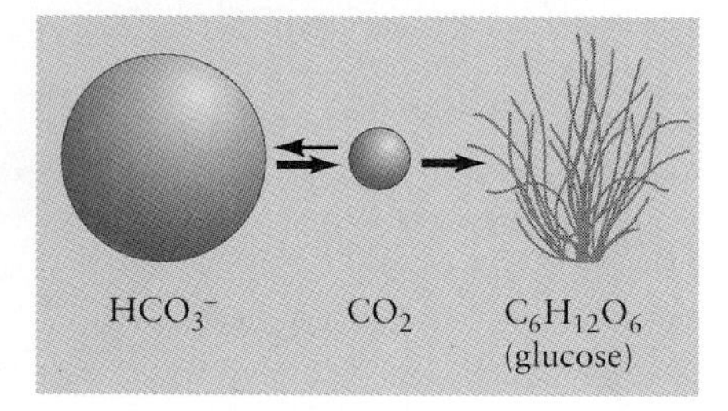

Figure 2.9 When aquatic plants and algae deplete carbon dioxide in their immediate vicinity, it is replenished from the pool of bicarbonate ions. (a) The equilibrium between bicarbonate and dissolved carbon dioxide in water. (b) Plants and algae remove carbon dioxide from the water during photosynthesis, and the reduced carbon dioxide concentration causes bicarbonate to release additional CO_2 into solution ($H^+ + HCO_3^- \rightarrow H_2O + CO_2$).

Inorganic nutrients

Organisms assimilate a wide variety of chemical elements. After hydrogen, carbon, and oxygen, the elements required in greatest quantity are nitrogen, phosphorus, sulfur, potassium, calcium, magnesium, and iron (Table 2.1). Certain groups of organisms need other elements in abundance. For example, diatoms construct their glassy shells of silicates; tunicates accumulate vanadium in high concentrations; and nitrogen-fixing bacteria require molybdenum as a part of the key enzyme in the process of nitrogen assimilation. Plants need elements such as boron and selenium in minute quantities, although the physiological functions of many of these micronutrients are not well understood.

Plants acquire mineral nutrients—other than oxygen, carbon, and some nitrogen—in soluble forms from the soil water around their roots. They obtain nitrogen in the form of ammonia ions (NH_4^+) or nitrate ions

Figure 2.10 The knees of these bald cypress trees in a fresh-water swamp in South Carolina conduct air from the atmosphere to roots growing in waterlogged, anaerobic sediments. Other marsh and swamp plants have air-conducting tissues (aerenchyma) in their stems. Courtesy of the U.S. Forest Service.

TABLE 2.1 Major nutrients required by organisms, and some of their primary functions

ELEMENT	FUNCTION
Nitrogen (N)	Structural component of proteins and nucleic acids
Phosphorus (P)	Structural component of nucleic acids, phospholipids, and bone
Sulfur (S)	Structural component of many proteins
Potassium (K)	Major solute in animal cells
Calcium (Ca)	Structural component of bone and of material between woody plant cells; regulator of cell permeability
Magnesium (Mg)	Structural component of chlorophyll; involved in the function of many enzymes
Iron (Fe)	Structural component of hemoglobin and many enzymes
Sodium (Na)	Major solute in extracellular fluids of animals

(NO_3^-), phosphorus in the form of phosphate ions (PO_4^{3-}), calcium and potassium as the elemental ions Ca^{2+} and K^+, and so on. The availability of these elements varies with their chemical form in the soil and with temperature, acidity, and the presence of other ions. Phosphorus, in particular, often limits plant production; even when it is abundant, most of the compounds it forms in the soil do not dissolve easily.

All natural waters contain some dissolved substances. Although nearly pure, rainwater acquires some minerals from dust particles and droplets of ocean spray in the atmosphere. Most lakes and rivers contain 0.01–0.02% dissolved minerals, which is roughly one-fortieth to one-twentieth of the average salt concentration of the oceans (3.4% by weight), in which salts and other minerals have accumulated over the millennia.

The minerals dissolved in fresh water and salt water differ in composition as well as in quantity. Seawater abounds in sodium and chlorine and has significant amounts of magnesium and sulfate. Fresh water contains a greater variety of ions, but calcium usually makes up most of the positively charged ions, or **cations,** and carbonate and sulfate most of the negatively charged ions, or **anions.** The composition of fresh water differs from that of salt water because of the different rates of solution and solubilities of different substances. Rarely do the concentrations of compounds in fresh water approach their maximum solubilities; the concentrations of minerals in fresh water usually reflect the composition and rates of solution of materials in the rock and soil that the water flows through. Limestone consists primarily of calcium carbonate, which dissolves readily in the presence of the ubiquitous hydrogen ion ($H^+ + CaCO_3 \rightarrow Ca^{2+} + HCO_3^-$). As carbonates are dissolved, hydrogen ions are bound up in bicarbonate ions, and the pH of the water increases. Thus, water in limestone areas contains abundant calcium ions, which make it "hard" and somewhat alkaline (pH greater than 7; Table 2.2). Granite is composed of minerals, including quartz and feldspar, that do not contain calcium and dissolve slowly; water flowing through granitic areas contains few dissolved substances and accordingly is "soft" and slightly acid. Nitrogen (mostly nitrate and dissolved organic nitrogen compounds) enters bodies of fresh water in relative abundance in the runoff from surrounding terrestrial systems. Typical values for fresh water are 0.015 mg of nitrogen per liter (mg l^{-1}) in ammonium (NH_4^+), 0.10 mg l^{-1} in nitrate (NO_3^-), and 0.26 mg l^{-1} in dissolved organic matter. In contrast, most of the phosphorus in streams, ponds, and lakes readily complexes with iron and precipitates out of the system, typically leaving about 0.01 mg l^{-1} in solution as phosphate (PO_4^{3-}). As a result, phosphorus, rather than nitrogen, usually limits plant production in freshwater systems.

The oceans function like large stills, concentrating minerals as nutrient-laden water arrives via streams and rivers and as pure water evaporates from the surface. Here the concentrations of some elements, particularly calcium, reach limits set by the maximum solubilities of the compounds they form. Because of the high concentration of sodium ions (Na^+) and the low concentration of hydrogen ions in the oceans, calcium readily forms calcium carbonate, which dissolves only to the extent of 0.014 g per liter of water. Its concentration in the oceans reached this level eons ago, and excess calcium ions entering the oceans each year from streams and rivers precipitate to form limestone sediments. At the other extreme, the solubilities of sodium compounds, such as sodium chloride (360 g l^{-1}) and sodium bicarbonate (69 g l^{-1}), far exceed the concentration of sodium in seawater (10 g l^{-1}); most of the sodium chloride washing into ocean basins remains dissolved.

TABLE 2.2 Average chemical composition of natural rivers and streams flowing through areas of granitic and carbonate bedrock

	UNDERLYING ROCK	
COMPONENT (mg l^-)	*Granite*	*Limestone*
Acidity (pH)	6.6	7.9
Silicon dioxide (SiO_2)	9.0	6.0
Calcium (Ca^{2+})	1.6	102.6
Magnesium (Mg^{2+})	0.8	15.6
Sodium (Na^+)	2.0	0.8
Potassium (K^+)	0.3	0.5
Chloride (Cl^-)	0.0	0.0
Sulfate (SO_4^{2-})	3.0	8.2
Bicarbonate (HCO_3^-)	7.8	194.9

Osmotic potential

Left to their own devices, ions diffuse across the surfaces of organisms and the membranes surrounding cells from the side where they are more abundant to the side where they are less abundant. This movement tends to equalize the concentrations of ions between the organism and its surroundings. Water also moves across membranes (the process is called **osmosis**) toward regions of higher ion concentrations (that is, lower water concentrations), tending to equalize the concentrations of dissolved substances on both sides of the membrane. This tendency of a solution to attract water is known as its **osmotic potential.** The osmotic potential of a solution usually is expressed as the pressure required to keep water from diffusing into a solution contained within a semipermeable membrane (Figure 2.11). It is the osmotic potential in the roots of trees that causes water to enter the roots from the soil against the attraction of soil particles (the water potential of the soil is also expressed as a pressure).

If the solute responsible for the osmotic potential of a solution also can diffuse across cell membranes, then its concentration within cells and its concentration in the surrounding water will eventually come into equilibrium. At this point, the osmotic potentials of the cell and its surroundings will be the same, and there will be no net movement of water across the cell membrane. This equalization of osmotic potential can be prevented by two mechanisms. First, a membrane can be **semipermeable,** by which it is meant that some small molecules and ions can diffuse across it, but others cannot. When the osmotic potential of a cell is generated by molecules and ions too large to cross a membrane (some carbohydrates and proteins, for example), the osmotic pressure of the cell is maintained. Membranes may also **transport** ions and small molecules actively against the diffusion gradient to maintain their concentrations within the cell.

A molar concentration of a substance in solution (1 mole of a substance per liter) exerts an osmotic pressure of 21 atmospheres. The osmotic pressure of seawater is about 12 atm, and that of fresh water is practically zero. The body fluids of vertebrate animals, which have an osmotic pressure 30–40% that of seawater (3–5 atm), occupy an intermediate position. The tissues of fresh-water fish have higher concentrations of salts than the surrounding water. Such organisms, which are referred to as **hyperosmotic,** tend to gain water from and lose solutes to their surroundings. Marine fish, which have lower concentrations of salts than the surrounding seawater, are referred to as **hypo-osmotic.** They tend to gain solutes and lose water (Figure 2.12). Fish solve these osmotic problems by using active transport mechanisms to pump ions in one direction or the other across various body surfaces (skin, kidney tubules, and gills), expending considerable energy in the process.

Certain environments pose special osmotic problems. Aquatic environments with salt concentrations greater than that of seawater occur in some landlocked basins, particularly in dry regions where evaporation considerably exceeds precipitation. The Great Salt Lake (20% salt) in Utah and the Dead Sea (23% salt), lying between Israel and Jordan, are well-known examples of such **hypersaline** environments. The osmotic potentials of

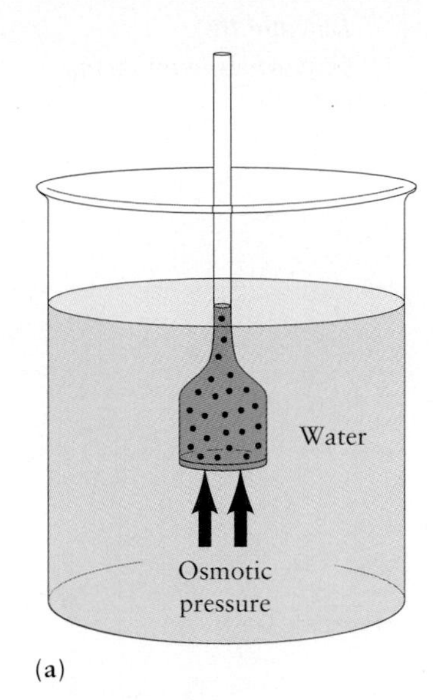

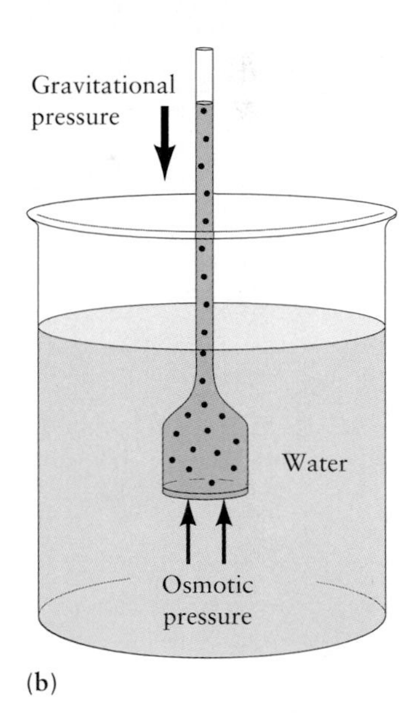

Figure 2.11 Schematic diagram of the osmotic potential developed by solutes enclosed within a membrane permeable to water, but not to the solutes (a semipermeable membrane). (a) Because the solutes are at high concentration, water tends to move across the membrane into the inverted funnel. (b) Within the funnel, the increasing volume pushes fluid up the stem. Eventually, the osmotic pressure of the fluid, which decreases as the solutes become more diluted, is balanced by the gravitational pressure exerted by the fluid in the stem.

these bodies of water—well over 100 atm—would suck the water from most organisms. However, a few aquatic creatures, such as brine shrimp *(Artemia),* can survive in salt water concentrated to the point of crystallization (300 g per liter, or 30%). Brine shrimp excrete salt at a prodigious rate to keep the salts in their body fluids less concentrated than those in their surroundings.

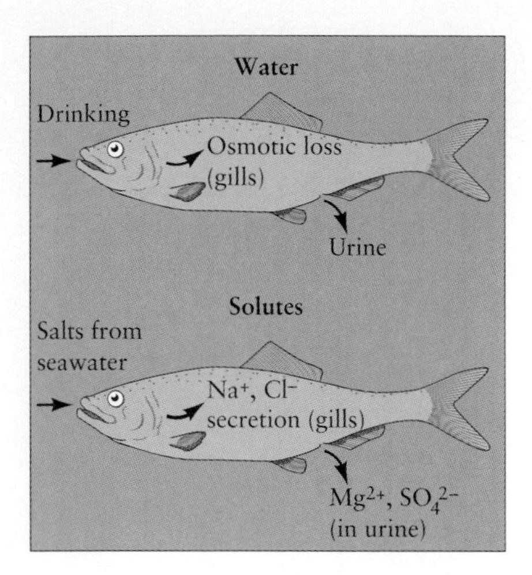

(a) Marine

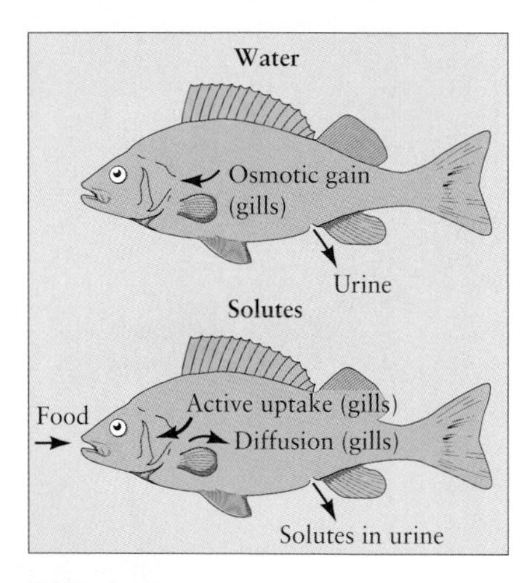

(b) Freshwater

Figure 2.12 Pathways of exchange of water and solutes by (a) marine fish, whose body fluids are hypo-osmotic (contain less concentrated salt than the surrounding water), and (b) fresh-water fish, whose body fluids are hyperosmotic. The gills and kidneys actively exclude or retain solutes to maintain salt balance. Marine fish must drink to acquire water. After K. Schmidt-Nielson, *Animal Physiology: Adaptation and Environment,* Cambridge University Press, Cambridge (1975).

Light

Light is the primary source of energy for the biosphere. Green plants, algae, and some bacteria absorb light and assimilate its energy by photosynthesis, but not all light striking the earth's surface can be used in this way. Rainbows and prisms show that light consists of a spectrum of wavelengths that we perceive as different colors. Wavelengths of light are generally expressed in micrometers (μm: one-millionth of a meter, 10^{-6} m), or nanometers (nm: one-billionth of a meter, 10^{-9} m). The visible portion of the spectrum, which corresponds to the wavelengths of light suitable for photosynthesis, ranges between about 400 nm (violet) and 700 nm (red). Light of wavelengths shorter than 400 nm makes up the **ultraviolet** part of the spectrum; we refer to light of wavelengths longer than 700 nm as **infrared.** The energy content of light varies with wavelength and hence with color; shorter-wavelength blue light has a higher energy level than longer-wavelength red light.

The light from the sun that reaches the upper part of the earth's atmosphere extends far beyond the visible range, through the ultraviolet region toward the short-wavelength, high-energy X rays at one end of the spectrum, and through the infrared region to extremely long-wavelength, low-energy radiation such as radio waves at the other end. Because of its high energy level, ultraviolet light can damage exposed cells and tissues. Fortunately, the atmosphere is completely transparent only to the visible range of the spectrum. As light passes through the atmosphere, most of its ultraviolet components are absorbed, primarily by a molecular form of oxygen known as ozone (O_3) that occurs in the upper atmosphere. The atmosphere thus shields life at the earth's surface from the most damaging wavelengths of light (Figure 2.13). Certain pollutants in the atmosphere, particularly the chlorofluorocarbons (CFCs) formerly used as refrigerants and as propellants in aerosol cans, chemically destroy ozone in the upper atmosphere. This degradation has progressed so far over some parts of the earth that ultraviolet radiation has increased to dangerous levels.

The atmosphere is also relatively opaque to the infrared portion of the spectrum. This means that much of the infrared portion of sunlight is absorbed by the atmosphere, which contributes to the warming of air. A more important effect of the infrared opacity of the atmosphere is the absorption of radiation from the surface of the earth. Most of the energy in the visible portion of the solar spectrum that reaches the earth's surface is absorbed by vegetation, soil, and surface waters and converted to heat energy. This heat is then reradiated from the warmed surface of the earth back toward space. Because the temperature of the earth's surface is much

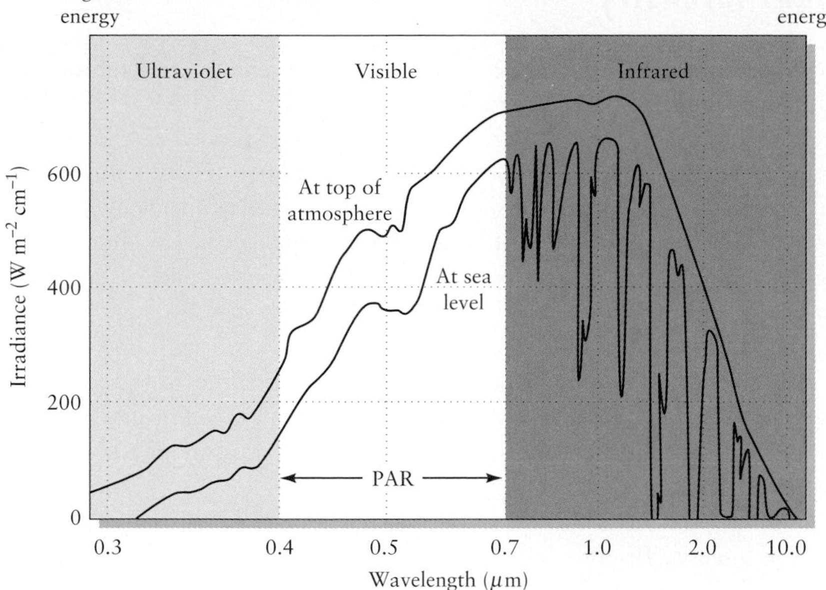

Figure 2.13 Spectral distribution of direct sunlight at the top of the atmosphere and at sea level. PAR indicates the photosynthetically active region of the spectrum. Ozone in the upper atmosphere absorbs light in the ultraviolet region of the spectrum, and water vapor and carbon dioxide absorb light in the infrared region. After D. M. Gates, *Biophysical Ecology,* Springer-Verlag, New York (1980).

lower than that of the sun, most of the heat energy is reradiated as low-intensity infrared radiation. Much of this radiation is absorbed by the atmosphere, which thereby acts as a blanket covering the earth and keeping its surface warm. Because this warming effect resembles the manner in which glass keeps a greenhouse warm, it is called the **greenhouse effect.** Eventually, this absorbed energy reaches the upper levels of the atmosphere and is lost to space, but at a much slower rate than would occur in the absence of water vapor, carbon dioxide, and other infrared-opaque components of air—the so-called greenhouse gases.

Vision and the photochemical conversion of light energy to chemical energy by photosynthetic organisms occur primarily within that portion of the solar spectrum at the earth's surface that contains the greatest amount of energy. The absorption of radiant energy depends on the nature of the absorbing substance. Water only weakly absorbs light in the visible region of the spectrum; as a result, a glass of water appears colorless. Dyes and pigments strongly absorb some wavelengths in the visible region, reflecting or transmitting light of a definite color that becomes an identifying characteristic. Leaves contain several kinds of pigments, particularly **chlorophylls** (green) and **carotenoids** (yellow), that absorb light and harness its energy (Figure 2.14). Carotenoids, which give carrots their orange color, absorb primarily blue and green and reflect light in the yellow and orange regions of the spectrum. Chlorophyll absorbs red and violet light while reflecting green and blue.

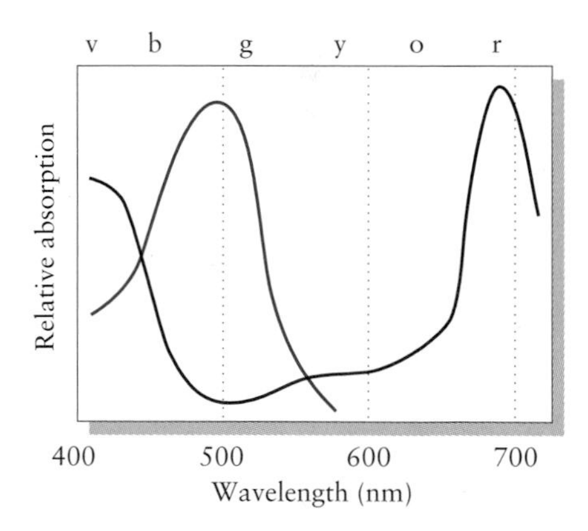

Figure 2.14 Absorption of light of different wavelengths by two groups of pigments—chlorophylls (black line) and carotenoids (colored line)—that capture light energy used in photosynthesis. The colors of the spectrum are violet (v), blue (b), green (g), yellow (y), orange (o), and red (r). After R. Emerson and C. M. Lewis, *J. Gen. Physiol.* 25:579–595 (1942).

Light intensity

Ecologists measure the intensity of light as the energy content of the **photosynthetically active region** of the spectrum **(PAR)**—between wavelengths of 400 and 700 nm. Intensity may be expressed in a variety of units, including the langley (ly), watt (W), and einstein (E), all named after major figures in physics. We'll stick with the watt, the familiar unit used to rate the power consumption of light bulbs and appliances. The intensity of light reaching the outer limit of the atmosphere—the so-called **solar constant**—is approximately 1,400 W m^{-2}. In reality, far less light energy reaches any area on the surface of the earth because of nighttime periods without light, the low incidence of light early and late in the day and at high latitudes, and cloud cover. A temperate habitat on a clear day in summer might typically receive 350 W m^{-2} of PAR (that is, about a quarter of the solar constant).

At low light levels, the rate of photosynthesis varies in direct proportion to light intensity. Brighter light saturates the photosynthetic pigments, however, and the rate of photosynthesis increases more slowly or levels off as intensity increases above a tenth or so of that of full sunlight. The response of photosynthesis to light intensity has two reference points (Figure 2.15). The first, called the **compensation point,** is the level of light intensity at which photosynthetic assimilation of energy just balances respiration. Above the compensation point, the energy balance of the plant is positive; below the compensation point, the energy balance is negative. The second reference point is the **saturation point,** above which the rate of photosynthesis no longer responds to increasing light intensity. Among terrestrial plants, the compensation points of species that normally grow in full sunlight (a maximum of about 500 W m^{-2}) occur between 1 and 2 W m^{-2}. The saturation points of such species usually are reached between 30 and 40 W m^{-2}—less than a tenth of the energy level of bright, direct sunlight. As one might expect, the compensation and saturation points of plants that typically grow in shade occur at lower light intensities.

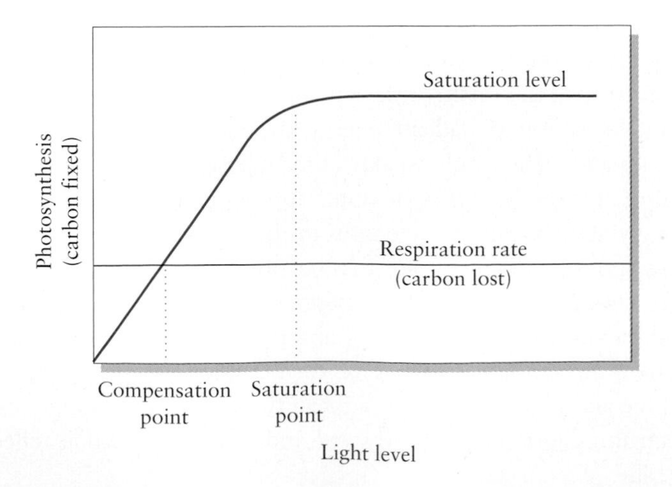

Figure 2.15 Relationship between carbon dioxide assimilation (photosynthesis, colored line) and light intensity, illustrating the compensation point, at which photosynthesis balances respiration (black line) and the saturation point of the photosynthetic process. After M. G. Barbour, J. H. Burk, and W. D. Pitts, *Terrestrial Plant Ecology,* Benjamin/Cummings, Menlo Park, Calif. (1980).

Water absorbs or scatters enough light to limit the depth of the sunlit zone of the sea. The transparency of a glass of water is deceptive. In pure seawater, the energy content of light in the visible part of the spectrum diminishes to 50% of the surface value within 10 m depth, and it drops to less than 7% within 100 m depth. Furthermore, water absorbs longer wavelengths more strongly than shorter ones; virtually all infrared radiation disappears within the topmost meter of water. Short wavelengths (violet and blue) tend to scatter when they strike water molecules, so they too fail to penetrate deeply. As a consequence of the absorption and scattering of light by water, green light predominates with increasing depth. The photosynthetic pigments of aquatic algae parallel this spectral shift. Algae that grow near the surface of the oceans, such as the green alga *Ulva* (sea lettuce), have pigments resembling those of terrestrial plants and best absorb blue and red light. The deep-water red alga *Porphyra* has additional pigments that enable it to use green light more effectively in photosynthesis (Figure 2.16).

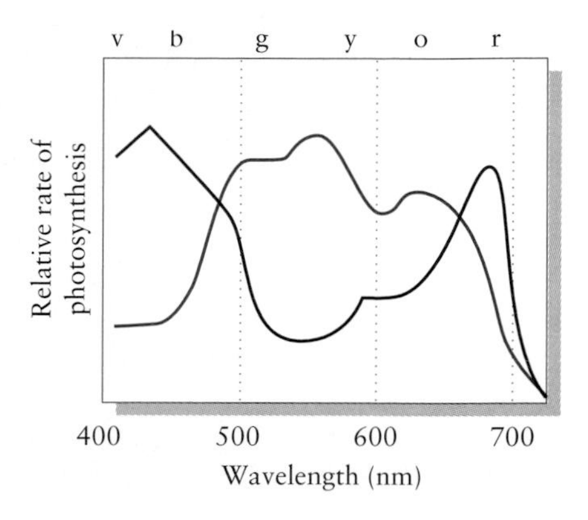

Figure 2.16 Relative rates of photosynthesis by the green alga *Ulva* (black line) and the red alga *Porphyra* (colored line) as a function of the color of light. After F. T. Haxo and L. R. Blinks, *J. Gen. Physiol.* 33:389–422 (1950).

The absorption of light by water limits the depth at which aquatic photosynthetic organisms can exist to a fairly narrow zone close to the surface. This area is called the **euphotic zone.** The lower limit of the euphotic zone, where photosynthesis just balances respiration, is the compensation point. This point may be defined in terms of either depth or light level. In some exceptionally clear ocean and lake waters, the compensation point may lie 100 meters below the surface, but this is a rare condition. In productive waters with dense phytoplankton or in water turbid with suspended silt particles, the euphotic zone may be as shallow as 1 meter.

The thermal environment

Much of the solar radiation absorbed by soil, plants, and animals is converted to heat. A particular spot on the surface of the earth warms each day and cools each night. As the days lengthen and the sun rises higher in the sky toward summer, the environment becomes warmer; more heat is added each day than is lost. The energy absorbed from the sun's and the earth's infrared radiation warms the atmosphere and drives the winds. Light absorbed by water is a source of heat for evaporation. Each object and each organism continually exchanges heat with its surroundings. When the temperature of the environment exceeds that of an organism, the organism gains heat and becomes warmer. When the environment is cooler, the organism loses heat and cools. An individual's heat budget includes several avenues of heat gain and heat loss (Figure 2.17).

Radiation is the absorption or emission of electromagnetic energy. Sources of radiation in the environment include the sun, the sky (scattered light), and the landscape. At night, objects that have warmed up in the sunlight radiate their stored heat to colder parts of the environment and, eventually, to space. Although we cannot see this infrared radiation, the bodies of organisms, especially warm-blooded birds and mammals, often are the "brightest" objects in the night (Figure 2.18). Because we are so much hotter than the black void of space, we radiate tremendous quantities of energy

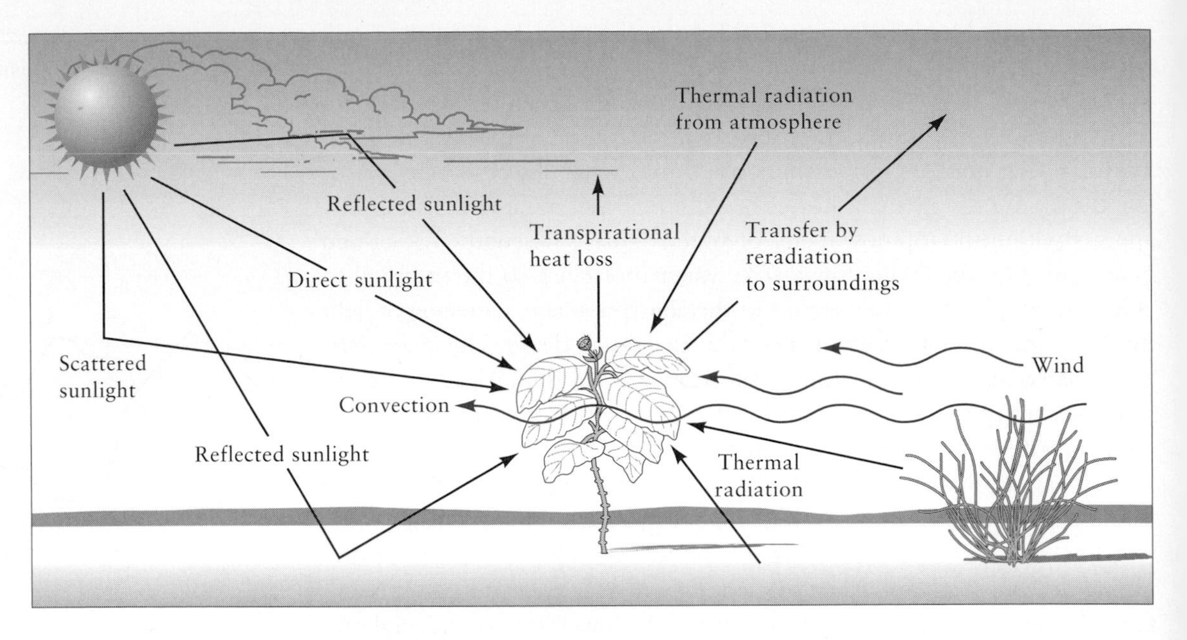

Figure 2.17 Pathways of heat exchange between a plant and its environment. After D. M. Gates, *Biophysical Ecology,* Springer-Verlag, New York (1980).

to the clear night sky. We can also receive radiation from atmospheric water vapor and from vegetation, which balances some of our nighttime radiation loss. That is why, at a particular temperature, one feels warmer at night in a humid environment, particularly if there is cloud cover.

Conduction is the transfer of the kinetic energy of heat between substances in contact. Thus, a vacuum conducts no heat. Water, because of its greater density, conducts heat more than 20 times faster than air. The rate of conductance between two objects, or between the inside and the outside of an organism, depends on the insulation of the surface (its resistance to heat transfer), the surface area, and the temperature gradient. An organism can either gain or lose heat by conduction, depending on its temperature relative to that of the environment.

Convection is the movement of liquids and gases of different temperatures, particularly over surfaces across which heat is transferred by conduction. Air conducts heat poorly. In still air, a boundary layer of air forms over a surface. A warm body tends to warm this boundary layer to its own temperature, effectively insulating itself against heat loss. A current of air flowing past a surface tends to disrupt the boundary layer and to increase the rate of heat exchange by conduction. This convection of heat away from the body surface is the basis of the "wind chill factor" we hear about on the evening weather report. On a cold day, air movement makes it feel the way it would on an even colder windless day. For example, a wind blowing 32 km per hour at an air temperature of –7°C has the cooling power of still air at –23°C.

Evaporation of water requires heat. The evaporation of 1 g of water from the body surface removes 2.43 kilojoules (kJ) of heat at 30°C. As plants and animals exchange gases with the environment, some water evaporates from their exposed surfaces. In plants, the evaporation of water from the surface of a leaf is referred to as **transpiration.** The rate of evaporative or transpirational heat loss depends on the permeability of the surface to water, the relative temperatures of the surface and the air, and the **vapor pressure** of the atmosphere. Vapor pressure is a measure of the capacity of the atmosphere to hold water. When vapor pressure is expressed in atmospheres, it represents the fractional weight of water vapor in saturated air. Thus, at 30°C, the vapor pressure of water is 0.042 atm, meaning that the air can hold 4.2% water by weight. When the temperature of air saturated with water drops from 30° to 20°C, its capacity to hold water decreases from 4.2 to 2.3%, and the difference—almost 2%—condenses to form clouds or precipitation.

Like heat, moisture can be trapped in the boundary layer of air that forms above surfaces. Convection tends to disrupt boundary layers and therefore increases evaporative heat loss as well as conductive heat loss. Because warm air holds more water than cold air, it has greater potential for evaporating water. In hot climates, water evaporating from the skin and respiratory surfaces cools many animals. For warm-blooded animals in cold climates, evaporation can become an unavoidable problem as cold inhaled air containing little water warms in contact with the body and respiratory surfaces, thereby speeding evaporation. We see evidence of such water loss on winter days when water evaporated from the warm surfaces of the lungs condenses as our breath mingles with the cold atmosphere.

Figure 2.18 Thermal images of Canada geese in an open meadow on a cool morning. It is clear that the geese lose more heat across their necks and legs than from their well-insulated bodies. Courtesy of R. Boonstra, from R. Boonstra, J. M. Eadie, C. J. Krebs, and S. Boutin, *J. Field Ornithol.* 66:192–198 (1995).

The buoyancy and viscosity of water and air

Because water is dense (800 times denser than air), it provides considerable support for organisms, which, after all, are themselves mostly water. But animals and plants also contain bone, proteins, dissolved salts, and other materials denser than salt water and much denser than fresh water. These materials would cause organisms to sink were it not for a variety of mechanisms that reduce their density or retard their rate of sinking. Many fish have a swim bladder, a small gas-filled structure whose size can be adjusted to make the density of the body equal to that of the surrounding water. Some large kelps, a type of seaweed found in shallow waters, have gas-filled bulbs that float their leaves to the sunlit surface waters (see Figure 1.12).

Most fats and oils have densities between 0.90 and 0.93 g cm^{-3} (90–93% of the density of pure water). Many of the microscopic, unicellular algae that float in great numbers in the surface waters of lakes and oceans (phytoplankton) contain droplets of oil that compensate for the natural tendency of cells to sink. The buoyancy of fish and other large marine organisms is also enhanced by accumulated lipids. Reduced skeletons, reduced musculature, and perhaps even the reduced salt concentration of their body fluids further lighten the bodies of aquatic organisms. It has been argued that aquatic vertebrates maintain low osmotic concentrations in their blood and body fluids (about one-third to one-half that of seawater) because such low concentrations reduce density. Unlike bony fishes, sharks and rays lack a swim bladder; to compensate for this, they reduce the densities of their bodies by not depositing mineral salts in most of the bones of their skeletons. Calcium carbonate and calcium phosphate, the principal components of mineralized bone, have densities close to three times that of water. The density of the cartilage skeleton of sharks and rays is much less—close to that of water.

The high viscosity of water helps to buoy up some organisms that would otherwise sink more rapidly, but it hampers the movement of others. Tiny marine animals often have long, filamentous appendages that retard sinking (Figure 2.19), just as a parachute slows the fall of a body through air.

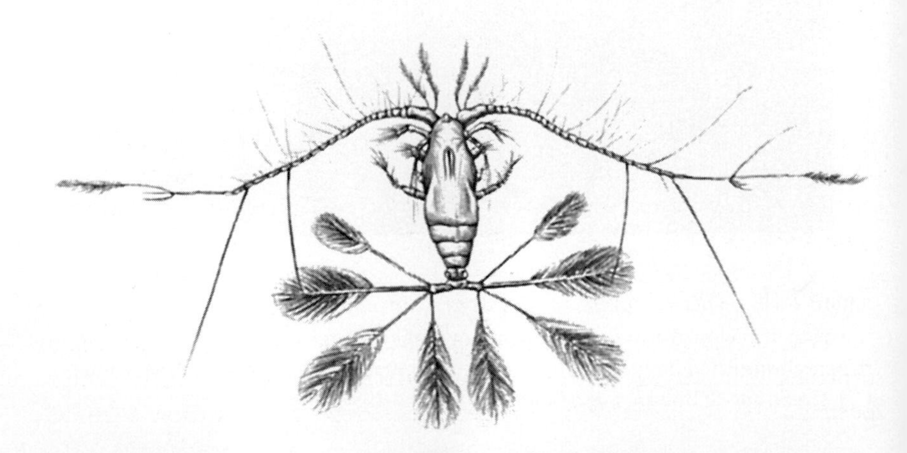

Figure 2.19 Filamentous and feathery projections from the body of the tropical marine planktonic crustacean *Calocalanus pavo.* Overall length is about 1.2 mm. After R. S. Wimpenny, *The Plankton of the Sea*, Faber and Faber, London (1966).

Figure 2.20 The streamlined shapes of young mackerel reduce the drag of water on the body and allow the fish to swim rapidly with minimum expenditure of energy. Courtesy of the U.S. Bureau of Commercial Fisheries.

The "wings" of maple seeds, the spider's silk thread, and the tufts on dandelion and milkweed seeds serve a similar function and increase the dispersal range of some terrestrial organisms. In contrast, fast-moving aquatic animals assume streamlined shapes to reduce the drag encountered in moving through a dense and viscous medium. Mackerel and other swift fishes of the open ocean closely approach the hydrodynamicist's body of ideal proportions (Figure 2.20). Of course, air offers far less resistance to movement, having less than one-fiftieth the viscosity of water. But the atmosphere offers little buoyancy. To provide lift against the pull of gravity, birds and other flying organisms expend prodigious amounts of energy.

Sensing the environment

To function in a complex and changing environment, organisms must be able to sense environmental change, detect and locate objects, and navigate the landscape. A predator must locate and recognize a suitable prey before it can capture it. Salmon must recognize the proper river for their spawning migration. Plants must sense the changing seasons to time their flowering properly. An organism's sensory modalities depend on the physical characteristics of information available in the environment, the way in which the organism relates to its environment (plants don't need the acute vision that some predators have, for example), and the evolutionary history of the species, which delimits the types of sensory organs that the organism can utilize.

That so many organisms rely on light to sense the environment is not surprising considering the high energy levels in the visible portion of the spectrum and the fact that light travels in a straight line, allowing accurate location and resolution of objects. We ourselves primarily use vision to locate food, particularly as it is now displayed on the shelves of supermarkets. Yet our vision is rather pathetic compared with that of hawks and falcons, and many insects can perceive ultraviolet light, which is invisible to us (Figure 2.21). Insects also can detect rapid movement, such as that of wings beating 300 times per second; we cannot distinguish individual movie frames flickering at 30 times per second. Thus, different organisms use the available information to different extents.

Many animals that are active during the night, when light levels are too low to be used effectively, rely on other sensory modalities—that is, they use other kinds of physical information about the environment. Among the more unusual sensory organs are the pit organs of pit vipers, a group of reptiles that includes the rattlesnakes. The pit organs, located on each side of the head in front of the eyes, detect the infrared (heat) radiation given off by the warm bodies of potential prey—a sort of "seeing in the dark" (Figure 2.22). Pit vipers are so sensitive to infrared radiation that they can detect a

Figure 2.21 The appearance of flowers to the human eye (left) and to eyes that are sensitive to ultraviolet light (right). Above: Marsh marigolds. Below: Five species of yellow-petaled Compositae from central Florida. Courtesy of T. Eisner. From T. Eisner, R. E. Silberglied, D. Aneshansley, J. E. Carrel, and H. C. Howland, *Science* 166:1172–1174 (1969).

small rodent several feet away in less than a second as the infrared radiation warms the tissues of pit organs. Moreover, because the pits are directionally sensitive, vipers can locate warm objects precisely enough to strike them.

What we perceive as sounds are pressure waves in air created by movements and impacts of objects, vibrating objects, or even turbulence in air flowing around objects. Pressure waves are propagated in all directions like ripples on the surface of water. This makes sounds easy to detect but more difficult to localize. The energy of pressure waves also decreases with distance, effectively limiting the range of detection. Nonetheless, sound may warn of an approaching predator, regardless of the direction from which it comes. Because the intensity of sound coming from a point source varies directionally, some nocturnal predators are able to use this information to locate prey accurately. Furthermore, direction may be sensed by the difference in the time of arrival of sound waves at paired hearing organs, such as our ears. Owls have such sensitive and directionally informative hearing that they can locate mice and other prey by the sounds they make as they move through the habitat.

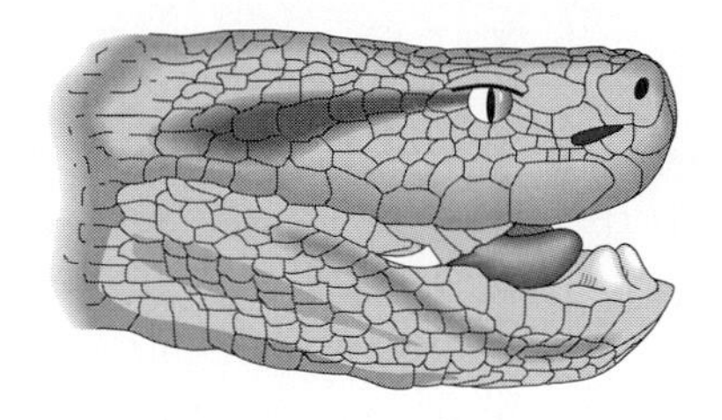

Figure 2.22 Head of the western rattlesnake *(Crotalis viridis),* showing the location of the infrared-sensitive pit between, and slightly lower than, the eye and the nostril. After D. Burkhardt, W. Schleidt, and H. Altner, *Signals in the Animal World* (trans. K. Morgan), McGraw-Hill, New York (1967).

Bats can use sound to find their way around the environment and locate prey in the absence of environmental sounds, because they produce the sound themselves by means of a biological sonar system. Bats emit very loud, high-pitched pulses of noise—generally beyond the range of our hearing—and sense the echoes that bounce back from objects in the environment, including such prey as moths in flight. The sound has to be produced in pulses so that the bat can listen for the fainter echoes during the quiet intervals between pulses. As a bat closes in on its prey, it emits pulses more frequently to increase the rate of incoming information. Both the projection of sound and hearing are highly directional, which increases the effective range of the sonar and its directional resolution.

Smell is the detection of molecules diffusing through air or water. This source of information has properties that differ considerably from those of electromagnetic waves (sight) and pressure waves in fluids (hearing). Smells are transported by air currents, so they are difficult to localize, and because they are persistent, the presence of a substance in the atmosphere can be detected long after its source has disappeared. Because volatile molecules are transported by wind currents, their sources sometimes can be localized by moving upcurrent. This property is the basis for a great deal of chemical communication, including the production of volatile mate attractants by many insects and the fragrances that many plants use to attract pollinators. Some predators use trails of volatile chemicals to follow potential prey. Snakes sense these chemicals by flicking their tongues against the substrate and transferring chemicals that adhere to the tongue to sensitive organs of smell (the vomeronasal organs) located in the roof of the mouth. The forked tongues of snakes and other reptiles allow them to simultaneously test for odors to the left and to the right to determine the correct direction of travel.

A few aquatic animals have developed the sensory ability to detect electrical fields. Some species of electric fish continuously discharge electricity from specialized muscle organs, creating a weak electrical field around them. Nearby objects distort the field, and these changes are picked

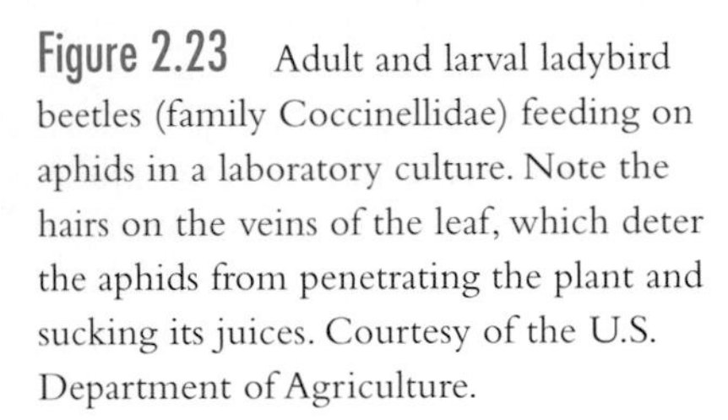

Figure 2.23 Adult and larval ladybird beetles (family Coccinellidae) feeding on aphids in a laboratory culture. Note the hairs on the veins of the leaf, which deter the aphids from penetrating the plant and sucking its juices. Courtesy of the U.S. Department of Agriculture.

up by receptors on the surface of the fish. Some species use electrical signals to communicate between individuals. The specialized electric ray *Torpedo* uses powerful electric currents (up to 50 volts at several amperes) to defend itself and to kill prey. As one might expect, the production and sensation of electrical fields are most highly developed in fishes that inhabit murky waters. In other habitats where visibility is poor, bottom-dwelling species such as catfish use elongated fins and barbels around the mouth as sensitive touch and taste receptors.

In contrast to the magnificent senses of many organisms, others perceive their surroundings only dimly and rely on chance to bump into things. The tactile sense has a very short range, of course, and if this tactic is to work for a predator, its prey must be equally oblivious. Touch can provide a tremendous amount of information not available through other senses because of the textural and structural richness of the environment. Consider, for example, the predatory larva of the ladybird beetle, which feeds on mites and aphids that infest the leaves of certain plants, and must physically contact its prey to recognize them (Figure 2.23). The movements of a beetle larva on a leaf are not oriented toward the prey, but neither are they random. The veins and rims of leaves make up a small percentage of the leaf surface, yet a larva will spend most of its search time on these structures, which it recognizes by touch; this is also where most of the aphids are.

We have barely touched on the ways in which organisms perceive and find their way around their environments. Sensory modalities are limited by the availability of particular kinds of information that can be interpreted to reveal structure and change in the environment. This information is a feature of the physical environment. The way in which an organism relates to its environment also depends on its size, as we shall see below.

Allometry and the consequences of body size

Size changes everything in ecology. In aquatic habitats, drag on the body depends on size and speed. Whereas a whale can coast on its momentum, a copepod stops dead in the water as soon as its power stroke ends; its momentum is insufficient to break through the viscosity of water! On land, the mechanical and supportive properties of appendages depend on their length and thickness; thus body size greatly affects locomotion. Such considerations influence the manner in which plants and animals adapt to the conditions of the environment.

Large organisms relate differently to their environments than do small ones because many physical and physiological processes vary out of proportion to size. The study of these relationships is referred to as **allometry.** In ecology, rates of processes and dimensions of objects are often related as follows:

$$Y = aX^b$$

where Y is being compared with X and a and b are constants pertaining to the relationship. Biologists refer to this equation as an **allometric relationship.** Y might be the heart rate of a mammal and X its body mass, or Y might be the average size of a leaf and X the diameter of the stem of a plant. The constant b is called the **allometric constant.** When b is 1, Y is directly proportional to X, and their relationship in this special case is referred to as **isometry.** When b exceeds 1, Y increases proportionally more rapidly than X, so the ratio of Y to X increases with larger X. When b is less than 1, the ratio of Y to X decreases with larger X; when b is less than 0, the absolute value of Y decreases with larger X.

The relationship between heart rate and body mass for several mammals ranging over many orders of magnitude in size is shown in Figure 2.24. The equation for the line that best fits the points has a slope, or allometric constant, of less than 0 ($b = -0.23$). The slower heart rates of larger

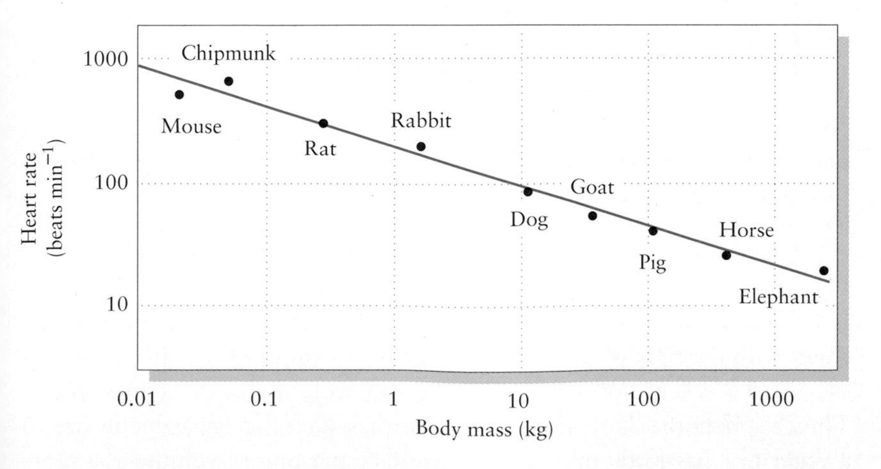

Figure 2.24 Allometric relationship between heart rate and body mass in a variety of mammals ranging in size from mouse to elephant. Data from P. L. Altman and D. S. Dittmer (eds.), *Biology Data Book*, Fed. Am. Soc. Exp. Biol., Washington, D.C. (1964).

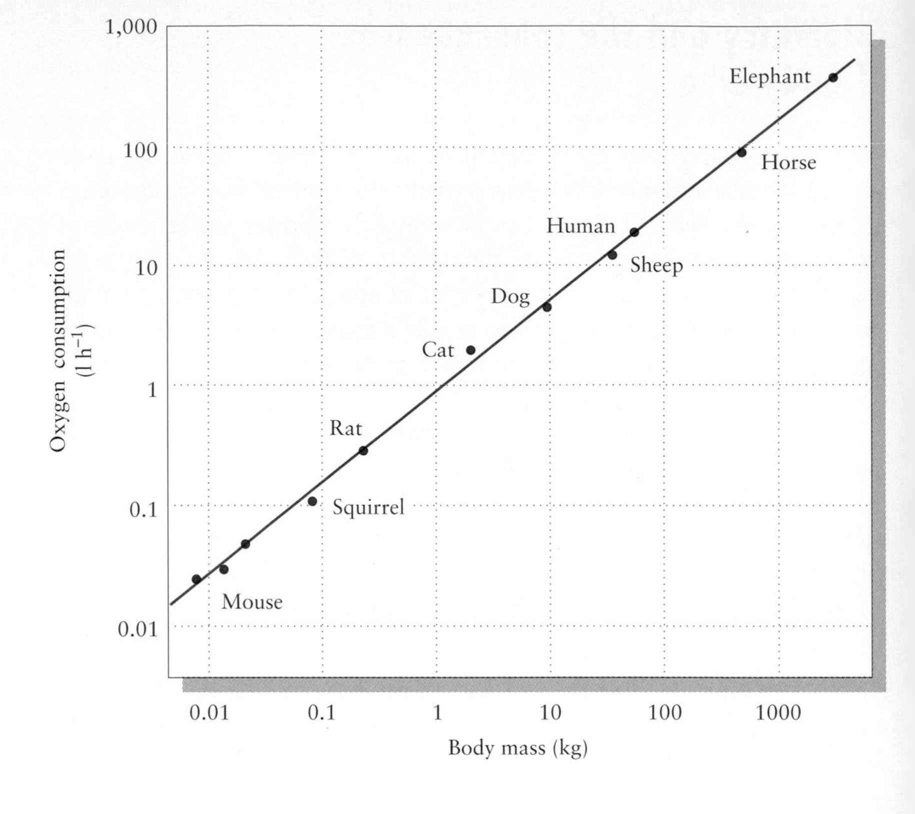

Figure 2.25 Allometric relationship between resting metabolic rate and body mass in several mammals. Data from P. L. Altman and D. S. Dittmer (eds.), *Biology Data Book,* Fed. Am. Soc. Exp. Biol., Washington, D.C. (1964).

species result from the mechanics of fluids moving through closed systems. If an elephant's heart were to beat as fast as that of a mouse, the pressure built up by the rapid contraction of such a large volume of blood couldn't be contained, and arteries would burst. In contrast to the heart rate–body size relationship, the resting metabolic rates (RMR) of mammals increase with larger size, but less rapidly than mass itself; the allometric constant of the relationship between RMR and body mass is 0.73 (Figure 2.25). An even higher allometric constant (2.05) is exhibited by the relationship between the average leaf area *(Y)* and the diameter of the stem *(X)* among species of the shrub *Leucadendron* of South Africa (Figure 2.26). Because the stem cross section is proportional to the square of stem diameter, leaf area increases in direct relation to cross-sectional area, probably because leaf area is related to transpiration and stem diameter to water transport.

The allometry of certain relationships arises from simple physical or geometric considerations. For example, the volume of a sphere increases in proportion to the cube of its diameter, but its surface area increases in proportion only to the square of the diameter. Hence the relationship between surface *(S)* and volume *(V)* has an allometric constant of 2/3, or

$$S = aV^{2/3}$$

Consequently, larger organisms have relatively smaller surfaces compared with the bulk of their bodies. A sphere of radius r has a surface area of $4\pi r^2$ and a volume of $4\pi r^3/3$; the ratio of surface area to volume is $3/r$. Thus, a sphere that has a diameter of 1 mm, which is comparable in size to a water flea, has 1,000 times as much surface per unit of volume as a bear-

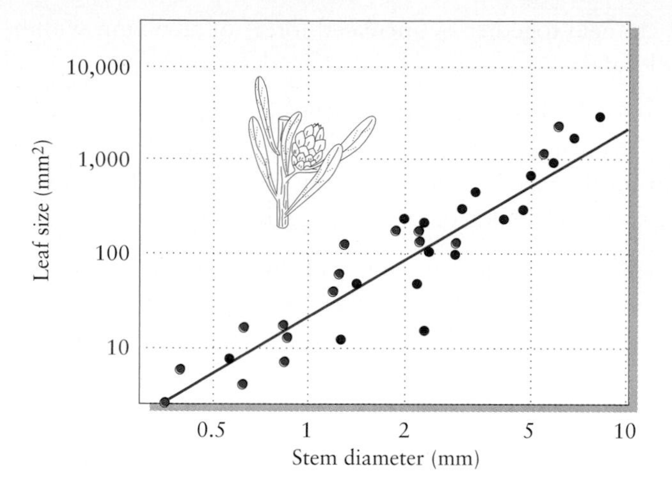

Figure 2.26 Allometric relationship between leaf area and stem diameter in male (colored symbols) and female (black symbols) plants of seventeen species of *Leucadendron*. From W. J. Bond and J. Midgley, *Am. Nat.* 131: 901–910 (1988).

sized sphere 1 m in diameter. This makes the uptake of oxygen relatively easy for the smaller organism, but a high rate of heat loss would make it impossible for such a small organism to maintain a high body temperature. This works both ways, however. A 2,000-kg elephant has only one forty-fifth as much surface area per gram of mass as a 20-g mouse, so avoiding sudden changes in body temperature and conserving water pose less severe problems for the elephant. But because of its small surface relative to its volume, the elephant could never have the metabolic rate of the mouse because it couldn't get rid of the heat fast enough across its body surface.

Organisms are physical systems, and they must obey physical laws and operate within limits set by the physical environment. By expending energy, however, plants and animals can maintain themselves out of equilibrium with their physical surroundings. Each type of organism devises, over evolutionary time, its own unique solutions to the problems posed by the physical environment. We shall consider some of these solutions in the next chapter.

Summary

1. Water is the basic medium of life. It is liquid within the range of temperatures encountered over most of the earth, and it has an immense capacity to dissolve inorganic compounds. These properties, and its abundance at the earth's surface, make water an ideal medium for living systems.

2. Because water clings tightly to soil surfaces, its availability depends in part on the physical structure of the soil. Soils made up of small particles—clays, for example—hold water more strongly than do sandy soils. Most plants cannot remove water from the soil when the water potential exceeds 15 atmospheres of pressure.

3. Most kinds of organisms cannot survive temperatures much greater than 45°C, but thermophilic bacteria grow in hot springs at up to 110°C. They apparently tolerate such temperatures because their proteins and

membranes are held together by increased forces of attraction within and between molecules.

4. Organisms in cold environments withstand freezing temperatures by metabolically maintaining elevated body temperatures, by lowering the freezing point of their body fluids with salts, glycerol, or glycoproteins, by supercooling their body fluids, or by tolerating freezing.

5. Acidity refers to the concentration of hydrogen (H^+) ions and is expressed as pH. Most natural waters have pH values between 6 (slightly acid) and 9 (slightly alkaline). Many organisms can tolerate high acidity (low pH) in the environment, but maintain their internal environments between pH 6 and 7, or close to neutral.

6. Biological energy transformations are based largely on the chemistry of carbon and oxygen. Energy is assimilated during photosynthesis as carbon is reduced from its state in carbon dioxide to its state in carbohydrates. In a coupled reaction, oxygen is oxidized from its form in water to molecular oxygen. The energy stored in carbohydrates is released by the oxidation of carbon to carbon dioxide (respiration).

7. Carbon dioxide is scarce in the atmosphere (0.03%) but is more abundantly distributed in aquatic systems, where it forms soluble bicarbonates. Oxygen, abundant in the atmosphere, is relatively scarce in water, where its solubility and rate of diffusion are low, and it may be depleted by bacterial respiration of organic matter (producing anoxic conditions) in deep, stagnant layers.

8. Organisms assimilate many elements necessary to life processes and biological structures. The availability of these elements relative to the amounts needed varies tremendously among environments. The scarcity (relative to need) of nitrogen and phosphorus often limits plant growth.

9. In aquatic habitats, differences in the concentrations of dissolved salts establish osmotic gradients between the organism and its surroundings. Hyperosmotic organisms, which have greater salt concentrations than their surroundings, tend to lose salt and gain water; hypo-osmotic organisms gain salt and lose water. Organisms achieve osmotic balance by altering the osmotic pressure of their body fluids and by actively pumping salts across membranes.

10. Virtually all of the energy for life ultimately comes from sunlight. Solar radiation varies over a spectrum of wavelengths. Plants extract energy primarily in the high-intensity, short-wavelength portion of the spectrum, which roughly coincides with visible light. Shorter wavelengths (ultraviolet) are absorbed in the atmosphere by ozone; longer wavelengths (infrared) are absorbed by carbon dioxide and water vapor.

11. Photosynthesis varies in proportion to light intensity at low light levels. Above the saturation point, usually 10–20% of direct sunlight, photosynthesis levels off and is limited by other factors. The level of light below which plant respiration exceeds photosynthesis, and there is no net growth, is the compensation point.

12. Light is attenuated by water. The depth of the euphotic zone, at the bottom of which photosynthesis balances respiration (the compensation point), varies from 100 meters in clear waters to a few tens of centimeters in turbid or polluted water.

13. The thermal environment of organisms, especially in terrestrial habitats, is determined by radiation, conduction, convection, and evaporation. In still air or water, organisms are surrounded by boundary layers, which impede exchange of heat, salts, and water vapor with the environment.

14. Water is denser than air and provides more buoyancy, but it is also more viscous and therefore impedes movement.

15. The senses of organisms depend on the availability of information in the physical environment. The nature of this information—whether it consists of light waves, sound (pressure) waves, or volatile or dissolved molecules—determines how and how well the organism can detect and localize sources of information.

16. Many factors bearing on the ecology of organisms scale disproportionately with respect to overall body size. Such relationships are described by the power (allometric constant) of the equation relating a measurement of some structure or function to body size. Surface area scales to the 0.67 power of body size when shape is held constant; hence larger organisms have proportionally smaller surface-to-volume ratios. This and other allometric relationships have important consequences for organisms of different sizes.

Suggested readings

Brock, T. D. 1985. Life at high temperatures. *Science* 230:132–138.

Fenchel, T., and B. J. Finlay. 1994. The evolution of life without oxygen. *American Scientist* 82:22–29.

Gates, D. M. 1965. Energy, plants, and ecology. *Ecology* 46:1–13.

Gates, D. M. 1971. *Man and His Environment: Climate.* Harper & Row, New York.

Hochachka, P. W., and G. N. Somero. 1984. *Biochemical Adaptation.* Princeton University Press, Princeton, N.J.

Knoll, A. H. 1991. End of the Proterozoic eon. *Scientific American* 265:64–73. (The role of primitive organisms in modifying the early environment of the earth.)

Lovelock, J. E. 1988. *Ages of Gaia.* W. W. Norton, New York. (Describes the interdependence of organisms and their environments and treats the entire biosphere as a superorganism; controversial.)

Peters, R. H. 1983. *The Ecological Implications of Body Size.* Cambridge University Press, Cambridge.

Schwenk, K. 1994. Why snakes have forked tongues. *Science* 263:1573–1577.

Vogel, S. 1981. *Life in Moving Fluids: The Physical Biology of Flow.* Princeton University Press, Princeton, N.J.

CHAPTER 3

ADAPTATION TO AQUATIC AND TERRESTRIAL ENVIRONMENTS

The physical environment includes many factors that are important to the well-being of living organisms, as we saw in the previous chapter. In this chapter, we shall explore a variety of the mechanisms that animals and plants have evolved so that they can function well as biological systems in the context of a physical environment. Many of these mechanisms enable plants and animals to control the movement, or **flux,** of heat and various substances across their surfaces. By regulating exchange with the physical environment, organisms can maintain their own internal environments in a state that is favorable for their life processes while obtaining necessary resources from the environment and ridding themselves of unnecessary, or even dangerous, waste products of their metabolism. Organisms can influence exchange with the environment by seeking appropriate external environments, adjusting the internal environment in response to the surrounding conditions, and modifying the qualities and area of the body surface. Such modifications, which better suit the organism to its particular environment, are called **adaptations.**

A common theme, then, is adjustment of the variables that affect flux: (1) surface area, (2) ease of movement across a surface, or **conductance,** and (3) the difference in the concentrations of a substance between the organism and the environment, or the **concentration gradient.** As we shall see, however, this theme manifests itself differently depending on which environmental factor is considered. Furthermore, an organism's relationship to any one environmental factor depends on its relationship to all the others. How one deals with salt balance, for example, influences the way in which water flux is regulated. Indeed, the coupled problems of regulating salt and water flux provide a good beginning for our discussion.

Salt balance and water balance

Left to their own devices, ions diffuse across cell membranes from regions of high concentration to regions of low concentration, thereby tending to equalize these concentrations. Water also moves across membranes by osmosis toward regions of high ion concentrations, tending to dilute dissolved substances. Maintaining an ionic imbalance between the organism and the surrounding environment against the physical forces of diffusion and osmosis requires energy and often is accomplished by organs specialized for salt retention or excretion.

Ion retention is critical to terrestrial and freshwater organisms. Freshwater fish, for example, continuously gain water by osmosis across surfaces of the mouth and gills, which are the most permeable of their tissues that are exposed to the surroundings, and in their food. To counter this influx, fish eliminate water as urine. If fish did not also selectively retain dissolved ions, they would soon become lifeless bags of water. The kidneys of freshwater fish retain salts by actively removing ions from the urine and infusing them back into the bloodstream. In addition, the gills can selectively absorb ions from the surrounding water and secrete them into the bloodstream.

Terrestrial animals acquire the mineral ions they need in the water they drink and the food they eat, although lack of sodium in some areas forces animals to obtain salt directly from such mineral sources as salt licks. Because they are not immersed continuously in fresh water, terrestrial animals have little trouble retaining ions. Plants absorb ions dissolved in soil water. Because plants lose water through evaporation from the leaves, ions are left behind and tend to concentrate in leaf tissues. To achieve salt balance, many plants regulate the flux of ions into the roots. Salts that do accumulate in leaves may be sequestered in vacuoles within cells where they cannot interfere with cell metabolism.

Marine fish are surrounded by water having a higher salt concentration than that of their bodies. As a result, they tend to lose water to the surrounding seawater and must drink seawater to replace this loss. The salt that comes in with the water and with food, as well as that which diffuses in across body surfaces, must be excreted at great metabolic cost from the gills and kidneys. Some sharks and rays have found a solution to one aspect of the problem of osmotic balance, that of water flux. Sharks retain urea [$CO(NH_2)_2$]—a common nitrogenous waste product of metabolism in vertebrates—in the bloodstream instead of excreting it from the body in the urine. The urea raises the osmotic potential of the blood to the level of seawater without any increase in the concentration of sodium and chloride ions. The high levels of urea in the blood effectively cancel the tendency of water to leave by osmosis, and there is consequently no net movement of water across a shark's surfaces. This makes it much easier to regulate the flux of ions such as sodium because sharks do not have to drink salt-laden water to replace water lost by osmosis. The fact that freshwater species of sharks and rays do not accumulate urea in their blood emphasizes the osmoregulatory role of urea in marine species.

The small copepod *Tigriopus* takes an approach to water balance similar to that of sharks. *Tigriopus* lives in pools high in the splash zone along rocky coasts. The pools receive seawater infrequently from the splash of high waves, and as the water in these pools evaporates, the salt concentration rises to high levels. Like the sharks, *Tigriopus* manages its water loss by increasing the osmotic potential of its body fluids. It accomplishes this by synthesizing large quantities of certain amino acids such as proline. These small molecules increase the osmotic potential of the body to match that of the habitat, without the deleterious physiological consequences of high levels of salt.

Water balance and salt balance are intimately related in terrestrial as well as aquatic organisms. Plants transpire hundreds of grams of water for every gram of dry matter they accumulate in tissue growth, and they inevitably take up salts along with the water that passes into their roots. In saline environments, plants actively pump excess salts back into the soil across the root surfaces, which therefore function as the plant's "kidneys." Mangrove plants grow in marine sediments that are inundated daily by high tides (Figure 3.1). Not only does this habitat present a high salt load, but the high osmotic potential of the root environment also makes it more difficult to take up water. To counter this problem, many mangrove plants have high levels of organic solutes, such as proline, sorbitol, and glycinebetaine, in their roots and leaves to increase their osmotic potential. In addition, salt glands secrete salt to the exterior surface of the leaves. The roots of many species exclude salts, apparently by means of semipermeable membranes that do not allow the salts to enter. We know that active transport is not involved, as it is in the salt glands of the leaves, because neither cooling nor metabolic inhibitors diminish salt exclusion by roots. Mangrove plants further reduce salt loads by decreasing the transpiration of water from leaves. Because many of these adaptations resemble those of plants from arid environments, where water is scarce, the mangrove habitat has been referred to as an osmotic desert.

Most terrestrial animals obtain more salts in their food than they need and eliminate the excess salts in their urine. Where water abounds, such animals can drink large quantities of water to flush out salts that would otherwise accumulate in the body. Where water is scarce, however, animals must produce a concentrated urine to conserve water. And so, as one would expect, desert animals have champion kidneys. For example, whereas humans can concentrate salt ions in their urine to about 4 times the level in blood plasma, the kangaroo rat's kidneys produce urine with a

(a)

(b)

salt concentration as high as 14 times that of its blood, and the Australian hopping mouse, a desert-adapted species, produces urine that has 25 times the salt concentration of its blood.

Figure 3.1 (a) The roots of mangrove vegetation are immersed in salt water at high tide. Some species exclude salt from their roots. (b) Specialized glands in the leaves of the button mangrove, *Conocarpus erecta,* excrete salt, which precipitates on their outer surfaces.

Nitrogen excretion

Most carnivores, regardless of whether they eat crustaceans, fish, insects, or mammals, consume excess nitrogen in their diets. This nitrogen, ingested in the form of proteins and nucleic acids, must be eliminated from the body when these compounds are metabolized (Figure 3.2). Most aquatic organisms produce the simple metabolic by-product ammonia (NH_3). Although ammonia is mildly poisonous to tissues, aquatic organisms eliminate it rapidly in a copious, dilute urine, or directly across the body surface, before it reaches a dangerous concentration within the body. Terrestrial animals cannot afford to use large quantities of water to excrete nitrogen. Instead, they produce protein metabolites that are less toxic than ammonia and which can therefore accumulate to higher levels in the blood and urine without danger. In mammals, this waste product is urea [$CO(NH_2)_2$], the same substance that sharks produce and retain to achieve osmotic balance in marine environments. Because urea dissolves in water, excreting it requires some urinary water loss—how much depends on the concentrating power of the kidneys. Birds and reptiles have carried adaptation to terrestrial life one step further: they excrete nitrogen in the form of uric acid ($C_5H_4N_4O_3$), which crystallizes out of solution and can then be excreted as a highly concentrated paste in the urine. Although water is saved by excreting urea and uric acid, there is also a cost, which is the energy lost in the

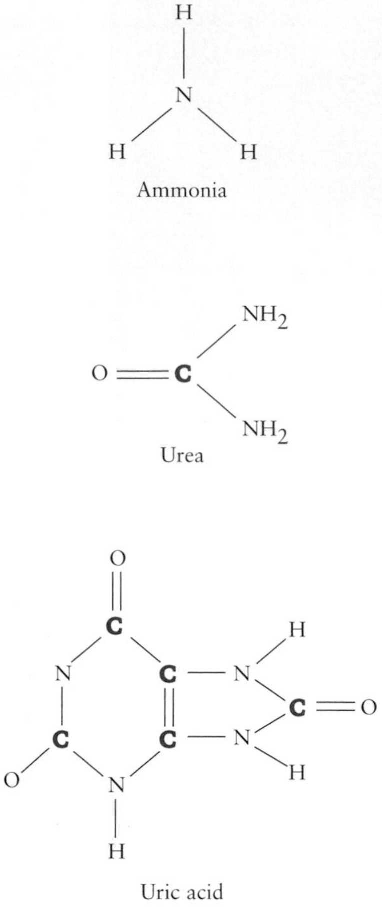

Ammonia

Urea

Uric acid

Figure 3.2 Chemical structures of common nitrogenous waste products. Ammonia lacks carbon, but is toxic to living tissues. Urea and uric acid are less harmful, but the considerable energy they contain in reduced carbon represents a loss to the organism.

organic carbon used to form these compounds. For each atom of nitrogen excreted, 0.5 and 1.25 atoms of organic carbon are lost in urea and uric acid respectively.

Temperature and water conservation

When air temperature approaches or exceeds the maximum permissible body temperature, individuals can dissipate heat only by evaporating water from their skin and respiratory surfaces. In deserts, the scarcity of water makes evaporative heat loss a costly mechanism; animals often must reduce their activity, seek cool microclimates (Figure 3.3), or undertake seasonal migrations to cooler regions. Many desert plants orient their leaves in such a way as to avoid the direct rays of the sun; others shed their leaves and become inactive during periods of combined heat and water stress.

Among mammals, the kangaroo rat is well suited to life in a nearly waterless environment (Figure 3.4). Its large intestine resorbs water from waste material so efficiently that it produces virtually dry feces. Kangaroo rats also recover much of the water that evaporates from their lungs by condensation in their enlarged nasal passages. When the kangaroo rat inhales dry air, moisture in its nasal passages evaporates, cooling the nose and saturating the inhaled air with water. When moist air is exhaled from the lungs, much of the water vapor condenses on the cool nasal surfaces. By alternating condensation with evaporation during breathing, the kangaroo rat minimizes its respiratory water loss.

Kangaroo rats avoid the desert's greatest heat by feeding only at night, when it is cool at the surface; during the blistering heat of the day, kangaroo rats remain comfortably below ground in their cool, humid burrows. In sharp contrast, ground squirrels remain active during the day. They also conserve water by restricting evaporative cooling. As you would expect, their body temperatures rise when they forage above ground, exposed to the hot sun. How do they manage? Before their body temperatures become dangerously high, they return to their burrows, where they cool down by conduction and radiation rather than by evaporation. By shuttling back and forth between their burrows and the surface, ground squirrels extend their activity into the heat of the day and pay a relatively small price in water loss. Like that of the ground squirrel, the camel's body temperature rises in the heat of the day—by as much as 6°C. Large body size gives the camel a distinct advantage, however. Because of its low surface-to-volume ratio, the camel heats up so slowly that it can remain in the sun most of the day. Excess heat is dumped at night by conduction and radiation to the cooler surroundings.

Plant-water relationships

Water is held in the soil surrounding roots by a variety of forces, whose strengths may be expressed in terms of water potential. The units of these forces are pressures—we are already familiar with the unit of sea-level

Figure 3.3 A jackrabbit seeking refuge from the hot sun of southern Arizona in the shade of a mesquite tree. The large ears and long legs of desert-inhabiting jackrabbits effectively radiate heat when the environment is cooler than the body. Courtesy of the U.S. Fish and Wildlife Service.

atmospheric pressure, or 1 atm. By convention, water potentials are given negative values: the larger the negative value, the greater the soil's attraction for water. Water within the spaces between soil particles is held by capillary attraction of water molecules to one another, which generates forces of about −0.15 atm. Water in large pores in the soil and at great distance from the surfaces of soil particles, held with pressures less than −0.15 atm, usually drains through the soil under the pull of gravity. Closer to the surfaces of soil particles, stronger forces between water molecules and the particles themselves hold water tightly in the soil with potentials of −50 atm or more. Plants first tap the soil water that is held with the least potential and is easiest to extract. As soils dry out, the remaining water is held with greater and greater force, eventually causing water stress. As a general rule,

Figure 3.4 The behavior and physiological features of the kangaroo rat adapt it exquisitely to desert environments. Courtesy of the U.S. Fish and Wildlife Service.

plants from moist environments cannot obtain water held with pressures more negative than −15 atm, the so-called permanent wilting point of the soil.

Plants obtain water from the soil by osmosis, so the ability of the roots to take up water depends on their osmotic potential. Osmotic potential, in turn, is a function of the concentration of dissolved molecules (including sugars) and ions in the cell sap. A 1-molar solution, for example, has an osmotic potential of about 21 atm (which is the same as a water potential of −21 atm). By manipulating the osmotic potential of their root cells, plants can alter their ability to remove water from soil. Plants growing in deserts and salty environments have been shown to increase the water potential of their roots to as much as −60 atm by increasing the concentrations of amino acids, carbohydrates, or organic acids in their root cells. They pay a high metabolic price, however, to maintain such concentrations of dissolved substances.

Plants conduct water to their leaves through xylem elements, which are the empty remains of xylem cells in the cores of the roots and stems, connected end-to-end to form the equivalent of water pipes leading from the roots to the leaves. For water to flow into these elements, their water potential must be more negative than that in the living cells of the roots, into which water enters from the soil. Then, for water to move through the xylem from the roots to the leaves, the water potential of the leaves must exceed that of the roots enough to draw water upward against the osmotic potential of the root cells, the pull of gravity, and the resistance of the xylem elements. Leaves generate water potential by transpiring water to the atmosphere. Dry air can have water potentials as great as −1,000 atm, which is more than enough to do the job, given reasonably low resistance to water flow in the stem.

C_3, C_4, and CAM photosynthesis

Plants require water primarily to replace losses that accompany gas exchange with the atmosphere. Because the atmosphere contains so little carbon dioxide (0.03%), the concentration gradient for water loss exceeds that for carbon dioxide assimilation by several orders of magnitude (see Figure 2.6). And because the vapor pressure of water increases with temperature, heat magnifies the problem of water loss.

Heat- and drought-adapted plants have anatomic and physiological modifications that reduce transpiration across plant surfaces, reduce heat loads, and enable plants to tolerate high temperatures. When plants absorb sunlight, they heat up. Plants can minimize overheating by increasing their surface area for heat dissipation and by protecting their surfaces from direct sunlight with dense hairs and spines (Figure 3.5). Spines and hairs also produce a still boundary layer of air that traps moisture and reduces evaporation. Because thick boundary layers retard heat loss as well, hairs are prevalent in cool, arid environments but less so in hot deserts. Plants may further reduce transpiration by covering their surfaces with a thick, waxy

(a) (b)

Figure 3.5 Cross section (a) and surface view (b) of the pubescent leaf of the desert perennial herb *Enceliopsis argophylla.* Courtesy of J. R. Ehleringer. From J. R. Ehleringer, in E. Rodriguez, P. Healy, and I. Mehta (eds.), *Biology and Chemistry of Plant Trichomes,* Plenum Press, New York (1984), pp. 113–132.

cuticle that is impervious to water and by recessing the stomates in deep pits, often themselves filled with hairs (Figure 3.6).

In some plants, the biochemical steps involved in the assimilation of carbon display modifications that help to conserve water. Most species of plants living in environments with adequate water assimilate the carbon in carbon dioxide into an organic molecule in a single biochemical step in a pathway known as the **Calvin cycle** (Figure 3.7). We can represent this step as

$$CO_2 + \text{RuBP} \rightarrow 2\ \text{PGA},$$

where RuBP (ribulose bisphosphate) is a 5-carbon compound and PGA (phosphoglycerate) is a 3-carbon compound. (Because the immediate product of carbon assimilation is a 3-carbon compound, biologists call this mechanism **C_3 photosynthesis.**) Several biochemical steps beyond the production of PGA, the Calvin cycle regenerates 1 molecule of RuBP while making 1 carbon atom available to synthesize glucose. Inasmuch as glucose contains 6 carbons, each molecule of glucose produced requires 6 turns of the Calvin cycle.

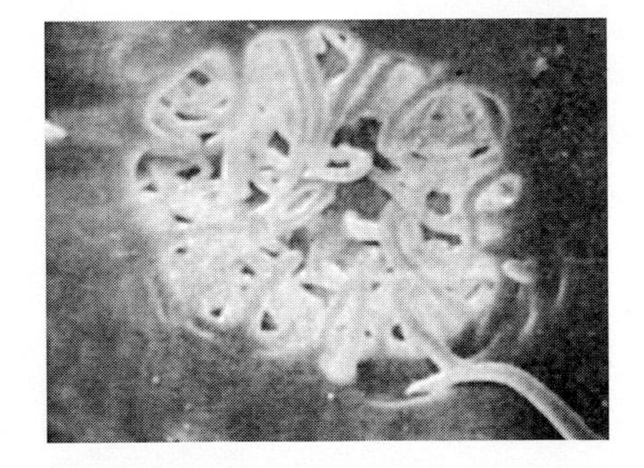

Figure 3.6 In oleander, a drought-resistant plant, the stomates lie deep within hair-filled pits on the leaf's undersurface (magnified about 500 times). The hairs reduce water loss by slowing air movement and trapping water. Courtesy of M.V. Parthasarathy.

The enzyme responsible for the assimilation of carbon, RuBP carboxylase, has a low affinity for carbon dioxide. As a result, at the low concentration of CO_2 found in the atmosphere and the resulting low concentration in the mesophyll cells of the leaves where photosynthesis takes place, plants assimilate carbon inefficiently. To achieve high rates of assimilation, plants must pack their cells with large amounts of RuBP carboxylase. This enzyme also facilitates the oxidation of RuBP in the presence of high oxygen and lowered carbon dioxide concentrations, especially at elevated leaf temperatures. That is, oxidation of RuBP partially undoes what RuBP carboxylase accomplishes when it chemically reduces carbon, making photosynthesis inefficient and self-limiting. Carbon assimilation tends to inhibit itself as levels of CO_2 decline and levels of oxygen produced by photosynthesis increase in the leaves. Plants can lessen this condition by keeping their stomates open—which, of course, leads to high water loss.

Many plants in hot climates exhibit a modification of C_3 photosynthesis that involves an additional step in the assimilation of carbon dioxide as well as spatial separation of that initial assimilation step from the Calvin cycle pathways within the leaf. Biologists call this modification **C_4 photosynthesis** because the assimilation of CO_2 initially results in a 4-carbon compound:

$$CO_2 + PEP \rightarrow OAA,$$

where PEP (phosphoenol pyruvate) contains 3 carbons and OAA (oxaloacetic acid) contains 4. The assimilatory reaction is catalyzed by the enzyme PEP carboxylase, which, unlike RuBP carboxylase, has a high affinity for CO_2. Assimilation occurs in the mesophyll cells of the leaf, but in most C_4 plants, photosynthesis (including the Calvin cycle) takes place in specialized cells surrounding the leaf veins called **bundle sheath cells** (Figure 3.8). Oxaloacetic acid diffuses into the bundle sheath cells, where it breaks down to produce CO_2 and pyruvate, a 3-carbon compound. The pyruvate moves back into the mesophyll cells, where enzymes convert it to PEP to complete the carbon assimilation cycle.

The CO_2 released by metabolism of OAA in the bundle sheath cells enters the Calvin cycle, as it does in C_3 plants. C_4 photosynthesis confers an advantage because CO_2 can be concentrated within the bundle sheath cells to a level, determined by the concentration of OAA, that far exceeds its equilibrium established by diffusion from the atmosphere. At this higher concentration, the Calvin cycle operates more efficiently. Also, because the enzyme PEP carboxylase has a high affinity for CO_2, it can bind CO_2 at a lower concentration in the cell, thereby allowing the plant to increase stomatal resistance and reduce water loss.

Certain succulent plants in desert environments use the same biochemical pathways as C_4 plants but segregate CO_2 assimilation and the Calvin cycle between day and night. The discovery of this arrangement in plants of the family Crassulaceae (the stonecrop family; sedum is one example) and the initial assimilation and storage of carbon dioxide as 4-carbon organic acids (malic acid and OAA) led to the name **crassulacean acid metabolism,** or **CAM.**

CAM plants open their stomates for gas exchange during the cool desert night, at which time transpiration of water is minimal. CAM plants initially assimilate CO_2 in the form of 4-carbon OAA and malic acid, which the leaf tissues store in high concentrations in vacuoles within the cells (Figure 3.9). During the day, the stomates close, and the stored organic acids are gradually recycled to release CO_2 to the Calvin cycle. The assimilation of CO_2 and the regeneration of PEP are regulated by different enzymes that have different temperature optima. CAM photosynthesis results in extremely high water use efficiencies and enables some types of plants to exist in habitats too hot and dry for other, more conventional species.

Submerged aquatic plants face problems entirely different from those that terrestrial plants confront. Water loss is not a problem, but many of

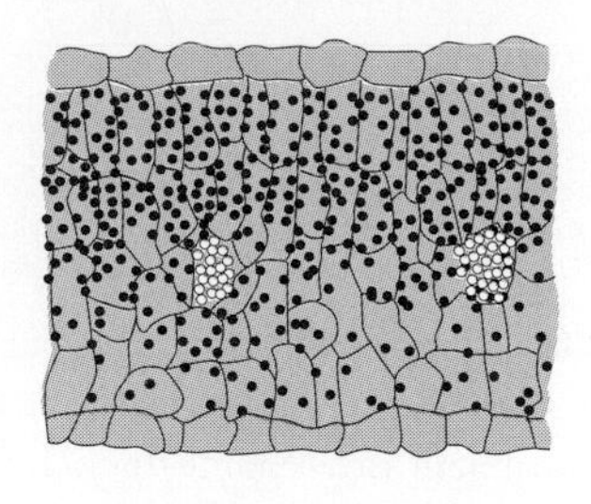

(a)

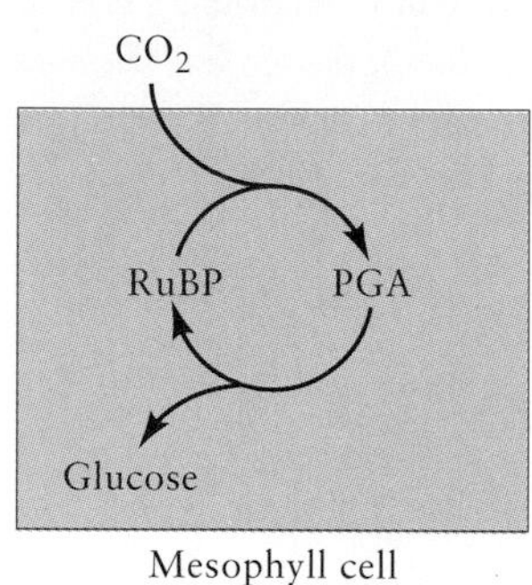

(b)

Figure 3.7 (a) An idealized cross section of a leaf of a C_3 plant, illustrating the general dispersion of chloroplasts (small black dots) throughout the mesophyll. (b) The major steps of the Calvin cycle of carbon dioxide assimilation. Courtesy of J. R. Ehleringer.

these plants nonetheless exhibit CAM and C_4 photosynthesis. *Myriophyllum* inhabits nutrient-rich ponds and lakes. It has several adaptations for obtaining nutrients from, and exchanging gases with, the surrounding water: carbon is assimilated from bicarbonate ions as well as from dissolved CO_2; the leaves are finely dissected to increase surface area for greater exchange; photosynthesis occurs in epidermal cells, which are at the surfaces of the leaves in contact with the surrounding water; the roots are greatly reduced and function only to anchor the plant to the bottom (Figure 3.10a). Photosynthetic rates are high, and the plant grows rapidly. One consequence of these adaptations is that oxygen produced by photosynthesis builds to high levels within the leaves, which potentially might limit photosynthesis. Plants such as *Myriophyllum* circumvent this problem by using C_4 pathways, which elevate the levels of carbon dioxide within leaf cells and help to maintain high rates of photosynthesis.

Unlike *Myriophyllum, Isoetes,* which is a primitive plant distantly related to ferns, grows in nutrient-poor ponds and lakes. It obtains nutrients and carbon dioxide primarily from sediments. Accordingly, its leaves have low surface-to-volume ratios and thick cuticles to prevent exchanges, particularly losses of cellular ions and carbon dioxide, with the surrounding water (Figure 3.10b). The root system is highly developed, and the entire interior of the plant, from the roots to the leaves, is filled with large air spaces. Most of the photosynthesis occurs in cells lining these air spaces in the leaves. The plant takes up carbon dioxide through the roots from the sediments, where it is produced by the respiration of bacteria and animals. CO_2 diffuses from the roots to the leaves through the interior air spaces, and is stored during the night by CAM assimilation. Oxygen produced by photosynthesis during the day diffuses to the roots, where it is used by the respiring root cells (remember that the water in the surrounding sediments may be depleted of oxygen).

Stylites, which is a relative of *Isoetes,* grows in peat deposits in seasonal bogs at over 4,000 m elevation in the Andes mountains of South America. The leaves of *Stylites* have a thick cuticle that is essentially impermeable to carbon dioxide and water vapor; this cuticle prevents water loss to the thin air during dry weather. Carbon dioxide is obtained from the decomposing peat through the roots, just as it is in aquatic relatives.

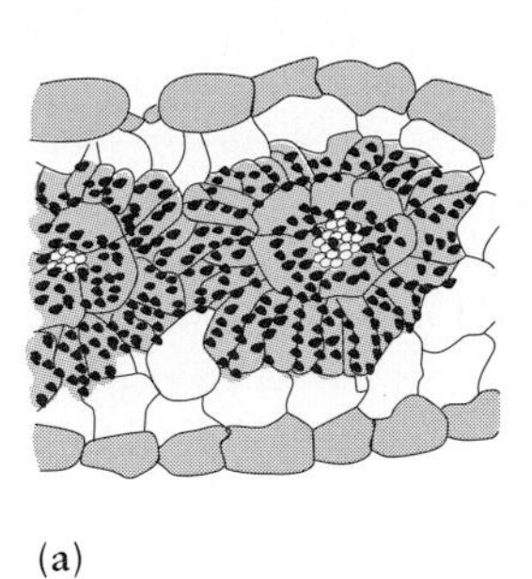

(a)

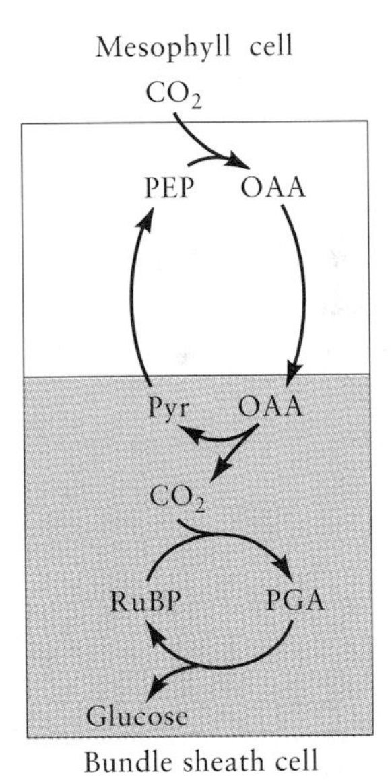

(b)

Figure 3.8 (a) An idealized cross section of a leaf of a C_4 plant, illustrating the concentration of chloroplasts in bundle sheath cells (color). (b) The major steps of carbon dioxide assimilation, including the transport of carbon from the mesophyll to the bundle sheath. Courtesy of J. R. Ehleringer.

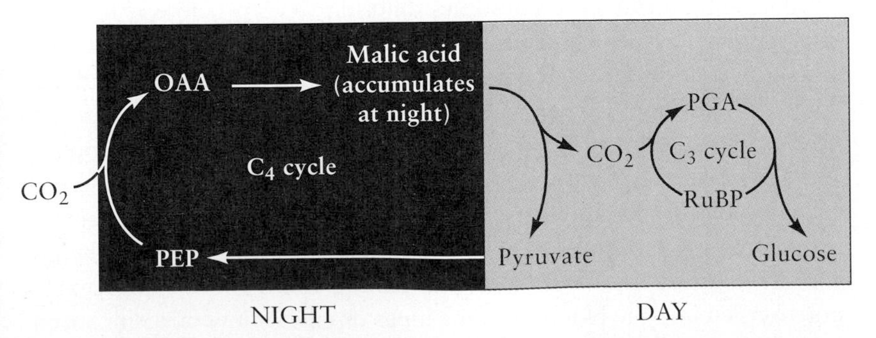

Figure 3.9 Diagram of the photosynthetic pathway in CAM plants. After J. B. Harbourne, *Introduction to Ecological Biochemistry,* Academic Press, New York (1982).

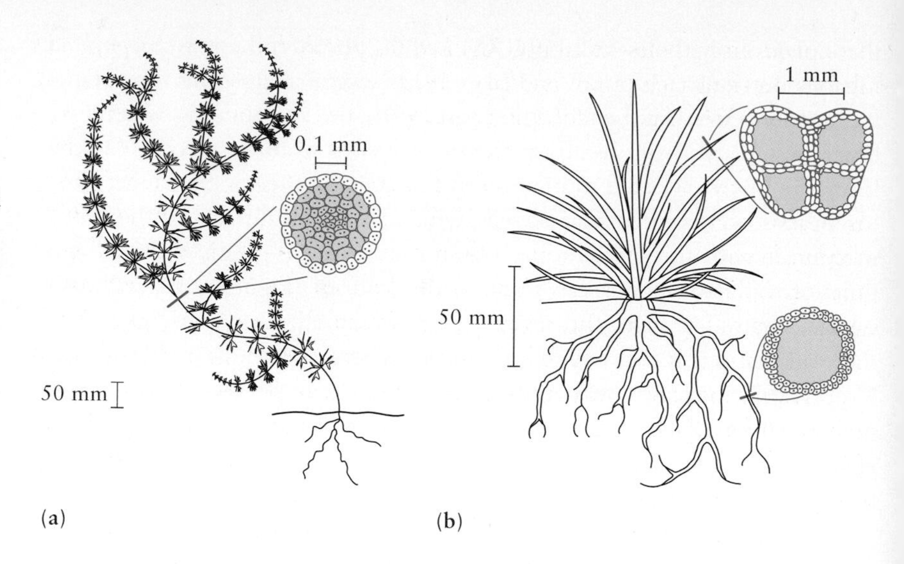

Figure 3.10 Structures of two submerged aquatic plants. (a) *Myriophyllum* is found in water with high nutrient concentrations. (b) *Isoetes* is characteristic of lakes and ponds with lower nutrient concentrations. Note the air-filled chambers, which allow internal gas exchange between the leaves and the roots. After J. E. Keeley, in H. A. Mooney, W. E. Winner, and E. J. Pell (eds.), *Response of Plants to Multiple Stresses,* Academic Press, San Diego, Calif. (1991).

In both cases, the leaves can have impermeable outer surfaces because they exchange gases internally with the roots and then with the root environment.

The procurement of oxygen by animals

Most animals release the chemical energy contained in organic compounds primarily by oxidative metabolism. The biochemical pathways involved—the converse of photosynthesis in many respects—collectively constitute respiration. Because oxygen plays such an important role in releasing energy, low availability of oxygen in the environment can restrict metabolic activity, particularly in aquatic habitats where low solubility and slow diffusion limit oxygen concentrations in the tissues of animals. Even for terrestrial organisms, which breathe an atmosphere containing abundant oxygen, the delivery of oxygen to tissues requires its transport as a dissolved gas through the aqueous medium of the body.

Active organisms require an abundant supply of oxygen for cell respiration. Diffusion can satisfy the oxygen needs of tiny aquatic organisms, but the centers of organisms larger than about 2 mm in diameter are too far from the external environment for diffusion to ensure a rapid supply of oxygen. Tissue metabolism consumes diffusing oxygen before it has gone much farther than a millimeter. In insects, systems of branching pipes (tracheae) carry air directly to the tissues. Other animals have blood circulatory systems to distribute oxygen from the respiratory surfaces to the body. To increase its oxygen-carrying capacity, the blood of most animals contains complex protein molecules, such as hemoglobin and hemocyanin, to which oxygen molecules readily attach for transport. When oxygen binds to hemoglobin—4 molecules of oxygen can bind to each molecule of hemoglobin—it leaves the blood solution, making room for the diffusion of more oxygen into the blood from the lungs or gills. At low oxygen concentrations, the binding process reverses, and oxygen is released to the tissues (Figure 3.11). Though plasma itself carries only limited oxygen in solution (about 1% by volume, or 14 ppm by weight), whole blood can transport up to 50 times more oxygen bound to oxygen-carrying molecules.

Hemoglobin molecules thus act as go-betweens for the uptake of oxygen from the surrounding environment and its release to the cells. The functions of uptake and release conflict, however, because the hemoglobin molecule that readily binds oxygen at the respiratory surfaces also holds oxygen tenaciously when it must be unloaded to supply active tissues. The compromise between the oxygen uptake and oxygen release functions of hemoglobin, or other blood pigments, may be seen in the oxygen dissociation curve (Figure 3.12). This curve portrays the amount of oxygen bound to hemoglobin, expressed as a percentage of the total binding possible (saturation), in relation to the concentration of oxygen in the blood plasma. Oxygen binds to hemoglobin (Hb) reversibly ($4O_2 + Hb \leftrightarrow Hb{=}[O_2]_4$), with the percentages of bound and unbound oxygen reaching an equilibrium. As oxygen diffuses into the blood plasma in the gills or lungs, the equilibrium shifts toward the right, and additional oxygen binds to hemoglobin. As active tissues deplete the blood of oxygen, the equilibrium shifts toward the left, and oxygen detaches from the oxyhemoglobin complex. In general, the proportion of hemoglobin that carries bound oxygen increases with greater concentration of dissolved oxygen in the blood plasma until the hemoglobin becomes saturated.

Tissues receive oxygen at a rate proportional to the difference between the concentrations of oxygen in the tissues and in the blood: the larger the difference, the higher the rate of diffusion. When dissolved oxygen leaves the bloodstream to enter the tissues, the concentration of oxygen in the plasma decreases. As the difference between oxygen concentrations in the blood and in the tissues decreases, diffusion slows. But as the concentration in the blood plasma decreases, hemoglobin releases some bound oxygen,

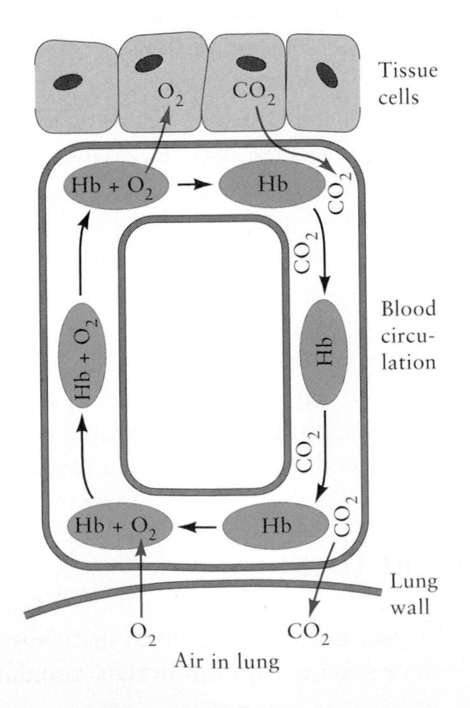

Figure 3.11 The binding of oxygen by hemoglobin in the red blood cells lowers the concentration of dissolved oxygen in the plasma and speeds the diffusion of oxygen from the lungs into the bloodstream. In the body's tissues, where the concentration of oxygen is low, the process is reversed, and oxygen is offloaded from hemoglobin and diffuses toward regions of high metabolic rate.

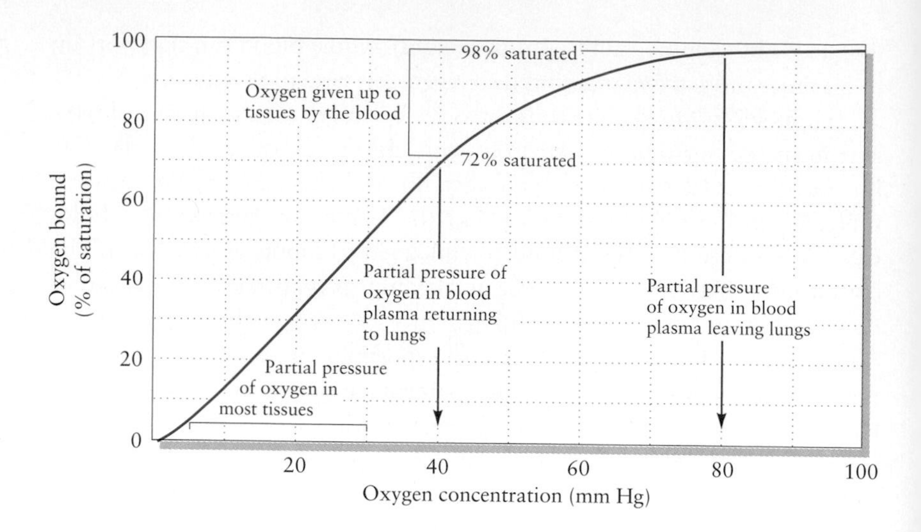

Figure 3.12 The oxygen dissociation curve for human hemoglobin. The blood releases about 25–30% of the oxygen bound to hemoglobin to the tissues, which have concentrations of oxygen equivalent to atmospheric pressures of 5–30 mm Hg.

thereby tending to restore the oxygen concentration. Thus, oxygen drawn from the blood plasma into the tissues is partially replenished by that bound to hemoglobin.

Small changes in the amino acid sequence of the hemoglobin molecule can modify its oxygen-binding capacity with respect to both the availability of oxygen in the environment and its requirement by the organism. Because small animals have high oxygen demands, their oxygen dissociation curves are generally shifted to the right so that oxygen is released more readily. Llamas, living at high elevations in the Andes, have oxygen dissociation curves shifted to the left so that they can obtain oxygen more readily from the thin air. The dissociation curve of a mammalian embryo is shifted to the left of that of its mother because its oxygen supply derives from the mother's blood, which has a lower concentration of oxygen than the air she breathes.

The total amount of hemoglobin in the bloodstream is also subject to adjustment. For example, the blood of the sedentary goosefish has a total oxygen capacity of 5% by volume; that of the more active mackerel is 16%. This difference reflects the hemoglobin concentration in the blood and parallels the sizes of the gills in the two species: the ratio of the surface area of the gills to body weight in the mackerel is 50 times that in the goosefish. Humans adapt to high elevations primarily by increasing the total amount of hemoglobin in their bloodstream; that is one reason why athletes often train at high elevations.

Adaptations for procuring oxygen illustrate a set of solutions to the problems that organisms confront at the interface between themselves and their environments. The design of the hemoglobin molecule, which simultaneously influences its oxygen-binding and its oxygen-releasing properties, further emphasizes the fact that adaptation often requires compromise.

Countercurrent circulation

A common mechanism in aquatic organisms that enhances the uptake of dissolved oxygen from water is **countercurrent circulation.** This mecha-

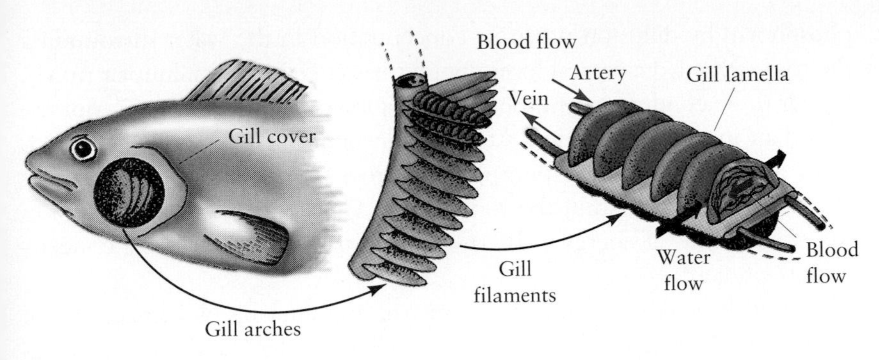

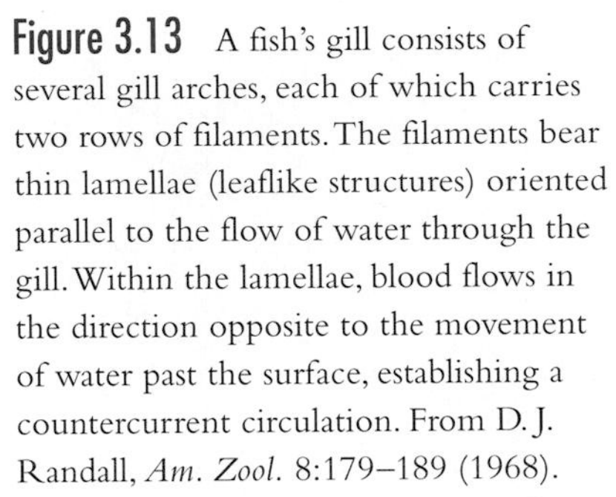

Figure 3.13 A fish's gill consists of several gill arches, each of which carries two rows of filaments. The filaments bear thin lamellae (leaflike structures) oriented parallel to the flow of water through the gill. Within the lamellae, blood flows in the direction opposite to the movement of water past the surface, establishing a countercurrent circulation. From D. J. Randall, *Am. Zool.* 8:179–189 (1968).

nism is illustrated by the structure of the gills of fish, which causes water and blood to flow in opposite directions (Figure 3.13). In a countercurrent system, as blood picks up oxygen from the water flowing past, it comes into contact with water that has progressively greater oxygen concentrations, because the water has flowed past a progressively shorter distance of the gill lamella (Figure 3.14). With this arrangement, the oxygen concentration of the blood plasma can approach the concentration in the surrounding water. If blood and water were to flow in the same direction through the gill (concurrent circulation), an equilibrium oxygen concentration would eventually be established, with equal, intermediate levels in the blood and water flowing past the gill. The countercurrent system keeps the blood and water out of equilibrium and maintains a constant gradient across which oxygen can flow.

The countercurrent principle appears frequently in adaptations that increase the flux of heat or materials across surfaces. Among terrestrial organisms, birds have a unique lung structure that, unlike the lungs of mammals, results in a one-way flow of air opposite to the flow of blood. This adaptation allows birds to achieve, with lungs whose weight and volume are small enough for flight, the high rates of oxygen delivery required by their active lives. The mammalian kidney employs countercurrent circulation to concentrate salts in the urine. The extremities of some birds and mammals have countercurrent blood circulation to reduce loss of heat to the surrounding environment. In all these examples, exchange with the environment is controlled by manipulating the gradient between the internal and external environments.

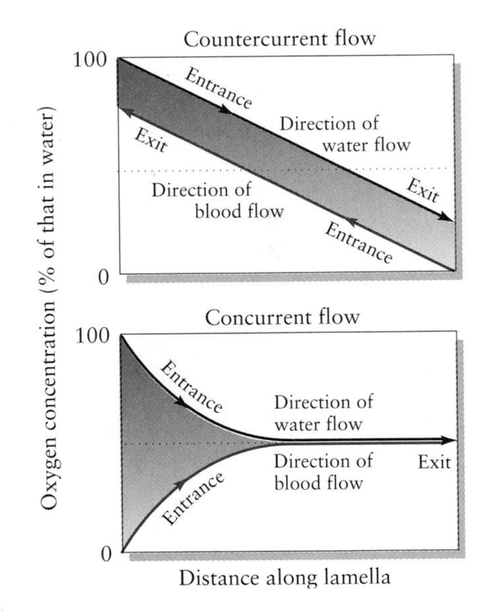

Figure 3.14 Changes in concentration of oxygen in blood and water in countercurrent and concurrent systems. The former maintains a constant gradient (color) for oxygen to diffuse across; as a result, the concentration of oxygen in the blood leaving the system approaches that of the incoming water. After K. Schmidt-Nielsen, *Animal Physiology,* Cambridge University Press, New York (1975).

Uptake of soil nutrients

Plants acquire mineral nutrients—nitrogen, phosphorus, potassium, calcium, and others—from dissolved ionic forms of these elements in soil water. In the case of abundant elements whose ions diffuse rapidly in the soil solution, such as calcium (Ca^{2+}) and magnesium (Mg^{2+}), uptake is limited primarily by the absorptive capacity of the root. Plants compensate for reduced levels of a nutrient in the soil by increasing the extent of the root system—that is, the absorptive surface of the roots—and by active uptake. In laboratory experiments, barley and beet roots passively took up phosphorus by diffusion when its concentration in the water surrounding the root exceeded a critical level, approximately 0.2–0.5 millimolar (mM). Under these conditions, phosphorus had a lower concentration in root tissues than in the soil solution. At soil concentrations below 0.2–0.5 mM, roots actively transported phosphorus across their surfaces and concentrated the element within the root cortex. This active absorption requires the expenditure of energy as the root tissue moves ions against a concentration gradient.

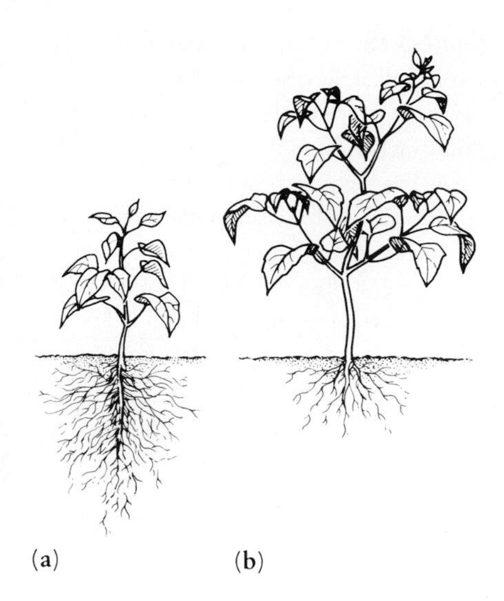

Figure 3.15 Different patterns of allocation between root tissue and shoot tissue during the development of a plant. (a) Allocation often favors roots when supplies of soil nutrients or water are limited. (b) In the absence of these restrictions, larger shoots allow more rapid growth.

Plants may also respond to a scarcity of soil nutrients by increasing root growth at the expense of shoot growth (Figure 3.15). This strategy brings the nutrient requirements of the plant into line with nutrient availability by reducing the nutrient demand created by the leaves, by increasing the absorptive surface area of the root system, and by sending roots into new areas of soil from which the plant has not already removed scarce minerals.

Crop plants and wild species growing on fertile soils have a great capacity to absorb nutrients across their root surfaces and to vary their growth rates in response to variations in soil nutrient levels. Species adapted to nutrient-poor soils are much more conservative. They cope with low nutrient availability by allocating a large fraction of their biomass to roots; by establishing symbiotic relationships with fungi, which enhance mineral absorption; and by growing slowly and retaining leaves for long periods, thereby reducing nutrient demand. Such species typically cannot respond to artificially increased nutrient levels by increasing their growth rates. Instead, their roots absorb more nutrients than the plant requires and store them for subsequent use when the soil nutrient availability declines.

Optimum environmental conditions

Unlike consumable resources, such as carbon dioxide, oxygen, and soil nutrients, conditions such as temperature and salt concentrations influence organisms through their effects on the rates at which physical and biochemical processes proceed and through their effects on the structures of such biologically important molecules as proteins and lipids. As a result of these interactions, each organism generally has a narrow range of conditions to which it is best suited, which define its **optimum.** The optimum is subject to natural selection, which acts on variations in the properties of enzymes and lipids, the structures of cells and tissues, and the form of the body to enable the organism to function well under the particular conditions of its environment. The breadth of conditions that organisms can tolerate is also quite variable. Organisms with wide tolerance ranges (physiological generalists, so to speak) are referred to as **eurytypic;** those with narrow tolerance ranges (physiological specialists) are **stenotypic.** Shifts in optimum environmental conditions may be seen by comparing the effects of salt concentration on the activities of various enzymes in bacteria adapted to low salt concentrations and in halophilic species that live in concentrated brine (Figure 3.16). Note, however, that although the optimum salt concentration for the halophilic *Halobacterium salinarium* is very high, not all of its individual enzymes have high salt optima. They are merely higher than those of species not adapted to high salt concentrations.

As a rule, within the temperature range between the freezing point of water (0°C) and the upper limit for most life forms (40°–50°C), higher temperature quickens the pace of life by increasing the rates of physical processes, such as diffusion and evaporation, as well as biochemical reactions. For example, increased temperature speeds the diffusion of gases through the shell of a bird's egg and that of mineral ions through the soil solution. Temperature also influences the shapes of enzymes and other proteins because the heat energy imparted to such molecules can break the weak forces holding the structure together. Each enzyme functions best over a narrow range of temperature determined by the amino acid sequences of protein chains and the way they affect the bonds that hold proteins in certain configurations. These sequences are, of course, subject to evolutionary modification.

Temperature adaptation is strikingly revealed by comparing organisms taken from environments with different characteristic temperatures. When we examine the performances of such organisms over a range of temperatures, each typically exhibits its maximum activity at a temperature corresponding to that of its normal environment. Many fish in the freezing oceans surrounding Antarctica swim actively and consume oxygen at a rate

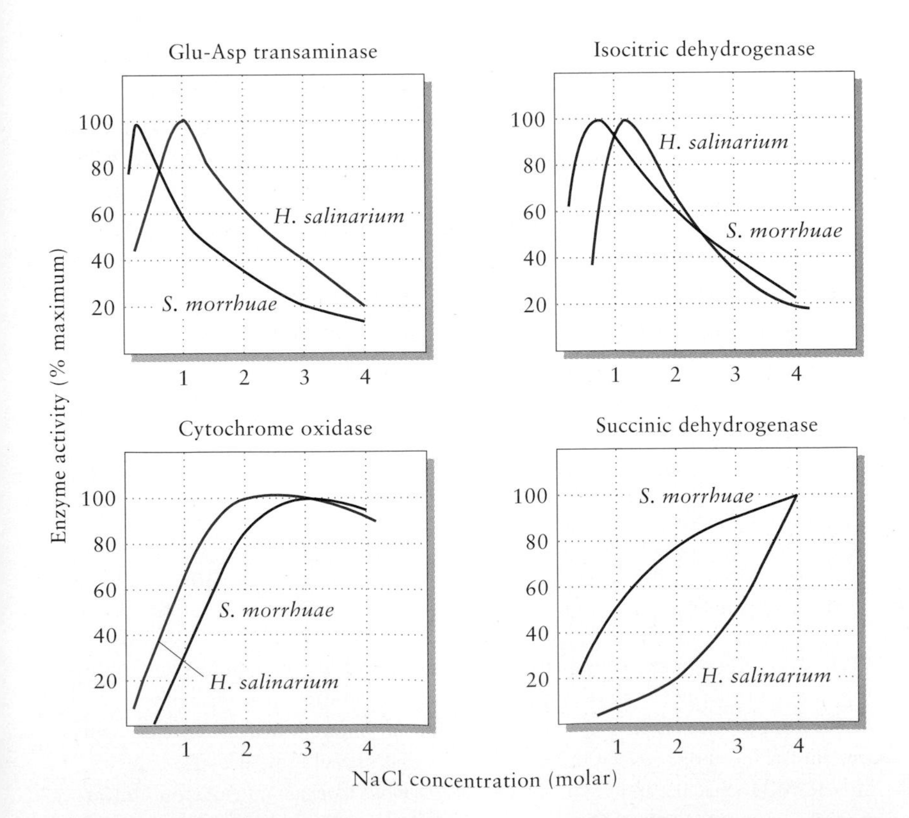

Figure 3.16 Activities of selected enzymes found in the halophilic bacterium *Halobacterium salinarium* (colored lines) at various salt concentrations, compared with those of the same enzymes in *S. morrhuae* (black lines), which cannot tolerate high salt concentrations. The concentration of salt in seawater is about 0.6 molar. After H. Larsen, in I. C. Gunsalus and R. Y. Stanier (eds.), *The Bacteria: A Treatise on Structure and Function*, Vol. 4, *The Physiology of Growth*, Academic Press, New York (1962), pp. 297–342.

comparable to fish living among tropical coral reefs (Figure 3.17). Put a tropical fish in cold water, however, and it becomes sluggish and soon dies; conversely, antarctic fish cannot tolerate temperatures warmer than 5°–10°C.

How can fish from cold environments swim as actively as fish from the Tropics? The metabolism that supports swimming consists of a series of biochemical transformations, most of which are catalyzed by enzymes. Because any given transformation occurs more rapidly at high temperatures than at low temperatures, the compensation observed in cold-adapted organisms must involve either a quantitative increase in the amount of substrate (substance catalyzed) or the amount of enzyme that catalyzes each step, or a qualitative change in the enzyme itself. Laboratory studies of the function of isolated enzymes have shown, in many cases, that a particular enzyme obtained from a variety of organisms exhibits different catalytic properties when tested over ranges of temperature, pH, salt concentration, and substrate abundance (see Figure 3.16). That differences in enzyme structure can help to make up for differences in the temperature of the environment is illustrated by the lactate dehydrogenase enzyme of barracudas. Different species of barracudas *(Sphyraena)* live in marine environments with markedly different ranges of temperature. The temperature dependence of the catalytic function of this enzyme in each species is shifted with respect to the others such that each has a similar level of activity within the range of temperatures normally encountered (Figure 3.18).

This picture of metabolic compensation is greatly simplified, of course. Adapting to changes in the environment requires a complete adjustment of metabolic pathways, which may involve changes in enzyme structure or concentration or the use of alternative metabolic pathways. Indeed, the examples of adaptation to the physical environment considered in this chapter emphasize the unity of organism structure and the interrelatedness of all facets of organism structure and function.

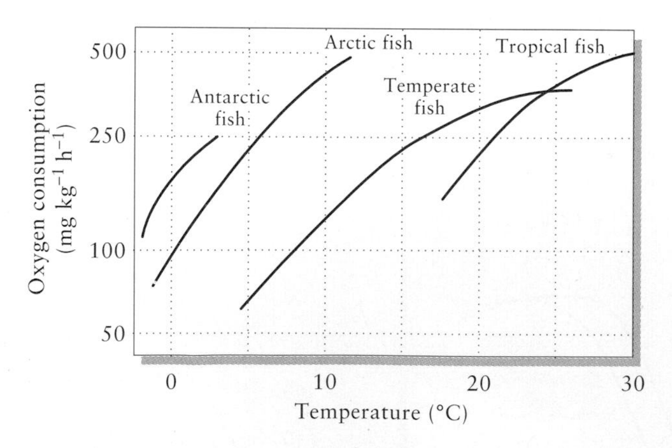

Figure 3.17 Temperature compensation in the rates of oxygen consumption of fish from different thermal regimes. Although metabolism in each increases with temperature, the curves are offset such that fish from different thermal environments have similar metabolic rates at the temperatures that prevail in their native habitats. After P. W. Hochachka and G. N. Somero, *Biochemical Adaptation,* Princeton University Press, Princeton, N.J. (1984).

Summary

1. Adaptations to the physical environment involve modifying the variables that regulate flux across an organism's surface and adjusting metabolic pathways with respect to conditions prevailing in the internal environment.

2. To maintain salt balance and water balance, marine organisms whose internal environments are hypo-osmotic actively exclude salts; freshwater organisms, which are hyperosmotic, retain salts while excreting the water that continuously diffuses into the body; terrestrial organisms minimize water loss in part by concentrating salts and nitrogenous waste products in their urine.

3. Nitrogenous waste products of protein metabolism are excreted as ammonia by most aquatic organisms, as urea by mammals, and as uric acid by birds and reptiles. Because uric acid crystallizes out of solution, birds and reptiles may excrete it at high concentrations and thereby gain considerable economy of water use.

4. Water stress increases with temperature. In dry environments, animals seek cool microclimates, and plants increase the stomatal resistance of their leaves. Such responses uniformly reduce productivity in return for enhanced survival.

5. Plants draw water from the soil by osmotic potential in their roots, and from the roots to the leaves by water potential generated by the evaporation of water from leaves. Resistance to water conduction from the soil through the root and xylem to the leaves limits the supply of water to the leaves.

6. During photosynthesis, plants assimilate carbon through a reaction (the C_3 pathway) catalyzed by the enzyme RuBP carboxylase. This enzyme has a low affinity for carbon dioxide and brings about oxidation at high temperatures, resulting in low efficiency. Plants adapted to high temperatures interpose a more efficient (C_4) carbon assimilation step, which is spatially separated from the C_3 reactions (Calvin cycle) in the leaf. In desert environments, some plants separate carbon assimilation and the Calvin cycle reactions into nighttime and daytime phases (CAM photosynthesis).

7. Some aquatic plants also use CAM and C_4 photosynthetic pathways, which act as mechanisms of storing CO_2 produced at night by respiration and of increasing CO_2 concentrations in leaf tissues to prevent inhibition of photosynthesis by high O_2 levels.

8. Oxygen diffuses too slowly to reach tissues directly that are more than about a millimeter from an organism's surface. Large animals overcome this problem either by conducting air directly to the tissues via a multibranched tracheal system (as in insects) or by transporting oxygen dissolved in circulating fluids. The low solubility of oxygen in water is compensated for by oxygen-binding proteins, such as hemoglobin, that take oxygen out of solution and transport it at high concentrations.

9. The uptake of oxygen by aquatic organisms is greatly facilitated by countercurrent circulation of blood through the gills in a direction opposite

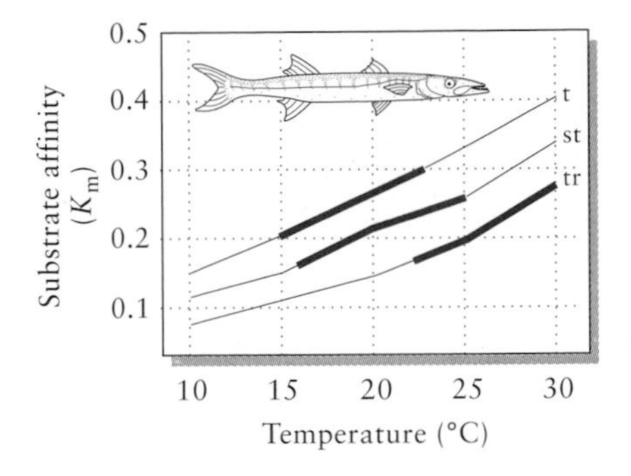

Figure 3.18 Relationship between substrate affinity (which is proportional to activity) and temperature for the lactate dehydrogenase enzymes of three species of barracudas *(Sphyraena)* from temperate (t), subtropical (st), and tropical (tr) waters. Because of temperature compensation, the substrate affinities of the three species are nearly identical within the normal temperature ranges of their environments (thick-colored portions of each curve). After P. W. Hochachka and G. N. Somero, *Biochemical Adaptation,* Princeton University Press, Princeton, N.J. (1984).

to that of water flowing over the gill surfaces. In this way, countercurrent circulation maintains high gradients of oxygen concentration, and the blood can achieve nearly the oxygen concentration of the surrounding water.

10. Plants obtain dissolved nutrients from soil water by passive diffusion when nutrients are abundant and by active uptake when they are scarce. When the concentrations of soil nutrients are low, plants may increase their total root surface area at the expense of shoot growth. Plants also may store nutrients when they are abundant.

11. Most organisms function best within a narrow range of environmental conditions. These optima may be shifted by evolution to more closely match the environmental conditions within which the organism lives. This is often accomplished by altering the structure and quantity of enzymes responsible for controlling metabolic processes.

12. Overall, adaptation to the physical environment depends on reaching compromises between opposing functions to both ensure the individual's survival and maximize its productivity in a particular environment.

Suggested readings

Chapin, F. S., III. 1991. Integrated responses of plants to stress. *BioScience* 41: 29–36.

Ehleringer, J. R., R. F. Sage, L. B. Flanagan, and R. W. Pearcy. 1991. Climate change and the evolution of C_4 photosynthesis. *Trends in Ecology and Evolution* 6:95–99.

Feldman, L. J. 1988. The habits of roots. *BioScience* 38:612–618.

Hochachka, P. W., and G. N. Somero. 1984. *Biochemical Adaptation.* Princeton University Press, Princeton, N.J.

Karov, A. 1991. Chemical cryoprotection of metazoan cells. *BioScience* 41: 155–160.

Kramer, P. J. 1983. *Water Relations of Plants.* Academic Press, New York.

Lee, R. E., Jr. 1989. Insect cold-hardiness: To freeze or not to freeze. *BioScience* 39:308–313.

Louw, G. 1982. *Ecology of Desert Organisms.* Longman, New York.

Schmidt-Nielsen, K. 1990. *Animal Physiology: Adaptations and Environment,* 4th ed. Cambridge University Press, London and New York.

Schulze, E.-D., R. H. Robichaux, J. Grace, P. W. Rundel, and J. R. Ehleringer. 1987. Plant water balance. *BioScience* 37:30–37.

Vogel, S. 1988. *Life's Devices.* Princeton University Press, Princeton, N.J.

CHAPTER 4

VARIATIONS IN THE PHYSICAL ENVIRONMENT

The physical environment varies widely over the surface of the earth. Conditions of temperature, light, substrate, moisture, salinity, soil nutrients, and other factors have shaped the distribution and adaptations of plants, animals, and microbes. The earth has many distinct climate zones whose expanses are broadly determined by the intensity of solar radiation and the redistribution of heat and moisture by wind and water currents. Within climate zones, such geologic factors as topography and composition of bedrock further differentiate the environment on a finer spatial scale. This chapter explores some important patterns of variation in the physical environment that underlie diversity in the biological components of ecosystems.

The surface of the earth, its waters, and the atmosphere above it behave like a giant heat-transforming machine. Climate patterns originate as the earth absorbs the energy in sunlight and as winds and ocean currents redistribute that energy over the globe as heat.

As the surface varies from bare rock to forested soil, open ocean, and frozen lake, its ability to absorb sunlight varies as well, thus creating differential heating and cooling. The heat energy absorbed by the earth eventually radiates back into space, after undergoing further transformations that perform the work of evaporating water and driving the circulation of the atmosphere and oceans. All these factors have created a great variety of physical conditions that, in turn, have fostered the diversification of ecosystems.

Global patterns in atmospheric temperature and precipitation

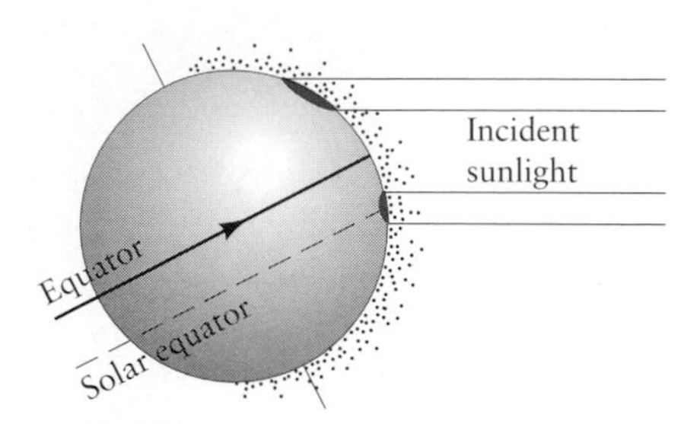

Figure 4.1 The warming effect of the sun is greatest at the solar equator because the sun is closer to the perpendicular there and shines directly down at the surface of the earth during the middle of the day. At higher latitudes, light strikes the earth's surface at a lower angle, and is therefore spread over a greater area; it must also pass through more of the earth's atmosphere.

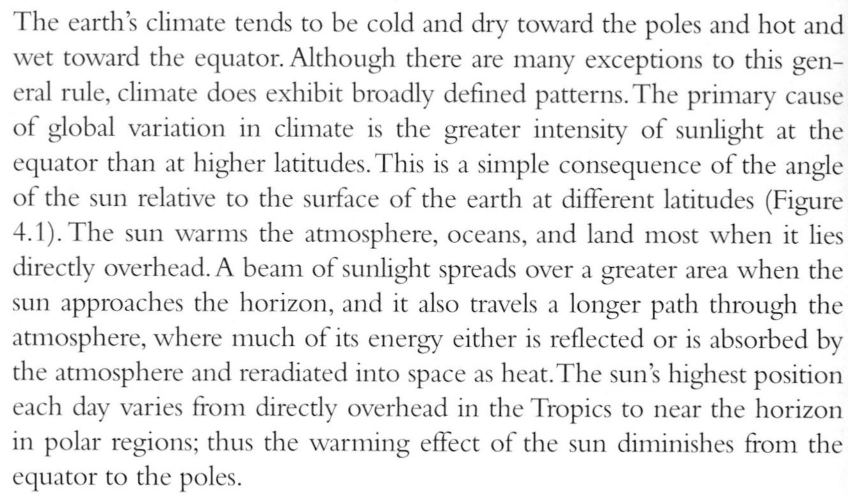

The earth's climate tends to be cold and dry toward the poles and hot and wet toward the equator. Although there are many exceptions to this general rule, climate does exhibit broadly defined patterns. The primary cause of global variation in climate is the greater intensity of sunlight at the equator than at higher latitudes. This is a simple consequence of the angle of the sun relative to the surface of the earth at different latitudes (Figure 4.1). The sun warms the atmosphere, oceans, and land most when it lies directly overhead. A beam of sunlight spreads over a greater area when the sun approaches the horizon, and it also travels a longer path through the atmosphere, where much of its energy either is reflected or is absorbed by the atmosphere and reradiated into space as heat. The sun's highest position each day varies from directly overhead in the Tropics to near the horizon in polar regions; thus the warming effect of the sun diminishes from the equator to the poles.

Warming air expands, becomes less dense, and tends to rise. As air heats up, its ability to hold water vapor increases, and evaporation quickens: the rate of evaporation from a wet surface nearly doubles with each 10°C rise in temperature. The mass of air that rises in the Tropics under the warming sun eventually spreads to the north and south in the upper layers of the atmosphere. It is replaced from below by surface-level air from subtropical latitudes. The rising tropical air mass cools as it radiates heat back into space. By the time this air has extended to about 30° north and south of the **solar equator** (the parallel of latitude that lies directly under the sun), the cooled air mass has become dense enough to sink back to the earth's surface, completing a cycling of air within the atmosphere (Figure 4.2). This type of circulation pattern is called a **Hadley cell.** One Hadley cell forms around the earth immediately to the north of the equator and another to the south. The sinking air of these tropical Hadley cells at about 30° north and south of the equator drives secondary Hadley cells in temperate regions, which circulate in the opposite direction. The circulation of these Hadley cells causes air to rise at about 60°N and 60°S, which leads to the formation of polar Hadley cells. All this circulation of air is driven by the differential heating of the atmosphere with respect to latitude.

The region within which surface currents of air from the northern and southern Subtropics meet in the equatorial region and begin to rise under

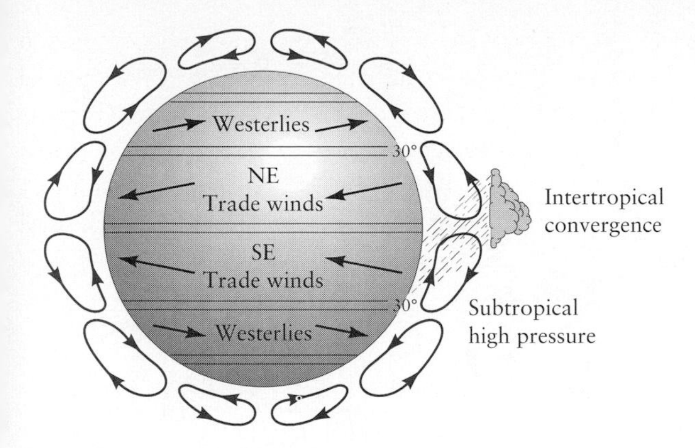

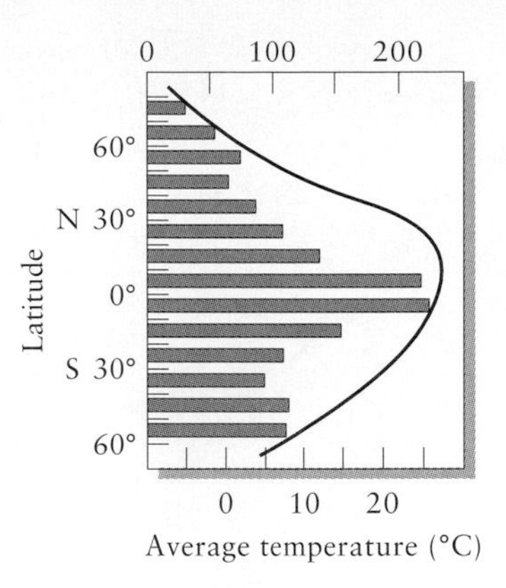

Figure 4.2 The illustration at left shows how differential warming of the earth's surface creates Hadley cells of air circulation. Warm, moist air rises in the Tropics, which results in abundant rainfall. Cool, dry air descends to the surface at subtropical latitudes. The intertropical convergence is the latitudinal belt at the solar equator within which surface winds converge from the north and south. At right, average annual precipitation (vertical bars) and temperature (line) are shown for 10° latitudinal belts within continental land masses. Because the chart presents averages for many localities, it obscures the great variation within each latitudinal belt.

the warming influence of the sun is referred to as the **intertropical convergence.** As moisture-laden tropical air rises and begins to cool, the moisture condenses to form clouds and precipitation. Thus, the Tropics are wet not because there is more water in tropical latitudes than elsewhere, but because water cycles more rapidly through the tropical atmosphere. The heating effect of the sun causes water to evaporate and warmed air masses to rise; the loss of heat as rising air expands causes precipitation.

The air mass moving high in the atmosphere to the north and south away from the intertropical convergence has already lost much of its water to precipitation in the Tropics. As it sinks and begins to warm at subtropical latitudes, its capacity to evaporate and hold water increases. As the air mass descends to ground level in the Subtropics and spreads to the north and south, it draws moisture from the land, creating zones of arid climate centered at latitudes of about 30° north and south of the equator (Figure 4.3). The great deserts of the world—the Arabian, Sahara, Kalahari, and Namib of Africa; the Atacama of South America; the Mojave, Sonoran, and Chihuahuan of North America; and the Australian—all fall within these belts.

The positions of continental land masses exert a secondary effect on the global pattern of precipitation. At any given latitude, rain falls more plentifully in the Southern Hemisphere because oceans and lakes cover a greater proportion of its surface (81% compared with 61% of the Northern Hemisphere). Water evaporates more readily from exposed surfaces of water than from soil and vegetation.

Energy from the sun drives the winds. The rotation of the earth deflects the surface flows in the Hadley cells to the west in the Tropics and to the east in the middle latitudes. The resulting wind patterns are called the trade winds and the westerlies, respectively (see Figure 4.2). Such air currents help to distribute water vapor through the atmosphere. Indeed, wind patterns interact with other features to create precipitation. Mountains force air upward, causing it to cool and lose its moisture as precipitation on the windward side of a mountain range. As the air descends the leeward slopes and travels across the lowlands beyond, it picks up moisture and creates arid environments called **rain shadows** (Figure 4.4). The Great Basin deserts of the western United States and the Gobi Desert of Asia lie in the rain shadows of extensive mountain ranges.

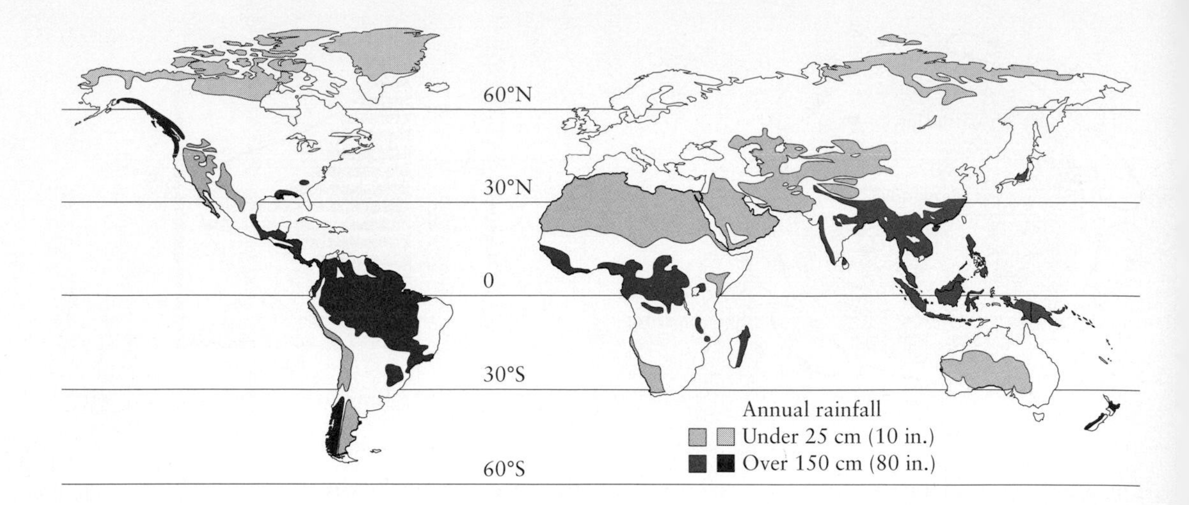

Figure 4.3 Distribution of the major deserts (regions with less than 25 cm annual precipitation within 50° of the equator) and wet areas (regions with more than 150 cm annual precipitation). Dark-colored areas are tropical rain forests; light-colored areas are subtropical deserts; black areas in western North America, Chile, and New Zealand are temperate rain forests; and gray areas are arctic deserts.

The interior of a continent usually experiences less precipitation than its coasts, simply because the interior lies farther from the major site of water evaporation, the surface of the ocean. Furthermore, coastal (maritime) climates vary less than interior (continental) climates because the great heat storage capacity and vertical mixing of ocean waters reduce temperature fluctuations. For example, the hottest and coldest mean monthly temperatures near the Pacific coast of the United States at Portland, Oregon, differ by only 16°C. Farther inland, this range increases to 18°C at Spokane, Washington; 26°C at Helena, Montana; and 33°C at Bismarck, North Dakota.

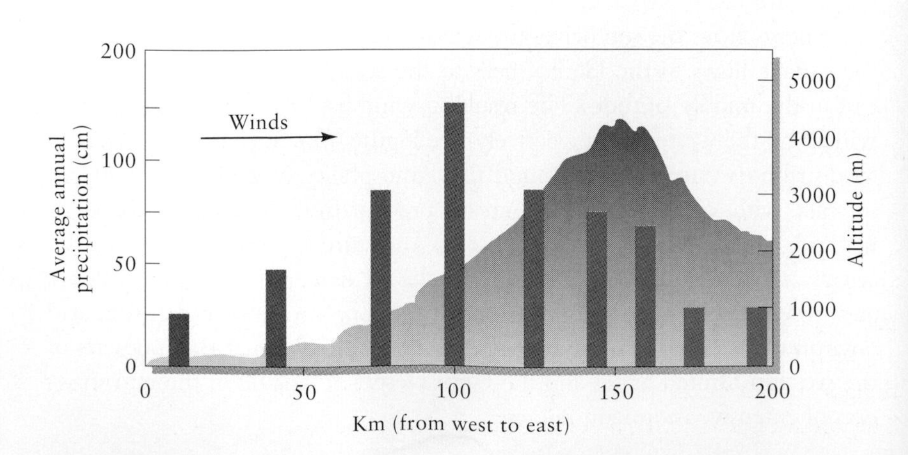

Figure 4.4 Influence of the Sierra Nevada mountain range on local precipitation and in causing a rain shadow to the east. Weather comes predominantly from the west across the Central Valley of California. As moisture-laden air is deflected upward by the mountains, it cools and its moisture condenses, resulting in heavy precipitation on the western slope of the mountains. As the air rushes down the eastern slope, it warms and begins to pick up moisture, creating arid conditions in the Great Basin. After E. R. Pianka, *Evolutionary Ecology,* 4th ed., Harper & Row, New York (1988).

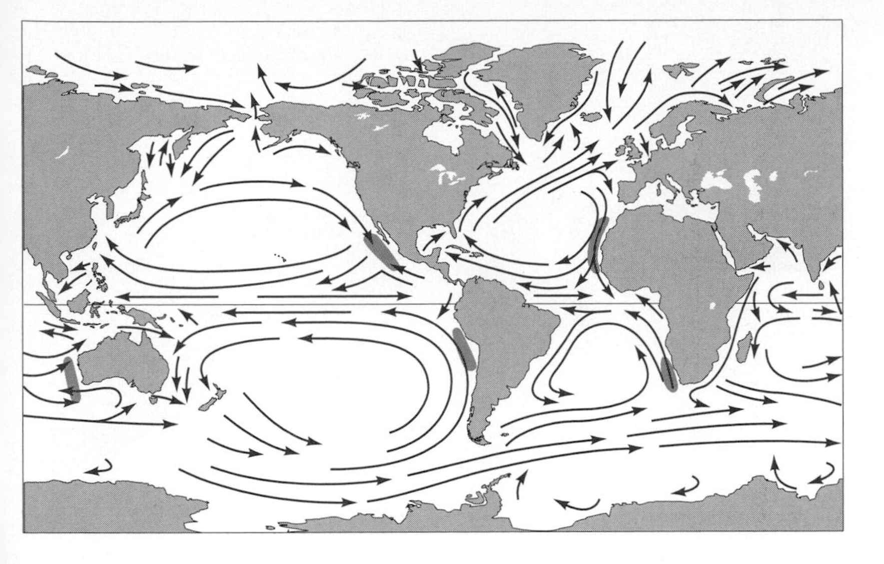

Figure 4.5 The major ocean currents. Water movement generally proceeds clockwise in the Northern Hemisphere and counterclockwise in the Southern Hemisphere. Zones of strong upwelling are indicated by light-colored shading. After A. C. Duxbury, *The Earth and Its Oceans,* Addison-Wesley, Reading, Mass. (1971).

The ocean environment

Ocean currents play a major role in moving heat over the surface of the earth and thereby influence the climates of the continents. In large ocean basins, cold water circulates toward the Tropics along the western coasts of the continents, and warm water circulates toward temperate latitudes along the eastern coasts of the continents (Figure 4.5). The cold Peru Current of the eastern Pacific Ocean, often called the Humboldt Current, which moves north from the Antarctic Ocean along the coasts of Chile and Peru, creates cool, dry environments along the west coast of South America right to the equator. Conversely, the warm Gulf Stream emanating from the Gulf of Mexico carries a mild climate far to the north into western Europe and the British Isles.

The physical conditions of the oceans themselves are as complex as those of the atmosphere. Variation in those conditions is caused by winds, which propel the major surface currents of the oceans, and by the underlying topography of the ocean basin. In addition, deeper currents are established by differences in the density of ocean water caused by variations in temperature and salinity.

Any upward movement of ocean water is referred to as **upwelling.** Vertical upwelling currents occur wherever surface currents diverge, as in the western tropical Pacific Ocean. As surface currents move apart, they tend to draw water upward from deeper layers. Strong upwelling zones are also established on the western coasts of continents where there are equatorward surface currents. A curious consequence of the rotation of the earth is the deflection of these currents away from the continental margins, which is aided by the trade winds. As this water moves away from the continents, it is replaced by water from greater depths. Because this water tends to be nutrient-rich, upwelling zones are often regions of very high biological productivity. The most famous of these support the rich fisheries of the Benguela Current along the western coast of southern Africa and the Peru Current along the western coast of South America.

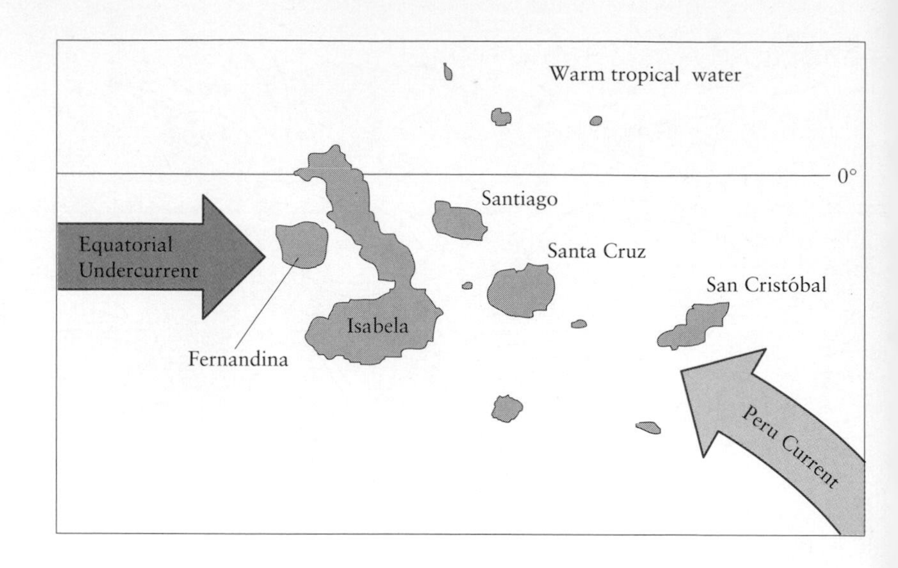

Figure 4.6 Map of the ocean currents in the region of the Galápagos archipelago. The primary influences are the deep, warm tropical waters to the north, the cold Peru Current, which approaches the islands from the southeast, and the cold, nutrient-laden Equatorial Undercurrent, which is forced to the surface as it encounters the western islands in the group. After G. T. Houvenaghel, in R. Perry (ed.), *Galápagos* (Key Environment Series), Pergamon Press, New York (1984), pp. 43–54

Cold temperature and high salinity at the ocean surface can result in vertical currents because cold, salty water is dense and tends to sink. These conditions are most pronounced around the continent of Antarctica. When water freezes each winter, salt is excluded from the crystalline ice. This causes an increase in the salinity of the water immediately underneath. This dense water descends and forms the Antarctic Bottom Water, which spreads north over the ocean basins. Close to the Antarctic continent, it is replaced at the surface by the so-called North Atlantic Deep Water, which has been slowly moving at great depth for hundreds of years. This water reaches the surface rather devoid of oxygen, but rich enough in accumulated nutrients to support tremendously high productivity. Just beyond this belt of upwelling around Antarctica is a region of downwelling caused by the increasing density of water as it warms toward 4°C. Because this pattern of downwelling and upwelling is driven by temperature and salinity differences in surface waters, it is referred to as **thermohaline circulation.**

Upwelling zones are extremely important to marine ecosystems. Because surface waters tend to be depleted of nutrients by the sedimentation of the remains of dead animals and plants, deep water rising to the surface usually creates a region of high productivity. Just as mountains force air currents upward, irregularities in the ocean bottom, especially around islands and archipelagoes, force water upward. Tropical coral reefs are productive in part because the islands around which they develop deflect deeper water to the surface. The Galápagos archipelago, which lies on the equator 1,000 km off the west coast of South America, has a particularly complex marine environment because the many islands intercept several ocean currents (Figure 4.6). To the north of the island group is deep ocean, warm at the surface and unproductive. From the coast of Peru to the southeast comes the Peru Current, cold and laden with nutrients received from coastal upwelling. These surface waters flow past the Galápagos to the west, bringing rich waters to the southernmost islands in the group. From the west, a strong current at mid-depth is forced to the surface by the westernmost islands of the archipelago, creating a zone exceptionally rich in ocean life and home to the only species of penguin living within the Tropics.

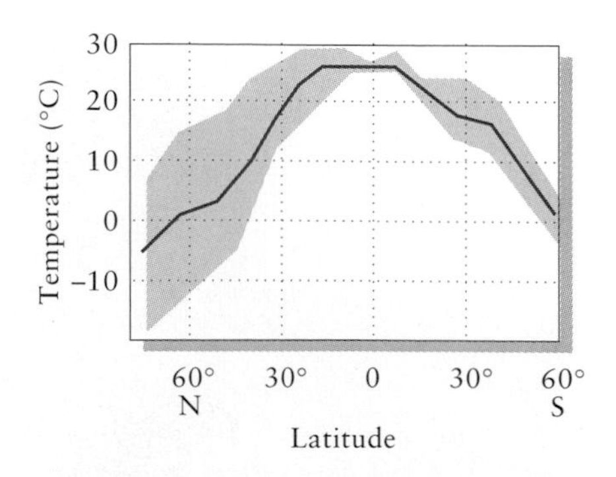

Figure 4.7 Annual range of mean monthly temperatures (shaded area) as a function of latitude. The mean annual temperature is indicated by the heavy line.

Seasonal variation in climate

Patterns of change in climate are as important to biological systems as the average temperature and precipitation. Periodic cycles in climate follow astronomical cycles: the rotation of the earth upon its axis causes daily periodicity; the revolution of the moon around the earth creates lunar cycles in the amplitude of the tides; and the revolution of the earth around the sun brings seasonal change.

The equator is tilted 23½° with respect to the path the earth follows in its orbit around the sun. As a result, the Northern Hemisphere receives more solar energy than the Southern Hemisphere during the northern summer, less during the northern winter. The seasonal range in temperature increases with distance from the equator (Figure 4.7). At high latitudes in the Northern Hemisphere, mean monthly temperatures vary by an average of 30°C, with extremes of more than 50°C annually; the mean temperatures of the warmest and coldest months in the Tropics differ by as little as 2° or 3°C.

Latitudinal patterns in the seasonality of rainfall result in part from the seasonal northward and southward movement of the solar equator and the resulting belts of wet and dry climate. Seasonality of rainfall is most pronounced in broad latitudinal belts lying about 20° north and south of the equator. As the seasons change, these regions alternately come under the influence of the solar equator, which brings heavy rains, and subtropical high-pressure belts, which bring clear skies (Figure 4.8).

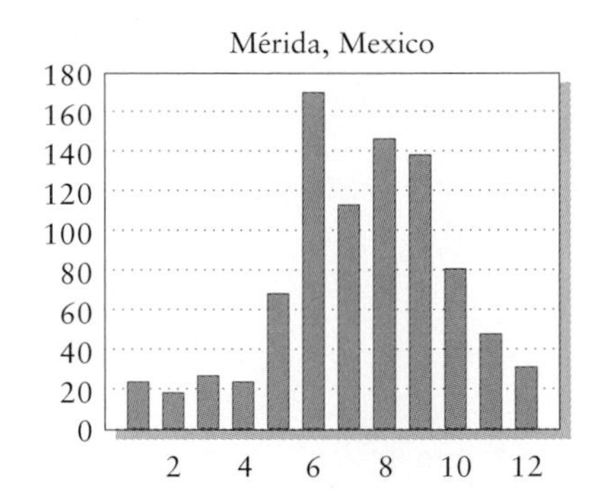

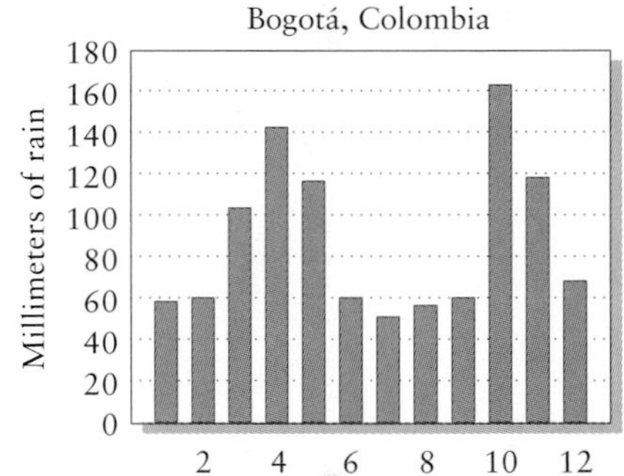

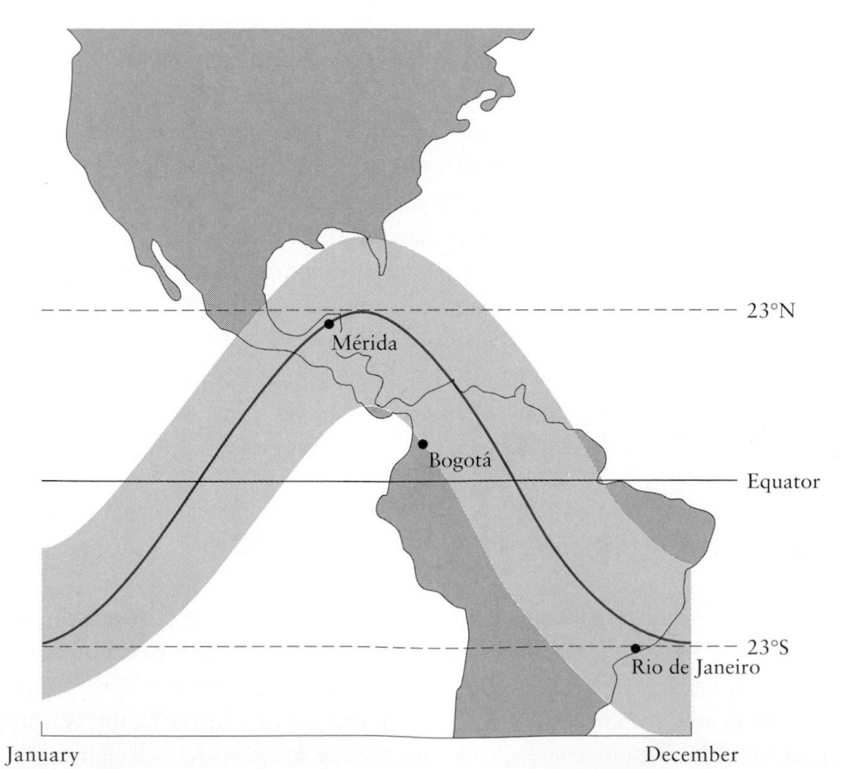

Figure 4.8 The seasonal latitudinal movement of the intertropical convergence (see Figure 4.2) results in two seasons of heavy precipitation at the equator and a single wet season alternating with a pronounced dry season at the edges of the Tropics.

Panama, at 10°N, lies within the wet Tropics, but even there the seasonal movement of the solar equator profoundly influences the climate. The major tropical belt of high rainfall remains south of Panama during most of the northern winter, but it lies directly overhead during the northern summer. Hence the winter is dry and windy, the summer humid and rainy. Panama's climate is wetter on the northern (Caribbean) side of the isthmus—the direction from which the prevailing trade winds come—than on the southern (Pacific) side; mountains intercept moisture coming from the Caribbean side and produce a rain shadow. The Pacific lowlands are so dry during the winter months that most trees lose their leaves (Figure 4.9). Tinder-dry forests and bare branches contrast sharply with the wet, lush, more typically tropical forest that abounds during the wet season.

Figure 4.9 Many trees on the Pacific slope of Panama shed their leaves during the dry season, which lasts from January through April. Photograph by M. A. Guerra, courtesy of the Smithsonian Tropical Research Institute.

Farther to the north, at 30°N in the Chihuahuan Desert of central Mexico, rainfall comes only during the summer, when the solar equator reaches its northward limit (Figure 4.10a). During the rest of the year this region falls within the dry subtropical high-pressure belt. Summer rainfall extends north into the Sonoran Desert of southern Arizona and New Mexico (Figure 4.10b). This area also receives moisture during the winter from the Pacific Ocean, carried by the southwesterly winds emanating from the subtropical high-pressure belt farther south. Thus the Sonoran Desert experiences both a winter and a summer peak of rainfall. Southern California lies beyond the summer rainfall belt and has a winter-rainfall, summer-drought climate (Figure 4.10c), often referred to as a **Mediterranean climate** because the Mediterranean region of Europe has the same seasonal pattern of temperature and rainfall. Mediterranean climates are also found in western South Africa, Chile, and Western Australia, all lying along the western sides of continents at about the same latitude north or south of the equator.

The sun warms the seas just as it does the continents and the atmosphere, but the ocean's great mass of water acts as a heat sink to dampen daily and seasonal fluctuations in temperature. Where ocean temperature does change seasonally, it reflects seasonal movements of water masses of different temperatures more often than it does local heating and cooling. During the Panamanian dry season, roughly January to April, steady trade winds blowing in a southwesterly direction create strong upwelling currents in the Pacific Ocean along the southern and western coasts of Central America. During these upwelling periods, winds blow warm surface water away from the coast, where cooler water moves upward from deeper regions to replace it. As a result, seawater temperature varies annually three times as much on the Pacific coast of Panama as on the Caribbean coast.

Seasonal cycles in temperate lakes

Small temperate zone lakes respond quickly to the changing seasons (Figure 4.11). In winter, a typical lake has an inverted **temperature profile;** that is, the coldest water (0°C) lies at the surface, just beneath the ice. (Because the density of water increases between the freezing point and 4°C, the warmer water within this range sinks, and temperature increases to as much as 4°C

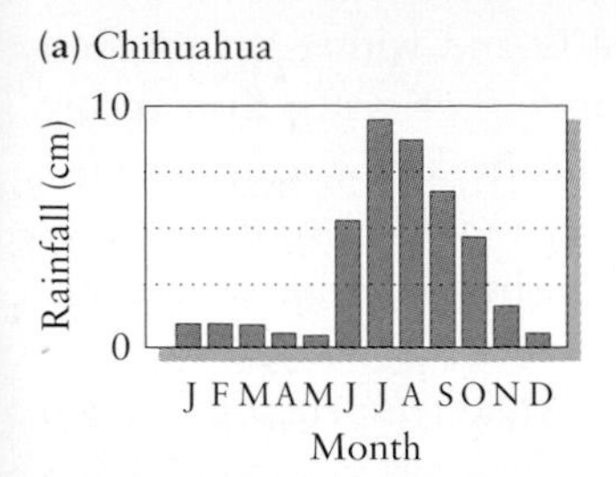

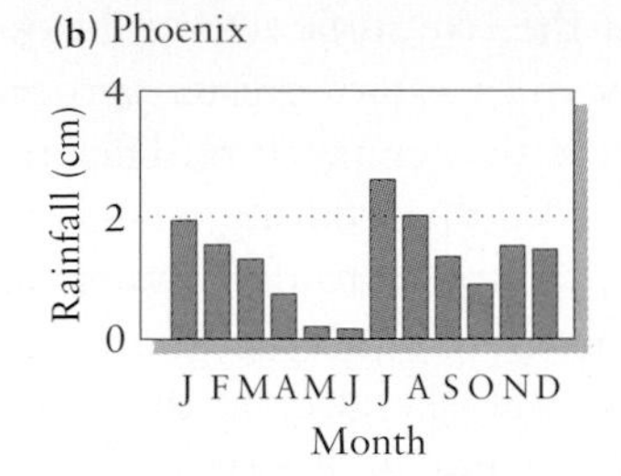

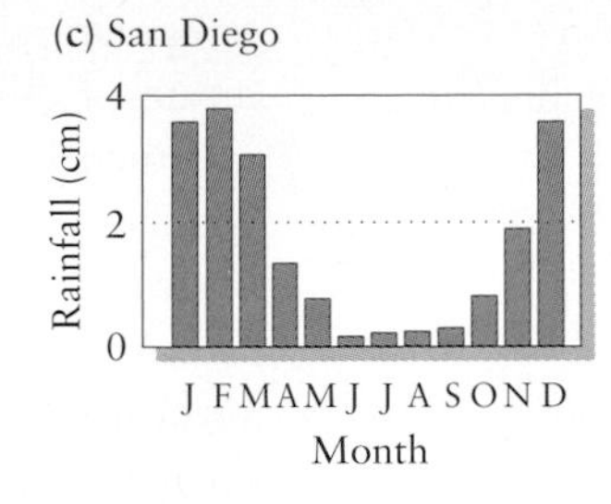

Figure 4.10 Seasonal occurrence of rainfall at three localities in western North America. (a) The summer rainy season of the Chihuahuan Desert in central Mexico. (b) The combined climate pattern of the Sonoran Desert. (c) The winter rain and summer drought of the Pacific coast (Mediterranean climate).

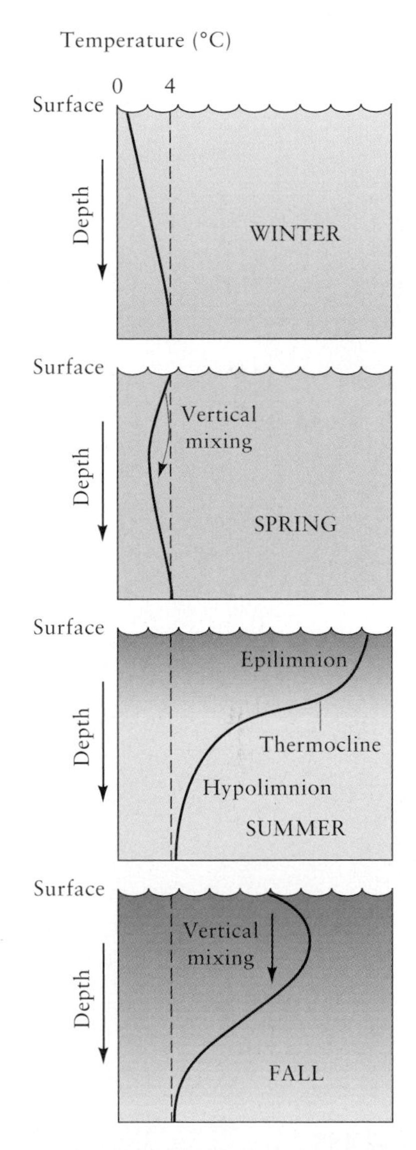

Figure 4.11 Seasonal changes in the temperature profile of a temperate lake result in vertical mixing of water in the spring and fall as surface layers become denser and begin to sink. Vertical mixing is also aided by wind-driven currents when the temperature of water is similar from the surface to the bottom of the lake.

toward the bottom of a lake.) In early spring the sun warms the lake surface gradually. But until the surface temperature exceeds 4°C, the sun-warmed surface water tends to sink into the cooler layers immediately below. This vertical mixing distributes heat throughout the water column from the surface to the bottom, resulting in a uniform temperature profile. Winds cause deep vertical movement of water in early spring **(spring overturn),** bringing nutrients to the surface from regions of nutrient release in bottom sediments and bringing oxygen from the surface to the depths.

Later in spring and in early summer, as the sun rises higher each day and the air above the lake warms, surface layers of water gain heat faster than deeper layers, creating a zone of rapid temperature change at intermediate depth, called the **thermocline.** Once the thermocline is well established, water does not mix across it. Now, at temperatures above 4°C, the warmer, less dense surface water literally floats on the cooler, denser water below, a condition known as **stratification.** The depth of the thermocline varies with local winds and with the depth and turbidity of the lake. It may occur anywhere between 5 and 20 m below the surface; lakes less than 5 m deep usually lack stratification.

The thermocline demarcates an upper layer of warm water (the **epilimnion**) and a deeper layer of cold water (the **hypolimnion**). Most of the primary production of the lake occurs in the epilimnion, where sunlight is most intense. Oxygen produced by photosynthesis supplements oxygen entering the lake at its surface, keeping the epilimnion well aerated and thus suitable for animal life, but plants and algae often deplete dissolved mineral nutrients and, in doing so, curtail their own production. The thermocline isolates the hypolimnion from the surface of the lake; animals and bacteria that remain below the euphotic zone of photosynthesis deplete the water of oxygen, creating anaerobic conditions. Thus, during late summer, the productivity of temperate lakes may become severely depressed, as nutrients needed to support plant growth in surface waters and oxygen needed to support animal life in the depths are both lacking.

During the fall, the surface layers of the lake cool more rapidly than deeper layers and, becoming heavier than the underlying water, begin to sink. This vertical mixing **(fall overturn)** persists into late fall, until the temperature at the lake surface drops below 4°C and winter stratification ensues. Fall overturn causes greater vertical mixing of water than spring overturn because the temperature differences in the lake during summer stratification exceed those during winter stratification. Fall overturn speeds the movement of oxygen to deep waters and pushes nutrients to the surface. Where the hypolimnion becomes fairly warm in midsummer, deep vertical mixing may take place in late summer while temperatures remain favorable for plant growth. Infusion of nutrients into surface waters at this time often causes an explosion in the population of phytoplankton—the **fall (autumn) bloom.** In deep, cold lakes, vertical mixing does not penetrate to all depths until late fall or early winter, when water temperatures are too cold to support phytoplankton growth.

Irregular fluctuations in climate

Everyone knows that weather is difficult to forecast far in advance. We often remark that a certain year was particularly dry or cold compared with others. The flooding in recent years in the Mississippi Valley and along many rivers in Europe drives home the capriciousness of nature. Most aspects of climate seem unpredictable. Rainfall varies most where it is sparsest: in deserts and, in wetter localities, during the driest season. Year-to-year variation in temperature on a particular date is greatest where temperature fluctuates most from season to season during the year. The most extreme conditions occur infrequently, but they may affect organisms disproportionately.

The rich Peruvian fishing industry, as well as some of the world's largest seabird colonies (Figure 4.12), thrives on the abundant fish in the rich waters of the Peru Current, a mass of cold water that flows up the western coast of South America and finally veers offshore at Ecuador,

Figure 4.12 Nesting colony of Peruvian boobies on an island off the coast of Peru. This dense population depends on the anchovy stocks in the rich Peru (Humboldt) Current. Photograph by R. C. Murphy, courtesy of the Department of Library Services, American Museum of Natural History.

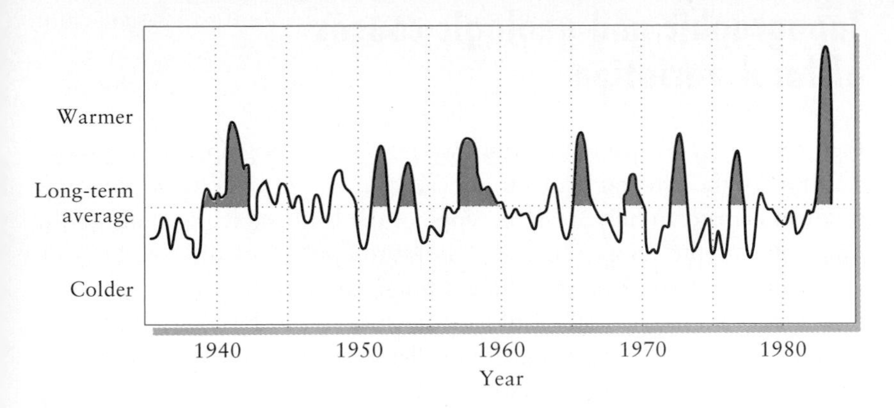

Figure 4.13 El Niño–Southern Oscillation (ENSO) events (shaded peaks) are marked by large positive anomalies in sea surface temperature in South American coastal waters. From E. M. Rasmussen, *Am. Sci.* 73:168–177 (1985).

toward the Galápagos archipelago. North of this, warm, tropical inshore waters prevail along the coast. Each year a warm countercurrent known as El Niño ("little boy" in Spanish, a name referring to the Christ child and chosen because this countercurrent appears around Christmastime) moves down the coast toward Peru. Infrequently, it flows strongly enough and far enough to force the cold Peru Current offshore, taking with it the food supply of millions of birds.

During "normal" years between El Niño "events," a steady wind blows across the equatorial central Pacific Ocean from an area of high atmospheric pressure centered over Tahiti to an area of low pressure centered over Darwin, Australia. An El Niño event appears to be triggered by a reversal of these pressure areas (the so-called Southern Oscillation) and of the winds that flow between them. As a result, the westward-flowing equatorial currents stop or even reverse; upwelling off the coast of South America weakens or ceases; and warm water—the El Niño current—piles up along the coast of South America. Historical records of atmospheric pressure at Tahiti and Darwin, and of sea surface temperatures on the Peruvian coast, reveal pronounced ENSO (El Niño–Southern Oscillation) events at irregular intervals of 2 to 10 years (Figure 4.13).

The climatic and oceanographic effects of an ENSO event extend over much of the world, affecting ecosystems in such distant areas as India, South Africa, Brazil, and western Canada. A major ENSO event in 1982–1983 disrupted fisheries and destroyed kelp beds in California, caused reproductive failure of seabirds in the central Pacific Ocean, and resulted in widespread mortality of coral in Panama. Precipitation was also dramatically affected in many terrestrial ecosystems, notably in the deserts of northern Chile, normally the driest places on earth, which received the first recorded rainfall in over a century. The conspicuous and well-studied example of the ENSO climate pattern is representative of what happens on a smaller scale in many parts of the world. No location, except perhaps in the depths of the oceans, escapes the vagaries of climate.

Topographic and geologic causes of local variation

Returning now to spatial variation, we note that topography and geology can modify the environment on a local scale within regions of otherwise uniform climate. In hilly areas, the slope of the land and its exposure to the sun influence the temperature and moisture content of the soil. Soils on steep slopes drain well, often causing drought stress for plants on the hillside at the same time that water saturates the soils of nearby lowlands. In arid regions, stream bottomlands and seasonally dry riverbeds may support well-developed forests, called **riparian** forests, which accentuate the contrasting bleakness of the surrounding desert. In the Northern Hemisphere, south-facing slopes directly face the sun, whose warmth and drying power limit vegetation to shrubby, drought-resistant **(xeric)** forms. The adjacent north-facing slopes remain relatively cool and wet and harbor moisture-requiring **(mesic)** vegetation (Figure 4.14).

Air temperature decreases with altitude by about 6°–10°C for each 1,000-m increase in elevation, depending on the region. This decrease in temperature, which is caused by the expansion of air with the lower atmospheric pressure at higher altitude, is referred to as **adiabatic cooling.** Climb high enough, even in the Tropics, and you will encounter freezing temperatures and perpetual snow. Where the temperature at sea level averages 30°C, freezing temperatures are reached at about 5,000 m, the approximate altitude of the snow line on tropical mountains.

In north-temperate latitudes, a 6°C drop in temperature corresponds to the temperature change encountered over an 800-km increase in latitude. In many respects, the climate and vegetation of high altitudes resemble those of sea-level localities at higher latitudes. But despite their similarities, alpine environments usually vary less from season to season than their low-elevation counterparts at higher latitudes. Temperatures in tropical montane environments remain nearly constant, and some of these areas remain frost-free over the year, which makes it possible for many tropical plants and animals to live in the cool environments found there.

In the mountains of the southwestern United States, changes in plant communities with elevation result in more or less distinct belts of vegetation, which the nineteenth-century naturalist C. H. Merriam referred to as **life zones.** Merriam's scheme of classification included five broad zones, which he named, from low to high elevation (or from south to north), the Lower Sonoran, Upper Sonoran, Transition, Canadian (or Hudsonian), and Alpine (or Arctic-Alpine) (Figure 4.15).

At low elevations in the North American Southwest, one encounters a cactus and desert shrub association characteristic of the Sonoran Desert of northern Mexico and southern Arizona. In the riparian forests along streambeds, the plants and animals have a distinctly tropical flavor. Many hummingbirds and flycatchers, ring-tailed cats, jaguars, and peccaries make their only temperate zone appearances in this area. In the Alpine zone, 2,600 m higher, one finds a landscape resembling the tundra of northern

Figure 4.14 The influence of exposure on the vegetation of a series of mountain ridges near Aspen, Colorado. The cool and moist north-facing (left-facing) slopes permit the development of spruce forest. Shrubby, drought-resistant vegetation grows on the south-facing slopes.

Canada and Alaska. Thus, by climbing 2,600 m, one experiences changes in climate and vegetation that would require a journey to the north of 2,000 km or more at sea level.

Local variation in the bedrock underlying a region promotes the differentiation of soil types and enhances biotic heterogeneity. In the northern Appalachian Mountains and in mountains near the Pacific coast of the United States, outcrops of **serpentine,** a kind of igneous rock, weather to form soils having so much magnesium that plants characteristic of surrounding soil types cannot grow. Serpentine **barrens,** as they are called, support little more than a sparse covering of grasses and herbs (Figure 4.16), many of which are distinct **endemics**—species found nowhere else—that have evolved a high tolerance for magnesium. Depending on the composition of the bedrock and on the rate of weathering, granite, shale, and sandstone also can produce barrens. The extensive pine barrens of southern New Jersey, where mature trees attain no more than waist height in some areas, occur on a large outcrop of sand, which produces a dry, acid, infertile soil.

The landscape concept

Variations in topography and soils within a region create a heterogeneous mosaic of habitat "patches" that make up the local **landscape.** The concept of landscape emphasizes environmental variation perceived over dimensions of tens of meters to kilometers, and stresses the importance of the movement of individuals and materials between habitat patches to the

Hudsonian Zone
Elevation 2500 m

Alpine Zone
Elevation 3500 m

Upper Sonoran Zone
Elevation 1500 m

Transition Zone
Elevation 2000 m

Lower Sonoran Zone
Elevation 900 m

Upper Sonoran Zone
Elevation 1200 m

Figure 4.15 Vegetation at different elevations in the mountains of southeastern Arizona. At the lowest elevations (bottom photographs), the Lower Sonoran Zone supports mostly saguaro cactus, small desert trees such as paloverde and mesquite, numerous annual and perennial herbs, and small succulent cacti; agave, ocotillo, and grasses are conspicuous elements of the Upper Sonoran Zone, and oaks appear toward its upper edge. Large trees predominate at higher elevations: ponderosa pine in the Transition Zone, spruce and fir in the Hudsonian Zone. These gradually give way to bushes, willows, herbs, and lichens in the Alpine Zone above the tree line. Courtesy of the U.S. Soil Conservation Service, the U.S. Forest Service, W. J. Smith, and R. H. Whittaker. From R. H. Whittaker and W. A. Niering, *Ecology* 46:429–452 (1965).

Figure 4.16 A small serpentine barren in eastern Pennsylvania. The soils surrounding the barren support a forest of oak, hickory, and beech.

maintenance of ecological processes over the entire landscape. Thus, populations are subdivided into local subpopulations residing in patches of suitable habitat. Their persistence depends not only on processes within each patch but also on the movement of individuals between patches. As a consequence, the organization of the landscape strongly influences population dynamics, a principle to which we shall return later.

The exchange of organisms and materials between habitats makes the landscape a functional level of organization; **landscape ecology** is dedicated to the study of the ways in which the mosaic nature of the environment influences the functioning of ecological systems. That the habitat mosaic functions as an integrated ecological system becomes clear when we consider that individuals can move between patches of a given habitat type and between habitat types (for example, breeding in one type of habitat and feeding in another), and that materials also find their way across habitat boundaries (for example, leaves falling into streams).

Human activities alter landscape patterns as well as individual habitats. Cutting trees not only reduces the area of forest but also creates a mosaic of forest and field patches. The sizes and proximity of these patches influence the dispersal of individuals between suitable areas and the maintenance of their populations within the region. If patches become small enough, a population may become extinct despite the presence of suitable habitat within the landscape. We shall consider landscape-related topics at many places in this book. At this point, it is enough to understand that habitat heterogeneity affects local ecosystems and populations. In this instance, the landscape is equal to more than the sum of its habitat parts.

Soils

Climate affects plants and animals indirectly through its influence on the development of soil, which provides the substrate within which plant roots grow and many animals burrow. The characteristics of soil determine its ability to hold water and to make available the minerals required for plant growth. Thus, its variation provides a key to understanding the distribution of plant species and the productivity of biological communities.

Soil defies simple definition, but we may describe it as the layer of chemically and biologically altered material that overlies rock or other unaltered materials at the surface of the earth. It includes minerals derived from the parent rock, modified minerals formed anew within the soil, organic material contributed by plants, air and water within the pores of the soil, living roots of plants, microorganisms, and the larger worms and arthropods that make the soil their home (see Figure 1.7). Five factors largely determine the characteristics of soils: climate, parent material (underlying rock), vegetation, local topography, and, to some extent, age.

Soils exist in a dynamic state, changing as they develop on newly exposed rock. And even after soils achieve stable properties, they remain in a constant state of flux. Groundwater removes some substances; other materials enter the soil from vegetation, in precipitation, as dust from above, and from the rock below. Where little rain falls, the parent material decomposes slowly and plant production adds little organic detritus to the soil. Thus arid regions typically have shallow soils, with bedrock lying close to the surface (Figure 4.17). Soils may not form at all where decomposed bedrock and detritus erode as rapidly as they form. Soil development also stops short on alluvial deposits, where fresh layers of silt deposited each year by floodwaters bury older material. At the other extreme, soil formation proceeds rapidly in parts of the humid Tropics, where chemical alteration of parent material may extend to depths of 100 meters. Most soils of temperate zones are intermediate in depth, extending to a rough average of about 1 meter.

Figure 4.17 Profile of a poorly developed soil (inceptisol) in Logan County, Kansas, illustrating shallow soil depth and absence of soil zonation. Courtesy of the U.S. Soil Conservation Service.

Soil horizons

Where a recent road cut or excavation exposes soil in cross section, one often notices distinct layers, which are called **horizons** (Figure 4.18). A generalized, and somewhat simplified, **soil profile** has four major divisions, the O, A, B, and C horizons; the A horizon has two subdivisions (A_1 and A_2). Arrayed in descending order from the surface of the soil, the horizons and their predominant characteristics are as follows:

O Primarily dead organic litter. Most soil organisms inhabit this layer.

A_1 A layer rich in humus, consisting of partly decomposed organic material mixed with mineral soil.

A_2 A region of extensive leaching of minerals from the soil. Because minerals are dissolved by water—that is to say, mobilized—in this layer, plant roots are concentrated here.

B A region of little organic material whose chemical composition resembles that of the underlying rock. Clay minerals and oxides of aluminum and iron leached out of the overlying A_2 horizon are sometimes deposited here.

C Primarily weakly altered material, similar to the parent rock. Calcium and magnesium carbonates accumulate in this layer, especially in dry regions, sometimes forming hard, impenetrable layers, or "pans."

Soil horizons reveal the decreasing influence of climate and biotic factors with increasing depth. Critical to soil formation is the movement of mineral elements upward and downward through the soil profile. But before considering these processes in detail, we shall examine the initial alteration of the bedrock and how it influences soil characteristics.

Weathering

Weathering—the physical and chemical alteration of rock material near the earth's surface—occurs wherever surface water penetrates. The repeated freezing and thawing of water in crevices physically breaks rock into smaller pieces and exposes a greater surface area to chemical action. Initial chemical alteration of the rock occurs when water dissolves some of the more soluble minerals, especially sodium chloride ($NaCl$) and calcium sulfate ($CaSO_4$). Other materials, such as the oxides of titanium, aluminum, iron, and silicon, dissolve less readily.

The weathering of granite exemplifies some basic processes of soil formation. The minerals making up the grainy texture of granite—feldspar, mica, and quartz—consist of various combinations of oxides of aluminum, iron, silicon, magnesium, calcium, and potassium, along with other, less abundant compounds. The key to weathering is the displacement of certain elements in these minerals—notably calcium, magnesium, sodium, and potassium—by hydrogen ions, followed by the reorganization of the remaining oxides of aluminum, iron, and silicon into new minerals.

Feldspar, which consists of aluminosilicates of potassium, weathers rapidly because of the displacement of potassium ions (K^+) by hydrogen ions (H^+) to form new, insoluble materials, particularly clay particles. Mica grains consist of aluminosilicates of potassium, magnesium, and iron. As in feldspar, the potassium and magnesium are displaced readily during weathering, and the remaining iron, aluminum, and silicon make new minerals. Hydrated iron and aluminum silicates form various kinds of **clay** particles, which contribute importantly to the water- and cation-holding capacity of soils. Quartz, a type of silica (SiO_2), is relatively insoluble and therefore remains more or less unaltered in the soil as grains of sand. Changes in chemical composition as granite weathers from rock to soil in different climate regions show that weathering is most severe under tropical conditions of high temperature and rainfall (Figure 4.19).

An important factor in the initial weathering of parent material, regardless of the chemical nature of the rock, is the presence of hydrogen ions in the water that percolates down to the bedrock. These ions derive from two sources. All precipitation contains dissolved carbon dioxide, which, as we have seen, forms carbonic acid. Some of the carbonic acid dissociates into hydrogen ions (H^+) and bicarbonate ions (HCO_3^-). In regions not affected by pollution-caused acidification, concentrations of hydrogen ions in rainwater produce a pH of about 5. This acidity is supplemented by hydrogen ions generated by the oxidation of organic material in the soil. The metabolism of carbohydrate, for example, produces carbon dioxide, and dissociation

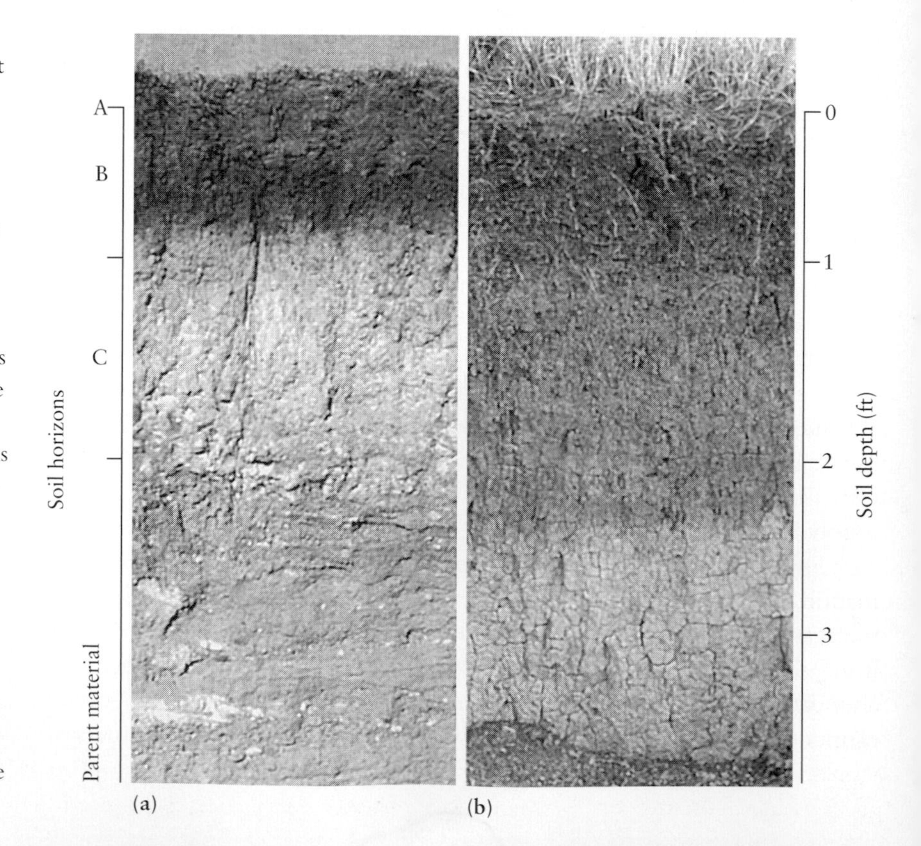

Figure 4.18 Soil profiles of mollisols (prairie soils) from the central United States, illustrating distinct layers, or horizons. (a) This profile, from eastern Colorado, is weathered to a depth of about 2 feet (0.6 m), where the subsoil contacts the parent material, which consists of loosely aggregated, calcium-rich, wind-deposited sediments (loess). A_1 and A_2 horizons are not clearly distinguished. The B horizon contains a dark band of redeposited organic materials that were leached from the uppermost layers of the soil. The C horizon is light colored and has been leached of much of its calcium. Some of the calcium has been redeposited at the base of the C horizon and at greater depths in the parent material. (b) A typical prairie soil from Nebraska. Rainfall here is sufficient to leach readily soluble ions completely from the soil; hence there are no B layers of redeposition, as in the drier Colorado soil, and the profile is more homogeneous. The A horizon is weakly subdivided into a darker upper layer and a lighter lower layer. The weathered soil lies on a parent material composed of loess, the wind-blown remnants of glacial activity.

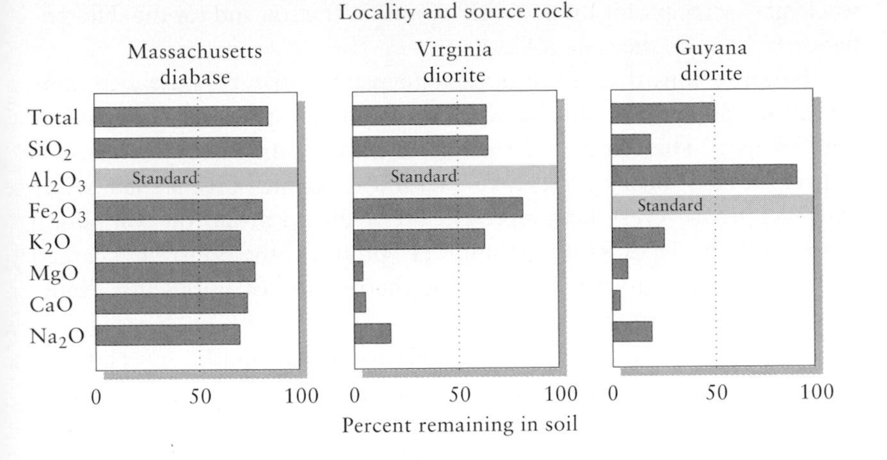

Figure 4.19 Differential removal of minerals from granitic rocks as a result of weathering in Massachusetts, Virginia, and Guyana. Values are compared with either aluminum or iron oxides (these standards = 100%), which are assumed to be the most stable components of the mineral soil. After E. W. Russell, *Soil Conditions and Plant Growth,* 9th ed., Wiley, New York (1961).

of the resulting carbonic acid generates additional hydrogen ions. In the Hubbard Brook Forest of New Hampshire, these internal processes account for about 30% of the hydrogen ions needed for the weathering of bedrock; the remainder comes from precipitation. In the Tropics, however, internal sources of hydrogen ions produced by biological oxidation of organic substrates in the soil assume greater importance and may lead to more rapid weathering.

Clay, humus, and the cation exchange capacity of soil

Plants obtain mineral nutrients from the soil in the form of dissolved ions, whose solubility derives from their electrostatic attraction to water molecules. Because ions are dissolved in water, those not immediately taken up by plants and fungi may wash out of the soil if they do not adhere strongly to stable soil particles. Clay particles and **humus** particles consisting of fine organic detritus, separately or associated in complexes, are large enough to form a stable component of the soil. These particles and complexes, known as **micelles,** have negative electric charges at their surfaces that hold the smaller, more mobile ions in the soil (Figure 4.20). The number of sites on soil particles available for binding positively charged ions (cations) is referred to as the soil's **cation exchange capacity.**

The bonds between soil particles and such ions as potassium (K^+) and calcium (Ca^{2+}) are relatively weak, so they constantly break and form anew. When a potassium ion dissociates from a micelle, its place may be taken by any other positive ion close by. Some ions cling more strongly to micelles than others—in order of decreasing tenacity: hydrogen, calcium, magnesium, potassium, and sodium. Hydrogen ions thus tend to displace calcium and all other cations in the soil. If cations were not added to or removed from the soil, the relative proportions of the various cations associated with clay-humus particles would achieve a steady state. But carbonic acid in rainwater and organic acids produced by the decomposition of organic detritus continuously add hydrogen ions to the upper layers of the soil; these readily supplant other cations, which are then washed out of the soil and into the groundwater. The influx of hydrogen ions in water percolating through the

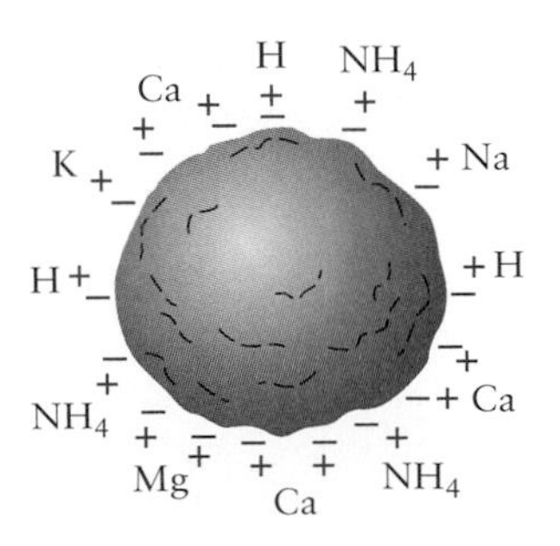

Figure 4.20 Schematic representation of a clay or humus particle (micelle) with hydrogen ions and mineral ions attracted by negative charges at its surface. After S. R. Eyre, *Vegetation and Soils,* 2d ed., Aldine, Chicago (1968).

soil largely accounts for the mobility of ions in the soil and for the differentiation of layers in the soil profile.

Negative ions that are important to plant nutrition, such as nitrate, phosphate, and sulfate, can be adsorbed onto clay particles by means of ion "bridges." These bridges form under acid conditions by the association of an additional hydrogen ion with a functional group such as the hydroxyl group (OH). Let's represent the hydroxyl group on compound R in a soil micelle as ROH. ROH + H^+ produces the positively charged ROH_2^+, which in turn makes possible the binding of anions (for example, $ROH_2^+ \cdots NO_3^-$).

Because of the cation-binding properties of clay and humus particles, the potential long-term fertility of soil—its capacity for storing nutrients—depends in large part on its clay content. Furthermore, because hydrogen ions displace others from soil particles and influence the electric charges on the surfaces of these particles, the retention of ions in the soil and the immediate availability of ions to plants depend to a large degree on the acidity of the soil.

Podsolization

The qualities of soils depend on the underlying parent rock and on the climate (Table 4.1). Under mild, temperate conditions of temperature and rainfall, sand grains and clay particles resist weathering and form stable components of the soil skeleton. In acid soils, however, clay particles break down in the A horizon of the soil profile, and their soluble ions are transported downward and deposited in lower horizons. This process, known as **podsolization,** reduces the ion exchange capacity, and therefore the fertility, of the upper layers of the soil.

Acid soils (spodosols) occur primarily in cold regions where needle-leaved trees dominate the forests. The slow decomposition of plant litter produces organic acids. In addition, rainfall usually exceeds evaporation in regions of podsolization. Under these moist conditions, because water continuously moves downward through the soil profile, little clay-forming material is transported upward from the weathered bedrock below.

In North America, podsolization advances furthest under spruce and fir forests in New England and the Great Lakes region and across a wide belt of southern and western Canada. A typical profile of a highly podsolized soil (Figure 4.21) reveals striking bands corresponding to the regions of leaching and redeposition. The topmost layer of the profile (A_1) is dark and rich in organic matter. This is underlain by a light-colored horizon (A_2) that has been leached of most of its clay content. As a result, the A_2 horizon consists mainly of sandy skeletal material that holds neither water nor nutrients well. One usually finds a dark band of deposition immediately below the A_2 horizon. This is the uppermost layer of the B horizon, where iron and aluminum oxides are redeposited. Other, more mobile minerals may accumulate to some extent in lower parts of the B horizon, which then grades almost imperceptibly into a C horizon and the parent material.

TABLE 4.1 **The major soil groups**

NAME	CHARACTER	DISTRIBUTION IN THE UNITED STATES
Alfisols	Moist, moderately weathered mineral soils	Ohio Valley, Great Lakes region, Rocky Mountains, central California
Aridosols	Dry mineral soils with little leaching and accumulations of calcium carbonate	Great Basin, southwestern deserts
Entisols	Recent mineral soils lacking development of soil horizons	On hard rock in the Rocky Mountains, on sands in the Southwest
Histosols	Organic soils of peat bogs; mucks	Northern Minnesota, Mississippi Delta, Florida Everglades
Inceptisols	Young, weakly weathered soils	New York, Pennsylvania, West Virginia; alluvial soils of Mississippi Valley, northwestern states, Alaska
Mollisols	Well-developed soils high in organic matter and calcium; very productive	Most of the prairie soils of the Great Plains
Oxisols	Deeply weathered, lateritic soils of moist tropics	Tropical South America and Africa; not present in the United States
Spodosols	Acid, podsolized soils of cool, moist climates with a shallow leached horizon and a deeper layer of deposition	Forested regions of New England; northern Michigan and Wisconsin
Ultisols	Highly weathered soils of warm, moist climates with abundant iron oxides	Most of the southeastern United States, the Pacific Northwest
Vertisols	High content of swelling-type clays developing deep cracks in dry seasons	Southern Texas

Laterization

The warm, wet climates of many tropical regions weather the soil to great depths. One of the most conspicuous features of weathering under these conditions is the breakdown of clay particles, which results in the leaching of silica from the soil and leaves oxides of iron and aluminum to predominate in the soil profile. This process is called **laterization,** and the iron and aluminum oxides give lateritic soils (oxisols) their characteristic reddish coloration. Even though the rapid decomposition of organic material in tropical soils contributes an abundance of hydrogen ions, these are quickly neutralized by the bases formed by the breakdown of clay minerals; consequently, oxisols usually are not very acidic. Laterization is enhanced in certain soils that develop on parent material deficient in quartz (SiO_2) but rich in iron and magnesium (basalt, for example); these soils contain little clay to

Figure 4.21 Profile of a podsolized soil in Plymouth County, Massachusetts. The light-colored A_2 horizon and the dark-colored B_1 horizon immediately below it form distinct bands. Compare the general absence of roots in the A_2 horizon with their presence in the lower B_1 horizon. Courtesy of the U.S. Soil Conservation Service.

begin with because they lack silicon. Regardless of the parent material, weathering reaches deepest and laterization proceeds furthest on low-lying soils, such as those of the Amazon Basin, where highly weathered surface layers are not eroded away and the soil profiles are very old.

One of the consequences of laterization in many parts of the Tropics is that the capacity of the soil to hold nutrients is very poor. Without clay and humus, cation exchange capacity can be low, in which case mineral nutrients are readily leached out of the soil. Where soils are weathered deeply, new minerals formed by the decomposition of the parent material are simply too far from the surface layers of the soil to contribute to soil fertility. Besides, heavy rainfall keeps water moving down through the soil profile, preventing the upward movement of nutrients. In general, the deeper the ultimate source of nutrients in the unaltered bedrock, the poorer the surface layers. Rich soils do, however, develop in many tropical regions, particularly in mountainous areas where erosion continually removes nutrient-depleted surface layers of the soil, and in volcanic areas where the parent material of ash and lava is often rich in such nutrients as potassium.

Soil formation emphasizes the role of the physical environment, particularly climate, geology, and landforms, in creating the tremendous variety of environments for life that exist at the surface of the earth and in its waters. In the next chapter, we shall see how this variety affects the distribution of life forms and the appearance of biological communities.

Summary

1. Global patterns of temperature and rainfall result from differential input of solar radiation in different regions and from the redistribution of heat energy by winds and ocean currents. Prominent features of terrestrial cli-

mates include a band of warm, moist climate over the equator and bands of dry climate at about 30° north and south latitude.

2. Variation in the marine environment is determined on a global scale by circulation of the major ocean currents. More locally, upwelling currents caused by winds, ocean basin topography, and variations in water density related to temperature and salinity bring cold, nutrient-rich water to the surface in some areas.

3. Seasonality in terrestrial environments is caused by the annual progression of the sun's path northward and southward and by the latitudinal movement of associated belts of temperature, wind, and precipitation. At high latitudes, the seasons primarily reflect annual cycles of temperature. Within the Tropics, seasonality of precipitation is more pronounced.

4. Seasonal warming and cooling profoundly change the characteristics of lakes in the temperate zones. During summer, such lakes are stratified, with a warm surface layer (epilimnion) separated from a cold bottom layer (hypolimnion) by a sharp thermocline. In spring and fall, the profile of temperature with depth becomes more uniform, allowing vertical mixing.

5. Irregular and unpredictable variations in climate, such as El Niño–Southern Oscillation events, may cause major disruptions of biological communities on a global scale.

6. Topography and geology superimpose local variation in environmental conditions on more general climate patterns. Mountains intercept rainfall, creating arid rain shadows in their lees. Conditions at higher altitudes resemble conditions at higher latitudes. Soil characteristics reflect the quality of the underlying bedrock and sometimes foster specialized floras, such as those of serpentine barrens.

7. Ecologists refer to mosaics of habitat patches as landscapes. The movement of individuals and materials between patches influences populations and ecosystems within the landscape as a whole.

8. Nutrient regeneration in terrestrial systems takes place in the soil. The characteristics of soil reflect the influences of the bedrock below and the climate and vegetation above. Weathering of bedrock results in the breakdown of some native minerals (feldspar and mica) and their re-formation into clay particles, which mix with organic detritus entering the soil from the surface. These vertically graded processes usually result in distinct soil horizons.

9. The clay and humus content of the soil determines its ability to retain nutrients needed by plants. Clay and humus particles (micelles) have negative charges on their surfaces that attract cations (Ca^{2+}, K^+, NH_4^+); anions (PO_4^{3-}, NO_3^-) may be held indirectly by ion bridges under acid conditions. Hydrogen ions tend to displace other cations on micelles and thereby reduce soil fertility.

10. In acid temperate zone (podsolized) soils and in deeply weathered (laterized) tropical soils, clay particles break down and the fertility of the soil is much reduced.

Suggested readings

Barber, R. T., and F. P. Chavez. 1983. Biological consequences of El Niño. *Science* 222:1203–1210.

Barry, R. G., and R. J. Chorley. 1976. *Atmosphere, Weather, and Climate.* 3d ed. Methuen, London.

Brady, N. C. 1974. *Nature and Property of Soils.* 8th ed. Macmillan, New York.

Forman, R. T. T., and M. Gordon. 1986. *Landscape Ecology.* John Wiley & Sons, New York.

Graedel, T. E., and P. J. Crutzen. 1995. *Atmosphere, Climate, and Change.* Scientific American Library, New York.

Hobbs, R. J., D. A. Saunders, and G. W. Arnold. 1993. Integrated landscape ecology: A Western Australian perspective. *Biological Conservation* 64: 231–238.

Jenny, H. 1980. *The Soil Resource: Origin and Behavior.* Springer-Verlag, New York.

Peterson, C. H. 1991. Intertidal zonation of marine invertebrates in sand and mud. *American Scientist* 79:236–249.

Philander, G. 1989. El Niño and La Niña. *American Scientist* 77:451–459.

Rasmussen, E. M. 1985. El Niño and variations in climate. *American Scientist* 73:168–177.

Shelford, V. E. 1963. *The Ecology of North America.* University of Illinois Press, Urbana.

Sherman, K., L. M. Alexander, and B. D. Gold (eds.) 1990. *Large Marine Ecosystems: Patterns, Processes, and Yields.* American Association for the Advancement of Science, Washington, D.C.

Waring, R. H., and J. Major. 1964. Some vegetation of the California coastal region in relation to gradients of moisture, nutrients, light, and temperature. *Ecological Monographs* 34:167–215.

Wiens, J. A., N. C. Stenseth, B. Van Horne, and R. A. Ims. 1993. Ecological mechanisms and landscape ecology. *Oikos* 66:369–380.

CHAPTER 5

BIOLOGICAL COMMUNITIES: THE BIOME CONCEPT

Climate, topography, and soil—and parallel influences in aquatic environments—determine the changing character of plant and animal life over the surface of the earth. Although no two points are inhabited by exactly the same assemblage of species, biological communities can be grouped into categories based on the dominant plant form, which gives the community its overall character. These categories are referred to as **biomes.** Important terrestrial biomes of North America are tundra, boreal forest, temperate deciduous forest, temperate evergreen forest, shrubland, grassland, and desert. As one would expect, these biomes show a close correspondence to the major climate zones of North America. Although each biome is immediately recognizable by its distinctive vegetation, it is important to realize that different systems of classification make coarser or finer distinctions among biomes and that the characteristics of one biome usually intergrade gradually into the next. Rather than referring to discrete units of biological organization, the biome concept is a useful tool for ecologists who wish to organize the tremendous variation of the natural world. Classifications such as the biome system enable ecologists to work together to understand the structure and functioning of large ecological systems.

Every biological system exchanges energy and materials with the physical world. Yet the expression of these processes varies according to the environmental conditions and particular assemblage of species found in each biome. For example, loss of water is of less consequence for plants and animals in tropical montane forests than for those in subtropical deserts; salt balance poses antithetical challenges to organisms in fresh and salt water; the different chemical composition of the leaf litter from needle-leaved and broad-leaved trees creates different soil acidity and nutrient status. Physical conditions largely determine which kinds of organisms can persist in a particular environment; therefore, physical conditions fashion the overall character of the biome. The combination of physical conditions and biological forms determines the productivity and other dynamic attributes of the ecosystem.

That biomes can be distinguished at all results from the simple fact that no one type of organism can endure the whole range of conditions at the surface of the earth. If organisms had such broad tolerance of physical conditions, then biological communities would be the same everywhere and the earth would be covered by a single biome. This reasoning can be taken one step further: no growth form can persist over the whole range of conditions on the earth. Simply because of their physical structure, or **growth form,** trees cannot grow under the dry conditions that shrubs and grasses can tolerate. The grassland biome exists because the grass growth form can thrive under the physical conditions found, for example, on the Great Plains of the United States, the steppes of Russia, and the pampas of Argentina.

This chapter emphasizes the general principle that the distribution of each species or growth form coincides with a particular range of conditions in the environment. We shall first see how small differences in conditions within a single biome can differentiate the distributions of species sharing the same growth form. Then we shall see how larger differences in conditions can differentiate the distributions of growth forms themselves and therefore define the limits of biomes. Finally, we will have a look at some important attributes of the major terrestrial and aquatic biomes.

The discussion in this chapter will focus on adaptations of organisms to their physical environments, building upon the background developed in the first part of this book. Certain structures and physiological adaptations of plants, animals, and microbes are better suited to some physical conditions than to others. This matching of growth form and environment allows us to make generalizations about the distributions of life forms and the extents of biomes. If that were the whole of it, the study of ecology could simply focus on the biological relationships of individual organisms to their physical environments, and everything else in ecology would emanate from that point. Life is not so simple, however. Two other kinds of factors influence the distributions of species and growth forms. The first of these includes the myriad interactions between species—competition, predation, disease, mutualism—that determine whether a species or growth form can persist in a particular place. For example, grasses grow perfectly well in eastern North America, as we see along roadsides and on abandoned agricultural lands, but trees are the dominant growth form, and in the absence of disturbance, they exclude grasses, which cannot grow and reproduce under the deep shade of trees.

The second kind of factor is that of chance and history. The present distributions of species and biomes have developed over long periods, during which the distributions of landmasses, ocean basins, and climate zones have changed continually. Most species fail to occupy some perfectly suitable environments simply because they have not had the opportunity to get to all ends of the earth. This fact is amply illustrated by the successful introduction by humans of such species as European starlings and Monterey pines to parts of the world having suitable environmental conditions but which are far outside their limited natural distributions. In addition, evolution has proceeded along independent lines in different parts of the world, leading in some cases to unique biomes.

Australia has been isolated from other continents for the past 40–50 million years, which accounts both for its unusual flora and fauna and for the absence of many of the kinds of plants and animals familiar to northerners. Because of its unique history, areas of Australia having a climate that in California would support a shrub vegetation biome, referred to as chaparral, instead are clothed with tall eucalyptus forest. Similarities between chaparral and eucalyptus forest include drought and fire resistance, but the predominant plant growth forms in these areas of Australia and California differ, primarily because of historical accident. We shall consider biological and historical factors later in this book. As we shall see in the present chapter, the physical environment sets the stage on which biological interactions take place and ultimately determines the character and distribution of the major biomes.

Climate and plant distribution

A single example—the distribution of several species of maples in eastern North America—will illustrate the point that species distributions are associated with physical conditions of the environment. The range of the sugar maple, a common forest tree in the northeastern United States and southern Canada, is limited by cold winter temperatures to the north, by hot summer temperatures to the south, and by summer drought to the west (Figure 5.1). Attempts to grow sugar maples outside their normal range have shown that they cannot tolerate average monthly temperatures above about 24°C or below about −18°C. The western limit of the sugar maple, determined by dryness, coincides with the western limit of forest in general. Because temperature and rainfall interact to control the availability of moisture, sugar maples tolerate lower annual precipitation at the northern edge of their range (about 50 cm) than at the southern edge (about 100 cm). To the east, the range of the sugar maple stops abruptly at the Atlantic Ocean.

Figure 5.1 The range of the sugar maple in eastern North America. After H. A. Fowells, *Silvics of Forest Trees of the United States,* U.S. Department of Agriculture, Washington, D.C. (1965).

Differences in the distributions of the sugar maple and other tree-sized species of maples—black, red, and silver—suggest differences in their ecological tolerances (Figure 5.2). Where their geographic ranges overlap, maples exhibit distinct preferences for certain local environmental conditions created by differences in soil and topography. Black maple frequently occurs together with the closely related sugar maple, but usually on drier, better-drained soils higher in calcium content (and therefore less acidic). Silver maple occurs widely in the eastern United States, but especially on the moist, well-drained soils of the Ohio and Mississippi river basins. Red maple grows best either under wet, swampy conditions or on dry, poorly developed soils—that is, under extreme conditions that limit the growth of the other species.

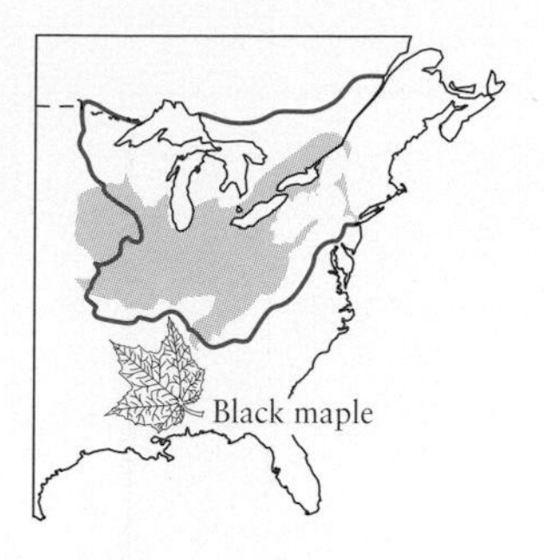

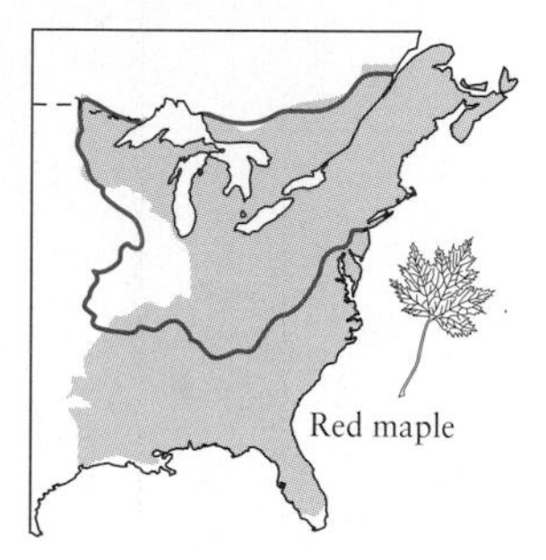

Figure 5.2 The ranges of black, red, and silver maples in eastern North America. The range of the sugar maple is outlined on each map to show the area of overlap. After H. A. Fowells, *Silvics of Forest Trees of the United States,* U.S. Department of Agriculture, Washington, D.C. (1965).

Topography, soils, and local distribution of plants

The distributions of plants clearly reveal the effects of many factors, which vary over different scales of distance. Climate, topography, soil chemistry, and soil texture exert progressively finer influences on geographic distribution. Elevation, slope, exposure, and underlying bedrock—factors that modify the plant environment—vary most in mountainous regions, and ecologists frequently turn to the varied habitats of mountains to study plant distribution.

Along the coast of northern California, mountains create conditions that support a variety of plant communities ranging from dry coastal chaparral to tall needle-leaved (coniferous) forests of Douglas fir and redwood. When localities are ranked on scales of available moisture, the distribution of each plant species among the localities exhibits a distinct optimum (Figure 5.3). The coast redwood dominates the central portion of the moisture gradient and frequently forms pure stands. Cedar, Douglas fir, and two

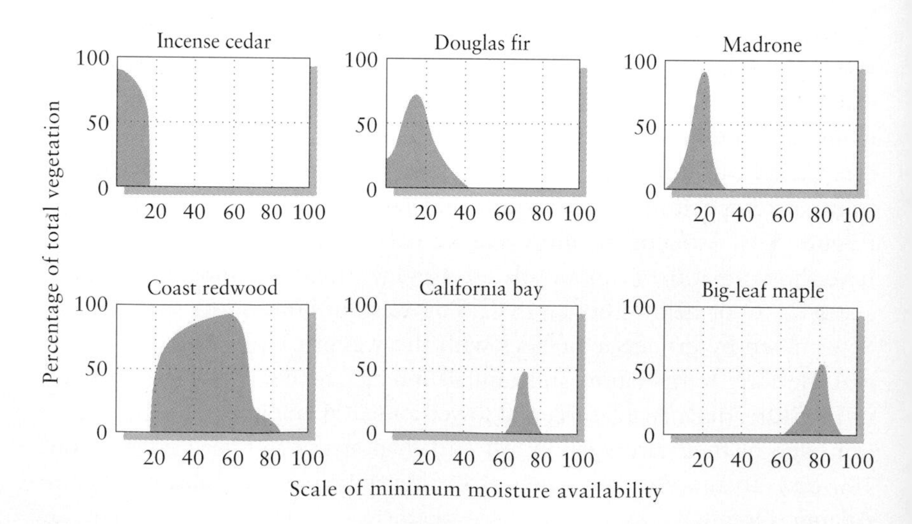

Figure 5.3 The distribution of tree species along a gradient of minimum available soil moisture in the northern coastal region of California. After R. H. Waring and J. Major, *Ecol. Monogr.* 34:167–215 (1964).

broad-leaved evergreen species with small, thick leaves—manzanita and madrone—occur at the drier end of the moisture gradient. Three deciduous species—alder, big-leaf maple, and black cottonwood—occupy the wetter end along with the broad-leaved, evergreen California bay tree.

Change in one environmental condition usually brings about changes in others. Increasing soil moisture alters the availability of nutrients. Variations in the amount and source of organic matter in the soil create parallel gradients of acidity, soil moisture, and available nitrogen. Such factors often interact in complex ways to determine the distributions of plants. Figure 5.4 relates the distributions of some forest floor shrubs, seedlings, and herbs in woodlands of eastern Indiana to levels of organic matter and calcium in the soil. The soils in this area contain between 2% and 8% organic matter and between 2% and 6% exchangeable calcium. Within the range of soil conditions in these woodlands, each species shows different preferences. Black cherry seedlings occur only within a narrow range of calcium, but tolerate considerable variation in the percentage of organic matter. Bloodroot is narrowly restricted by the percentage of organic matter in the soil, but its distribution is relatively insensitive to variation in calcium. The distributions of yellow violets and cream violets extend more broadly over varying levels of organic matter and calcium in the soil, but the two species overlap very little. Cream violets grow on soils higher in calcium and lower in organic matter than do yellow violets; where one occurs, the other usually does not.

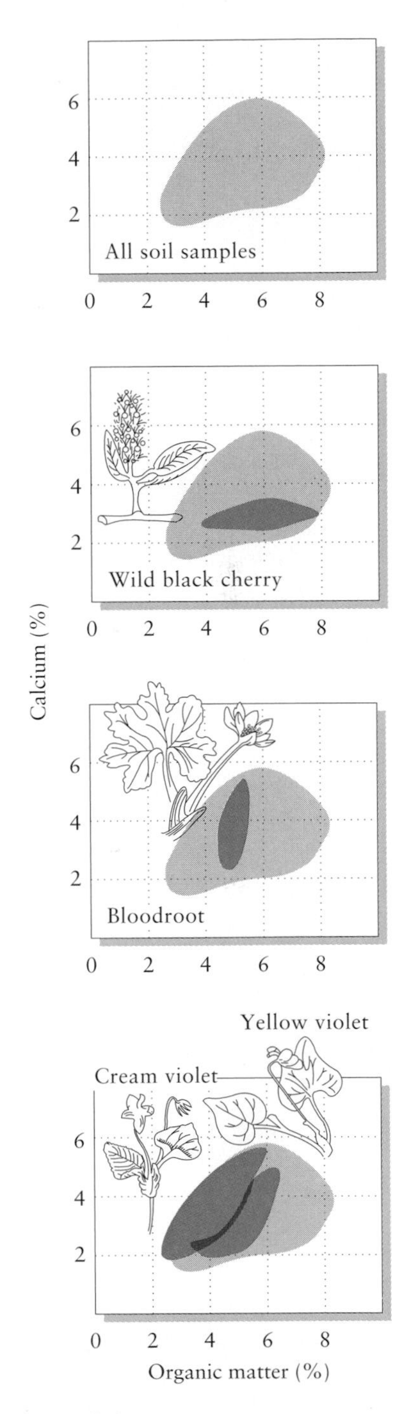

Figure 5.4 The occurrence of four forest floor plants with respect to the calcium and organic matter content of the soil in woodlands of eastern Indiana. After E. W. Beals and J. B. Cope, *Ecology* 45:777–792 (1964).

In the coastal mountain ranges of northern California, several species of pines and cypresses are restricted to serpentine soils, whereas others occur only on extremely acid soils. When grown on soils from different localities, seedlings of these endemics often do best when planted in soil from their native habitats. Thus lodgepole pine grows only on acid soil, and Sargent cypress, a serpentine endemic, grows somewhat better on serpentine soil than on "normal" soil, and not at all on acid soil (Figure 5.5). Not all endemics perform best on their home soil, however. When given the chance in an experimental garden, pygmy cypress, normally restricted to acid soils, grows much better on "normal" and serpentine soils. How does it tolerate serpentine soils? What factors exclude mature pygmy cypress from soils on which its seedlings grow vigorously? Clearly, factors other than soil conditions, such as competitors and pathogens, influence the distributions of these species. Such biological agents might, for example, exclude pygmy cypress from richer soils, restricting its distribution by default to soils in which superior competitors, such as Sargent cypress, or disease organisms cannot grow.

Environment, form, and function

The adaptations of an organism cannot easily be separated from the environment in which it lives. Insect larvae from stagnant aquatic environments in ditches and sloughs can survive longer without oxygen than related species from well-aerated streams and rivers; species of marine snails that occur high in the intertidal zone, where they are frequently exposed to air,

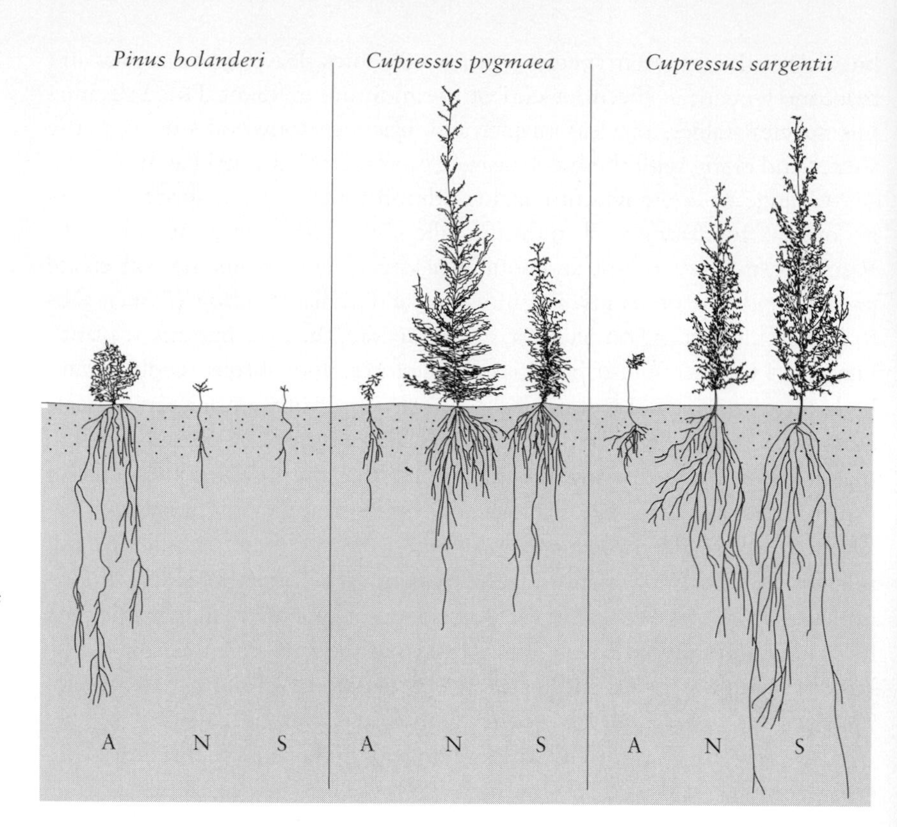

Figure 5.5 Seedling growth of lodgepole pine *(Pinus bolanderi),* pygmy cypress *(Cupressus pygmaea),* and Sargent cypress *(Cupressus sargentii)* in acid (A), "normal" (N), and serpentine (S) soils. After C. McMillan, *Ecol. Monogr.* 26:177–212 (1956).

tolerate desiccation better than do species from lower levels. These are examples of **specializations,** by which are meant adaptations that suit organisms to particular, restricted ranges of environmental conditions.

Compare the leaves of deciduous forest trees with those of desert species. The former are typically broad and thin, providing a large surface area for light absorption and also, unavoidably, for water loss. Desert trees have small, finely divided leaves (Figure 5.6)—or sometimes none at all

Figure 5.6 Leaves of some plants from the Sonoran Desert in Arizona. (a) Mesquite *(Prosopis)* leaves are subdivided into numerous small leaflets, which facilitate the dissipation of heat. (b) The paloverde *(Cercidium)* carries this adaptation even further; its leaves are tiny, and the thick stems, which contain chlorophyll, are responsible for much of the plant's photosynthesis (hence the name paloverde, which is Spanish for "green stick"). (c) Unlike most desert plants, limberbush *(Jatropha)* has broad, succulent leaves, but it produces them for only a few weeks during the summer rainy season.

(a)

(b)

(c)

(cacti rely entirely on their stems for photosynthesis; their leaves are modified into thorns for protection). Leaves heat up in the desert sun. Structures lose heat by convection most rapidly at their edges, where wind currents disrupt the insulating boundary layers of still air. The more edges, the cooler the leaf, and the lower the water loss; small leaf size means that a large proportion of each leaf is close to its edge. Even on a single plant, leaves exposed to full sun may be smaller and have more edges than shade leaves (Figure 5.7).

The water relations of coastal sage and chaparral plants in southern California illustrate the divergent forms and lifestyles of these plants in relation to the levels of water stress in their respective environments (Table 5.1). Chaparral habitat ranges over higher elevations than that of the coastal sage and thus is cooler and moister. Both vegetation types are exposed to prolonged summer drought, but the soils present greater water stress in the sage habitat. Coastal sage plants typically have shallow roots and small, delicate, deciduous leaves (Figure 5.8). Chaparral species have deep roots that often extend through tiny cracks and fissures far into the bedrock; their thick leaves have a waxy outer covering (cuticle) that reduces water loss. Most coastal sage species shed their delicate leaves during the summer drought period; the tougher leaves of chaparral plants persist.

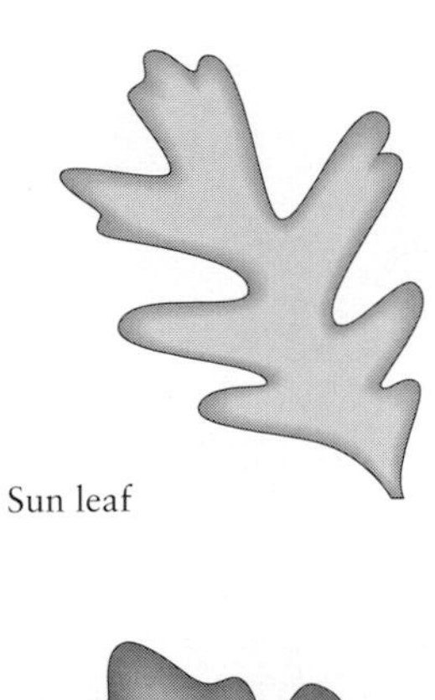

Figure 5.7 Silhouettes of a sun leaf and a shade leaf of white oak. The sun leaves have more edge per unit of surface area and therefore dissipate heat more rapidly. After S. Vogel, *J. Exp. Bot.* 21:91–101 (1970).

TABLE 5.1 Characteristics of chaparral and coastal sage vegetation in southern California

	VEGETATION TYPE	
CHARACTERISTIC	*Chaparral*	*Coastal sage*
Roots	Deep	Shallow
Leaves	Evergreen	Summer deciduous
Average leaf duration (months)	12	6
Average leaf size (cm^2)	12.6	4.5
Leaf weight (g dry weight dm^{-2})	1.8	1.0
Maximum transpiration (g H_2O dm^{-2} h^{-1})	0.34	0.94
Maximum photosynthetic rate (mg C dm^{-2} h^{-1})	3.9	8.3
Relative annual CO_2 fixation	49.8	46.8

Source: A. T. Harrison, E. Small, and H. A. Mooney, *Ecology* 52:869–875 (1971); H. A. Mooney and E. L. Dunn, *Am. Nat.* 104:447–453 (1970).

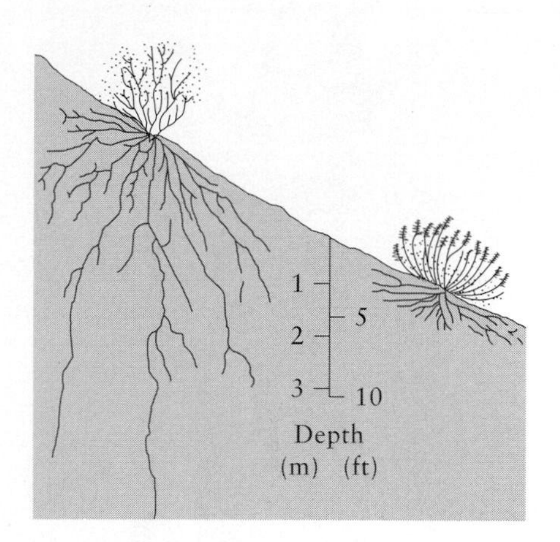

Figure 5.8 Profiles of the root systems of chamise *(Adenostoma fasciculatum),* a chaparral species (left), and black sage *(Salvia mellifera),* a member of the coastal sage community (right). After H. Hellmers et al., *Ecology* 36:667–678 (1955).

The influence of leaf morphology on photosynthetic rate parallels its influence on transpiration. The leaves of coastal sage species, with their numerous stomates, are designed for rapid gas exchange with the surrounding air. This means that they lose water rapidly, but they can also assimilate carbon rapidly from the atmosphere when water is available in the soil to replace water lost by transpiration. The relationship between transpiration and carbon assimilation can be demonstrated readily in the laboratory. In one set of experiments, scientists clipped leaves from plants and placed them in a chamber within which they could monitor transpiration and photosynthesis. Both functions declined as the leaves dried out and their stomates closed to prevent further water loss. Rates of both photosynthesis and transpiration for such coastal species as the black sage (a member of the mint family) were high initially, but decreased rapidly (Figure 5.9). However, even when the stomates were fully closed and carbon assimilation had ceased, the leaves continued to lose water across their thinly protected outer surfaces. Photosynthetic rates for such chaparral species as the toyon (a member of the rose family) were at most only one-fourth to one-third those of the black sage, but the leaves resisted desiccation better and continued to be active under drying conditions for longer periods. Furthermore, the outer surfaces of the leaves of the toyon have a thick waxy cuticle to minimize water loss when the stomates are fully closed.

Sage and chaparral plants are differently specialized. Black sage is active only during the rainy season of winter and early spring. Its leaves are designed for high rates of photosynthesis and high rates of growth, but they are dropped and the plant becomes dormant as soon as water becomes scarce in the soil. Black sage is thus specialized for the transient moist conditions of the Mediterranean climate winter. Toyon and other chaparral species make use of the more limited water that lies deeper in the soil, but nonetheless persists through a longer part of the year; they cannot utilize the winter water bonanza in the upper layers of the soil as efficiently as coastal sage species.

Where chaparral and coastal sage species grow together near the overlapping edges of their ranges, they exploit different parts of the environment: deep, perennial sources of water versus shallow, ephemeral sources. In spite of these differences and the corresponding adaptations of leaf morphology and drought response, these species are equally productive at intermediate levels of water availability. This more or less balanced competitive footing enables the two growth forms to coexist in intermediate environments. In drier habitats, the prolonged seasonal absence of deep water tilts the outcome toward the deciduous coastal sage vegetation; the increasing availability of deep water at higher elevations favors the evergreen chaparral vegetation.

These examples drive home the point that growth form is closely related to the physical conditions of the environment. With respect to terrestrial plants, those with large growth forms are often competitively superior to those with smaller growth forms, but they require moister soils. Thus, we should not be surprised that the availability of water is the single predominating factor determining the character and distribution of terrestrial biomes. Because temperature influences moisture stress and moisture availability, it also makes an important contribution.

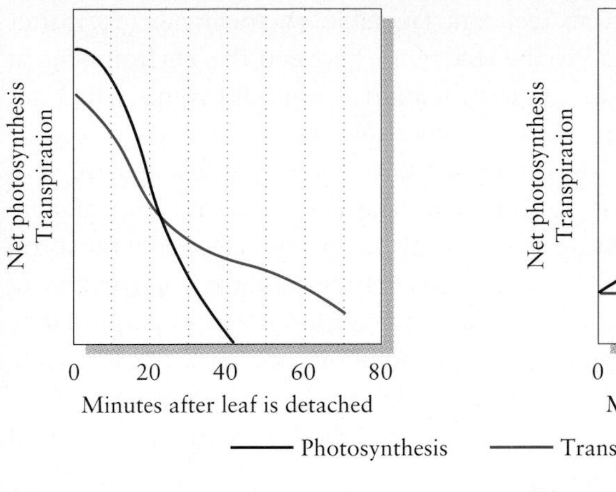

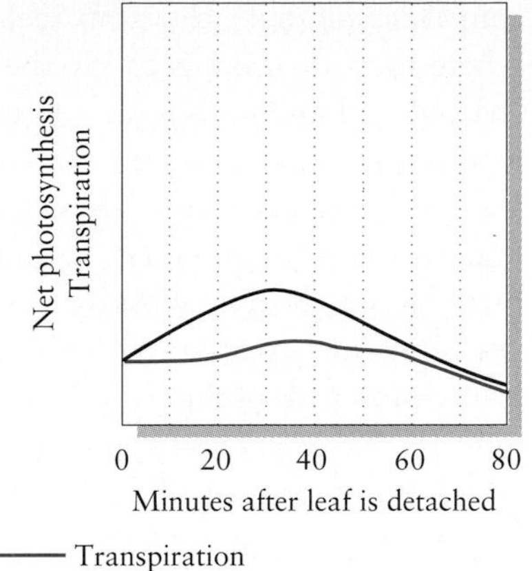

(a) Black sage (b) Toyon

Figure 5.9 Time courses of photosynthesis and transpiration for detached leaves under standard drying conditions. (a) A coastal sage species (black sage, *Salvia mellifera*). Note that transpiration continues well after photosynthesis has been shut off; hence leaf dormancy is an ineffective long-term solution to drought. (b) A chaparral species (toyon, *Heteromeles arbutifolia*). After A. T. Harrison, E. Small, and H. A. Mooney, *Ecology* 52:869–875 (1971).

We shall look at the interaction of these factors in their effects on biomes below, but we should keep in mind a number of other effects. One of these is fire, whose influence is greatest where moisture availability is intermediate in level and highly seasonal. Deserts and moist forests burn infrequently: deserts rarely accumulate enough plant debris to fuel a fire, and moist forests rarely dry out enough to be highly flammable. Grassland and shrub biomes have the combination of abundant fuel and seasonal drought that makes fire a frequent visitor. In these biomes, fire is a predominating factor to which all members must be adapted and, indeed, for which many are specialized: for some species, fire is necessary for germination of seeds and growth of seedlings. Another effect is conditions in the soil, referred to as **edaphic** factors, such as nutrient status, water availability, and toxic mineral content, which can also modify biome formations. We have seen in the last chapter how the dry, toxic soils produced over serpentine rock can restrict vegetation to grasses and herbs, whereas normal soils in the same region support forests.

Finally, before continuing with the theme of climate and vegetation, which has been the focus of studies on the distributions of biomes, we should say a few words about animals. Historically, the biome concept grew up around variation in terrestrial plants. Animals do not figure in the distinctions between biomes. In general, the life forms of animals are less sensitive to climate than are those of plants. This does not mean that animals are not adapted to physical conditions and do not specialize on narrow ranges of these conditions, only that such adaptations are not expressed as major differences in life form. Birds of deserts and birds of forests have the same basic body plan; so do mammals, reptiles, amphibians, and insects.

Why the different responses of plants and animals? A facile answer—which cannot fully explain the plant–animal dichotomy—is that because animals are mobile, they can seek out favorable microclimates no matter where they are. That is to say, the difference between the environment in Arizona and in Illinois is not so great for animals as it is for plants, which are firmly rooted in place and must therefore tolerate all extremes of conditions. Three other factors seem important, however. First, plants have very high surface-to-volume ratios, both above and below ground level, and are therefore much more sensitive than animals to changes in, and extremes of, environmental conditions. Second, because of the modular construction of plants, most cells of the plant "body" are more or less independent, and they must survive extremes of environmental conditions on their own, with little help from other parts of the body—with the conspicuous exception of the delivery of water from the roots. The animal body has many specialized organs, such as the kidneys, lungs, lymphoid system, and muscles, that support other tissues, particularly by ameliorating the cell environment. Third, all plants play basically the same ecological role: they draw water and minerals through their root systems and fix carbon by means of photosynthesis in their leaves. As a result, their entire existence is dominated by water balance and the problems of delivering an adequate water supply to the aboveground portion of the plant organism. Again because of the modular design of plants, the only effective ways of manipulating water balance are through the design of the water delivery systems and leaves, and through the overall size of the plant and the ratio of its aboveground to belowground parts. These are precisely the characteristics of growth form that define the major terrestrial biomes.

Variations in animal form, as there are between worms, insects, and vertebrates, reflect the different roles that animals play within biomes. Thus, differences in animal form are associated more with different ecological roles within biomes (soil dwelling, seed gathering, and predation, for instance) than with differences in the physical environment between biomes. Also, in the case of animals, behavior can compensate for differences in growth form. For example, in the deserts of Arizona, seeds are gathered from the soil surface by rodents, birds, and ants; other rodents, birds, and ants gather seeds in the forests of North Carolina and in the savannas of East Africa. In spite of their different body plans, these animals' ecological roles are the same. It is important to remember that in spite of their similarity, plants can also occupy different positions within the same biome. Tropical forests exhibit the greatest diversity of plant growth form, which includes trees (both deciduous and evergreen in some localities), vines (both woody and herbaceous), epiphytes, understory treelets and shrubs, and herbaceous plants on the forest floor. Each of these growth forms lives in a very different environment with respect to light and moisture stress.

Climate and terrestrial biomes

One of the most widely adopted biome classification schemes is the climate zone system of the German ecologist Heinrich Walter. This system, which has nine major divisions, is based on the annual course of temperature and precipitation. The important attributes of climate and the characteristic veg-

TABLE 5.2 **H. Walter's classification of the climate zones of the world**

CLIMATE ZONE	CORRESPONDING VEGETATION
I **Equatorial** Always moist and lacking temperature seasonality	Evergreen tropical rain forest
II **Tropical** Summer rainy season and cooler "winter" dry season	Seasonal forest, scrub, or savanna
III **Subtropical** Highly seasonal, arid climate	Desert vegetation with considerable exposed surface
IV **Mediterranean** Winter rainy season and summer draught	Sclerophyllous (drought-adapted), frost-sensitive shrublands and woodlands
V **Warm temperate** Occasional frost, often with summer rainfall maximum	Temperate evergreen forest, somewhat frost-sensitive
VI **Nemoral** Moderate climate with winter freezing	Frost-resistant, deciduous, temperate forest
VII **Continental** Arid, with warm or hot summers and cold winters	Grasslands and temperate deserts
VIII **Boreal** Cold temperate with cool summers and long winters	Evergreen, frost-hardy needle-leaved forest (taiga)
IX **Polar** Very short, cool summers and long, very cold winters	Low, evergreen vegetation, without trees, growing over permanently frozen soils

etation in each of these zones is presented in Table 5.2. The boundaries of the climate zones correspond to conditions of moisture and cold stress that are particularly important determinants of plant form. For example, within the Tropics, tropical climates are distinguished from equatorial climates by periods of water stress during a pronounced dry season. Subtropical climate zones are perpetually water-stressed. The typical vegetation of each of these climate zones is evergreen rain forest, deciduous forest or savanna, and desert scrub, respectively.

Another approach to the relationship between terrestrial biomes and climate variables was that of Cornell University ecologist Robert H. Whittaker. Whittaker defined biomes on the basis of vegetation structure, then devised a simple climate diagram on which he plotted the approximate boundaries of the major biomes with respect to average temperature and precipitation. When one plots a sample of terrestrial localities on a graph according to their mean annual temperature and precipitation, most points fall within a triangular area whose three corners represent warm-moist, warm-dry, and cool-dry climates (Figure 5.10). Cold regions with high rainfall are rare because water does not evaporate rapidly at low

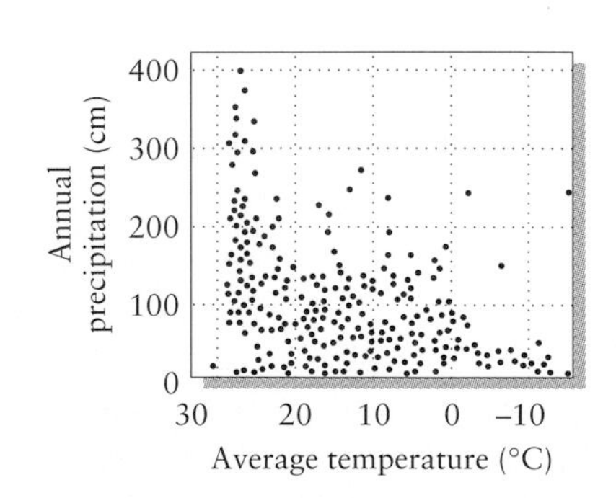

Figure 5.10 Average annual temperature and precipitation for a sample of localities more or less evenly distributed over the land area of the earth. Most of the points fall within a triangular region that includes the full range of climates, excluding those of high mountains.

temperatures and because the atmosphere in cold regions holds little water vapor. The major biomes defined by Whittaker are superimposed on the same graph in Figure 5.11.

Within the tropical and subtropical realms, with mean temperatures between 20° and 30°C, vegetation ranges from rain forest, which is wet throughout the year and generally receives more than 250 cm (about 100 inches) of rain annually (Walter's equatorial climate zone), to desert, which generally receives less than 50 cm of rain, depending on the amount of rainfall received. Intermediate climates support seasonal forests (150–250 cm rainfall), in which some or all trees lose their leaves during the dry season, or dry forests, scrublands with many thorn trees, and savannas (50–150 cm rainfall).

Plant communities in temperate areas follow the pattern of tropical communities with respect to rainfall, with corresponding vegetation types distinguishable in both. In colder climates, however, precipitation varies so little from one locality to another that vegetation types are poorly differentiated by climate. Where mean annual temperatures are below –5°C, all plant associations may be lumped into one type: tundra.

Toward the drier end of the precipitation spectrum within each temperature range, fire plays a distinct role in shaping plant communities. In many areas of African savanna and North American prairie, frequent fires kill the seedlings of trees and prevent the encroachment of forests, which could be sustained by the local precipitation if it were not for fire. Burning favors perennial grasses with extensive root systems that can survive underground. After an area has burned over, grass roots sprout fresh shoots and quickly revegetate the surface. In the absence of frequent fires,

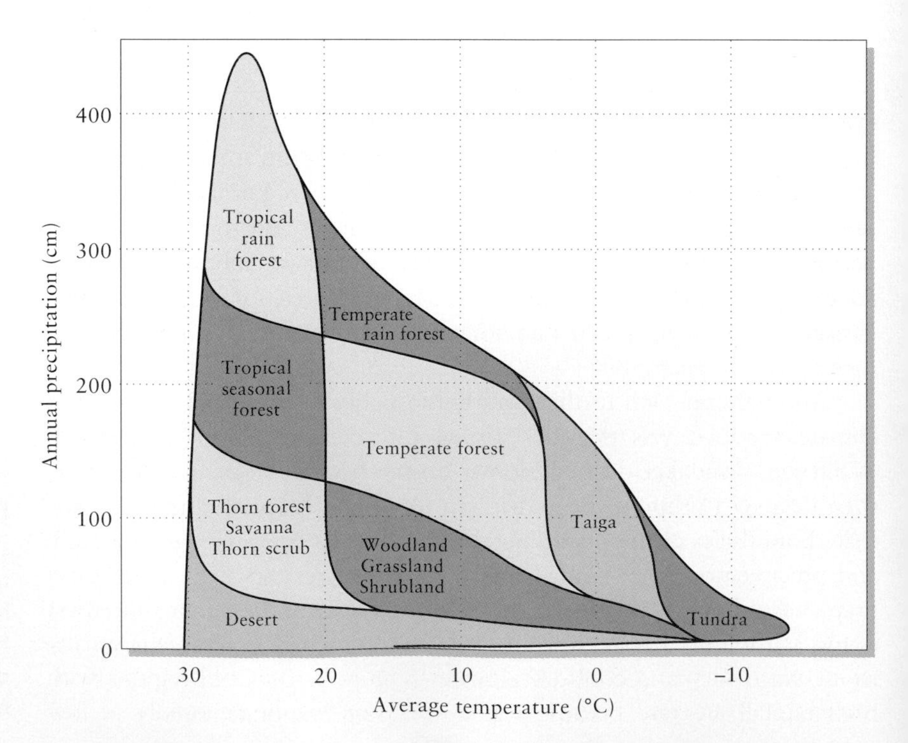

Figure 5.11 Whittaker's classification of vegetation types superimposed on the range of terrestrial climates. In climates intermediate between those of forested and desert regions, fire, soil, and climate seasonality determine whether woodland, grassland, or shrubland develops. From R. H. Whittaker, *Communities and Ecosystems,* 2d ed., Macmillan, New York (1975).

tree seedlings can become established and eventually shade out savanna and prairie vegetation.

As in all classification systems, exceptions appear frequently, and boundaries between biomes are fuzzy. Moreover, not all plant growth forms correspond with climate in the same way; for example, Australian eucalyptus trees form forests under climatic conditions that support only shrubland or grassland on other continents. Finally, plant communities reflect factors other than temperature and rainfall. Topography, soils, fire, seasonal variations in climate, and herbivory all leave their mark. The following brief overview of the major biomes emphasizes the distinguishing features of the physical environment and how these are reflected in the form of the dominant plants.

Terrestrial biomes

Temperate climate zones

Because most readers of this book live in the so-called temperate zone, it is a good place to start. Temperate climates are characterized by average annual temperatures in the range of 5°–20°C at low elevations. Such climates are distributed approximately between 30° and 45°N in North America and between 40° and 60°N in Europe, which is warmed by the Gulf Stream current. Frost is an important factor throughout the temperate zone, perhaps even a defining one. Within the temperate zone, biomes are differentiated primarily by total amounts and seasonal patterns of precipitation, although the length of the frost-free season, which is referred to as the **growing season,** and the severity of frost are also important.

Deciduous forest biome. The deciduous forest biome is found in North America principally in the eastern United States and southern Canada, but is also widely distributed in Europe and eastern Asia (Figure 5.12). It is poorly developed in the Southern Hemisphere (New Zealand and southern Chile) because of the milder winter temperatures at moderate latitudes. The length of the growing season varies from 130 days at high latitudes to 180 days at lower latitudes. Precipitation usually exceeds the potential evaporation and transpiration; as a result, water tends to move downward through soils and to drain from the landscape as groundwater and as surface streams and rivers. Soils are podsolized, tend to be slightly acid and moderately leached, and are brown in color owing to abundant organic humus. Vegetation is dominated by deciduous trees, often with a subcanopy layer of small trees and shrubs, and herbaceous plants on the forest floor, many of which complete their growth and flower early in spring, before trees have fully leafed out.

Figure 5.12 A stand of hardwoods in Indiana dominated by white oak and having a well-developed understory of small shrubs. Courtesy of the U.S. Forest Service.

Temperate needle-leaved biome. Temperate needle-leaved forests are dominated by pines and exist under conditions of water and nutrient stress, often on sandy soils. The most important of these formations in North America are the pine forests of the coastal plains of the Atlantic and Gulf states of the United States, the jack pine forests of the northern parts of the Great Lakes states and central Canada, and the montane pine forests of the American West. In the Southeast, because of the warm climate, soils are usually lateritic (oxisols) and have low nutrient levels. The low availability of nutrients and water favors evergreen, needle-leaved trees, which resist desiccation and give up nutrients slowly because they retain their needles for several years. Because soils tend to be dry, fires are frequent, and most species are able to resist fire damage.

Temperate rain forest biome. Near the coast in the northwestern United States and British Columbia, and also in southern Chile, New Zealand, and Tasmania, mild winters, heavy winter rains, and summer fog create conditions that presently support extremely tall evergreen forests. In North America these forests are dominated toward the south by coast redwood and toward the north by Douglas fir (Figure 5.13). Trees are typically 60–70 meters high and may grow to over 100 meters. It is not well understood why these sites are dominated by needle-leaved trees, but the fossil record shows that these plant formations are very old and that they are mere remnants of forests that were vastly more extensive during the Mesozoic era, as recently as 70 million years ago. In contrast to rain forests in the Tropics, the diversity of temperate rain forests typically is very low.

Figure 5.13 Redwood trees growing in a temperate rain forest in northern California. Courtesy of the U.S. Forest Service.

Temperate grassland biome. In North America, grasslands develop where the rainfall is between 30 and 85 cm per year, depending on the average temperature. Summers are hot and wet; winters are cold. The growing season increases from north to south from about 120 to 300 days. These grassland biomes are often called **prairies.** Extensive grasslands are also found in central Asia, where they are called **steppes.** Because precipitation is low, organic detritus does not decompose rapidly, and the soils are rich in organic matter. Because of the low acidity, the soils, which belong to the mollisol group, are not heavily leached and tend also to be rich in nutrients. The vegetation is dominated by grasses, which grow to over 2 meters in the moister parts of the grassland biome and to less than 0.2 meters in more arid regions. There are also abundant nongrass herbaceous species, which are called **forbs.** Fire is a dominant influence in grasslands, particularly where the habitat dries out during the late summer. Most grassland species have fire-resistant underground stems, or **rhizomes,** from which shoots resprout, or they have fire-resistant seeds.

Temperate shrubland biome. Where precipitation ranges between 25 and 50 cm per year, and winters are cold and summers are hot, grasslands grade into shrublands. The shrubland biome covers most of the Great Basin of the western United States. In the northern part of the region, sagebrush is the dominant plant, whereas toward the south and on somewhat moister soils, widely spaced juniper and piñon trees predominate, forming open woodlands of less than 10 meters stature with sparse coverings of grass (Figure 5.14). In these shrublands, the evaporation and transpiration potential of the habitat exceeds precipitation during most of the year, so soils are dry and little water percolates through them to form streams and rivers. The soils grade from mollisols to aridosols, and they tend to accumulate, at depths to which water usually penetrates, calcium carbonate leached from the surface layers. Fires occur infrequently in shrublands because the habitat produces little fuel. However, because of the low productivity, grazing can exert strong pressure on vegetation and may even favor the persistence of shrubs, which are not good forage. Indeed, many dry grasslands in the western United States and elsewhere in the world have been converted to shrublands by overgrazing.

Mediterranean woodland biome. The Mediterranean climate zone is distributed at 30–40° north and south of the equator—somewhat higher in Europe—on the western sides of continental landmasses: in southern Europe and southern California in the Northern Hemisphere, and in central Chile, the Cape region of South Africa, and southwestern Australia in the Southern Hemisphere. Mediterranean climates are characterized by mild winter temperatures, winter rain, and summer drought. These climates support thick, evergreen, shrubby vegetation 1–3 m in height, with deep roots and drought-resistant foliage. The small, durable leaves of typical Mediterranean-climate plants give them the label **sclerophyllous** (hard-leaved) vegetation. Fires are frequent in Mediterranean biomes, and most plants have either fire-resistant seeds or root crowns that resprout soon after a fire.

Subtropical and temperate desert biomes. What people call deserts varies tremendously, which is a danger in using everyday terms to name biomes. Many people refer to the dry areas of central Asia as deserts—the Mongolian Desert and the Gobi Desert are names familiar to most of us—but the climate and vegetation of these "deserts" differ utterly from those of the arid areas located within the subtropical belts of high pressure that girdle the earth. The Gobi Desert falls within Walter's temperate continental climate zone, characterized by low precipitation and cold winters. This climate, which is similar to that of the Great Basin and high western plains of North America, produces dry grasslands—steppes—and shrublands. Where precipitation dwindles to near zero, the vegetation dwindles accordingly, leaving a landscape that is more rock and sand than vegetation.

Subtropical deserts develop at latitudes of 20°–30° north and south of the equator in areas of very sparse rainfall (less than 25 cm) and generally long growing seasons. Because of the low rainfall, the soils of subtropical deserts (aridosols) are shallow, virtually devoid of organic matter, and neutral in pH. Impermeable hardpans of calcium carbonate often develop at the limits of water penetration—depths of a meter or less. Whereas sagebrush dominates Great Basin (continental climate) deserts, creosote bush takes its place in the subtropical deserts of the Americas. Wetter sites support a profusion of succulent cacti, shrubs, and small trees, such as mesquite and paloverde. Most subtropical deserts receive summer rainfall, during which many herbaceous plants sprout from dormant seeds, and quickly grow and reproduce before the soils dry out again. Many of the plants in subtropical deserts are not frost-tolerant. Species diversity is usually much higher than it is in temperate arid lands.

Figure 5.14 An open juniper-piñon pine woodland at about 2,000 meters elevation in the Coconino National Forest of northern Arizona. Courtesy of the U.S. Forest Service.

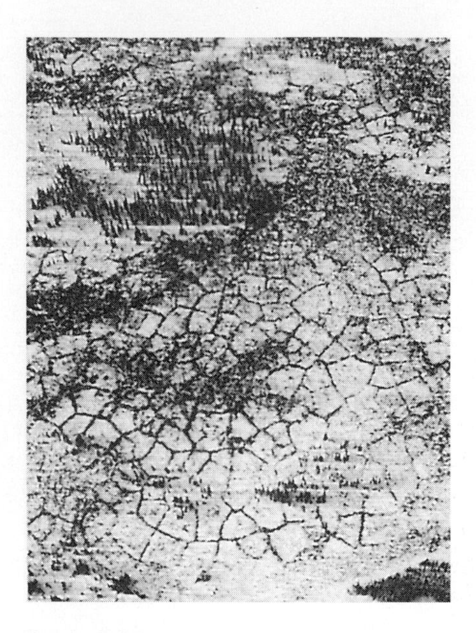

Figure 5.15 Alaskan tundra showing the characteristic polygonal patterns in the ground surface created by freezing and thawing of the surface layers of the soil. Courtesy of the U.S. Soil Conservation Service.

Figure 5.16 Wind-driven ice strips bark and branches from trees near the forest-alpine tundra border (timberline) in the Rocky Mountains of Colorado. Courtesy of the U.S. Forest Service.

Boreal and polar climate zones

Boreal forest biome. Stretching in a broad belt centered at about 50°N in North America and about 60°N in Europe and Asia lies the boreal forest biome, often called **taiga.** The average annual temperature is below 5°C, and winters are severe. Precipitation is in the range of 40–100 cm, and because evaporation is low, soils are moist throughout most of the growing season. The vegetation consists of unendingly dense stands of 10–20 m high evergreen needle-leaved trees, mostly spruce and fir. Because of the low temperatures, litter decomposes very slowly and accumulates at the soil surface. The needle-leaf litter does, however, produce high levels of organic acids, so the soils are acid, strongly podsolized, and generally of low fertility. Growing seasons are rarely 100 days, and more often half that. The vegetation is extremely frost-tolerant, as temperatures may reach –60°C during the winter. Species diversity is very low.

Tundra biome. To the north of the boreal forest, in the so-called polar climate zone, lies the arctic tundra, a treeless expanse underlain by permanently frozen soil, or **permafrost** (Figure 5.15). The soils thaw to a depth of 0.5–1 m during the brief summer growing season. Precipitation is generally less than 60 cm, and often much less, but in low-lying areas where drainage is prevented by the permafrost, soils may remain saturated with water throughout most of the growing season. Soils tend to be acid because of their high organic matter content, and they are very low in nutrients. In this nutrient-poor environment, plants hold their foliage for years. Most plants are dwarf, prostrate, woody shrubs, which grow low to the ground to gain protection under the winter blanket of snow and ice. Anything protruding above the surface of the snow is sheared off by blowing ice crystals. For most of the year, the tundra is an exceedingly harsh environment, but during the 24-hour-long summer days, the rush of activity in the tundra biome exuberantly testifies to the remarkable adaptability of life.

Alpine tundra. At high elevations in temperate climate zones, and even within the Tropics, one finds vegetation resembling that of the arctic tundra and containing some of the same species, or their close relatives. These areas above the tree line occur most broadly in the Rocky Mountains of North America and, especially, on the Tibetan Plateau of central Asia (Figure 5.16). In spite of their similarities, alpine and arctic tundra have important points of dissimilarity as well. Areas of alpine tundra generally have warmer and longer growing seasons, higher precipitation, less severe winters, greater productivity, better-drained soils, and higher species diversity than arctic tundra. Still, as in the high-latitude tundra, it is the harsh winter conditions that ultimately limit the growth of trees.

Equatorial and tropical climate zones

Within 20° of the equator, temperature varies more throughout the day than average monthly temperatures vary throughout the year. Average temperatures at sea level generally exceed 20°C. Environments within tropical latitudes are distinguished by differences in the seasonal pattern of rainfall, which create a continuous gradient of vegetation from wet, aseasonal rain forest through seasonal forest and scrub, savanna, and desert. Frost is not a factor in tropical biomes, even at high elevations, and tropical plants and animals generally do not tolerate freezing when investigators expose them to cold temperatures.

Tropical rain forest biome. Climates under which rain forests develop are always warm and receive at least 200 cm of precipitation throughout the year, with not less than 10 cm during any one month. These conditions prevail in three important regions within the Tropics: The Amazon and Orinoco basins of South America, with additional areas in Central America and along the Atlantic coast of Brazil, constitute the American rain forest; the area from southernmost West Africa and extending eastward through the Congo River basin constitutes the African rain forest; and the Indo-Malayan rain forest covers parts of Southeast Asia (Vietnam, Thailand, and the Malay Peninsula); the islands between Asia and Australia, including the Philippines, Borneo, and New Guinea; and the Queensland coast of Australia.

The tropical rain forest climate often exhibits two peaks of rainfall centered around the equinoxes, corresponding to the periods when the intertropical convergence overlies the equatorial region. Rain forest soils are typically old and deeply weathered oxisols. Because they are relatively devoid of humus and clay, they take on the reddish color of aluminum and iron oxides and have poor ability to retain nutrients. In spite of the low nutrient status of the soils, rain forest vegetation is dominated by a continuous canopy of tall evergreen trees rising to 30–40 meters, with occasional **emergent trees** rising above the canopy to heights of 55 meters. Because water stress on emergents is great due to their height and exposure, they are often deciduous, even in an evergreen rain forest. Tropical rain forests have understory tree, shrub, and herb layers, but these are usually quite sparse because so little light penetrates the canopy. Climbing **lianas,** or woody vines, and **epiphytes,** plants that grow on the branches of other plants and are not rooted in soil (also called air plants), are prominent in the forest canopy itself (Figure 5.17). Species diversity is higher than anywhere else on earth.

Figure 5.17 Vines and epiphytes drape the trees in a lowland tropical rain forest in the Republic of Panama. Courtesy of W. J. Smith.

The productivity of rain forests is greater than that of any other terrestrial biome, and their standing biomass exceeds that of all other biomes except temperate rain forests. Because of the continuously high temperatures and abundant moisture, plant litter decomposes quickly, and the nutrients released are immediately taken up by the vegetation. This rapid nutrient cycling supports the high productivity of the rain forest, but also makes the rain forest ecosystem extremely vulnerable to disturbance. When tropical rain forests are cut, many of the nutrients are carted off in logs or go up in smoke. The vulnerable soils erode rapidly and fill the streams with silt. In many cases, the environment degrades rapidly and the landscape becomes unproductive.

Tropical seasonal forest biome. Within the Tropics, but beyond 10° from the equator, tropical climates often exhibit a pronounced dry season, corresponding to winter at higher latitudes. Seasonal forests in the Tropics have a preponderance of deciduous trees that shed their leaves during the season of water stress. Increasingly longer and more severe dry seasons generally result in vegetation with lower stature and more thorns to protect leaves from grazing, leading to dry forest, thorn scrub, and finally true deserts under the extremely dry conditions that occur in the rain shadows of mountain ranges or along coasts with cold ocean currents running alongside.

Tropical savanna biome. Savanna may be defined as grassland with scattered trees, and it typifies large areas of the dry Tropics, especially in Africa. Rainfall is typically 90–150 cm per year, but the driest three or four months receive less than 5 cm each (Figure 5.18). Fire and grazing undoubtedly play important roles in maintaining the character of the savanna biome, particularly in wetter regions, as grasses can persist better than other forms of vegetation under both influences. Often when grazing and fire are controlled within a savanna habitat, dry forest begins to develop. It is possible that vast areas of African savanna owe their character to the influence of human activity, including burning, over many millennia.

Figure 5.18 Savanna vegetation, typified by drought-adapted trees interspersed by grassland, develops in hot tropical climates with highly seasonal rainfall, as in the Samburu district of Kenya.

The biome concept in aquatic systems

The biome concept was developed for terrestrial ecosystems, and biomes are distinguished principally by the growth form of their dominant vegetation. Throughout most of the development of ecology as a science, terrestrial and aquatic ecologists generated concepts and descriptive terms for ecological systems independently. As a consequence, aquatic "biomes" do not exist in the sense in which the term is applied to terrestrial systems. Indeed, employing a vegetation concept would be impossible in aquatic systems, because the primary producers in many aquatic systems are single-celled algae, which do not form "vegetation" with a characteristic structure, as in terrestrial systems. As a result, classifications of aquatic systems have been based primarily on physical characteristics: salinity, water movement, depth, and so on.

The major kinds of aquatic environments are streams, lakes, estuaries, and oceans, and each of these can be subdivided further with respect to many factors. Streams form wherever precipitation exceeds evaporation and excess water drains from the surface of the land. Within small streams, ecologists distinguish areas of **riffles,** where water runs rapidly over a rocky substrate, and **pools,** which are deeper stretches of slowly moving water. Water is well oxygenated in riffles; pools tend to accumulate silt and organic matter. Production in small streams is often dominated by **allochthonous** material, that is to say, organic material such as leaves that enters the aquatic system from the outside. Streams grow with distance as they join together to form rivers. The larger a river, the more of its production is home-grown, or **autochthonous.** A **"river continuum"** concept has grown up around the continuous change in environment and ecological system between the headwaters and the mouth of a river drainage. As one moves downstream, water is warmer, more slowly flowing, and richer in nutrients; ecosystems are more complex and generally more productive. **Fluvial** systems, as river systems are called, are also distinguished by the fact that material, including animals and plants, is continually moved downstream by currents. Inasmuch as fluvial systems exist in steady states, this so-called **downstream drift** is balanced by active movement of animals upstream, by the productivity of the upstream portions of the system, and by input of materials from outside.

Lakes form in any kind of depression. For the most part, such bodies of water are the products of recent glaciation, which leaves behind gouged-out basins and blocks of ice buried in glacial deposits, which eventually melt and form lakes. Lakes are also formed in geologically active regions, such as the Rift Valley of Africa, where vertical shifting of blocks of the earth's crust creates basins within which water accumulates. Broad river valleys, such as those of the Mississippi and Amazon, may have oxbow lakes, which are broad bends of the former river cut off by shifts in the main channel. An entire lake could be considered a biome, but it is usually subdivided into regions, each of which has its own character. The **littoral** zone is the shallow zone around the edge of a lake or pond within which one finds rooted vegetation, such as water lilies and pickerel weed (Figure 5.19). The open water beyond the littoral zone is the **limnetic** zone, where primary production is accomplished by floating single-celled algae, or phytoplankton. Lakes may also be subdivided vertically on the basis of light penetration and the formation of thermally stratified layers of water. The sediments at the bottoms of lakes and ponds form a special **benthic** habitat for burrowing animals and microorganisms.

Figure 5.19 Cattails growing in a shallow lake in New York State.

Estuaries are special environments found at the mouths of rivers, especially where the outflow is partially enclosed by landforms or barrier islands. The unique character of estuaries derives from the mixture of fresh and salt water, within which the larvae of many species of marine organisms grow in great profusion. In addition, the nutrients carried by rivers and the rapid exchange between surface waters and sediments contribute to extremely high biological productivity. Because estuaries tend to be shallow areas within which sediments are deposited, they are often edged by extensive tidal marshes with **emergent vegetation.** Indeed, the marshes that surround many estuaries are among the most productive habitats on earth, owing to a combination of high nutrient levels and freedom from water stress. These marshes then contribute abundant additional organic matter to the estuarine ecosystem, which in turn supports abundant populations of estuarine and marine species.

The largest portion of the surface of the earth is covered by **oceans.** Beneath the surface of the water lies an immensely complex realm harboring a great variety of ecological conditions and ecosystems. Variation in marine systems comes from temperature, depth, current, substrate, and, at the edge of the seas, tides. Many marine ecologists have recognized several oceanic zones according to depth. The **littoral** zone extends between the highest and lowest tidal levels and so, to a varying extent depending on position within the intertidal range, is exposed periodically to air (Figure 5.20). The rapid changes in ecological conditions within the intertidal range often create sharp **zonation** of organisms according to their ability to tolerate the stresses of terrestrial conditions. Beyond the range of the

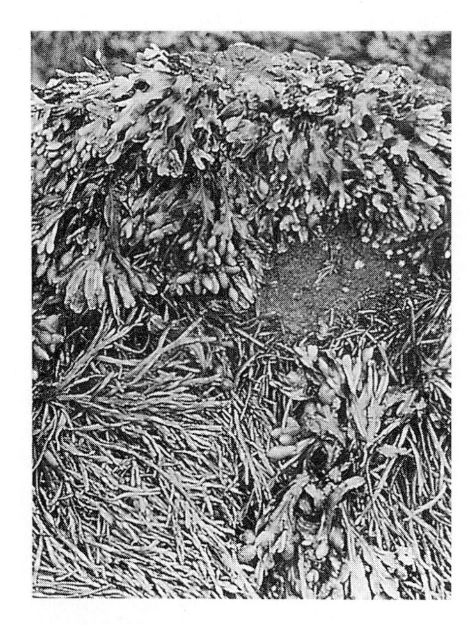

Figure 5.20 The intertidal zone, which becomes exposed to air twice each day, may support prolific growth of algae and a variety of animals, as in this area of the New Brunswick coast in Canada.

lowest tide level, the so-called **neritic** zone extends to depths of about 200 meters, which corresponds to the edge of the continental shelf. Beyond this, looking from the shore, the seafloor drops rapidly to the great depths of the **oceanic** zone, thousands of meters below. Often the continental shelves are regions of high productivity because the sunlit surface layers of water are not far removed from the regeneration of nutrients in the sediments below. Even strong waves can move suspended materials from depths of 100–200 meters to the surface. In the oceanic zone, production usually is strictly limited by low availability of nutrients. Both the neritic and the oceanic zones may be subdivided vertically into a superficial **photic,** or euphotic, zone in which there is sufficient light for photosynthesis, and an **aphotic** zone without light, in which organisms depend mostly on organic material raining down from above.

Figure 5.21 The high productivity of coral reefs in warm, tropical waters, provides abundant food for a diverse biological community. Courtesy of J. W. Porter.

Whereas the open ocean has been compared to a desert, coral reefs are like tropical rain forests, both in the richness of their biological production and the diversity of their inhabitants (Figure 5.21). Reef-building corals occur in shallow waters of warm oceans, usually where water temperatures remain above 20°C year round. Many coral reefs develop around volcanoes, which are widely distributed throughout the western Pacific and Indian Oceans. The volcanoes themselves may gradually disappear through erosion or subsidence under their own weight, but as long as the rate of coral growth exceeds the rate of subsidence, the reef continues to build. Eventually, all that may be left is a ring of coral, an **atoll,** outlining the position of the former volcanic island. The high production of the reef is fed by nutrients eroding off the encircled volcano and by deep-water currents forced upward by the profile of the island. Corals are doubly productive because they contain symbiotic photosynthetic algae within their tissues, which generate the carbohydrate energy base for their phenomenal rates of growth.

The unique qualities that characterize each type of biome or aquatic system are manifested in every aspect of ecosystem structure and function. The most direct way to evaluate these attributes is to measure the flux of energy through the ecosystem and the cycling of nutrients within the ecosystem. These aspects of ecological structure and function, and how they differ among the terrestrial biomes and aquatic ecosystems, are the subject of the next part of this book.

Summary

1. The geographic distributions of plants on continental scales are determined primarily by climate, whereas local distributions within regions may vary according to topography and soils.

2. Climate profoundly affects the evolution of plants and animals, which become specialized to particular conditions of the physical environment. As a consequence, each climatic region has characteristic types of vegetation that differ in growth form, leaf morphology, and seasonality of foliage.

3. Because plant growth form is directly related to climate, we can match major types of vegetation, or biomes, to temperature and precipitation. This relationship emphasizes the way in which temperature and precipitation interact to determine water availability; soil, climate seasonality, fire, and grazing additionally influence the character of biomes.

4. Two major approaches to the classification of biomes are the climate-zone approach of Walter and the vegetation approach exemplified by Whittaker. The first classifies regions on the basis of climate within which a characteristic vegetation normally develops. The second classifies regions according to vegetation, which generally reflects the local climate.

5. Climate zones and biomes are broadly divided into tropical, temperate, boreal, and polar zones according to their latitudes north and south of the equator. These latitudinal bands are distinguished principally by temperature and the adaptations of plants to temperature. Within latitudinal zones, biomes are further distinguished by the annual level of precipitation and its seasonality.

6. Within temperate climate zones, the major biomes are deciduous forest, needle-leaved forest, temperate rain forest, grassland, and shrubland. At lower latitudes within temperate regions, one also finds Mediterranean-climate woodlands. Subtropical deserts lie between temperate and tropical climate zones.

7. At high latitudes, one encounters boreal forests, usually consisting of needle-leaved trees with persistent foliage and low growth rates on nutrient-poor, acid soils, and tundra, a treeless biome that develops on permanently frozen soils, or permafrost.

8. Tropical climate zones are dominated by evergreen rain forest, seasonal forest, which ranges from partly to fully deciduous and becomes thorn scrub in drier climates, and savanna, a grassland with scattered trees that is maintained by fire and grazing pressure.

9. Aquatic systems are not classified as biomes because they lack the equivalent of terrestrial vegetation. One may, however, distinguish streams, lakes, estuaries, and oceans, and each of these systems can be further subdivided based on salinity, current, and depth of water.

Suggested readings

Eyre, S. R. 1968. *Vegetation and Soils: A World Picture.* 2d ed. Aldine, Chicago.

Forman, R. T. T. (ed.). 1979. *Pine Barrens: Ecosystem and Landscape.* Academic Press, New York.

Jaeger, E. C. 1957. *The North American Deserts.* Stanford University Press, Stanford, Calif.

Jeffree, E. P., and C. E. Jeffree. 1994. Temperature and biogeographical distributions of species. *Functional Ecology* 8:640–650.

Levinton, J. S. 1982. *Marine Ecology.* Prentice-Hall, Englewood Cliffs, N.J.

McMillan, C. 1956. Edaphic restriction of *Cupressus* and *Pinus* in the coast ranges of central California. *Ecological Monographs* 26:177–212.

Prentice, I. C., et al. 1992. A global biome model based on plant physiology and dominance, soil properties and climate. *Journal of Biogeography* 19:117–134.

Teal, J., and M. Teal. 1969. *Life and Death of a Salt Marsh.* Little Brown, Boston.

Terborgh, J. 1992. *Diversity and the Tropical Rain Forest.* Scientific American Library, New York.

Weaver, J. E. 1956. *Grasslands of the Great Plains.* Johnsen, Lincoln, Nebraska.

Whitmore, T. C. 1990. *An Introduction to Tropical Rain Forests.* Oxford University Press, New York.

Whittaker, R. H., and W. A. Niering. 1965. Vegetation of the Santa Catalina Mountains, Arizona: A gradient analysis of the south slope. *Ecology* 46: 429–452.

PART II

ECOSYSTEMS

CHAPTER 6

ENERGY IN THE ECOSYSTEM

From its infancy, ecology has sought to understand the relationships of organisms to the physical environment. As we have seen in the first part of this book, these relationships explain much of the structure and functioning of organisms, as well as the geographic distributions of species. The relationship of plant form to climate underlies the biome concept. The biome comprises an assemblage of plants, animals, and microorganisms, having similar tolerances of physical conditions, living together in the same place.

As ecologists developed an understanding of the organism–environment relationship during the early part of the twentieth century, several new concepts emerged that led the study of ecology in novel directions. One of these was the realization that feeding relationships link organisms into a single functional entity. Foremost among the proponents of this new ecological viewpoint during the 1920s was the English ecologist Charles Elton. Elton argued that organisms living in the same place not only had similar tolerances of physical factors in the environment, but also interacted with each other, most importantly in a system of feeding relationships that he called the **food web.** Of course, every organism must feed in some manner to gain nourishment, and each may be fed upon by some other. But that these feeding relationships define an ecological unit was a novel idea early in the twentieth century.

A decade later, the English plant ecologist A. G. Tansley took Elton's idea an important step further by regarding animals and plants, together with the physical factors of their surroundings, as ecological systems. Tansley called this concept the **ecosystem,** which he regarded as the fundamental unit of ecological organization. Tansley envisioned the biological and physical parts of nature together, unified by the dependence of animals and plants on their physical surroundings and by their contributions to the maintenance of the physical world.

Thermodynamic views of the ecosystem

Working independently of the ecologists of his day, Alfred J. Lotka, a chemist by training, developed ecosystem concepts from considerations of energetics. Lotka was the first to consider populations and communities as **thermodynamic** systems. In principle, he said, each system can be described by a set of equations that represents exchanges of mass and energy among its components. Such exchanges include the assimilation of carbon dioxide into organic carbon compounds by green plants, the consumption of plants by herbivores, and the consumption of animals by carnivores.

Lotka believed that the size of a system and the rates of energy and material transformations within it were determined in accordance with certain thermodynamic principles. Just as heavy machines and fast machines require more fuel to operate than lighter and slower ones, and efficient machines require less fuel than inefficient ones, the energy transformations of ecosystems grow in direct proportion to their size (roughly, the total mass of their constituent organisms), productivity (rate of transformations), and inefficiency. The earth itself is a giant thermodynamic machine in which the circulation of winds and ocean currents and the evaporation of water are driven by the energy in sunlight. Part of that energy is assimilated by the photosynthesis of plants, and this energy ultimately fuels all biological processes.

Lotka's ideas were not widely appreciated by ecologists of his time. His mathematical representations were difficult and unfamiliar, and he did little to promote his ideas at scientific meetings or through further publications. The idea of the ecosystem as an energy-transforming system was brought to the attention of many ecologists for the first time in a paper published in 1942 by Raymond Lindeman, a young aquatic ecologist from the University of Minnesota. Lindeman's framework for understanding ecological systems on the basis of sound thermodynamic principles made a deep impression. He adopted Tansley's notion of the ecosystem as the fundamental unit in ecology and Elton's concept of the food web, including inorganic nutrients at the base, as the most useful expression of ecosystem structure.

The sequence of feeding relationships through which energy passes in the ecosystem is referred to as a **food chain.** A food chain has many links—plant, herbivore, and carnivore, for example—which Lindeman referred to as **trophic levels.** (The Greek root of the word *trophic* means "food.") Furthermore, Lindeman visualized a **pyramid of energy** within the ecosystem. He argued that less energy reaches each successively higher trophic level; energy is lost at each level because of the work performed by organisms and because of the inefficiency of biological energy transformations, which result in energy being lost from the system as heat. Thus, of the light energy impinging on a field, plants gather only a portion. Herbivores harvest even less of this energy because plants use a portion of the energy they assimilate to maintain themselves, and that energy is not available to herbivores as plant biomass.

By the 1950s, the ecosystem concept had fully pervaded ecological thinking and had spawned a new branch of ecology, called **ecosystem ecology,** in which the cycling of matter and the associated flux of energy through an ecosystem provided a basis for characterizing that system's structure and function. Energy and the masses of elements, such as carbon, provided a common "currency" for ecological description. These units of measurement make it possible to compare directly the structure and functioning of different ecosystems in terms of the energy and materials residing in, and transferred among, the plants, animals, microbes, and abiotic sources of energy and nutrients. Measurements of energy assimilation and energetic efficiencies became the tools for exploring this new thermodynamic concept of the ecosystem.

With a clear conceptual framework for the ecosystem and a currency of energy and the masses of elements to describe its structure, ecologists began to measure energy flow and the cycling of nutrients in the ecosystem. One of the strongest proponents of this approach has been Eugene P. Odum of the University of Georgia, whose text *Fundamentals of Ecology,* first published in 1953, influenced a generation of ecologists. Odum depicted ecosystems as energy flow diagrams (Figure 6.1). For any one trophic level, such a diagram consists of a box representing the biomass (or its energy equivalent) at any given time and pathways through the box representing the flow of energy. These diagrams simplify nature, but nonetheless convey the important principle that energy passes from one link in the consumer food chain to the next, diminished by respiration and the shunting of unused foodstuffs to detritus-based food chains. Feeding relationships are depicted as two or more energy flow diagrams linked into food chains.

Unlike energy, which ultimately comes from sunlight and leaves the ecosystem as heat, nutrients are regenerated and retained within the system. Odum elaborated his energy flow diagrams to include this cycling of elements (Figure 6.2). In the development of ecosystem ecology, the cycling of elements has assumed equal standing with the flow of energy. One reason for this prominence is that the amounts of elements and their movement among ecosystem components can provide a convenient index to the flow of energy, which is difficult to measure directly. Carbon, in particular, bears a close relationship to energy content because of its intimate association with the assimilation of energy via photosynthesis.

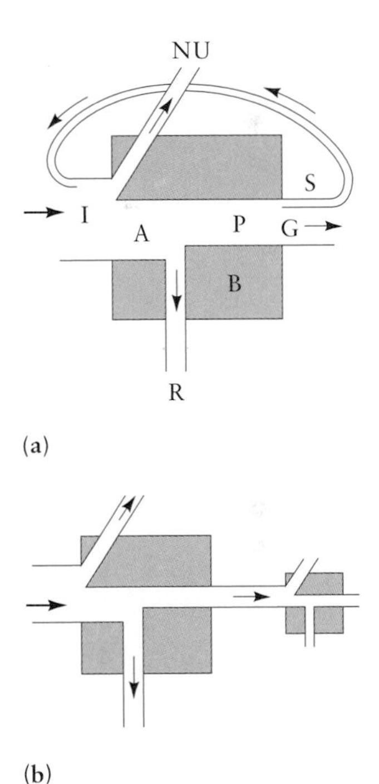

Figure 6.1 (a) E. P. Odum's "universal" model of ecological energy flow, which can be applied to any organism. I = ingestion; A = assimilation; P = production; NU = not used (excreta); R = respiration; G = growth; S = storage, as in the form of fat, for future use; B = biomass. (b) Representation of a food chain by Odum's energy flow models. The net production of one trophic level becomes the ingested energy of the next higher level. After E. P. Odum, *Am. Zool.* 8:11–18 (1968).

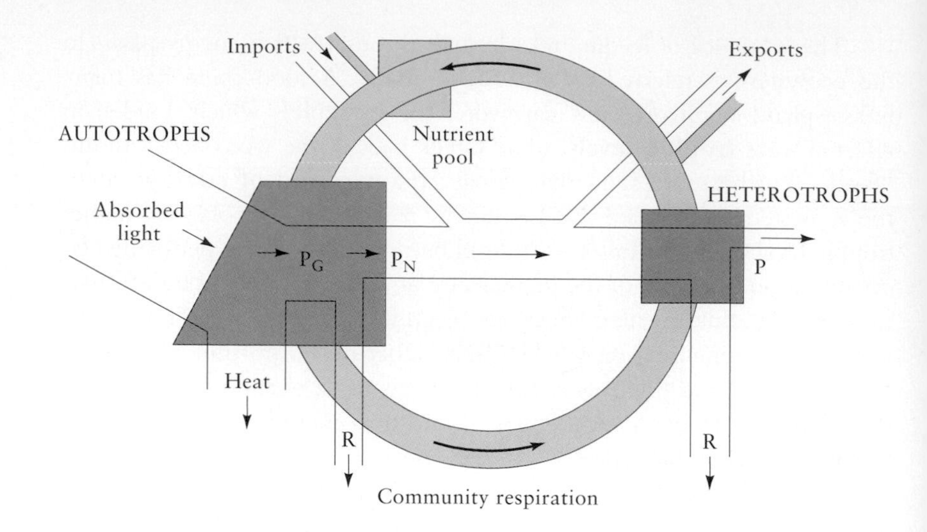

Figure 6.2 E. P. Odum's flow diagram of an ecosystem showing the one-way flow of energy and the recycling of materials. P_G = gross production; P_N = net production; P = heterotrophic production; R = respiration.

A second reason for the prominence of nutrient cycling in ecosystem ecology is the fact that in many circumstances the quantities of certain nutrients regulate primary production. The productivity of desert plants reflects the amount of water available rather than sunlight or minerals in the soil. The open oceans are deserts by virtue of their scarce nutrients, particularly nitrogen. Understanding how elements cycle among components of the ecosystem seems crucial to understanding the regulation of ecosystem structure and function.

Primary production

Plants capture light energy and transform it into the energy of chemical bonds in carbohydrates. Photosynthesis chemically unites two common inorganic compounds, carbon dioxide (CO_2) and water (H_2O), to form glucose ($C_6H_{12}O_6$), with the release of oxygen (O_2). The overall chemical balance of the photosynthetic reaction is

$$6CO_2 + 6H_2O \rightarrow C_6H_{12}O_6 + 6O_2.$$

Photosynthesis transforms carbon from an oxidized (low-energy) state in CO_2 to a reduced (high-energy) state in carbohydrates. Because work is performed on carbon atoms to increase their energy level, photosynthesis requires energy. This energy is provided by the energy of visible light. For each gram of carbon assimilated, a plant transfers 39 kJ of energy from sunlight to the chemical energy of carbon in carbohydrates. But because of inefficiencies in the many biochemical steps of photosynthesis, no more than a third (and usually much less) of the light energy absorbed by photosynthetic pigments eventually appears in carbohydrate molecules.

Photosynthesis supplies the carbohydrate building blocks and energy that a plant needs to synthesize tissues and grow. Rearranged and joined together, glucose molecules become fats, starches, oils, and cellulose. Glucose and other organic compounds (starches and oils, for example) may be transported throughout the plant or stored as a source of energy for future needs. Combined with nitrogen, phosphorus, sulfur, and magnesium, simple carbohydrates derived ultimately from glucose produce an array of proteins, nucleic acids, and pigments. Plants cannot grow unless they have all these basic building materials. Chlorophyll contains an atom of magnesium, and so even when all other necessary elements are present in abundance, a plant lacking sufficient magnesium cannot produce chlorophyll and thus cannot grow.

Plants lie at the base of the food chain, and for this reason they are referred to as the **primary producers** of the ecosystem. The total energy assimilated by photosynthesis is referred to as the **gross primary production** of the ecosystem. Plants use some of this energy to build and maintain themselves, and so their biomass contains substantially less energy than the total assimilated. The energy accumulated in plants, and which is therefore available to consumers, is referred to as **net primary production.** The difference between gross and net production is the energy of respiration, the amount used for maintenance and biosynthesis (Figure 6.3).

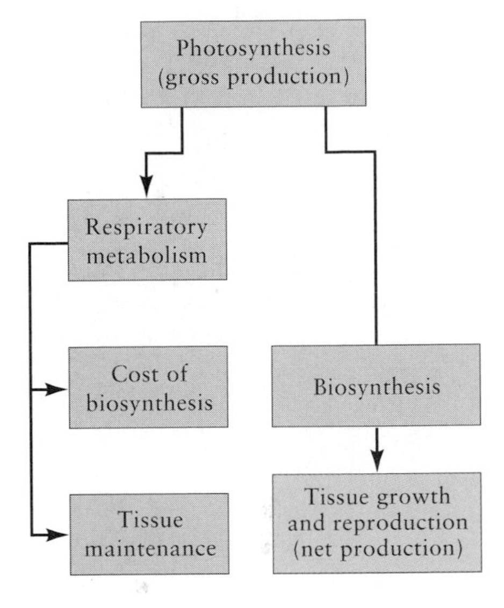

Figure 6.3 A diagram of the allocation of energy by plants. The metabolic costs of biosynthesis and tissue maintenance represent energy lost by way of respiration and thus unavailable to higher trophic levels.

Measurement of primary production

Plant production involves fluxes of carbon dioxide, oxygen, minerals, and water on one hand, and the accumulation of biomass on the other (Figure 6.4). In principle, the rates of any of these flows could provide an index to the overall rate of primary production. In practice, which measure is appropriate depends on habitat and growth form. Furthermore, ecologists use different measures to estimate gross and net production, as well as different measures to estimate the production of an entire system and that of a small part of a single plant.

The units of production are energy per unit of area per unit of time. For comparing ecosystems, kilojoules per square meter per year (kJ m^{-2} yr^{-1}) are convenient units; for comparing the photosynthetic rates of the leaves of two species of plants, it would be more meaningful to use joules per square centimeter of leaf area per second (J cm^{-2} s^{-1}). Production need not be expressed only in terms of energy, however. Net production can be expressed conveniently as grams of carbon assimilated, dry weight of plant tissues, or their energy equivalents. Ecologists use such indices interchangeably because they have found a high degree of correlation among them. The energy content of an organic compound depends primarily on its carbon content, which represents approximately 39 kilojoules per gram of carbon, with some energy added or subtracted during various biochemical transformations.

In terrestrial ecosystems, ecologists often estimate plant production by the annual increase in plant biomass. In areas of seasonal production, annual growth may be determined by cutting, drying, and weighing plants at the end of the growing season. Root growth is often ignored because roots are difficult to remove from most soils; thus harvesting measures the **annual aboveground net productivity (AANP),** the most common basis for comparing terrestrial communities.

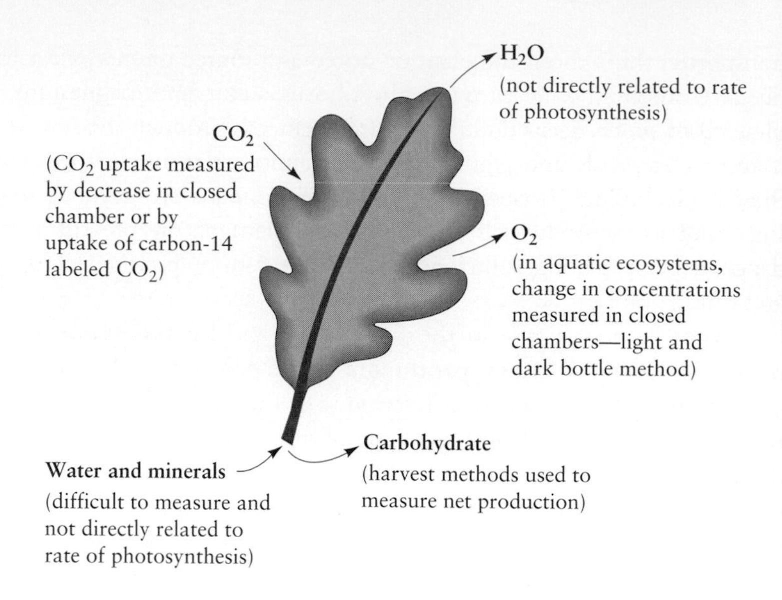

Figure 6.4 Mass balances of photosynthesis and plant production, and how these fluxes are measured and used to estimate primary productivity.

Because the atmosphere contains so little carbon dioxide (0.03%), uptake by plants can measurably reduce its concentration in an enclosed chamber within a short period. This concentration can provide a direct measure of photosynthetic rate under some circumstances. The most convenient application of this method is to enclose samples of vegetation (whole herbaceous plants or branches of trees) in a clear chamber (light must penetrate for photosynthesis) and measure the change in the concentration of CO_2 in air passed through the chamber. CO_2 uptake per gram of dry weight or per square centimeter of leaf surface area in the chamber is then extrapolated to the entire tree or forest. Carbon dioxide flux during the light period of the day includes both assimilation (uptake) and respiration (output), and thus measures net production. Respiration can be estimated separately by measuring carbon dioxide production during the night, when photosynthesis shuts down. Gross production during the day can then be estimated by adding the respiration rate—determined at night—to the daytime rate of net production.

The radioactive isotope carbon-14 (^{14}C) provides a useful variation on the gas exchange method of measuring productivity. When a known amount of ^{14}C-carbon dioxide is added to an airtight chamber, plants assimilate the radioactive carbon atoms roughly in proportion to their occurrence in the air inside the chamber. Thus, we can calculate the rate of carbon fixation by dividing the amount of ^{14}C in the plant by the proportion of ^{14}C in the chamber at the beginning of the experiment. For example, if a plant assimilates 10 mg of ^{14}C in an hour, and the proportion of isotopic carbon in the chamber is 0.05, we can conclude that the plant assimilates carbon at a rate of about 200 mg h^{-1} (10 divided by 0.05).

In aquatic systems, harvesting provides a convenient method for estimating the primary production of large photosynthetic organisms, such as kelps, but this technique is not practical for phytoplankton. The low natural concentration of oxygen dissolved in water does, however, enable ecologists to measure small changes in oxygen concentration in most aquatic systems. To estimate production, samples of water containing phytoplankton are suspended in pairs of sealed bottles at desired depths beneath the surface of a lake or the ocean; one of each pair (the "light bottle") is clear and allows sunlight to enter; the other (the "dark bottle") is opaque. In the light bottles, photosynthesis and respiration occur together, and part of the oxygen produced by the first process is consumed by the second. In the dark bottles, respiration consumes oxygen without its being replenished by photosynthesis. Thus, gross production can be estimated by adding the change in oxygen concentration in the dark bottle (respiration alone) to that in the light bottle (photosynthesis and respiration).

In unproductive waters, such as those of deep lakes and the open ocean, the uptake of ^{14}C by plants and algae is used to estimate carbon assimilation. The principle of the ^{14}C method is the same in aquatic systems as in terrestrial ones, except that the isotope is usually provided in the form of bicarbonate ion (HCO_3^-).

Light, temperature, and photosynthesis

Light levels during the growing season usually exceed the saturation points of photosynthesis in most plants; therefore the productivity of terrestrial habitats generally is not restricted by the availability of light, but rather by the availability of water or nutrients. Temperature, however, does have an effect on production, and temperature may depend on light intensity. Like most other physiological processes, photosynthesis proceeds most rapidly within a narrow range of temperatures; the optimum varies with the prevailing temperature of the environment to which plants acclimate—from about 16°C in many temperate species to as high as 38°C in tropical species. Net production depends on the rate of respiration as well as on the rate of photosynthesis, and respiration generally increases with increasing leaf temperature.

Photosynthetic efficiency is the percentage of the energy in incident radiation that is converted to net primary production during the growing season, and it provides a useful index to rates of primary production under natural conditions. Where water and nutrients do not severely limit plant production, photosynthetic efficiency varies between 1% and 2%. What happens to the remaining 98–99% of the energy? Leaves and other surfaces reflect anywhere from one-quarter to three-quarters of it. Molecules other than photosynthetic pigments absorb most of the remainder, which is converted to heat and either radiated or conducted across the leaf surface or dissipated by the evaporation of water from the leaf (transpiration).

Water and primary production in terrestrial habitats

The tiny openings (stomates) in leaves through which carbon dioxide and oxygen are exchanged with the atmosphere also allow the passage of water vapor (transpiration). As the moisture content of soil decreases, plants have increasing difficulty in obtaining water. As soil moisture approaches the wilting point, the stomates close to reduce water loss. This prevents uptake of CO_2, and photosynthesis slows to a standstill. Consequently, the rate of photosynthesis depends on the availability of soil moisture, the plant's ability to tolerate water loss, and the influence of air temperature and solar radiation on the rate of transpiration.

Agronomists rate the drought resistance of crop plants in terms of **transpiration efficiency,** also called **water use efficiency,** which is the number of grams of dry matter produced (net production) per kilogram of water transpired. In most plants, transpiration efficiencies are less than 2 grams of production per kilogram of water, but they may be as high as 4 g kg^{-1} in drought-tolerant crops. Because transpiration efficiency varies little among a wide variety of plant species, production is directly related to water availability in the environment (Figure 6.5).

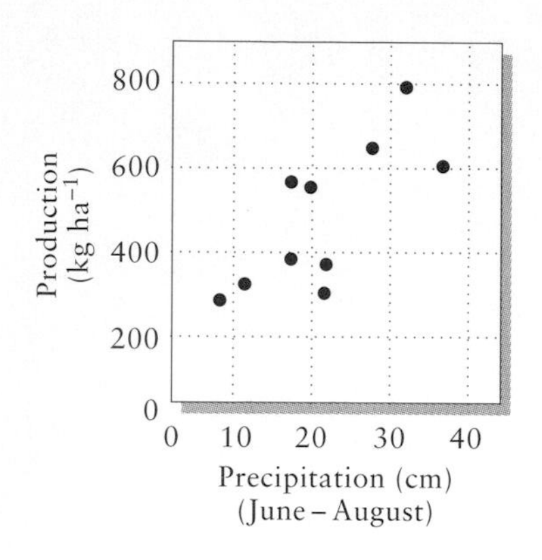

Figure 6.5 Relationship between production and summer precipitation for perennial grassland in southern Arizona. After D. R. Cable, *Ecology* 56:981–986 (1975).

Nutrients and plant production

Fertilizers stimulate plant growth in most habitats. When nitrogen and phosphorus fertilizers were applied singly and in combination to chaparral habitat in southern California, most species responded to the application of nitrogen, but not to that of phosphorus, with increased production (Figure 6.6). This result suggests that production in most chaparral species is limited by the availability of nitrogen. However, the growth of California lilac *(Ceanothus greggii)* bushes, which harbor nitrogen-fixing bacteria in their root systems, responded to the application of phosphorus, but not to that of nitrogen. The productivity of annual plants (forbs and grasses) in the same habitat increased when nitrogen was applied, but was depressed somewhat by the application of phosphorus alone. When equal amounts of nitrogen and phosphorus were applied together, however, production soared. Evidently, the plants could take advantage of increased phosphorus only in the presence of high levels of nitrogen.

Plants and algae generally suffer nutrient limitation most strongly in aquatic habitats, particularly in the open ocean, where the scarcity of dissolved minerals reduces production far below terrestrial levels. Even in shallow coastal waters, where vertical mixing, upwelling currents, and runoff from the land maintain nutrients at high concentrations, the addition of fertilizers (as often occurs inadvertently through pollution) may greatly enhance aquatic production. In a classic study conducted along the southern coast of Long Island, New York, observations showed that phytoplankton abundance closely paralleled the level of inorganic phosphorus, the latter being a general indicator of pollution. But the addition of nutrients to standard cultures of the marine alga *Nannochloris* in water samples taken from different locations demonstrated that nitrogen, rather than phosphorus, limited primary production in both polluted and relatively clean waters (Figure 6.7). Thus, phosphorus occurred in sufficient abundance to promote algal growth, even in the unpolluted sections of coastline.

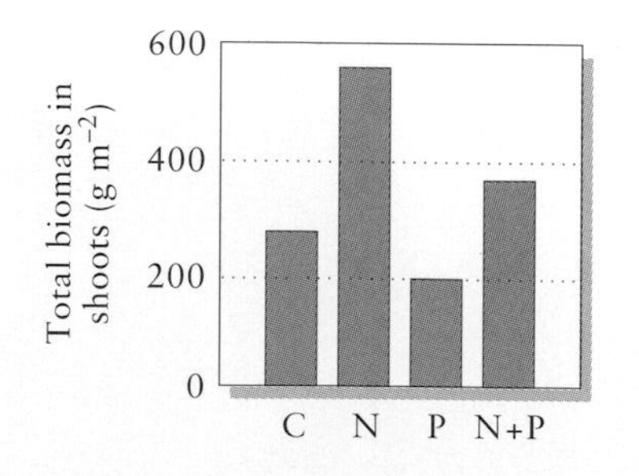

Figure 6.6 Response of the chaparral shrub *Adenostema* to nitrogen (N) and phosphorus (P) fertilization (C = control). After G. S. McMaster, W. M. Jow, and J. Kummerow, *J. Ecol.* 70:745–756 (1982).

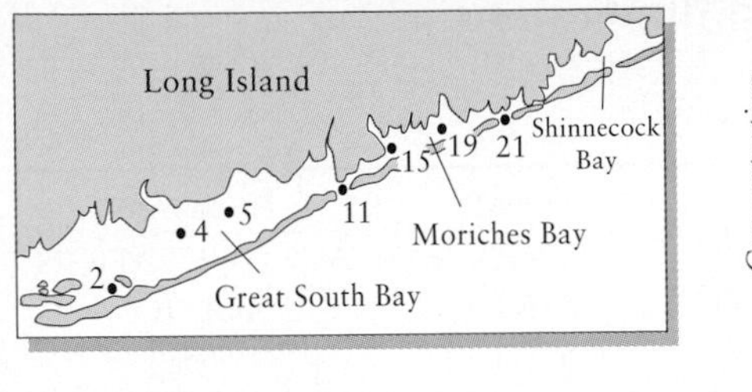

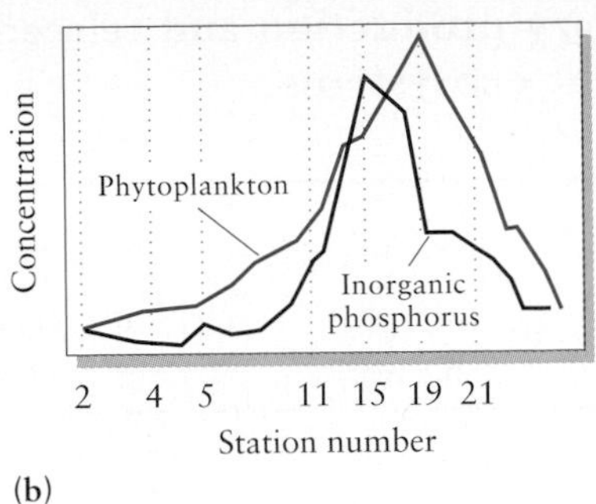

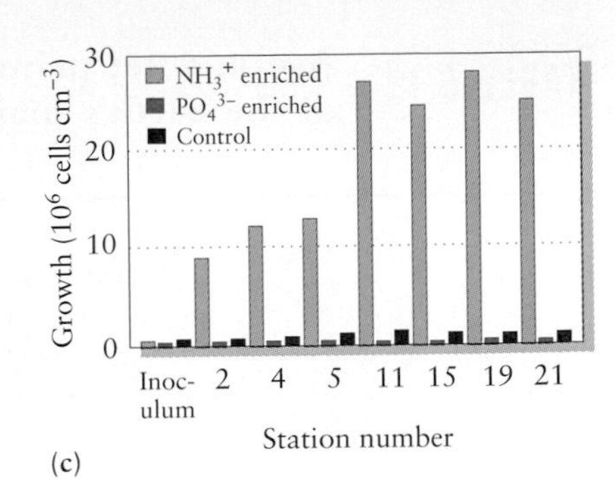

Figure 6.7 (a) Map of water collection stations in Great South Bay and Moriches Bay, Long Island. (b) Phytoplankton and inorganic phosphorus concentrations in water collected in the summer of 1962 from the stations shown in (a). (c) Growth of standard cultures of the marine alga *Nannochloris* in water, taken from each of the stations, to which ammonium or phosphate had been added. From J. H. Ryther and W. M. Dunstan, *Science* 171: 1008–1013 (1971).

Global patterns of primary production

The favorable combination of intense sunlight, warm temperature, and abundant rainfall in the humid Tropics results in the highest terrestrial productivity on earth. In temperate and arctic ecosystems, low winter temperatures and long winter nights curtail production. Within a given latitude belt, where light and temperature do not vary appreciably from one locality to the next, net production is directly related to annual precipitation. Above a certain threshold of water availability, net production increases by 0.4 gram of dry mass per kilogram of water in hot deserts and by 1.1 g kg^{-1} in short-grass prairies and cold (Great Basin) deserts. Thus, a given amount of water supports almost three times as much plant production in cold climates as in hot climates. The water use efficiencies of forest trees under drought stress vary between about 0.9 and 1.8 g kg^{-1}, but under normal conditions—at least 50 cm of precipitation annually in the cooler regions of the United States and perhaps 100 cm in the warmer eastern and southeastern regions—the production of forest ecosystems is generally not limited by water, but rather by light and nutrients.

Global patterns of net primary production are summarized in Table 6.1. These values come from many studies employing a wide variety of techniques, but their lack of strict comparability does not override the general patterns they show. Production of terrestrial vegetation is highest in the wet Tropics and lowest in tundra and desert habitats. Swamp and marsh ecosystems, which occupy the interface between terrestrial and aquatic habitats, can produce as much biomass annually as tropical forests because of the continuous abundance of water and the rapid regeneration of nutrients in the mucky sediments surrounding the plant roots.

In the open ocean, scarcity of mineral nutrients limits productivity to a tenth that of temperate forests, or even less. Upwelling zones (where

TABLE 6.1 **Average net primary production and related dimensions of the earth's major ecosystems**

HABITAT	NET PRIMARY PRODUCTION ($g\ m^{-2}\ yr^{-1}$)	BIOMASS ($kg\ m^{-2}$)	CHLOROPHYLL ($g\ m^{-2}$)	LEAF SURFACE AREA ($m^2\ m^{-2}$)	BIOMASS ACCUMULATION RATIO (yr)
Terrestrial					
Tropical forest	1,800	42	2.8	7	23
Temperate forest	1,250	32	2.6	8	26
Boreal forest	800	20	3.0	12	25
Shrubland	600	6	1.6	4	10
Savanna	700	4	1.5	4	6
Temperate grassland	500	1.5	1.3	4	3
Tundra and alpine	140	0.6	0.5	2	4
Desert scrub	70	0.7	0.5	1	10
Cultivated land	650	1	1.5	4	1.5
Swamp and marsh	2,500	15	3.0	7	6
Aquatic					
Open ocean	125	0.003	0.03	—	0.02
Continental shelf	360	0.01	0.2	—	0.03
Algal beds and reefs	2,000	2	2.0	—	1.00
Estuaries	1,800	1	1.0	—	0.56
Lakes and streams	500	0.02	0.2	—	0.04

Source: R. H. Whittaker and G. E. Likens, *Human Ecol.* 1:357–369 (1973).

nutrients reach the surface from deeper waters) and continental shelf areas (where bottom sediments in shallow water rapidly exchange nutrients with surface waters) support greater production. In estuaries, coral reefs, and coastal algal beds, production approaches the levels observed in adjacent terrestrial habitats. Primary production in freshwater habitats compares favorably with that in marine habitats, achieving the highest levels in rivers, shallow lakes, and ponds and the lowest levels in clear streams and deep lakes. In general, phosphorus limits production in freshwater systems, nitrogen in marine systems.

Food chain energetics

Plants, algae, and some bacteria manufacture their own food from raw inorganic materials. Ecologists refer to such organisms, which form the bases of ecological food chains, as **autotrophs,** which means, literally, "self-nourishers." Animals, fungi, and most microorganisms, which obtain their energy and most of their nutrients by eating plants or animals or the dead remains of either, are called **heterotrophs,** which means "nourished from others." The dual roles of living organisms as food producers and food consumers give the ecosystem a trophic structure, determined by feeding relationships, through which energy flows and nutrients cycle. The food chain from grass to caterpillar to sparrow to snake to hawk traces one particular path that energy follows through the trophic structure. With each link in a food chain, biochemical transformations dissipate much of the energy of gross production before organisms feeding at the next higher trophic level can consume it. All the grass in Africa piled together would dwarf a mound of all the grasshoppers, gazelles, zebras, wildebeests, and other animals that eat grass. That mound of herbivores, in turn, would overwhelm the pitiful heap of all the lions, hyenas, and other carnivores that feed on them.

As Lindeman pointed out in 1942, the amount of energy reaching each trophic level depends on the net primary production at the base of the food chain and on the efficiencies at each higher trophic level with which animals convert food energy into their own biomass energy through growth and reproduction. Of the light energy assimilated by photosynthesis, plants use between 15% and 70% for maintenance, thereby making that portion unavailable to consumers. Herbivores and carnivores are more active than plants and expend correspondingly more of their assimilated energy on maintenance. As a result, the productivity of each trophic level is typically only 5–20% that of the level below it (Figure 6.8). Ecologists refer to the percentage of energy transferred from one trophic level to the next as the **ecological efficiency** or the **food chain efficiency.**

Once consumed, food energy follows a variety of paths through the organism (Figure 6.9). Regardless of the source of its food, what the organism digests and absorbs constitutes its **assimilated energy,** which supports maintenance, builds tissues, or may be excreted in the form of unusable metabolic by-products. The portion of the assimilated energy used to fulfill metabolic needs, most of which escapes the organism as heat, makes up the **respired energy.** Animals excrete another, usually smaller, portion of the

Figure 6.8 An ecological pyramid of energy in which the breadth of each bar represents the net productivity of each trophic level in the ecosystem. For this particular system, ecological efficiencies are 20%, 15%, and 10% between trophic levels, but these values vary widely in different communities.

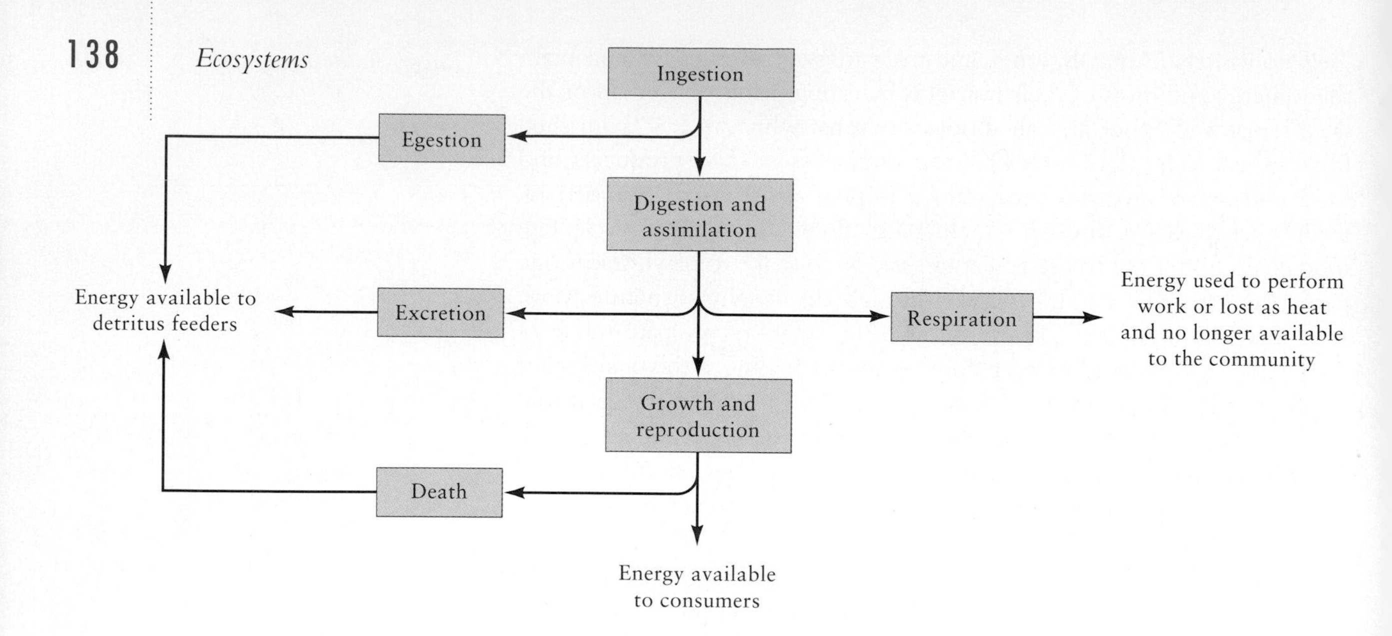

Figure 6.9 Allocation of energy within one link of the food chain.

assimilated energy in the form of nitrogen-containing organic wastes (primarily ammonia, urea, or uric acid) produced when the diet contains an excess of nitrogen; this is called the **excreted energy.** The assimilated energy retained by the organism becomes available for the synthesis of new biomass (production) through growth and reproduction, which animals feeding at the next higher trophic level may then consume. Many components of food resist digestion and assimilation: hair, feathers, insect exoskeletons, cartilage, and bone in animal foods, and cellulose and lignin in plant foods (Figure 6.10). These substances may be defecated or regurgitated, and the energy they contain is referred to as **egested energy.** Although rejected as unsuitable by one consumer, egested materials may be food for other consumers that are specialized to digest the more recalcitrant materials, as we will see below.

Assimilation efficiency

Ecological efficiency is the product of the efficiencies with which organisms exploit their food resources and convert them into biomass: the exploitation, assimilation, and net production efficiencies. Because most biological production is consumed by one organism or another, the **exploitation efficiency** of an entire trophic level approaches 100% overall. Therefore, ecological efficiency depends primarily on **assimilation efficiency,** which is the proportion of consumed energy that is assimilated, and **net production efficiency,** which is the proportion of assimilated energy that is incorporated in growth, storage, and reproduction. Box 6.1 summarizes the energetic efficiencies discussed in this chapter.

For plants, net production efficiency is the ratio of net production to gross production. This index varies between 30% and 85%, depending on habitat and growth form. Rapidly growing plants in temperate zones—whether trees, old-field herbs, crop species, or aquatic plants—have uni-

Figure 6.10 A photograph of elephant dung, showing the poorly digested fibrous plant material it contains.

formly high net production efficiencies (75–85%). Similar types of vegetation in the Tropics exhibit lower net production efficiencies (perhaps 40–60%). As we should expect because of the higher temperature, respiration increases relative to photosynthesis in tropical latitudes.

The energy value of plants to their consumers depends on the amount of cellulose, lignin, and other indigestible materials they contain. Herbivores assimilate as much as 80% of the energy in seeds and 60–70% of that in young vegetation. Most grazers and browsers (elephants, cattle, grasshoppers) extract 30–40% of the energy in their food for themselves. Millipedes, which eat decaying wood composed mostly of cellulose and lignin (and the microorganisms that occur in decaying wood), assimilate only 15% of the energy in their diet.

The effect of food quality on assimilation can be seen in the efficiency with which a single animal retains energy from different portions of its diet. In studies with mountain hares *(Lepus timidus),* the efficiency of energy assimilation on a diet of small willow twigs came to 39%, of which the hares lost an additional 5% through excretion. Lower assimilation efficiencies characterized diets of larger willow twigs (31%), presumably because of their thicker, less digestible bark, and of birch twigs (23–35%), on which the hares could not maintain a constant weight. By measuring the content of fiber (cellulose and lignin) in the food and feces, investigators determined that the digestibility of fiber varied between only 15% and 25%.

Food of animal origin is more easily digested than food of plant origin; assimilation efficiencies of predatory species vary between 60% and 90%. Vertebrate prey are digested more efficiently than insect prey because the indigestible exoskeletons of insects constitute a larger proportion of the body than the hair, feathers, and scales of vertebrates. Assimilation efficiencies of insectivores vary between 70% and 80%, whereas those of most carnivores are about 90%.

BOX 6.1 Definitions of several energetic efficiencies

$$\text{Exploitation efficiency} = \frac{\text{Ingestion of food}}{\text{Prey production}}$$

$$\text{Assimilation efficiency} = \frac{\text{Assimilation}}{\text{Ingestion}}$$

$$\text{Net production efficiency} = \frac{\text{Production (growth and reproduction)}}{\text{Assimilation}}$$

$$\text{Gross production efficiency} = \text{Assimilation efficiency} \times \text{Net production efficiency} = \frac{\text{Production}}{\text{Ingestion}}$$

$$\text{Ecological efficiency} = \text{Exploitation efficiency} \times \text{Assimilation efficiency} \times \text{Net production efficiency} = \frac{\text{Consumer production}}{\text{Prey production}}$$

Net and gross production efficiencies

Maintenance, movement, and heat production require energy that animals otherwise could use for growth and reproduction. Active warm-blooded animals exhibit low net production efficiencies: birds less than 1% and small mammals with high reproductive rates up to 6%. More sedentary cold-blooded animals, particularly aquatic species, channel as much as 75% of their assimilated energy into growth and reproduction. The **gross production efficiency,** which is the product of assimilation efficiency and net production efficiency, represents the overall efficiency of biomass production within a trophic level. The gross production efficiencies of warm-blooded terrestrial animals rarely exceed 5%, and those of some birds and large mammals fall below 1%. For insects, these efficiencies lie within the range of 5–15%, and for some aquatic animals they exceed 30% (Figure 6.11).

Detritus feeding

Terrestrial plants, especially woody species, allocate much of their production to structures that are difficult to ingest, let alone digest. As a result, most terrestrial plant production is consumed as **detritus**—dead remains

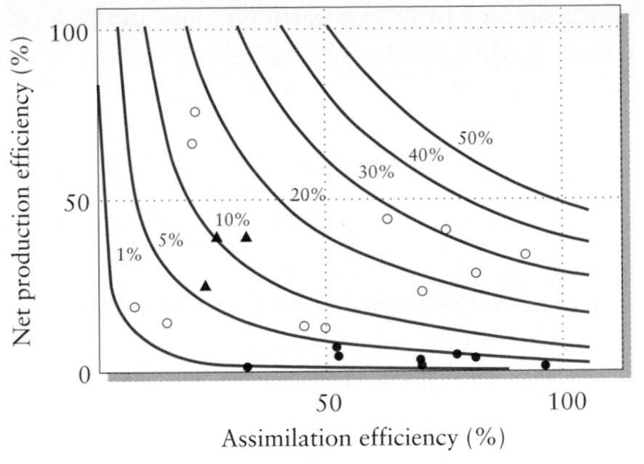

Figure 6.11 Gross production efficiencies for a variety of animal species. Gross production efficiency levels are indicated by the curved lines on the graph. From R. E. Ricklefs, *Ecology,* 2d ed., Chiron Press, New York (1979).

of plants and undigestible excreta of herbivores—by organisms specialized to attack wood, leaf litter, and fibrous plant egesta. This partitioning between herbivory and detritus feeding establishes two parallel food chains in terrestrial communities. The first originates when relatively large animals feed on leafy vegetation, fruits, and seeds; the second originates when relatively small animals and microorganisms consume detritus in the litter and soil layer. These separate food chains sometimes mingle considerably at higher trophic levels, but the energy of detritus tends to move into the food chain much more slowly than the energy consumed by herbivores.

The relative importance of herbivore-based and detritivore-based food chains varies greatly among communities. Herbivores predominate in plankton communities, detritivores in terrestrial communities. The proportion of net production that enters herbivore–predator food chains depends on the relative allocation of plant tissue between structural and supportive functions on one hand and on growth and photosynthetic functions on the other. Herbivores consume 1.5–2.5% of the net production in temperate deciduous forests, 12% of that in old-field habitats, and 60–99% of that in plankton communities.

Rates of energy flow through the ecosystem

Food chain efficiencies indicate the amount of energy that eventually reaches each trophic level of the community. The rate of transfer of energy or, inversely, its **residence time** in each trophic level provides a second index to the energy dynamics of the ecosystem. For a given rate of production, the residence time (also called the **transit time**) of energy in the community and the storage of energy in living biomass and detritus are directly related: the longer the residence time, the greater the accumulation of energy.

The average residence time in a particular link in the food chain equals the energy stored divided by the rate at which energy is converted into biomass:

$$\text{Residence time (yr)} = \frac{\text{Energy stored in biomass (kJ m}^{-2})}{\text{Net productivity (kJ m}^{-2}\text{ yr}^{-1})}.$$

We may also calculate the residence time defined by this equation in terms of mass rather than energy, in which case it expresses the **biomass accumulation ratio.** Wet tropical forests produce dry matter at an average rate of 1,800 g m^{-2} yr^{-1} and have an average living biomass of 42,000 g m^{-2}. Inserting these values into the above equation, we obtain 23 years (42,000/1,800). Biomass accumulation ratios for representative ecosystems may be more than 20 years in forested terrestrial environments and less than 20 days in aquatic plankton-based communities (see Table 6.1). These figures are only averages for the system as a whole. Some energy remains in the system longer; much of it disappears more quickly. For example, leaf eaters and root feeders consume much of the energy assimilated by forest trees during the year of its production, some of it within days of assimilation by the plant.

The figures in Table 6.1 also underestimate the average residence time of energy in plant biomass because they do not include the accumulation of dead organic matter in the litter. The residence time of energy in accumulated litter can be determined by an equation analogous to that for the biomass accumulation ratio:

$$\text{Residence time (yr)} = \frac{\text{Litter accumulation (g m}^{-2})}{\text{Rate of litter fall (g m}^{-2}\text{ yr}^{-1})}.$$

In forests, this value varies from 3 months in the wet Tropics to 1–2 years in dry and montane tropical habitats, 4–16 years in the southeastern United States, and more than 100 years in temperate mountains and boreal regions. Warm temperatures and the abundance of moisture in lowland tropical regions create optimal conditions for rapid decomposition of litter.

Ecosystem energetics

The flux of energy and its efficiency of transfer are used to summarize certain aspects of the structure of an ecosystem: the number of trophic levels, the relative importance of detritus feeding and herbivory, the steady-state values for biomass and accumulated detritus, and the turnover rates of organic matter in the community. The importance of these measures was argued by Lindeman, who constructed the first energy budget for an entire biological community—that of Cedar Bog Lake in Minnesota. The proliferation of energy flow studies during the 1950s and 1960s clearly reflected energy's acceptance as a universal currency, a common denominator to which all populations and their acts of consumption could be reduced.

The overall energy budget of an ecosystem reflects a balance between income and expenditure, just as in a bank account. The ecosystem gains energy through the photosynthetic assimilation of light by green plants and the transport of organic matter into the system from external sources. Organic materials produced outside the system are referred to as **allochthonous** inputs (from the Greek *chthonos,* "of the earth," and *allos,* "other"); photosynthesis that occurs within the system is referred to as **autochthonous** production. In Root Spring, near Concord, Massachusetts, herbivores assimilated energy at a rate of 0.31 W m^{-2} (J m^{-2} s^{-1}), but the net productivity of aquatic plants and algae was only 0.09 W m^{-2}, the balance being transported into the spring in the form of leaves from nearby vegetation. In general, autochthonous production predominates in large rivers, lakes, and most marine ecosystems; allochthonous imports make up the largest part of energy flux in small streams and springs under the closed canopies of forests; and life in caves and the abyssal depths of the oceans, to which no light penetrates, subsists entirely on energy transported in from outside.

Lindeman constructed the Cedar Bog Lake energy budget (Table 6.2) from measurements of the harvestable net production at each of three trophic levels—plants and algae, herbivores, and carnivores—and from laboratory determinations of respiration and assimilation efficiencies. Lindeman's findings are somewhat startling in that herbivores consumed only 20% of the net primary production, and carnivores consumed only 33% of the net production of herbivores. These are extremely low exploitation efficiencies. The majority of the plant and herbivore biomass, which is not consumed, ends up as organic sediments at the bottom of the lake.

TABLE 6.2 An energy flow model for Cedar Bog Lake, Minnesota

	ENERGY (kcal m^{-2} yr^{-1})		
ENERGY PRODUCTION OR REMOVAL	*Primary producers*	*Primary consumers*	*Secondary consumers*
Harvestable production*	704	70	13
Respiration	234	44	18
Removal by consumers			
Assimilated	148	31	0
Unassimilated	28	3	0
Gross production (totals)	1,114	148	31

*Does not include net production removed by consumers. Actual net production, including removal by consumers, was 879 kcal m^{-2} yr^{-1} for primary producers, 104 kcal m^{-2} yr^{-1} for primary consumers, and 13 kcal m^{-2} yr^{-1} for secondary consumers.

Source: R. Lindeman, *Ecology* 23:399–418 (1942).

BOX 6.2

Estimating the average number of trophic levels in various ecosystems from primary production, consumer energy flux, and ecological efficiencies

We can estimate the average length of food chains in a community from net primary production, average ecological efficiency, and average energy flux of a top predator population. The energy available [$E(n)$] to a predator at a given trophic level n (plants being level 1) is equal to the product of the net primary production (NPP) and the intervening ecological efficiencies (Eff). Thus

$$E(n) = \text{NPP Eff}^{n-1},$$

where Eff is the geometric mean of the efficiencies of transfer between each level. This expression can be rearranged to give

$$n = 1 + \frac{\log[E(n)] - \log(\text{NPP})}{\log(\text{Eff})}.$$

Using this equation and some rough estimates for the values on its right-hand side, we can calculate the average number of trophic levels to be about 7 for marine plankton-based systems, 5 for inshore aquatic communities, 4 for grasslands, and 3 for wet tropical forests, as shown in the following table.

COMMUNITY	NET PRIMARY PRODUCTION (kcal m^{-2} yr^{-1})	PREDATOR INGESTION (kcal m^{-2} yr^{-1})	ECOLOGICAL EFFICIENCY (%)	NUMBER OF TROPHIC LEVELS
Open ocean	500	0.1	25	7.1
Coastal marine	8,000	10.0	20	5.1
Temperate grassland	2,000	1.0	10	4.3
Tropical forest	8,000	10.0	5	3.2

Values are approximations based on many studies.

These estimates should be taken with a grain of salt, to be sure, but they do indicate the general size of the pyramid of energy built upon a base of primary production within an ecosystem.

Even with sedimentation, the Cedar Bog Lake ecosystem achieved a 12% overall ecological efficiency of energy transfer between trophic levels. After comparing similar analyses for five aquatic communities, ecologist D. G. Kozlovski concluded that (1) assimilation efficiency increases at higher trophic levels; (2) net production efficiency decreases at higher trophic levels; (3) gross production efficiency also decreases at higher trophic levels; and (4) ecological efficiency (assimilation or gross production at level n divided by that at level $n - 1$) averages about 10%.

Food chain length

Kozlovski's 10% generalization is not a fixed law of ecological thermodynamics. In Silver Springs, Florida, investigators measured ecological efficiencies of 17% between the plant and herbivore levels, but only 5% between herbivores and carnivores. Ecological efficiencies are usually lower in terrestrial than in aquatic habitats. A useful rule of thumb is that the top carnivores in terrestrial communities, on average, can feed no higher than the third trophic level, whereas aquatic carnivores may feed as high as the fourth or fifth level. Of course, a tiny fraction of the total energy may travel through a dozen links before it is dissipated by respiration. But such high trophic levels certainly do not contain enough energy to fully support a predator population (Box 6.2). The key determinant of food chain length is the average ecological efficiency of the links of the food chain.

Summary

1. An ecosystem is the whole complex of organisms and the physical environments they inhabit. It is also a giant thermodynamic machine that continuously dissipates energy in the form of heat. This energy initially enters the biological realm of the ecosystem via photosynthesis and plant production, which provide energy for animals and nonphotosynthetic microorganisms.

2. Charles Elton described biological communities in terms of feeding relationships, which he emphasized as a dominant organizing principle in community structure.

3. In 1935, A. G. Tansley coined the term *ecosystem* to include the organisms and all the abiotic factors in a habitat.

4. A. J. Lotka, in 1925, provided a thermodynamic perspective on ecosystem function, showing that the movement and transformations of mass and energy conform to thermodynamic laws.

5. Raymond Lindeman, in 1942, popularized the idea of the ecosystem as an energy-transforming system, providing a formal notation for energy flux in trophic levels and for ecological efficiency.

6. The study of ecosystem energetics dominated ecology during the 1950s and 1960s, due largely to the influence of Eugene P. Odum, who championed energy as a common currency for describing ecosystem structure and function.

7. Gross primary production is the total energy fixed by photosynthesis. Net primary production is the accumulation of energy in plant biomass; hence it is gross production minus respiration.

8. Primary production can be measured by one or some combination of a variety of methods, such as harvest, gas exchange (carbon dioxide in terrestrial habitats, oxygen in aquatic habitats), or assimilation of radioactive carbon (^{14}C).

9. The rate of primary production in most terrestrial habitats is not limited by light, which generally exceeds the saturation points of most plant species. The efficiency of photosynthesis (gross production divided by total incident light energy) during daylight periods in the growing season is 1–2% in most habitats.

10. Because plants lose water in direct proportion to the amount of carbon dioxide assimilated, plant production in dry environments is limited by, and varies in direct proportion to, the availability of water. Transpiration efficiency, also called water use efficiency, is the ratio of production (in grams of dry mass) to water transpired (in kilograms). Transpiration efficiency typically ranges between 1 and 2; it occasionally reaches 4 in drought-adapted species.

11. Production in both terrestrial and aquatic environments can be enhanced by the application of various nutrients, especially nitrogen and phosphorus, which indicates that nutrient availability limits production. This effect is greatest in the open ocean and deep lakes, where nutrient inputs from sediments and runoff are lowest.

12. The movement of energy and materials through a food chain can be characterized by the assimilation efficiency (the ratio of assimilation to digestion) and the net production efficiency (the ratio of production to assimilation). Unassimilated material enters detritus-based food chains.

13. Assimilation efficiency depends on the quality of the diet, particularly the amount of digestion-resistant structural material (cellulose, lignin, chitin, keratin) it contains. Assimilation efficiency varies from about 15% to 90%.

14. Net production efficiency is lowest in animals whose costs of maintenance and activity are greatest, especially warm-blooded vertebrates. Typical net production efficiencies of 1–5% for warm-blooded vertebrates contrast with the values of 15–45% that are typical of invertebrates.

15. Gross production efficiency (the ratio of production to ingestion) varies between about 5% and 20% in most studies.

16. The average residence time of biomass or energy in a single link of the food chain is the ratio of biomass or the energy stored in it to the rate of net production. Average residence times for primary production vary from 20 years in some forests to 20 days or less in aquatic, plankton-based communities.

17. Considerations of energy flux and ecological efficiencies suggest that the highest trophic level at which a consumer population can be maintained ranges from the third level in terrestrial food chains to the seventh level in plankton-based communities of the open ocean.

Suggested readings

Clarkson, D. T., and J. B. Hanson. 1980. The mineral nutrition of higher plants. *Annual Review of Plant Physiology* 31:239–298.

Cook, R. E. 1977. Raymond Lindeman and the trophic-dynamic concept in ecology. *Science* 198:22–26.

Fenchel, T. 1988. Marine plankton food chains. *Annual Review of Ecology and Systematics* 19:19–38.

Golley, F. B. 1994. *A History of the Ecosystem Concept in Ecology.* Yale University Press, New Haven, Conn.

Griffin, D. H. 1981. *Fungal Physiology.* Wiley, New York.

Howarth, R. W. 1988. Nutrient limitation of net primary production in marine ecosystems. *Annual Review of Ecology and Systematics* 19:89–110.

Laws, R. M. 1985. The ecology of the Southern Ocean. *American Scientist* 73: 26–40.

Lawton, J. H. 1994. What do species do in ecosystems? *Oikos* 71:367–374.

Lindeman, R. 1942. The trophic-dynamic aspect of ecology. *Ecology* 23: 399–418.

Odum, E. P. 1968. Energy flow in ecosystems: A historical review. *American Zoologist* 8:11–18.

Pauly, D., and V. Christensen. 1995. Primary production required to sustain global fisheries. *Nature* 374:255–257.

Webb, W., S. Szarek, W. Lauenroth, R. Kinerson, and M. Smith. 1978. Primary productivity and water use in native forest, grassland, and desert ecosystems. *Ecology* 59:1239–1247.

Whittaker, R. H., and G. E. Likens. 1973. Primary production: The biosphere and man. *Human Ecology* 1:357–369.

Wiegert, R. G. 1988. The past, present, and future of ecological energetics. In L. R. Pomeroy and J. J. Albert (eds.), *Concepts of Ecosystem Ecology: A Comparative View,* pp. 29–55. Springer-Verlag, New York.

CHAPTER 7

PATHWAYS OF ELEMENTS IN THE ECOSYSTEM

Nutrients, unlike energy, remain within the ecosystem, where they continually cycle between organisms and the physical environment. Most nutrients originate in rocks of the earth's crust or in the earth's atmosphere, but within the ecosystem they are reused over and over by plants and animals before being lost in sediments, streams, and groundwater or escaping to the atmosphere as gases. Though all of the energy assimilated by green plants is "new" energy received from outside the ecosystem, most of the nutrients assimilated by plants have been used before. The ammonia absorbed from the soil by roots might have been leached out of decaying leaves on the forest floor that same day. The carbon dioxide assimilated by a green plant might have been produced recently by animal, plant, or microbial respiration.

Each element follows a unique route, determined by its particular biochemical transformations, in its wanderings through the ecosystem. Living systems transform chemical compounds to obtain nutrients for building their structures and to mobilize the energy required by their metabolic processes. This chapter shows how physical, chemical, and biological processes result in the cycling of elements within ecosystems. We shall see that many aspects of element cycling make sense only when one understands that chemical transformations and energy transformations go hand in hand.

Energy transformation and element cycling

Transformations that change elements to biological forms are referred to as **assimilatory** processes, referring to the assimilation of elements into the molecules of plants, animals, and microbes. Photosynthesis, in which plants use energy to change an inorganic form of carbon (carbon dioxide) to the organic form of carbon found in carbohydrates, is one example of an assimilatory transformation of an element. In the overall cycling of carbon, photosynthesis is balanced by respiration, a complementary **dissimilatory** process that involves the transformation of organic carbon back to the inorganic form with an accompanying release of energy. Transformations involving nitrogen and sulfur are equally important, as we shall see.

Not all transformations of elements in the ecosystem are biologically mediated, nor do all involve the net assimilation or release of useful quantities of energy. Many chemical reactions take place in the air, soil, and water. Some of these, such as the weathering of bedrock, release certain elements (potassium, phosphorus, and silicon, for example) to the ecosystem. Lightning storms produce small amounts of reduced nitrogen from the molecular nitrogen in the atmosphere, and these products can be assimilated by plants. Such reactions are thought to have been involved in the origin of life itself. Other physical and chemical processes, such as the sedimentation of calcium carbonate in the oceans, remove elements from circulation and incorporate them into rocks in the earth's crust, where they can remain locked away for eons.

Most energy transformations are associated with the biochemical **oxidation** and **reduction** of carbon, oxygen, nitrogen, and sulfur. An atom is oxidized when it gives up electrons, and it is reduced when it accepts electrons. In a sense, these electrons carry with them a portion of the energy content of the atom, which can be used in biological transformations. In each case, an energy-releasing transformation (oxidation) pairs with an energy-requiring transformation (reduction), and energy shifts from the reactants in the first transformation to the products in the second transformation (Figure 7.1). Such coupled transformations work—that is, they are thermodynamically possible—only when the oxidation side releases at least as much energy as the reduction side requires. The energy changes associated with various reactions vary widely depending on the compounds involved and the number of electrons exchanged. Unfortunately, the energies of the two transformations rarely match, and the excess energy supplied by an oxidation over that required by a reduction cannot be used; the balance is lost in the form of heat. These imbalances are responsible for the thermodynamic inefficiency of life processes.

A typical coupling of transformations might involve the oxidation of carbon in carbohydrate (perhaps glucose), which releases energy, and the reduction of nitrate-nitrogen to amino-nitrogen (the building blocks of

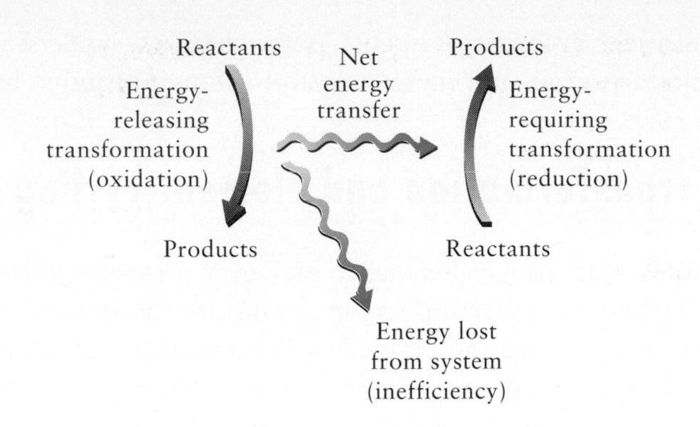

Figure 7.1 The coupling of energy-releasing and energy-requiring transformations is the basis of energy flow in the ecosystem.

proteins), which requires energy. Some transformations involve many intermediate steps of the type shown in Figure 7.1, linked together into a biochemical, or metabolic, pathway (Figure 7.2). Plants accomplish the initial input of energy into the ecosystem by an assimilatory transformation—the reduction of carbon—in which light, rather than a coupled dissimilatory process, serves as the source of energy. A portion of that energy escapes biological systems with each subsequent transformation. Some of these transformations involve the assimilation of other elements required for growth and reproduction; in animals, most are biochemical transformations required to maintain the cell environment and to effect movement. The cycling of elements between the living and nonliving parts of the ecosystem is connected to energy flow by the coupling of the dissimilatory part of one cycle to the assimilatory part of another.

Compartment models of the ecosystem

A **compartment model** is a model in which the various parts of a system are represented as units (compartments) that receive inputs from, and provide outputs to, other such units. The food web is a compartment model of the biological community, in which species are compartments and materials and energy move between compartments when one organism feeds on another. Applying this concept to the cycling of elements within the ecosystem, each form of an element can be thought of as occupying a separate compartment, like a room in a house. In this analogy, biochemical transformations result in exchanges of the elements between the compart-

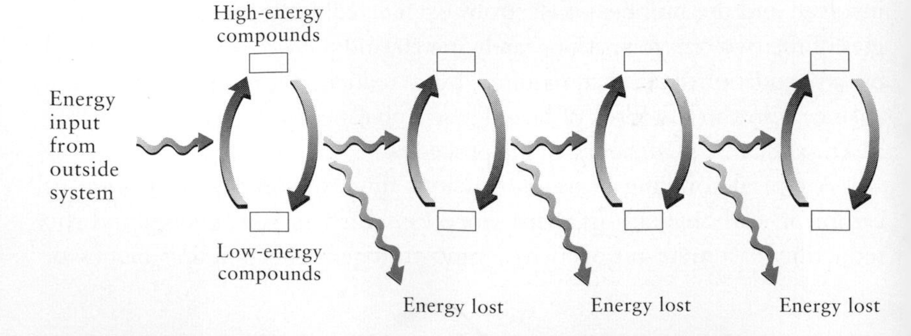

Figure 7.2 As energy flows through the ecosystem, elements alternate between assimilatory and dissimilatory transformations, thus going through cycles.

ments—that is, their movement between rooms (Figure 7.3). For example, photosynthesis moves carbon from the carbon dioxide compartment to that containing organic forms of carbon (assimilation); respiration brings it back (dissimilation). Compartment models can be hierarchically organized, having subcompartments within compartments. The inorganic carbon compartment includes carbon dioxide both in the atmosphere and dissolved in water, carbonate and bicarbonate ions dissolved in water, and calcium carbonate, mostly as a precipitate in the water column and in sediments. The organic carbon compartment also has many subcompartments: animals, plants, microorganisms, and detritus. Herbivory, predation, and detritus feeding move carbon among these subcompartments.

Some transformations involve changes in energy state. Photosynthesis adds energy to carbon, which we may think of as lifting the element to the second floor of the house. In descending the respiration "staircase," carbon releases this stored chemical energy, which the organism can then use for other purposes.

Elements cycle rapidly among some compartments of the ecosystem and much more slowly among others. The movement of an element between living organisms and the inorganic forms that organisms both produce and use occurs over periods ranging from a few minutes to the life span of the organism and its subsequent existence as organic detritus. Both organic and inorganic forms of elements may leave rapid circulation within the ecosystem for compartments not readily accessible to transforming agents. For example, coal, oil, and peat contain vast quantities of organic carbon that has been removed from circulation in the ecosystem. Sedimentation removes inorganic carbon from circulation in the ecosystem by the precipitation of calcium carbonate, which forms thick layers of marine sediments that may eventually turn to limestone. These forms of carbon are returned to the rapidly cycling compartments of the ecosystem only by the slow geologic processes of uplift and erosion.

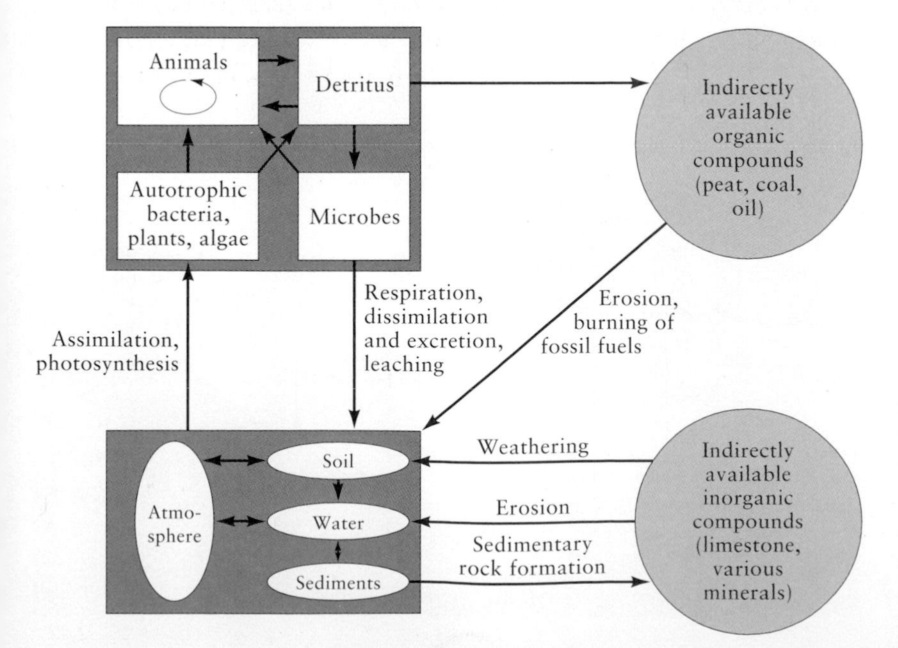

Figure 7.3 A generalized compartment model of the cycling of elements within the ecosystem. Within each compartment, we can recognize subcompartments; for example, the compartment that represents available organic forms of nutrients is further subdivided into compartments occupied by plants, algae, and autotrophic bacteria; animals; detritus; and microbes.

The water cycle

Although water is chemically involved in photosynthesis, most of the movement of water through the ecosystem takes place by the physical processes of evaporation, transpiration, and precipitation (Figure 7.4). Light energy absorbed by water performs the work of evaporation. The condensation of atmospheric water vapor to form clouds releases the potential energy in water vapor as heat. Thus evaporation and condensation resemble photosynthesis and respiration thermodynamically, and they provide an instructive model for the cycling of nutrients within the ecosystem more generally.

Most of the water that presently circulates within the biosphere originated through volcanic outpourings of steam from deep within the earth. The total available water at the earth's surface amounts to about 1.4 billion cubic kilometers, or $1{,}400{,}000 \times 10^{18}$ g. More than 97% of all available water resides in the oceans. Other reservoirs of available water are ice caps and glaciers ($29{,}000 \times 10^{18}$ g), underground aquifers ($8{,}000 \times 10^{18}$ g), lakes and rivers (100×10^{18} g), soil moisture (100×10^{18} g), water vapor in the atmosphere (13×10^{18} g), and all the water in living organisms (1×10^{18} g).

Over land surfaces, precipitation (111×10^{18} yr^{-1}, which is 22% of the global total) exceeds evaporation and transpiration (71×10^{18} g yr^{-1}; 16% of the global total). Over the oceans, evaporation exceeds precipitation by a similar amount. Much of the water that evaporates from the surface of the oceans is carried by winds to the continents, where it is captured as precipitation by the land. This net flow of atmospheric water vapor from ocean to land (40×10^{18} g yr^{-1}) is balanced by runoff from the land back into the ocean basins.

We can calculate the energy that drives the global hydrologic cycle by multiplying the total weight of water evaporated (456×10^{18} g yr^{-1}) by the energy required to evaporate 1 g of water (2.24 kJ). The product, approximately 10^{21} kJ yr^{-1}, represents about one-quarter of the total

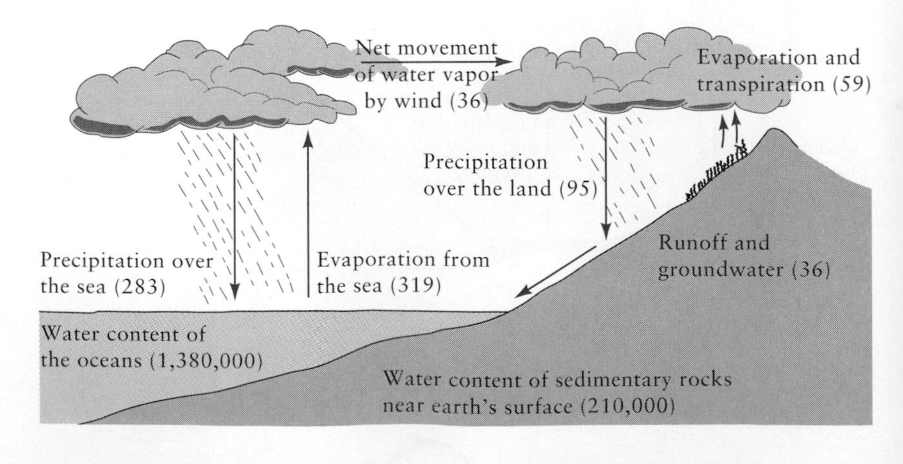

Figure 7.4 The water cycle, with its major components expressed on a global scale. The sizes of compartments and transfers between compartments (shown in parentheses) are expressed as billion billion (10^{18}) grams and billion billion grams per year. After R. G. Barry and R. J. Chorley, *Atmosphere, Weather, and Climate,* Holt, New York (1970); G. E. Hutchinson, *A Treatise on Limnology,* Vol. 1: *Geography, Physics, and Chemistry,* Wiley, New York (1957).

energy of the sun's radiation striking the earth. Evaporation, not precipitation, determines the rate of movement of water through the ecosystem. The absorption of radiant energy by liquid water couples an energy source to the hydrologic cycle. Evaporation and precipitation are closely linked because the atmosphere has a limited capacity to hold water vapor; any increase in the evaporation of water into the atmosphere creates an excess of vapor and causes an equal increase in precipitation.

The water vapor in the atmosphere at any one time (the compartment size) corresponds to an average of 2.5 cm (1 in) of water spread evenly over the surface of the earth. An average of 65 cm (26 in) of rain or snow falls each year (the water flux), which is 26 times the average amount of water vapor. Thus the steady-state content of water in the atmosphere—the atmospheric compartment—replaces itself 26 times each year on average. (Conversely, water has an average residence time in the atmosphere of 1/26th of a year, or 2 weeks.) The soils, rivers, lakes, and oceans contain more than 100,000 times the amount of water in the atmosphere. Fluxes through both pools are the same, however, because evaporation balances precipitation. Thus the average residence time of water in its liquid form at the earth's surface (about 2,800 years) is about 100,000 times longer than its residence time in the atmosphere.

Redox potential

Just as the energy transformation of evaporation drives the water cycle, chemical energy transformations propel the cycling of elements through the biosphere. The energy potential of chemical transformations is measured by the **redox potential** of a reaction, which reflects the capacity of an atom to accept electrons—that is, to become reduced. Redox potentials are expressed as volts (electric potential). A high value indicates that an atom accepts electrons readily and therefore that the reaction has the potential to oxidize some other substance; lower values indicate more powerful reducing potential (Box 7.1).

Molecular oxygen (O_2) is an example of a strong oxidizer. In its organic form, carbon is a strong reducer. Giving electrons and taking them are opposite sides of the same coin; each substance may serve as an oxidizing or a reducing agent. But because strong oxidizers hold onto electrons tightly, they always perform as weak reducers, and vice versa. To put it another way, a strong oxidizer, like molecular oxygen, can always oxidize (be reduced by) a weaker oxidizer, like organic carbon.

Oxygen and carbon are only two of the electron acceptors and donors that are important to biological reactions and whose redox reactions drive the cycling of elements in the ecosystem. Nitrate (NO_3^-) is a powerful oxidizer; hydrogen sulfide (H_2S) and ammonium (NH_3) are good reducers. Indeed, the synthesis and metabolism of many organic compounds, including amino acids, involve assimilatory and dissimilatory redox reactions of nitrogen and sulfur. In general, oxygen is the oxidizer of choice because of its ubiquity and because its common reduced form (H_2O) is innocuous. In contrast, the reduction of nitrate produces nitrite (NO_2^-),

BOX 7.1 An introduction to oxidation and reduction

Every oxidation-reduction (redox) reaction can be characterized as a pair of half-reactions, one for the reduction (electron-accepting) step and the other for the oxidation (electron-donating) step. For example, when molecular oxygen acts as an oxidizing agent, each atom accepts 2 electrons (an electron is designated e^-): $O_2 + 4e^- \rightarrow 2O^{2-}$. In this reduced form, oxygen readily combines with any of a variety of positive ions, such as H^+, giving

$$O_2 + 4e^- + 4H^+ \rightarrow 2H_2O,$$

or C^{4+}, giving

$$O_2 + 4e^- + C^{4+} \rightarrow CO_2.$$

These half-reactions include only the reduction step. The electrons must come from an oxidation half-reaction. An example is

$$CH_2O \rightarrow C^{4+} + H_2O + 4e^-.$$

The carbon atom donates electrons (it is oxidized), and the electrons are available to reduce an oxidizer. The overall reaction combining the last two half-reactions has the form

$$CH_2O + O_2 \rightarrow CO_2 + H_2O,$$

which is the familiar equation for respiration.

The redox potentials, or relative strengths, of oxidizers and reducers are measured by their electric potentials (Eh), which are expressed in volts (V). An electric current represents the movement of electrons; a chemical battery generates an electric potential by a pair of redox reactions, one at each electrode. Oxygen has a high redox potential (+0.81 V at pH 7 and 25°C). Opposites attract, and oxygen has a positive redox potential because it attracts electrons. Oxygen is an excellent oxidizer. So are nitrate (NO_3^-; +0.42 V), ferric iron (Fe^{3+}; +0.36 V), and, to a lesser extent, sulfate (SO_4^{2-}; –0.20 V). Near the bottom of the redox scale are carbon dioxide (CO_2; –0.43 V) and molecular nitrogen (N_2; –0.28 V). Atoms of these elements in their oxidized forms tend to give up rather than accept electrons, and so considerable energy is required to reduce them. The electric potentials of the half-reactions that reduce CO_2-carbon (C^{4+}) to organic carbon (C^0; –0.43 V) and N_2 to organic (ammonia) nitrogen (N^{3-}; –0.25 V) are unfavorable thermodynamically under most conditions, particularly in the presence of oxygen.

Any oxidization reaction having a high redox potential coupled with a reduction reaction having a low redox potential can proceed with the release of energy. Thus the oxidation of carbon by oxygen (O_2) releases energy, because oxygen is a stronger oxidizing agent than carbon dioxide. For the reaction to proceed in the opposite direction (for example, fixation of carbon by photosynthesis), energy must be added.

which most organisms can tolerate only in minute quantities. Oxidizers such as nitrate and ferric iron can nonetheless be important in soils and aquatic sediments in which oxygen has been depleted (anaerobic conditions), as we shall soon see.

The carbon cycle

Three major classes of processes cause the cycling of carbon in aquatic and terrestrial systems (Figure 7.5). The first includes the assimilatory and dissimilatory reactions of carbon in photosynthesis and respiration. These are the major energy-transforming reactions of life. Approximately 10^{17} g (10^{11} metric tons) of carbon enter into such reactions worldwide each year. During photosynthesis, the oxidation state (electric charge) of carbon decreases from +4 to 0, indicating a gain of 4 electrons, each with charge of −1 (Figure 7.6). This gain of electrons is accompanied by a gain in chemical energy. The energy is released by respiration, which results in a gain in the oxidation state of carbon from 0 to +4. Under anaerobic conditions, some bacteria can transform carbon dioxide and organic carbon into methane, which has an oxidation state of −4. **Methanogenesis** results in a net gain of energy because some of the organic carbon is oxidized to carbon dioxide while the rest is reduced to methane (CH_4). The overall reaction, starting with methanol (CH_3OH), is

$$4CH_3OH \rightarrow 3CH_4 + CO_2 + 2H_2O,$$

in which case carbon acts both as an electron acceptor to produce methane and an electron donor to produce carbon dioxide. When hydrogen gas is present under anaerobic conditions, CO_2 can be used as an oxidizer, giving

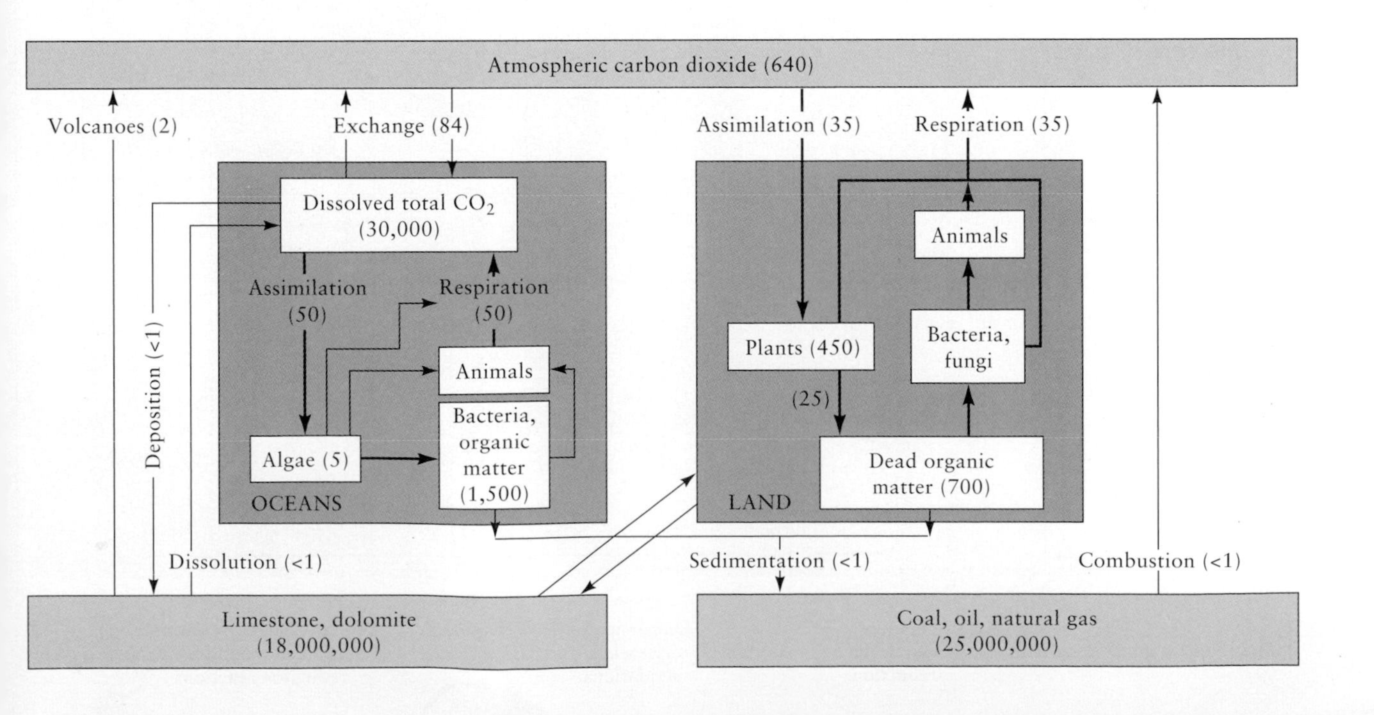

Figure 7.5 The global carbon cycle. The sizes of compartments and transfers between compartments (shown in parentheses) are in billions of metric tons (10^{15} g) and billions of metric tons per year. After T. Fenchel and T. H. Blackburn, *Bacteria and Mineral Cycling,* Academic Press, New York (1979); W. D. Grant and P. E. Long, *Environmental Microbiology,* Wiley, New York (1981).

$$CO_2 + 4H_2 \rightarrow CH_4 + 2H_2O,$$

with a net release of energy. These methanogenic reactions are very restricted biologically, but they have assumed great importance recently because methane is an important greenhouse gas that helps to trap heat in the atmosphere and is contributing to the warming of the earth's climate. Atmospheric methane has been increasing in recent years because it is produced in great quantity by the gut bacteria of cattle and by anaerobic bacteria in rice paddies, both of which are important sources of food for humans.

The second class of carbon cycling processes involves the physical exchange of carbon dioxide between the atmosphere and oceans, lakes, and streams. Carbon dioxide dissolves readily in water; indeed, the oceans contain about 50 times as much CO_2 as the atmosphere. Exchange across the air–water boundary links the carbon cycles of terrestrial and aquatic ecosystems.

The third class of carbon cycling processes consists of the dissolution and precipitation (deposition) of carbonate compounds as sediments, particularly limestone and dolomite. On a global scale, dissolution and precipitation approximately balance each other, although certain conditions favoring precipitation have led to the deposition of extensive layers of calcium carbonate (limestone) sediments in the past. In aquatic systems, dissolution and deposition occur about two orders of magnitude (100 times) more slowly than assimilation and dissimilation by biological systems. Thus, the exchange between sediments and the water column is relatively unimportant to the short-term cycling of carbon in the ecosystem. Locally and over long peri-

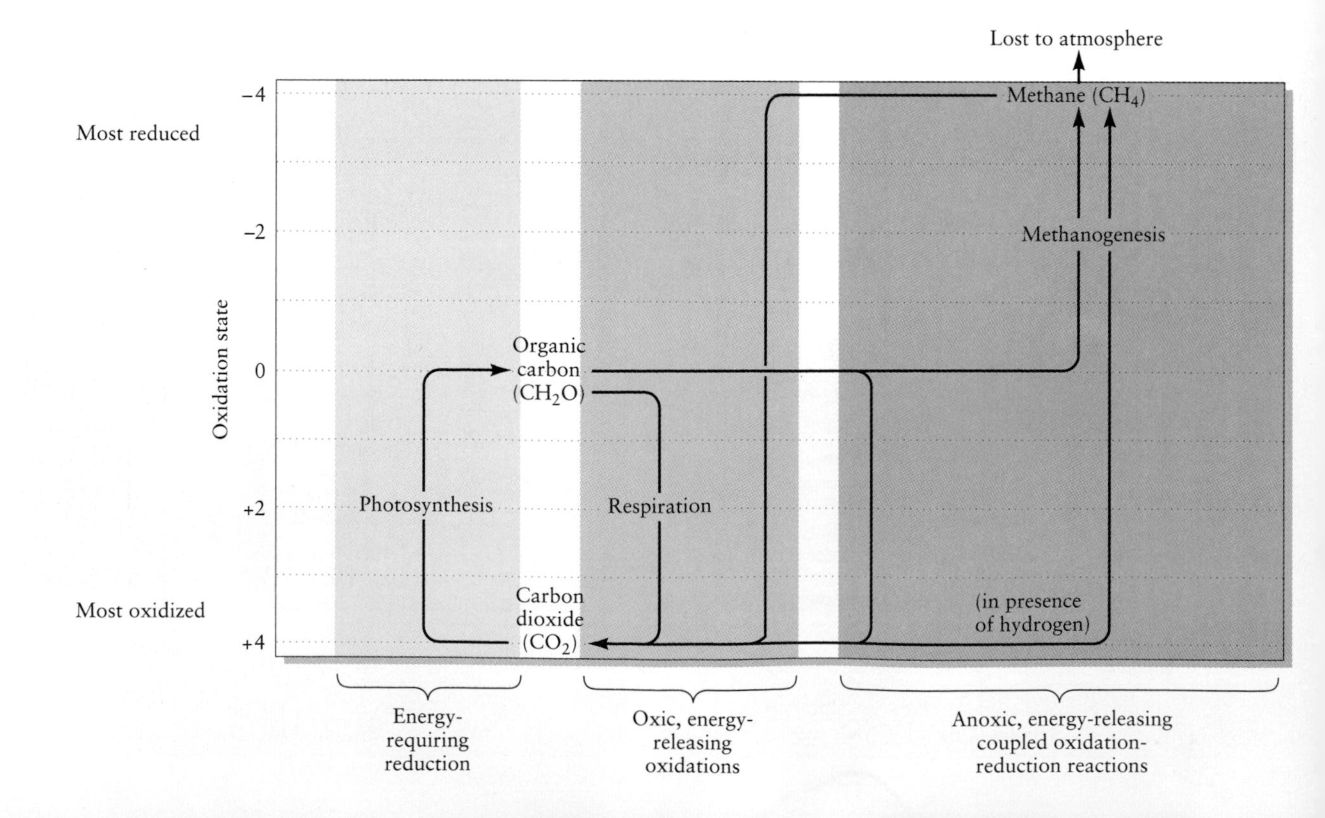

Figure 7.6 Schematic diagram of transformations and oxidation states of compounds in the carbon cycle. Most of the carbon cycles between organic forms (C^0) and carbon dioxide (C^{4+}). Methane (C^{4-}) is produced only by certain bacteria under anaerobic conditions.

Figure 7.7 Sedimentary deposits of limestone in mountains of southern Texas represent calcium carbonate precipitated out of solution in the shallow seas that once covered the area. Courtesy of the U.S. National Park Service.

ods, however, it can assume much greater importance; in fact, almost half of the ecosystem's carbon is locked up in sedimentary rocks (Figure 7.7).

When carbon dioxide dissolves in water, it forms carbonic acid,

$$CO_2 + H_2O \rightleftharpoons H_2CO_3,$$

which readily dissociates into hydrogen, bicarbonate, and carbonate ions:

$$H_2CO_3 \rightleftharpoons H^+ + HCO_3^- \rightleftharpoons 2H^+ + CO_3^{2-}.$$

Calcium, when present, also equilibrates with the carbonate ions:

$$Ca^{2+} + CO_3^{2-} \rightleftharpoons CaCO_3.$$

Calcium carbonate ($CaCO_3$) has low solubility under most conditions and readily precipitates out of the water column to form sediments that eventually can become limestone. This sedimentation effectively removes carbon from aquatic ecosystems, but the rate of removal is less than 1% of the annual cycling of carbon in these ecosystems, and this amount is added back by input from rivers, which are naturally somewhat acid and tend to dissolve limestone (carbonate) sediments.

Dissolution and dissociation may be affected locally by the activities of organisms. In the marine system, under approximately neutral conditions, carbonate and bicarbonate are in chemical equilibrium:

$$CaCO_3 \text{ (insoluble)} + H_2O + CO_2 \rightleftharpoons Ca^{2+} + 2HCO_3^- \text{ (soluble)}.$$

Uptake of CO_2 by photosynthesis shifts the equilibrium to the left, resulting in the formation and precipitation of calcium carbonate. Many algae excrete this calcium carbonate to the surrounding water, but reef-building algae and coralline algae incorporate the substance into hard body structures (Figure 7.8). In the system as a whole, when photosynthesis exceeds respiration (as it does during algal blooms), calcium tends to precipitate out of the system.

Figure 7.8 The "skeleton" of coralline algae is made of calcium carbonate precipitated in conjunction with the uptake of dissolved carbon dioxide during photosynthesis.

The nitrogen cycle

The ultimate source of nitrogen for the ecosystem is molecular nitrogen (N_2) in the atmosphere. This form of nitrogen dissolves to some extent in water, but no nitrogen of any kind is found in native rock. Some molecular nitrogen is converted to forms that plants can use by lightning discharges, but most enters the biological pathways of the nitrogen cycle (Figure 7.9) through its assimilation by certain microorganisms in a process referred to as **nitrogen fixation.** Although this pathway ($N_2 \rightarrow NH_3$) constitutes a small fraction of the earth's annual nitrogen flux, most biologically cycled nitrogen can be traced back to nitrogen fixation.

Once in the biological realm, nitrogen follows pathways more complicated than those of carbon because it has more numerous oxidation states. Beginning arbitrarily with reduced (organic) nitrogen, the first step in the nitrogen cycle is **ammonification**: the hydrolysis of protein and the oxidation of the carbon in amino acids, resulting in the production of ammonia (NH_3), which is accomplished by all organisms. During the initial breakdown of amino acids, although some carbon is oxidized, releasing energy, the energy potential of the nitrogen atom does not change from its oxidation state of –3 (N^{3-}).

Nitrification involves the oxidation of nitrogen, first from ammonia to nitrite, then from nitrite to nitrate, during which transformations the nitrogen atom releases much of its potential chemical energy. Both steps are carried out only by specialized bacteria: $NH_3 \rightarrow NO_2^-$ ($N^{3-} \rightarrow N^{3+}$) by *Nitrosomonas* in the soil and by *Nitrosococcus* in marine systems; $NO_2^- \rightarrow NO_3^-$ (N^{5+}) by *Nitrobacter* in the soil and *Nitrococcus* in the oceans.

Because nitrification steps are oxidations, they require the presence of oxygen, which can act as an electron acceptor. In waterlogged, anoxic soils and sediments and in oxygen-depleted bottom waters, nitrate and nitrite can act as electron acceptors (oxidizers), and the nitrification reactions reverse:

$$NO_3 \rightarrow NO_2 \rightarrow NO\ (N^{5+} \rightarrow N^{3+} \rightarrow N^{2+}).$$

This **denitrification** is accomplished by such bacteria as *Pseudomonas denitrificans.* Additional chemical reactions under anaerobic conditions in soils and water can produce molecular nitrogen,

$$NO \rightarrow N_2O \rightarrow N_2\ (N^{2+} \rightarrow N^{+} \rightarrow N^{0}),$$

with the consequent loss of nitrogen from general biological circulation. Denitrification may be one of the major causes of the low availability of nitrogen in marine systems. When organic remains of plants and animals sink to the depths of the oceans, their oxidation by bacteria in deep waters and bottom sediments often is accomplished anaerobically using nitrate as an oxidizer. This results in the conversion of nitrate and nitrite to the dissolved gases NO and N_2, which cannot be used by algae.

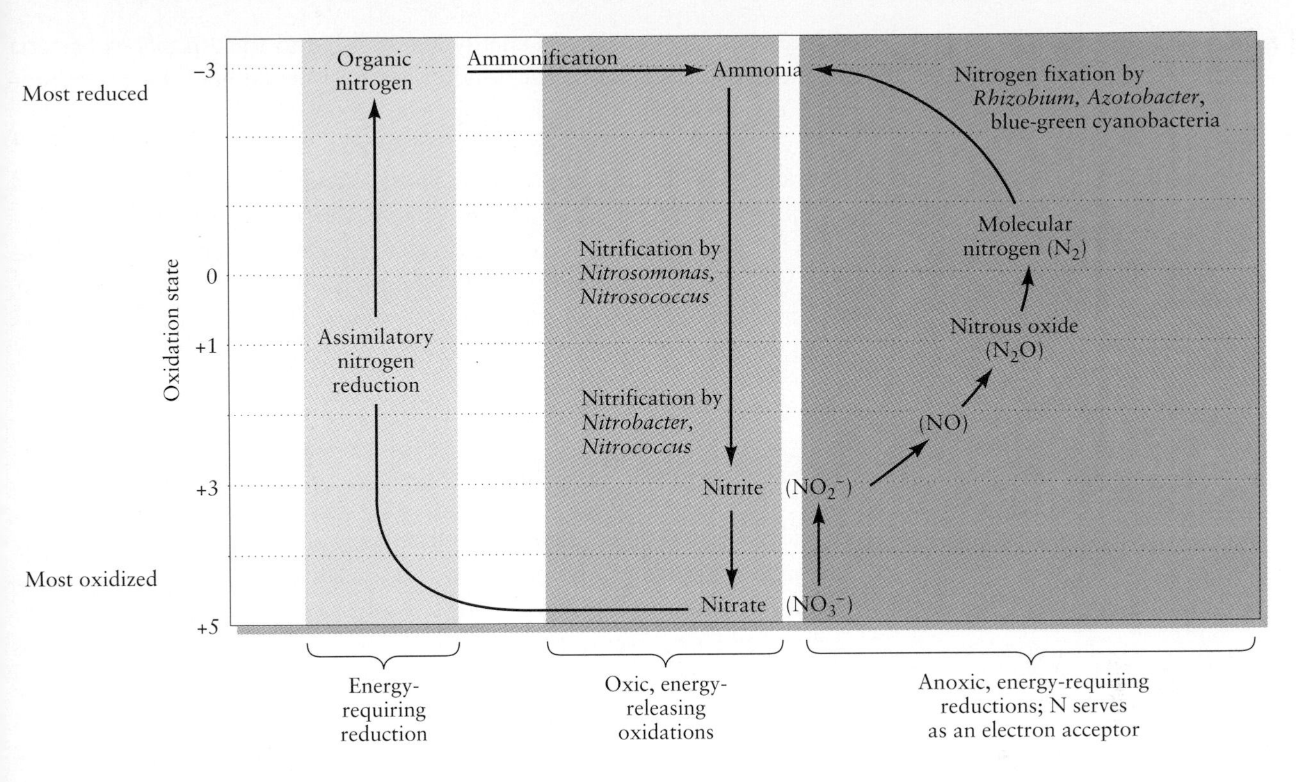

Figure 7.9 Schematic diagram of transformations and oxidation states of compounds in the nitrogen cycle. The most reduced state of the atom, having an electric charge of –3, has the highest chemical energy potential.

Denitrification is balanced in terrestrial and aquatic systems by nitrogen fixation. This assimilatory reduction of nitrogen is accomplished by bacteria such as *Azotobacter,* which is a free-living species; *Rhizobium,* which occurs in symbiotic association with the roots of some legumes (members of the pea family) and other plants (Figure 7.10); and the blue-green cyanobacteria. The enzyme responsible for nitrogen fixation—nitrogenase—consists of two protein subunits, one containing one atom each of iron and molybdenum and the other containing an atom of iron. The enzyme is extremely sensitive to oxygen and works efficiently only under extremely low oxygen tensions. This explains why *Azotobacter* bacteria, living freely in the soil, exhibit only a small fraction of the nitrogen-fixing capacity of *Rhizobium* bacteria, which are sequestered in the relatively anoxic cores of root nodules.

Nitrogen fixation uses energy, though no more than the conversion of an equivalent amount of nitrate to ammonia by plants. The reduction of one atom of molecular nitrogen to ammonia requires approximately the amount of energy released by the oxidation of an atom of organic carbon to carbon dioxide. Nitrogen-fixing microorganisms obtain the energy and reducing power they need to reduce N_2 to NH_3 by oxidizing sugars or other organic compounds. Free-living bacteria must obtain these resources by metabolizing organic detritus in the soil, sediments, or water column.

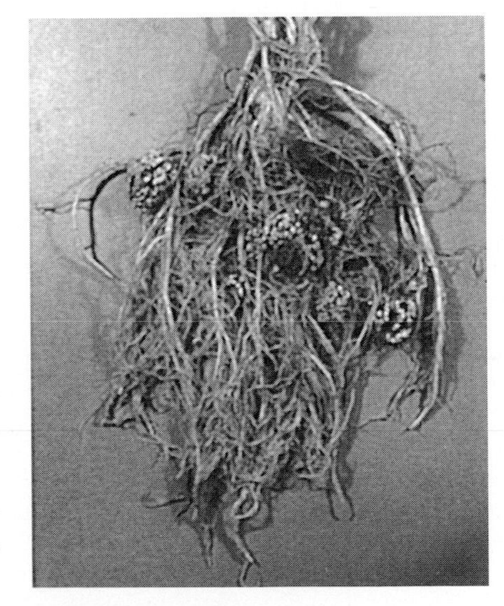

Figure 7.10 The root system of an Austrian winter pea plant, showing the clusters of nodules that harbor symbiotic nitrogen-fixing bacteria. Courtesy of the U.S. Soil Conservation Service.

More abundant supplies of energy are available to the *Rhizobium* bacteria that enter into symbiotic relationships with plants because the plants provide them with sugars produced by photosynthesis.

On a global scale, nitrogen fixation approximately balances the production of N_2 by denitrification. These fluxes amount to about 2% of the total cycling of nitrogen through the ecosystem. On a local scale, nitrogen fixation can assume much greater importance, especially in nitrogen-poor habitats. When land is first exposed to colonization by plants—as, for example, are areas left bare by receding glaciers or newly formed lava flows—species with nitrogen-fixing capabilities dominate the colonizing vegetation.

The phosphorus cycle

Ecologists have studied the role of phosphorus in the ecosystem intensively because organisms require phosphorus at a relatively high level (about one-tenth that of nitrogen). Phosphorus is a major constituent of nucleic acids, cell membranes, energy transfer systems, bones, and teeth. Phosphorus is thought to limit plant productivity in many aquatic habitats, and influxes of phosphorus into rivers and lakes in the form of sewage and runoff from fertilized agricultural lands can artificially stimulate production in aquatic habitats, which can upset natural ecosystem balances and greatly alter the quality of aquatic habitats.

The phosphorus cycle has fewer steps than the nitrogen cycle: plants assimilate phosphorus as phosphate ions (PO_4^{3-}) directly from the soil or water and incorporate it into various organic compounds in the form of phosphate esters. Animals eliminate excess organic phosphorus in their diets by excreting phosphorus salts in urine; phosphatizing bacteria also convert organic phosphorus in detritus to phosphate ions. Phosphorus does not enter the atmosphere in any form other than dust, so the phosphorus cycle involves only the soil and aquatic compartments of the ecosystem. Except in very limited microbial transformations, phosphorus does not undergo oxidation-reduction reactions in its cycling through the ecosystem.

The acidity of the environment greatly affects the availability of phosphorus and consequently its uptake and assimilation by plants. In general, at low pH (acid conditions), phosphorus binds tightly to clay particles in soil and forms relatively insoluble compounds with ferric iron and aluminum. At high pH, it forms other insoluble compounds (for example, with calcium). When both calcium and ferric iron or aluminum are present under aerobic conditions, the highest concentration of dissolved phosphate—that is, the greatest availability of phosphorus—occurs at a pH of between 6 and 7. Nonetheless, marine and freshwater sediments act as a phosphorus sink, continually removing precipitated phosphorus from rapid circulation in the ecosystem. Under anoxic conditions in aquatic sediments and bottom waters, phosphorus compounds readily dissolve and enter the water column. For example, when iron is reduced from the ferric (Fe^{3+}) to the ferrous (Fe^{2+}) state, it tends to form sulfides rather than phosphate compounds.

The sulfur cycle

Sulfur occurs in the amino acids cysteine and methionine and hence is required by plants and animals, but the importance of sulfur in the ecosystem goes far beyond this role. Like nitrogen, sulfur has many oxidation states, and so it follows complex chemical pathways and affects the cycling of other elements (Figure 7.11).

The most oxidized form of sulfur is sulfate (SO_4^{2-}; that is, oxidation state S^{6+}); the most reduced forms are sulfide (S^{2-}) and the organic (thiol) form of sulfur (also S^{2-}). Under well-oxygenated (oxic) conditions, assimilatory sulfur reduction ($SO_4^{2-} \rightarrow$ organic S) balances the oxidation of organic sulfur back to sulfate, which occurs either directly or with sulfite (SO_3^{2-}) as an intermediate step. This oxidation occurs when animals excrete excess dietary organic sulfur and when microorganisms decompose plant and animal detritus.

Under anoxic conditions, sulfate may function as an oxidizer. In sediments, the bacteria *Desulfovibrio* and *Desulfomonas* couple dissimilatory sulfate reduction ($SO_4^{2-} \rightarrow S^{2-}$) to the oxidation of organic carbon to make energy available. The reduced S^{2-} may then be used as a reducing agent ($S^{2-} \rightarrow S^0 + 2e^-$) by photoautotrophic bacteria, which assimilate carbon by pathways analogous to photosynthesis in green plants. In these reactions, sulfur takes the place of the oxygen atom in water as an electron donor. The elemental sulfur (S) accumulates unless the sediments are exposed to aeration or oxygenated water, at which point the sulfur may be further oxidized to sulfite (SO_3^{2-}) and sulfate (SO_4^{2-}).

Figure 7.11 Schematic diagram of transformations and oxidation states of compounds in the sulfur cycle.

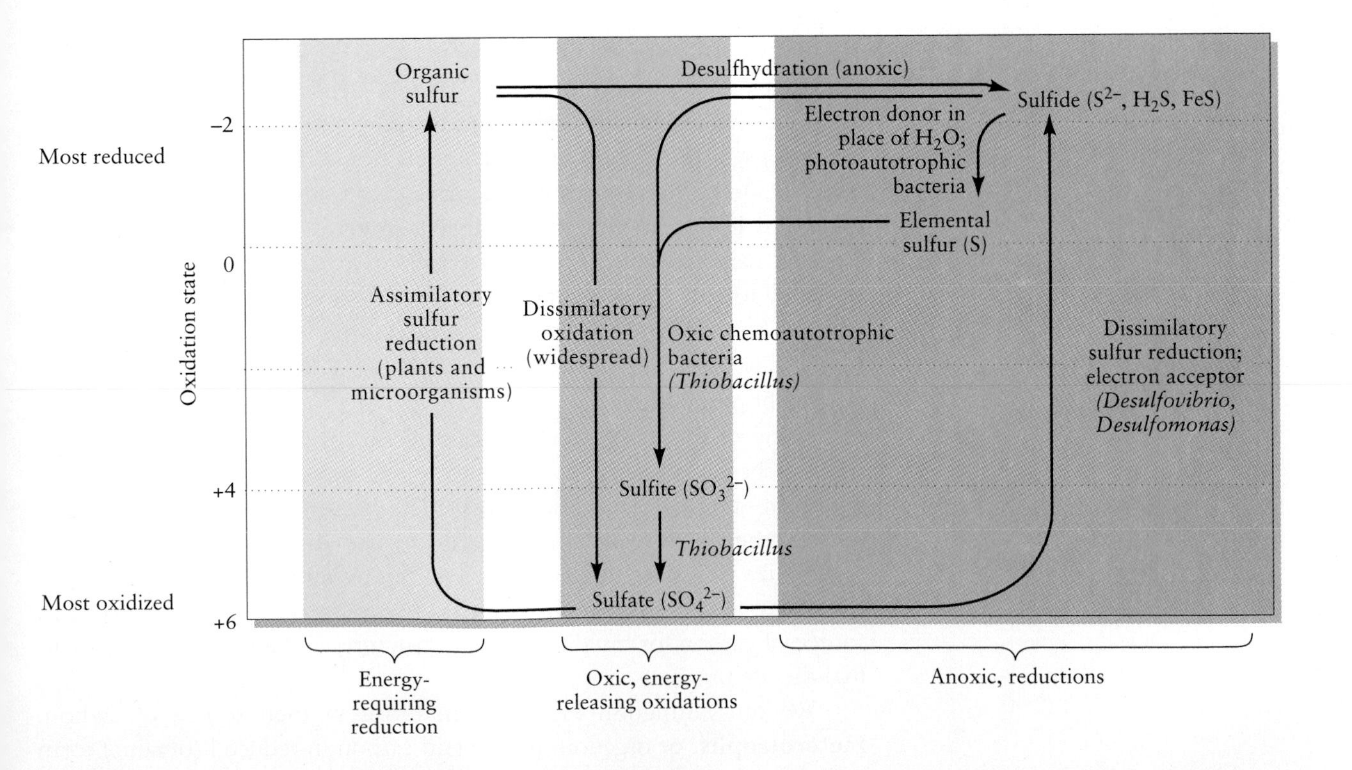

The fate of thiol sulfur (S^{2-}) produced under oxic conditions depends on the availability of positive ions. Frequently, hydrogen sulfide forms; it escapes from shallow sediments and mucky soils as a gas having the familiar smell of rotten eggs. Because anoxic conditions generally favor the reduction of ferric iron (Fe^{3+}) to ferrous iron (Fe^{2+}), the presence of iron in sediments leads to the formation of iron sulfide (FeS). Sulfides are commonly associated with coal and oil. When these materials are exposed to the atmosphere in mine wastes or burned for energy, the reduced sulfur oxidizes (with the help of *Thiobacillus* bacteria in mine wastes) and the oxidized forms combine with water to produce the sulfuric acid of acid mine drainage and acid rain.

The roles of microorganisms in element cycles

Many of the transformations discussed in this chapter are accomplished mainly or entirely by bacteria. In fact, were it not for the activities of bacteria, many element cycles would be drastically altered and the productivity of the ecosystem much reduced. For example, without the capacity of some microbes to use nitrogen, sulfur, and iron as electron acceptors, little decomposition would occur in anoxic organic sediments, and their resulting accumulation would reduce the amount of inorganic carbon in the ecosystem.

The origin of forms of nitrogen that can be assimilated by plants depends mostly on nitrogen fixation by microorganisms, although some useful nitrogen fixation is also produced by lightning discharges in the atmosphere. Without nitrogen-fixing bacteria, however, denitrification under anaerobic conditions would slowly deplete ecosystems of useful nitrogen and reduce biological productivity in proportion. Under oxic conditions, plants and bacteria compete for ammonia in the soil and water column. Because plants and algae can assimilate ammonia as easily as, and at less energetic expense than, nitrate, the activities of nitrifying bacteria may reduce the primary productivity of terrestrial and aquatic plants by forcing them to allocate more of their energy to nitrogen assimilation. Nonetheless, without microorganisms, much organic detritus would never decompose, and its minerals would not be released to support plant production.

Many of the transformations carried out by microorganisms, such as the metabolism of sugars and other organic molecules, are accomplished in similar ways by plants and animals. The bacteria and cyanobacteria are distinguished physiologically primarily by the ability of many species to metabolize substrates under anoxic conditions and to use substrates other than organic carbon as energy sources. Every organism needs, above all, a source of carbon for building organic structures and a source of energy to fuel the life processes.

We can distinguish organisms in terms of their source of carbon. **Heterotrophs,** or organotrophs, obtain carbon in reduced (organic) form by consuming other organisms or organic detritus. All animals and fungi, and many bacteria, are heterotrophs. **Autotrophs** assimilate carbon as carbon dioxide and expend energy to reduce it to an organic form. **Photoautotrophs** use sunlight as their source of energy for photosynthesis. Photoautotrophs include all green plants and algae; the cyanobacteria, which use H_2O as an electron donor (reducing agent) and are aerobic; and the purple and green bacteria, which have light-absorbing pigments different from those of green plants, use H_2S or organic compounds as electron donors, and are anaerobic.

Chemoautotrophs, which are sometimes called **chemolithoautotrophs,** all use CO_2 as a carbon source, but they obtain energy for its reduction by the oxic oxidation of inorganic substrates: methane (for example, *Methanosomonas* and *Methylomonas*); hydrogen *(Hydrogenomonas* and *Micrococcus);* ammonia (the nitrifying bacteria *Nitrosomonas* and *Nitrosococcus*); nitrite (the nitrifying bacteria *Nitrobacter* and *Nitrococcus*); hydrogen sulfide, sulfur, and sulfite *(Thiobacillus);* or ferrous iron salts *(Ferrobacillus and Gallionella)*. The chemoautotrophs are almost exclusively bacteria, which apparently are the only organisms that can become so specialized biochemically as to make efficient use of inorganic substrates and efficiently dispose of the products of chemoautotrophic metabolism.

Organisms thus obtain energy from three sources: sunlight (photoautotrophs), the oxidation of organic compounds (heterotrophs), and the oxidation of inorganic compounds (chemoautotrophs). Oxidation of glucose combines the oxidation of carbon, which releases energy, and the reduction of oxygen. Heterotrophs all use organic carbon as a reduced substrate. Under oxic conditions, they may use O_2 as the oxidizing agent. But many bacteria (such as sulfate-reducing bacteria and methanogens) use sulfate (SO_4^{2-}), ferric iron (Fe^{3+}), or carbon dioxide (CO_2) as oxidizing agents under anoxic conditions.

The special role of microorganisms in ecosystem function is illustrated nicely by the highly productive communities of marine organisms that develop around deep-sea hydrothermal vents. These miniature ecosystems were first discovered in deep water off the Galápagos archipelago in 1977 and have since been found to be distributed more widely in the ocean basins of the world. The most conspicuous members of the community are giant white-shelled clams and tube worms (pogonophorans) that grow to 3 meters, but numerous crustaceans, annelids, mollusks, and fish also cluster at great density around hydrothermal vents. The high productivity of the vent communities contrasts strikingly with the desertlike appearance of the surrounding ocean floor. As you might suspect, it is based on the unique qualities of the water issuing from the vents themselves, which is hot and loaded with a reduced form of sulfur, hydrogen sulfide (H_2S). Where vent water and seawater mix, conditions are ideal for chemoautotrophic sulfur bacteria. These bacteria use the oxygen in seawater to oxidize the hydrogen sulfide in vent water as a source of energy for assimilatory reduction of inorganic carbon and nitrogen in seawater. All the other members of the vent community feed on the bacteria, which thus form the base of the food chain. The pogonophoran worms have gone so far as to house symbiotic colonies of the bacteria within the tissues of a specialized organ, the trophosome, trading a protected place to live for a share of the carbohydrate and organic nitrogen produced by the bacteria.

In this chapter we have examined the cycling of several important elements from the standpoint of their chemical and biochemical reactions. Elements are cycled through the ecosystem primarily because the metabolic activities of organisms result in chemical transformations of the elements. The kinds of transformations that predominate depend on the physical and chemical conditions of the system. Each type of habitat presents a different chemical environment, particularly with respect to the presence or absence of oxygen and possible sources of energy. It stands to reason, therefore, that the patterns of element cycling should differ greatly among habitats and ecosystems. In the next chapter, we shall contrast element cycling in aquatic habitats and in terrestrial habitats by focusing on how some of the unique physical features of each of these environments affect the chemical and biochemical transformations involved in organic production and recycling of elements.

Summary

1. Unlike energy, nutrients are retained within the ecosystem and are cycled between its physical and biotic components. The paths that elements follow through the ecosystem are the result of chemical and biological transformations, which themselves depend on the chemistry of each element, the physical and chemical conditions of the environment, and the ways in which each element is used by various organisms.

2. The movement of energy through the ecosystem parallels the paths of several elements, particularly carbon, whose transformations either require or release energy.

3. The cycling of each element may be thought of as the result of movement between compartments of the ecosystem. The major compartments are living organisms, organic detritus, immediately available inorganic forms, and unavailable organic and inorganic forms, for the most part locked away in sediments.

4. The water cycle, or hydrologic cycle, provides a physical analogy for element cycling in the ecosystem. Energy is required to evaporate water because molecules of water vapor have a higher energy content than molecules of liquid water. This energy is released as heat when water vapor condenses in the atmosphere to produce precipitation.

5. Energy transformations in biological systems occur primarily in the course of oxidation-reduction (redox) reactions. An oxidizer is a substance that readily accepts electrons (O_2, NO_3^-); a reducer is one that readily donates electrons (H_2, organic carbon). Upon being reduced, an atom gains energy along with the electrons it accepts; upon being oxidized, an atom releases energy along with the electrons it gives up. Because elements such as carbon, nitrogen, and sulfur in organic compounds tend to be in reduced forms, biological assimilation of these elements requires energy.

6. All organisms require organic carbon as the primary substance of life. Organic carbon is also the major source of energy for most animals and microorganisms. Carbon shuttles between organic forms (C^0) and the carbon dioxide (C^{4+}) compartment of the ecosystem by way of photosynthesis and respiration. Some anaerobic bacteria produce methane (C^{4-}), in which case carbon acts as an electron acceptor.

7. The circulation of carbon in the biosphere also involves nonbiological processes such as the solution of carbon dioxide in waters at the surface of the earth. Dissolved carbon dioxide enters into a chemical equilibrium with bicarbonate ions and with carbonate ions, which in the presence of calcium in the neutral acidity of the oceans, tend to precipitate and form sediments. Thick accumulations of these marine sediments can form limestone rock.

8. Nitrogen has many oxidation states and consequently follows many pathways through the ecosystem. Quantitatively, most of the flux follows the cycle leading from nitrate through organic nitrogen (following assimilation by plants), ammonia, nitrite (following nitrification by bacteria), and then back to nitrate (following further nitrification). The last two steps are accomplished by certain bacteria in the presence of oxygen.

9. Under anoxic conditions in soils and sediments, certain bacteria can use nitrate in place of oxygen as an oxidizing agent (denitrification), and the reactions reverse: nitrate leads to nitrite and (eventually) to molecular nitrogen (N_2). This loss of nitrogen from the general biological cycling is balanced by nitrogen fixation by some microorganisms.

10. Plants assimilate phosphorus in the form of phosphate (PO_4^{3-}), whose availability varies with the acidity and oxidation level of the soil or water. The energy potential of the phosphorus atom does not change during its cycling through the ecosystem.

11. Sulfur is an important redox element in anoxic habitats, where it may serve as an oxidizer in the form of sulfate (SO_4^{2-}) or as a reducing agent (for photoautotrophic bacteria) in the forms of elemental sulfur (S^0) and sulfide (S^{2-}).

12. Many elemental transformations, particularly under anoxic conditions, are accomplished by biochemically specialized microorganisms (bacteria, cyanobacteria). The activities of these organisms therefore play important roles in the cycling of elements through the ecosystem.

Suggested readings

Berner, R. A., and A. C. Lasaga. 1989. Modeling the geochemical carbon cycle. *Scientific American* 260:74–81.

Coleman, D. C., C. P. P. Reid, and C.V. Cole. 1983. Biological strategies of nutrient cycling in soil systems. *Advances in Ecological Research* 13:1–55.

Fenchel, T., and T. H. Blackburn. 1979. *Bacteria and Mineral Cycling.* Academic Press, New York.

Fenchel, T., and B. J. Finlay. 1995. *Ecology and Evolution in Anoxic Worlds.* Oxford University Press, Oxford.

Grant, W. D., and P. E. Long. 1981. *Environmental Microbiology.* Wiley (Halsted Press), New York.

Grassle, J. F. 1985. Hydrothermal vent animals: Distribution and biology. *Science* 229:713–717.

Grassle, J. F. 1986. The ecology of deep sea hydrothermal vent communities. *Advances in Marine Biology and Ecology* 23:301–362.

Howarth, R. W. 1993. Microbial processes in salt-marsh sediments. In T. E. Ford (ed.), *Aquatic Microbiology,* pp. 239–259. Blackwell Scientific Publications, Oxford.

Jannasch, H. W., and M. J. Mottl. 1985. Geomicrobiology of deep-sea hydrothermal vents. *Science* 229:717–725.

Schlesinger, W. H. 1991. *Biogeochemistry: An Analysis of Global Change.* Academic Press, San Diego.

Sprent, J. I. 1987. *The Ecology of the Nitrogen Cycle.* Cambridge University Press, New York.

Stacey, G., R. H. Burris, and H. J. Evans (eds.). 1992. *Biological Nitrogen Fixation.* Chapman & Hall, New York.

CHAPTER 8

NUTRIENT REGENERATION IN TERRESTRIAL AND AQUATIC ECOSYSTEMS

Elements cycle through the ecosystem along paths established by their chemical properties, which in turn determine chemical and biochemical reactions in the biosphere. The regeneration of nutrients into forms that can be used by other organisms provides a key to understanding the regulation of ecosystem function. These processes of regeneration are different in terrestrial and aquatic systems. To be sure, both systems exhibit similar chemical and biochemical transformations: oxidation of carbohydrates, nitrification, and chemoautotrophic oxidation of sulfur, among many others. But terrestrial and aquatic systems differ in the material basis for nutrient regeneration. In terrestrial habitats, most elements cycle through detritus at the soil surface, where plant roots have ready access to nutrients. In aquatic habitats, particularly in lakes and oceans, sediments are the ultimate source of regenerated nutrients; these sediments are often far removed from the sites of primary production. In this chapter, we shall discuss how biochemical processes in soils, water, and sediments influence the productivity of the ecosystem and the cycling of elements within it.

Nutrient regeneration in terrestrial ecosystems

One of the major sources of new nutrients in terrestrial systems is the formation of soil by the decomposition, or weathering, of bedrock and other parent materials. How rapidly does this process occur? Weathering normally takes place under deep layers of soil, where it is impossible to measure directly. Soil scientists can estimate the rate of weathering indirectly, however, by measuring the net efflux (loss) of certain elements from a system. When a soil attains a steady state, as it may in undisturbed areas, the loss of an element from a system equals the weathering input of that element from the parent material plus any gains from other sources, such as precipitation. Thus, it is possible to estimate weathering input from information about precipitation input and total loss. The basic cations—calcium, potassium, sodium, and magnesium—are good candidates for the study of such balances because they dissolve readily in water and leave the soil as ions in streams, where they can be measured easily.

Cation budgets of watersheds

It is most practical to study the cation budget of a large area making up a **watershed,** the entire drainage area of a stream or river, from which all surface water and groundwater leaves at a single point. Scientists have obtained detailed cation budgets for several small watersheds by measuring inputs in rainwater collected at various locations in the watersheds (Figure 8.1) and outputs in the streams that drain them (Figure 8.2). If we assume that the soil is in a steady state, then the net loss (precipitation input minus stream outflow) equals the input to the soil from weathering.

The best-known watershed study comes from the Hubbard Brook Forest of New Hampshire. There, the annual input of calcium in precipitation averaged 2 kilograms per hectare (kg ha^{-1}), while loss of dissolved Ca^{2+} in stream flow was 14 kg ha^{-1}. Therefore, the net loss to the system equaled 12 kg ha^{-1}. The investigators also determined that living and dead plant biomass increased in the watershed during the study period because the forest was recovering from earlier clearing; net assimilation of calcium in vegetation and detritus brought its overall removal from the mineral soil to 21 kg ha^{-1} yr^{-1}. Because calcium constitutes about 1.4% of the weight of the bedrock in the area, its annual loss equaled the weathering of about 1,500 kg (21/0.014) of bedrock per hectare, or approximately 1 mm of depth per year. The major lesson of this and other studies has been how small the weathering input is relative to the annual uptake of nutrients by vegetation. The bulk of these nutrients is made available to plants by

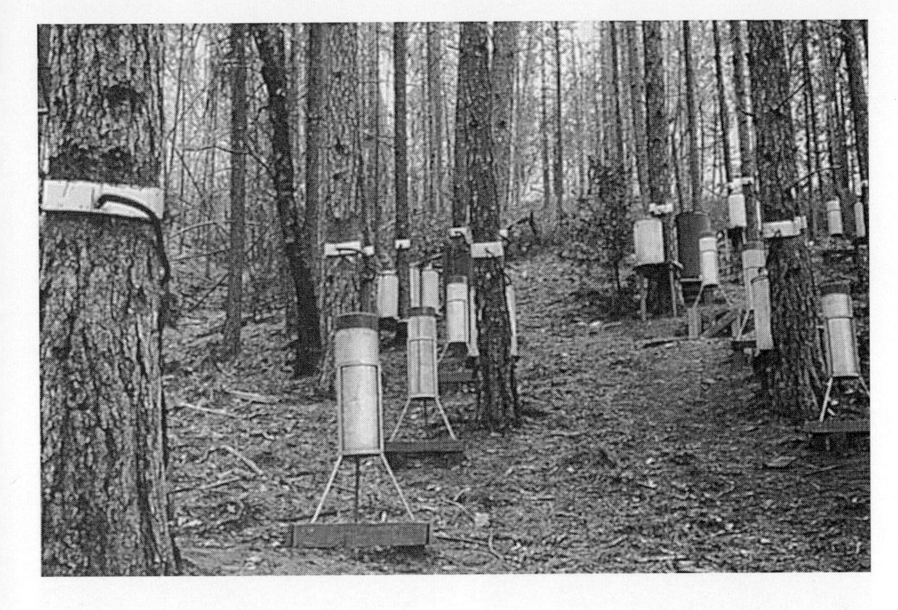

Figure 8.1 Rain gauges installed in a ponderosa pine stand in California to intercept precipitation falling through the canopy of the forest and running down the trunks of trees. Analyses of the nutrient content of water collected in sampling programs like this one help to determine the overall cation budget of the forest and the specific routes of mineral cycles. Courtesy of the U.S. Forest Service.

regenerative processes within the soil profile; typically, only 10% are provided by weathering of bedrock.

Detritus and nutrient regeneration

Plants assimilate elements from the soil far more rapidly than weathering generates them from the parent material. Most of the basic cations (Ca^{2+}, Mg^{2+}, K^+, and Na^+) do not figure prominently in biochemical transformations, and for the most part, plants merely assimilate these cations with the water they need in quantity. Plants sequester excesses of these elements, whose depletion from the soil probably has little direct effect on plant production. This is not the case with supplies of such important nutrients as nitrogen, phosphorus, and sulfur, which are poorly represented in parent material. Igneous rocks—granite and basalt, for example—contain no nitrogen and only 0.3% phosphate and 0.1% sulfate by mass. Hence weathering adds little of these nutrients to the soil; inputs from precipitation and nitrogen fixation are also small. Plant production therefore depends on the rapid regeneration of these nutrients from detritus and their retention within the ecosystem.

Organic detritus occurs everywhere, most conspicuously in terrestrial habitats, where the parts of plants not consumed by herbivores accumulate at the soil surface along with animal excreta and other organic remains. Ninety percent or more of the plant biomass produced in forested habitats passes through this detritus reservoir. The nutrients locked up in detritus are regenerated into forms that can be reused by plants by the activities of the countless worms, snails, insects, mites, bacteria, and fungi that consume detritus—their primary source of carbon and energy—as food.

The breakdown of leaf litter on the forest floor occurs in four ways: (1) leaching of soluble minerals and small organic compounds by water; (2) consumption by large detritus-feeding organisms (millipedes, earthworms, wood lice, and other invertebrates); (3) breakdown of the woody

Figure 8.2 A stream gauge at the lower end of a watershed at the Coweeta Hydrological Laboratory, North Carolina. The V-shaped notch is engineered so that the flow of water through the weir can be estimated from the water level in the basin. Courtesy of the U.S. Forest Service.

components of leaves by fungi; and (4) decomposition of almost everything by bacteria. Between 10% and 30% of the substances in newly fallen leaves dissolve in cold water. Leaching rapidly removes most of these (salts, sugars, amino acids) from the litter, making them available to microorganisms and plant roots in the soil; complex carbohydrates, such as cellulose, and other large organic compounds remain behind. Large detritus feeders typically assimilate only 30–45% of the energy available in leaf litter, and even less from wood, but they nonetheless speed the decay of litter beyond what they themselves extract because they macerate plant detritus in their digestive tracts, and the finer particles in their egested wastes expose new surfaces to microbial feeding.

Leaves of different species of trees decompose at different rates depending on their composition, particularly under cold or dry conditions. For example, in eastern Tennessee, the weight loss of fallen leaves during the first year after leaf fall was found to range from 64% for mulberry to 39% for oak, 32% for sugar maple, and 21% for beech. The needles of pines and other conifers also decompose slowly. The differences among species depend to a large extent on the lignin content of the leaves. **Lignins** are a heterogeneous class of **phenolic** polymers (long chains of phenolic subunits, which are six-carbon rings with attached hydroxyl [OH] groups). Lignins lend wood many of its structural qualities and are even more difficult to digest than cellulose; in fact, only the "white rot" fungi can break down lignins. The decomposition rate of detritus also depends on its content of nitrogen, phosphorus, and other nutrients required by bacteria and fungi for their own growth.

The resistance of some types of litter to degradation highlights the unique role of fungi in the regeneration of nutrients. Most fungi consist of a network, or **mycelium,** of **hyphae,** threadlike structures that can pene-

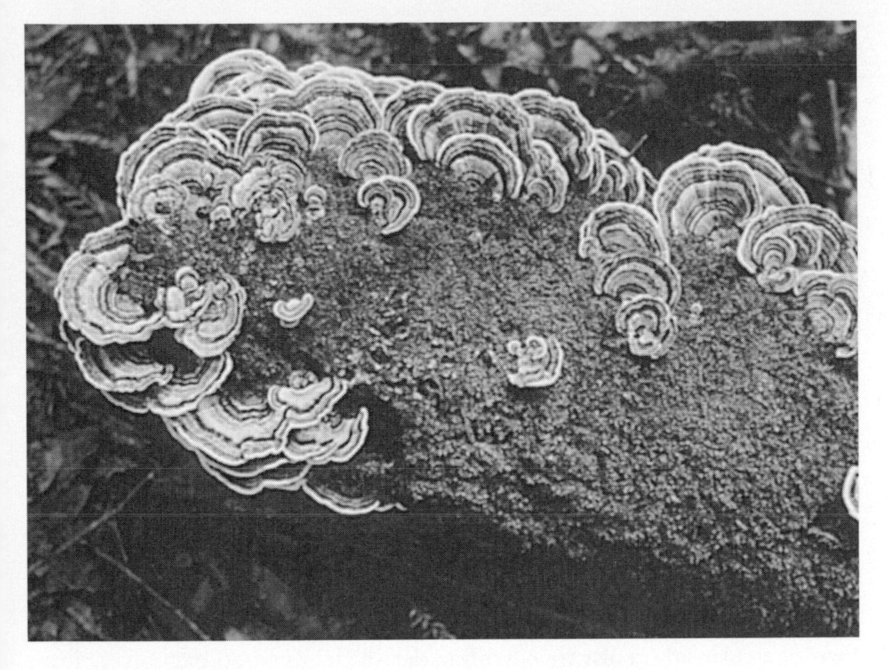

Figure 8.3 Shelf fungi speed the decomposition of a fallen log. The visible fruiting structures are produced by the fungal hyphae, together called the mycelium, that grow throughout the interior of the log, slowly digesting its structure. Courtesy of the U.S. National Park Service.

trate to the depths of plant litter and wood that bacteria cannot reach. The familiar mushrooms and shelf fungi are merely fruiting structures produced by the mass of hyphae deep within the litter or wood (Figure 8.3). Like bacteria, fungi secrete enzymes into the substrate and absorb the simple sugars and amino acids produced by this extracellular digestion. Fungi differ from bacteria in being able to digest cellulose (which a few bacteria, the protozoans in the guts of termites, and snails also can accomplish) and, especially, lignin. Cellulose digestion begins with the hydrolysis of polysaccharides (long chains of sugar subunits) into simple sugars that can be absorbed into the hyphae. The breakdown of lignin by fungi apparently is initiated by an oxidation that cleaves the phenolic ring structure.

Mycorrhizae

In addition to their role in decomposing detritus, some kinds of fungi grow on the surfaces of, or inside, the roots of many types of plants, especially woody species. This association, which is called a **mycorrhiza** (plural, *mycorrhizae;* literally, "fungus root"), enhances the plant's ability to extract mineral nutrients from the soil. Although many forms of mycorrhizae are recognized, they are classified as **endomycorrhizae** when the fungus penetrates the root tissue and as **ectomychorrhizae** when it forms a sheath over the root surface (Figure 8.4).

Mycorrhizae occur everywhere, but their importance is best demonstrated experimentally by growing plants on sterile soil to which spores of mycorrhiza-forming fungi are either added or not added. In an experiment with *Pinus strobus* seedlings, for example, ectomycorrhizae increased by two to three times the uptake of nitrogen, phosphorus, and potassium per unit of root mass and greatly improved the growth of the plant. Mycorrhizae

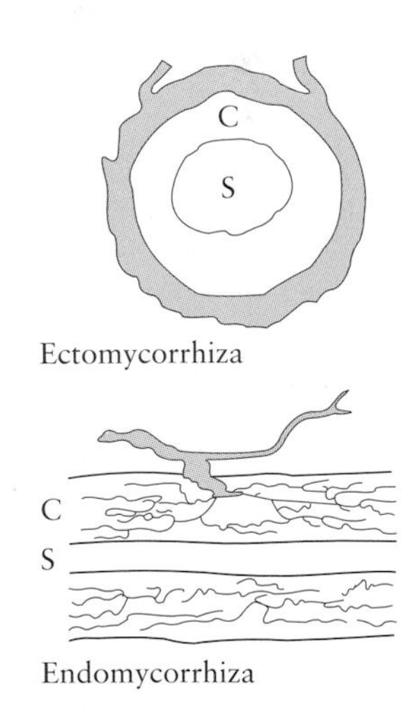

Figure 8.4 Generalized structure of an ectomycorrhiza and an endomycorrhiza, represented by the shaded areas on the diagrams. C = root cortex; S = vascular tissue of stele. After W. D. Grant and P. E. Long, *Environmental Microbiology,* Wiley, New York (1981).

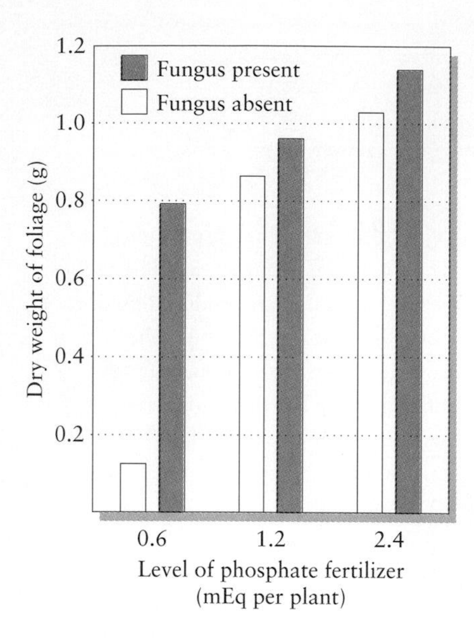

Figure 8.5 The effects of phosphate fertilizer and inoculation with the mycorrhizal fungus *Endogone macrocarpa* on the growth of tomato plants *(Lycopersicon esculentum)*. The mycorrhizae stimulated growth under low phosphate concentrations, but not under high phosphate concentrations, at which point other nutrients become limiting. After J. L. Harley and S. E. Smith, *Mycorrhizal Symbiosis,* Academic Press, London (1983).

promote growth the most in soils that are relatively depleted of nutrients, as shown by the experiments with tomatoes summarized in Figure 8.5.

Mycorrhizae increase a plant's uptake of minerals by penetrating a greater volume of soil than the roots could alone and increasing the total surface area available for nutrient assimilation. In addition, because mycorrhizae secrete enzymes and acid (hydrogen ions) into the surrounding soil, they are more effective than plant roots at extracting certain mineral nutrients, such as phosphorus. Mycorrhizae, especially ectomycorrhizal forms, may also protect the plant root from infection by pathogens by physically excluding the pathogens or by producing antibiotics (antibacterial toxins). What do the fungi derive from the association? The main advantage appears to be a reliable source of carbon in the form of photosynthates transported to the roots of the plant.

The commonest type of endomycorrhiza is the **vesicular-arbuscular** mycorrhiza, so called because of the structures developed within the host tissue. These fungi do not grow freely in the soil, but rather infect plant roots from spores left behind from dead, previously infected roots. (Presumably, spores blow into virgin soils along with other dust.) The fungi that form vesicular-arbuscular mycorrhizae derive their carbon exclusively from the plant root. Their mineral nutrients are derived from the soil, into which the fungi send long hyphae. Apparently, they can use sources of phosphorus not available to plants, such as the highly insoluble "rock phosphorus" $Ca_3(PO_4)_2$, which dissolves only under acid conditions; without the help of mycorrhizae, plants can use only the more soluble forms $CaHPO_4$ and KH_2PO_4. Over and above their unique ability to secrete hydrogen ions and organic acids, fungi may be aided in this task when their hyphae grow in association with phosphate-solubilizing bacteria.

Ectomycorrhizae are widespread, especially among the roots of trees and shrubs. In addition to its hyphae, which extend out into the soil, the fungus forms a tough sheath around the root that may account for as much as 40% of the weight of the mycorrhizal association (root plus fungus). The sheath stores soil-derived nutrients and carbon compounds, which may be one of the chief advantages that the mycorrhizae offer to the plant. Regardless of how they function, ectomycorrhizal fungi can account for a major part of the energy and nutrient budgets of some habitats. For example, the carbon assimilated by ectomycorrhizae in one fir *(Abies)* forest accounted for 15% of the total net primary production. The fungus was undoubtedly responsible for a much greater share of the mineral nutrient uptake.

Nutrient regeneration in tropical and temperate forests

Nutrient cycling differs in tropical and temperate ecosystems because of the effects of climate on weathering, soil properties, and the decomposition of detritus. Tropical soils tend to be deeply weathered and low in clay content, which results in a poor ability to retain nutrients (low cation exchange capacity). The high productivity of tropical forests is supported by rapid regeneration of nutrients from detritus under warm, humid conditions, rapid uptake of nutrients by plants from the top layers of the soil, and efficient retention of nutrients by plants. Ecologists have concluded that in typical tropical ecosystems, most of the nutrients occur in the living biomass, and elements are regenerated and assimilated very rapidly. This pattern has important implications for tropical agriculture and conservation.

Over extensive regions of old, deeply weathered soils in the Tropics, planting crops such as corn on clear-cut land has predictable consequences for soil fertility (Figure 8.6). The practice of cutting the forest and burning the felled trees releases many mineral nutrients that may support a year or two of crop growth, but these nutrients are quickly leached out of the soil when the natural vegetation is no longer there to assimilate them, and levels of mineral nutrients in the soil decline rapidly. Furthermore, as the exposed soil dries, the upward movement of water draws iron and aluminum oxides to the surface, where they form a concretelike substance called **laterite.** Surface runoff over the impenetrable laterite accelerates erosion, further depleting nutrients and choking streams with sediment.

A comparison of forested ecosystems cleared for crops in Canada, Venezuela, and Brazil shows the importance of soil organic matter in sustaining soil fertility under intensive agriculture. Carbon contents of undisturbed soils were 8.8 kg m^{-1} at a prairie site in Canada, 3.4 kg m^{-1} at a semiarid thorn forest site in Brazil, and 5.1 kg m^{-1} under a Venezuelan rain forest. After 65 years of cultivation, the carbon content of the Canadian soil had been reduced by 51%, an exponential decline at a rate of about 1% per year. In marked contrast, the carbon content of the Brazilian soil had decreased by 40% after 6 years of cultivation (9% per year), and that of the Venezuelan soil had decreased by 29% after 3 years of slash-and-burn agriculture (11% per year). These results suggest that cultivated temperate zone soils retain organic matter ten times longer than tropical soils, and thereby provide a more persistent store of mineral nutrients that can be released slowly by decomposition.

Figure 8.6 An area of about 2 hectares on a steep slope in Costa Rica that has been cut and burned for agriculture. The close-up reveals small corn plants emerging among the debris. This practice exposes soil to erosion and promotes leaching of nutrients and laterization. Such clearings produce crops for 2 or 3 years at most; decades of forest regeneration are required to restore the fertility of the soil. Courtesy of D. H. Janzen.

Figure 8.7 A clear-cut watershed at the Coweeta Hydrological Laboratory, North Carolina, employed in studies of evapotranspiration and runoff in forest ecosystems. Courtesy of the U.S. Forest Service.

Experience has taught us that vegetation is critical to the development and maintenance of soil fertility in many tropical systems. Even in temperate zones, the removal of vegetation reveals its important role in the retention of soil nutrients (Figure 8.7). For example, clear-cutting of small watersheds in the Hubbard Brook Forest increased stream flow severalfold because transpiring leaf surfaces were removed; losses of cations increased 3–20 times over cation losses in comparable undisturbed systems. The nitrogen budgets of the clear-cut watersheds sustained the most striking change. Plants assimilate available soil nitrogen so rapidly that the undisturbed forest gained nitrogen at the rate of only 1–3 kg ha^{-1} yr^{-1} from precipitation and nitrogen fixation. In the clear-cut watershed, net loss of nitrogen as nitrate soared to 54 kg ha^{-1} yr^{-1}, a value comparable to the annual assimilation of nitrogen by vegetation and many times the precipitation input (7 kg ha^{-1} yr^{-1}). The loss of nitrate resulted from nitrification of organic nitrogen by soil microorganisms at the normal annual rate without simultaneous rapid uptake by plants. (Remember that the negatively charged nitrate ions [NO_3^-] do not bind well to soil particles.)

Comparative studies of element dynamics in temperate and tropical forests shed further light on their differences. Litter on the forest floor constitutes an average of about 20% of the total biomass of vegetation (including trunks and branches) and detritus in temperate needle-leaved forests, 5% in temperate hardwood forests, and only 1–2% in tropical rain forests. The ratio of litter to the biomass of only the living leaves is between 5 and

10 to 1 in temperate forests, but less than 1 to 1 in tropical forests. Of the total organic carbon in the system as a whole, more than 50% occurs in soil and litter in northern forests, but less than 25% in tropical rain forests. Clearly, dead organic material decomposes rapidly in the Tropics and does not form a substantial nutrient reservoir relative to plant production, as it does in temperate regions.

Fewer data exist on the relative proportions of nutrient elements in the soil and in the living vegetation. The distributions of potassium, phosphorus, and nitrogen in a temperate and a tropical forest with similar living biomass are compared in Table 8.1. Two points stand out: First, the accumulation of nutrients in vegetation, on a weight-for-weight basis, was somewhat greater in the tropical forest. For example, the total dry weight of living vegetation in the Belgian ash-oak forest exceeded that in the tropical deciduous forest of Ghana by 14%, but the accumulation of the three elements per gram of dried vegetation was 32–38% lower in the temperate forest. Second, the ratio of each element in soil to its level in biomass was much lower in the Tropics. In the temperate forest, 96% of the phosphorus occurred in the soil; in the tropical forest, more than 90% of the much smaller amount of phosphorus was found in the living biomass.

While recognizing the general nutrient poverty of many tropical soils, we must also distinguish between nutrient-rich and nutrient-poor soils within the Tropics. **Eutrophic** ("well-nourished") soils develop in geologically active areas where natural erosion is high and soils are relatively young. With the bedrock closer to the surface, weathering adds nutrients more

TABLE 8.1 **Distribution of mineral nutrients in the soil and living biomass of a temperate and a tropical forest ecosystem**

		NUTRIENTS (kg ha^{-1})		
FOREST (LOCALITY)	BIOMASS (T ha^{-1})*	*Potassium*	*Phosphorus*	*Nitrogen*
Ash and oak (Belgium)	380			
Living vegetation		624	95	1,260
Soil		767	2,200	14,000
Ratio of soil to biomass		1.2	23.1	11.1
Tropical deciduous (Ghana)	333			
Living vegetation		808	124	1,794
Soil		649	13	4,587
Ratio of soil to biomass		0.8	0.1	2.0

* T = metric tons.

Source: P. Duvigneaud and S. Denayer-de-Smet, in D. E. Reichle (ed.), *Analysis of Tropical Forest Ecosystems,* Springer-Verlag, New York (1970), pp. 199–225; D. J. Greenland and J. M. Kowal, *Plant Soil* 12:154–174 (1960); J. D. Ovington, *J. Biol. Rev.* 40:295–336 (1965).

rapidly, and soils retain nutrients more effectively. In the Neotropics, such eutrophic soils occur widely in the Andes mountains, in Central America, and in the Caribbean. By contrast, nutrient-poor **(oligotrophic)** soils develop in old, geologically stable areas, particularly on sandy alluvial deposits (as in much of the Amazon Basin), where intense weathering removes clay and reduces nutrient retention.

Comparison of an oligotrophic forest (in the Amazon Basin) and a eutrophic forest (in Puerto Rico) illustrates that although production and biomass are similar in the two areas, the distribution of nutrients (calcium, in this case) in the oligotrophic forest favors biomass over soil compared with their distribution in the eutrophic forest (Table 8.2). Moreover, the oligotrophic forest holds nutrients more tightly; that is, loss of calcium from the eutrophic forest is equivalent to more than half the annual flux through vegetation, whereas the oligotrophic system appears to gain calcium through precipitation input.

Especially in nutrient-poor areas, nutrient retention by vegetation is crucial to the high productivity of tropical ecosystems. Plants retain nutrients by holding on to their leaves for long periods and by withdrawing nutrients from them before they are dropped. They also grow dense mats of roots (and associated fungi) that remain close to the surface where litter decomposes and even extend up the trunks of trees to intercept nutrients washing down from the canopy. Data from Africa reveal that between 68%

TABLE 8.2 **Standing crops and fluxes of dry biomass and calcium in a nutrient-rich and a nutrient-poor tropical rain forest**

CHARACTERISTIC	EUTROPHIC*	OLIGOTROPHIC*
Soil		
Exchangeable calcium† (kg ha^{-1})	1,900	306
Standing crop		
Living vegetation (T ha^{-1})	263	298
Production		
Living vegetation (g m^{-2} yr^{-1})	1,033	1,012
Calcium in standing crop		
Living vegetation (kg ha^{-1})	760	529
Calcium flux		
Precipitation (kg ha^{-1} yr^{-1})	21.8	16.0
Subsurface runoff (kg ha^{-1} yr^{-1})	43.1	13.2
Net change (kg ha^{-1} yr^{-1})	–21.3	+2.8

*Eutrophic: montane tropical rain forest, Puerto Rico; oligotrophic: Amazonian rain forest, Venezuela.

†Soil calcium measured to a depth of 40 cm.

Source: C. F. Jordan and R. Herrera, *Am. Nat.* 117:167–180 (1981).

and 85% of the root biomass of forests is concentrated within the top 25–30 cm of the soil. In other tropical areas, the application of radioactively labeled compounds has shown that root mats intercept nutrients regenerated by the leaching and decomposition of detritus before they can penetrate into the mineral soil and be washed out of the system.

Nutrient regeneration in aquatic ecosystems

Because most chemical and biochemical processes involved in the cycling of elements take place in an aqueous medium, the processes themselves do not differ markedly in terrestrial and aquatic systems. What is distinctive about most rivers, lakes, and oceans is the sedimentation of nutrients into bottom deposits, from which they are regenerated and returned to zones of productivity very slowly.

The sediments in aquatic systems are the counterparts of terrestrial soils, but they differ from soils in two important ways. First, the regeneration of nutrients from terrestrial detritus takes place near plant roots. In contrast, algae and aquatic plants assimilate nutrients directly from the water column in the uppermost sunlit (photic) zones, often far removed from sediments at the bottom. Second, decomposition of terrestrial detritus occurs, for the most part, aerobically and hence relatively rapidly. Aquatic sediments often become anoxic; the lack of oxygen greatly slows most biochemical transformations and changes the pathways of some nutrient cycles.

The maintenance of high aquatic production depends in part on the proximity of bottom sediments to the photic zone at the surface or on the presence of upwelling currents bringing nutrients regenerated from sediments back to the surface. A map of productivity of the oceans (Figure 8.8) reveals high rates of carbon fixation in shallow seas—both in the Tropics (for example, the Coral Sea and the waters surrounding Indonesia) and at high latitudes (the Baltic Sea, the Sea of Japan)—and in zones of upwelling. These zones occur along the western coasts of Africa and the Americas, where winds blow surface waters away from shore, thus establishing a vertical current to replace them (see Figure 4.5).

Excretion and microbial decomposition regenerate some nutrients in the water column, where assimilation and production take place, just as they do in terrestrial soils. For some elements, rates of regeneration in the water column itself match rates of production. For example, in a study conducted off the western coast of North America, phytoplankton assimilated ammonia as rapidly as it was produced by zooplankton. About half the nitrogen was assimilated directly as ammonia, and about half was first nitrified ($NH_4^+ \rightarrow NO_3^-$) by bacteria. Studies in temperate freshwater systems indicate turnover rates of organic nitrogen on the order of 0.2–2 days; in temperate marine systems, turnover varies between 1 and 10 days, suggesting lower productivity overall.

Plants and algae assimilate regenerated nitrogen rapidly, especially in nutrient-depleted waters. Under conditions of high nutrient concentration, phytoplankton can take up more nitrogen than they need for growth. This

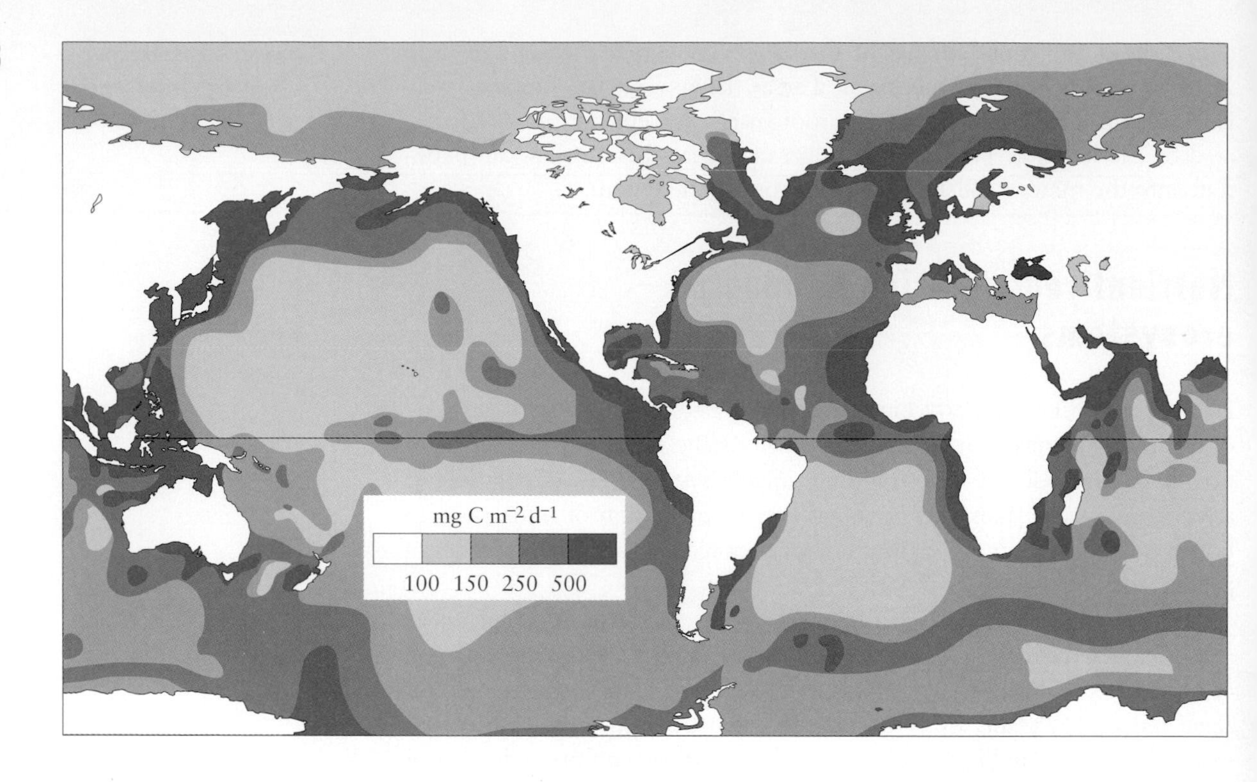

Figure 8.8 Primary production in the world's oceans, in milligrams of carbon fixed per square meter per day. Productivity is greatest on the continental shelves and in regions of upwelling on the west coasts of Africa and South and Central America. After R. K. Barnes and K. H. Mann, *Fundamentals of Aquatic Ecosystems,* Blackwell, Oxford (1980).

ability to store nitrogen, called **luxury consumption,** enables algae to take advantage of short-term abundances or local "pockets" of nutrients made available by discrete bursts of excretion or decomposition. We may think water is homogeneous because it all looks the same to our eyes, but the concentration of nutrients in organisms makes the distribution of elements such as nitrogen and phosphorus extremely heterogeneous. Excretion and decomposition produce transient local abundances of the nutrients that algae rely on. Once turbulent mixing disperses these concentrations, the nutrients may become too sparse to be assimilated.

Nitrogen budgets measured in the Bay of Quinte, Lake Ontario, illustrate the relative magnitudes of assimilation and regeneration in a freshwater aquatic ecosystem. The studies were conducted within columns of water enclosed by "limnocorrals," which are triangular or circular in cross section and are formed by sheets of plastic suspended by floats at the surface and entrenched in sediment at the bottom (in this case, 4 meters beneath the surface). Such enclosures make it possible to study the fluxes of elements by adding isotopically labeled compounds. The limnocorrals were large enough that vertical mixing could take place between surface waters and water immediately over the sediments at the bottom.

Although nitrate accounted for 30–40% of the regenerated nitrogen, algae preferentially assimilated ammonia by ratios of 4 to 1 early in the growing season (June 5) and 30 to 1 late in the growing season (September 4) (Table 8.3). Although the levels of available nitrogen were similar, gross production in September exceeded that in June by nearly 7 times, probably because of a combination of differences in water temperature and some other, limiting nutrient. Short-term uptake of nitrogen by plants amounted to about one-tenth the level of carbon fixation.

TABLE 8.3 **Estimates of release and uptake of nitrogen in limnocorrals in the Bay of Quinte, Lake Ontario**

CHARACTERISTIC	JUNE 5, 1974	SEPTEMBER 4, 1974
Concentration (μg N l^{-1} d^{-1})		
Ammonia (NH_4^-)	120	176
Nitrate (NO_3^-)	84	72
Primary production (μg C l^{-1} d^{-1})		
Gross	185	1,281
Net	−139	+807
Uptake (μg N l^{-1} d^{-1})		
Ammonia	20	117
Nitrate	4.5	4.5
Nitrogen fixation	1.2	2.7
Total	26	124
Release (μg N l^{-1} d^{-1})		
Zooplankton grazing	9.2	27
Sedimentation	3.7	63
Total	13	90

Source: C. F. H. Liao and D. R. S. Lean, *J. Fish. Res. Bd. Can.* 35:1102–1108 (1978).

Sedimentation of nitrogen in sinking particulate matter averaged 14% as much as uptake in June and 28% as much as uptake in September. Accordingly, during the September sample period, when the total nitrogen in the system (NH_4^+, NO_3^-, and particulate) was 586 g l^{-1} and sedimentation was 36 g l^{-1} d^{-1}, physical removal of nitrogen from the system could have depleted the resource quickly. But sedimentation of particulate nitrogen was approximately balanced throughout the season as vertical mixing of the water column returned ammonia from bottom sediments; thus total nitrogen in the water column varied little relative to internal cycling.

Vertical mixing and thermal stratification

The vertical mixing of water requires an input of energy to accelerate water masses and keep them moving. Winds supply most of this energy, causing turbulent mixing of shallow water and upwelling currents along some seacoasts, although variations in water density related to temperature and salinity establish vertical currents (thermohaline circulation) in some marine ecosystems. Vertical movement of water can be hindered in aquatic systems when fresh water floats over denser salt water, as in an estuary, or when the sun heats surface water, establishing a warm layer of lower density overlying a cooler layer of higher density. Other processes promote vertical mixing. In marine systems, when evaporation exceeds freshwater input, the surface layers become more saline, hence denser, and literally fall through the lighter water below. The same phenomenon occurs when ice forms and salt is excluded from the crystallized water. When surface layers of temperate zone lakes cool during the fall, the denser surface water tends to descend through the warmer, less dense layers below.

The vertical mixing of water affects production in two opposing ways. On one hand, mixing can bring nutrient-rich water from the depths to the sunlit surface and thereby promote production. On the other hand, mixing can carry phytoplankton below the zone of photosynthesis and thereby reduce production. Indeed, when vertical mixing extends far below the photic zone, phytoplankton cannot maintain themselves, much less reproduce. Under such conditions, primary production may shut down altogether, resulting in the seeming contradiction of nutrient-rich water without primary production.

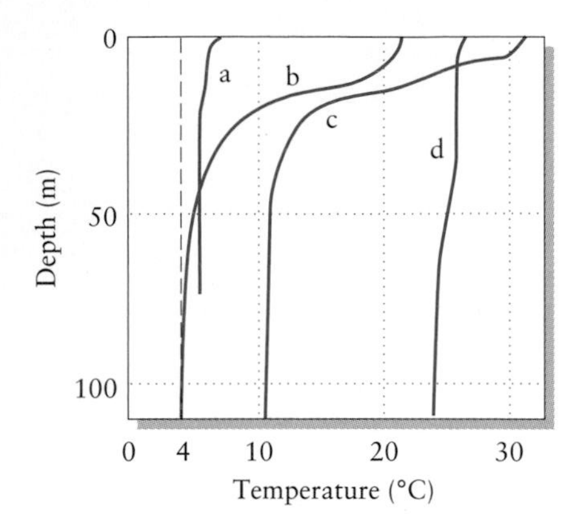

Figure 8.9 Temperature profiles of four lakes at different latitudes at the height of summer stratification: a = Lake Flakevatn, Norway; b = Lake Cayuga, New York; c = Lake Ikedako, Japan; d = Lake Edward, Uganda. At high latitudes (Norway), the summer sun is not strong enough to warm the surface layers. At the equator (Uganda), the strong sun and year-round high temperatures warm lakes uniformly to their bottoms. From G. E. Hutchinson, *A Treatise on Limnology,* Vol. 1, Wiley, New York (1957).

The more typical situation in many temperate zone lakes and ponds is one in which thermal stratification during the summer prevents vertical mixing; then, as sedimentation removes nutrients from the surface layers, production decreases. Nutrients may be regenerated in the deeper layers of the lake, but they cannot reach the surface until stratification breaks down and vertical mixing ensues with cool fall temperatures.

Thermal stratification develops only weakly, if at all, at high and low latitudes (Figure 8.9). In arctic and subarctic regions, the heat input into lakes is not sufficient to counter turbulent mixing, and the water column warms uniformly to the extent that water temperature rises at all. In the Tropics, the lack of a pronounced seasonal temperature cycle reduces the sharpness of thermal stratification because the sun and constant high air temperatures warm the water uniformly to the bottom of the lake.

In marine systems, two very different water masses may meet at a **front,** and here intermixing may create excellent conditions for high production. Sometimes at the boundary of a shallow-water system and a deep-water system, mixed (deep) and stratified (shallow) water masses are brought together. On the mixed side, nutrients may be abundant, but phytoplankton may not remain within the photic zone. On the stratified side, nutrients may have been depleted from the surface waters. Where the two systems meet, some of the nutrient-laden mixed water may enter the stratified layer, creating ideal conditions for photosynthesis and nutrient assimilation (Figure 8.10).

Nutrient limitation of production in the oceans

On the whole, the production of marine ecosystems is closely related to the supply of nutrients, particularly nitrogen, in the surface layers of water. As a result, the highest levels of production are observed in shallow seas, where vertical mixing reaches to the bottom, and in areas of strong upwelling. However, there appear to be areas of the oceans where nitrogen and phosphorus are present in abundance, but phytoplankton concentrations and primary production are low. These conditions suggest limitation by other elements, among which iron and silicon are strong candidates. Iron is an important component of many electron transport and catalytic systems. Silicate is the primary material in the glass shells of diatoms, which are the predominant kind of phytoplankton in the oceans, and is lost from the photic zone when the organisms die and the shells fall to the bottom.

High phytoplankton densities in the Southern Ocean are clearly associated with the proximity of continental sources of nutrients: plankton blooms are concentrated in waters downcurrent of Australia and New Zealand, South America and the Antarctic Peninsula, and southern Africa (Figure 8.11). However, concentrations of nitrogen and phosphorus are high enough to sustain high phytoplankton densities throughout the entire area due to the upwelling currents established by thermohaline circulation in the oceans around Antarctica. The absence of dense phytoplankton concentrations in much of the region suggests limitation by other nutrients. In particular, the area to the west of southern South America between 40° and 50°S appears to have too little silicon.

Over 20% of the open oceans appear to have abundant nitrogen and phosphorus but low densities of phytoplankton. These regions are referred to as high-nutrient low-chlorophyll (HNLC) areas, and they have puzzled marine biologists for years. One hypothesis is that phytoplankton populations in these areas are kept low by zooplankton grazers, although it is unclear why this would happen in some regions of the sea and not others. In the late 1980s, John H. Martin, of the Moss Landing Marine Laboratories in California, proposed that production in these areas is limited by iron. In well-aerated surface waters, iron complexes with other elements, including phosphorus, forms precipitates, and sediments out of the system.

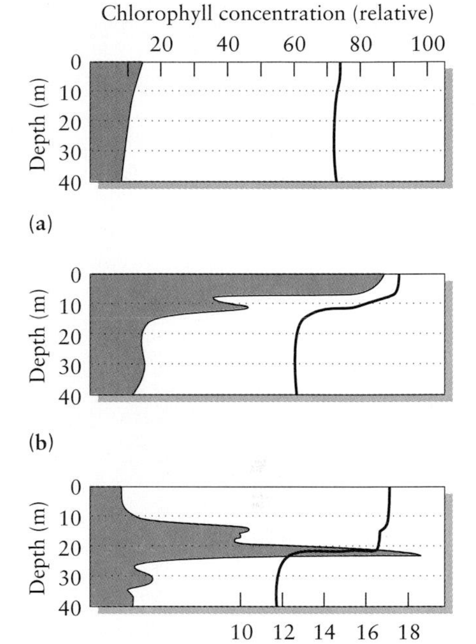

Figure 8.10 Vertical profiles of chlorophyll concentration (represented by the shaded areas) and temperature (represented by the lines) in the western English Channel in July 1975. (a) A well-mixed water mass. (b) A "front" in the region of mixing between water masses a and c. (c) A stratified water mass. After R. K. Barnes and K. H. Mann, *Fundamentals of Aquatic Ecosystems,* Blackwell, Oxford (1980).

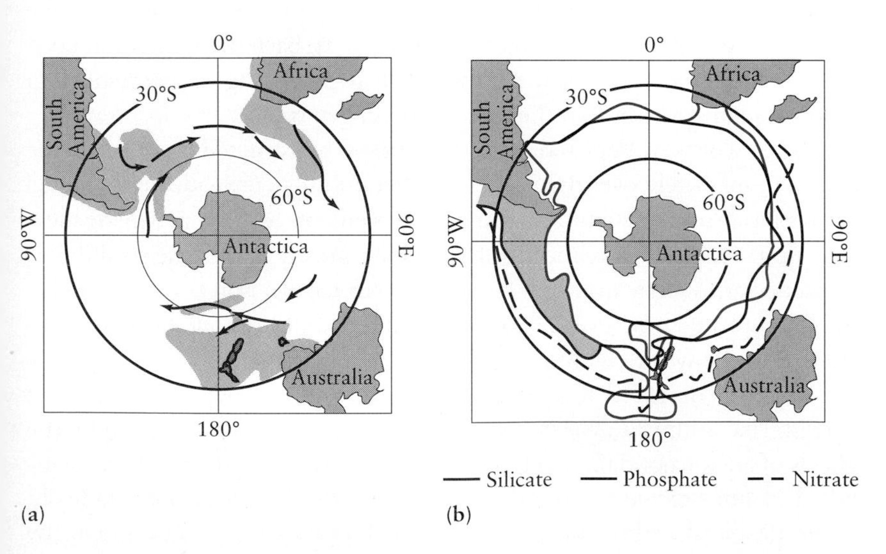

Figure 8.11 Maps of the Southern Ocean showing (a) areas of high phytoplankton concentration, with predominant surface current directions indicated, and (b) regions within which nutrients are sufficient for abundant phytoplankton growth (nitrogen > 10 μM nitrate; phosphorus > 1 μM phosphate; silicon > 5 μM silicate). The shaded region has sufficient nitrogen but insufficient silicon, suggesting silicon limitation. From C. W. Sullivan et al., *Science* 262:1832–1837 (1993).

Inputs of iron to remote parts of the oceans come almost exclusively from windblown dust.

In an enormous experiment conducted in 1993 off the Pacific coast of South America, about 5° south of the equator, scientists fertilized a target area by distributing 450 kg of dissolved iron over 64 km^2, increasing the concentration of iron in that area almost a hundredfold. Within a few days, phytoplankton production inside the fertilized patch, as measured by the concentration of chlorophyll, tripled. This result clearly demonstrated iron limitation in natural surface waters. The original motivation for conducting the experiment was to determine whether stimulation of marine production could quickly sequester large amounts of carbon in biomass and help to reduce the carbon dioxide concentration of the atmosphere, which is increasing rapidly because of the burning of fossil fuels and the clearing of forests. In this respect, however, the experiment was a failure, probably because zooplankton populations increased along with phytoplankton populations and regenerated much of the assimilated carbon dioxide by respiration. Nonetheless, the point that production might be limited by particular essential nutrients, heterogeneously distributed throughout the oceans, was well made.

Regeneration of nutrients in deep, oxygen-depleted waters

In lakes, the layer of water lying above the sharp change (thermocline) in a stratified temperature profile is called the epilimnion; the zone below the thermocline is the hypolimnion. During prolonged periods of stratification, bacterial respiration in the hypolimnion depletes the oxygen supply in that layer (Figure 8.12), provided that abundant organic matter exists for the bacteria to oxidize. In such anoxic bottom waters, bacterial respiration continues using sulfate as an oxidizer and results in increasing concentrations of reduced sulfur, primarily in the form of hydrogen sulfide.

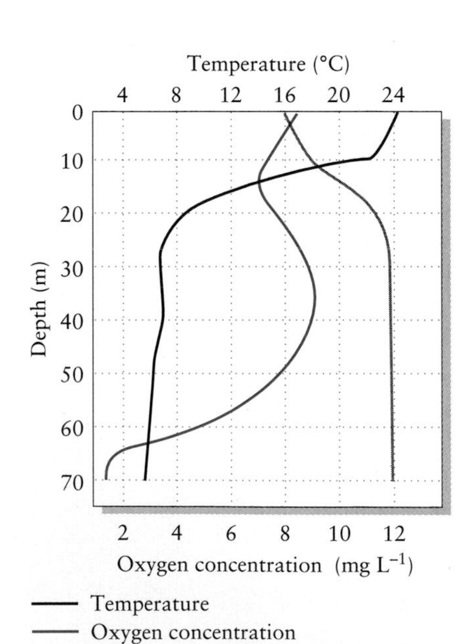

Figure 8.12 Profiles of temperature and oxygen concentration in Green Lake, Wisconsin, during summer stratification, illustrating oxygen depletion in the hypolimnion. From G. E. Hutchinson, *A Treatise on Limnology*, Vol. 1, Wiley, New York (1957).

In the oxygen-depleted environment of bottom sediments and the waters immediately over them, there is often insufficient oxidizing potential for bacteria to nitrify ammonia, and such elements as iron and manganese shift from oxidized to reduced forms, which greatly affects their solubility. In particular, as ferric iron (Fe^{3+}) is reduced to ferrous iron (Fe^{2+}), insoluble iron-phosphate complexes become solubilized, and both elements tend to move into the water column.

Changes observed in the water chemistry of the hypolimnion of an English lake, Esthwaite Water, during the course of a single season show the effects of anoxic conditions (Figure 8.13). After stratification becomes established in June, oxygen at the deepest level of the lake decreases gradually, while dissolved carbon dioxide increases. The water becomes anoxic by early July and remains so until the end of stratification and the onset of vertical mixing in late September. During the period of anoxia, levels of ferrous iron, phosphate, and ammonia increase dramatically as reduced forms in the sediments, and at the sediment–water boundary these materials solubilize and enter the water column. The return of oxidizing conditions in the fall reverses the chemistry of the bottom water, initially because of the

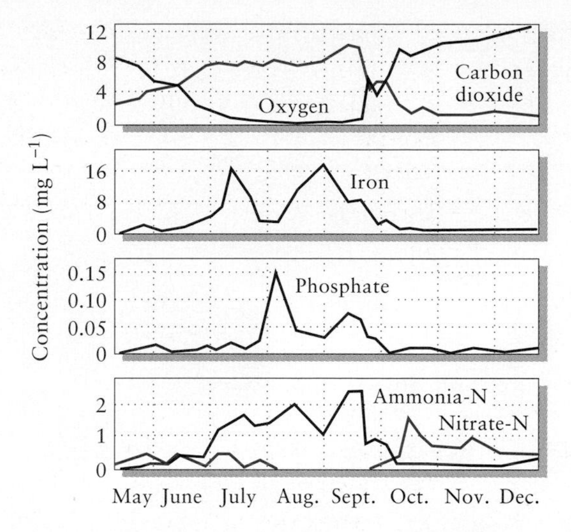

Figure 8.13 The seasonal course of water chemistry in the hypolimnion of Esthwaite Water, England, showing the solubilizing of reduced phosphorus and iron compounds during the period of summer anoxia. From G. E. Hutchinson, *A Treatise on Limnology,* Vol. 1, Wiley, New York (1957).

replacement of bottom water by surface water, but ultimately because the oxidized forms of several redox elements produce insoluble compounds, which precipitate out of the water column. Nitrogen is a conspicuous exception: under oxic conditions, nitrifying bacteria convert ammonia to nitrate, which generally remains in solution.

Life in the anoxic world can often be surprising, as is the case of the white sulfur bacterium *Thioploca,* which forms dense, thick mats on the surface of sediments at 40–280 m water depth off the coast of Peru and Chile (Figure 8.14). The habitat of *Thioploca* occurs within an upwelling zone that supports a rich fishery, but the water chemistry there is unusual in that there is little oxygen in the deep waters, but relatively high concentrations of nitrate. The sediments themselves are anoxic, but other bacteria oxidize organic carbon in a reaction coupled with the reduction of sulfate to hydrogen sulfide. *Thioploca* is chemolithotrophic and gains energy by oxidizing the hydrogen sulfide back to sulfate, using nitrate as an oxidizer ($NO_3^- \rightarrow N_2$). The only problem is that the nitrate is in the water column above the sediment. Filaments of *Thioploca* cells live within tubular sheaths, which extend from the water-sediment interface 5–10 cm into the sediment. The bacterial filaments slowly glide up and down in the sheaths, shuttling between the water and sediment environments. At the top of the sheath the bacterium takes up nitrate, which it stores at high concentration in a large vacuole in its center. Armed with a supply of oxidizer, *Thioploca* slides down the sheath into the sediment, where it converts H_2S to elemental sulfur, which forms globules of pure sulfur in the bacterium's cytoplasm, and then further oxidizes the sulfur to sulfate. When its nitrate supply runs low, it slides back up to the sediment surface to restock.

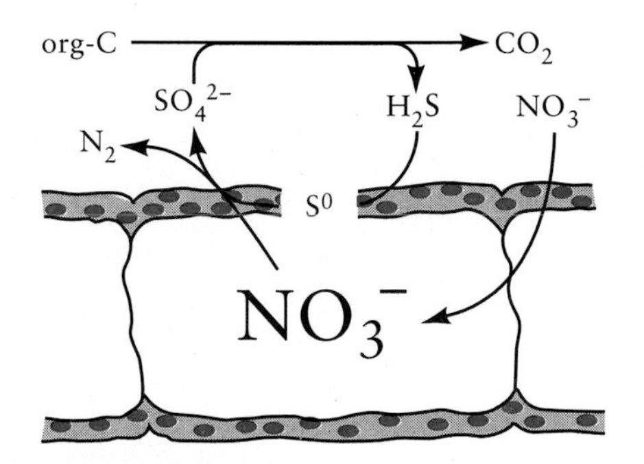

Figure 8.14 A schematic diagram of a filament of a *Thioploca* bacterium, indicating biochemical transformations involving sulfur and nitrogen. The center of the cell is a large vacuole within which nitrate is stored at high concentration. Globules of pure elemental sulfur embedded in the cytoplasm are represented as colored dots. From H. Fossing et al., *Nature* 374:713–715 (1995).

Phosphorus and eutrophication

Phosphorus is often scarce in the well-oxygenated surface waters of lakes, and low levels of phosphorus limit the production of freshwater systems. In small lakes on the Canadian Shield, productivity increased dramatically in response to the addition of phosphorus, but not nitrogen or carbon

Figure 8.15 An experiment in a natural lake demonstrating the crucial role of phosphorus in eutrophication. The near basin, fertilized with carbon (in sucrose) and nitrogen (in nitrates), exhibited no change in organic production. The far basin, separated from the first by a plastic curtain, received phosphate in addition to carbon and nitrogen, and was covered by a heavy bloom of cyanobacteria within 2 months. Courtesy of D. W. Schindler, from D. W. Schindler, *Science* 184:897–899 (1974).

(Figure 8.15). Natural lakes exhibit a wide range of productivity, depending on inputs of nutrients from outside (**external loading:** rainfall, streams) and the regeneration of nutrients within the lake **(internal loading).** In shallow lakes lacking a hypolimnion, internal loading occurs continuously through resuspension of bottom sediments. In somewhat deeper lakes where the thermocline develops only weakly, vertical mixing may occur periodically as a result of occasional strong winds or during unusual periods of summer cold. Such mixing returns regenerated nutrients to the surface and stimulates production. In very deep lakes, bottom waters rarely mix with surface waters, and production depends almost entirely on external loading.

Aquatic ecologists classify lakes on a continuum ranging from poorly nourished (oligotrophic) to well-nourished (eutrophic), depending on their nutrient status and production. Naturally eutrophic lakes have characteristic temporal patterns of production and nutrient cycling that maintain the system in a steady state. The addition of nutrients in the form of sewage and drainage from fertilized agricultural lands can cause inappropriate nutrient loading and greatly alter natural cycles in lakes.

Increased production is not bad in and of itself; indeed, many lakes and ponds are artificially fertilized to increase commercial fish production. But overproduction can lead to imbalance when natural regeneration processes cannot handle the increased demands on cycling. Heavy organic pollution, such as that which results from dumping raw sewage into rivers and lakes, creates **biological oxygen demand (BOD),** resulting from the oxidative breakdown of the detritus by microorganisms. Inorganic nutrients, including runoff from fertilized agricultural land, stimulate the production of organic detritus, adding to the BOD. The problem is heightened in winter, when photosynthesis rates are low and little oxygen is generated within the water column. In its worst manifestations, this type of pollution can deplete the surface water of oxygen, causing fish and other obligately aerobic organisms to suffocate.

Despite the frequently devastating effects of external nutrient loading, artificially eutrophic lakes can recover their original condition when inputs are shut off. Eventually, oxic conditions return, phosphorus precipitates out of the water column, and the normal cycles of assimilation and regeneration are restored. A spectacular and convincing demonstration of such a recovery occurred after the diversion of sewage from Lake Washington, in Seattle, where an advanced and rather ugly case of eutrophication quickly reversed itself.

Estuaries and marshes

Shallow estuaries—semi-enclosed coastal regions subject to both freshwater inputs from rivers and tidal inputs from the sea—are among the most productive ecosystems on earth. Salt marshes, which are intertidal areas with emergent vegetation (Figure 8.16), exhibit similarly high productivity. The high production in these ecosystems results from rapid and local regeneration of nutrients and external loading in the form of nutrients brought into the system by rivers. The effects of high production in estuaries and coastal

Figure 8.16 Salt marshes are a common feature of protected bays along most temperate coasts.

marshes extend to marine ecosystems in many areas through net export of organic matter. A Georgia salt marsh exports nearly 10% of its gross primary production and almost half of its net primary production to surrounding marine systems in the form of organisms, particulate detritus, and dissolved organic material carried out with the tides (Figure 8.17). Because of their high productivity and the hiding places they offer prey organisms, coastal marshes and estuaries are important feeding areas for the larvae and immature stages of many fishes and invertebrates that later complete their life cycles in the sea.

The high production of coastal systems reflects high nutrient levels. Because such a large fraction of the production of coastal habitats is carried out to sea, exported nutrients must be replaced by imports. Studies of the nitrogen budget of Great Sippewissett Marsh on Cape Cod have revealed nutrient inputs to the marsh through precipitation (minor), groundwater flow from surrounding terrestrial systems, and local fixation of atmospheric nitrogen. These inputs approximately balance losses through denitrification, local accumulation of sediments, and export in tidal water. A high rate of denitrification, which occurs primarily in the anoxic sediments at the bottom of the creek draining the salt marsh, underscores the role of chemolithotrophic bacteria and photoautotrophic bacteria in salt marsh metabolism. Rich organic sediments are anoxic just below their surfaces due to rapid microbial decomposition of organic matter. As a result, oxidations based on denitrification ($N^{5+} \rightarrow N^0$) and sulfate reduction become important for the regeneration of nutrients.

As we have seen, the basic chemical and biochemical transformations of the element cycles are uniquely modified by the physical and chemical conditions created in each type of terrestrial and aquatic ecosystem. The paths of elements discussed in this chapter reflect patterns of nutrient cycling. In terrestrial systems, most ecosystem metabolism is aerobic, and production appears to be limited by the regeneration of nutrients in soil where it is not limited by availability of water. In aquatic systems, anoxic regeneration of nutrients in sediments, often far removed from sites of primary production, controls much of the overall productivity and cycling of nutrients within the system. In the next chapter, we shall examine the

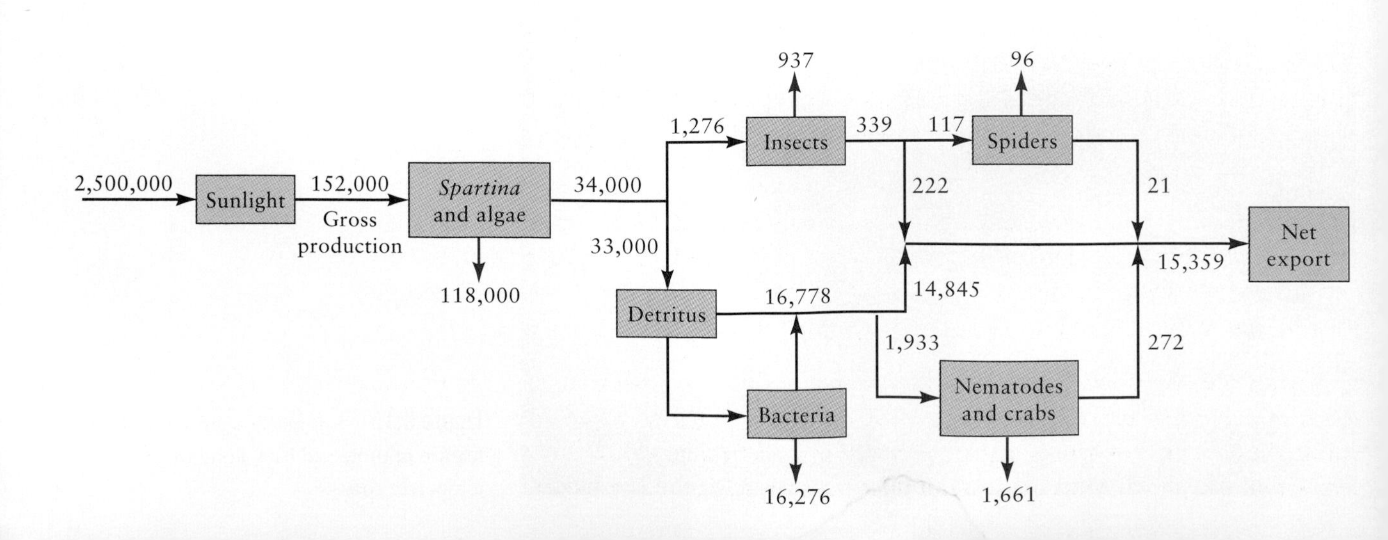

Figure 8.17 Energy flow diagram for a Georgia salt marsh; units are kJ m^{-2} yr^{-1}. The large, striated arrowheads represent respired energy lost from the system. From D. S. McLusky, *The Estuarine Ecosystem*, Wiley, New York (1981).

regulation of the fluxes of nutrients through their cycles—and hence the control of the overall productivity of the ecosystem.

Summary

1. Nutrient cycles in terrestrial and aquatic ecosystems result from similar chemical and biochemical reactions expressed in different physical and chemical environments.

2. Nutrient regeneration in terrestrial ecosystems takes place in the soil. The weathering of bedrock and the associated release of new nutrients proceed slowly compared with the assimilation of nutrients from the soil by plants. Therefore, the productivity of vegetation depends on the regeneration of nutrients from plant litter and other organic detritus.

3. Nutrients are regenerated from litter by the leaching of soluble substances; consumption by large detritus feeders; fungi that break down cellulose and lignin; and the eventual mineralization of phosphorus, nitrogen, and sulfur, primarily by bacteria.

4. Mycorrhizae are symbiotic associations of certain types of fungi with the roots of plants. The fungi, which may either penetrate the root tissue or form a dense sheath around the root, enhance the plants' uptake of soil nutrients. They do this primarily by enlarging the volume of soil accessible to the roots and by secreting acid into the surrounding soil to increase the solubility of such nutrients as phosphorus. In return, they obtain a reliable source of carbon from the plant.

5. In many lowland tropical habitats, deeply weathered soils retain nutrients poorly. In such habitats, regeneration and assimilation of nutrients proceed rapidly, and most of the nutrients, especially phosphorus, occur in the living vegetation. When such soils are clear-cut for agriculture, they soon lose their fertility because nutrients are removed along with the native vegetation and crops, organic matter in the soil decomposes rapidly, and the nutrients that are released and not assimilated are washed out of the soil.

6. The sediments at the bottoms of lakes and oceans resemble terrestrial soils but differ from them in two important respects: aquatic sediments are spatially removed from the site of nutrient assimilation by aquatic plants and algae, and sediments often develop anoxic conditions that retard the regeneration of some nutrients.

7. The productivity of aquatic systems is maintained by the transport of nutrients from bottom sediments to the surface, as occurs in shallow waters and areas of upwelling, by the recycling of nutrients regenerated within the photic zone, and by the import of nutrients from other systems.

8. Vertical mixing is inhibited by stratification, which is the formation of layers of water that differ in density because they differ in temperature or salinity. Stratification enhances aquatic production by retaining phytoplankton within the photic zone, but it diminishes production to the extent that

the sedimentation of detritus carries nutrients below the depth where light is sufficient for photosynthesis.

9. The productivity of marine systems is generally limited by availability of nutrients. The limiting nutrients may be silicon or iron in the open ocean, where both elements tend to sediment out of the system, silicon in the shells of diatoms and iron in precipitated complexes with other elements, such as phosphorus. Large-scale fertilization experiments have demonstrated iron limitation of photosynthesis in parts of the ocean.

10. The annual cycles of temperate zone lakes include a period of temperature stratification during the summer intervening between spring and fall periods of vertical mixing. Nutrients are brought to the photic zone near the surface during the periods of mixing. During the summer, nutrients are depleted by the sedimentation of organic material.

11. Nutrients are regenerated in aquatic sediments by bacterial decomposition. The anaerobic conditions that develop beneath the thermocline because of the consumption of oxygen by bacteria result in the chemical reduction of iron, magnesium, and sulfur and the solubilizing of phosphate compounds. Thus, phosphorus is released from anoxic sediments and may reach the photic zone by vertical mixing. The sediment-water interfaces of lakes and oceans are extremely complex systems, often with rapid changes in nutrient concentrations and redox potentials across the boundary. These systems are dominated by chemolithotrophic and chemoautotrophic bacteria, which form the base of local food chains.

12. Because phosphorus forms insoluble compounds with iron and precipitates readily under the oxic conditions of surface waters, it frequently is in short supply and limits aquatic production in fresh waters. Sewage and agricultural runoff add phosphorus and other nutrients to streams and lakes and may greatly alter natural patterns of production and nutrient cycling, upsetting natural balances in freshwater ecosystems.

13. Shallow-water communities, particularly estuaries and salt marshes, are extremely productive because of rapid local regeneration of nutrients and the external loading of additional nutrients from nearby terrestrial habitats. Studies of nutrient budgets indicate that marshes and estuaries are major exporters of both organic carbon and mineral nutrients to surrounding marine systems and are therefore an indispensable component of marine production in some areas.

Suggested readings

Barnes, R. K., and K. H. Mann (eds.). 1980. *Fundamentals of Aquatic Ecosystems.* Blackwell, Oxford.

Baskin, Y. 1995. Can iron supplementation make the equatorial Pacific bloom? *BioScience* 45:314–316.

Bertness, M. D. 1992. The ecology of a New England salt marsh. *American Scientist* 80:260–268.

Binkley, D., and D. Richter. 1987. Nutrient cycles and H^+ budgets of forested ecosystems. *Advances in Ecological Research* 16:1–51.

Coleman, D. C., C. P. P. Reid, and C.V. Cole. 1983. Biological strategies of nutrient cycling in soil systems. *Advances in Ecological Research* 13:1–55.

Edmondson, W. T. 1970. Phosphorus, nitrogen, and algae in Lake Washington after diversion of sewage. *Science* 169:690–691.

Gage, J. D., and P. A. Tyler. 1991. *Deep-Sea Biology: A Natural History of Organisms at the Deep-Sea Floor.* Cambridge University Press, Cambridge.

Jordan, C. F. 1982. Amazon rain forests. *American Scientist* 70:394–401.

Jordan, C. F., and R. Herrera. 1981. Tropical rain forests: Are nutrients really critical? *American Naturalist* 117:167–180.

Libes, S. M. 1992. *An Introduction to Marine Biogeochemistry.* Wiley, New York.

Mann, K. H., and J. R. N. Lazier. 1991. *Dynamics of Marine Ecosystems: Biological-Physical Interactions in the Oceans.* Blackwell Scientific Publications, Boston.

McLusky, D. S. 1989. *The Estuarine Ecosystem.* 2d ed. Chapman & Hall, New York.

McNaughton, S. J., R. W. Ruess, and S. W. Seagle. 1988. Large mammals and process dynamics in African ecosystems. *BioScience* 38:794–800.

Richards, B. N. 1987. *The Microbiology of Terrestrial Ecosystems.* Wiley, New York.

Stevenson, F. J. 1986. *Cycles of Soil: Carbon, Nitrogen, Phosphorus, Sulfur, Micronutrients.* Wiley, New York.

Tunnicliffe, V. 1992. Hydrothermal-vent communities of the deep sea. *American Scientist* 80:336–349.

Van Cleve, K., F. S. Chapin III, C. T. Dyrness, and L. A. Viereck. 1991. Element cycling in taiga forest: State-factor control. *BioScience* 41:78–83.

Warning, R. H., and W. H. Schlesinger. 1985. *Forest Ecosystems: Concepts and Management.* Academic Press, Orlando, Fla.

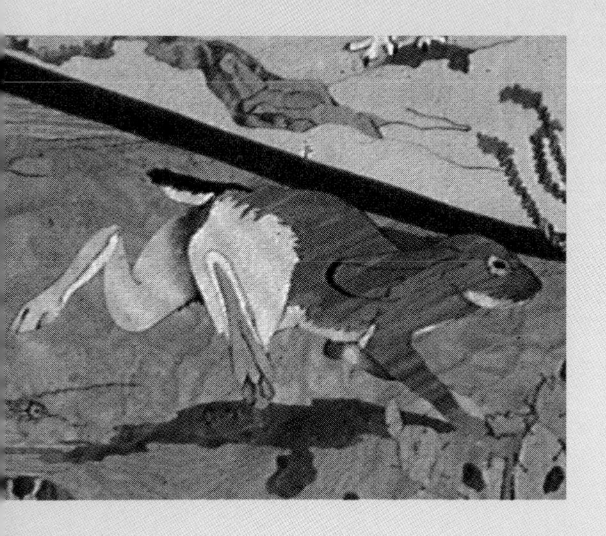

CHAPTER 9

REGULATION OF ECOSYSTEM FUNCTION

Primary production drives energy flux and cycling of elements within ecosystems. The most productive systems are tropical rain forests, coral reefs, and estuaries, where favorable combinations of high temperature, abundant water, and intense sunlight promote rapid photosynthesis. However, plant growth also depends on the assimilation of nutrients, which are in large measure regenerated within ecosystems by bacteria, fungi, and animals. Because nutrient regeneration depends on biological processes, it too is sensitive to variations in temperature, moisture, and other conditions. Thus, the physical environment may control ecosystem function by influencing energy assimilation, nutrient assimilation, or both.

Ecologists do not fully understand how the productivity of most ecosystems is controlled. Broad comparisons among ecosystems reveal that primary production is correlated with temperature, precipitation, nutrient availability, and other physical factors, but such comparisons do not show how or where these factors act. Experiments provide another means of studying the regulation of ecosystem function, but rarely can ecologists conduct experiments on the scale of ecosystems. How, for example, could one eliminate nitrifying bacteria from soil and leave all other factors unaltered? In those experiments that have been performed (for example, by watering an area of dry habitat or by adding iron to a patch of ocean), it is often unclear which step or steps in the cycling of critical elements responded to the treatment and how those responses might change over time as the system established a new equilibrium. One may also question whether experimental manipulation of any one factor can be informative about differences among natural systems that may result from the influences of other factors.

Another approach to understanding ecosystems is to develop mathematical models of their functioning and to examine how the outputs of these models vary with respect to different inputs and conditions. Modeling requires detailed knowledge of the processes that are critical to regulating ecosystem function (which could be interpreted as everything that happens in the system!). It is also important to validate each mathematical equation by field observation and experimentation. These criteria are difficult to meet; however, even simple models can provide a concise and rigorous statement of our understanding of a system. To the extent that the behavior of a model parallels the behavior of natural systems, it confirms that our thinking is on the right track. Even when the correspondence between the model and the system cannot be judged, a model may provide insights into what is possible in natural systems and may guide observation and experimental investigation accordingly. Modeling efforts are central to the discipline of **systems ecology,** which is the study of how interdependent parts contribute to the functioning of an entire system.

In this chapter, we shall develop some simple models of systems, with the modest goal of gaining some insights into the control of ecosystem function. We shall then use these insights to interpret particular observations and experimental results. The approach is mathematical, but requires only simple algebra. Modeling can help us to organize what we have learned about ecosystem function and to see how different processes influence ecosystem function as a whole.

External factors and internal controls

From the standpoint of systems ecology, a system may be defined as a group of interdependent items making up a unified whole. As such, a system has a well-defined boundary in the sense that some items are included within the system and others are not. For example, one does not normally include the sun as a part of the salt marsh ecosystem, although the sun's light provides the energy to drive the system. Items like the sun are external to the system. Such **external factors** exclude material inputs from outside the system (light, precipitation, weathering of parent rock) and physical conditions of the environment that influence the system's structure and function (temperature, salinity, acidity).

The **internal controls** within an ecosystem result from the various transformations of energy and elements within the compartments of the ecosystem and exchanges between them. These transformations and exchanges depend in part on physical processes and in part on the activities of organisms, both of which are influenced by the availability of energy and resources and by the conditions of the environment. The important point is that internal controls are part of the structure of the system itself. As we saw in Chapter 8, biological production depends on the assimilation of energy and nutrients, primarily by plants or algae, and the regeneration of nutrients, primarily by microorganisms; these activities exert internal controls on ecosystem function. Because external factors influence both assimilatory and regenerative processes, it is impossible to understand how ecosystem function is controlled without appreciating the influence of external factors on all the internal controls within the system.

An analogy may clarify the interaction of external factors and internal controls. Consider as a simple system a bucket that holds water. The rate at which the bucket fills with water depends on an external factor, which is the flow of water from a tap. If the bottom of the bucket has a hole in it, the hole, as a property of the system, imposes an internal control on the amount of water in the bucket. When the system achieves a steady state—when water entering the bucket equals the amount leaving through the hole—the flow of water through the bucket depends only on the flow from the tap (the external factor), although the level of water in the bucket depends on both the flow from the tap and the size of the hole (the external factor and the internal control). If one were to observe several similarly porous buckets sitting under flowing taps, and each bucket had a different amount of water in it, one could not know whether the amount of water in any bucket was determined by the flow from the tap or by the size and number of holes in the bucket. A systems model would tell us, however, that it is important to measure both aspects of the system in order to understand its structure and functioning.

When we extrapolate the bucket analogy to the level of ecosystem function, we are confronted by two issues. The first is the degree to which variations in ecosystem structure and function result from variations in external factors as opposed to unique internal properties of the system—essentially, the difference between variation in the flow of water from the tap and variation in the number and size of holes in the bucket. The second issue concerns the particular means by which external factors influence the internal controls of the system.

Energy and element fluxes in ecosystem function

Energy provides the most generalizable currency of ecosystem structure and function, and assimilated energy ultimately drives virtually all ecosystem processes. However, the availability of energy appears to cause little of the variation observed among ecosystems. Ecologists agree that basic aspects of ecosystem structure and function respond to external factors, particularly temperature and precipitation in the case of terrestrial ecosystems. What remains to be discovered is how these factors act to regulate ecosystems.

Peter Vitousek, a plant ecologist at Stanford University, attempted to pinpoint the influence of external factors by comparing the cycling of individual elements to the flow of energy through the system as a whole. Vitousek reasoned that the processes responsible for regenerating the element whose cycling shows the strongest correlation with primary production exercise predominant control. When a forest achieves a steady state, net aboveground primary production approximately equals litter production. Similarly, the cycling of each element approximately equals the amount of that element that falls each year in litter. Vitousek compared the total dry matter of litter produced each year (which is proportional to carbon and hence to energy content) with the amounts of several elements in the litter; the data for nitrogen and phosphorus are given in Figure 9.1. As can be seen, production more closely parallels the cycling of nitrogen than that of phosphorus (or calcium, which is not shown). Vitousek concluded that factors regulating the cycling of nitrogen predominantly control primary production in forests.

Vitousek's study also showed that production (energy flow) does not vary in direct proportion to nutrient cycling. Rather, forests with lower rates of cycling exhibit relatively greater production per unit of nutrient cycled **(nutrient use efficiency, NUE)** than do those with high fluxes. Two factors can result in higher nutrient use efficiency: first, trees may assimilate more energy per unit of nutrient assimilated, and second, trees may retain nutrients for reuse by drawing them back into their stems before they drop their leaves. For both nitrogen and phosphorus, nutrient use efficiency clearly decreases as nutrient cycling increases, and for phosphorus, NUE in the Tropics greatly exceeds that in temperate latitudes; tropical trees evidently retain phosphorus to a greater extent than do temperate trees.

Nutrient use efficiency may reflect certain adaptations of the plant that enable it to retain nutrients or reduce nutrient requirements for production. Thus, while the relationship between nutrient cycling and production may indicate which nutrients limit ecosystem function, the picture is somewhat blurred in this case by adaptations of plants to manage nutrient use. Furthermore, even when a limiting nutrient has been identified, one still does not know at what point the cycling of that nutrient is controlled. Here, we shall use some simple systems models to attempt to sort out the factors responsible for the regulation of ecosystem function. These models consist of mathematical expressions representing the internal controls that govern the movement of nutrients between compartments within the

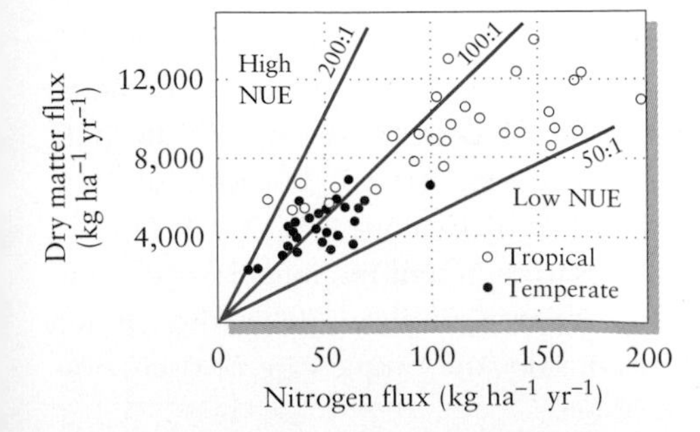

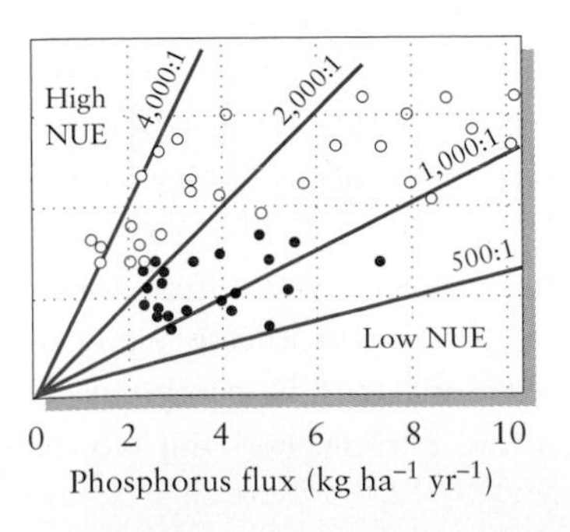

Figure 9.1 The relationship of dry matter flux to nitrogen and phosphorus fluxes in the litter fall of temperate and tropical forests. NUE = nutrient use efficiency. After P. M. Vitousek, *Am. Nat.* 119:553–572 (1982).

ecosystem. Various parameters of the equations incorporate the effects of external factors, such as temperature, moisture, and light. We shall then use these models to determine whether variations in external factors and internal controls produce diagnostic, measurable variations in ecosystem properties from which we can infer the mechanisms that regulate ecosystem function. First, however, some basics.

Development of systems models

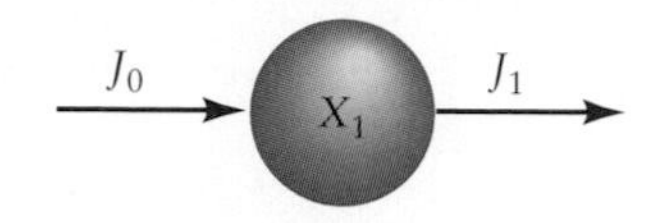

Figure 9.2 Diagram of a single compartment in an ecosystem model, showing input (J_0), output (J_1), and compartment size (X_1).

Let us suppose that each form of a nutrient or of the energy within a system makes up a distinct compartment, which we shall designate X_i for the ith form (organic nitrogen, ammonia, or nitrate, for example). Each compartment has inflows and outflows, which we shall designate as J. A schematic diagram of a single compartment (X_1) with one input (J_0) and one output (J_1) is shown in Figure 9.2. If compartment X_1 represented the water in our bucket, then J_0 would be the flow from the tap and J_1 would be what leaks out through the hole in the bottom. The rate of change with respect to time (t) in the amount of water in the bucket ($\mathrm{d}X_1/\mathrm{d}t$) is equal to the difference between the input and the output—algebraically,

$$\frac{\mathrm{d}X_1}{\mathrm{d}t} = J_0 - J_1$$

The amount of water in the bucket achieves a steady state (that is, $\mathrm{d}X_1/\mathrm{d}t = 0$) when inflow equals outflow ($J_1 = J_0$).

Each flux J may be a constant, or it may vary depending on other factors, including the value of X and the values of external factors. We may express such variable fluxes in symbolic notation as $J = f(X,S,P)$, where f denotes that J is a function of the values inside the parentheses; S represents the state of the system (for example, the number and size of holes in the bucket, the leaf area available for photosynthesis, the population density of nitrifying bacteria); and P stands for various parameters that are the external factors (such as the flow of water from the tap, the temperature of the environment, the amount of precipitation).

In the bucket example, suppose that J_0 is a constant ($J_0 = k_0$) but that J_1 increases in direct proportion to the volume of water in the bucket (that is, $J_1 = k_1X_1$). As the bucket fills, the water pressure at the bottom increases the flow of water through the hole. The amount of water in the bucket (X_1) reaches a steady state when $J_0 = J_1$, and therefore $k_0 = k_1X_1$. We may rearrange this equation to show that the steady-state amount of water (X_1) equals k_0/k_1. Thus, when the value of the external factor (k_0) increases, water reaches a higher level in the bucket. Reducing the size of the hole (the variable k_1) has the same effect on X_1 (Figure 9.3), but note that the flux through the system is always controlled by k_0 alone.

We need at least two compartments to represent the internal cycling of elements within ecosystems. Realistic systems models can become much more complicated, but we shall first consider the simple case of two com-

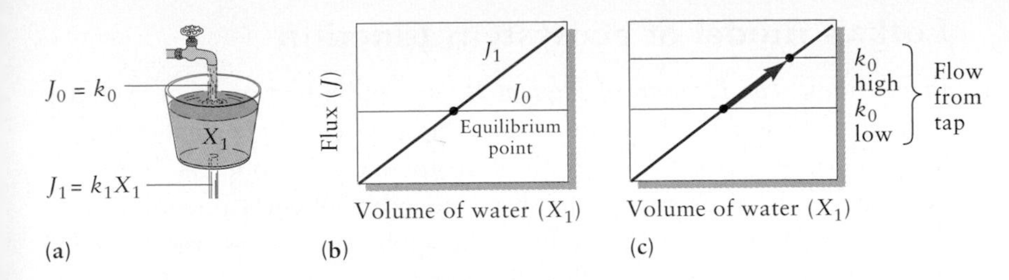

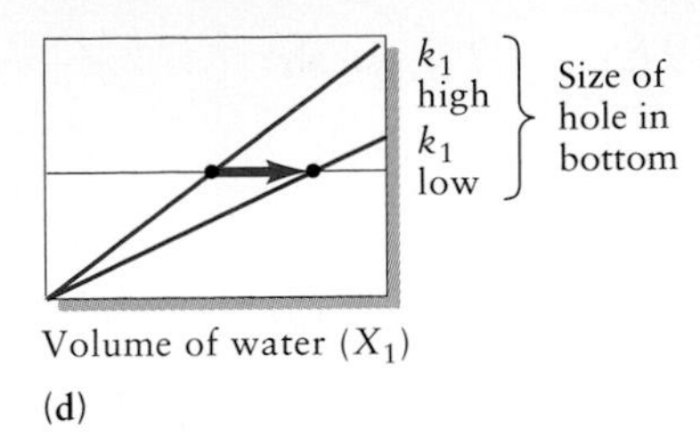

Figure 9.3 The effect of changing input (J_0) and output (J_1) rates on the volume of water (X_1) in a bucket. (a) Because water leaves the bucket through a hole in the bottom, the rate of output is proportional to the water pressure at the bottom and thus to the depth and volume of water. (b, c) Water depth and flux increase as water input (J_0) increases. (d) Water depth (X_1) increases, but flux (J) does not change, as rate of emptying (k_1) decreases.

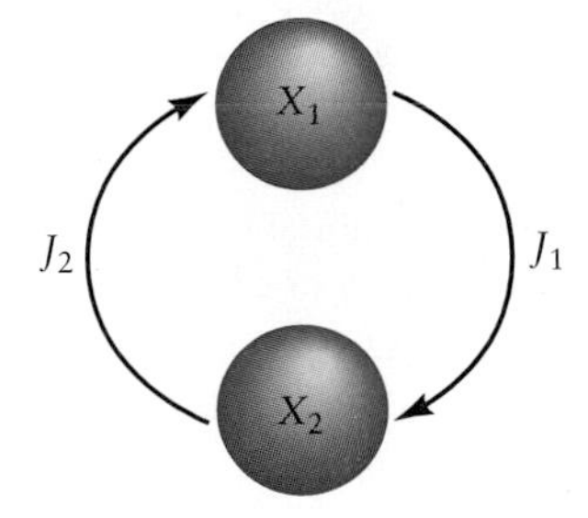

Figure 9.4 Diagram of a closed, two-compartment system.

partments that cycle an element between them (Figure 9.4). Compartment sizes are X_1 and X_2, and fluxes are J_1 and J_2. As we have seen, the Js represent functions. These may be of zero order, in which case J is a constant ($J = c$); of first order, in which case J_1 is a function of X_1; or of second order, in which case J_1 is a function of both X_1 and X_2.

We can say little about how a particular system functions without knowing the details of the flux functions. Simple models may, however, provide insight into general features of system function. The global water cycle, for example, can be thought of as two compartments, vapor (X_1) and liquid (X_2). What we know about the change of water between vapor and liquid phases allows us to infer what the functions (f) for the fluxes (J) might look like. Precipitation (J_1) is a first-order equation depending only on the water vapor content of the atmosphere (X_1) and on various external factors, such as air temperature; the amount of water at the earth's surface (X_2) has no direct effect on precipitation. Air has a limited capacity to hold water vapor at a given temperature, and so condensation probably increases more rapidly with increasing atmospheric vapor pressure as this ceiling is approached.

Evaporation (J_2) is a second-order equation depending on the surface area of water (some function of X_2), the vapor pressure of water in the atmosphere (proportional to X_1), and external factors: temperature and insolation (intensity of sunlight). On a global scale, changes in the amount of water in the oceans resulting from evaporation and precipitation have little effect on the total surface area of water, so we can safely ignore X_2 in the function J_2. Accordingly, we may portray the general features of the hydrologic cycle on a graph relating J_1 and J_2 to X_1 (Figure 9.5). The model shows that the system assumes a steady state with $J_1 = J_2$ at some intermediate value of X_1; changes in external factors that affect J_1 or J_2 (change in temperature, for example) would adjust the equilibrium point for the entire system. Clearly, however, this model can't be used to predict the local weather.

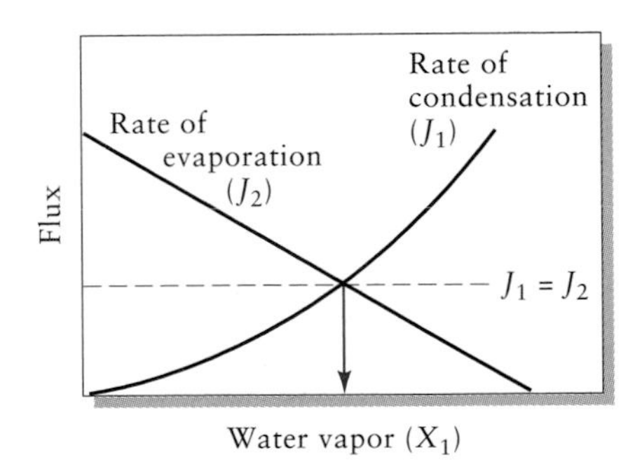

Figure 9.5 Rate of evaporation decreases and rate of condensation increases with increasing water vapor in the atmosphere; this creates a steady-state level (arrow) of water vapor in the atmosphere, at which point evaporation equals condensation, at the level indicated by the dashed line.

Lotka's model of ecosystem function

In his book *The Elements of Physical Biology,* published in 1925, Alfred J. Lotka investigated the behavior of biological systems by applying insights from thermodynamics and tools of mathematical modeling borrowed from the study of chemical equilibria and other physical phenomena. In the space of a few pages, he outlined the application of systems modeling to the problem of nutrient cycling in ecosystems. Lotka treated the specific case of three compartments, X_1, X_2, and X_3, through which some element or other material cycles with fluxes J_1, J_2, and J_3 (Figure 9.6).

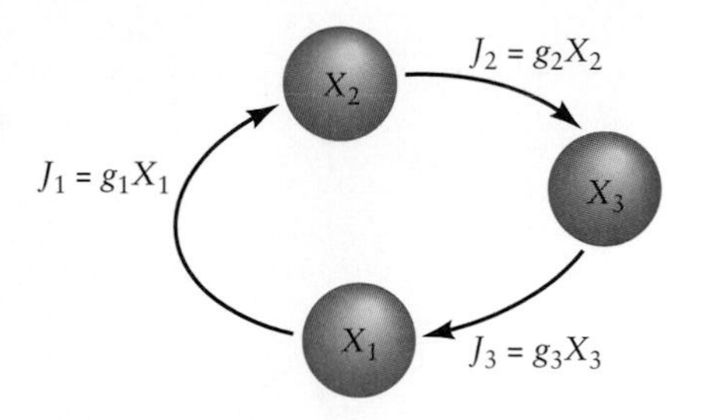

Figure 9.6 A diagram of Lotka's system of three compartments in which the transfer between one compartment and the next is a function of the size of only the first compartment.

Change in any compartment—for example, dX_1/dt—equals the difference between the fluxes into and out of the compartment (J_3 and J_1 in the case of compartment X_1). To illustrate the behavior of such a system, Lotka described each flux as a first-order term $f_i(X_i)$ arbitrarily having the form g_iX_i; g_i is the rate at which material in compartment X_i is transferred to the next compartment, thus $dX_1/dt = g_3X_3 - g_1X_1$. When the system is in a steady state, all the fluxes are equal; hence $J_1 = J_2 = J_3$, or $g_1X_1 = g_2X_2 = g_3X_3$. Because $X_i = J_i/g_i$ and all the Js are equal, the sizes of the compartments are in the relative proportions

$$X_1 : X_2 : X_3 = \frac{1}{g_1} : \frac{1}{g_2} : \frac{1}{g_3}.$$

All the compartments of the system together contain a total amount M (= $X_1 + X_2 + X_3$) of the cycling material. Lotka showed that the compartment sizes can be described by equations of the form

$$X_i = \frac{M}{g_i}\left(\frac{g_1g_2g_3}{g_1 + g_2 + g_3}\right)$$

and the fluxes by

$$J_i = g_iX_i = M\left(\frac{g_1g_2g_3}{g_1 + g_2 + g_3}\right).$$

These equations tell us that the structure (Xs) and function (Js) of the system are defined completely by the transfer functions (gs), which incorporate the external factors and internal controls. In particular, the size of compartment X_i is inversely related to transfer rate g_i.

Lotka further showed that differences between systems in structure and function probably derive from differences in the lowest of the transfer rates. Low transfer rates place bottlenecks in the path of material flow through a system, causing material to accumulate in the preceding compartment. Therefore, the underlying cause of variation in flux through a system should be apparent in the shifts of materials between compartments in the system. For example, if an increase in J were accompanied by a shift of material from compartment X_1 to compartments X_2 and X_3, we could infer that an increase in the function g_1 was responsible. Distinguishing among the roles of state variables, internal controls, and external factors in causing this change would require additional study of the system, but research efforts would be focused by the insights of the systems model. The practical lesson to be learned is that a change in one segment of a nutrient cycle can alter the function of the entire system: a chain is only as strong as its weakest link.

An aquatic systems model

The cycling of nitrogen within a water column by and large follows the path (ammonia → nitrate) → particulate nitrogen (organisms + detritus) → dissolved organic nitrogen (DON) → (ammonia → nitrate). Algae may assimilate either ammonia or nitrate. In a study of nitrogen transformations in the water column of the Bay of Quinte, Lake Ontario, the sizes of some compartments were found to change dramatically with the seasons (Figure 9.7). In particular, between winter and summer, particulate and dissolved organic nitrogen increased while nitrate decreased. This shift implies that the key difference between the cycling of nitrogen in summer and in winter is the rate of nitrate uptake by phytoplankton. A simple first-order systems model will illustrate how we can elaborate on this idea.

When the system in Figure 9.7 has achieved a steady state, fluxes into and out of each compartment must be balanced: for example, $J_2 = J_3$, $J_3 = J_4 + J_5$, and so on. Hence, under steady-state conditions,

$$g_2X_2 = g_3X_3 = (g_4 + g_5)X_4 = g_1X_1 + g_4X_4.$$

From the last two quantities, we can show algebraically that $X_1/X_4 = g_5/g_1$, and by making the appropriate substitutions, we obtain the relationship

$$X_1{:}X_2{:}X_3{:}X_4 = \frac{g_5}{g_1\,(g_4 + g_5)} : \frac{1}{g_2} : \frac{1}{g_3} : \frac{1}{(g_4 + g_5)}$$

Setting each X_i at its proportion of the total nitrogen enables us to estimate relative values for the g_is during the winter and summer (Table 9.1). These estimates indicate, quite dramatically, that the ratio g_1/g_5 (the ratio of algal assimilation to nitrification) is an order of magnitude lower during the winter than it is during the summer. The sum $g_4 + g_5$ differs little between the seasons, so we may assume that g_5 is relatively constant. Evidently, it is the

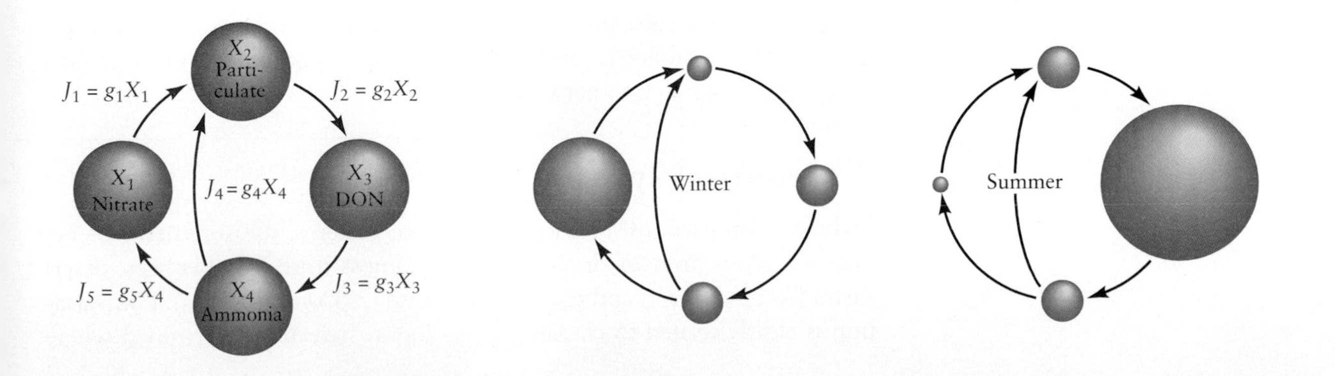

Figure 9.7 A first-order systems model of nitrogen cycling in the Bay of Quinte, Lake Ontario, showing the changes in compartment sizes between summer and winter.

TABLE 9.1 **Standing crops and fluxes of dry biomass and calcium in a nutrient-rich and a nutrient-poor tropical rain forest**

TRANSFER RATE	DESCRIPTION	WINTER	SUMMER
g_2	Production of dissolved organic nitrogen by grazers, excretion, and leakage	8.0	4.1
g_3	Ammonification	4.4	2.0
$g_4 + g_5$	Assimilation of ammonium plus nitrification	4.2	4.7
$\frac{g_1}{g_5}$	Ratio of nitrate assimilation to nitrification	0.6	5.1
$\frac{g_1}{g_5}\ (g_4 + g_5)$		2.5	24

Source: From data in C. F. H. Liao and D. R. S. Lean, *J. Fish. Res. Bd. Can.* 35:1102–1108 (1978).

value of g_1 that decreases so much between summer and winter (g_1 describes the rate of assimilation of nitrate by algae). During the winter, assimilation decreases markedly compared with nitrification (g_5) and ammonification (g_3), suggesting that some factor such as light limits production. Alternatively, g_1 and g_4 (assimilation of nitrate and ammonia, respectively) might both decrease in winter, in which case nitrification rate (g_5) would have to increase at the same time to maintain the sum $g_4 + g_5$ constant.

The preceding model may oversimplify or even misrepresent nitrogen transformations in the water column. It was intended only to illustrate an approach. Most systems models are much more complicated, often including dozens of compartments and equations of much higher order than unity. In our simple model, for example, fluxes J_1 and J_4 almost certainly should have been represented by second-order equations involving X_2 (the populations of phytoplankton that accomplish the assimilation; no nitrogen would be assimilated in the absence of algae). More realistic equations can be developed for particular systems, but simplified models, such as this one and the one that follows, can also serve to direct inquiry into more general comparisons of function between ecosystems.

A terrestrial systems model

In broad comparisons among terrestrial ecosystems, measurements of net primary production of dry matter vary almost thirtyfold between desert shrub (70 g m^{-2} yr^{-1}) and tropical rain forest (2,000 g m^{-2} yr^{-1}). This variation is clearly related to climate, yet ecologists have not determined where external factors exert their influence within the system. A systems approach can provide clues in this case.

Let us consider the cycling of nitrogen again. We shall represent an ecosystem as three compartments—mineral soil (X_1), living plant biomass (X_2), and organic detritus (X_3)—with fluxes J_1, J_2, and J_3 between them (Figure 9.8). (Animals are trivial in this model; the activities of microorganisms are implicit in flux J_3). J_2 (the annual shedding of leaves and other detritus) and J_3 (the mineralization of detritus by microorganisms) can be considered first-order processes. Nitrogen assimilation (J_1) is modeled as a second-order process, incorporating both nutrient availability and plant activity. Under steady-state conditions, we obtain the relationships

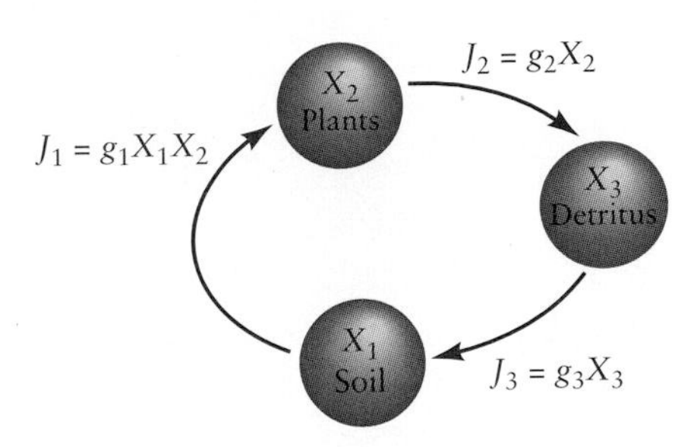

Figure 9.8 A second-order systems model of nitrogen cycling in a forest ecosystem. Note that the assimilation of nitrogen by plants (J_1) is a function of both the availability of nitrogen in the soil and the amount of living plant biomass.

$$g_1X_1X_2 = g_2X_2 = g_3X_3X_2$$

and

$$X_1X_2 : X_2 : X_3 = \frac{1}{g_1} : \frac{1}{g_2} : \frac{1}{g_3}$$

This simple model offers some surprises. For example, the level of inorganic nitrogen in the soil, mostly nitrate (X_1), equals the ratio of g_2 (rate of detritus production) to g_1 (rate of nitrogen assimilation) and is independent of g_3 (rate of microbial regeneration of inorganic nitrogen). Nitrogen flux does depend on g_3, of course. Algebra shows that the relationship of flux to the rates g_i follows

$$\mathrm{J} = \left[\frac{g_2g_3(g_1 - g_2)}{g_1(g_2 + g_3)}\right]M,$$

where M equals the total nitrogen in the system. Fluxes of elements are difficult to measure directly, but we may estimate relative values of g_i from the sizes of compartments. Finally, differences among systems in the relative rates of transfer between compartments can provide insights into points of internal control.

To illustrate these points, we shall compare the cycles of nitrogen and phosphorus in a 47-year-old Scots pine (*Pinus sylvestris*) plantation in England and a 50-year-old mixed tropical forest in Ghana (Figure 9.9). The differences are striking. In England, for both nitrogen and phosphorus, the values of g_1, g_2, and g_3 are of the same magnitude. With respect to nitrogen in the Ghanian forest, g_1 and g_2 are similar, but g_3 is almost two orders of magnitude greater. Thus the regeneration of mineral nutrients from organic detritus apparently proceeds much more rapidly in the Tropics than in temperate areas, confirming direct measurements of litter decomposition rates. External factors appear to influence ecosystem function through their effects on soil microorganisms rather than their effects on plants.

The relative transfer rates of phosphorus behave similarly, but they reveal another difference between the tropical and temperate forests. In the forest in Ghana, the assimilation coefficient ($g_1 = 11.9$) greatly exceeds the rate at which vegetation gives up nutrients ($g_2 = 1.1$) (for comparison, the values of g_1 and g_2 in the English pine plantation are 6.7 and 4.0, respectively). These values of g imply that, compared with *Pinus sylvestris,* tropical

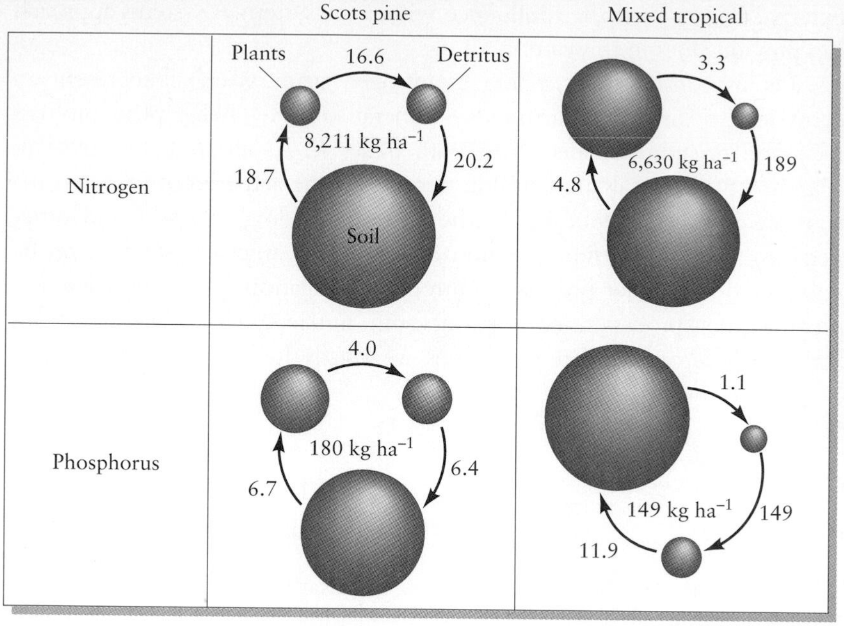

Figure 9.9 Compartment sizes and values of g_i for nitrogen and phosphorus estimated from compartment sizes in a Scots pine plantation in England and a tropical forest in Ghana. After data in J. D. Ovington, *Adv. Ecol. Res.* 1:103–192 (1962); D. J. Greenland and J. M. L. Kowal, *Plant Soil* 12:154–174 (1960).

forest trees assimilate phosphorus more efficiently and hold onto it more tightly (for example, by withdrawing phosphorus into the stem before shedding leaves). Therefore, the systems model is consistent with the higher nutrient use efficiency observed in tropical vegetation.

Energy and materials flow through ecosystems because plants, animals, and microbes must acquire energy and nutrients to grow and reproduce. As the models in this chapter have shown, variations in ecosystem function are influenced by the ways in which the activities of these organisms respond to variations in the physical environment and by the particular ways in which organisms are adapted to manage their energy and nutrient resources. In the preceding chapters, we have examined these aspects of ecology from the perspective of the ecosystem. In the next section of this book (Part 3), we will begin to develop an idea of ecology from the point of view of the organism, in which the prime considerations are survival and reproduction. The particular lifestyle of each kind of organism has implications for the flow of energy through the ecosystem and the cycling of nutrients within it, but from the organism's perspective, these considerations are secondary to the production of offspring. The principle of evolution by natural selection tells us that the criterion for success is number of descendants, not energy, although, all other things being equal, the two probably are closely related.

Summary

1. Ecologists use three methods to study the regulation of ecosystem function: comparison, experimentation, and modeling. We can interpret comparisons and experiments with greater confidence when their results are

compared with the predictions of models based on mechanisms of ecosystem function.

2. Systems models incorporate external factors, which influence the system from outside, and internal controls, which result from interactions among the components of a system.

3. Examination of terrestrial systems shows that primary production is strongly correlated with such external factors as temperature and precipitation. Hence, ecosystem function responds to the physical environment.

4. The strength of the correlation between ecosystem production and the cycling of a particular element indicates the degree to which transformation of that element limits ecosystem function overall. The concept of nutrient use efficiency (NUE) describes the relationship between production and nutrient assimilation.

5. Systems models based on the cycling of a particular element or energy consist of compartments (*X*s) and fluxes (*J*s) between those compartments. When a compartment achieves a steady state, input equals output. For an element cycling within the ecosystem in a steady state, fluxes through each compartment are equal. Each output from a compartment equals a transfer rate (g) times the compartment size. Each transfer rate may be a function of the attributes of one or more compartments, other properties of the system, and factors external to the system.

6. Alfred J. Lotka first applied systems models to ecological systems. He showed that the flux of a cycling element responds most sensitively to variation in the smallest transfer rate.

7. When a system exists in a steady state, relative transfer rates can be estimated from compartment sizes. In a simple model of nitrogen cycling within the water column of a lake, seasonal changes in the sizes of nitrogen compartments indicated that the rate of assimilation of nitrate decreased dramatically during the winter, perhaps as colder temperatures and lower light levels slowed algal growth.

8. A three-compartment model (soil, plants, and detritus) has been developed for forest ecosystems. Comparisons of compartment sizes in temperate and tropical forests indicate that nitrogen and phosphorus are regenerated from litter much more rapidly in the Tropics and that tropical trees, to a greater extent than temperate trees, withdraw phosphorus from leaves about to be shed.

Suggested readings

Carpenter, S. R., and J. F. Kitchell. 1988. Consumer control of lake productivity. *BioScience* 38:764–769.

Jeffers, J. N. R. 1978. *An Introduction to Systems Analysis, with Ecological Applications.* Edward Arnold, London, and University Park Press, Baltimore.

Kitching, R. L. 1983. *Systems Ecology: An Introduction to Ecological Modelling.* University of Queensland Press, St. Lucia, London, and New York.

Mann, K. H. 1982. *Ecology of Coastal Waters: A Systems Approach.* University of California Press, Berkeley.

Miller, P. C., W. A. Stoner, and L. L. Tieszen. 1976. A model of stand photosynthesis for the wet meadow tundra at Barrow, Alaska. *Ecology* 57:411–430.

Odum, H. T. 1983. *Systems Ecology: An Introduction.* Wiley, New York.

Perry, D. A., M. P. Amaranthus, J. G. Borchers, S. L. Borchers, and R. E. Brainerd. 1989. Bootstrapping in ecosystems. *BioScience* 39:230–237.

Vitousek, P. M. 1982. Nutrient cycling and nutrient use efficiency. *American Naturalist* 119:553–572.

Vitousek, P. M. 1984. Litterfall, nutrient cycling, and nutrient limitation in tropical forests. *Ecology* 65:285–298.

PART III

ORGANISMS

CHAPTER 10

HOMEOSTASIS AND LIFE IN VARYING ENVIRONMENTS

The relationships of organisms to their physical environments are the cornerstones of ecology. As we have seen, these relationships govern many pathways of energy and elements in ecosystems, as well as overall rates of ecosystem function. How well an organism functions within its physical environment also influences its probability of survival and rate of reproduction. These attributes of individuals—survival and reproduction—underlie the dynamics of populations, including natural selection and interactions among species. We shall consider these topics in detail later in this book. Here, our task is to understand how organisms live in environments that are both stressful and varied.

To varying degrees, organisms maintain themselves out of equilibrium with their physical environments with respect to heat, water, salts, and chemical energy. Keeping this state of imbalance—upholding gradients in concentrations of energy and substances between the body and the surroundings—requires metabolic work. It is costly in terms of energy. The degree to which an organism controls its internal environment is thus an economic decision, with costs and benefits measured in currencies of energy income and expenditure and, ultimately, survival and reproduction. Each organism's success in a particular environment depends on its adaptations of form and function. Any particular organism will perform better in some environments than in others. By the same reasoning, some organisms perform better than others in particular environments. Thus, each is specialized to function most efficiently and productively within a narrow range of environmental conditions. We have seen in the first part of this book how specialization results in restricted ecological and geographic distribution. Each type of organism has an **activity space,** which comprises habitats within which conditions and resources can sustain its population.

Complicating the relationship of every organism to its environment is the fact that conditions vary. Except perhaps for environments at great depths in the seas and in the farthest reaches of caves, the world is constantly changing. Witness the annual cycle of the seasons, the daily periods of light and dark, and the frequent unpredictable turns of climate familiar to us all. To the extent possible, organisms adjust their morphology and physiology to changes in their surroundings. Humans can be conscious of their own responses to change. When we step from a warm room into the outdoors on a cold day, we soon shiver to generate heat. A few days on the beach and our skin darkens, blocking some of the damaging radiation from the sun. Like other organisms, we respond to environmental change in such a way as to maintain our internal conditions within a suitable range for proper functioning. When changes are too rapid or too extreme, there may be no alternative to suffering the consequences of impaired function. But what determines the best internal conditions? At what rate should individuals function? How can organisms respond to environmental change most effectively?

Responses to environmental change, like a problem in economics, can be evaluated in terms of costs and benefits. When the environment cools, shivering helps to maintain our body temperature and ensure our survival. But shivering also uses metabolic energy, and thus may deplete fat reserves and render life more precarious in the face of a sudden food shortage. Some warm-blooded organisms reduce the costs of temperature regulation by lowering their internal temperatures, just as we turn down our thermostats to save fuel. But turning down the fire of an organism's life also renders it less active, and less able to gather food and avoid enemies. Other organisms, such as lizards, rely on external sources of heat to warm their bodies. Most animals and plants don't regulate their temperatures at all.

When we consider the many factors that affect the costs and benefits of a particular response, we begin to understand why different organisms regulate their internal conditions at different levels, or not at all, and why they employ different means of response to environmental change.

Homeostasis and negative feedback

Homeostasis is the ability of an individual to maintain constant internal conditions in the face of a varying external environment. All organisms exhibit homeostasis to some degree with respect to some environmental conditions, though the occurrence and effectiveness of homeostatic mechanisms vary. Regardless of how organisms regulate their internal environments, all homeostatic systems exhibit **negative feedback,** meaning that they counteract change and tend to return to a stably maintained state. Those of you who live in cold climates can readily understand how a negative feedback system works by considering the thermostat in your living room. When the room is cold, a temperature-sensitive switch turns on a heater, which restores the temperature to its desired setting. In accordance with the same general principle, when we move from a dark room into bright sunlight, the pupils of our eyes contract rapidly, restricting the amount of light entering the eye. Walking outdoors from an air-conditioned office on a hot summer afternoon may bring on sweating; evaporation of sweat draws heat away from the skin and helps to maintain body temperature at its normal level, just as the air conditioner draws heat from the office.

Such responses are forms of negative feedback: when a system deviates from its norm, or desired state, internal response mechanisms act to restore that state. The system may be changed from its norm either by external influences—that is, environmental conditions—or by internal influences resulting from changes in the physiological state of the organism. For example, activity is accompanied by increased metabolic production of heat, which must be dissipated by sweating or some other physical mechanism. In the blood, concentrations of oxygen decrease and carbon dioxide increases; these changes are countered by an increase in the rate of breathing, which increases gas exchange between blood and air in the lungs.

The essential elements of such negative feedback systems are (1) a mechanism that senses the internal condition of the organism; (2) a means of comparing the actual internal state with the desired state, or **set point;** and (3) an effector apparatus that alters the organism's internal condition in the direction of the preferred condition (Figure 10.1). In some cases, a negative feedback system exists within a single cell or is even a property of a single molecule. For example, as metabolic rate increases, oxygen concentrations in blood and tissues decrease; at the same time, carbon dioxide and hydrogen ion concentrations (from the formation of carbonic acid)

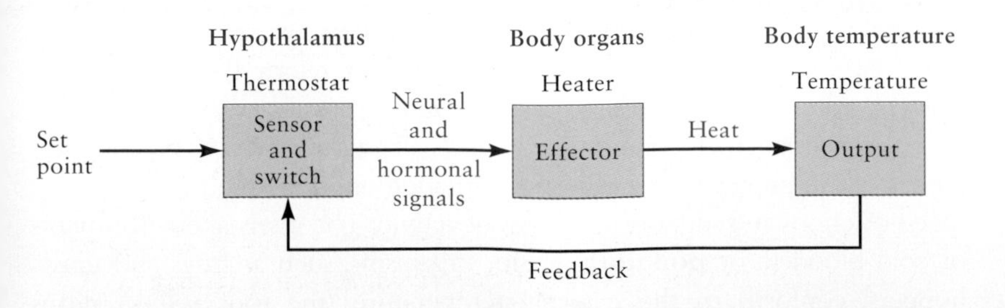

Figure 10.1 The major features of a negative feedback system, with particular components for the regulation of body temperature indicated in color. The thermostat compares body condition with a set point; when the two differ, it signals the effector organs to bring body condition back into line with the set point. In the case of body temperature regulation, the thermostat resides in the hypothalamus of the brain.

increase. The structure of the hemoglobin molecules in our red blood cells responds to increasing hydrogen ion concentration in such a way that the molecules hold oxygen less tightly and release it more readily to the blood plasma and tissues. Thus, hemoglobin has a mechanism built into its structure for increasing delivery of oxygen to tissues where it is needed to support metabolism.

In other cases, many parts of the body are involved in a single negative feedback system. For example, animals measure their body temperatures by taking the temperature of the blood flowing through the hypothalamus of the brain. Appropriate signals are sent out to various parts of the body by the nervous system, which stimulates the muscles to shiver and the skin to produce sweat, and by hormones secreted into the blood, which may increase the metabolic heat production of the visceral organs. Some control over heat exchange is retained at the local level, however. When an area of skin is exposed to cold, the blood vessels contract and restrict the flow of blood—and heat—to the surface. Responses to temperature changes also include behaviors: many small birds and mammals huddle together in cold weather; dogs curl up in a ball to reduce exposed surface area; we humans put on sweaters and coats.

Regulators and conformers

Regulators are organisms that maintain constant internal environments, whether we refer to temperature, pH, or solutes; **conformers** allow their internal environments to follow external changes (Table 10.1). Few organisms fit either ideal in all respects. Frogs regulate the salt concentration of their blood, but conform to external temperature. Even warm-blooded animals conform partially to ambient temperature; in cold weather our hands, feet, noses, and ears—our exposed extremities—become noticeably cool.

Regulation of body temperature

The ecological aspect of homeostasis can be simply and amply illustrated by regulation of body temperature. Temperature is one of the most important determinants of organism function because the rates of most biochemical processes increase rapidly with increasing temperature. Furthermore, changes in temperature represent changes in the amount of heat energy in a system, and therefore can be expressed in the currency of energy as metabolism and heat lost from surfaces of organisms when water evaporates.

Most mammals and birds maintain their body temperatures between 36° and 41°C, even though the temperature of their surroundings may vary from –50° to +50°C. Such regulation, which is referred to as **homeothermy** (the Greek root *homos* means "same"), creates constant temperature (homeothermic) conditions within cells, under which biochemical processes can proceed most efficiently. In addition, these high temperatures speed reactions and support high rates of activity. The internal environments of cold-blooded, or **poikilothermic,** organisms, such as frogs and grasshoppers, conform to the external temperature (the root *poikilo* means

TABLE 10.1 Terms associated with homeostasis, particularly with respect to body temperatures of organisms

Regulation	Maintenance of constant internal conditions in the face of varying environmental conditions (such organisms are regulators)
Conformance	Allowing internal conditions to vary in parallel with external conditions (such organisms are conformers)
Homeothermy	Maintenance of a constant body temperature, usually warmer than that of the environment (hence "warm-blooded")*
Endothermy	Use of elevated metabolism in response to body cooling to maintain homeothermy
Ectothermy	Reliance on external sources of heat (solar radiation, conduction of heat from warm surfaces) to maintain an elevated body temperature
Poikilothermy	Failure to regulate body temperature; hence, conformance to environmental temperature ("cold-blooded")*

*Note that homeothermic organisms may have to dissipate heat, usually by evaporative cooling, when environmental temperature exceeds preferred body temperature; also, although called "cold-blooded," poikilotherms may become very warm at high environmental temperatures.

"varying"). Of course, frogs cannot function at either high or low temperature extremes, so they are active only within a narrow part of the range of environmental conditions over which mammals and birds thrive.

Many so-called cold-blooded organisms, including reptiles, insects, and plants, adjust their heat balance behaviorally simply by moving into or out of shade or by changing their orientation with respect to the sun. When horned lizards are cold, they increase the surface area of their bodies exposed to the sun by lying flat against the ground; when hot, they decrease their exposure by standing erect upon their legs. By lying flat against the ground, horned lizards also gain heat by conduction from the sun-warmed surface. Such basking behavior, widespread among reptiles and insects, effectively regulates body temperature within a narrow range. The temperatures of these animals may rise considerably above that of surrounding air, well into the range of the "warm-blooded" birds and mammals. So long as lizards and snakes can gain heat from the sun, they can act as homeotherms. Because their source of heat lies outside the body, biologists refer to them as **ectotherms** (external heat); warm-blooded animals are referred to as **endotherms** (internal heat).

Energy costs of endothermy

Sustaining internal conditions that differ significantly from the external environment requires work and energy. Let us consider the costs to birds and mammals of maintaining constant high body temperatures in cold

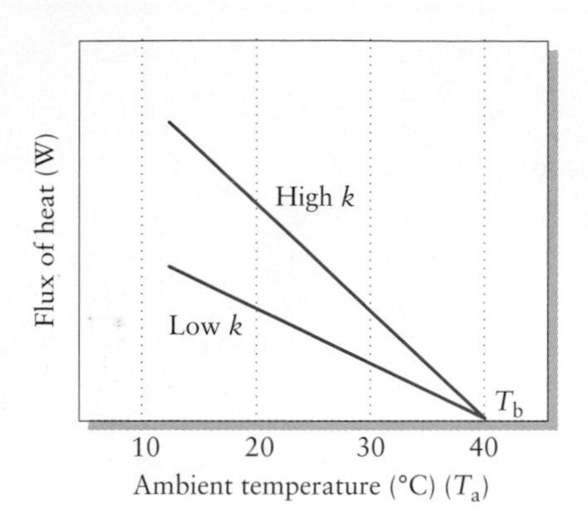

Figure 10.2 According to Newton's law of cooling, rate of loss of heat from an object increases in direct proportion to the difference in temperature between the object (T_b) and its surroundings (T_a). Thus, heat loss = $k(T_b - T_a)$, where k is a constant that depends on the conductance across the object's surface. Heat loss curves for high and low conductances are compared. When T_a exceeds T_b, the object gains heat from the environment (not shown).

environments. As air temperature decreases, the gradient (difference) between internal and external environments increases. Heat is lost across body surfaces in direct proportion to this gradient. An animal that maintains its body temperature at 40°C loses heat twice as fast at an **ambient** (surrounding) temperature of 20°C (a gradient of 20°C) as at an ambient temperature of 30°C (a gradient of only 10°C). The principle that heat loss varies in direct proportion to the gradient between body and ambient temperature is called **Newton's law of cooling** because it was first formulated by the physicist Isaac Newton. To maintain a constant body temperature, endothermic organisms replace heat lost to their environment by generating heat metabolically. Thus, the rate of metabolism required to maintain body temperature increases in direct proportion to the difference between body and ambient temperature, all other things being equal (Figure 10.2).

If the only purpose of metabolism were temperature regulation, individuals would require no metabolic heat production when body temperatures equaled ambient temperature, at which point no heat would flow between organisms and their surroundings. But organisms generate metabolic heat while sustaining basic physiological processes, such as heartbeat, breathing, muscle tone, and kidney function, regardless of ambient temperature. The metabolism of an organism resting quietly and without food in its digestive tract (postabsorptive condition) achieves a **basal,** or **resting, metabolic rate (BMR,** or **RMR).** BMR represents the lowest level of energy use under normal conditions. At this basal rate, the individual produces sufficient heat to maintain its body temperature when ambient temperatures remain above a certain point, referred to as the **lower critical temperature** (T_{lc}). Below the lower critical temperature, the individual must increase its metabolism to balance increased heat loss. The lower critical temperature depends on BMR and thermal conductance. As body size increases, basal metabolic rate increases more rapidly than body surface area, and thicker fur and feathers tend to reduce thermal conductances. As a result, T_{lc} decreases with increasing size—for example, from about 30°C in sparrow-sized birds to below 0°C in penguins and other large species.

An organism's ability to sustain a high body temperature while exposed to extremely low ambient temperatures reaches limits set over the short term by its physiological capacity to generate heat and over the long term by its ability to gather food or metabolize nutrients to satisfy the energy requirements of generating heat (Figure 10.3). The maximum rate at which an organism can perform work generally does not exceed ten to fifteen times its basal metabolism even during the most strenuous exercise, such as that of a bird in powered flight. Such rates of metabolism are rarely maintained for more than a few minutes or hours. Over the course of a day, few organisms—even migrating birds—expend energy at a rate exceeding four or five times BMR.

When the environment becomes so cold that heat loss exceeds an organism's physiological capacity to produce heat or its capacity to make metabolic energy available for heat-producing organs (point *c* in Figure 10.3), body temperature begins to drop, a condition that is fatal to most homeotherms. The lowest environmental temperatures that homeotherms

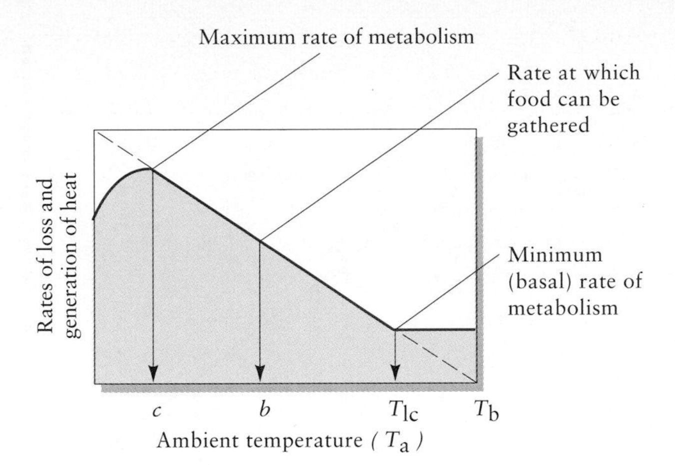

Figure 10.3 Relationship between metabolic rate and ambient temperature for a homeothermic bird or mammal whose body temperature is maintained at T_b. T_{lc} is the lower critical temperature, below which metabolism must increase to maintain body temperature. Point *c* is the lowest temperature at which an organism can sustain metabolic energy production for temperature regulation, and point *b* is the lowest temperature at which an organism can maintain itself indefinitely in its natural environment, limited by its ability to procure food.

can survive for long periods often depend on their ability to gather food (point *b*) rather than on their ability to assimilate and metabolize the energy in food. At low temperatures, animals may starve rather than freeze to death when they metabolize food energy to maintain body temperature more rapidly than they can gather food.

When homeothermy requires more energy than an individual can provide, certain "economy measures" are available. For example, the regulated temperatures of portions of the body may be lowered, reducing the temperature difference between air and body. Because the legs and feet of most birds do not have feathers, they would be major avenues of heat loss in cold regions were they not held at a lower temperature than the rest of the body (Figure 10.4). Gulls conserve heat by using countercurrent heat exchange in their legs. Warm blood in arteries leading to the feet cools as it passes close to veins that return cold blood to the body. In this way, heat is transferred from arterial to venous blood and transported back into the body rather than lost to the environment.

Because they are small, hummingbirds expose a large surface area relative to their weight and consequently lose heat rapidly compared with their ability to produce it. As a result, hummingbirds must sustain very high

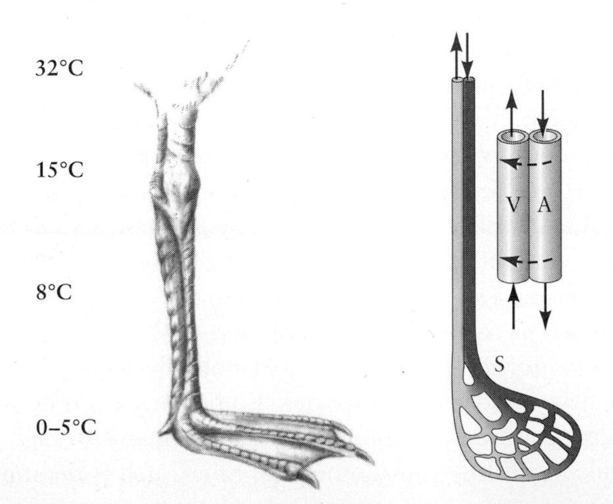

Figure 10.4 Skin temperatures of the leg and foot of a gull standing on ice. The anatomic arrangement of blood vessels and countercurrent heat exchange between arterial blood (A) and venous blood (V) are diagrammed at right. Arrows indicate direction of blood flow, and dashed arrows indicate heat transfer. A shunt at point S between the artery and vein in the leg allows blood vessels in the feet to constrict, thereby reducing blood flow and heat loss further, with no increase in blood pressure. After L. Irving, *Sci. Am.* 214:93–101 (1966); K. Schmidt-Nielsen, *Animal Physiology,* Cambridge University Press, London and New York (1983).

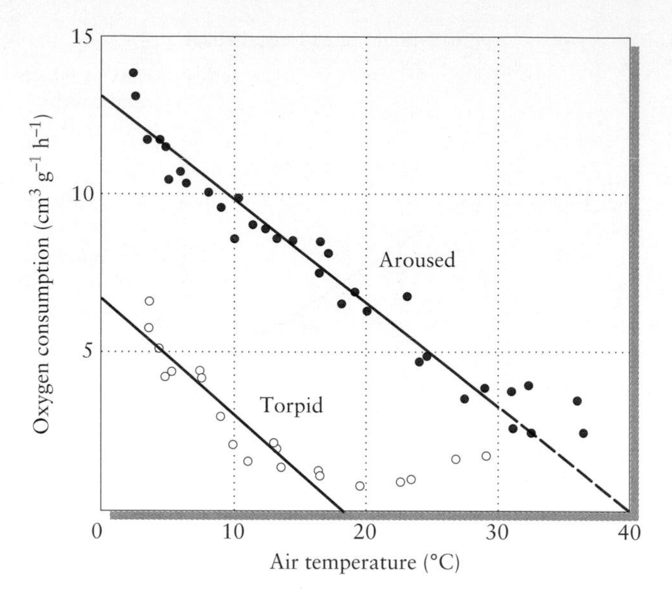

Figure 10.5 Relationship between energy metabolism and air temperature in the hummingbird *Eulampis jugularis* during periods of torpor and normal arousal, illustrating endothermy in each case, but with body temperature regulated at different set points. After F. R. Hainsworth and L. L. Wolf, *Science* 168:368–369 (1970).

metabolic rates to maintain their at-rest body temperatures near 40°C. Species inhabiting cool climates would risk starving overnight if they did not enter **torpor,** a voluntary, reversible condition of lowered body temperature and inactivity. The West Indian hummingbird, *Eulampis jugularis,* drops its body temperature to between 18° and 20°C when resting at night. It does not cease to regulate body temperature; it merely changes the setting on its thermostat to reduce the difference between ambient and body temperature, and thereby reduces the energy expenditure needed to maintain its temperature at the regulated set point (Figure 10.5).

Partial homeostasis

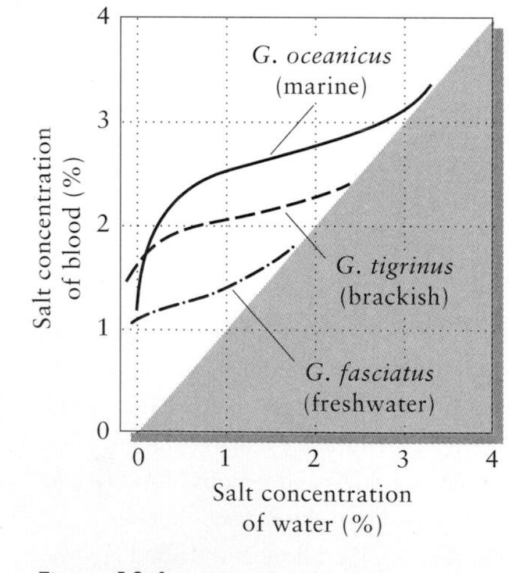

Figure 10.6 Salt concentrations in the blood of three gammarid crustaceans from different habitats as a function of the salt concentration of their external environment. The normal salt concentration of seawater is 3.5%. After C. L. Prosser and F. A. Brown, *Comparative Animal Physiology,* 2d ed., Saunders, Philadelphia (1961).

Some organisms regulate their internal environments over moderate ranges of external conditions but conform under extremes. Small aquatic amphipods of the genus *Gammarus* control the salt concentrations of their body fluids when placed in water with a higher salt concentration than their blood, but not when placed in water with a lower concentration (Figure 10.6). The freshwater species *G. fasciatus* regulates the salt concentration of its blood at a lower level than the saltwater species *G. oceanicus* and thus begins to conform to concentrated salt solutions at a lower level. In their natural habitat, however, neither the freshwater species nor the saltwater species encounters water with a salt concentration higher than that in its blood, so the absence of regulation in this range is not surprising. Animals that inhabit salt lakes and brine pools, by contrast, actively maintain the salt concentrations of their blood below that of the surrounding water. The brine shrimp *Artemia* maintains its internal salt concentration below 3%, even when placed in a 30% salt solution.

All birds and mammals generate heat metabolically to regulate body temperature, but many cold-blooded species also become endothermic or partially endothermic at times. Pythons, for example, maintain high body temperatures while incubating eggs. Some large fishes, such as the tuna, use a countercurrent arrangement of blood vessels to maintain temperatures up to 40°C in the center of their metabolically active muscle masses; swordfish employ specialized metabolic heaters, derived from muscle tissue, to keep their brains warm. Large moths and bees often require a preflight warm-up period during which the flight muscles shiver to generate heat. Even among plants, temperature regulation based on metabolic heat production has been discovered in the floral structures of philodendron and skunk cabbage.

Clearly, organisms other than birds and mammals can generate heat and maintain elevated body temperatures, and many do so under certain conditions. Why, then, is endothermy so rare throughout the animal and plant kingdoms? Part of the answer certainly lies in body size. Birds and mammals are relatively large as animals go. As body size increases, volume increases more rapidly than the surface area, across which heat leaves the body. In general, the lower the ratio of surface area to volume, the more comprehensive and precise temperature regulation can be.

Although body size may explain why mammals are hot and insects are not—large moths that exhibit preflight warm-up approach the size of small mammals and birds—it does not explain why most large fishes and reptiles have not made the shift to homeothermy. Water contains too little oxygen to support a high rate of metabolism, and it conducts heat away from the body too rapidly for endothermy to be practical for most fishes. Reptiles have evolved endothermy several times—that is to say, birds and mammals evolved from reptilian ancestors—but poikilothermic and ectothermic reptiles persist. Why wasn't the transition to homeothermy complete? After all, many contemporary reptiles are comparable in body size to small mammals, and they breathe air. Also, many reptiles are effective ectothermic homeotherms, so they possess the necessary feedback control mechanisms. Undoubtedly, the lower energy requirements of ectothermy confer advantages in certain environments, particularly where food supplies are highly seasonal or erratic, or too sparse or of too low a quality to sustain endothermy. The coexistence of reptiles with birds and mammals illustrates that there may be many different solutions to problems posed by the environment.

Activity space and microhabitat selection

Each organism functions best within a limited range of conditions, which we may refer to as its **activity space,** whether it is active or not. This concept applies to all aspects of an individual's life. Seeds, for example, require specific combinations of light, temperature, and moisture for proper germination, and these conditions vary even among closely related plant species. When seeds arrive at suitable sites, the seedlings survive and grow; elsewhere, they die. Irregularities in the surfaces of natural soils provide the variety of conditions needed to allow the germination of many species. An experiment in which investigators created a heterogeneous soil environment dramatized the differences in germination requirements between closely related species. Plantains are common lawn and roadside weeds in the genus *Plantago.* Three species sown in experimental seedbeds responded differently to modifications in the environment produced by slight depressions, by squares of glass placed on the soil surface, and by vertical walls of glass or open-topped wooden boxes (Figure 10.7). Relatively few seeds germinated on the smooth surface of soil that had not been modified in some way by one of these treatments.

Unlike plants, most animals can choose where they live. In many environments, the diurnal behavior cycles of lizards respond to the varying temperatures of habitat patches. Although lizards do not regulate their body temperatures by generating heat metabolically, they do take advantage of solar radiation and warm surfaces to maintain their temperatures within a suitable range. At night, these sources of heat disappear, and the lizard's body temperature gradually drops to that of the surrounding air.

The mallee dragon *(Amphibolurus fordi),* a medium-sized (50 mm snout-vent length) Australian lizard that is by no means a dragon, is fully active only when its body temperature ranges between 33° and 39°C. In the early morning, before its temperature has risen above 25°C and when it still moves sluggishly, a dragon will bask within a large clump of grass of the genus *Triodia,* where it enjoys protection from predators (Figure 10.8). At this time of day, the *Triodia* clump is part of the dragon's activity space. When the dragon's temperature rises above 25°C, it leaves the *Triodia* clump and basks in the sunshine nearby, with its head and body in direct contact with the ground surface, from which it absorbs additional heat.

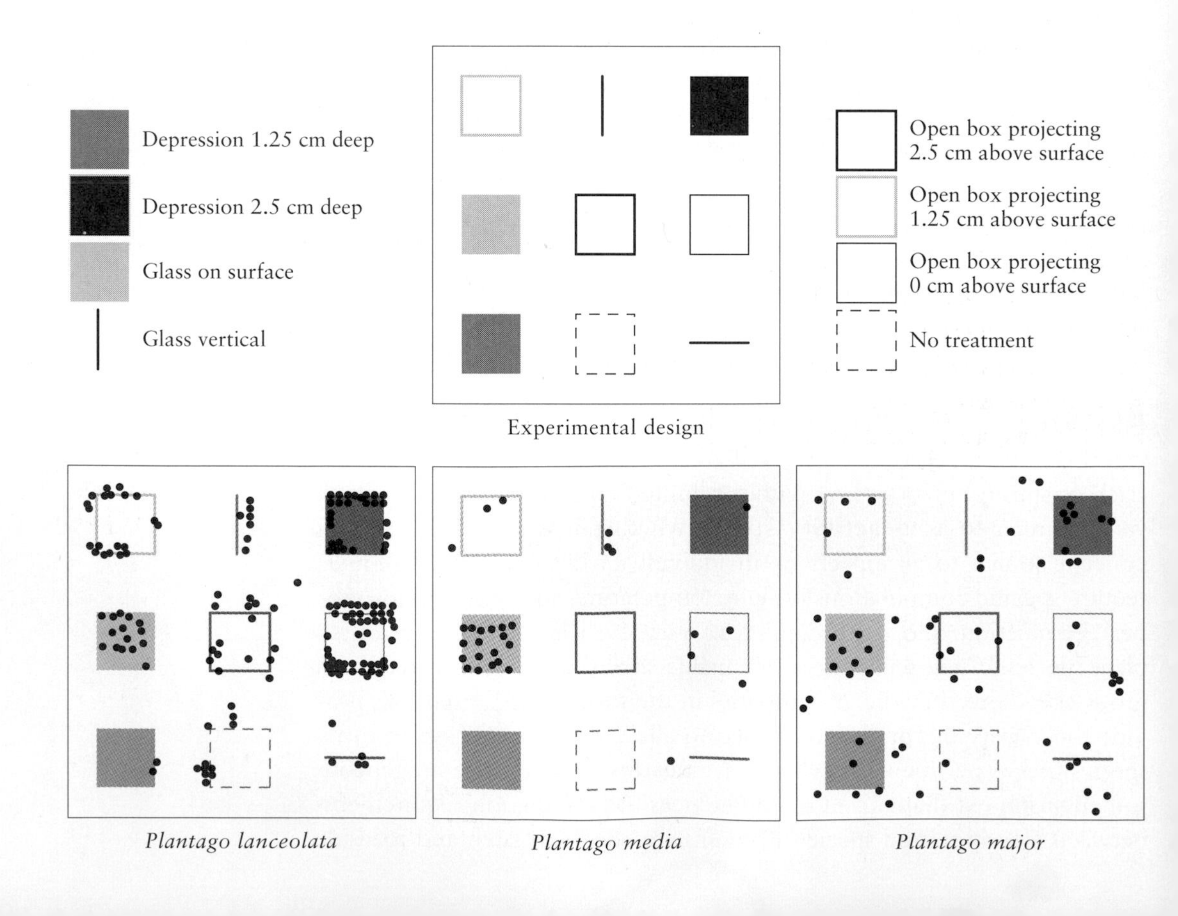

Figure 10.7 Germination of seedlings of three species of plantains (genus *Plantago*) with respect to artificially produced variation in conditions at the soil surface. The activity spaces for seed germination include different soil surface conditions for each species. After J. L. Harper, J. T. Williams, and G. R. Sagar, *J. Ecol.* 51:273–286 (1965).

(a)

(c)

(b)

Figure 10.8 The mallee dragon *(Amphibolurus fordi)* at different times during its activity cycle. (a) Early morning basking in a *Triodia* grass clump. (b) Midmorning basking on the ground (note that the body is flattened against the surface to increase the exposed profile as well as the animal's contact with warm soil). (c) Normal foraging attitude. Courtesy of H. Cogger. From H. Cogger, *Aust. J. Zool.* 22:319–339 (1974).

When its body temperature enters the range for normal activity, the dragon ventures farther from *Triodia* clumps to forage, its head and body normally raised above the ground as it moves. When its body temperature exceeds 39°C, it moves less rapidly to reduce metabolically produced heat and seeks the shade of a small *Triodia* clump. Above 41°C, it reenters a large *Triodia* clump, where it finds cooler temperatures and deeper shade at the center. It may also pant to dissipate heat by evaporation. If heat stress continues unabated, the dragon will lose locomotor ability above 44°C, and it will die if its body temperature exceeds 46°C. Thus, a dragon's activity space at a particular time depends on its own body temperature; this example emphasizes the fact that activity spaces are determined by the interactions of organisms with their environments and thus are sensitive to the tolerance ranges of organisms.

On a typical summer day, during which air temperature varies from about 23°C at dawn to 34°C at midday, the mallee dragon does not begin to forage until about 8:30 A.M. By 11:30 the habitat has become too hot for normal activity, and most individuals seek shade and become inactive.

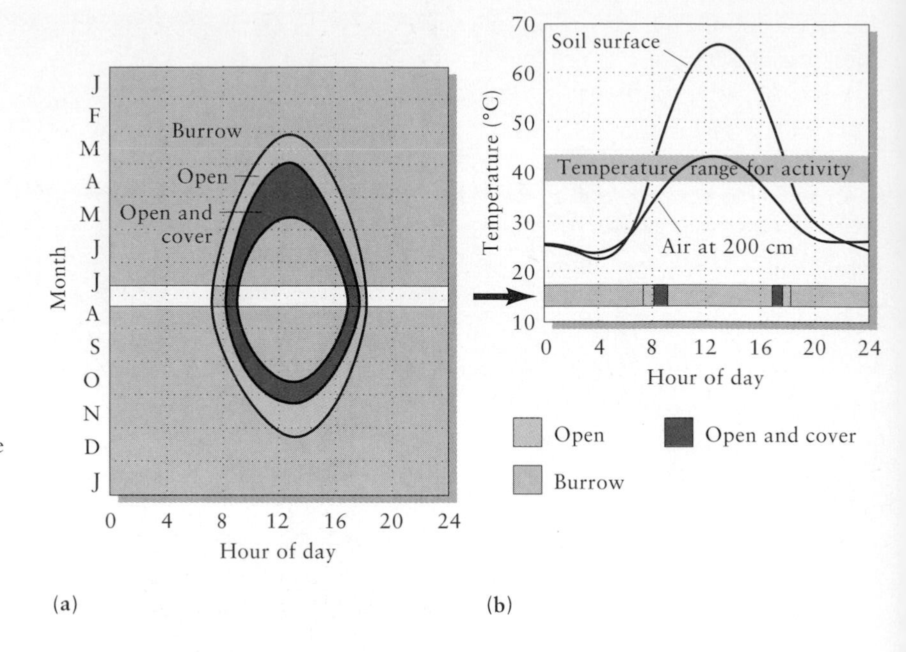

Figure 10.9 Seasonal activity space of the desert iguana *(Dipsosaurus dorsalis)* in southern California. (a) The daily activity budget for an entire seasonal cycle. (b) The activity budget for July 15 is shown with the time course of environmental temperature. After W. A. Beckman, J. W. Mitchell, and W. P. Porter, *J. Heat Transfer* (May 1973):257–262.

By 2:30 P.M. the habitat has cooled off enough for the dragons to resume foraging, but by 6:00, declining evening temperatures force them back into *Triodia* clumps before their bodies cool too much. Any individuals that remain in the open after this time of day move sluggishly and are easily caught by warm-blooded predators.

The desert iguana *(Dipsosaurus dorsalis)* of the southwestern United States faces an environment more severe than that of the mallee dragon, with a greater annual temperature fluctuation. Shade temperatures can reach 45°C in summer and plunge below freezing in winter. During mid-July, the thermal environment cycles so rapidly between day and night extremes that desert iguanas can move about within their preferred body temperature range of 39°–43°C for only about 45 minutes in mid-morning and a similar period in the early evening (Figure 10.9). During the remainder of the day, they seek the shade of plants or the coolness of their burrows, where temperatures rarely rise above the preferred range. At night desert iguanas enter their burrows, where they find safety from predators; at dawn, the burrows are warmer than the desert surface, so the early morning warm-up period is correspondingly brief.

In summer, desert iguanas restrict their activity to two brief bouts separated by inactivity through midday to avoid heat stress. Spring offers more favorable temperatures for activity. The thermal environment in May does not exceed the desert iguana's preferred range, and individuals forage actively on the ground surface from a businesslike 9:00 A.M. to 5:00 P.M., only occasionally seeking the cool shade of plants. Winter cold restricts *Dipsosaurus* to brief periods of activity in the middle of the day when body temperatures rise to the point at which individuals can come aboveground and forage. Between early December and the end of

February, most days are so cold that desert iguanas cannot venture from their burrows.

Subdivisions of the habitat that exhibit distinctive attributes of structure, temperature, salinity, and so on are referred to as **microhabitats** or **microenvironments.** In deserts, for example, the shaded ground under a shrub offers cooler, moister conditions than surrounding areas exposed to direct sunlight. As we have seen in the case of the mallee dragon, an individual may include various microhabitats within its activity space or not, depending on conditions in the microhabitats and its own body condition. Unlike the dragon, the cactus wren, an insectivorous bird that lives in deserts of the southwestern United States and northern Mexico, maintains a constant body temperature. Thus, its activity space reflects changes in environmental conditions in various microhabitats as these change throughout the day and season. Cactus wrens seek favorable conditions within which to feed as the thermal environment changes throughout the day **(microhabitat selection).** In the cool temperatures of the early morning, wrens forage throughout most of the environment, searching for food among foliage and on the ground. As the day brings warmer temperatures, wrens select cooler parts of their habitat, particularly the shade of small trees and large shrubs, always managing to avoid feeding where the temperature of the microhabitat exceeds 35°C (Figure 10.10). When the minimum temperature in the environment rises above 35°C, at which point birds must use evaporative cooling to maintain their body temperatures even when inactive, the wrens stop feeding and perch quietly in deep shade.

Many desert birds build enclosed nests or place their nests in holes in the stems of large cacti, where the young are protected from the sun and from extremes of temperature. Cactus wrens build untidy nests—bulky,

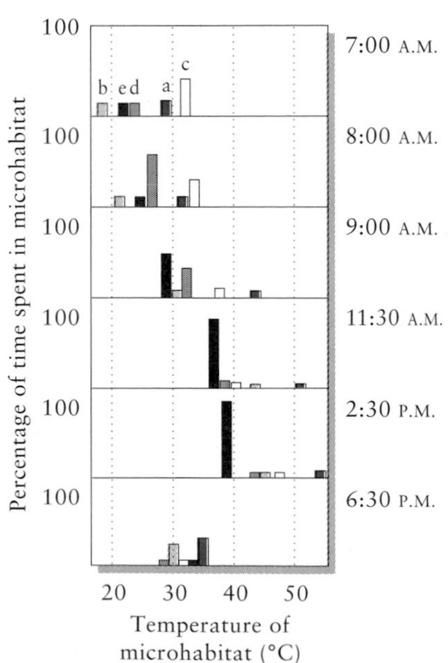

Figure 10.10 Microhabitat use by cactus wrens *(Campylorhynchus bruneicapillus)* in southeastern Arizona during the course of a day in late spring. Microhabitats vary in degree of thermal stress between exposed ground (a) and the deep shade of trees (e). Wrens distribute their activity among all microhabitats in the cool hours of early morning (7:00 A.M.) but restrict their activity to cool shade (e) during the hottest part of the day (2:30 P.M.), when other microhabitats are above 40°C. From R. E. Ricklefs and F. R. Hainsworth, *Ecology* 49:227–233 (1968).

somewhat haphazardly constructed balls of grass—with side entrances. Once built, of course, the position and orientation of a nest cannot be changed. For a month and a half, from the laying of the first egg until the young fly off, a nest must provide a suitable environment day and night, in hot and cool weather. During the long breeding period (March through September) in southern Arizona, cactus wrens usually rear several broods of young. They build their early nests so that the entrances face away from the direction of the cold winds of early spring; during the hot summer months, they orient their nests to face prevailing afternoon breezes, which circulate air through the nest chamber and facilitate heat loss (Figure 10.11). It makes a difference! Nests oriented properly for the season are consistently more successful (82% produce viable offspring) than nests facing in the wrong direction (45% do so).

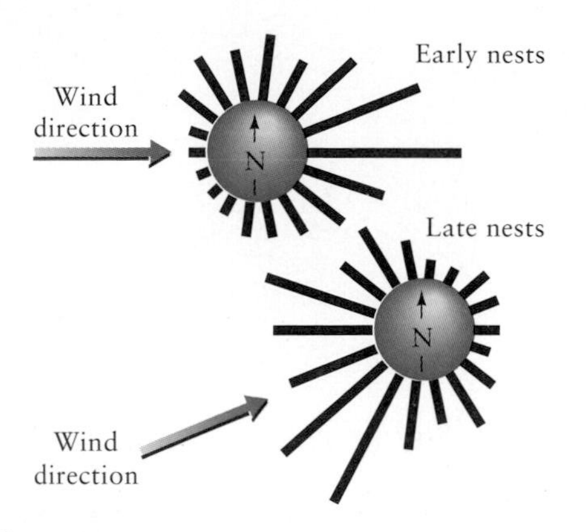

Figure 10.11 Orientation of nest entrances of the verdin *(Auriparus flaviceps)* during early (cool) and late (hot) parts of the breeding season in Nevada and Arizona. The orientation of cactus wren nest entrances is similar. Lengths of bars represent relative numbers of nests with each orientation. Data courtesy of G. T. Austin.

Structural responses to environmental change

The suitability of the environment for each organism varies over time and space. At any given moment, cactus wrens may feed within any of several parts of their habitat, each offering different thermal characteristics and access to food. These distinctive patches of microhabitat shift continuously throughout the daily cycle of solar radiation and temperature, and they change seasonally as both the physical environment and populations of suitable prey vary. Behaviors such as microhabitat selection enable animals to respond quickly to changing conditions. Nest placement and orientation undergo "change" much more slowly; the cactus wren can make such structural changes only between broods. The cactus wren's body also undergoes structural changes in response to environmental variation: a new plumage is produced twice each year, thicker in winter, thinner in summer; birds fatten up a bit in winter, just in case food supplies decrease; the size of the gut adjusts to changes in diet. Such changes, which characterize all plants and animals, involve modifications of the body's structure and metabolic machinery, and thus take time.

Appropriate responses to change depend on the rate and persistence of change and on the rate and cost of response. Responses must be substantially quicker than environmental change; otherwise, today's form and function may be well suited only to yesterday's conditions. The most rapid responses involve behavior and changes in rates of physiological processes; examples include the homeotherm's shade-seeking behavior under heat stress and its elevation of metabolism in response to cold stress. Such responses do not require modification of existing morphology or biochemical pathways. Slower changes in the environment may be accommodated by slower, structural changes in the organism. Biologists recognize two types of structural responses: Acclimation (often referred to as acclimatization) is the faster of the two and is reversible. Developmental responses are slower, resulting from modifications of growth and development, and generally are not reversible.

Acclimation

Acclimation is a reversible change in structure that helps to maintain homeostasis in response to environmental change. Growing thicker fur in winter, increasing the number of red cells in blood at high altitude, and producing enzymes with different temperature optima or lipids that remain fluid at different temperatures are all forms of acclimation. Such changes take days to weeks, so acclimation is a strategy restricted to seasonal and other persistent variations in conditions. These changes may be thought of as shifts in the ranges of behavioral and physiological responses of the individual. Acclimation is reversible, as it must be to follow the ups and downs of the environment.

By producing enzymes and other molecules having temperature optima that correspond to the temperatures experienced, a cold-blooded animal can adjust its tolerance range, or activity space, in response to prevailing environmental conditions. The relationship between the swimming speed of goldfish and water temperature shows the advantages and the limitations of acclimation. Goldfish swim most rapidly when acclimated to 25°C and placed in water between 25° and 30°C, conditions that closely resemble their natural habitat (Figure 10.12). Lowering the acclimation temperature to 5°C increases the swimming speed at 15°C but reduces it at 25°C. Increased tolerance of one extreme often brings reduced tolerance of the other. Acclimation takes many days, but as long as temperature change is persistent, it is a good tactic.

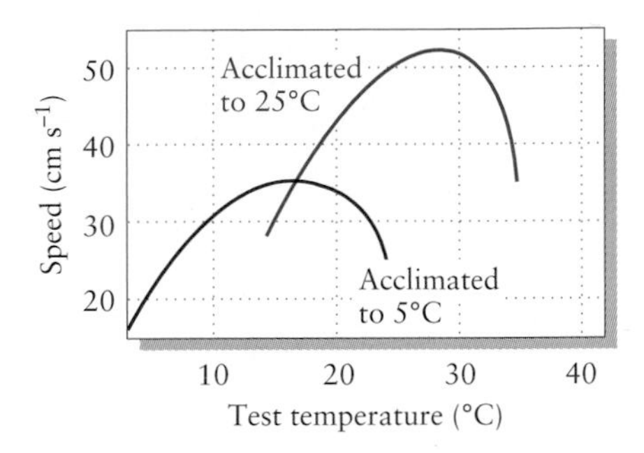

Figure 10.12 Swimming speed of goldfish as a function of temperature. Separate curves are shown for individuals acclimated to 5°C and to 25°C. After F. E. J. Fry and J. S. Hart, *J. Fish. Res. Bd. Can.* 7:169–175 (1948).

When goldfish are exposed to temperatures approaching their upper lethal limits, they exhibit progressive stages of behavioral abnormality, passing from hyperexcitability to loss of equilibrium and, finally, coma. The temperatures at which these syndromes appear can be shifted by acclimation over periods of 2 to 6 weeks, as shown in Figure 10.13. These neurological syndromes depend upon the fluidity of lipids in the membranes of nerve synapses, which apparently affects transmission of nerve impulses to muscles. Acclimation involves the rate of exchange of lipid components of the membrane, which, as one would expect, is lower at 5°C than at the higher acclimation temperature of 25°C because biological processes, including lipid biosynthesis, run more slowly at colder temperatures.

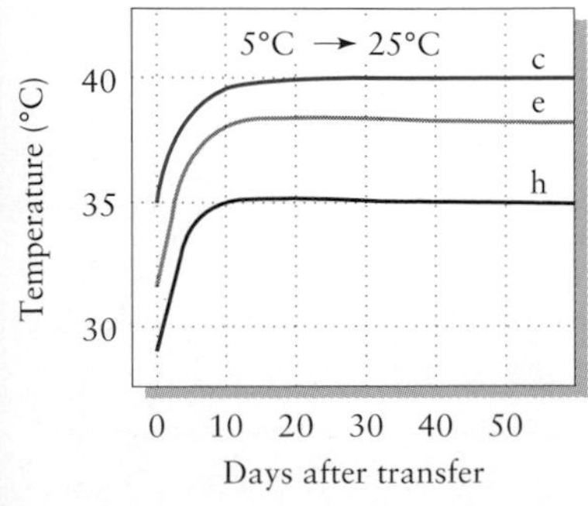

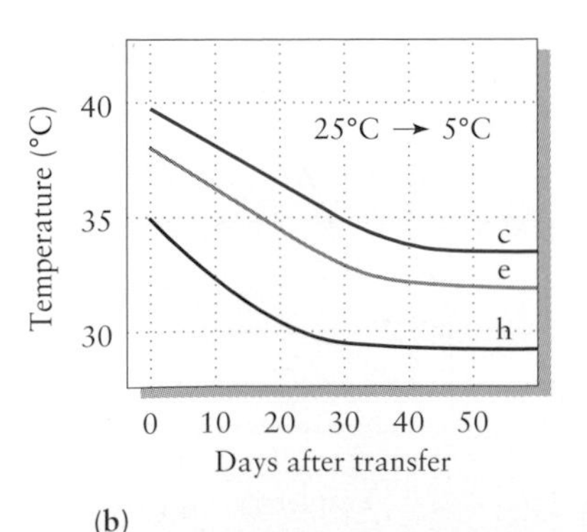

Figure 10.13 Time courses of acclimation by goldfish to high-temperature stress for three behavioral indices: hyperexcitability (h), loss of equilibrium (e), and coma (c). Individuals were first acclimated to either (a) 5°C or (b) 25°C and then transferred to the other temperature. To test temperature tolerance, individual fish were briefly exposed to high temperatures after differing periods of acclimation to the new temperature. From A. R. Cossins, M. J. Friedlander, and C. L. Prosser, *J. Comp. Physiol.* 120:109–121 (1977).

Acclimation of photosynthetic rate to temperature shows that a plant's capacity for acclimation often reflects the range of temperatures experienced in the natural environment (Figure 10.14). *Atriplex glabriuscula,* a species of saltbush native to cool coastal regions of California, does not increase its photosynthetic rate at high temperatures when acclimated to 40°C; individuals acclimated to 16°C, within the range of temperature the species normally experiences, are uniformly more productive at all temperatures. Presumably, changes stimulated by growth at 40°C may benefit the plant in ways other than photosynthetic rate. The heat-loving **(thermophilic)** species *Tidestromia oblongifolia* cannot acclimate to cool temperatures. Its temperature optimum shifts to a lower temperature, but photosynthesis declines over a wide range of leaf temperatures from 10° to 40°C. *Larrea divaricata* (creosote bush), a species that inhabits interior deserts but maintains photosynthetic activity during the cool winters as well as the hot summers, shows the shift in temperature optimum characteristic of thermal acclimation. The basis for this acclimation appears to be associated with changes in the viscosity of membranes directly related to photosynthetic pathways. That some plant species lack this capability suggests that mechanisms to acclimate photosynthetic rate to temperature entail a cost to individuals and that these mechanisms have been dispensed with when plants experience only narrow ranges of temperatures.

Developmental responses

When the environment changes slowly relative to life span, an individual may experience relatively uniform conditions throughout its life, although they may differ from the conditions experienced by individuals in earlier or later generations. Spring and summer generations of insects are exposed to consistently different environmental conditions. In this case, the conditions experienced by immature individuals are predictive of their environment as adults, and an individual may alter its development accordingly so

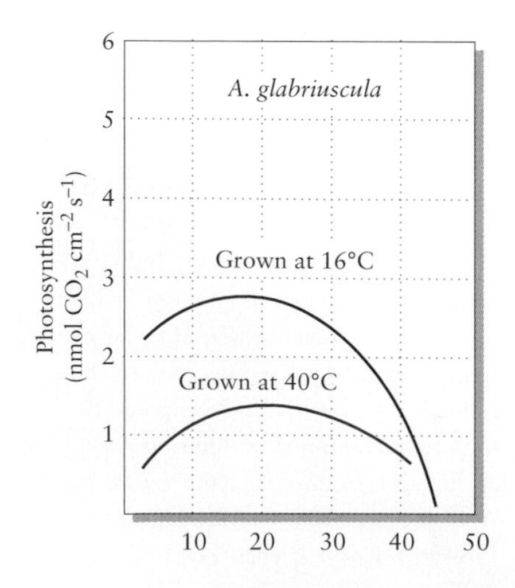

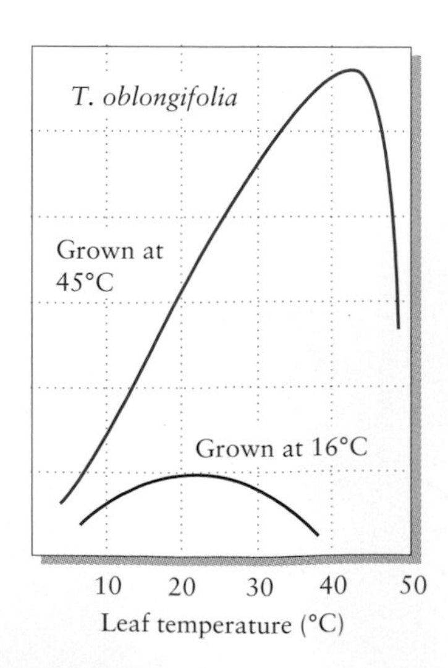

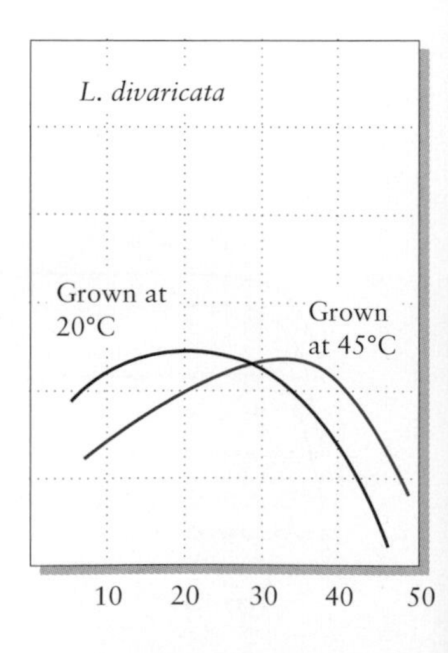

Figure 10.14 Light-saturated photosynthetic rate as a function of leaf temperature in three species of plants (genera *Atriplex, Tidestromia,* and *Larrea*) grown under moderate and hot temperatures. From P.W. Hochachka and G. N. Somero, *Biochemical Adaptation,* Princeton University Press, Princeton, N.J. (1984); after O. Bjorkman, M. R. Badger, and P.A. Arnold, in N. C. Turner and P.J. Kramer (eds.), *Adaptation of Plants to Water and High Temperature Stress,* Wiley, New York (1980), pp. 231–249.

TABLE 10.2 Distribution of dry matter and rates of photosynthesis in loblolly pine *(Pinus taeda)* seedlings grown under shade and in full sunlight

CHARACTERISTIC	SHADE-GROWN	SUN-GROWN
Percentage of dry weight, 6-month-old seedlings		
Roots	35	52
Needles	47	37
Stems	18	11
Photosynthetic rate, 4-month-old seedlings (mg CO_2 h^{-1} g^{-1})		
Low light intensity (500 fc)*	1.9	1.0
Moderate light intensity (1,500 fc)	4.6	4.0
High light intensity	7.2	6.6

*fc = footcandles.

Source: After F. H. Bormann, in D.V. Thimann (ed.), *The Physiology of Forest Trees,* Ronald Press, New York (1958), pp. 197–215.

that as an adult, its structure and function are most suitable to prevailing conditions. Such **developmental responses** require long periods relative to the life spans of individuals and generally do not reverse themselves; once fixed during development, they remain unchanged for the rest of an individual's life. Because of their long response times and irreversibility, developmental responses cannot accommodate short-term environmental changes. As a general rule, therefore, only plants and animals in environments with persistent variation in the conditions experienced by individuals exhibit developmental flexibility. Such organisms include plant species whose seeds may settle in many different kinds of habitats. Thus, spatial heterogeneity in the environment may create the kind of persistent environmental variation for individuals that favors developmental responses.

Light intensity, among many other factors, influences the course of development in plants. We saw in an earlier chapter how white oak leaves that grow in sun develop a highly lobed leaf outline that helps them to dissipate heat. Loblolly pine seedlings grown in shade have smaller root systems and more foliage than seedlings grown in full sunlight. Because shaded environments tax a plant's water economy less, shade-grown seedlings can allocate more of their production to stem and needles; sun-grown seedlings must develop more extensive root systems to obtain sufficient water. The larger proportion of foliage in shade-grown seedlings results in a higher rate of photosynthesis per unit of plant mass under given light conditions, particularly under low light intensities (Table 10.2).

A striking example of developmental response is the coloration of several species of locusts and grasshoppers. In tropical habitats with seasonal precipitation, the onset of the wet season brings growth of lush, green vegetation. During the early part of the dry season this vegetation browns and dies, often exposing red-brown earth. As the seasonal drought intensifies, natural fires and those set by humans blacken the ground over vast areas. As a result, there is a regular seasonal progression of color in the dry Tropics from green to brown to black and back to green again. Grasshoppers develop so rapidly that the life span of each individual may coincide largely with one background color. And many species of grasshoppers do, in fact, match the background coloration of their environment closely, even as it changes with season.

Coloration in the grasshopper *Gastrimargus africanus* responds to environmental conditions, particularly quality and intensity of light and humidity. The epidermis (outer skin) of a grasshopper has a pigment system that permits any given area of skin to be either green or brown; both colors may occur on a single animal, but not in the same area of the body. Green and brown colors represent small biochemical variations on a single pigment molecule. Where brown pigment occurs in the epidermis, additional pigments may produce colors ranging from yellow through orange and red to black. Furthermore, black pigment (melanin) may be deposited in the cuticle that covers the epidermis.

The color response in *Gastrimargus* is most pronounced in the early developmental stages; adults have a more characteristic and unvarying pattern of color distribution. Between developmental stages a grasshopper sheds its epidermis, discarding its pattern of camouflaging coloration with it. A new layer of epidermis develops underneath, and thus a grasshopper can change its color with each molt. High-intensity light, a characteristic of the dry season, during which much of the vegetation has died back, leads to predominance of brown and black pigment systems. The low-intensity sunlight and high humidity that prevail during the rainy season increase the proportion of green coloration. When black ash covers a recently burned area, the soil reflects little incident light, a condition that stimulates the black pigment system in grasshopper nymphs feeding on newly sprouting vegetation. These environmental signals are perceived by the eye and transmitted to the epidermis by hormones produced in the brain. The sensitivity of pigment systems in the developing epidermis to prevailing conditions of light and humidity thus enables a grasshopper to match its coloration to the conditions experienced during its growth period.

Strategies and mechanisms in response to environmental change

Temporal variation in the environment requires a flexible response, the nature of which depends on the time course, predictability, and amplitude of the variation. Observed responses generally match change rather well, because individuals that respond inappropriately don't survive to reproduce themselves. To the extent that mechanisms of response are genetically determined, genes responsible for appropriate responses are propagated in a population at the expense of others. That is to say, the history of environmental changes experienced by every population has selected those individuals that have best responded to the changes to survive and reproduce. Populations that lack appropriate mechanisms for coping with environmental change die out. As humans are causing the environments of many species to change at unprecedented rates, it is becoming ever more important to understand the mechanisms by which organisms accommodate change. The case of wing development in water striders illustrates the complexity of response to temporal variation and the variety of mechanisms by which this may occur. Differences in the responses of different species show how mechanisms of response are also keyed to the nature of environmental change.

Water striders are freshwater bugs of the genus *Gerris.* They spend most of their lives moving about the surfaces of streams and ponds, where they feed on other small insects. They fly only occasionally: to disperse between feeding areas, and at the end of the summer, when individuals of many species leave lakes and ponds to find protected sites in forests in which to spend the winter. In Europe, some species of water striders have short, nonfunctional wings; others have long, functional wings; and still others may produce individuals with either wing type.

The length to which wings grow appears to be correlated with the seasonal variability and predictability of the habitat (Table 10.3). At one extreme, species inhabiting large, permanent lakes have short wings, or none at all, and do not disperse between lakes. At the other extreme, species living in temporary ponds usually have long, functional wings and disperse to find suitable sites for breeding each year. Between these extremes, species that are characteristic of small ponds, which are more or less persistent from year to year but tend to dry up during summers, frequently have both long-winged and short-winged forms. In these cases, the length of an individual's wings is determined genetically and does not respond to environmental conditions. Even where both long-winged and short-winged individuals occur within a single population, genes control the differences between these individuals. In none of the above cases is the wing length of an individual influenced by the environment. Each population has evolved a different genetic strategy to deal with its environment, although that strategy may involve a mixture of two forms where the environment is unpredictable.

The life cycles of most *Gerris* species in central Europe, England, and southern Scandinavia include two generations per year. The first (summer) generation hatches during spring, reproduces during summer, and then dies. The second, hatched from eggs laid by females of the summer generation, develops to the adult stage during late summer, then overwinters before breeding in early spring the following year. In species that inhabit seasonal ponds, the summer generation is **dimorphic** (literally, "of two forms"); both long-winged and short-winged forms are represented

TABLE 10.3 Wing lengths of water strider species *(Gerris)* inhabiting bodies of fresh water with different levels of permanence and predictability

CHARACTERISTIC OF HABITAT	CHARACTERISTIC WING LENGTH*	CHARACTERISTIC DETERMINATION
Permanent	Short	Genetic
Fairly persistent but unpredictable	Both short and long	Genetic dimorphism
Seasonal	Seasonally dimorphic (summer generation)	Developmental switch
Very unpredictable	Long	Genetic

*Short wings are not functional and prevent dispersal.

Source: K.Vepsalainen, *Acta Zool. Fenn.* 141:1–73 (1974).

(Figure 10.15). All the individuals in the winter generation have long wings and can fly. They leave their ponds in late summer and move into nearby woodlands for the winter. In spring, the same individuals return to small bodies of water to lay eggs.

Dimorphism in the summer generation reflects two extreme strategies. The long-winged forms can fly to other habitats if their pond dries up, a particular advantage when this happens early in the season. The short-winged forms gamble that their pond will persist. The advantage of having short wings may be higher fecundity, because nutrients that would have been devoted to wings and flight muscles can be converted into eggs.

Because all members of the winter generation, including those produced by short-winged parents, have long wings, seasonal dimorphism must be controlled by responsiveness to particular environmental conditions during development, rather than by genetic factors. In fact, studies have shown that whether an individual grows long or short wings depends on the length of the day during the early developmental period. When day length increases continuously during the developmental period and, in southern Finland, exceeds 18 hours during the last nymphal (immature) stage prior to adulthood, individuals grow short wings. When day length begins to decrease before the end of the nymphal stage (as it does when larval development extends beyond June 21, the summer solstice), long-winged adults are produced. Thus, summer generation individuals become short-winged or long-winged depending on their hatching date and rate of nymphal development. Because ponds dry up in late summer, late hatching and slow development place the individual in jeopardy more often than not. The switch between long and short wings is also influenced by temperature: high temperatures, which cause ponds to dry quickly, favor the development of long-winged forms.

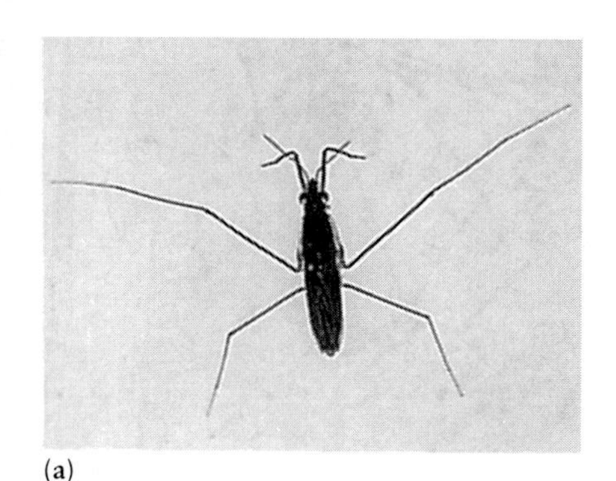

(a)

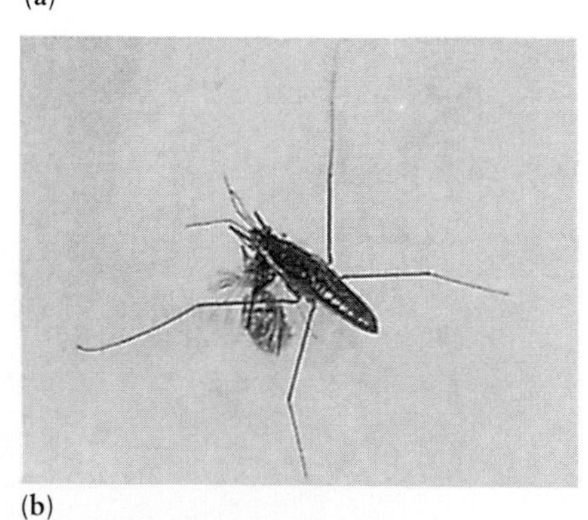

(b)

Figure 10.15 Alary (wing) polymorphism in the water strider *Gerris odontogaster:* (a) the long-winged form; (b) the short-winged form. Courtesy of K.Vepsalainen.

As in the case of coloration in *Gastrimargus,* particular environmental conditions trigger a switch from one developmental pathway to another in some species of *Gerris.* The switch point itself is selected to match the development of individuals to the expected conditions of their environments. Of course, evolution cannot predict whether a particular year will be dry or wet, but the history of variation experienced by a population during its evolution provides an estimate of the best date to switch, which is judged by day length. This estimate is further refined by sensitivity to temperature, which also influences the rate at which ponds dry up.

Migration, storage, and dormancy

In many parts of the world, extremes of temperature, drought, low light, and other adverse conditions are so severe that individuals cannot change enough to maintain normal activities, or, if they could, it would not be worth the cost. Under such conditions, organisms resort to a number of extreme responses. These include **migration** (moving to another region where conditions are more suitable), **storage** (relying on resources accumulated during bountiful seasons), and **dormancy** (becoming inactive).

Many animals, particularly among those that fly or swim, undertake extensive migrations. Arctic terns probably hold the record for long-distance migration. Individuals make yearly round trips of 30,000 km between their North Atlantic breeding grounds and antarctic wintering grounds. Many antarctic breeders, such as the tiny Wilson's storm-petrel, migrate to oceans of the Northern Hemisphere during the nonbreeding season. Each fall hundreds of species of land birds leave temperate and arctic North America, Europe, and Asia for the south in anticipation of cold winter weather and dwindling supplies of their invertebrate food. In East Africa, many large ungulates, such as wildebeests, migrate long distances, following the geographic pattern of seasonal rainfall and fresh vegetation (Figure 10.16).

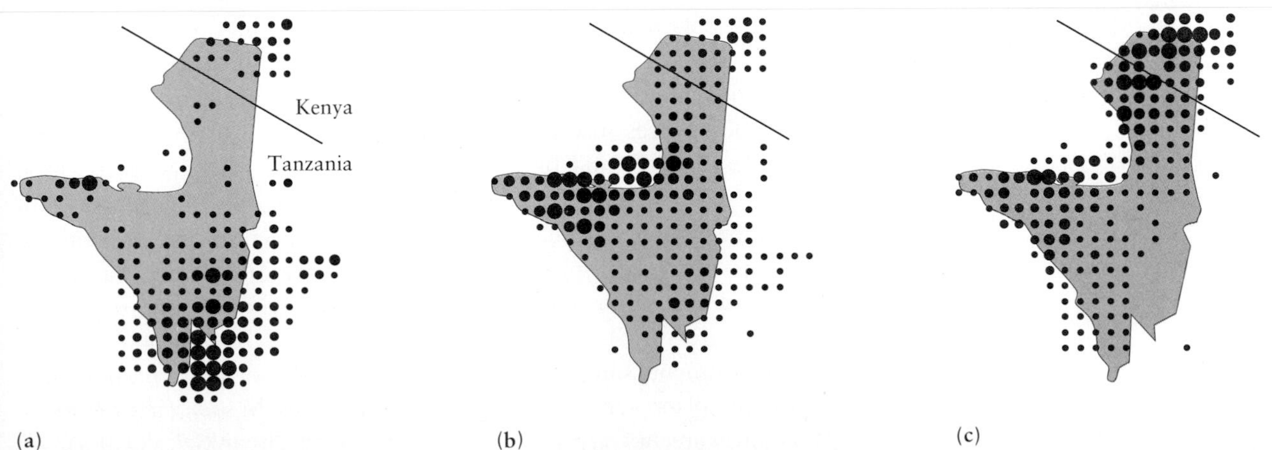

Figure 10.16 Distribution of wildebeest populations of the Serengeti ecosystem (shaded area) of northern Tanzania and southern Kenya throughout the annual cycle during 1969–1972: (a) December to April; (b) May to July; (c) August to November. Migrations follow lush growth of grasses following seasonal rains in each area. The size of each dot indicates the relative size of the population in that area. Adapted from L. Pennycuick, in A. R. E. Sinclair and M. Norton-Griffiths (eds.), *Serengeti: Dynamics of an Ecosystem,* University of Chicago Press, Chicago (1979), pp. 65–87.

Figure 10.17 A dense swarm of migratory locusts in Somalia, Africa, in 1962. Courtesy of the U.S. Department of Agriculture.

Some migratory movements occur in response to occasional local depletion of food supplies, which forces individuals to move out of an area in search of new feeding places. Such movements, which are often called population **irruptions,** are perhaps best known from outbreaks of migratory locusts from areas of high local density where food has been depleted. These migrations can reach immense proportions and cause extensive crop damage over wide areas (Figure 10.17). Irruptive behavior in locusts is a developmental response to population density. When locusts occur in sparse populations, they become solitary and sedentary as adults. In dense populations, however, frequent contact with other locusts stimulates young individuals to develop gregarious, highly mobile behavior, which can develop into a mass migration.

When environmental changes plunge organisms from feast into famine, and migration is not a possibility, storage of resources during periods of abundance for use in times of scarcity may be a way to cope. During infrequent rainy periods, desert cacti store water in their succulent stems. Plants growing on infertile soils absorb, in times of abundance, more nutrients than they require and use them when soil nutrients are depleted. Many temperate and arctic animals accumulate fat during mild weather in winter as a reserve of energy for periods when snow and ice make food sources inaccessible. Some winter-active mammals (beavers, squirrels, and pikas) and birds (acorn woodpeckers and jays) cache food supplies underground or under the bark of trees for later retrieval. In habitats that frequently burn severely—such as the chaparral of southern California—perennial plants store food reserves in fire-resistant root crowns, which sprout and send up new shoots shortly after a fire has passed (Figure 10.18).

Environments sometimes become so cold, dry, or nutrient, poor that animals and plants can no longer function normally. In such circumstances, species that are not capable of migration enter physiologically dormant states. Many tropical and subtropical trees shed their leaves during seasonal periods of drought; temperate and arctic broad-leaved trees shed theirs in autumn in anticipation of winter cold and darkness. Many mammals, such as ground squirrels and polar bears, **hibernate** (spend winter in a dormant state) because they cannot find food in winter, not because they are physiologically unable to cope with the harsh physical environment.

In most species, conditions requiring dormancy are anticipated by a series of physiological changes in the individual (for example, production of antifreezes, dehydration, and fat storage) that prepare it for a partial or complete shutdown of activity. Before winter, some insects enter a resting state known as **diapause,** in which water is chemically bound or reduced in quantity to prevent freezing and metabolism drops so low that it is barely detectable. Drought-resistant insects that enter a summer diapause either dehydrate themselves and tolerate the desiccated condition of their bodies or secrete an impermeable outer covering to prevent drying. Plant seeds and spores of bacteria and fungi exhibit similar dormancy mechanisms. Indeed, there are many cases of seeds stored in burial chambers or recovered in other archeological settings that have sprouted after hundreds of years of dormancy. By whatever mechanism it occurs, dormancy reduces exchange between organisms and their environments, enabling animals and plants to "ride out" unfavorable conditions.

Proximate and ultimate cues to environmental change

What stimulus indicates to birds wintering in the Tropics that spring is approaching in northern forests? What urges salmon to leave the seas and migrate upstream to their spawning grounds? How do aquatic invertebrates in the Arctic sense that if they delay entering diapause, a quick freeze may catch them unprepared for winter? In 1938, J. R. Baker made an important distinction between **proximate factors**—cues, such as day length, by which organisms can assess the state of the environment but that do not directly affect its well-being—and **ultimate factors**—features of the environment, such as food supplies, that bear directly on the well-being of the organism.

(a)

(b)

Figure 10.18 Root-crown sprouting by chamise *(Adenostoma fasciculatum)* following a fire in chaparral habitat in southern California. (a) Photograph taken on May 4, 1939, 6 months after the burn. (b) Photograph taken on July 16, 1940, showing extensive regeneration. Courtesy of the U.S. Forest Service.

Virtually all plants and animals sense the length of the day **(photoperiod)** as a proximate factor that indicates season, and many can distinguish periods of lengthening and shortening days. We have seen how wing development in water striders responds to both day length and changes in day length. Different populations of a single species may differ strikingly in their responses to photoperiod in different areas, reflecting different relationships of environmental changes to day length. Under controlled cycles of light and dark, southern populations (at 30°N) of side oats grama grass flower when day length is 13 hours, whereas more northerly populations (at 47°N) flower only when the light period exceeds 16 hours each day. In Michigan, at 45°N, populations of small freshwater crustaceans known as water fleas *(Daphnia)* form diapausing broods at photoperiods of 12 hours (mid-September) or less. In Alaska, at 71°N, related species enter diapause when the light period decreases to fewer than 20 hours per day, which happens in mid-August. Warm temperatures and low population densities tend to shorten the day length that triggers diapause (and hence delay the inception of diapause in autumn), suggesting that these factors portend more favorable environmental conditions for *Daphnia.* Differences in response to day length are under genetic control.

When day length does not accurately predict sporadically changing conditions, animals and plants must take their cues more directly from changes in ultimate factors in the environment. Annual cycles in equatorial regions, where day length is nearly constant, follow upon seasonal cycles of rainfall and their effects on humidity and vegetation. In highly unpredictable environments such as deserts, many organisms adopt a conservative strategy of readiness during the entire period during which sporadic rains are likely to occur. In some desert birds, for example, the lengthening photoperiod in spring stimulates development of reproductive organs to a point just short of breeding. The gonads maintain this state of readiness throughout the time when rains might occur, but only rainfall itself finally triggers completion of their physiological development and initiation of breeding.

Summary

1. In varying environments, organisms must be able to respond to changing external conditions to maintain suitable internal environments, obtain appropriate food items, and avoid predation. Responses include a variety of physiological changes and structural modifications, and each has a characteristic time period.

2. Maintenance of constant internal conditions, called homeostasis, depends on negative feedback responses. Organisms sense changes in their internal environments and respond in such a manner as to return those conditions to their optimum.

3. Homeostasis requires energy when a gradient must be maintained between internal and external conditions. Endotherms (birds and mammals), for example, maintain elevated body temperatures by generating heat metabolically to balance loss of heat to their cooler surroundings.

4. The magnitude of the body-environment gradient and the constancy of regulated internal conditions reflect the balance an organism strikes between the costs of homeostasis (maintaining the physiological apparatus and sustaining gradients) and the benefits of a narrow range of internal conditions.

5. A behavioral component of homeostasis is selection of microhabitats that minimize the body-environment gradient. This strategy is illustrated by the temperature-dependent foraging behavior of desert birds and by their construction of nests in such a way as to protect their chicks from environmental extremes.

6. The activity space of an individual—that is, its selection of habitat and microhabitat—depends on the suitability of conditions and availability of resources at any given time. This principle is illustrated by the influence of changing thermal conditions on habitat utilization by desert lizards.

7. Habitat selection cannot fully compensate for environmental change, and organisms must bring a variety of homeostatic responses into play. The most rapid of these are behavioral changes in orientation and activity and changes in rates of metabolic processes. Shivering in response to cold stress and contraction of the pupil in response to bright light are examples. Such responses are rapid and reversible.

8. Acclimation involves reversible changes in structure (for example, fur thickness) or biochemical pathways (induction of enzymes or changes in the amounts of enzymes and their products). Such changes require longer periods than behavioral or metabolic changes (usually days or weeks). Acclimation plays a prominent role in responses of long-lived organisms to seasonal change.

9. Developmental responses express the interaction between an organism and its environment during the growth period. Different environmental conditions lead to characteristic, irreversible structures and appearances. Developmental responses include those of plant seedlings to sun or shade and the expression of pigment systems in grasshoppers depending on light quality and intensity.

10. When conditions exceed the range of tolerance, organisms may migrate elsewhere, rely on materials stored during periods of abundance, or enter inactive states (such as torpor, hibernation, or diapause).

11. In many cases, the individual must anticipate environmental changes in order to respond successfully. Organisms rely on proximate cues, such as day length and seasonal changes in climate, to predict changes in ultimate factors, such as food supply and temperature, that directly affect their well-being.

Suggested readings

Alerstam, T. 1990. *Bird Migration.* Cambridge University Press, Cambridge.

French, A. R. 1988. The patterns of mammalian hibernation. *American Scientist* 76:568–575.

Gunn, D. L. 1960. The biological background of locust control. *Annual Review of Entomology* 5:279–300.

Hardy, R. N. 1983. *Homeostasis.* 2d ed. Edward Arnold, London.

Heinrich, B., and H. Esch. 1994. Thermoregulation in bees. *American Scientist* 82:164–170.

Hunter, A. F. 1995. The ecology and evolution of reduced wings in forest macrolepidoptera. *Functional Ecology* 9:275–287.

Johnston, I. A., J. D. Fleming, and T. Crockford. 1990. Thermal acclimation and muscle contractile properties in cyprinid fish. *American Journal of Physiology* 259:R231–R236.

Lyman, C., et al. 1982. *Hibernation and Torpor in Mammals and Birds.* Academic Press, New York.

Pough, F. H. 1980. The advantages of ectothermy for tetrapods. *American Naturalist* 115:92–112.

Prosser, C. L. (ed.). 1991. *Environmental and Metabolic Animal Physiology.* Wiley-Liss, New York.

Ricklefs, R. E., and F. R. Hainsworth. 1968. Temperature dependent behavior of the cactus wren. *Ecology* 49:227–233.

Roff, D. A. 1994. Habitat persistence and the evolution of wing dimorphism in insects. *American Naturalist* 144:772–798.

Schmidt-Nielsen, K. 1972. *How Animals Work.* Cambridge University Press, London and New York.

Schmidt-Nielsen, K. 1990. *Animal Physiology: Adaptations and Environment.* 4th ed. Cambridge University Press, London and New York.

CHAPTER 11

LIFE HISTORIES

Organisms are generally well suited to the conditions of their environments. Form and function vary in parallel with the ranges of temperature, water availability, salinity, oxygen, and other factors encountered by each species. We have seen how homeostatic mechanisms enable individuals to respond to temporal and spatial variation in their environments. Whether the differences we observe between populations and species are evolved or result from individual responses to different environments, we presume that modification of form and function improves either survival or reproduction, or both. That is, we assume that evolutionary and individual modifications are adaptive and that they increase the fitness of individuals. When structure and function respond to prominent physical conditions in the environment, simple engineering principles often suggest how these modifications can improve fitness. From an engineering perspective, we can see why desert plants have small leaves with thick cuticles to reduce water loss, or why swift runners have long legs. Many other adaptations are equally straightforward. The close color matching of grasshoppers to their backgrounds, for example, makes sense when one understands that they are eaten by visually hunting predators.

Organisms have limited time, energy, and nutrients at their disposal. Adaptive modifications of form and function serve two purposes in this regard. One is to increase the resources available to individuals. The other is to use those resources to their best advantage, that is, in a manner that maximizes the survival and reproduction of individuals in their particular environmental settings. Every modification involves a **trade-off,** meaning that an increase in any one thing implies a decrease in another. If resources are limited, then the time, energy, or materials devoted to one structure or function cannot be allotted to another. Therefore, each individual is faced with the problem of **allocation:** given that resources are limited, how can the organism best use its time and resources?

Practical solutions to the allocation problem depend on how changes in any given structure or function affect fitness. When modification of a trait influences several components of survival and reproduction, as is often the case, the evolution of that trait can be understood only by considering the entire life strategy. For example, an increase in the number of seeds produced by a plant may contribute to fitness by increasing fecundity (number of offspring), but it may also reduce survival of seedlings (if seed size is reduced to make more of them), survival of adult plants (if resources are shifted from root growth to support increased seed production), or subsequent fecundity (if seed production in one year reduces the growth of a plant and therefore its size in subsequent years).

In this chapter, we shall consider some general rules governing the allocation of time and resources in life strategies of plants and animals. From an evolutionary point of view, the object of life is to produce successful progeny—as many as possible. Reproduction involves choosing among options: when to begin to breed, how many offspring to have at one time, how much care to bestow upon them. The set of rules and choices pertaining to an individual's schedule of reproduction is referred to as its **life history.** Each life history has many components, the most important of which concern **maturity** (age at first reproduction), **parity** (number of episodes of reproduction), **fecundity** (number of offspring produced per reproductive episode), and termination of life (senescence and programmed death). Each of these choices affects other aspects of an individual's life. Because breeding takes time and resources from other activities and entails risks, investment in offspring generally diminishes survival of parents. In many cases, rearing offspring drains the parent's resources so much that fewer offspring are produced later. Thus, an optimized life history represents a resolution of conflicts between the competing demands of survival and reproduction to the best advantage of the individual in terms of perpetuating its lineage.

Before we discuss life histories in detail, we shall consider briefly an issue that will help us to distinguish between evolutionary adaptations of populations and responses of individuals to the range of environmental conditions normally encountered. Both are governed by sets of decision rules concerning allocation of time and energy. Individual responses are nongenetic, in the sense that an individual has one set of genetic material that does not change during its life, but the ways in which an individual responds to its environment may be under genetic control and thus subject to evolutionary change by natural selection.

Phenotypic plasticity

Virtually all attributes of an individual are affected by environmental conditions, at least to some extent. The relationship between the form or function of an individual—its **phenotype**—and the environment is referred to as a **reaction norm** (Figure 11.1). Many of the responses discussed in Chapter 10 are examples of reaction norms. The general responsiveness of the phenotype to the surroundings is called **phenotypic plasticity.**

Some reaction norms are a simple consequence of the influence of the physical environment on life processes. Heat energy accelerates most life processes. Therefore, we should not be surprised that caterpillars of the swallowtail butterfly *Papilio canadensis* grow faster at higher temperatures. However, the fact that individuals of the same species from Michigan and from Alaska exhibit different relationships between growth rate and temperature indicates that reaction norms may be modified by evolution. In one experiment, larvae from Alaskan populations grew more rapidly at low temperatures and larvae from Michigan grew more rapidly at high temperatures, as one might have predicted from the range of temperatures found in each location during the growing season (Figure 11.2).

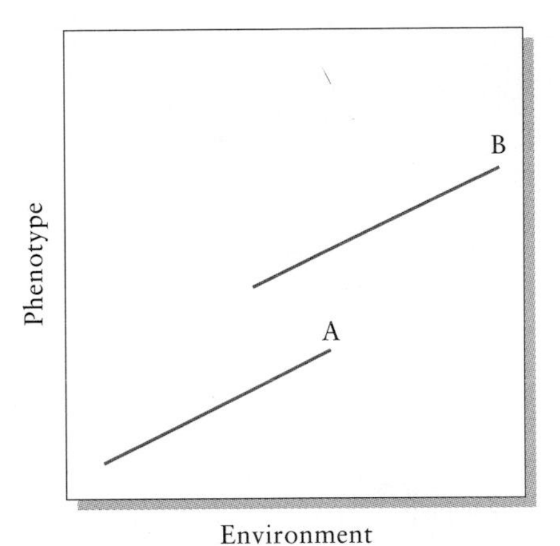

Figure 11.1 The relationship between phenotype and environment, called the reaction norm, is a property of the individual organism. The responsiveness of individuals to their environment is under genetic control and may be modified by evolution, as indicated by a shift in the reaction norm of species A compared with that of related species B.

As swallowtail growth rates show, genotype and environment interact to determine phenotypic traits. **Genotype-environment interaction** means simply that the slope of the reaction norm differs between genotypes. When two reaction norms cross, as they do in the case of the swallowtail butterfly, then each genotype (or population) performs better in one environment and worse in another. The effects of such interactions between genetic factors and environmental factors on performance are the basis for the evolution of specialization. Over time, when two populations are exposed to different ranges of environmental conditions, genotype-environment interactions will cause different genotypes to predominate in each population. The populations will therefore become differentiated and will have different reaction norms, each of which enables individual organisms to perform better in their own environments.

Whether differences between populations are due to evolutionary differentiation or to phenotypic responses of individuals to different environments often can be revealed by **reciprocal transplant experiments.** Transplant studies compare the phenotypes of individuals kept in their native environment to those of individuals transplanted to a different environment. Reciprocal transplants involve the switching of individuals between two localities. The traits of interest are assumed to be genetically determined when phenotypic values of native and transplanted individuals do not vary between the two environments. When trait values reflect where an individual is living (environment) rather then where it came from (genotype), then the results of the experiment are consistent with phenotypic flexibility as a cause of differences between populations. Of course, intermediate results are possible, in which case one might conclude that the reaction norm has been subject to evolutionary modification.

In one reciprocal transplant experiment, fence lizards *(Sceloporus undulatus)* from New Jersey and Nebraska were switched between those locations. The effect of the transplants on growth rate revealed both genetic determination and phenotypic plasticity (Figure 11.3). The growth rates of Nebraska lizards, about twice those of New Jersey lizards in their native environments, decreased by half—to the New Jersey level—in individuals transplanted to New Jersey. In contrast, New Jersey lizards did not grow faster in Nebraska. A simple interpretation of these results is that resources available for growth are consistently fewer in New Jersey than in Nebraska and that Nebraska lizards transplanted to New Jersey cannot gather resources fast enough to support their natural growth rates. Apparently, New Jersey lizards have a genetically regulated growth rate that is adjusted to a low resource level. That is, they have lost the ability to modify individual growth rates in response to higher resource levels—levels that they probably experience rarely, if ever.

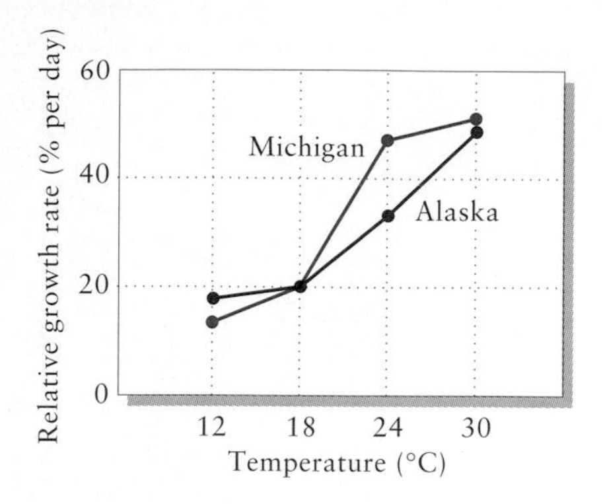

Figure 11.2 Growth rates of fourth instar larvae of the swallowtail butterfly *Papilio canadensis* as a function of temperature. Each of the two relationships shown is a reaction norm for a population. Larvae were obtained from populations in Alaska and Michigan and reared on balsam poplar *(Populus balsamifera)*. Note that Alaskan butterflies grew more rapidly than Michigan individuals at cold temperatures; the reverse was true at warm temperatures, even though individuals from both populations grew faster as temperatures increased. These results are fairly typical of genotype-environment interactions in that differences between populations are small. Modified from M. P. Ayres and J. M. Scriber, *Ecol. Monogr.* 64:465–482 (1994).

It is a fair question whether the slower growth of Nebraska lizards transplanted to New Jersey is adaptive or merely a consequence of reduced resources. If phenotypic plasticity is adaptive, a change in the form or function of an organism should reduce the negative effects of environmental change on fitness. A simple experiment conducted by Bill Schew at the University of Pennsylvania illustrates the difference between adaptive and nonadaptive phenotypic plasticity. Schew restricted the food supplies of hand-reared chicks of Japanese quail and European starlings, for periods of 10 and 3 days respectively, to a level that was just sufficient to maintain a constant weight. Thus, the experiment simulated the effects of poor feeding conditions during the period of rapid chick growth.

Although chicks of neither species could gain body weight, the responses of other aspects of growth and development to food restriction differed markedly between the two species. The quail chicks quickly decreased their body temperatures and metabolic rates (thereby saving energy); their feathers and extremities, particularly the long bones of the legs and wings, ceased growing almost immediately. In effect, the quail stopped their development "clock" and remained at a physiological age equivalent to that at the beginning of the experiment. When food restrictions were lifted, normal growth and development resumed, and the quail subsequently reached normal adult size.

The starlings showed none of these compensating responses. Their metabolism and body temperatures remained high, and their bones and feathers continued to grow throughout the restriction period, at the expense of the size of their internal organs. As a result, physiological age kept pace with chronological age, and the chicks became progressively more undernourished. When food restrictions were lifted, growth resumed at a rapid rate, but the chicks did not reach adult size before their developmental program told them to stop growing. As a result, the restricted chicks became stunted adults, with bones and feathers significantly shorter than those of adults that were well nourished during the growth period.

The story of the quail and starlings shows that although organisms may have little control over their rate of growth, other aspects of their lives can be modified in response to growth performance. Many types of organisms undergo dramatic changes during the course of their growth. Metamorphosis from larval to adult forms and sexual maturation are the most prominent of these changes. Optimal timing of such developmental events depends on resources and natural enemies, and is made more complicated by variations in the rate of growth due to food supply, temperature, and other environmental factors.

Imagine two growth curves resulting from two levels of food supply (Figure 11.4). Let us suppose that under a good nutritional regime resulting in rapid growth, an individual matures at a given weight and age. Poorly nourished individuals clearly cannot reach the same weight at a given age, and therefore must use a different decision point for maturation. Faced with such environmental variation, an individual may adopt one of two rules for the switch, or some intermediate between them. First, the individual may mature at a predetermined weight; with poor nourishment, this will take longer to achieve and will therefore expose the individual to a longer period of risk prior to reproduction. Alternatively, the individual may mature at a predetermined age; with poor nourishment, this will result in smaller size at maturity and perhaps a reduced reproductive rate as an adult. The optimum solution is usually somewhere in between, depending in part on the risk of death as a juvenile (high risk favors earlier maturation at a smaller size) and the slope of the relationship between fecundity and size at maturity (higher values favor delayed maturation at a larger size because the fecundity payoff is greater).

Tadpoles raised under high and low food availabilities exhibit different growth rates, as one would expect. In one experiment, the tadpoles given the poorer diet metamorphosed into adult frogs at a smaller size, but a later age, than those reared with abundant food (Figure 11.5). This finding supports the theoretical conclusion that timing of metamorphosis should be

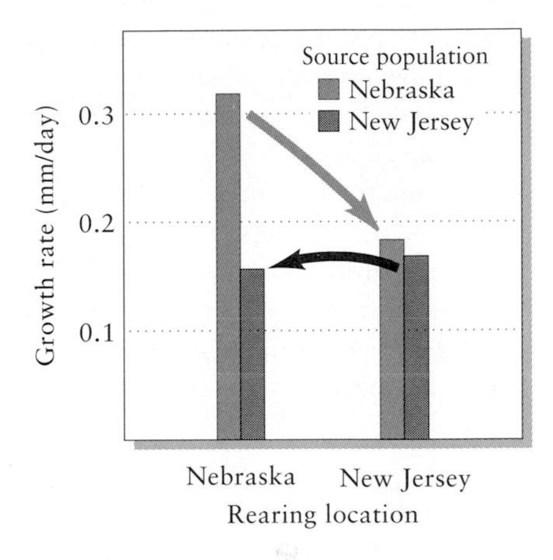

Figure 11.3 Growth rates of juvenile eastern fence lizards *(Sceloporus undulatus)* from populations in Nebraska and New Jersey exchanged in a reciprocal transplant experiment. Arrows indicate transplanted populations. From data in P. H. Niewiarowski and W. Roosenburg, *Ecology* 74:1992–2002 (1993).

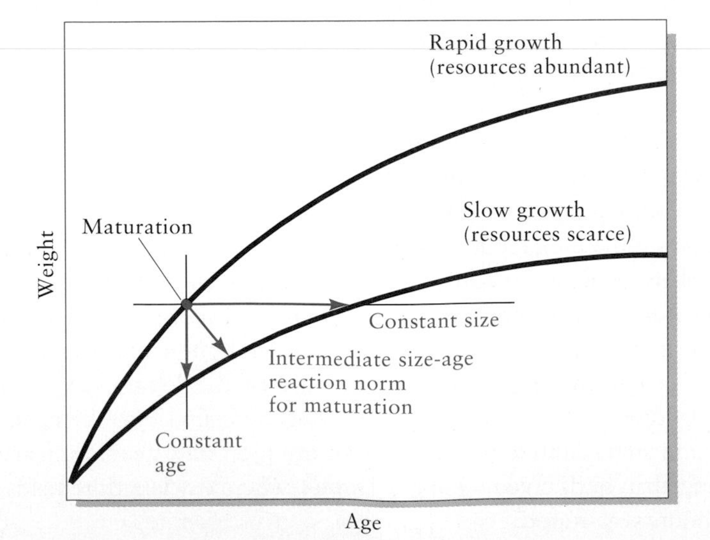

Figure 11.4 Possible relationships between age and size at maturation or metamorphosis when growth rates differ. Individuals may switch at a constant age, a constant size, or some intermediate between the two. From S. C. Stearns, *Am. Zool.* 23:65–76 (1983), and S. C. Stearns and J. Koella, *Evolution* 40:893–913 (1986).

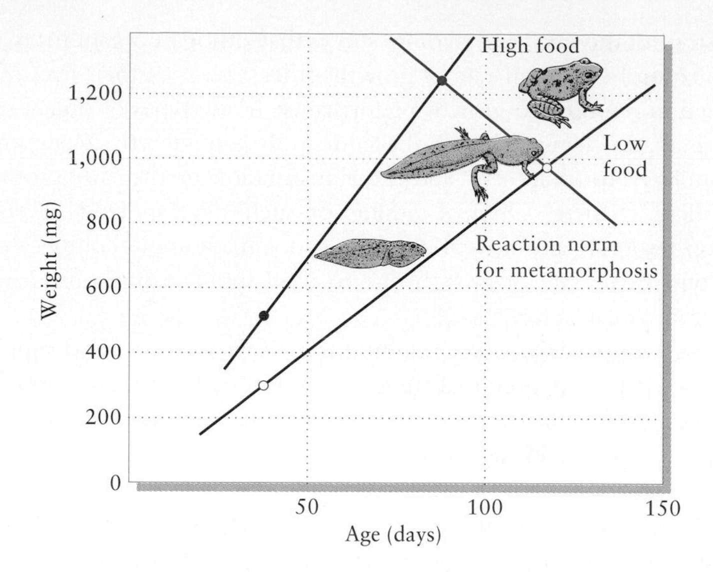

Figure 11.5 The relationship between age and size at metamorphosis in samples of frogs grown at high and low food availabilities. The metamorphosis reaction norm (colored line) lies between constant size and constant age extremes. Symbols represent weights of tadpoles at 40 days and at metamorphosis, which were used to obtain the growth relationship (black lines) of each treatment group. After data in J. Travis, *Ecology* 65:1155–1160 (1984).

sensitive to both age and size: poor nutrition slows the developmental program in frogs, but does not stop it altogether, as it seems to in Japanese quail. The relationship between age and size at metamorphosis under different feeding regimes is the reaction norm of metamorphosis with respect to age and size. To illustrate the generality of this pattern, we note a similar reaction norm for maturity in human females in the United States and other developed countries, who are, on average, much better nourished now then they were a century ago. In response to a shifted weight-age relationship, the age at maturity of young women has decreased by about 4 years, and weight at maturity has increased by about 2 kilograms, since the beginning of the twentieth century.

Variation in life histories

During the early 1940s Reginald Moreau, a British ornithologist who had worked in Africa for many years, called attention to the fact that songbirds in the Tropics lay fewer eggs (two or three, on average) than their counterparts at higher latitudes (generally four to ten, depending on the species). The study of life histories is mostly a legacy of one of Moreau's colleagues, David Lack, whose influence is ubiquitous in population biology and evolutionary ecology. Lack recognized that an increase in clutch size (number of eggs per nest) would increase the overall reproductive success of the parents, unless something reduced survival of offspring in large broods. He presumed that the ability of adults to gather food for their young was limited and, accordingly, that in broods having too many mouths to feed relative to the availability of food, chicks would be undernourished and survive poorly. Lack further suggested that because of the longer day length at higher latitudes in the season when offspring are reared, birds living at temperate and arctic latitudes could gather more food, and therefore rear more offspring, than birds breeding in the Tropics, where day length remains close to 12 hours year-round.

Lack made three important points. First, he related life history traits, such as fecundity, to reproductive success and thus to evolutionary fitness. Second, he demonstrated that life histories vary consistently with respect to factors in the environment. This suggested the possibility that life history traits are molded by natural selection. Third, he proposed a hypothesis that could be subjected to experimentation. In the particular case of clutch size, Lack suggested that food supply limits the number of offspring that parents can rear. To test this idea, one could add eggs to nests to create enlarged clutches and broods. According to Lack's hypothesis, parents should be unable to rear added chicks because they cannot gather the additional food required. We shall return to this problem below.

Life history traits vary with respect to environmental conditions and to one another. Consistent patterns of variation, such as the general decrease in reproductive rates of animals from high latitudes to the equator, have aroused curiosity and at the same time have suggested factors in the environment that may govern the pattern. Lack noted that clutch size and day length vary together; additional explanations for the increase in clutch size with latitude have been based on average temperature, seasonal variation in temperature and food supplies, and rates of predation on bird nests, all of which vary consistently with latitude.

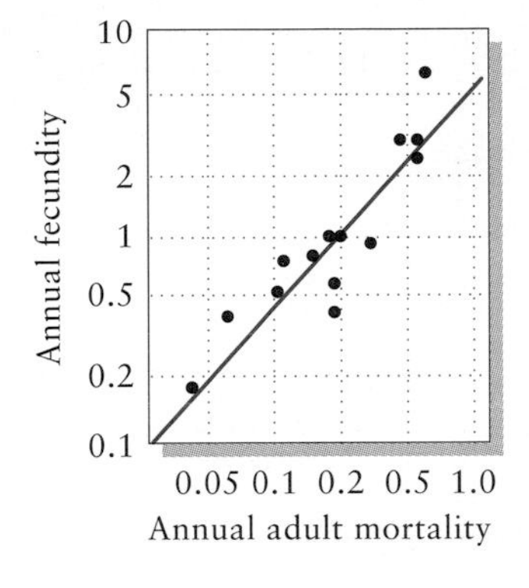

Figure 11.6 Relationship between annual fecundity and adult mortality in several populations of birds ranging from albatrosses (low values) to sparrows (high values). From data in R. E. Ricklefs, *Am. Nat.* 111:453–478 (1977).

Some life history traits, such as fecundity and mortality, tend to vary in close association (Figure 11.6). At one extreme, elephants, albatrosses, giant tortoises, and oak trees exhibit long life, slow development, delayed maturity, high parental investment, and low reproductive rates; at the other extreme are mice, fruit flies, and weedy plants. In broad comparisons within the plant and animal kingdoms, such associations of traits vary with body size and undoubtedly reflect the relative slowness of all life processes in large organisms. But even among organisms of similar size and body plan, different environments produce widely divergent life histories. Storm-petrels, which are seabirds the size of thrushes, do not begin to reproduce until they are 4 or 5 years old, rear at most a single chick each year, and may live to the age of 30 or 40 years. Thrushes may produce several broods of three or four young each year, beginning with their first birthday, but they rarely live beyond 3 or 4 years. Similarly varied life histories may be found even among different populations of the same species, as illustrated in the case of the fence lizard in Table 11.1. The correlation between mortality and fecundity across species must in part reflect the fact that in persistent populations, births and deaths must balance on average. In addition, however, these life history traits may be modified by evolution.

The English plant ecologist J. P. Grime emphasized the relationship between the life history traits of plants and certain conditions of the environment. He envisioned variation in life history traits between three extreme apexes, like points of a triangle, and called plants with life histories at these extremes "stress tolerators," **"ruderals,"** and "competitors" (Table 11.2). As the name implies, stress tolerators grow under extreme environmental conditions. They grow slowly and conserve resources. Because seedling establishment is difficult in stressful environments, vegetative spread is emphasized. Where conditions for plant growth are more favorable, ruderals and competitors occupy opposite ends of a spectrum of disturbance.

TABLE 11.1 **Life history traits of several populations of the fence lizard *Sceloporus undulatus***

TRAIT	*Arizona*	*Utah*	*Colorado*	*New Mexico*	*Kansas*	*Texas*	*Ohio*	*South Carolina*
	LOCATION							
Clutch size	8.3	6.3	7.9	9.9	7.0	9.5	11.8	7.4
Clutches per year	3	3	2	4	1–2	3	2	3
Egg weight (g)	0.29	0.36	0.42	0.24	0.26	0.22	0.35	0.33
Relative clutch mass	0.22	0.21	0.23	0.21	0.28	0.27	0.25	0.23
Age at maturity (months)	12	23	21	12	12	12	20	12
Survival to breeding	0.07	0.05	0.11	0.03	0.10	0.06	0.03	0.11
Annual adult survival	0.24	0.48	0.37	0.20	0.27	0.11	0.44	0.49

Source: D. W. Tinkle and A. E. Dunham, *Copeia* (1986):1–18.

Ruderals colonize disturbed patches of habitat, where they exhibit rapid growth, early maturation, high reproductive rates, and easily dispersed seeds, which enable them to reproduce quickly and disperse their progeny to other disturbed sites before being overgrown by superior competitors. Plants with the "competitor" suite of life history traits tend to grow to large stature, mature at large size, and exhibit long life spans. The competitor life history therefore requires stable, favorable conditions for its success.

Allocation of limited time and resources

Ecologists believe that observed life histories represent the best resolutions of conflicting demands on organisms. Therefore, a critical effort in the study of life histories has been to understand the allocation of limited time and resources to competing functions. An increase in any one function requires a shift in the allocation of energy and nutrients. Because these are limited in supply, any increase in one component of demand, such as that created by reproduction, results in a decrease in delivery to some other component of demand. Time spent searching for food, for example, cannot be applied to caring for offspring or watching for predators.

Energy and nutrients used for growth cannot be earmarked for reproduction. In many species, individuals produce eggs in direct proportion to the size of their gonads, or produce seeds in direct proportion to number of flowers. Therefore, when growth is shifted from reproductive structures to other parts of the body, fecundity also decreases. Photosynthetic rate depends in part on how much of a plant's production ends up in photosynthetic tissue at the expense of root, support, and reproductive tissue (Figure 11.7). Because of the modular growth form of plants, an individual bud at a

TABLE 11.2 **Typical life histories of plants in environments with different selective factors**

COMPETITORS	RUDERALS	STRESS TOLERATORS
Herbs, shrubs, or trees	Herbs, usually annuals	Lichens, herbs, shrubs, or trees; usually evergreen
Large, with a fast potential growth rate	High potential growth rate	Potential growth rate slow
Reproduction at a relatively early age	Reproduction at an early age	Reproduction at a relatively late age
Small proportion of production to seeds	Large proportion of production to seeds	Small proportion of production to seeds
Seed bank sometimes, vegetative spread often important	Seed bank and/or highly vagile seeds	
Vegetative spread important		

Source: J. P. Grime, *Plant Strategies and Vegetation Processes,* Wiley, Chichester (1979).

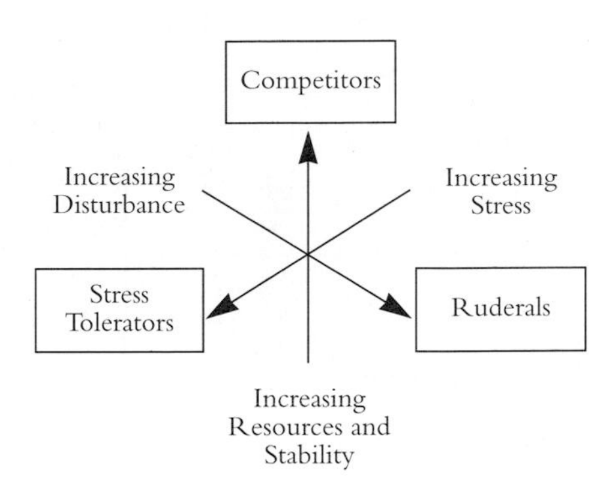

point of leaf attachment may produce either a lateral branch or a flower, but not both. Thus, in some species, shoot growth is traded off against reproduction through allocation of growth meristems.

In spite of the presumed pervasiveness of trade-offs in life histories, demonstrating their existence has proved difficult because this requires controlled experimental manipulation of individual components of the phenotype. Adding and subtracting eggs in the nests of birds has, in many

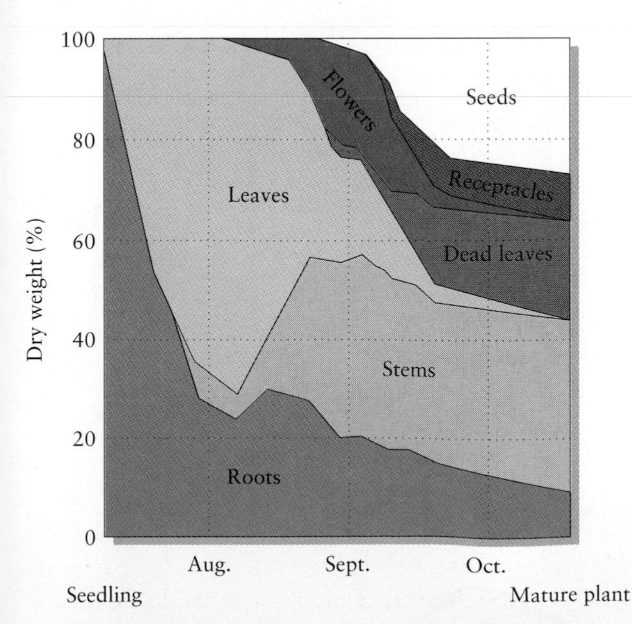

Figure 11.7 Proportional distribution of dry weight among different plant parts of the groundsel, *Senecio vulgaris* (Compositae), during its life cycle. Note the development of reproductive parts at the expense of leaves and roots toward the end of the growing season. Based on J. Ogden, in J. L. Harper, *J. Ecol.* 55: 247–270 (1967).

cases, revealed an inverse relationship between the number of chicks in a nest and their survival. As a result, production of offspring is often greatest from clutches of intermediate size, as predicted by David Lack. For example, magpies (relatives of crows) lay a clutch that corresponds to the maximum number of offspring that a pair can rear. Either adding or subtracting eggs results in fewer offspring fledged (Figure 11.8). In some species, however, the most productive clutch size is larger than the commonest clutch size observed in a population (Figure 11.9). To explain such results, we may hypothesize that rearing an enlarged clutch has bad effects on the parents. Brood manipulation experiments with blue tits and European kestrels (a small hawk) revealed that parents that reared enlarged broods had lower survival to the following breeding season, and those that did survive produced fewer offspring.

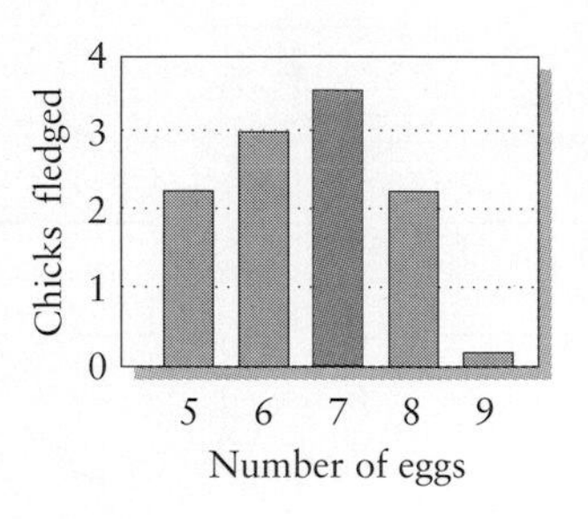

Figure 11.8 Number of chicks fledged from nests of European magpies in which seven eggs were laid, but in which eggs were added or removed by the experimenter to make up manipulated clutches of five to nine eggs. The most productive clutch size was seven. After G. Hogsted, *Science* 210:1148–1150 (1980).

Other manipulations have failed to demonstrate trade-offs between life history traits. For example, in one study on guppies, the resources allocated to reproduction were experimentally reduced by preventing females from mating with males. Now, if growth and reproduction compete for allocation of assimilated resources, the experimental fish should have attained a larger size than controls by the end of the study. In fact, little of the difference in accumulated reproductive tissue between mated and unmated females was converted to growth (Figure 11.10). These results suggest several interpretations. Possibly, food availability does not limit reproduction in guppies. Alternatively, genetic factors may determine the growth of females at a rate that corresponds to the expected availability of resources (guppies normally do not abstain from mating), just as New Jersey fence lizards appear to grow at a low, genetically determined rate commensurate with their food supply.

Figure 11.9 The frequencies of clutch sizes (black bars, left-hand scale) among 4,489 clutches of the great tit *(Parus major)* near Oxford, England, between 1960 and 1982, and number of young per clutch surviving at least to the next season (colored bars, right-hand scale) as a function of clutch size. Note that the most common clutch size is not the most productive. From M. S. Boyce and C. M. Perrins, *Ecology* 68:142–153.

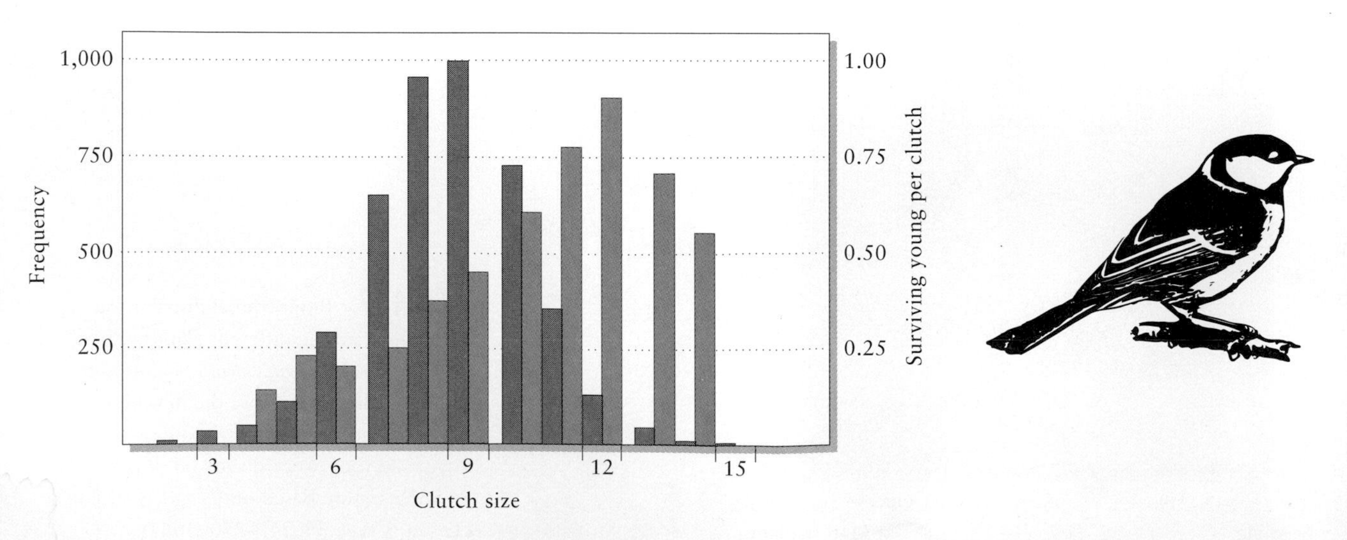

Evolution of life histories

Most issues concerning life histories can be phrased in terms of three questions: When should an individual begin to produce offspring? How often should it breed? How many offspring should it attempt to produce in each breeding episode? The answers that each species has provided to these questions express in a different way the fundamental trade-off between fecundity and adult growth and survival, that is, between present and future reproduction.

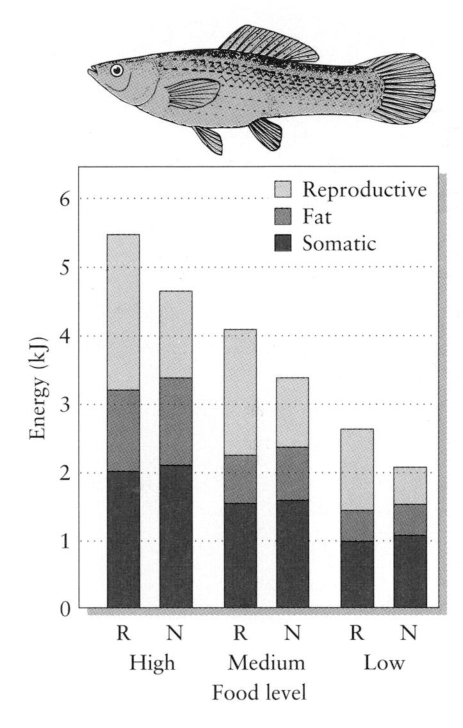

Figure 11.10 Contents of energy in somatic (body) tissue, fat deposits, and reproductive tissue (including eggs) in female guppies that were raised at three food levels and either were permitted access to males (R) or were not (N). After D. Reznick, *Ecology* 64:862–873 (1983).

Age at first reproduction

When should an animal or plant begin to breed? Long-lived organisms typically begin to reproduce at an older age than short-lived ones (Figure 11.11). Why should this be so? At every age, an individual must choose, whether consciously or not, between attempting to reproduce and abstaining from breeding. When young individuals resolve this choice in favor of abstention, they effectively delay the onset of sexual maturity. Thus, we may understand age at first reproduction in terms of the benefits and costs of breeding at a particular age. The benefit appears as an increase in fecundity at that age. Costs may appear as reduced survival to older ages or reduced fecundity at older ages, or both.

Consider the following hypothetical example. A type of lizard continues to grow only until it reaches sexual maturity. Its fecundity varies in direct proportion to body size at maturity. Suppose that number of eggs laid per year increases by 10 for each year that an individual delays reproduction, so that individuals that begin to breed in their first year produce 10 eggs that year and the same number each year thereafter; individuals first breeding in their second year produce 20 eggs per year; individuals maturing in their third year produce 30 eggs; and so on. Comparing the cumulative egg production of early-maturing and late-maturing individuals (Table 11.3) reveals that the age at maturity that maximizes lifetime reproduction varies in direct proportion to the life span. For example, for a life span of 3 years, maturing at 2 years results in the greatest lifetime reproduction. When the life span is 7 years, 4 years is the best age to mature.

For organisms that do not grow after their first year (most birds, for example), the choice between breeding and not breeding comes down to balancing current reproduction against survival. Nonbreeding individuals avoid the risks of preparations for reproduction: courtship, nest building, migration to breeding areas. Presumably, life experience gained with age also reduces the risks of breeding or increases realized fecundity from a certain level of parental investment, or both, and thereby favors delayed reproduction. Among birds, age at maturity varies directly with annual survival rates of adults, up to about 10 years in certain long-lived seabirds, as we have seen in Figure 11.11. Tending to offset the advantages of delayed reproduction are many factors that reduce the expectation of future reproduction. These factors include high predation rates, encroaching senescence at old

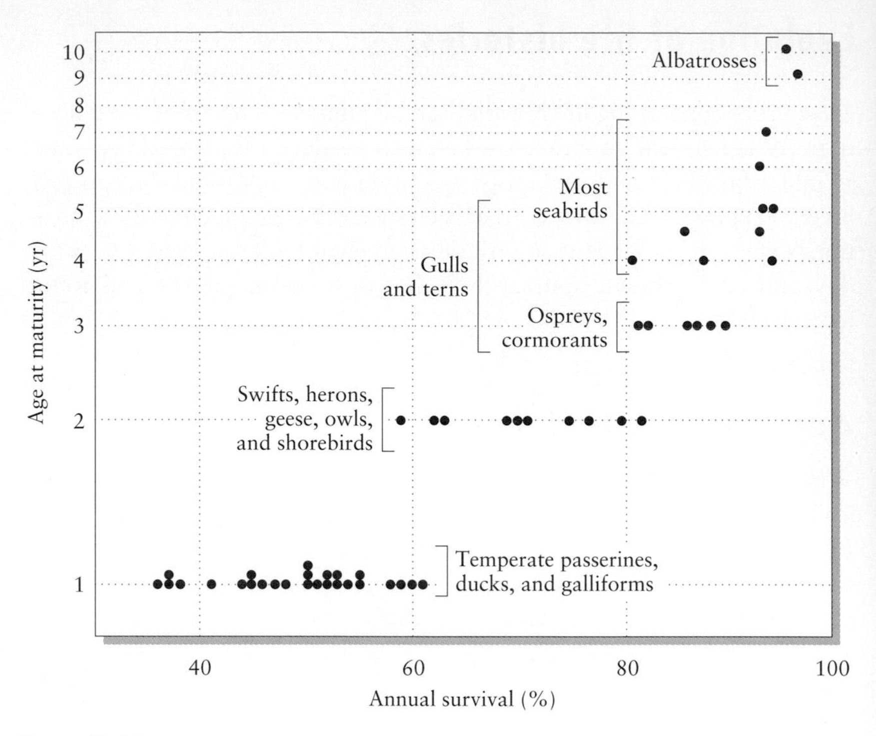

Figure 11.11 Relationship between age at maturity and annual adult survival rate, which is directly proportional to life span, in a variety of birds. From R. E. Ricklefs, in D. S. Farner (ed.), *Breeding Biology of Birds,* National Academy of Sciences, Washington, D.C. (1973), pp. 366–435.

age, and, for organisms that live a single year or less in seasonal environments, the end of the reproductive season.

Annual and perennial life histories

Plants and animals either reproduce during a single season and die (**annual** reproduction) or have the potential to reproduce over a span of many seasons (**perennial** reproduction). Population biologists have pondered the relative advantages of each habit in terms of the trade-off between survival probability and fecundity. To survive the nonreproductive "winter" period, a perennial plant must store materials in roots and form freeze-resistant or drought-resistant buds, presumably at the expense of production. In perennial plants, the advantages of the perennial habit must outweigh the costs of reduced fecundity relative to that of the annual habit. Theory shows that the pertinent consideration is the ratio of survival of the adult plant to that of the immature plant (up to the point of reproduction; see Box 11.1). When few perennials survive from one breeding season to the next, or when perennials produce relatively few seeds, the annual habit is favored. Where individuals have a high probability of survival once they are established, but seedlings survive poorly (a high ratio of adult to immature survival), an annual must have many seeds in order to outproduce a perennial. Thus, annuals tend to dominate the floras of deserts, where few adult plants can

TABLE 11.3 Total eggs produced by individuals in a hypothetical population as a function of life span and age at first reproduction

AGE AT FIRST REPRODUCTION (YEARS)	LIFE SPAN (YEARS)							
	1	*2*	*3*	*4*	*5*	*6*	*7*	*8*
1	**10***	**20**	30	40	50	60	70	80
2	0	**20**	**40**	**60**	80	100	120	140
3	0	0	30	**60**	**90**	**120**	150	180
4	0	0	0	40	80	**120**	**160**	**200**
5	0	0	0	0	50	100	150	**200**
6	0	0	0	0	0	60	120	180

*Bold type indicates most productive ages at first reproduction for a given life span.

survive long drought periods, and perennials tend to dominate tropical floras, where competition and predator pressure make establishment of seedlings difficult.

Reproductive effort

Annual plants have little life expectancy beyond their first breeding season and therefore devote as many of their resources as possible to current reproduction. Perennials, however, must allocate resources between current reproduction and functions that prolong life or increase size. When a particular life history attribute affects both fecundity and growth or survival, the individual must optimize the trade-off between the two. Intuitively, high mortality rates for adults should tip the balance in favor of current fecundity. Conversely, when life span is potentially long, individuals should not increase current fecundity so much as to unduly jeopardize future reproduction. This insight has a simple algebraic proof, which is shown in Box 11.2.

Many plants and invertebrates, as well as some fishes, reptiles, and amphibians, do not have a characteristic adult size. They grow, at a continually decreasing rate, throughout their adult lives, a condition referred to as **indeterminate growth.** Fecundity is directly related to body size in most species with indeterminate growth. Because egg production and growth draw on the same resources of assimilated energy and nutrients, increased fecundity during one year must be weighed against reduced fecundity in subsequent years. Accordingly, a long life expectancy should favor growth over fecundity during each year. For organisms with less chance of living to reproduce in future years, allocating limited resources to growth rather than eggs wastes potential fecundity.

BOX 11.1 A theoretical comparison of annual and perennial reproduction

Suppose a population of plants contains some individuals that produce a large number of seeds at the end of the first growing season and then die (annual) and others that produce fewer seeds but survive through the winter to reproduce in subsequent growing seasons (perennial). Which has the greater fitness? For simplicity, let us assume that annual and perennial plants have the same probability of survival during their first (in the annual's case, only) growing season (S_0) and that perennials have a constant probability of survival thereafter (S).

The factor by which a population of an annual plant grows (λ) equals the number of seeds each individual produces (B_a) times their survival to reproductive age (S_0):

$$\lambda_a = B_a S_0$$

The increase in a population of a perennial plant equals the number of seeds (B_p) times their survival (S_0) plus the probability of survival of the parent (S); hence

$$\lambda_p = B_p S_0 + S$$

The population growth rate of the annual exceeds that of the perennial ($\lambda_a > \lambda_p$) when $B_a S_0 > B_p S_0 + S$. Dividing both sides of the inequality by S_0, we obtain $B_a > B_p + S/S_0$, or, rearranging,

$$\beta_a - B_p > \frac{S}{S_0}$$

Accordingly, an annual life history leaves more offspring than a perennial life history when the number of seeds produced by the annual exceeds the fecundity of the perennial by the ratio S/S_0.

This algebraic model comes from E. L. Charnov and W. M. Schaffer, *Am. Nat.* 107:791–793 (1973).

Consider two hypothetical fish, both of which weigh 10 g at sexual maturity, but which allocate resources to growth and reproduction differently. Both gather enough food each year to reproduce their weight in new tissue or eggs. Fish A allocates 20% of its production to growth and 80% to eggs, whereas fish B allocates half of its production to growth and half to eggs. Calculations of growth, fecundity, and accumulated fecundity (Table 11.4) show that for fish living 4 or fewer years, on average, high fecundity

BOX 11.2 The balance between fecundity and survival

We use a model similar to the one presented for perennials in Box 11.1, in which the rate of population growth (λ) is equal to $S_0B + S$. However, we partition adult survival into two components, one directly related to reproduction (S_R) and the other independent of reproduction (S). Now, the number of descendants may be expressed as

$$\lambda = S_0B + SS_R$$

Certain reproductive traits that cause small changes in the values of survival (ΔS_R) and fecundity (ΔB) will influence the number of descendants as follows:

$$\Delta\lambda = S_0\Delta B + S\Delta S_R$$

When changes that enhance fecundity (ΔB positive) also reduce survival (ΔS_R negative), their effects on $\Delta\lambda$ depend on the relative values of S and S_0. In general, when S is large compared with S_0, the criterion of number of descendants favors traits that increase adult survival at the expense of fecundity, and vice versa. Thus one expects parental investment in offspring to decrease with increasing adult life span.

and slow growth result in greater overall productivity (cumulative weight of eggs), whereas for fish living longer than 4 years, low fecundity and rapid growth are more productive. Adult mortality, therefore, determines the optimal allocation of resources between growth and reproduction.

Life histories in variable environments

Several theoretical models suggest that low adult survival relative to prereproductive survival favors high fecundity, and vice versa. Further theoretical studies have indicated that for given average values of survival, increased fluctuation in adult survival rate about the average value favors increased fecundity, and increased fluctuation in prereproductive survival favors reduced fecundity. The reasoning is simple: Variations above the mean convey smaller advantage than variations below the mean convey disadvantage. Suppose the average adult survival rate is 50% per year. In one population this survival rate is constant, while in another it alternates with equal frequency between 0% and 100%. In the case of the extreme fluctuation in

TABLE 11.4 Numerical comparisons of the strategies of slow growth/high fecundity and rapid growth/low fecundity in two hypothetical fish

	YEARS					
CHARACTERISTIC	*1*	*2*	*3*	*4*	*5*	*6*
Slow growth/high fecundity						
Body weight	10	12	14.4	17.3	20.8	25.0
Growth increment	2	2.4	2.9	3.5	4.2	5.0
Weight of eggs	8	9.6	11.5	13.8	16.6	20.0
Cumulative weight of eggs	8	17.6	29.1	42.9	59.5	79.5
Rapid growth/low fecundity						
Body weight	10	15	22.5	33.8	50.7	76.1
Growth increment	5	7.5	11.3	16.9	25.4	38.1
Weight of eggs	5	7.5	11.3	16.9	25.4	38.1
Cumulative weight of eggs	5	12.5	23.8	40.7	66.1	104.2

Note: All weights in grams. Body weight + growth increment = next year's body weight. Cumulative weight of eggs to last year + weight of eggs = cumulative weight of eggs to this year. Growth increment and weight of eggs in each year are equal to the body weight.

survival rate, an adult has a high and unpredictable chance of dying in any given year and is therefore best served by increasing its current reproductive output. As a general rule, selection shifts allocation away from stages of the life cycle with the greatest uncertainty. This strategy is sometimes referred to as **bet hedging.**

Mosquitofish *(Gambusia affinis)* were introduced to the Hawaiian Islands early in the twentieth century, and they have maintained populations in dozens of reservoirs ever since. In some of these reservoirs, water depth is kept at stable levels. In others, water levels fluctuate markedly in response to rainfall and demand for irrigation water. Mosquitofish from stable and fluctuating reservoirs differ consistently in life history traits (Figure 11.12). Mature females from fluctuating reservoirs tend to be smaller, allocate more of their body mass to reproduction, and produce larger numbers of offspring, each having smaller size, than females from stable reservoirs. Although variations in the mortality rates of adults and juveniles have not been measured directly in stable and fluctuating reservoirs, the differences in reproductive allocation are consistent with more variable adult survival, and perhaps lower average adult survival, in the fluctuating reservoirs. Smaller offspring size also indicates a reduced commitment to each individual offspring, perhaps because juveniles in fluctuating reservoirs also have highly variable survival rates.

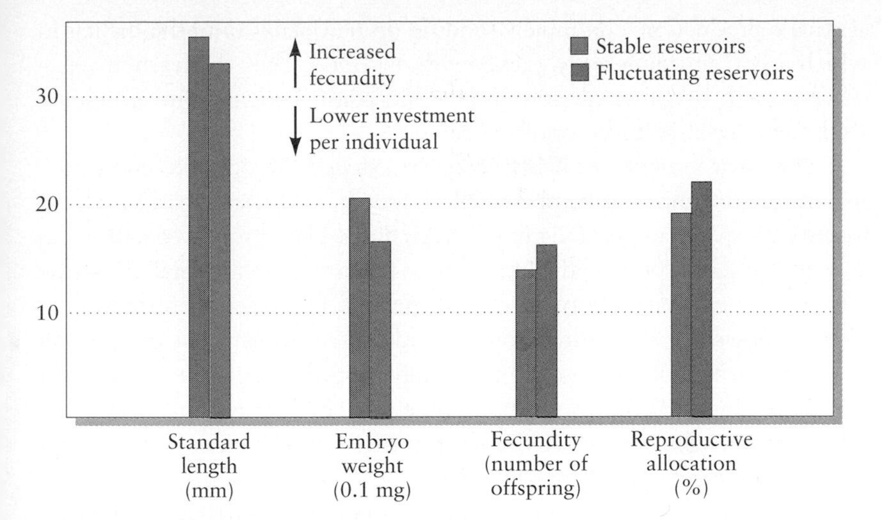

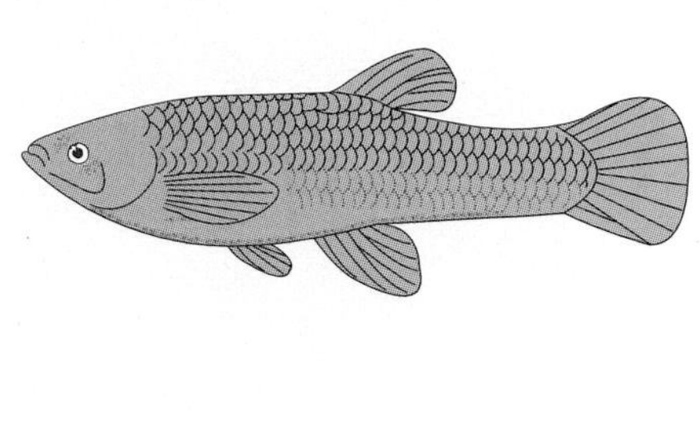

Figure 11.12 Life history characteristics of populations of mosquitofish *(Gambusia affinis)* introduced into reservoirs in Hawaii indicate greater total reproductive investment, but lower individual investment, in fluctuating environments than in stable environments. Reproductive allocation is the proportion of the dry mass comprising embryos. From data in K. T. Scribner et al., *J. Evol. Biol.* 5:267–288 (1992).

Semelparity

Unlike mosquitofish, which breed repeatedly, some species of salmon grow rapidly for several years, then undertake a single episode of breeding. During this one burst of reproduction, females convert a large portion of their body tissues into eggs, and then die shortly after spawning. Because salmon make so great an effort to migrate upriver to reach their spawning grounds, it may be to their advantage to make the trip just once, at which time they should produce as many eggs as possible, even if this supreme reproductive effort results in the wastage of most body tissues and ensures death. This pattern is called **programmed death.**

The salmon life history is sometimes called "big-bang" reproduction, but is more properly referred to as **semelparity.** This term comes from the Latin *semel* ("once") and *pario* ("to beget"); the opposite of semelparity is **iteroparity,** from *itero* ("to repeat"). Semelparity is not the same as annual reproduction. For one thing, annuals may have more than one episode of reproduction, or prolonged continuous reproduction, within a season; for another, like perennials, semelparous individuals must survive at least one nonbreeding season—and usually survive many—before maturing sexually. Semelparity is rare among animals and plants that live for more than 1 or 2 years. So many resources are allocated to survival between reproductive seasons, as compared with those used to prepare for breeding, that once a perennial life form is adopted, reproduction every year seems the most productive pattern.

The best-known cases of semelparous reproduction in plants occur in agaves and bamboos, two distinctly different groups, although this life history pattern has been reported even for some tropical forest trees. Most bamboos are tropical or warm temperate zone plants that form dense stands in disturbed habitats. Reproduction in bamboos does not appear to require substantial preparation or resources, as would be needed to grow a heavy flowering stalk. But there are probably few opportunities for successful seed germination. Once established, a bamboo plant increases by asexual reproduction, continually sending up new stalks until the habitat in which it germinated is fairly packed with bamboo. Only then, when vegetative growth becomes severely limited, do plants benefit from producing seeds, which can colonize disturbed sites.

The environment and habits of agaves occupy the opposite end of the spectrum from those of bamboos. Most species of agaves inhabit arid climates with sparse and erratic rainfall. Each agave plant grows vegetatively as a rosette of leaves produced from a single meristem over several years (the number of years varies from species to species). The plant then sends up a gigantic flowering stalk. Physiological studies have shown that the growth of the flowering stalk is too rapid to be fully supported by photosynthesis or uptake of water by the roots. As a consequence, the nutrients and water necessary for stalk growth are drawn from the leaves, which die soon after the seeds are produced (Figure 11.13).

Agaves frequently live side by side with yuccas, a closely related group of plants that have a similar growth form but which flower year after year. Yuccas typically are branched and have many terminal rosettes of leaves (see Figure 1.2, for example). The root systems of agaves and yuccas also differ markedly. Yucca roots descend deeply to tap persistent sources of groundwater; agaves have shallow, fibrous roots that catch water percolating through the surface layers of desert soils after rain showers, but are left high and dry during drought periods.

Figure 11.13 Stages in the life cycle of the Kaibab agave *(Agave kaibabensis)* in the Grand Canyon of Arizona. An individual plant grows as a rosette of thick, fleshy leaves (a) for up to 15 years. Then it rapidly sends up its flowering stalk (b) and sets fruit, after which the entire plant dies (left-hand rosette in a).

(a)

(b)

TABLE 11.5 **Ecological, life history, demographic, and reproductive traits of *Lobelia telekii* and *Lobelia keniensis* on Mount Kenya**

TRAIT	*LOBELIA TELEKII*	*LOBELIA KENIENSIS*
Life history	Semelparous	Iteroparous
Habitat	Dry rocky slopes	Moist valley bottoms
Growth form	Unbranched	Branched
Reproductive output	Larger inflorescences, more seeds	Smaller inflorescences, fewer seeds
Variation in inflorescence size	Highly variable, increases with soil moisture	Relatively invariable, independent of soil moisture
Demography	Virtually no adult survivorship	Populations in drier sites have lower adult survivorship and less frequent reproduction
Variation in number of seeds per pod	Strongly positively correlated with inflorescence size	Independent of inflorescence size, positively correlated with number of rosettes
Effects of pollinators	Increased seed quality, but not seed quantity	Increased seed quality, but not seed quantity

Source: T. P. Young, *Evol. Ecol.* 4:157–171 (1990).

Several explanations have been proposed for the occurrence of semelparous and iteroparous reproduction in plants. First, variable environments might favor iteroparity, which would reduce the variation in lifetime reproductive success by spreading reproduction over both good and bad years. This hypothesis can be rejected because semelparous plants tend to occur in more variable (usually drier) environments than their iteroparous relatives. Second, variable environments might favor semelparity when a plant can time its reproduction to occur during a very favorable year. Storing resources for the big event makes sense, just as not holding back resources for an uncertain episode of future reproduction also makes sense. *Carpe diem:* seize the day. This tactic is particularly favored when adult survival is relatively low and the interval between good years is long. Finally, attraction of pollinators to massive floral displays might favor plants that put all their effort into one reproductive episode. The few observations on this point are equivocal to mildly supportive. For example, in the semelparous rosette plant *Lobelia telekii,* which grows high on the slopes of Mount Kenya in Africa, a doubling of inflorescence size was seen to result in a fourfold increase in seed production.

Comparison of *Lobelia telekii* with its iteroparous relative *L. keniensis* (Table 11.5), like the comparison between agaves and yuccas, suggests that semelparity is associated with dry habitats that are highly variable in both space and time. Presumably, infrequent conditions that are highly favorable for the establishment of seedlings trigger the massive flowering episodes in these plants. In summary, semelparity appears to arise either when preparation for reproduction is extremely costly, as in the undertaking of long migrations to breeding grounds, or when the payoff for reproduction is highly variable and predictable.

Senescence

Although few organisms exhibit programmed death associated with reproduction, most do experience a gradual increase in mortality and a decline in fecundity resulting from the deterioration of physiological function. This phenomenon is known as **senescence,** and humans are no exception to the general pattern seen in virtually all animals. The rates of most physiological functions in humans decrease in a roughly linear fashion between the ages of 30 and 85 years, by 15–20% for nerve conduction and basal metabolism, 55–60% for volume of blood circulated through the kidneys, and 63% for maximum breathing capacity. Birth defects in offspring and infertility generally occur with increasing prevalence in women progressively older than 30 years (Table 11.6). Reproductive decline and death in old age do not result from abrupt physiological changes. Rather, these consequences of senescence follow upon a gradual decrease in physiological function with age.

Why does senescence exist, when survival and reproduction presumably confer advantages on an individual at any age? Senescence may reflect the accumulation of molecular defects that fail to be repaired, just as an automobile eventually deteriorates and has to be junked. Ionizing radiation and highly reactive free radicals break chemical bonds; macromolecules

TABLE 11.6 **Number of defects per 10,000 human births according to the age of the mother**

BIRTH DEFECT	AGE OF MOTHER (YEARS)						
	15–19	*20–24*	*24–29*	*30–34*	*35–39*	*40–44*	*45+*
Anencephalus	10	9	7	8	9	10	6
Spina bifida	12	11	10	10	13	13	16
Hydrocephalus	18	7	7	8	10	15	25
Microcephalus	1	1	1	1	1	13	—
Heart defects	17	18	17	17	24	28	50
Cleft lip, palate	10	11	11	10	12	16	—
Down syndrome	2	2	2	3	11	33	89
Clubfoot	15	14	11	11	12	24	18
Total	85	73	66	68	92	142	202

Source: S. Milham, Jr., and A. M. Gittelsohn, *Human Biol.* 3:13–22 (1965).

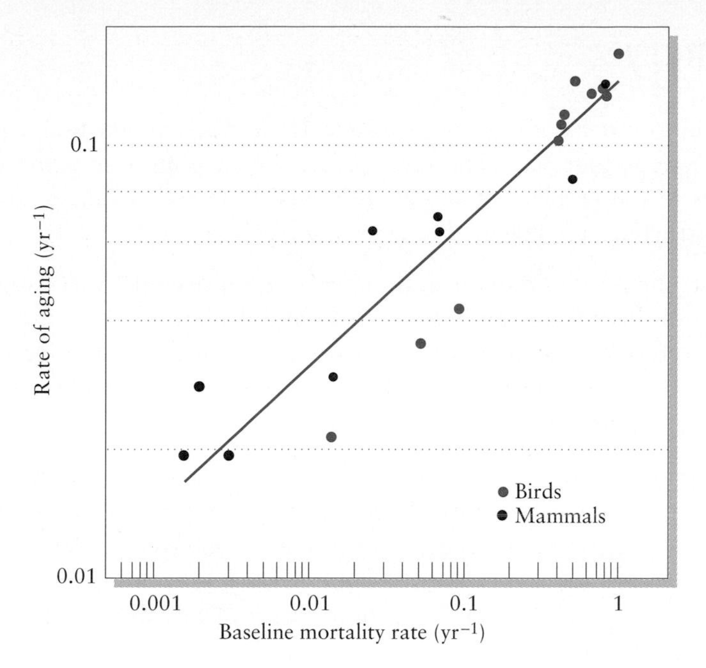

Figure 11.14 Species of birds and mammals with a high baseline mortality rate appear to have a high rate of aging as well. The rate of aging is the inverse of the number of years that would be required for 90% of adult individuals to die from causes of aging only. The baseline mortality rate is the rate for young adults. From R. E. Ricklefs and C. E. Finch, *Aging: A Natural History,* Scientific American Library, New York (1995).

become cross-linked; DNA accumulates mutations. This wear and tear cannot be the entire explanation for patterns of aging, however, because maximum longevity varies widely even among species of similar size and physiology. Many small insectivorous bats achieve ages in captivity of 10–20 years, whereas mice of similar size rarely live beyond 3–5 years. In addition, cellular mechanisms to repair damaged DNA and protein molecules appear to be better developed in long-lived animals than in their short-lived relatives. These observations suggest that rates of senescence may be under the influence of natural selection and evolutionary modification. Like any other life history trait, postponement of senescence may exact costs in terms of reduced reproduction at younger ages. For example, if repair processes require both time and resources, and if mortality is so high that few individuals live to old age, it may be more productive to allocate resources to early reproduction and let the body fall apart eventually.

In a population without senescence, accidental causes of death affect young and old equally. Even so, progressively fewer individuals survive as age increases. With few individuals living to old age, progressive physiological deterioration has little effect, on average, on lifetime reproduction. By this reasoning, the strength of selection on changes in survival or fecundity declines as these changes are confined to older ages. Thus, selection tends to favor improvements in reproductive success at a young age over those later in life. The lower the survival rate of adults, on the whole, the weaker is selection for improvement in reproductive success at older ages, and the faster senescence should progress. This relationship can be tested by comparing the rate of senescence with the "baseline" mortality rate experienced by young adults in a population prior to the onset of aging. As you can see in Figure 11.14, data for birds and mammals seem to bear out this prediction.

Summary

1. Life history traits of organisms include age at first reproduction, reproductive rate, and life span. The values of these traits can be interpreted as solutions to the problem of allocating limited time and resources among various structures, physiological functions, and behaviors.

2. Most phenotypic traits of individuals are sensitive to variations in the environment. This response of form and function to the environment is referred to as phenotypic plasticity, and the quantitative relationship between phenotypic values and environmental variables is called the reaction norm.

3. Phenotypic plasticity itself is under genetic control. The variation in the sensitivities of individuals with different genotypes to variation in the environment is referred to as genotype-environment interaction. When each genotype is superior to others over some range of environmental conditions, genotype-environment interactions can lead to habitat and microhabitat specialization.

4. Variations in different life history traits are often correlated among species. Delayed reproduction, long life, and low reproductive rates—and their opposites—are frequently associated.

5. Plant ecologists have recognized clusters of life history attributes, including relative allocation of resources to reproduction and size of seeds, associated with ruderal, stressful, and highly competitive environments.

6. Theories concerning life history variation among species, including correlations among life history traits, are based on the principle that limited time and resources are allocated among competing activities in such a way as to maximize lifetime reproductive success.

7. Delayed reproduction is favored when the life span is relatively long and when immature individuals benefit from increased growth or accumulation of experience by having greater fecundity later in life.

8. When survival between breeding seasons is low or requires large sacrifices in fecundity, annual life histories are favored over perennial life histories.

9. High prereproductive survival as compared with adult survival favors increased reproductive effort, or investment in offspring, at the expense of adult survival and future reproduction.

10. In a variable environment, allocation of resources toward components of the life history showing the least variation tends to increase. Thus, highly variable adult survival favors earlier maturation and increased reproductive investment.

11. When reproduction requires costly preparation, selection may favor a single all-consuming reproductive event followed by death, as in salmon. This pattern of reproduction, called semelparity, is the converse of iteroparity, or repeated reproduction. Some cases of semelparity in plants, including

agaves and lobelias, appear to be associated with dry habitats and marked temporal variation in conditions for seedling establishment.

12. Senescence, the progressive deterioration of physiological function with age, causes declines in fecundity and probability of survival. Senescence is caused by the wear and tear and the detrimental biochemical changes brought about just by living. Senescence is also subject to evolutionary modification, and rates of aging vary considerably among animals. Owing to accidental deaths, few individuals survive to old age under any circumstances. As a result, the strength of selection diminishes on traits expressed at progressively later ages, all the more so in populations subjected to high overall mortality rates. Therefore, individuals in populations subjected to higher extrinsic mortality age faster.

Suggested readings

Bazzaz, F. A., N. R. Chiarello, P. D. Coley, and L. F. Pitelka. 1987. Allocating resources to reproduction and defense. *BioScience* 37:58–67.

Dijkstra, C., A. Bult, S. Bijlsma, S. Daan, T. Meijer, and M. Zijlstra. 1990. Brood size manipulations in the kestrel *(Falco tinnunculus)*: Effects on offspring and parental survival. *Journal of Animal Ecology* 59:269–286.

Fleming, I. A., and M. R. Gross. 1989. Evolution of adult female life history and morphology in a Pacific salmon (Coho: *Oncorhynchus kisutch*). *Evolution* 43: 141–157.

Gross, M. R. 1996. Alternative reproductive strategies and tactics: Diversity within sexes. *Trends in Ecology and Evolution* 11:92–98.

Janzen, D. H. 1976. Why bamboos wait so long to flower. *Annual Review of Ecology and Systematics* 7:347–391.

Reznick, D. 1985. Costs of reproduction: An evaluation of the empirical evidence. *Oikos* 44:257–267.

Reznick, D. N., H. Bryga, and J. A. Endler. 1990. Experimentally induced life-history evolution in a natural population. *Nature* 346:357–359.

Ricklefs, R. E., and C. E. Finch. 1995. *Aging: A Natural History.* Scientific American Library, New York.

Rose, M. R. 1991. *Evolutionary Biology of Aging.* Oxford University Press, Oxford.

Schlichting, C. D. 1989. Phenotypic integration and environmental change. *BioScience* 39:460–464.

Sibly, R. M., and P. Calow. 1986. *Physiological Ecology of Animals: An Evolutionary Approach.* Blackwell Scientific Publications, Oxford.

Stearns, S. C. 1992. *The Evolution of Life Histories.* Oxford University Press, Oxford.

Strathmann, R. R. 1990. Why life histories evolve differently in the sea. *American Zoologist* 30:197–207.

Williams, G. C. 1966. Natural selection, the costs of reproduction, and a refinement of Lack's principle. *American Naturalist* 100:687–690.

Young, T. P., and C. K. Augspurger. 1991. Ecology and evolution of long-lived semelparous plants. *Trends in Ecology and Evolution* 6:285–289.

CHAPTER 12

SEX

Reproduction is the ultimate goal of an individual's life strategy. In most animals and plants, it is accomplished by the production of **gametes:** eggs or ovules are made by female sexual organs, and sperm or pollen grains by male sexual organs. Male and female gametes join together in fertilization to form the next generation. Some of the most important and fascinating attributes of life concern sexual function. Among these are gender, the gender ratio of an individual's offspring (when it can control this), and the various devices and behaviors that enhance the success of an individual's gametes, which often means improving its own success in mating. The peacock's glorious tail, whose purpose is to make its bearer more attractive to females, is one of nature's most fantastic productions. Indeed, sex underlies much of what we see in nature. In this chapter, we shall consider how sexual function influences the evolutionary modification of organisms and many of their behaviors as individuals. A good place to start is with sex itself.

Sexual function

Sexual reproduction unites two **haploid** gametes to form a **diploid** cell, the **zygote.** Sexual reproduction mixes the genetic material of two individuals, resulting in new combinations of genes in their offspring. As a result of this **recombination,** siblings differ from one another genetically. Thus, in a variable environment, at least some offspring of a sexual union are likely to have a genetic constitution that enables them to survive and reproduce, regardless of the particular conditions. Sexual reproduction may also produce new combinations of genetic factors (genotypes) previously absent from a population. Because the expression of individual genes is influenced by other genes, new combinations of old genes may provide new variation in phenotypic expression for natural selection to work on. In the view of many biologists, sexual reproduction evolved very early in the history of life as a means of generating the genetic diversity necessary to cope with a varied and changing environment. In contrast, progeny produced by asexual reproduction are identical to one another and to their parent, and thus none of them are likely to be well adapted to novel conditions.

The sexual gametes themselves are formed by a special type of cell division, called **meiosis,** which occurs in germ cells, specialized cells within the primary sex organs—the gonads. The more usual sort of cell division, which is called **mitosis,** is responsible for producing all the other cells in the body. Mitosis involves replication of the genetic material, or DNA, of a cell and its division into two daughter cells, each receiving identical copies of all of the genetic material of the parent cell (Figure 12.1). Each of an individual's cells contains a set of chromosomes inherited from its mother (maternal chromosomes) and a corresponding set (except for the so-called sex chromosomes) inherited from its father (paternal chromosomes). During mitosis, each daughter cell receives one copy of the maternal chromosomes and one copy of the paternal chromosomes, which make up a set identical to that of the parent cell.

Meiosis features two major modifications of mitotic cell division. First, after replication, the homologous maternal and paternal chromosomes exchange parts. As a result of these so-called crossing-over events, each chromosome may harbor new combinations of maternal and paternal genes. Second, the cell then undergoes two divisions, so that each of the four cell products of meiosis is haploid—that is, it contains only one member of each of the chromosome pairs present in diploid cells. Each of these haploid cell products of meiosis contains a single full set of chromosomes, but whether a particular chromosome has a paternal or maternal origin is generally up to random chance. These haploid cells are the ones that eventually develop into gametes. As a consequence of meiosis, the genetic makeup of each zygote is a unique, random combination of the genetic material of each of the individual's four grandparents.

The alternative to sex is **asexual reproduction,** by which an individual reproduces genetically exact copies of itself (except for the occasional mutation), either by vegetative means or by producing a diploid, egglike propagule that develops directly without fertilization. Vegetative reproduction occurs in many plants, most of whose cells retain the ability to produce an entire new individual. Thus, shoots may sprout from roots or even from the margins of leaves, and then separate from the parent plant to become new individuals (Figure 12.2). Many simple animals, such as hydras, corals, and their relatives, can form buds in their body walls that develop into new individuals. When these remain attached to the parent individual, a colony forms, as in the case of hydroids, corals, bryozoans, and many other aquatic animals; when buds detach, independent new individuals are formed.

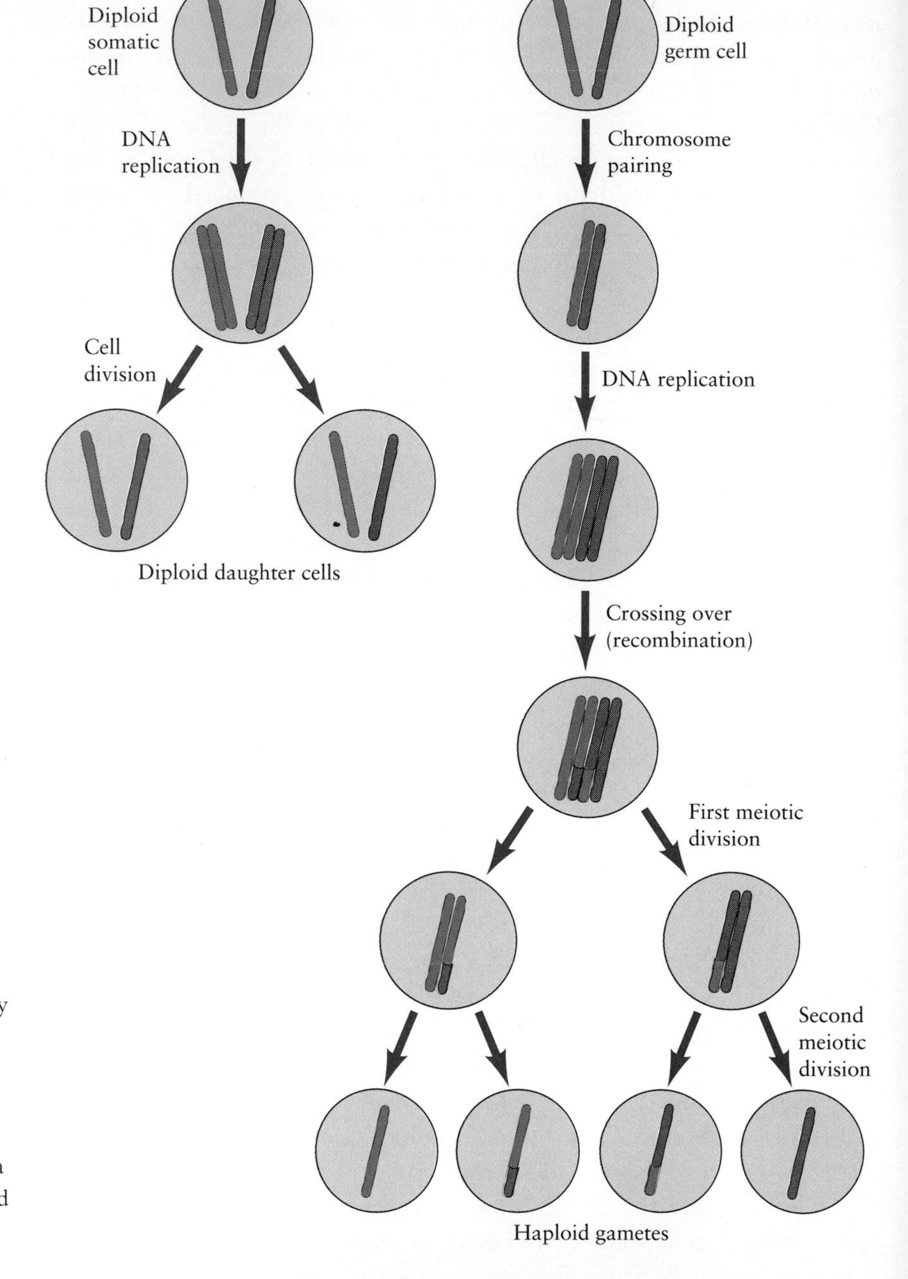

Figure 12.1 The steps in mitosis and meiosis, showing the fates of the maternal (black) and paternal (color) chromosome sets. During mitosis, each chromosome of the parental cell, or a genetically exact copy of it, is passed intact to each daughter cell. During meiosis, the maternal and paternal chromosomes pair, replicate, and then exchange segments during the process called crossing over. Each gamete receives a single chromosome, containing recombined segments exchanged during crossing over.

Some animals produce diploid eggs asexually by modifying the usual sexual production of gametes by meiosis. We see such alterations in all-female populations of fish, lizards, and some insects, to name a few. They may occur by direct transformation of diploid germ line cells into egg cells, in which case all eggs are genetically identical. Alternatively, meiosis may proceed through the replication and crossing-over stages and the first meiotic division; suppression of the second meiotic division at this point will result in diploid egg cells that nonetheless differ from one another genetically because of recombination. In another variation, meiosis may proceed to completion, but gametes may then fuse to form diploid egg

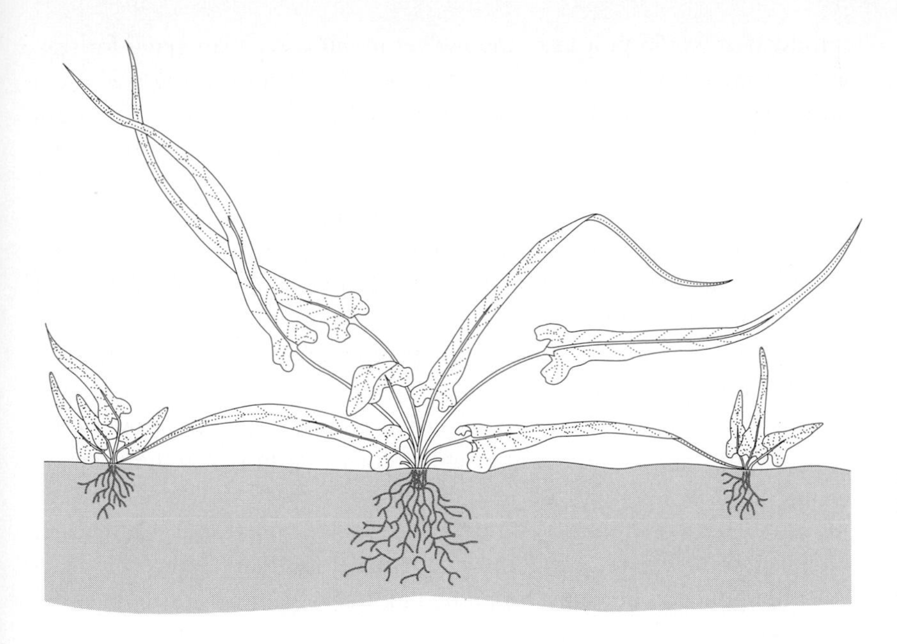

Figure 12.2 The walking fern propagates asexually by sprouting a fully formed plant from the tip of a leaf. After V. A. Greulach and J. E. Adams, *Plants: An Introduction to Modern Botany,* John Wiley & Sons, New York (1962).

cells. This process is a type of self-fertilization, and its products also vary genetically. Finally, individuals that have both male and female sexual organs may form male and female gametes and then fertilize themselves. This method of reproduction is frequently encountered in plants. It is sexual in that both types of gametes are made, but asexual in that progeny are produced by a single individual.

Sexual and asexual reproduction would appear to be viable alternative life histories. Asexual reproduction is widespread among plants and is found in all major groups of animals, with the exception of birds and mammals. And, indeed, sex has its costs. Gonads are expensive organs that are of little direct benefit to the individual and require resources that could be devoted to other purposes. Mating itself is a major production for animals and plants, involving floral displays to attract pollinators and elaborate courtship rituals to please mates. These costs presumably are offset by the advantages of producing genetically varied offspring when the environment itself varies in time or space. Genetic variation among offspring assures that at least some will be well adapted to conditions that differ from the parental environment.

When the sexes are separate—that is, when individuals are either male or female—sexual reproduction has a much higher cost arising from the fact that only half the genetic material of each individual offspring comes from a given parent. Thus, compared with asexually produced offspring, which contain only the genes of their single parent, the progeny of a sexual union contribute only half as much to the evolutionary fitness—that is, the contribution of its genes to future generations—of either parent. This 50% cost of sexual reproduction to the individual parent is sometimes referred to as the **cost of meiosis.**

Evolution favors genetically determined traits that reproduce the greatest number of copies of their genes in future generations. If sexual and asexual reproduction were alternative genetic traits, genes for asexual reproduction would propagate themselves much faster than genes for sexual reproduction whenever the production of varied offspring was not a consideration. A female can produce a limited number of eggs. Thus, from the standpoint of an individual female, asexually produced offspring would have twice as many copies of her genes as sexually produced offspring. Under this scenario, males would not only be superfluous, but mating with them would reduce the personal fitness of an individual female because only half of her genes would be transmitted to her progeny, the other half coming from male gametes.

In light of the high genetic (fitness) cost of sexual reproduction to many kinds of animals and plants, one might wonder why sex is so pervasive. Are the advantages of producing varied offspring enough to overcome the 50% disadvantage of sex in species like ourselves, in which the sexes are separate individuals? It is true that the cost of meiosis does not apply to individuals having both male and female sexual function because each individual produces offspring through female function. However, once sexual reproduction becomes established in such organisms, it may be difficult to get rid of if the sexes are later separated. The cost of meiosis also does not apply when the sexes are separate, but males contribute to the number of offspring produced through male parental care, because males effectively produce offspring by means of their mate's sexual function, but their own parental investment.

Asexual reproduction does occur, of course, but only rarely in animals. Once it does arise, asexual reproduction may initially have a high success rate. However, the long-term evolutionary potential of asexual populations may be low because of their greatly reduced genetic variability, and their lines may die out over time. A combination of infrequent origination from sexual ancestors and a high rate of extinction could limit the occurrence of asexual reproduction. Indeed, most cases of asexual reproduction appear in species that belong to genera, such as *Ambystoma* (salamanders), *Poeciliopsis* (fishes), and *Cnemidophorus* (lizards), in which other species are sexual. This indicates that, for the most part, asexual forms do not have a long evolutionary history. If they did, one would expect to see whole families or even orders of animals sharing this derived trait.

Another intriguing possibility is that sexual reproduction serves as a mechanism to weed out harmful genetic mutations that inevitably accumulate in the germ line of each individual. As the cells of the body proliferate, spontaneous changes in DNA and damage to DNA caused by radiation and reactive chemicals cause mutations. These mutations are passed on to each daughter cell produced by mitosis. With continued asexual reproduction, each cell in a clone will eventually accumulate so many mutations that it may cease to function and die. Gamete formation and sexual reproduction may stave off this deterioration of the genotype in two ways. First, recombination of genes during meiosis produces, just by chance, a few gametes that lack the harmful genetic mutations that may have arisen in the germ line of an individual. Second, harmful genetic factors present in one gamete may be masked by their unmutated, properly functioning counterparts in the gamete with which they fuse to form a zygote. Therefore, among the varied gametes and varied offspring produced, some may be relatively free of genetic contamination. Thus, the advantage to sexual reproduction may be independent of variation in the environment, but related more intimately to the intrinsic variation produced by dysfunction of the genetic mechanism itself.

Sex remains one of the most challenging problems that face biologists. At this point, however, we shall accept the fact that sex is with us, and turn to explore some of the consequences of sexual reproduction in the lives of organisms.

Gender

As with many issues, botanists and zoologists have divergent perspectives on sex. We humans are used to thinking in terms of two genders, female and male. But female and male sexual functions may be combined in the same individual, or an individual may change its gender during its lifetime. In most species of plants, individuals have both male and female functions (Figure 12.3). When both functions occur in the same individual, biologists label the individual a **hermaphrodite,** after the mythological Hermaphroditus, son of Hermes and Aphrodite, who while bathing became joined in one body with a nymph. Hermaphrodites may be **simultaneous,** as in the case of many snails and most worms, or they may be **sequential:** male first in some mollusks and echinoderms, female first in some fishes.

Figure 12.3 A flower of the hibiscus plant, showing the receptive structure (stigma) of the female part of the plant (F) and the anthers of the male part of the plant (M), which produce the pollen.

Botanists refer to plants exhibiting separate sexes as **dioecious,** from the Greek *di-* ("two") and *oikos* ("dwelling," also the root of the word *ecology*). They apply an additional term, **monoecious,** to individual plants that bear separate male and female flowers, rather than so-called **perfect flowers** that include both male and female parts. Though perfect-flowered hermaphrodites are the rule among plants (by one estimate they account for 72% of plant species), nearly all imaginable combinations of sexual patterns are known, including populations with hermaphrodites and either male or female individuals; populations with male, female, and monoecious individuals; and hermaphroditic individuals with both perfect flowers and also either male or female flowers.

Which sexual arrangement occurs in a given population depends on the relative fitness costs and benefits that would accrue to individuals having either or both male and female sexual function. One can measure the fitness contributions of male and female sexual function by the number of sets of genes transmitted to offspring through either male or female gametes. When females can achieve a certain amount of male function by giving up a smaller amount of their female function, selection favors individuals that shift resources to male function. Similarly, males that add female function when this change does not cut deeply into their male productivity are also selected. It would seem that both male and female flowers could add the other sexual function with little cost. After all, the basic flower structure, and the floral display necessary to attract pollinators, are already in place in unisexual flowers. Thus, we would expect hermaphroditism to arise frequently, as it has among plants and most simple forms of animal life.

Sequential hermaphroditism reflects changes in the costs and benefits of male and female sexual function as an organism grows. In some marine gastropods having internal fertilization, such as the slipper shell *Crepidula,* insemination requires the production of only small amounts of sperm. Hence male function consumes few resources and has little effect on growth. As a consequence, individuals of many such species are male when they are small and become female when they are large and thus able to produce correspondingly large clutches of eggs. Female-to-male sequential hermaphroditism occurs when males compete aggressively for opportunities to mate with females, and large size confers an advantage in these contests. This circumstance seems to occur in some highly territorial reef-inhabiting fishes, such as the wrasses. In these species, there would be no point to a small individual becoming male because it would be excluded from mating. Female function at this stage gives an individual an increment of fitness that it could not gain otherwise. When an individual wrasse has grown to large size, however, it can achieve greater reproductive success as a male, and so the switch occurs.

Separate sexes provide the best compromise when gains from adding the function of one sex bring about even greater losses in the function of the other. This may occur when establishing new sexual function in an individual entails a substantial fixed cost before any gametes can be produced. Sexual function requires gonads, ducts, and other structures for transmitting gametes as well as secondary sexual characteristics for attracting mates and competing with individuals of the same sex. In many animals in which maleness requires specializations for mate attraction and antagonistic interaction with other males, or in which femaleness requires specializations for egg production or brood care, such fixed costs may put hermaphroditism at a disadvantage compared with sexual specialization. In fact, hermaphroditism occurs rarely among animal species that actively seek mates and that engage in brood care, but it is much more common among sedentary aquatic forms that simply shed their gametes into water.

Sex ratio

When the sexes are separate, one may define a **sex ratio** among the progeny of an individual, or within a population, as number of males relative to number of females. Two facts about sex ratio stand out. First, many populations have very nearly equal numbers of females and males. Second, many populations present exceptions to this rule. Because females and males occur in the human population in a ratio of approximately 1:1, and because roughly equal numbers of females and males typify populations of most species, we consider the 1:1 sex ratio the usual condition and regard deviations from this ratio as special cases.

Biologists explain the predominant 1:1 sex ratio by the following simple reasoning. Every product of a sexual union has exactly one mother and one father. Consequently, if the sex ratio of a population is not 1:1, individuals of the rarer sex will enjoy greater reproductive success because they will compete for matings with fewer others of the same sex. For example, if a population of 5 males and 10 females produced 100 offspring, each male would contribute 20 sets of genes, but each female would contribute only 10 sets of genes. Thus, individuals of the rarer male sex would contribute more sets of their chromosomes to subsequent generations. Therefore, when there are more females than males in a population, any genetic tendency on the part of a parent to produce a larger proportion of male offspring will be favored, and the frequency of males in the population will increase, bringing the sex ratio closer to 1:1. Similarly, when females are the rarer sex, genotypes that increase the proportion of female progeny will be favored, and the frequency of females will increase. When males and females are equally numerous, individuals of both sexes contribute equally to future generations, on average, and different frequencies of males and females among the progeny of individuals are of no consequence to their relative long-term reproductive success.

Some situations exist, however, in which a parent may benefit from a **skewed sex ratio** among its progeny, meaning that a preponderance of either males or females is produced. As we have already noted, competition for matings among individuals of one sex (usually males) can create tremendous variation in reproductive success: when competition is keen, some males may achieve many matings, others none. The largest males may win all the contests over access to females. In mammals, a mother cares directly for her offspring through gestation and lactation, and her condition is likely to influence the condition and reproductive success of her male offspring. As a result, females in good condition ideally should produce male offspring, which will grow large and fare well in male-male competition for mates. Females in poor condition should invest more in female offspring, which are likely to mate successfully regardless of the parental care they receive.

Experimental confirmation of this idea emerged from a laboratory study of female wood rats *(Neotoma floridana).* In this species, as in most mammals, females normally invest equally in male and female offspring. But when investigators restricted food intake during the first 3 weeks of lactation to below the maintenance level of a nonreproductive female, mothers actively rejected the attempts of male offspring to nurse. As a result, males starved, and the sex ratio of the offspring at 3 weeks shifted to about one male for every two females. Thus, faced with the likelihood that their young would be poorly nourished and that some of them would probably die before they were weaned, the mothers favored their female offspring.

Local mate competition and sex ratio in wasps

The usual explanation for a 1:1 sex ratio depends on individuals having the opportunity to mate with unrelated individuals within a large population. Otherwise, individuals of the rarer sex would not benefit from their potentially greater reproductive potential. When individuals disperse only short distances or not at all, or when they mate prior to dispersal, mating often takes place among close relatives. At the extreme, mating may occur among the progeny of an individual parent. In this situation, **local mate competition** among males would take place among brothers. From the standpoint of the parent of these siblings, one son would serve just as well as many to fertilize his sisters and propagate the parent's genes. In the case of brother-sister (sib) mating, the number of copies of her genes that a mother passes on to her grandoffspring depends only on the number of daughters she produces, because each son's contribution to the daughters' offspring will also come from the mother. Thus, females that produce daughters at the expense of sons will have more grandoffspring and greater evolutionary fitness.

Sib mating occurs commonly in certain wasps that parasitize other insects (see, for example, Figure 20.1) and whose progeny mate on their hosts before dispersing. These wasps can alter the sex ratio among their progeny, and they do so in a manner predicted by the degree of inbreeding. Hymenopterans (bees, ants, and wasps) have an unusual sex-determining mechanism by which fertilized eggs produce females and unfertilized eggs produce males (Figure 12.4). As a result, females are diploid and males haploid, giving rise to the term **haplodiploidy.** Reproductive females can control the sex ratio of their offspring simply by using sperm stored when they mate to fertilize eggs, thereby producing females, or not, thereby producing males.

Many wasps parasitize other insects or complete their larval development within the fruits of certain plants. For some of these species, hosts are so scarce and mates so difficult to find that females mate where they hatch before dispersing to find new hosts on which to lay their own eggs. When a host is parasitized by a single female wasp, the female offspring within each brood are limited to mating with their brothers. Under this circum-

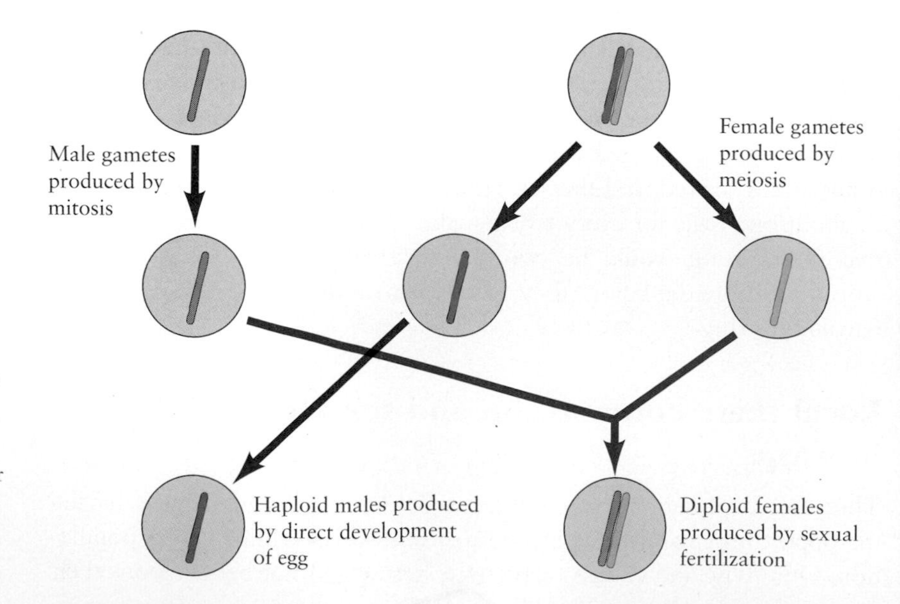

Figure 12.4 A diagram of haplodiploidy in the Hymenoptera, showing the fate of maternal and paternal sets of genes and how a female can determine the sex of her offspring by using stored sperm to fertilize eggs to produce diploid females or by not fertilizing them to produce haploid males.

stance, male offspring make a reduced contribution to the reproductive success of their mothers, as described above, and therefore wasps skew the sex ratios of their progeny greatly in favor of females—to the point of producing only one male per brood in some species. Males of many of these species lack wings and, in extreme cases, fertilize females as larvae within the host. When two or more females lay their eggs in the same host, however, their male offspring can engage in local competition for mates with the progeny of unrelated wasps. As the possibility that sons might inseminate the female offspring of another wasp increases, the proportion of males in broods increases, just as one would expect.

Mating systems

The **mating system** of a population is the pattern of matings between males and females—specifically, the number of simultaneous or sequential mates and the permanence of the pair bond. Like its sex ratio, the mating system of a population is based on behavioral tendencies of individuals that are subject to natural selection and evolutionary modification. As a consequence, mating systems are usually correlated with attributes of the species and the environment.

It is a basic asymmetry of life that a female's evolutionary fitness depends on her ability to make eggs and otherwise provide for her offspring, whereas a male's fitness usually depends on the number of matings he can procure. Large female gametes individually require more resources than male gametes, so a female's ability to gather resources to make eggs determines her fecundity. Males of many species neither contribute resources to their mates (for example, by defending territories within which their mates can feed, or by feeding their mates directly) nor care for their young. In such cases, a male can have more offspring only by mating with additional females.

When a male mates with as many females as he can locate and persuade, and provides his offspring with nothing more than a set of genes, he is said to be **promiscuous.** Promiscuity usually precludes a lasting pair bond. It also tends to increase the variation in mating success among males: some individuals may obtain dozens of matings while others get none. Among animal taxa as a whole, promiscuous mating is by far the commonest system, and it is universal among outcrossing plants.

When a male can contribute to the fecundity of his mate by procuring resources for her or by caring for their offspring directly, pair bonds may outlast the copulatory rapture of promiscuous species. As males can increasingly augment the realized fecundity of a single mating through caring for their mates and progeny, mating systems progress from promiscuity (least care) through polygamy to strict monogamy (most care, relative to the female).

In **polygamy,** a single individual of one sex forms long-term bonds with more than one individual of the opposite sex. Normally males mate with more than one female, in which case the system is referred to as **polygyny** (literally, "many females"). Polygyny may involve defense of several females against mating attempts by other males (a harem), or defense of territories or nesting sites to which more than one female gravitates to raise her young. Thus, polygyny may arise because a male can prevent access by other males to more than one female residing within his territory, in which case his contribution to his progeny may be primarily genetic, or because he can control or provide resources that a female needs to reproduce. The rare cases of a single female having many mates (usually sequentially) are referred to as **polyandry.** Among birds, polyandry has been identified only among a few species of wading birds—sandpipers and jacanas.

Monogamy is the formation of a pair bond between one male and one female that persists through the period required to rear offspring, and which may endure even until one of the pair dies. Monogamy arises primarily when males can contribute substantially to the number and survival of their offspring by providing parental care. Hence it is most common in species with dependent offspring that can be cared for equally well by either sex. Monogamy is not common in mammals because males neither carry the developing embryo nor produce milk. But it is common among birds, especially those in which parents feed their offspring. Males and females can incubate eggs and feed young equally well.

Recent genetic surveys of monogamous bird populations have revealed that many offspring are fathered by males other than a female's mate, a result of so-called **extra-pair copulations,** or **EPCs.** As many as a third of the broods produced by some species contain one or more offspring sired by a different male. Most EPCs involve males on neighboring territories, indicating considerable opportunism and infidelity in natural populations. It is not known whether this behavior benefits a mated female, but it surely increases, at relatively little cost, the fitness of neighboring males. The constant threat of EPCs also has selected strongly for **mate guarding** behaviors on the part of males during their mates' periods of fertility. It is also reasonable to suppose that the intensity of male parental care should decrease with increasing uncertainty of paternity, and so frequent EPCs might even lead to the breakdown of a monogamous mating system.

A species' mating system is strongly influenced by its reproductive biology, particularly the care of offspring. Mating system seems to go hand in hand with ecology as well. In African weaverbirds, for example, monogamy predominates among insectivorous species inhabiting forest and savanna, whereas polygyny is the rule among seed-eating species of open savannas and grasslands. Furthermore, most monogamous species breed as solitary pairs on widely dispersed territories, whereas polygynous species are invariably colonial. But why is it that some species adopt polygynous habits and others do not? And why should mating system be related to habitat?

The polygyny threshold model

As long as his territory holds sufficient resources, a male gains by increasing the number of his mates. A female benefits from choosing a territory or a mate of high quality. Thus, polygyny arises when a female can obtain greater reproductive success by sharing a male with one or more other females than she can by forming a monogamous relationship. Suppose that the quality of two males' territories differs so much that a female could rear as many offspring on the better territory, sharing it with other females and having little or no help from her mate, as she could on the poorer territory with help from her monogamous mate. The difference in territory quality at which polygynous and monogamous females do equally well is referred to as the **polygyny threshold** (Figure 12.5). According to the polygyny threshold model, polygyny should occur only when the quality of male territories varies so much that some females will have higher reproductive success mated to a polygynous male on a high-quality territory than they would mated monogamously to a male on a poor-quality territory.

In cattail marshes throughout North America, male red-winged blackbirds establish territories in early spring. Marsh habitat is heterogeneous with respect to vegetation cover and water depth, which affect food supply and the safety of nests, and therefore territories vary greatly in their intrinsic quality. Females return to the breeding grounds after the males, at which time they settle in and pair with established males. Female blackbirds appear to assess the quality of male territories, and the first individuals to arrive pair monogamously with the best males—that is, those holding the best territories. Latecomers are faced with the choice between pairing monogamously with a low-quality male or pairing polygynously with a high-quality male, but sharing his territory's resources with one or more other females. In contrast to blackbirds, birds of forests live in habitats that are more homogeneous than marshes. Bird territories vary less in quality, and most species of birds are primarily monogamous; few territories drop below the polygyny threshold.

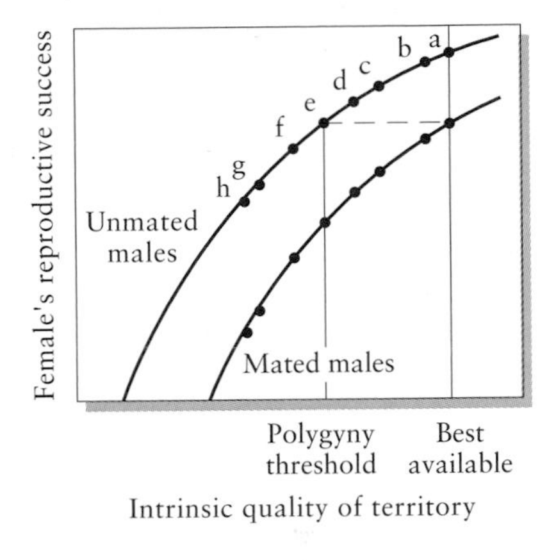

Figure 12.5 The polygyny threshold model. Female fitness (reproductive success) varies with the intrinsic quality of a male's territory. The presence of a mated female reduces that quality to subsequent females. Females should select unmated males (a through e) until the quality of the territory of any remaining unmated male (f, g, and h) drops below the "net" quality (presence of another female "factored in") of the best territory controlled by a mated male (a). At this point, called the polygyny threshold, female choice will alternate between mated males (a through e, which then become bigamists) and unmated males (f through h). After G. H. Orians, *Am. Nat.* 103:589–603 (1969).

Promiscuous mating

Although promiscuous mating is the rule in the animal and plant kingdoms, it appears rarely and sporadically among birds, being predominant only among pheasants and grouse and a few groups of tropical, frugivorous species—manakins, cotingas, and birds of paradise. Because such mating systems are so unusual among birds, ecologists have paid considerable attention to those cases in which promiscuity has surfaced on the calm sea of avian monogamy. Males of most bird species contribute substantially to the fecundity of their mates (and therefore to their own fecundity) either by defending territories or by directly providing for their young. The key to the evolution of promiscuity in birds therefore appears to be the emancipation of males from parental investment without substantially cutting into the reproductive success of their mates.

This shift can happen in several situations, which seem to explain the distribution of promiscuity among birds. First, because grouse chicks feed themselves, males can do relatively little to increase the fecundity of their mates. Females of such species (Figure 12.6) can rear nearly as many young by themselves as they could with the help of males. The same is true of mammals, in which the female alone nurses the young, and potential male contributions (defending resources, protecting the litter, and providing food directly to the female) pale by comparison. Second, the resources of some fruit-eating birds may be conspicuous and of fixed quantity, in which case a female alone can gather as much as a male and a female working together. Clearly, when males can desert their mates without diminishing the number of their progeny, they can increase their reproductive success by attempting to attract additional mates.

Mixed reproductive strategies

In many fishes with promiscuous mating systems, males defend territories to which they attract females to spawn with them. The males then take care of the eggs until they have hatched. Indeed, fertilized eggs do not survive without the constant attention and protection of a territorial male. The bluegill sunfish is typical of such species. Because males compete vigorously for good spawning sites, they usually do not mature until they are at least 7 years old and large enough to fend off other males. However, a small number of males adopt a different tactic: they mature sexually at 2 years and a small body size, and they sneak matings with females attracted to territorial males. Sneaker males are reproductive parasites in that the survival of their offspring depends on the care of territorial males. Whether a male becomes a territory holder or a sneaker is probably genetically based.

The mating system of the bluegill sunfish is a good example of a **mixed reproductive strategy,** in which two or more phenotypes are stably maintained in a population because each has a higher fitness when rare. We have already seen this principle of **frequency-dependent selection** operating on the proportion of males and females among progeny to maintain a balanced 1:1 sex ratio. In the case of the bluegill sunfish, opportunities for sneak matings are limited by vigilance of territory holders, so as more and more sneakers compete for these opportunities, their reproductive success goes down.

(a)

(b)

Figure 12.6 (a) Two male sharp-tailed grouse displaying on a communal courting area (lek) in southern Michigan. (b) The female has a dull plumage compared with that of the male. Courtesy of U.S. Soil Conservation Service.

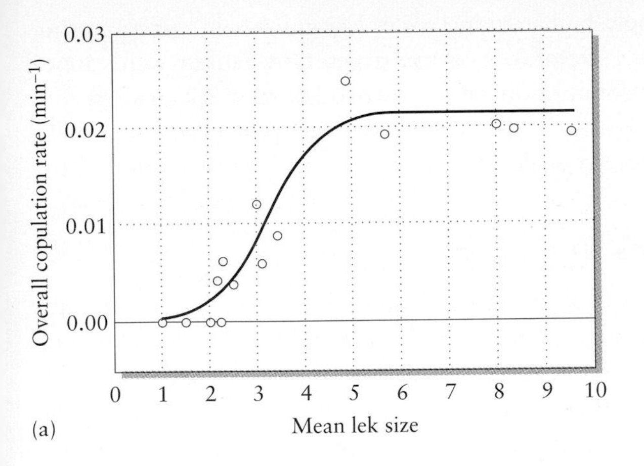

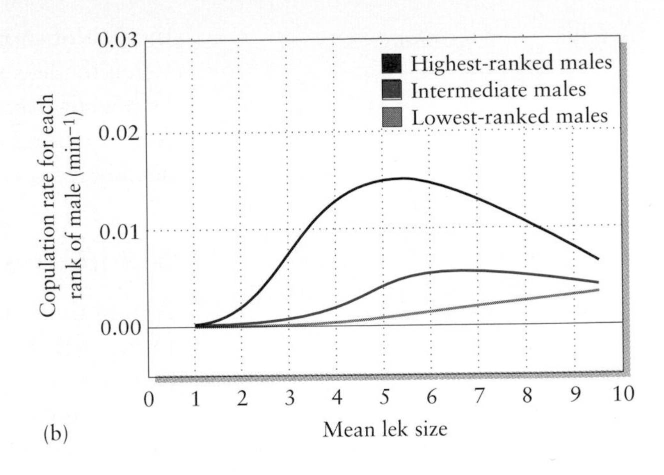

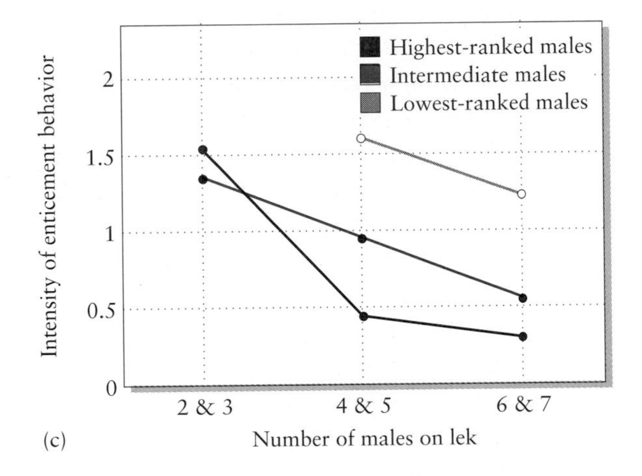

Figure 12.7 The influence of lek size on male mating success and enticement behavior in ruffs *(Philomachus pugnax)*. (a) The rate of copulation on a lek increases with lek size up to a size of about five males. (b) Most of the increase is enjoyed by the highest-ranking male, but subordinate males benefit by obtaining matings when leks exceed five individuals. (c) High-ranking males try to entice other males to join the lek only when it is small, but subordinate males maintain enticement behaviors even after a lek has grown to a large size. After F. Widemo and I. P. F. Owens, *Nature* 373:148–151 (1995).

A conspicuous feature of the mating systems of many promiscuous birds, as well as other promiscuous animals such as frogs, is the formation of communal mating areas, called **leks,** where males congregate to perform elaborate displays to attract females to mate with them. Leks appear to form because several males displaying together attract more females than do single males displaying alone. In a species of sandpiper called the ruff, males form leks of up to ten individuals. Within the lek, each male has a courtship area to which a female may come to mate, and the socially dominant males occupy the best locations within the lek.

Detailed observations on ruffs in Sweden showed that the overall frequency of matings increased with lek size up to a mean lek size of five males. The maximum copulation rate was about three per hour. Most of these matings were gained by the highest-ranking male within the lek. Only after lek size had increased to more than five males did the second- and third-ranking males gain a significant share of the matings, at the expense of the highest-ranking male (Figure 12.7). Because lower-ranking males fare so poorly, one wonders why they join leks at all. It is just not possible for us to think like ruffs, but we can easily observe that residents on a lek engage in a variety of behaviors—wing-flapping, jumping up and down, and hovering over the lek—to entice other males flying by to join them. Not surprisingly, high-ranking males lose interest in enticing others when the leks reach a size of four or five males. Low-ranking birds appear to benefit from the commotion on a crowded lek to sneak matings from time to time, and so they solicit other males to join even when a lek already has more than five males.

Mating systems in plants

Almost three-quarters of all species of flowering plants are hermaphroditic, meaning that an individual plant has both male and female sexual function. Most of these species have mechanisms that prevent self-fertilization, resulting in a fully outcrossed population. In the terms applied to animal populations, such plants would be referred to as promiscuous in their mating behavior. In fact, individual plants have relatively little control over where their pollen will land, or from whom they will receive pollen. However, self-compatibility and inbreeding (selfing) are apparently favored, and do occur, in habitats where pollinators are scarce (windswept coastal landscapes, for example), and in species that are good colonizers of remote locations. Often, only a single colonizing individual gains an initial roothold, so that selfing is the only option that can lead to a new population.

The other one-quarter of plant species have adopted a variety of mating systems. Two of the commonest of these are gynodioecy, in which populations contain a mixture of female and hermaphroditic individuals, and dioecy, in which populations are made up of only male and female individuals. The ecological conditions that favor one or another mating system in plants are poorly understood. Several possible factors have been suggested, including specialization of male and female individuals to different ecological roles, and interference between male and female function when these occur on the same plant. There is little evidence for ecological segregation between the sexes in most dioecious species, so the first hypothesis has little empirical support. However, male and female sexual function may interfere with each other in several ways. Wind-pollinated species produce prodigious amounts of pollen because their pollen is scattered indiscriminately by a physical factor in the environment, rather than being delivered more precisely by an animal pollinator. High pollen production may result in the female parts of hermaphroditic flowers becoming clogged with pollen from the same plant, which may reduce the success of the smaller amount of pollen that arrives from other individuals. It is perhaps a consequence of this fact that dioecy is particularly common in wind-pollinated species.

Individual plants may be selected to give up male function in order to increase their allocation of energy and nutrients to female function. Because male function depends on attracting pollinators, it requires the shifting of resources into showy displays of flowers, typically many more than are needed to produce the fruits that depend on female function. In general, female function is more severely limited than male function by the availability of resources for seed and fruit production. Of course, dioecy also guarantees outcrossing, but so many other mechanisms exist to prevent selfing that it is unlikely that this factor is responsible for the evolution of dioecy.

Sexual selection

Regardless of mating system, the initial stages of reproduction involve choosing a mate. In polygynous and promiscuous mating systems, males gain by mating with as many females as they can, and choice usually is the prerogative of females. How should a female choose among the males that court her attention? If males differ in obvious features that could affect a female's reproductive success, and if her offspring could inherit those features, she should choose to mate with the male of highest quality. Of course, males should do everything in their power to advertise their quality—that is, they should strut their stuff. This phenomenon of **mate choice,** also known as **sexual selection,** sets the stage for intense competition among males for mates. This competition has resulted in male attributes evolved for use in combat with other males or in attracting females. A usual result of sexual selection is strong **sexual dimorphism,** meaning a difference in the outward appearance of male and female individuals of the same species. Sexual selection tends to produce dimorphism especially in ornamentation, coloration, and courtship behavior. Such traits, which distinguish gender over and above the primary sexual organs, are known as **secondary sexual characteristics.**

Charles Darwin, in his book *The Descent of Man and Selection in Relation to Sex,* published in 1871, was the first to propose that sexual dimorphism could be explained by selection applied uniquely to one sex. Sexual dimorphism can arise in three ways. First, the dissimilar sexual functions of males and females emphasize different considerations in the evolution of their life histories and ecological relationships. For example, because females produce large gametes, the number of their offspring often increases in direct relation to body size; this may explain why females are larger than males in many species (Figure 12.8). Furthermore, females must acquire additional nutrients to make eggs, and the duties of protecting the eggs and young usually fall upon the female. These special requirements may lead females to use the environment differently than males. Simply because they have to find suitable nest sites, females may exploit different habitats than males during the nesting season.

Second, sexual dimorphism may result from contests between males, which may favor the evolution of elaborate weapons for combat, such as the antlers of deer (Figure 12.9) and the horns of mountain sheep. Whichever males win such contests are more likely to gain access to females.

Third, sexual dimorphism may arise through the direct exercise of choice by individuals of the opposite sex. With few exceptions, females do the choosing, and males attempt to influence their choices with magnificent courtship displays. That females choose, and males compete among themselves for the opportunity to mate, is a consequence of the asymmetry of parental investment that defines the male and female conditions.

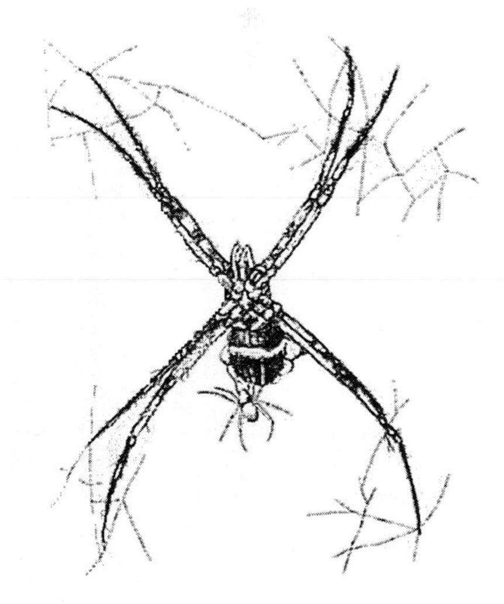

Figure 12.8 Extreme sexual dimorphism in size in the garden spider *Argiope argentata.* The male is much smaller than the female, which is portrayed in a normal resting position at the hub of her web.

Figure 12.9 Elk have immense antlers that are used during contests between males to establish control over harems of females. Courtesy of U.S. Department of the Interior.

Males enhance their fecundity in direct proportion to the number of matings they obtain; females are limited in number of offspring by the number of eggs they can produce, but they stand to improve the quality of their offspring by choosing to mate with males that have superior genotypes.

Female choice

Female choice is experienced at some level by most males. One of the first demonstrations of female choice came from an experimental study of tail length in male long-tailed widowbirds *(Euplectes progne)*. This polygynous species inhabits open grasslands of central Africa. The females, which are about the size of a sparrow, are mottled brown, short-tailed, and altogether ordinary in appearance. During the breeding season, the males are jet black, with a red shoulder patch, and sport a half-meter-long tail that is conspicuously displayed in courtship flights. Males may attract up to a half dozen females to nest in their territories, but they provide no care for their offspring. The tremendous variation in male reproductive success in this species provides classic conditions for sexual selection. In a simple yet elegant experiment, the tail feathers of some males were cut to shorten them, and the clipped feather ends were glued onto the feathers of other males' tails to lengthen them. Length of tail had no effect on a male's ability to maintain a territory, but males with experimentally elongated tails attracted significantly more mates than those with shortened or unaltered tails (Figure 12.10). This result strongly suggests female choice of mates on the basis of tail length.

Many subsequent studies have demonstrated that females choose their mates on the basis of conspicuous differences among males. There are nonetheless many issues regarding male traits and female preferences that have not been fully resolved. Which came first, female choice or male traits that indicated intrinsic quality? How are the various ornaments of males related to fitness attributes? Why don't low-quality males cheat by taking on a high-quality appearance?

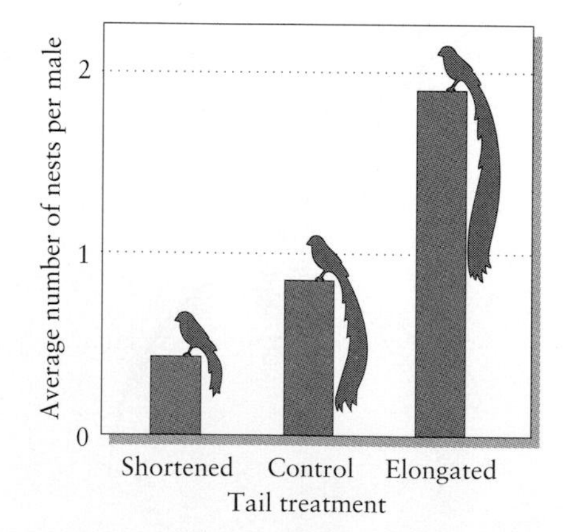

Figure 12.10 Relative reproductive success of male long-tailed widowbirds *(Euplectes progne)* with artificially shortened and artificially elongated tails. After M. Andersson, *Nature* 299:818–820 (1982).

There are two principal hypotheses for the origin of female choice. These hypotheses differ regarding whether choice arises first and drives evolution of male traits or whether choice evolves secondarily in response to manifest variation in male quality. The first, called the **sensory exploitation hypothesis,** assumes that females have intrinsic preferences for certain appearances or behaviors because of the ways in which their sensory systems (which are designed for many things besides mate choice) receive and process sensory information. Two groups of animals have provided support for this hypothesis. Swordtails are small fish of the genus *Xiphophorus* in which the lower part of the tail fin is greatly elongated. It can be shown that female swordtails prefer males with longer tail fins (Figure 12.11). It has been further demonstrated, however, that in close relatives of swordtails that lack the tail fin, females also prefer experimentally produced males that have a swordlike tail. The argument is simply that the female preference was present in the ancestors of swordtails and that males having genetic factors which produced elongated tail fins were greatly favored. The absence of the long tail in the relatives of swordtails could be attributed either to selection against the trait by other factors, such as predators, or to lack of genetic variability for tail length in the population. Why long tails tickle the fancy of female swordtails is not known.

Figure 12.11 A male green swordtail *Xiphophorus helleri* showing the greatly elongated lower portion of the tail fin. Courtesy of A. Basolo.

Another species that supports the sensory exploitation hypothesis is a tropical frog of the genus *Physalaemus,* in which males court females by vocalizations. One can experiment with this system easily because artificial courtship calls can be constructed on sound synthesizers, and females will respond to these calls by hopping toward speakers that play them. One can also measure the neural sensitivity of females to sounds of different frequencies. Experiments have shown that females prefer *chuck* calls whose sound frequencies are closer to their maximum auditory sensitivity than the *chucks* that males of the species actually give. This finding suggests that there is ongoing directional selection for deeper-voiced frogs in this species. Moreover, females of related species of *Physalaemus* that do not use the *chuck* call prefer it to the calls of males of their own species. In order to know whether female preference preceded the evolution of matching male vocalizations, it is essential to know which is the ancestral species. After all, a lineage of frogs could lose the *chuck* call if there was strong selection by predators to suppress it, in spite of the fact that females retained a preference for it. In both frogs and swordtails, other morphological and genetic information suggests that the preference arose prior to the appearance of the male trait, but more work is needed to resolve this issue.

The second hypothesis proposed to explain sexual selection is that female preferences are based on perceptible variations in the quality of male genotypes: that is, females prefer to mate with males that confer higher fitness on their offspring. In this scenario, female preference evolves because those females that make the best choices of mates leave the most descendants. Any genetic trait that influences preference by females thereby comes under strong selection.

Regardless of whether female preference arises by sensory bias or in response to variation in male quality, once female choice is established in a population, it exaggerates fitness differences among males and may create what is known as **runaway sexual selection.** Whether females intrinsically preferred males with longer tails, or tail length indicated fitness and females therefore evolved a preference for long tails, their mating preference would give long-tailed males a fitness advantage solely because of this endowment. After all, a large component of a male's fitness is his mating success, which females can influence through their choice of mates. If longer tails are what females prefer, then the tail length of males will respond. If females choose by comparing among males, rather than by comparing males to some idealized standard of beauty, then sexual selection will continually apply pressure for further elaboration of male traits. The peacock's tail, as well as the other outlandish (to our eyes) sexual ornaments and behaviors liberally spread throughout the animal kingdom, provides convincing evidence that some sort of runaway process must be at work.

If sensory exploitation is the basis for female preference, then sexually selected male traits need not have any direct connection to the quality of males; "quality" is determined solely by what females prefer. If selected traits indicate, at least initially before runaway selection takes hold, intrinsic attributes of male quality, we are then faced with a paradox. Presumably, such outlandish traits as the tail of the long-tailed widowbird burden males by making them more conspicuous to predators and by requiring energy and resources to maintain. How can such traits indicate, let alone contribute to, male quality?

One intriguing possibility, suggested by Israeli biologist Amotz Zahavi, is that male secondary sexual characteristics act as handicaps. That a male can survive while bearing such a handicap indicates to a female that he has an otherwise superior genotype. This idea is known as the **handicap principle.** It may sound crazy, but if you wanted to demonstrate your strength to someone, you might make your point by carrying around a large set of weights. A weaker individual couldn't do it and thus could not falsely advertise strength. Accordingly, the greater the handicap borne,

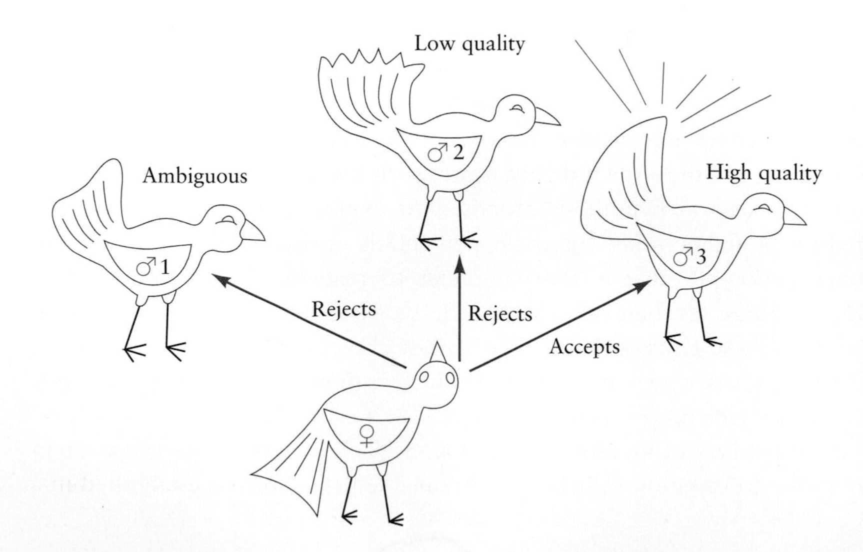

Figure 12.12 Parasite-mediated mate choice. Given a choice of three males, a female rejects the one with a short tail, which is too inconspicuous to reveal his parasite load. She also rejects the male whose long tail is obviously damaged by parasites, and chooses the male whose well-kept long tail reveals that he is parasite-free. After D. H. Clayton, *Parasitology Today* 7:329–334 (1991).

(a)

(b)

Figure 12.13 (a) Scanning electron microscopic view of a louse on a host's feather. The louse is about 1 mm long, and is seen from a dorsal view. (b) Average (center) and heavy (right) damage to abdominal contour feathers by feather lice. A normal feather is at left. Photographs courtesy of D. H. Clayton from D. H. Clayton, *Am. Zool.* 30:251–262 (1990).

the greater the ability of the individual to offset the handicap by other virtues—and to pass genes for those virtues on to his offspring. One small European songbird, the wheatear, takes the iron-pumping analogy literally and festoons its nesting ledge with up to 2 kilograms of small stones carried from a distance in its beak.

One virtue that males might possess, and which might be demonstrated by producing a showy plumage, is resistance to parasites and other disease-causing organisms. This idea was first proposed by William D. Hamilton and Marlene Zuk in 1982, who suggested that only individuals having genetic factors to resist parasite infection could produce or maintain a bright and showy plumage. Thus, an elaborate and well-maintained sexual display may provide a convincing demonstration of high male fitness, even if the display itself is an encumbrance (Figure 12.12). Indeed, the fact that a male survives with such an encumbrance may be evidence enough of his superior constitution. The importance of parasites to this theory is that they evolve rapidly and thereby continually apply selection for genetic resistance factors.

The **Hamilton-Zuk hypothesis,** along with its subsequent modifications, comes under the general heading of **parasite-mediated sexual selection.** Its general assumptions—that parasites reduce host fitness, that parasites alter male showiness, that parasite resistance is inherited, and that females choose less parasitized males—are generally supported by experiments and field observations. For example, feather lice produce obvious damage by eating the downy portions of feathers and the barbules of feather vanes (Figure 12.13). In feral rock doves, females preferred clean to lousy males by a ratio of three to one; furthermore, highly infested males had higher metabolic requirements in cold weather because of the reduced insulative quality of their plumage, and they were lighter in weight.

Sexual selection remains an active area of research, and much has yet to be learned. Studies of sexual displays show quite clearly, however, the power of natural selection to produce evolutionary modification of structures and behaviors, and how these changes can be directed by the asymmetry of sexual function in males and females.

Summary

1. In most species, reproductive function is divided between two sexes. Sexual reproduction involves the production of male and female gametes with haploid chromosome numbers. Male and female gametes unite in fertilization to form the zygotes that start a new generation. Haploid gametes are formed by a special kind of cell division, called meiosis, in which chromosome number is halved and maternal and paternal sets of genes are mixed. In mitotic cell division, which produces all the cells of the body of the organism, each daughter cell receives a full, identical set of the parental genes.

2. The origin and maintenance of sexual reproduction remain controversial topics. Sex is thought to benefit individuals by increasing genetic variation among their progeny, which increases the probability that at least some may be well suited to changed conditions. Balancing this potential advantage in species with separate sexes is the so-called cost of meiosis: female sexual parents pass only half as many genes on to their progeny as asexually reproducing individuals.

3. Sexual reproduction creates conflicts over the allocation of resources between male and female sexual function, and it affects interactions between individuals of the same and different sexes.

4. Most organisms are hermaphrodites, meaning that they have both male and female sexual organs. Separation of the sexual functions between individuals (dioecy) occurs infrequently among plants, but commonly among animals. It is favored when either sexual function imposes large fixed costs and limited resources must be allocated between the requirements of male and female function.

5. The sex ratio in a population balances contributions of genes to progeny through male and female function. In general, because the rarer sex is favored, most populations at evolutionary equilibrium have equal numbers of males and females.

6. In some parasitic wasps, males compete with siblings for matings, and sex ratio is shifted in favor of producing female offspring. In wasps and other hymenopterans, sex is determined by whether an egg is fertilized or not, and thus is under direct control of the mother.

7. Mating systems may be monogamous (a lasting bond is formed between one male and one female), polygamous (more than one mate,

usually female, per individual), or promiscuous (mating at large within the population, without lasting pair bonds). Monogamy usually occurs in species in which males can increase their individual fitness more by caring for their offspring than by seeking additional matings. Monogamy is most frequent in birds whose offspring are fed by their parents.

8. Polygyny arises when males can monopolize either resources or mates through intrasexual competition. In polygynous birds, some females gain greater fitness by joining an already mated male that holds a superior territory than by joining an unmated male on an inferior territory. Many mammals are polygynous; males defend harems of females to which they have exclusive mating access.

9. Promiscuity may arise when males contribute little, other than their genes, to the number or survival of their offspring; this is the common condition in all plants and most animals. Males of promiscuous species often defend territories to which females come to mate. Sometimes an alternative male mating strategy may be maintained within a population, such as that of nonterritorial, so-called sneaker males that gain matings with females attracted to male territories.

10. Promiscuous males may join together in leks where communal displays, sometimes involving cooperation among males, increase their attractiveness to females.

11. When males compete among themselves for mates, females can choose among them. Female choice leads to sexual selection of traits in males that indicate fitness. Female preferences may be intrinsic and based on attributes of the sensory system, or they may evolve in response to the success of choices made among males that vary in appearance. Sexually selected structures or behaviors may function as "handicaps" that only the more fit males in a population can bear without encumbrance.

12. Because parasites can evolve rapidly, and because they may directly affect the appearance or survival of males with elaborate ornaments or displays, resistance to parasites may be one factor that female preferences judge.

Suggested readings

Alcock, J. 1980. Natural selection and the mating systems of solitary bees. *American Scientist* 68:146–153.

Andersson, M., and Y. Iwasa. 1996. Sexual selection. *Trends in Ecology and Evolution* 11:53–58.

Barrett, S. C. H., and L. D. Harder. 1996. Ecology and evolution of plant mating. *Trends in Ecology and Evolution* 11:73–79.

Borgia, G. 1995. Why do bowerbirds build bowers? *American Scientist* 83:542–547.

Bradbury, J. W., and M. B. Andersson (eds.). 1987. *Sexual Selection: Testing the Alternatives.* Wiley-Interscience, New York.

Charnov, E. L. 1982. *The Theory of Sex Allocation.* Princeton University Press, Princeton, N.J.

Clayton, D. H. 1991. The influence of parasites on host sexual selection. *Parasitology Today* 7:329–334.

Emlen, S. T., and L. W. Oring. 1977. Ecology, sexual selection, and the evolution of mating systems. *Science* 197:215–223.

Godfray, H. C. J., and J. H. Werren. 1996. Recent developments in sex ratio studies. *Trends in Ecology and Evolution* 11:59–63.

Møller, A. P. 1994. *Sexual Selection and the Barn Swallow.* Oxford University Press, Oxford.

Ryan, M. J., and A. S. Rand. 1995. Female responses to ancestral advertisement calls in Túngara frogs. *Science* 269:390–392.

Shaw, K. 1995. Phylogenetic tests of the sensory exploitation model of sexual selection. *Trends in Ecology and Evolution* 10:117–120.

Slater, P. J. B., and T. R. Halliday (eds.). 1994. *Behaviour and Evolution.* Cambridge University Press, Cambridge.

Small, M. F. 1992. Female choice in mating. *American Scientist* 80:142–151.

Stevens, G., and R. Bellig (eds.). 1988. *The Evolution of Sex.* Harper & Row, San Francisco.

Warner, R. R. 1984. Mating behavior and hermaphroditism in coral reef fishes. *American Scientist* 72:128–136.

Werren, J. H. 1987. Labile sex ratios in wasps and bees. *BioScience* 37:498–506.

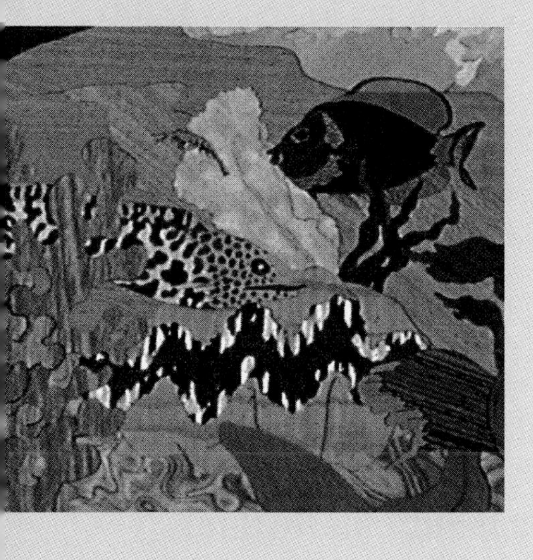

CHAPTER 13

FAMILY AND SOCIETY

During the course of its life, each individual interacts with many others of the same species: mates, offspring, nondescendant relatives, and unrelated members of social groups. Each interaction requires the individual to perceive the behavior of others and make appropriate responses. In general, interactions between close relatives are more likely to be mutually supportive than interactions between unrelated individuals because of the shared genetic factors the relatives have inherited from their ancestors. Mates too must cooperate if they are to raise offspring successfully. Nonetheless, social behaviors among unrelated individuals emphasize that all interactions between members of the same species delicately balance conflicting tendencies of cooperation and competition, altruism and selfishness. In this chapter we shall explore some of the consequences for individuals of interactions within social and family groups, and we shall describe various ways in which these relationships are managed behaviorally.

Humans are the most social of all animals. Our societies are sustained by role specialization among members, the interdependence attendant upon specialization, and the cooperation that interdependence requires. Yet humans also are competitive, to the point of violence, within this mutually supportive structure. Social life balances contrasting tendencies toward mutual help and conflict. Some animal populations exhibit much of the complexity of human societies. The social insects—ants, bees, wasps, and termites—are remarkable for their division of labor and behavioral integration within the hive or nest. Similar subtlety of social interaction, including role specialization and altruistic behavior, is being discovered increasingly among other animals, especially mammals and birds.

Social behavior includes all types of interactions between individuals, from cooperation to antagonism. Sometimes social behavior provides a means of organizing and ritualizing the expression of competition within populations and social groups: for example, outright conflict is averted when it is channeled into the posturing of males for social rank or access to mates. The defense of territories and the establishment of dominance hierarchies also may serve this purpose with a minimum of social strife.

Territoriality and dominance hierarchies

Any area defended against the intrusion of others may be regarded as a **territory.** Territories may be transient or more or less permanent, depending on the stability of resources and an individual's need of them. Territoriality is most conspicuously displayed by birds, which may actively defend areas throughout the year or only during the breeding season. Many migratory species establish territories on both the breeding and wintering grounds; shorebirds defend feeding areas for a few hours or days on stopover points on their long migrations. Hummingbirds defend individual flowering bushes and abandon them when their flowering periods are over. Male ruffs and grouse defend a few square meters of space on a communal lek. During the egg-laying period, males of many species accompany their mates and chase off would-be cuckolders.

In some situations, dispersion of individuals on territories may not be practical because of the pressures of high population density, the transience of critical resources, or the overriding benefits of living in groups. In such circumstances, conflict among individuals may be resolved by contest, but social rank rather than space is the winner's prize. Once individuals order themselves into a **dominance hierarchy** of social status, subsequent contests between them are resolved quickly in favor of higher-ranking individuals. When a social hierarchy is linearly ordered, the first-ranked member of a group dominates all others, the second-ranked dominates all but the first-ranked, and so on down the line to the last-ranked individual, who dominates none.

Occupation of space and social rank are opposite sides of the same coin, and often are directly related. For example, low-ranking individuals—those that win few contests—usually occupy territories of low quality or small size, to which it is hard to attract mates. Even within a social group, the position of an individual in a dominance hierarchy is sometimes reflected by its spatial position within the flock or herd. In large foraging flocks of wood pigeons, for example, individuals low in dominance tend to be at the periphery, where they are more vulnerable to predators than the

ENVIRONMENTS WITHOUT LIFE

The Antarctic ice cap is too cold for life. Water freezes and life processes come to a standstill. In contrast, new land is formed when lava reaches the surface in volcanic eruptions, as at Kilauea volcano in Hawaii. Several decades will pass after the lava has cooled before the first plants begin to colonize. Hot temperatures and noxious gases keep plants from establishing in the crater of Poas volcano in Costa Rica.

Photo by Bill Johnson, courtesy of U. S. Dept of Agriculture, Soil Conservation.

Photo by R. E. Ricklefs.

Photo by R. E. Ricklefs.

THE TROPICAL RAIN FOREST

Warm temperatures and abundant rainfall in many parts of the tropics support the highest levels of biological production and diversity of life on earth. Tropical rain forest biome.

'hoto by R. E. Ricklefs.

The unbroken canopy of a Panamanian rain forest may include 200 species of trees. Some of these may reach 50 to 60 meters in height. In the monotonous green of the forest, plants use bright colors such as the red bracts of *Heliconia,* to attract pollinators. Birds such as the wire-tailed manakin from Ecuador adopt similar colors to enhance their sexual displays. In many tropical plants, such as *Clusia grandiflora,* the flowers of male and female plants differ in the rewards they offer to pollinators. The white-petaled *Miconia mirabilis* is harvested by insects for both its pollen and nectar rewards.

Photo by Carl C. Hansen, courtesy of Smithsonian Tropical Research Institute.

Photo by Volker Bittrich.

Photo by R. E. Ricklefs.

Photo by Volker Bittrich.

Photo by R. E. Ricklefs.

Photo by R. E. Ricklefs.

Most tropical animals, such as the katydid at right, are cryptically colored or resemble inedible objects such as dead leaves to avoid notice by predators. Others are distasteful because of painful stings or noxious chemicals that they acquire from the plants they eat. These animals often have bright colors that warn predators that they are unpalatable. The predominantly green color of some tree frogs is cryptic, but bright colors on the legs and ventral surfaces may be displayed during behavioral encounters. Conspicuous banding in a caterpillar often warns of noxious chemicals. The coloration and antenna-like tails on the wings of this tropical butterfly confuse predators into mistaking its hind end for its head.

Photo by R. E. Ricklefs.

Photo by Marcos A. Guerra, courtesy of Smithsonian Tropical Research Institute.

Photo by J. Burgett, courtesy of Smithsonian Tropical Research Institute.

Photo by Marcos A. Guerra, courtesy of Smithsonian Tropical Research Institute.

Photo by Carl C. Hansen, courtesy of Smithsonian Tropical Research Institute.

Conspicuously marked noxious animals often aggregate to enhance the effects of their coloration, as in the case of these Panamanian caterpillars and true bugs (hemipterans), shown below. Tropical wasps show various levels of sociality from the solitary *Polistes* to the highly social *Polybia* portrayed at right. Parental care as a defense against predators extends to many other insects, such as the bright yellow coccinelid beetle guarding her eggs. Ants provide protection for many herbivores, such as these homopteran nymphs, from which they obtain a concentrated excreted solution of sugars. This is one of the many kinds of mutualistic interactions in tropical ecosystems.

Photo by James H. Hunt.

Photo by James H. Hunt.

Photo by Carl C. Hansen, courtesy of Smithsonian Tropical Research Institute.

Photo by Carl C. Hansen, courtesy of Smithsonian Tropical Research Institute.

Photo by Carl C. Hansen, courtesy of Smithsonian Tropical Research Institute.

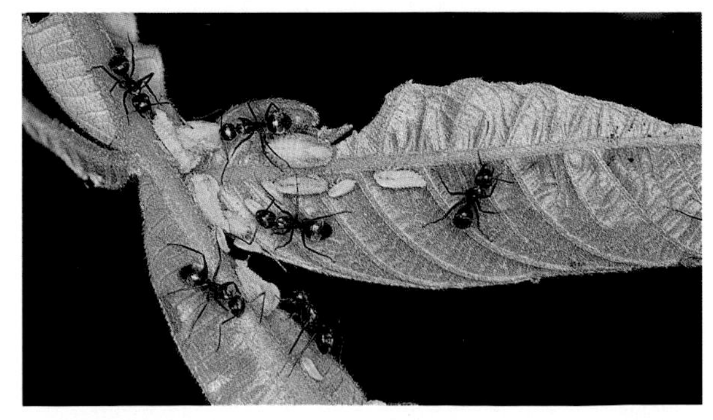

Photo by R. E. Ricklefs.

With the high productivity of rain forest plants, herbivores are also abundant and varied, including thousands of species of insects. Among vertebrate herbivores, the sloth may be thought of as the sheep of the rain forest. Even reptiles get into the act: the green iguana is a major consumer of leaves. Because birds have high energy requirements and could not fly with their stomachs filled with fermenting vegetation, they concentrate on nectar and fruits.

Photo by Marcos A. Guerra, courtesy of Smithsonian Tropical Research Institute.

Photo by R. E. Ricklefs.

Photo by Marcos A. Guerra, courtesy of Smithsonian Tropical Research Institute.

Photo by Carl C. Hansen, courtesy of Smithsonian Tropical Research Institute.

Within the tropics, high mountains and other features create rain shadows and seasonally dry conditions (tropical seasonal forest biome). The false-color satellite image of western Panama shows heavy forest (brown) to the north of the continental divide where the prevailing winds blow humid air from the Caribbean Sea. On the Pacific side of the isthmus, the green color in this dry-season image indicates pasture and dry forest, illustrated in the accompanying scene from western Costa Rica. Over much of the tropics, pastures are burned during the dry season to kill tree seedlings. Forest clearing is extending the acreage under pasture and cultivation in the tropics, for which most forest soils are poorly suited.

Photo by R. E. Ricklefs.

Photo by Marcos A. Guerra, courtesy of Smithsonian Tropical Research Institute.

Photo by Marcos A. Guerra, courtesy of Smithsonian Tropical Research Institute.

Photo by Marcos A. Guerra, courtesy of Smithsonian Tropical Research Institute.

Photo by R. E. Ricklefs.

Throughout much of tropical Africa, seasonal rainfall supports mixed grassland and woodland vegetation, called savanna, with vast herds of grazing herbivores; Amboseli National Park, Kenya, with Mt. Kilamanjaro in the background (tropical savanna biome).

dominant individuals at the flock's center. Peripheral birds appear nervous, and because they spend much of their time looking up from feeding, they are often undernourished. Birds in the center of the flock remain calm and feed more because they are protected from surprise attack by the vigilance of individuals at the periphery.

Social contests

Whether an individual lives within a territorial system or in a group setting, its social rank is determined by its ability to win contests. The outcomes of these contests are all-important to the individual because they determine the quality and amount of space it can defend and its access to food and mates. Each contest between two individuals can be resolved only through behavioral decisions taken by each participant. A spider confronting another over a particularly good place to build a web assesses the situation and decides either to back down or to escalate the contest. Sometimes the projected outcome of a physical contest is obvious, and the smaller individual retreats. When the outcome is more difficult to judge in advance, the two spiders may engage in a series of elaborate displays that help them to weigh each other's fighting abilities, each hoping (though perhaps not consciously) that the other will be duly impressed and back down. If the match appears to be close and the outcome uncertain, the contest may then escalate to actual fighting, with the risk of serious injury or death to one or both participants.

Optimal behavior in a contest depends on assessment by each contestant of the likely outcome of the contest and the payoffs of winning or losing. What actually happens—that is, how the contest plays out—also depends on the decisions made by each contestant. These decisions are taken to maximize the net benefit to the individual, but they also depend on the behavior of the other contestant, over which the first individual has no control. In these circumstances, optimal behavior is contingent upon the behavior of others engaged in the same contest. Humans are faced with such decision making all the time, not only in social behavior, but also in business, war, and other competitive enterprises. Optimal behaviors in these situations are the subject of **game theory,** which analyzes the outcomes of behavioral decisions when outcomes depend on the behavior of other players.

Game theory analysis is based on outcomes of behaviors. Consider a spider's decision whether to escalate a contest or not. If the other contestant backs down, the payoff to the first spider is the territory, and the cost is nil. If the other contestant meets the challenge, then the payoff depends on the chance of winning the contest, and the cost—win or lose—is much higher. Without making a quantitative analysis, it is still easy to see that an individual's behavior should depend on its best estimate of the other contestant's response and on the reward for winning. When the first spider is much larger than the second, it is likely that the second will back down from any confrontation, so escalation carries little risk of harmful conflict—an easy win, so to speak. When the two are evenly matched, both the response of the second spider and the outcome of a conflict are more difficult to predict, and the probability of getting hurt is higher. Under such circumstances, both escalation and meeting the challenge are likely to occur only when the potential rewards for winning a contest are large. It is no surprise, then, that spiders are actually observed to fight only over the best web sites, and only when the two contestants are similar in size. We'll use a game theoretical approach again a bit later to analyze the evolution of cooperation within social groups.

Ritualized antagonistic behavior

Most conflict associated with social rank or territorial defense involves **ritualized behaviors** that rarely lead to risky physical struggle. These behaviors allow individuals to ascertain their chances of winning a contest, and presumably they have evolved because they allow contests to be resolved without bloodshed. Certain appearances or behaviors seem to signal higher status than others. If such ritualized behavior represents a means of signaling an individual's prowess, and if most members of a group or population agree on the meaning of the signals, it is reasonable to ask, why don't some less dominant individuals cheat by assuming the behavior of their betters? Why don't they try to deceive other members of a population into thinking they are more aggressive or stronger than they really are? Part of the answer is that some ritualized behaviors allow contestants to judge each other's size, which is difficult to fake and which often determines the outcome of earnest combat. Moreover, for a ritualized system to work, status signaling and ability to back up the signaled status must go together, because closely matched individuals will always be tempted to escalate a conflict. Consider the following experiments that demonstrate this fact.

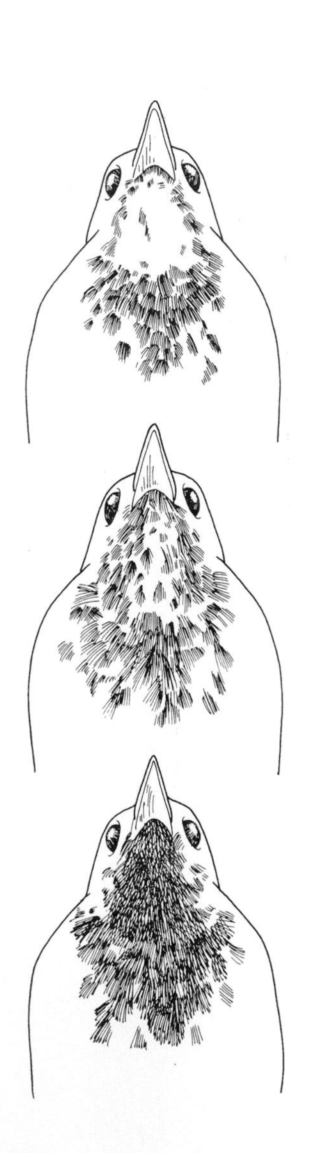

Figure 13.1 Examples of variation in the throat plumage of Harris's sparrows. After D. Morse, *Behavioral Mechanisms in Ecology*, Harvard University Press, Cambridge, Mass. (1980).

Harris's sparrows breed in the Canadian Arctic and winter in small flocks in the central United States. Social status within a flock is correlated with the amount of dark coloration in the plumage of the throat and upper breast (Figure 13.1). Upon first encounters, lighter birds generally avoid darker individuals, so dark coloration meaningfully signals status. When the plumage of some light individuals was experimentally dyed dark, they were mercilessly attacked by individuals of truly high status who easily saw through the ruse. When the tables were turned, dark individuals bleached to a lighter color found themselves constantly having to attack naturally light birds to regain their status. The system works because dark plumage and aggression go together—there is little cheating. When lighter birds were implanted with testosterone to raise their level of aggression, such individuals that were also dyed darker rose in the dominance hierarchy, whereas those left with their light plumage were persecuted by naturally dark birds and could not rise in status despite their higher level of aggression. The researchers concluded that behavior and plumage are probably affected by common physiological conditions associated with hormone levels. When the two do not match, birds get into trouble because they either don't get the respect that is due them or are treated as impostors.

These results raise a difficult question: If social status depends on the level of a hormone whose production imposes little physiological cost,

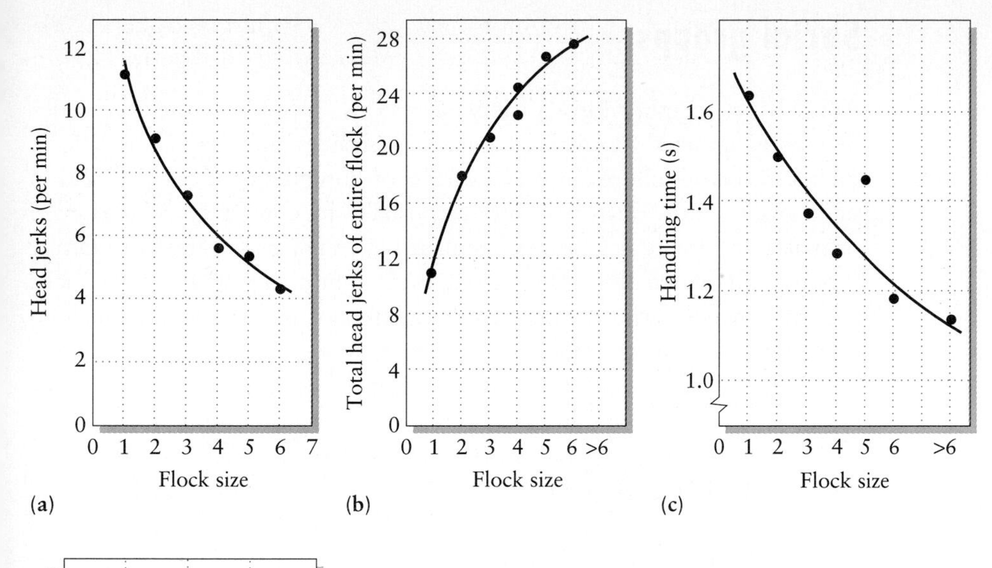

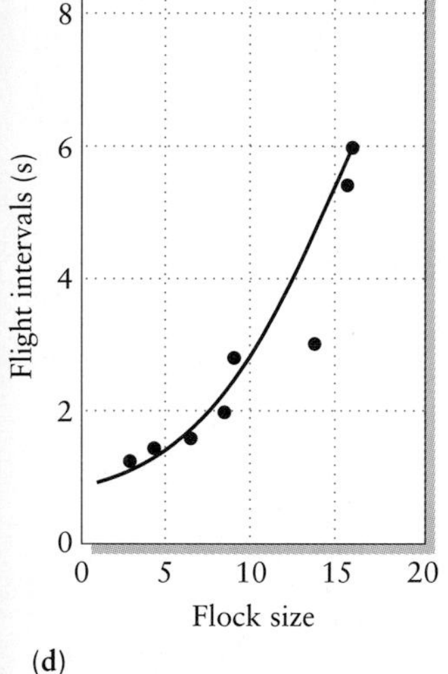

Figure 13.2 Increased flock size leads to increased security from predation, but reduced efficiency of foraging, in the European goldfinch *(Carduelis carduelis)*. (a) Mean rates of looking up from foraging, (b) total vigilance rate for the entire group, (c) time required to husk each seed of sorrel *(Rumex acetosa)*, and (d) time required to move from one plant to the next by foraging individuals, at different flock sizes. After E. Glück, *Ethology* 74:65–79 (1987).

why don't all members of a population have a high level of aggression? Experimental hormone implants have turned monogamous song sparrows into bigamists and nonterritorial red grouse into landowners. Perhaps status is linked to age and experience, on the basis of which hormone-mediated levels of aggression are adjusted to serve other purposes. As an individual grows older, it rises in status; younger individuals await their turns. Alternatively, variation in rank may indicate that dominance has its costs—such as spending too much time fighting and not enough feeding and watching out for predators—and that all members of the hierarchy have, in effect, roughly equivalent fitness.

Given that status truly represents an individual's potential performance in behavioral conflicts, one might also ask why low-ranking individuals, which are excluded from food and mates by dominant individuals and are exposed to higher risks, remain with a group. The answer must be simply that it is better to be a low-ranking individual in a group, perhaps rising in rank with age, than to be a loner.

Social groups

Animals get together for a variety of reasons. Sometimes they are independently attracted to suitable habitat or resources and form aggregations, such as those of vultures around a carcass or dungflies on a cowpat. Within such groups, individuals may interact, usually to contest space, resources, or mates. In other cases, progeny remain with their parents to form family groups, and aggregation results from their failure to disperse. True social groups, however, arise through the attraction of unrelated individuals to one another—that is, through a purposeful joining together.

Animals form groups to increase their chances of surviving, feeding, or finding mates. In groups, individuals tend to spend more time feeding and less time looking out for predators. Consider the data presented in Figure 13.2 for the European goldfinch *(Carduelis carduelis),* which feeds on seed heads of plants in open fields and hedgerows. Two factors control optimal flock size in these birds. As flock size increases, each individual spends less time looking out for predators. If you watch closely as birds feed, you will notice that they raise their heads and look around from time to time. In a larger group, an individual goldfinch can spend more time going about the business of eating, and can gather and husk seeds more rapidly, because the total vigilance of the flock is higher. Balancing this advantage of reduced individual vigilance time, a larger flock depresses a local food supply faster, and individuals are forced to fly farther between suitable foraging patches, using valuable feeding time and energy and perhaps increasing their vulnerability to predators. Thus, joining a flock is a good choice for an individual as long as the flock is not too large. This situation is reminiscent of the problem of optimal lek size in the ruff discussed in the previous chapter.

Individuals in flocks may learn the location of food from others, cooperate in obtaining food, or obtain information about their environment. Some researchers have argued that groups serve as **information centers,** meaning that individuals can learn about places to feed, for example, from the behavior of others, perhaps by following them from a roost or breeding colony to a food source. Some colonially nesting seabirds appear to gain information about the quality of a nesting location from the reproductive success of their neighbors. Kittiwakes are small seabirds, related to gulls, that nest on cliffs (Figure 13.3). Nesting areas vary in quality due to variation in the abundance of ticks, which infest the chicks. When a pair of kittiwakes successfully rears a family in one year, it usually returns to the same cliff to breed in the next. When a pair is unsuccessful, its decision to return to the same cliff or move elsewhere depends on the success of others breeding in the same area. Working on the coast of Brittany in France, behavioral ecologist Etienne Danchin observed kittiwake pairs that failed in the egg stage of the nesting cycle. Nearly all pairs nonetheless returned to the same site the following year when the failure rate of others around them was less than 25%, but more than one-third abandoned the cliff when the failure rate of their neighbors exceeded 50%. Success or failure of an individual in any given year is largely a matter of chance. By observing what is happening around them, individuals can gain a better estimate of their own probability of success in the future.

Figure 13.3 Black-legged kittiwakes on a nesting ledge on the coast of Newfoundland, Canada. The chicks have a black collar marking.

Social groups open the door to cooperative behavior. Cooperation involves the coordination of individual behavior to achieve a common goal, such as defense of the herd by male musk oxen or pack hunting by wolves and killer whales. Such instances of cooperation are not common, but where they do occur, social behavior fosters mutualism between individuals more than it organizes competition between them: individual behavior becomes subservient to the group, and groups assume a characteristic behavior of their own. Considering our own position as the most social of all species, it is natural that humans have devoted considerable effort to learning about the evolution and organization of animal societies.

Social behavior

Any social interaction other than mutual display can be dissected into a series of behavioral acts by one individual (the **donor** of the behavior) directed toward another (the **recipient**). One individual delivers food, the other receives it; one threatens, the other is threatened. When one individual attacks another, the attacker may be thought of as the donor of a behavior. The attacked individual (the recipient in this case) may respond by standing its ground or by fleeing; in either case, it thereby becomes the donor of a subsequent behavior. The donor-recipient distinction is useful because each act has the potential to affect the reproductive success of both the donor and the recipient of the behavior. These increments of reproductive success, or fitness, may be positive or negative, depending on the interaction.

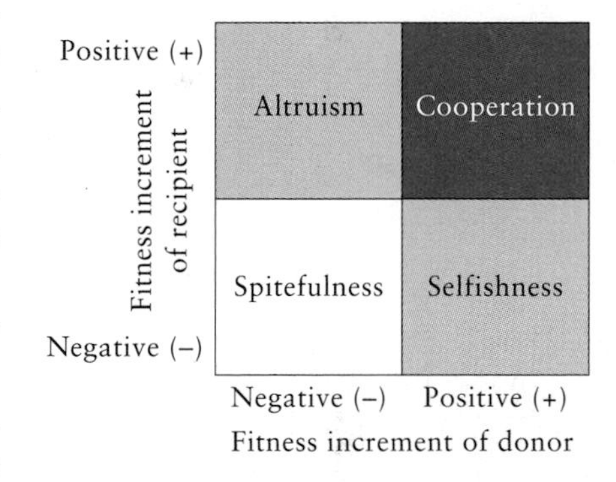

Figure 13.4 Four types of social behavior classified according to their effects on the fitness of donors and recipients of the actions.

Four combinations of cost and benefit to donor and recipient can be used to organize social interactions into four categories (Figure 13.4). **Cooperation** and **selfishness** both benefit the donor of the behavior, and therefore should occur frequently. **Spitefulness**—behavior that reduces the fitness of both donor and recipient—cannot be favored by natural selection under any circumstance, and presumably does not occur in natural populations. The fourth type of behavior, **altruism,** benefits the recipient at a cost to the donor. Altruism presents a difficult problem because it requires the evolution of behaviors that reduce the fitness of the individuals performing them. We would expect selfish behaviors to prevail to the exclusion of altruism because they increase the reproductive success of the donor. However, altruism appears to have arisen in colonies of social insects, in which workers forgo personal reproduction to rear the offspring of the queen, their mother. We humans also like to think that we are not only capable of altruistic behavior, but that such interactions hold together the fabric of our society.

Kin selection

The evolutionary dilemma posed by the apparent altruism of social insects is resolved when one realizes that their colonies are discrete family units, containing mostly the offspring of a single female (the queen). Therefore, behavioral interactions within an ant colony or beehive occur between

TABLE 13.1 Probabilities of identity by descent between one individual and others having various degrees of relationship

RELATIONSHIP	PROBABILITY OF IDENTITY BY DESCENT
Parent	0.50
Offspring	0.50
Full sibling	0.50
Half-sibling	0.25
Grandparent	0.25
Grandchild	0.25
Uncle or aunt	0.25
Nephew or niece	0.25
First cousin	0.125

close relatives, in this case siblings. When an individual directs a behavior toward a sibling or other close relative, it influences the reproductive success of an individual with which it shares more of its own genetic makeup than it does with an individual drawn at random from the population. This special outcome of social behavior among close relatives is referred to as **kin selection.**

Close relatives have a certain probability of inheriting copies of the same gene from a particular ancestor. The likelihood that two individuals share copies of any particular gene is the probability of **identity by descent,** the value of which varies with degree of relationship (Table 13.1). For example, two siblings have a 50% probability of inheriting copies of the same gene from one parent. This probability is also called their **coefficient of relationship.** Two cousins have a probability of one in eight (0.125) of inheriting copies of the same gene from one of their grandparents, which are their closest shared ancestors.

When an individual behaves in a particular way toward a close relative, that act influences not only its own personal fitness, but also the fitness of an individual that shares a portion of its genes. Now suppose that an act of altruism is caused by a genetic factor that was inherited from one parent. When this act is directed toward a sibling, the probability that the recipient of the behavior will also have a copy of that gene is 50%. Therefore, the occurrence of the gene within the population as a whole will be determined both by its influence on the fitness of the donor of the behavior and by its influence on the fitness of the recipient, discounted by the coefficient of relationship.

Biologists refer to the total fitness of a gene responsible for a particular behavior as its **inclusive fitness,** which equals the contribution to the reproductive success of the donor resulting from its own behavior *plus* the product arrived at by multiplying the change in reproductive success of the recipient times the probability that it carries a copy of the same gene. Therefore, the inclusive fitness of an altruistic gene would exceed that of its selfish alternative as long as the cost to the altruist was less than the benefit to the recipient multiplied by the average genetic relationship of donor and recipient. Thus, a genetic factor for altruism will have a positive fitness and increase in the population when the cost *(C)* of a single altruistic act is less than the benefit *(B)* to the recipient times the coefficient of relationship *(r)* of the recipient; that is, $C < Br$. When this equation is rearranged, the condition for the evolution of altruism becomes $C/B < r$; that is, the cost-benefit ratio, which is a measure of how altruistic the behavior is, must be less than the average coefficient of relationship of the recipient of the altruistic behavior.

At the same time that inclusive fitness makes possible the evolution of altruism among close relatives, the same considerations constrain the evolution of selfish behavior. With B now representing the benefit to the donor of a behavior and C the cost to the recipient, selfish behavior among close relatives can evolve only when $B > Cr$, or $C/B < 1/r$. The cost-benefit ratio (C/B) is, in this case, a measure of the selfishness of the behavior. The higher the coefficient of relationship *(r)* between donor and recipient, the lower the level of selfishness that can evolve.

The maintenance of altruistic behavior by kin selection requires that such behaviors have a low cost to the donor and be restricted to close relatives. Individuals of many species tend to associate in family groups, and limited dispersal often keeps close relatives together. Moreover, individuals of many species can sense their degree of relationship to others by chemical or behavioral cues even when they have had no family experience. Thus, the opportunity for altruistic behavior to evolve by kin selection seems to exist.

Many behaviors have been identified as being altruistic. One of the most conspicuous of these is alarm calling. By uttering an alarm in the presence of a predator, an individual warns others of danger, but places itself at increased risk by calling attention to its position. One species in which the costs and benefits of alarm calling have been studied is Belding's ground squirrel, which lives in high meadows in the Sierra Nevada mountains of California. Because ground squirrels are social creatures and are exposed to view in their montane meadow habitat (Figure 13.5), predators pose a considerable threat. Alarm calls reduce this threat, but also impose a cost on the caller when directed toward terrestrial predators, such as weasels. Acts of predation are infrequent, and many hours of observation are needed to discern any pattern to their occurrence. After several years of study, Paul Sherman, of Cornell University, and his coworkers tallied instances when predators approached groups of ground squirrels, at least one of which gave an alarm call, and found that 13% of calling individuals were attacked, compared with only 5% of silent individuals.

Figure 13.5 Belding's ground squirrel at Tioga Pass, California. Courtesy of P. W. Sherman. From P. W. Sherman, *Science* 197:1246–1253 (1977).

Ground squirrels normally give alarm calls only when close relatives can hear them. Individuals sounding alarms in the presence of terrestrial predators usually are adult and yearling females; males and juvenile females do not exhibit this behavior (Figure 13.6). This makes sense from the standpoint of kin selection because males disperse widely from their place of birth, whereas females remain close to home. Moreover, females called more frequently when they had occupied the same territory for many years (and hence were likely to have female relatives nearby) or when relatives were known to be present in the population.

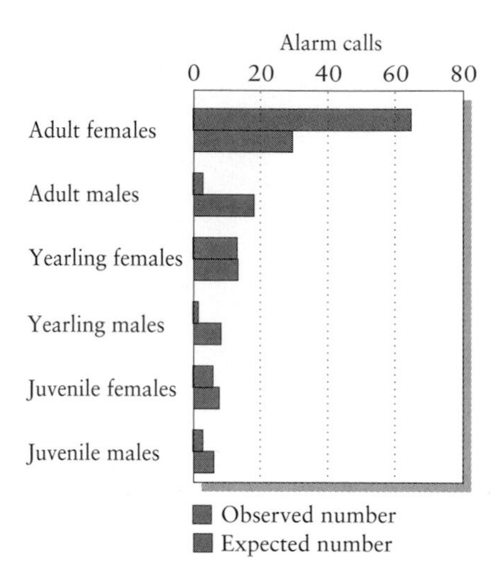

Figure 13.6 Expected and observed frequencies of alarm calling in response to the presence of a predatory mammal by sex and age classes of Belding's ground squirrels. Expected values were computed by assuming that individuals call in direct proportion to the number of times they are present when a predatory mammal appears. From P. W. Sherman, *Science* 197:1246–1253 (1977).

Life in the family

Extended families in humans include the nuclear family of a mated pair and their young progeny as well as, to varying degrees, grandparents, uncles and aunts, cousins, nephews and nieces, and sometimes individuals of uncertain relationship to the rest. These are complex social units within which occur a tremendous variety of social interactions, most of them cooperative but many competitive enough to stress the bonds that hold a family unit together. Rarely do human extended families include more than one child-producing pair, and at least a portion of the behavior of nonnuclear members of the family is directed toward supporting the well-being and upbringing of the children.

For the most part, family life in animals, and in plants for that matter, is not so complex. In most groups of animals, offspring receive no support from their parents or other relatives beyond nutrients placed in the egg to nourish the embryo. Even when parents care for their young, as is the case in all mammals, most birds, and sporadically in other groups, the family is a transient phenomenon limited to the period of growth and early development, after which offspring disperse to begin their own independent lives.

Occasionally, however, offspring stay with their parents for extended periods, and may even remain permanently associated with them in what become extended families. The Florida scrub jay is typical of several hundred species of birds in which offspring remain with their parents for a year or more. The family life of these birds has been studied in detail by Glen Woolfenden and John Fitzpatrick at the Archbold Biological Station in Florida. They found that the stay-at-home jays do not themselves breed, but rather, help to care for the offspring produced by their parents during the next breeding season, which are their younger brothers and sisters. Such conduct, now documented in many species, prompted many researchers to include this "helping at the nest" in the category of altruistic behaviors. In many such cases, including that of the scrub jay, it has been shown that helpers increase the reproductive success of their parents. That is reason enough for parents to tolerate their offspring staying at home.

Whether helping is a purely altruistic behavior, however, depends on the alternatives available. Scrub jay habitat is densely occupied, and young birds have little chance of setting up territories of their own. One-year-olds can increase their inclusive fitness by helping their parents while gaining experience that may later improve their own personal fitness as

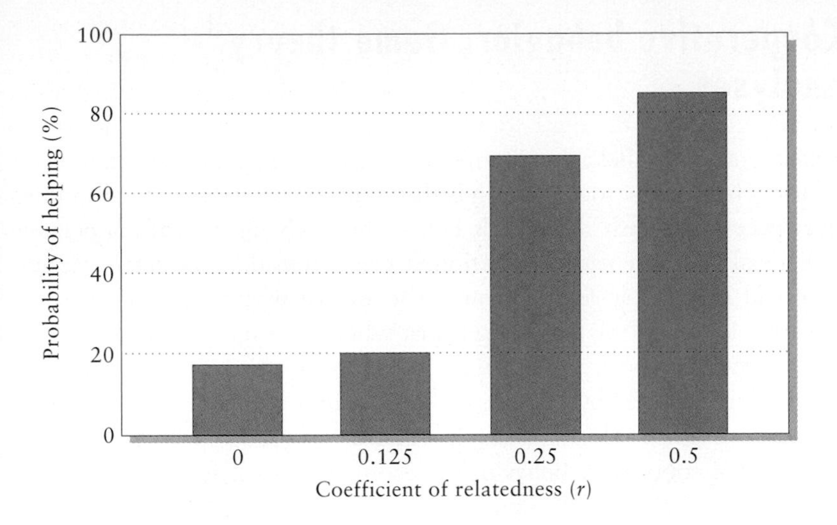

Figure 13.7 Distribution of altruistic behaviors as a function of degree of relationship within family groups of the white-fronted bee-eater. From S. T. Emlen, P. H. Wrege, and N. J. Demong, *Am. Sci.* 83:148–157 (1995).

independent breeders. Staying at home provides another immense benefit: the possibility of inheriting the family territory when the father dies.

Studies of the white-fronted bee-eater in East Africa by Stephen Emlen, Peter Wrege, and Natalie Demong, of Cornell University, have revealed complex extended families. These families are typically multigenerational groups of three to seventeen individuals, often including two or three mated pairs plus assorted single birds—unpaired young and widowed older individuals. Careful observations of individually marked birds over several years have shown that these social groups are truly extended families, comprising only related individuals and their mates, which normally come from other families. Although relationships within extended families tend to be cooperative, bee-eater family groups are hardly models of harmonious behavior; one sees the usual squabbling over food, nest sites, and who mates with whom typical of society at large. Remarkably, however, selfless and selfish acts appear to be directed toward other individuals very much in accordance with their degree of relationship: brothers and sisters are treated better than half-siblings and uncles, for example, and cousins fare almost as badly as nonrelatives outside the family group (Figure 13.7). Through their behavior, bee-eaters tell us that individuals know who their relatives are and can distinguish subtle differences in degree of relationship. We can also conclude from the distribution of helpful and harmful behaviors in this species that inclusive fitness is the appropriate measure of selection on social behavior. Altruistic behaviors can indeed evolve among close relatives by kin selection.

We will return to interactions between family members later in this chapter. Before doing so, however, we shall consider whether or not cooperative or altruistic behavior can evolve among nonrelatives within a society. Clearly, social groups can form out of the self-interest of group members seeking protection from predators or perhaps some efficiency gained by foraging or hunting in groups. Whether groups of unrelated individuals can take the next step toward true cooperation is a fundamental issue in the evolution of social behavior. We shall address this issue by means of a simple game-theory analysis.

Cooperative behavior: Game theory analyses

Self-interest rules behavior among nonrelated individuals. A paradox of selfish behavior in a social setting is that conflict can reduce the reproductive success of selfish individuals below the likely success of cooperative individuals. Because natural selection favors increased reproductive success, it should be possible for cooperation to evolve within societies at large. The problem with this reasoning is that when most of a society consists of cooperative members, a selfish individual can greatly increase its personal reproductive success by "cheating." Thus, selfish behavior will always be favored by natural selection, which will prevent groups from crossing the threshold of cooperative behavior to become true societies.

The hawk-dove game

The logic of this somewhat pessimistic argument can be shown by a simple game theory analysis. The analysis we will use is called the hawk-dove game (it is also known in a different context as the prisoner's dilemma). Let's assume that one type of individual will always behave selfishly in conflict situations, always being willing to fight over contested resources and taking all the reward when it wins: this is hawk behavior (H). In contrast, doves (D) never contest a potential reward, but share it evenly with other doves. Each contest between two individuals has a potential reward, or benefit *(B),* and a cost *(C)* when a contest results in physical conflict. The payoff, both to hawks and to doves, depends on the behavior of the second contestant—that is, whether it is a hawk or a dove (Table 13.2). For example, two hawks always fight and, on average, get half the reward, so that the payoff is $\frac{1}{2}B - C$. When a hawk confronts a dove, the hawk gains the entire uncontested benefit without a cost; thus, the payoff is B.

The average payoff (fitness) to hawks and doves depends on the relative proportions of the two kinds of individuals in a population. Let p be the proportion of hawks and $(1 - p)$ the proportion of doves. The payoffs are now as follows: hawks receive $p(\frac{1}{2}B - C) + (1 - p)B$, and doves receive $\frac{1}{2}(1 - p)B$. One can see that a population consisting only of hawks ($p = 1$) has an average payoff of $\frac{1}{2}B - C$, which is less than the average payoff of $\frac{1}{2}B$ in a population consisting only of doves ($p = 0$). Clearly, the dove strategy would be the best all around from a social point of view.

The problem is that dove behavior is not an **evolutionarily stable strategy:** that is, it cannot resist evolutionary invasion by an alternative strategy (a genetic mutation, if you like)—namely, hawkish behavior. A single hawk in a population of doves (p close to 0) receives twice the average payoff that doves do (B versus $\frac{1}{2}B$) because it never encounters another hawk, and conflicts are never contested. Thus, in a world of doves, the hawk strategy increases rapidly. Not only can hawkish behavior invade a dove population, but a pure hawk population is also resistant to invasion by doves, except when the cost of conflict is very high relative to the benefit. In that case, doves can survive in a hawk population because the hawks

TABLE 13.2 Costs and benefits in the hawk–dove game

BEHAVIOR OF FIRST PLAYER	RESPONSE OF SECOND PLAYER	
	Hawk	*Dove*
Hawk	Share both the benefit and the cost of conflict $\frac{1}{2}B - C$	Gains entire benefit B
Dove	Neither gains benefit nor assumes cost 0	Shares benefit without cost of conflict $\frac{1}{2}B$

fight so much among themselves, but when two doves meet, they share. When p is close to 1 (a pure hawk population), the payoff to hawks is $\frac{1}{2}B - C$, and that to doves is 0. Thus, hawkish behavior is an evolutionarily stable strategy as long as $B > 2C$. When the benefit is less than twice the cost of conflict, doves can invade the hawk population, and the eventual outcome is a mixed population of hawks and doves with the proportion of hawks (p) equal to $\frac{1}{2}B/C$. The hawk-dove game, as well as more complex game theory analyses, demonstrates how difficult it is for cooperative behavior to evolve among unrelated individuals in a population.

Reciprocal altruism

One way around the evolutionary constraint of selfishness is the strategy of **reciprocal altruism,** in which an individual is cooperative (altruistic) toward doves but fights back against hawks. In this case, the behavior of the individual is contingent on the behavior of others in the population. This strategy is also referred to as **tit-for-tat,** that is to say, giving back in kind. The average payoff to a reciprocal altruist in the hawk-dove game is $p(\frac{1}{2}B - C) + \frac{1}{2}(1 - p)B$, or $\frac{1}{2}B - pC$. This payoff is always as good or better than that of a dove, and never worse than that of a hawk. Thus, reciprocal altruism is a superior strategy that could invade and predominate in any mixed population of doves and hawks.

How do reciprocal altruists behave toward each other? In a tit-for-tat world, the first encounter colors all future interactions between two individuals. Thus, a benefit-of-the-doubt strategy is the only one that will lead to a cooperative society: Cooperate initially, until one finds out that another individual's intentions are not altruistic. Under this scenario, reciprocal altruism will lead to cooperation generally within the population—reciprocal altruists are nice guys who are willing to fight against cheaters.

Although reciprocal altruism provides a model for the evolution of altruistic behavior, it relies on long-term associations among individuals during which they can learn individual behavior patterns and act toward others in their group accordingly. Individual recognition and a high probability of return of altruistic acts in the future are both necessary ingredients for the evolution of reciprocal altruism. These conditions appear to have been met in a number of carefully studied organisms, including the dwarf mongoose and impala in Africa. Perhaps the best-documented case of reciprocal altruism is in vampire bats, in which successfully foraging individuals share their blood meals with others that fail to feed on a particular night. Vampires roost in small groups whose composition may be stable for many months or years, so individuals get to know each other well and could easily refuse others who do not cooperate. Because fewer than 10% of bats are unsuccessful on a particular night, sharing meals is not overly costly (less than 5% of total food intake when sharing is equally spread), but the survival benefit of receiving part of a meal may be substantial.

Clearly, mechanisms exist for the evolution of cooperative and altruistic behavior in small family groups and in social groups. Whether these behaviors can balance the universal self-interest of the individual in such large groups as human societies remains to be seen. Mechanisms of reciprocal altruism and cooperation are firmly embedded in our culture and social institutions, but cheating and selfish behavior are also widespread. Regardless of how this issue is resolved, the study of social behavior emphasizes the conflicting values of self-interest and group interest. Nowhere is this conflict stronger (and perhaps more personally felt) than in relations between parents and their offspring.

Parent–offspring conflict

Rather than passively accept whatever their parents offer, most offspring actively solicit care: young animals beg for food and solicit brooding; eggs actively take up yolk from the ovarian tissues or bloodstream of the mother; developing seeds take up nutrients from the ovary of a flower. For the most part, the interests of parent and offspring are compatible: when progeny thrive, so do their parents' genes. But when the selfish accumulation of resources by one offspring reduces the overall fecundity of its parents, parent and offspring can come into conflict. We may define **parent–offspring conflict** as a situation that arises when the optimal level of parental investment in a particular offspring differs from the standpoints of the parent and that offspring.

Each act of parental care and each unit of parental investment benefits an individual offspring by enhancing its survival, but costs the parent by decreasing the number of its other contemporary or future offspring. Resources allocated to one child cannot be delivered to others, prolonged care delays the birth of subsequent children, and the risks of caring for today's children decrease the probability that parents will survive to rear tomorrow's. Thus, there is always a conflict between present and future reproductive success. Offspring try to resolve that conflict in favor of present reproductive success (that is, themselves); parents benefit from a more balanced distribution of their investment.

From the standpoint of the parent, its offspring are genetically equivalent, and a parent should not exhibit preferences in delivering care to them. From the standpoint of an individual offspring, however, the self has twice the genetic value of a sibling because a sibling shares only half of an individual's genes. Therefore, when an individual possesses a genetic factor that increases the care it receives from its parents, that trait is favored as long as the cost to the parents, in terms of number of siblings, is less than twice the benefit to the individual. This is the limit to selfish behavior under kin selection that we discussed above.

As offspring develop and become more self-sufficient, the ratio of benefit to cost of continuing to care for them decreases (Figure 13.8). The cost of a particular act of parental investment may change with increasing offspring age, but as young mature and become better able to care for themselves, the benefits of parental care dwindle. When the benefit-cost ratio drops below 1, at the latest, a parent should cease to provide care to those offspring in favor of producing additional ones. Suppose, however, that a child has a gene that increases its solicitation of parental care. Because the inclusive fitness of the child discounts by one-half the cost of not delivering care to siblings, it "prefers" that parental care should continue until the benefit-cost ratio is 0.5 when its parents' future offspring are full siblings, and less when they are not. Thus, the period of time between the age at which B/C is 1 and the age at which it is 0.5 is one of conflict between parent and offspring.

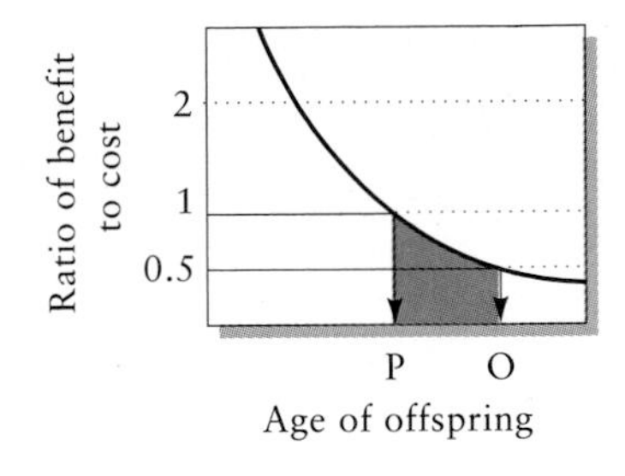

Figure 13.8 The ratio of benefit to cost of an act of parental care toward an offspring decreases with the offspring's age as it grows and becomes more self-sufficient. Because all offspring are of equal genetic value to an individual parent, parents should shift care to succeeding offspring when the benefit-cost ratio falls below 1 (age P). Because siblings have a coefficient of relationship of one-half, however, inclusive fitness arguments dictate that they should solicit parental care until the benefit-cost ratio equals 0.5 (age O). This difference establishes a region of parent-offspring conflict (shaded). After R. L. Trivers, *Am. Zool.* 14:249–264 (1972).

Postadolescent humans may encounter this region of conflict if they express a wish to live with their parents after finishing college (unless they also engage in helping behavior). Biologists believe that similar conflicts over time of weaning can be seen in many species of mammals and birds that exhibit extensive postnatal care. Young, often perfectly capable of taking care of themselves, sometimes hound their parents mercilessly for food. One would think that parents would have the upper hand in any conflict with their offspring, but one must remember that parents are adapted to respond positively to solicitations by their offspring when they are growing and not yet independent. By prolonging juvenile appearance and dependent behavior, offspring may be able to take advantage of their parents' responsiveness and prolong parental care.

The most extreme manifestation of family living is seen in the social insects, in which offspring stay with parents and help them to rear siblings to the exclusion of their own personal reproduction. There is still some debate over the origins of insect societies—particularly over whether these extended families serve the purposes of the parents by means of despotic domination or the purposes of the sterile offspring (workers) through their inclusive fitness. Regardless of their origins, the structure of insect societies and the peculiar sex-determining mechanism of the bees, ants, and wasps create ample conditions for parent-offspring conflict, as well as plenty of fascinating questions to attract the attention of biologists.

Insect societies

The complex societies of termites, ants, bees, and wasps have presented a formidable challenge to evolutionary ecologists primarily because of the existence of nonreproductive castes. How can natural selection produce individuals with no reproductive output—that is to say, with no individual fitness? Before considering the evolutionary issues raised by insect societies, let's have a quick look at their natural history. There are several grades of sociality in the animal world, the highest of which is **eusociality.** This grade is characterized by (1) several adults living together in groups; (2) overlapping generations—that is, parents and offspring living together in the same colony; (3) cooperation in nest building and brood care; and (4) reproductive dominance, including the presence of sterile **castes.** Thus defined, eusociality is limited among insects to the termites (Isoptera) and the ants, bees, and wasps (Hymenoptera); elements of eusociality are present in at least one mammal, the naked mole rat of Africa.

The complex organization of insect societies is dominated by one or a few egg-laying females, which are referred to as **queens.** Nonreproductive progeny of a queen gather food and care for developing brothers and sisters, some of which become sexually mature, leave the colony to mate, and establish new colonies. Most insect societies are huge extended families.

From its distribution across taxonomic groups, it is clear that eusociality has evolved independently many times in bees, wasps, and ants. It is less clear what route was followed. The most widely accepted sequence of evolutionary steps includes a lengthened period of parental care of the developing brood, with parents either guarding their nests or continuously provisioning their larvae in a manner similar to birds feeding their young. If parents lived and continued to produce eggs after their first progeny emerged as adults, then their offspring would be in a position to help raise subsequent broods consisting of their younger siblings. This overlapping of generations, which is not common among insects, and extensive parental care are necessary ingredients in the recipe for eusociality. Once progeny remain with their mother after they attain adulthood, the way is open to relinquishing their own reproductive function solely to support hers.

Bee societies are organized simply: the offspring of a queen are divided among a sterile worker caste, which is all genetically female, and a reproductive caste, consisting of both males and females, that is produced seasonally. Whether an individual will become a sterile worker or a fertile reproductive is controlled by the quality of nutrition it receives as a developing larva. In general, differentiation of sterile castes is stimulated by environmental (usually nutritional) factors. The development of sexual forms can be inhibited by substances produced by a queen and fed to her larvae. In bees, the worker caste represents an arrested stage in the development of reproductive females, stopped short of sexual maturity.

Ant and termite colonies often have a continuous gradation of worker castes, ranging from very small individuals that are primarily responsible for the nutrition of the colony to larger individuals specialized morphologically to defend it against intruders (Figure 13.9). In the leaf-cutting ant

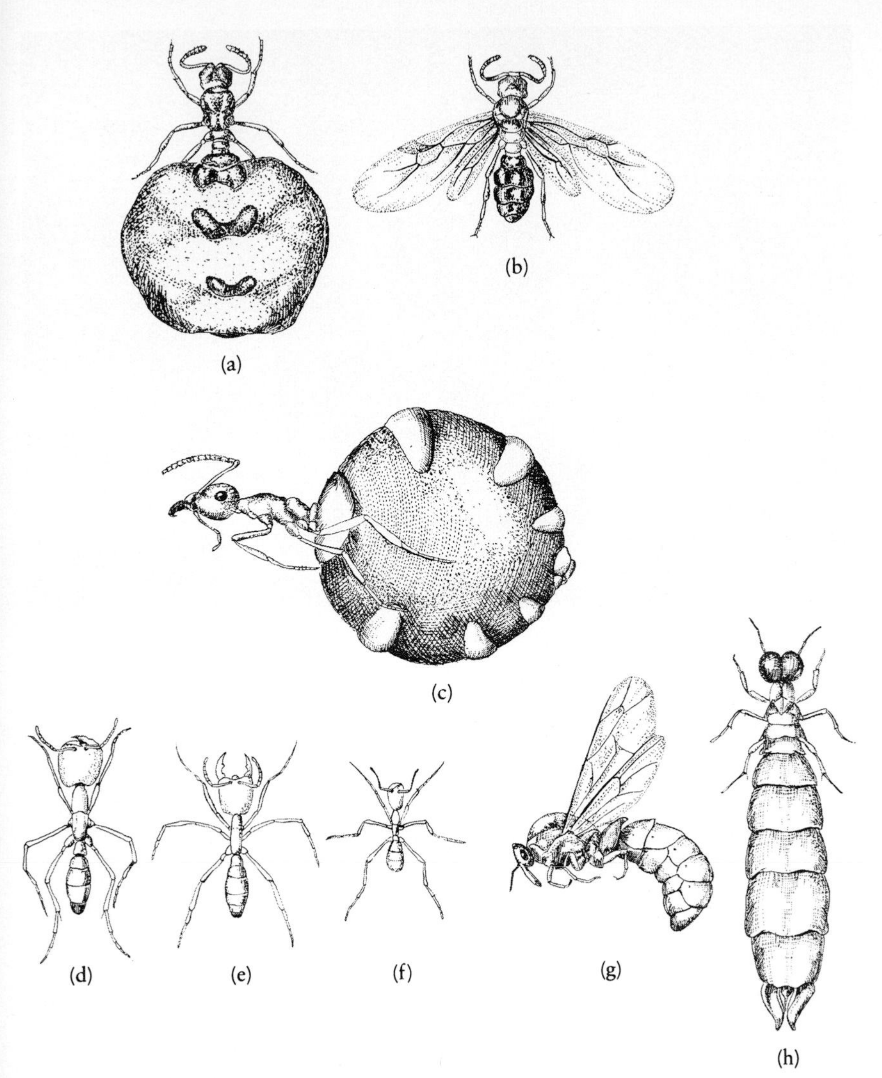

Figure 13.9 Castes of several species of ants: (a) a virgin queen and (b) an old egg-laying queen of the workerless social parasite *Anergates atralulus* of Europe; (c) a so-called replete worker, which stores plant sap in its gut, of the honey ant *Myrmecosystus melliger* from Mexico; (d, e, and f) three sizes of blind workers, (g) a winged male, and (h) a queen, blind and wingless, of the African visiting ant *Dorylus nigricans.* After W. M. Wheeler, *Social Life Among the Insects,* Harcourt, New York (1923); P. P. Grassé, *Traité de zoologie,* Vol. 10, *Insects superieurs et hemipteroides,* Part 2, Masson, Paris (1951).

(Atta), small workers sometimes ride on sections of leaves carried by large workers from which they ward off parasitic flies, like the sidekick riding shotgun on a stagecoach (Figure 13.10). In most insect societies, workers assume specialized roles at any given time, although these may change with the age of a worker or conditions in the colony. Colonies of harvester ants in the southwestern deserts of the United States have five distinct worker roles, each with a characteristic pattern of behavior: nest interior worker (brood care, nest construction, seed storage), midden (refuse pile) worker, forager, patrolling ant, and nest surface maintenance worker (Figure 13.11). Changes in conditions at a nest cause certain ants to switch to different tasks, and once a switch is made, it is permanent. For example, increased food availability causes midden workers, nest surface maintenance workers, and patrolling ants to switch to foraging behavior; threats of intrusion by foreign ants cause nest surface maintenance workers to become patrolling ants, which defend the colony.

Figure 13.10 Leaf-cutting ants returning to their nest. Note the small worker riding on one of the leaf cutouts.

Unlike ant, bee, and wasp societies, termite colonies are headed by a mated pair—the king and queen—which produce all the workers by sexual reproduction. Workers are both male and female, but neither sex matures sexually unless either the king or the queen dies. The queens that produce colonies of ants, bees, and wasps mate only once during their lives and store enough sperm to produce all their offspring, up to a million or more over 10–15 years in some army ants. As we saw in the last chapter, hymenopterans have a haplodiploid sex-determining mechanism: workers are all females produced from fertilized eggs. Males, which develop from unfertilized eggs, appear in colonies only as reproductives (drones) that leave to seek mates.

As mentioned above, the caste structure of insect societies is maintained primarily by means of substances produced by the queens that arrest sexual development in the worker castes. Workers acquiesce to this system—that is, they respond appropriately to chemical signals and nutrition—partly because they would be killed by their queen or by other workers if they embarked on sexual development, and partly because their inclusive fitness differs little whether they rear siblings or their own offspring. The haplodiploid mating system creates strong asymmetries in coefficients of genetic relationship within insect societies (Table 13.3). In particular, a female worker's coefficient of relationship to a female sibling is 0.75, whereas to a male sibling it is 0.25. The queen herself has the same genetic relatedness to sons and to daughters (0.50), so she can be relatively ambivalent about the sex of her offspring, especially when the sex ratio among reproductives in the population as a whole is near equality. The skewed genetic relatedness among siblings means that cooperation is likely to be greater among all-female castes than among male castes or, especially, among mixed castes. This may explain why workers in hymenopteran societies are all female, and why broods of reproductives usually favor females, by about 3:1 on a weight basis. Furthermore, when a female

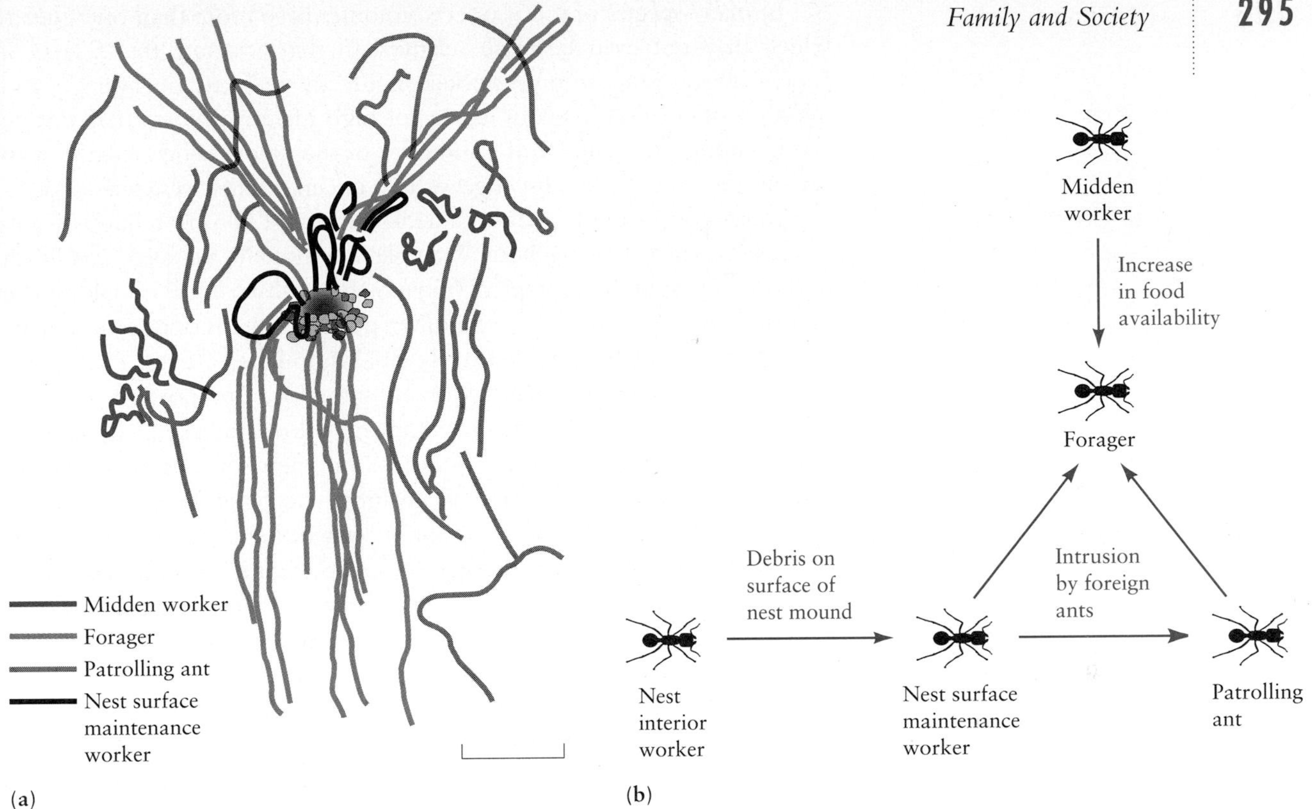

Figure 13.11 (a) Characteristic patterns of paths taken by four types of harvester ant *(Pogonomyrmex barbatus)* workers outside the nest, and (b) the directions of switching between these roles when conditions at the nest change. After D. M. Gordon, *Am. Sci.* 83:50–57 (1995).

worker can produce more female than male reproductives, her own inclusive fitness may actually be higher than it would be if she raised a brood of her own consisting of an equal number of males and females. Under this bizarre circumstance, it is not surprising that sterile castes might have evolved.

TABLE 13.3 Probabilities of identity by descent between male and female individuals and their relatives in eusocial hymenopterans

PROBABILITY OF IDENTITY BY DESCENT WITH	MALE	FEMALE
Mother	0.50	0.50
Father	0.00	0.50
Brother	0.50	0.25
Sister	0.25	0.75
Son	0.00	0.50
Daughter	1.00	0.50

In many species of social insects, colonies have more than one queen, which may not even be close relatives. In this situation, the degree of genetic relationship among workers is greatly reduced on average, and selection of altruistic traits by means of high inclusive fitness may not be strong enough to maintain the integrity of the colony. The fact that such multiqueen social units exist suggests that queens themselves exercise decisive control over caste formation. The balance must be very tenuous, however, as shown by the following example. The fire ant *Solenopsis invicta,* an introduced pest in the southern United States, has two kinds of colonies: in one, there is a single queen; in the other, up to hundreds of queens may be sexually active, each laying relatively small numbers of eggs. The multiqueen colonies produce smaller workers and a lower proportion of reproductives, many of which remain within the colony to breed; colonies are founded by multiple, unrelated queens, and new queens may also be adopted by a nest from outside the colony. The *Solenopsis* population in the United States is polymorphic for an enzyme-encoding genetic locus known as *Pgm-3.* Allele *a* of this gene, when present in homozygous form, causes queens in multiqueen colonies to have a higher reproductive rate than *a/b* heterozygotes and *b/b* homozygotes. Apparently, the *a/a* genotype upsets the delicate balance of relationships within a colony, and *a/a* queens are quickly killed by workers. How they recognize *a/a* queens isn't known, but probably involves a chemical cue.

Behavioral relationships among the social insects represent one extreme along a continuum of social organization from animals that live alone except to breed to those that aggregate in large groups organized by complex behavior. Regardless of its complexity, however, all behavior balances costs and benefits to the individual and to close relatives affected by the behavior. Like morphology and physiology, behavior is strongly influenced by genetic factors and thus is subject to evolutionary modification by natural selection. The evolution of behavior becomes complicated when individuals interact within a social setting and the interests of individuals within a population may either coincide or conflict. Understanding the evolutionary resolution of social conflict in animal societies continues to be one of the most challenging and important concerns of biology.

Summary

1. Selection imposed by behavioral interactions with members of one's family and with unrelated individuals within one's population provides the basis for the evolutionary modification of social behavior.

2. Territoriality is the defense of an area or resource from intrusion by other individuals. Animals maintain territories when the resources they gain by doing so are rewarding and defensible.

3. Dominance hierarchies order individuals within social groups by rank, which is established by direct confrontation. Because rank is generally respected, dominance relationships reduce conflict within the group.

4. Living in large social groups may benefit individuals by enabling them to better detect and defend against predators or to obtain food more efficiently. Groups form to the extent that these advantages outweigh the negative effects of competition among group members.

5. Isolated acts of social behavior involve a donor and a recipient. When both benefit, the behavior is termed cooperation or mutualism; when the donor benefits at a cost to the recipient, the behavior is selfish; when the recipient benefits at a cost to the donor, the behavior is altruistic.

6. The presence of altruistic behavior in populations has been explained in terms of kin selection. Kin selection arises because, when an individual interacts with a relative, it affects the fitness of that portion of its own genotype that is also inherited by the relative directly from a common ancestor.

7. Inclusive fitness expresses the benefit (or cost) of a behavior to the donor plus the benefit or cost to the recipient, adjusted by the coefficient of relationship. In the case of interactions between siblings, which have a coefficient of relationship of 0.50, selection will favor any altruistic behavior whose cost to the donor is less than one-half the benefit it confers on the recipient.

8. Inclusive fitness arguments have been invoked to explain such behaviors as alarm calling and delaying one's own sexual maturation to help one's parents rear younger siblings. In general, the distribution of cooperation and altruism within social groups is sensitive to degree of genetic relatedness between individuals.

9. Game theory analyses, such as the hawk-dove game, indicate that cooperative behavior cannot evolve among nonrelatives even though the average benefit to individuals in a purely cooperative social group exceeds that gained by individuals through confrontation and conflict. The reason is that cooperative behavior is not an evolutionarily stable strategy, but can be invaded by selfish cheaters.

10. When individuals live in close association over long periods, the choice of cooperation or conflict by one individual may be made contingent on the experienced behavior of others in the group. In this way, cooperation may pervade through the mechanism of reciprocal altruism, in which members of the social group withhold cooperation from cheaters.

11. Conflict may arise between parents and offspring over the optimal level of parental investment. All siblings are genetically equal in the eyes of their parents, but siblings are genetically related to each other by only ½. Therefore, individual offspring should prefer unequal parental investment in themselves, even when parental fitness is reduced as a result.

12. Social insects (termites, ants, wasps, and bees) live in extended family groups in which most offspring are retained in a colony as sterile workers, increasing their mother's fitness by rearing reproductive siblings.

13. The haplodiploid sex-determining mechanism of the Hymenoptera results in females having a ¾ relationship to sisters, but only a ¼ relation-

ship to brothers. This skew probably has contributed to the facts that workers in ant, bee, and wasp colonies are all female and that more female reproductives are produced than males.

Suggested readings

Emlen, S.T., P. H. Wrege, and N. J. Demong. 1995. Making decisions in the family: An evolutionary perspective. *American Scientist* 83:148–157.

Gordon, D. M. 1995. The development of organization in an ant colony. *American Scientist* 83:50–57.

Heinrich, B., and J. Marzluff. 1995. Why ravens share. *American Scientist* 83: 342–349.

Honeycutt, R. L. 1992. Naked mole-rats. *American Scientist* 80:43–53.

Keller, L., and K. G. Ross. 1993. Phenotypic plasticity and "cultural" transmission of alternative social organizations in the fire ant *Solenopsis invicta. Behavioral Ecology and Sociobiology* 33:121–129.

Lima, S. L. 1995. Back to the basics of anti-predatory vigilance: The group-size effect. *Animal Behavior* 49:11–20.

Rohwer, S. 1982. The evolution of reliable and unreliable badges of fighting ability. *American Zoologist* 22:531–546.

Seeley, T. D. 1989. The honey bee colony as a superorganism. *American Scientist* 77:546–553.

Sherman, P. W., J. U. M. Jarvis, and S. H. Braude. 1992. Naked mole rats. *Scientific American* 267:72–78.

Trivers, R. L. 1971. The evolution of reciprocal altruism. *Quarterly Review of Biology* 46:35–57.

Trivers, R. L. 1974. Parent-offspring conflict. *American Zoologist* 14:249–264.

Trivers, R. L. 1985. *Social Evolution.* Benjamin/Cummings, Menlo Park, Calif.

Wilkinson, G. S. 1984. Reciprocal food sharing in the vampire bat. *Nature* 308:181–184.

Wilkinson, G. S. 1988. Social organization and behavior. In A. M. Greenhall and U. Schmidt, eds., *Natural History of Vampire Bats,* pp. 85–97. CRC Press, Boca Raton, Fla.

Wilson, E. O. 1975. *Sociobiology.* Harvard University Press, Cambridge, Mass.

Winston, M. L., and K. N. Slesser. 1992. The essence of royalty: Honey bee queen pheromone. *American Scientist* 80:374–385.

Woolfenden, G. E., and J. W. Fitzpatrick. 1984. *The Florida Scrub Jay: Demography of a Cooperatively Breeding Bird.* Princeton University Press, Princeton, N.J.

PART IV

POPULATIONS

CHAPTER 14

POPULATION STRUCTURES

A **population** can be defined as the individuals of a species within a given area. A population's boundaries may be natural ones imposed by the geographic limits of suitable habitat, or they may be defined arbitrarily for the purposes of scientific study. In either case, a population has a spatial structure, which means that within its geographic boundary, individuals live primarily within patches of suitable habitat, and their abundances may vary with food supplies, predators, nest sites, and other ecological factors within that habitat.

Population structure, which includes density and spacing of individuals, proportions of individuals in each age class, mating system, and genetic variation, provides us with a snapshot at an instant in time. Populations also exhibit dynamic behavior, continuously changing over time because of births, deaths, and movements of individuals. The regulation of these processes depends on the varied interactions of individuals with their environments and with one another. Moreover, evolution by natural selection and the regulation of both community structure and ecosystem function become apparent when these issues are cast in terms of population processes. These processes determine whether a gene will spread within a population and whether a population will persist within a habitat; they also govern the flux of energy and contribute to the cycling of elements within ecosystems. Hence much of ecology focuses on processes at the population level.

Populations have temporal continuity because the individuals alive at any one time have descended from others that were alive at an earlier time. Populations also have spatial continuity because individuals in different parts of a population's range have common ancestors: as a rule, the greater the distance between two individuals, the more remote in time is their common ancestor and the more likely they are to differ genetically. Individuals within a population derive their genes from a common source, the so-called **gene pool,** and thus also share a common history of adaptation to their environments, but populations also may be subdivided spatially and may be adapted to local conditions.

In this chapter, we consider the distribution and genetic organization of individuals within populations. Distribution and movements of individuals determine the spatial coherence of a population—the degree to which different parts of a population share the same dynamics or, alternatively, exhibit independence from one another. As populations become fragmented when humans transform the landscape, issues of population coherence and, particularly, the vulnerability of small subpopulations to extinction or drastic genetic change become important management and conservation concerns. Populations may also be subdivided as a result of natural habitat heterogeneity. Most island populations pass through initial stages of colonization characterized by small numbers of individuals. Because chance events have a greater impact on number of individuals and genetic diversity in small populations than in large ones, ecologists must pay close attention to the history of a population's size, no matter how large it is at present, to interpret its present structure and prospects for the future.

Within populations, individuals may vary with respect to gender, age, experience, social position, genotype, and the accumulated effects of accident and chance. Evolution occurs when genetic differences between individuals result in different birth rates or death rates, and hence different contributions of progeny to future generations. Sex ratio and age structure influence population dynamics through sex-related and age-related variations in birth and death rates. These aspects of population structure are therefore an important part of our understanding of how populations change over time. We shall consider many of these issues in the chapters that follow.

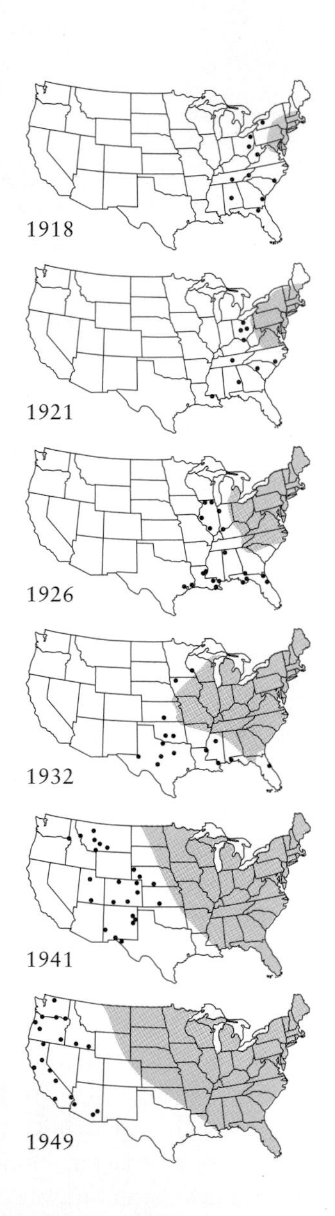

Figure 14.1 Western expansion of the range of the European starling *(Sturnus vulgaris)* in the United States. The shaded areas represent the breeding range; dots indicate records of birds in preceding winters. The population now inhabits the entire country. After B. Kessel, *Condor* 55:49–67 (1953).

Habitat and the distribution of populations

The **distribution** of a population is its geographic range, which often is correlated with the variety of habitats that the population occupies, a measure of its **ecological range.** Distributions are determined primarily by the presence or absence of suitable habitat. Thus, the natural range of the sugar maple in the United States and Canada stops abruptly to the east at the Atlantic Ocean, but is limited more gradually to the west by low precipitation, to the north by cold winters, and to the south by hot summers (see Figure 5.1). Undoubtedly, much suitable habitat for the sugar maple exists throughout the world, especially in Europe and Asia, where it has been transplanted successfully, but the species evolved in North America and has not had the opportunity to colonize these areas on its own.

Distributional limits imposed by barriers to long-distance dispersal reveal themselves dramatically when introduced species expand successfully into new regions. For example, in 1890 and 1891, 160 European starlings were released in the vicinity of New York City, evidently because of someone's wish to introduce all the birds mentioned in Shakespeare's works to the New World. Within 60 years, the population had expanded to cover more than 3 million square miles and stretched from coast to coast (Figure 14.1).

Within the geographic range of a population, individuals are not equally numerous in all regions. Individuals generally live only in suitable habitat. During the summer months, starlings feed by probing the ground for insect larvae and other food items. Because starlings have difficulty feeding when soil is dry and hard, populations are most dense in the cooler and moister parts of the species' distribution, and otherwise where parks, golf courses, crops, and pastures are well irrigated (Figure 14.2). Returning to an earlier example, sugar maples do not grow in marshes, on serpentine barrens, on newly formed sand dunes, in recently burned areas, or in a variety

Figure 14.2 Relative densities, indicated by density of shadings, of European starlings in North America averaged over 1966–1989, as determined by the Breeding Bird Survey. From B. A. Maurer and M. Villard, *Nat. Geogr. Res. Explor.* 10:306–317 (1994).

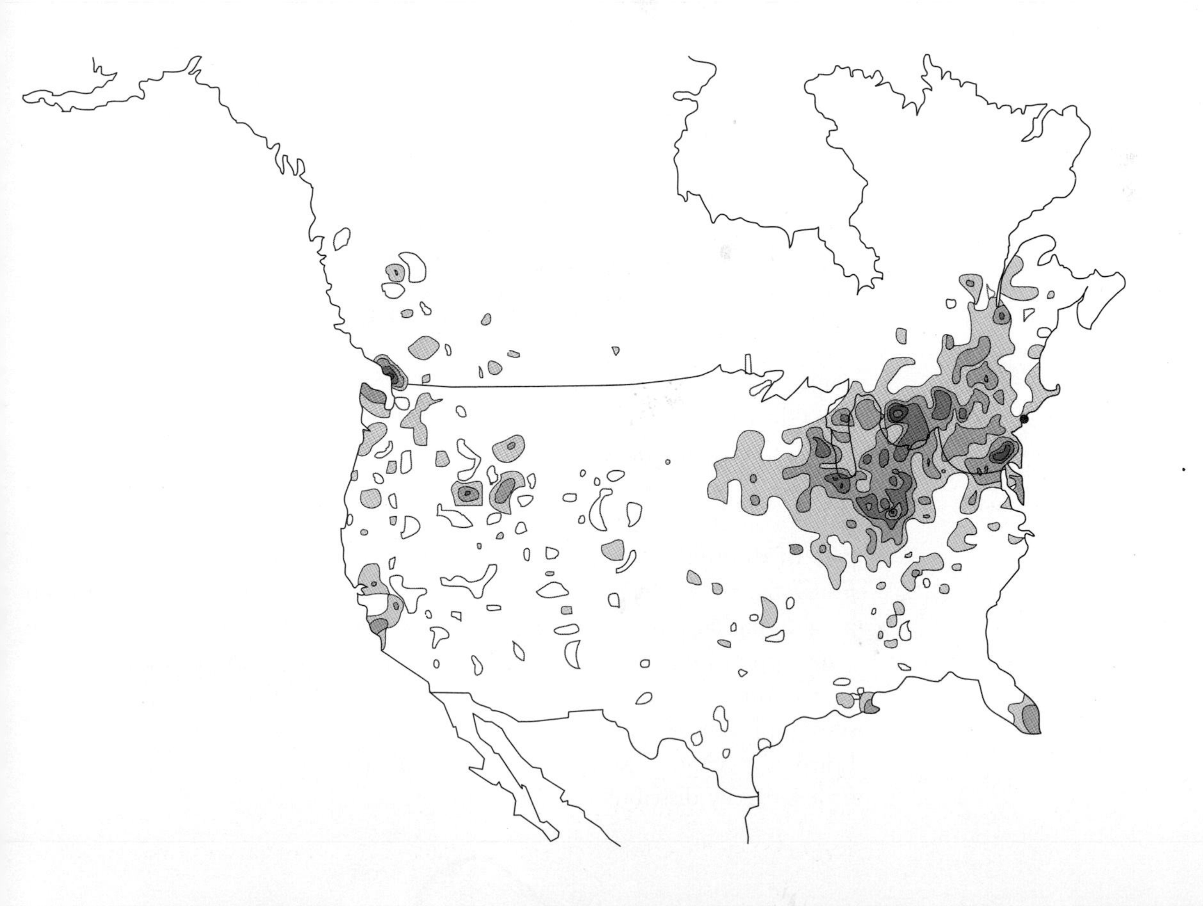

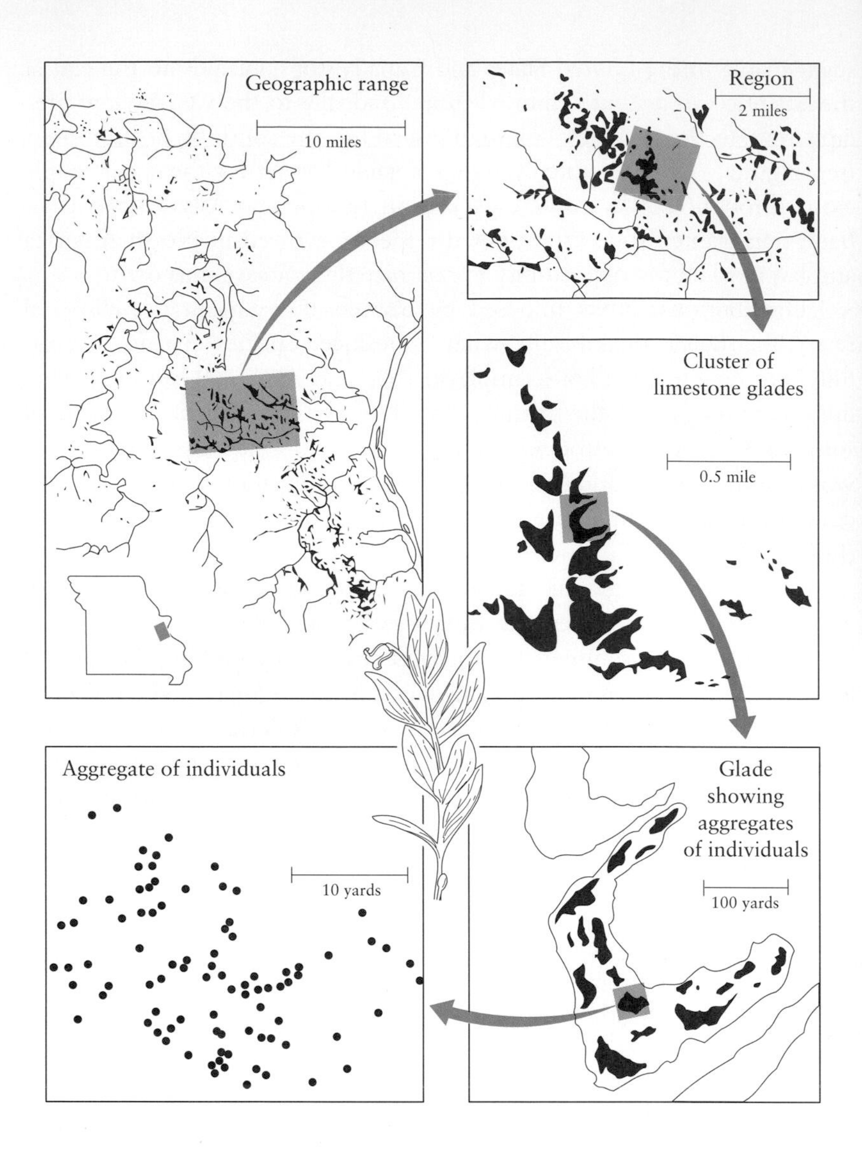

Figure 14.3 Hierarchy of patterns in the geographic distribution of *Clematis fremontii,* variety *riehlii,* in Missouri. After R. O. Erickson, *Ann. Mo. Bot. Gard.* 32:416–460 (1945).

of other habitats that simply lie outside their range of ecological tolerance. Hence, as in the case of the starling, the geographic range of the sugar maple is a patchwork of occupied and unoccupied areas.

Climate, topography, soil chemistry, and soil texture exert progressively finer influences on the geographic distribution of the perennial shrub *Clematis fremontii* (Figure 14.3). Climate and perhaps interactions with ecologically similar species restrict this species of *Clematis* to a small part of the midwestern United States. The distinctive variety of *Clematis fremontii* named *riehlii* occurs only in Jefferson County, Missouri. Within its geographic range, *Clematis fremontii* is restricted to dry, rocky soils on outcroppings of limestone. Small variations in relief and soil quality further confine the distribution of *Clematis* within each limestone glade to sites with suitable conditions of moisture, nutrients, and soil structure. Local aggregations occurring on each of these sites consist of many more or less evenly distributed individuals.

The horned lark *(Eremophila alpestris),* a small songbird of short grasslands, occurs more widely in North America than *Clematis fremontii,* but its local distribution also reflects subtle variations in habitat. In Colorado, populations of horned larks are denser on heavily grazed rangeland than on lightly grazed land because the larks prefer open spaces with low vegetation. Even within their territories, individuals use some parts more than others; for example, the best display sites on small shrubs are often not in the best areas for foraging (Figure 14.4).

The distribution of a population includes all of the areas its members occupy during their life cycle. Thus the geographic range of salmon includes not only the rivers that are their spawning grounds, but also vast areas of the sea where individuals grow to maturity before making the long migration back to their birthplace (Figure 14.5). Many birds sensibly make annual migrations to warm climates during the winter months. Distributions of the golden plover and blackburnian warbler, for example, lie entirely within North America during the summer but entirely within Central and South America during the winter (Figure 14.6). The size and year-to-year variations of populations of such migratory species depend on the interactions of individuals with their environments throughout the year. Thus, ecological changes in Mexico, the Amazon Basin, or central Africa may affect populations of migratory species that breed in the forests of eastern North America or Europe.

Figure 14.4 Hierarchy of patterns in the distribution of the horned lark *(Eremophila alpestris)* during the breeding season. Although the species is widely distributed on a continental scale, areas of high population density are more localized, particularly in the western Great Plains. Within a local area, density depends on habitat quality and land use practices, with heavier grazing actually favoring the larks. Even in the best habitats, territories (bounded by black lines) may not fill all the space, and within an individual's territory, certain parts receive heavier use than others. From J. A. Wiens, *Ecol. Monogr.* 43:237–270 (1973).

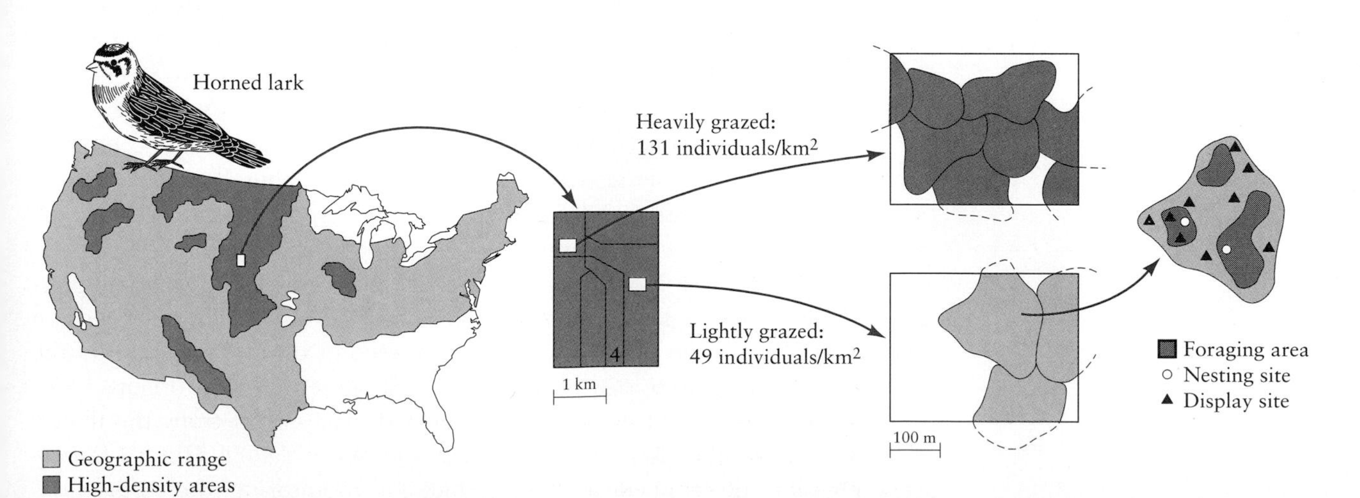

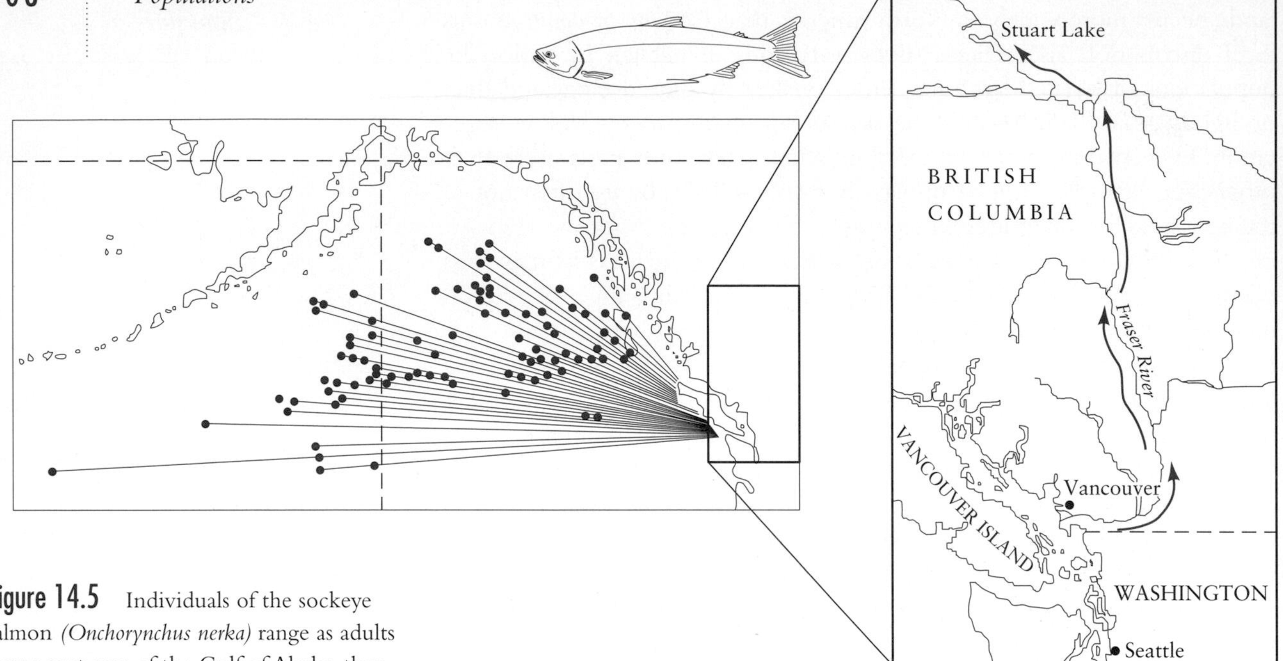

Figure 14.5 Individuals of the sockeye salmon *(Onchorynchus nerka)* range as adults over a vast area of the Gulf of Alaska, then migrate up rivers to their birthplaces to breed. The map shows the distribution of individual salmon tagged as adults at sea and then subsequently recaught as breeders at Stuart Lake, more than 1,000 km upriver from the mouth of the Fraser River. Adapted from C. Groot and T. P. Quinn, *Fishery Bull.* 85 (1967).

Population density

The ultimate measure of a population is number of individuals. From a management and conservation standpoint, it is important to understand the factors that cause population size to change and the processes that govern the regulation of population size. This understanding must begin with an empirical knowledge of the numbers of individuals in populations. Total population size has two components, local density of individuals and total range of the population. **Density** is defined as the number of individuals per unit of area. The density of individuals in a particular habitat depends on the intrinsic quality of that habitat for the species of concern and on the net movement of individuals from other habitats. From the standpoint of understanding the ecological relationship of a population to its environment, local density is more revealing than total population size because it is more directly connected to local ecological interactions. As a rule, individuals are most numerous where resources are most abundant. Thus, density provides information about the relationship of a population to its environment, and changes in density reflect changing local conditions.

Because densities of populations change over time and space, no population has a single structure; one's perception of a population depends on where and when one looks. Long-term records of the population of the chinch bug *(Blissus leucopterus)* in Illinois illustrate this point (Figure 14.7). These records exist because chinch bugs damage cereal crops; the Illinois State Entomologist's Office and later the State Natural History Survey Division determined the importance of monitoring the population,

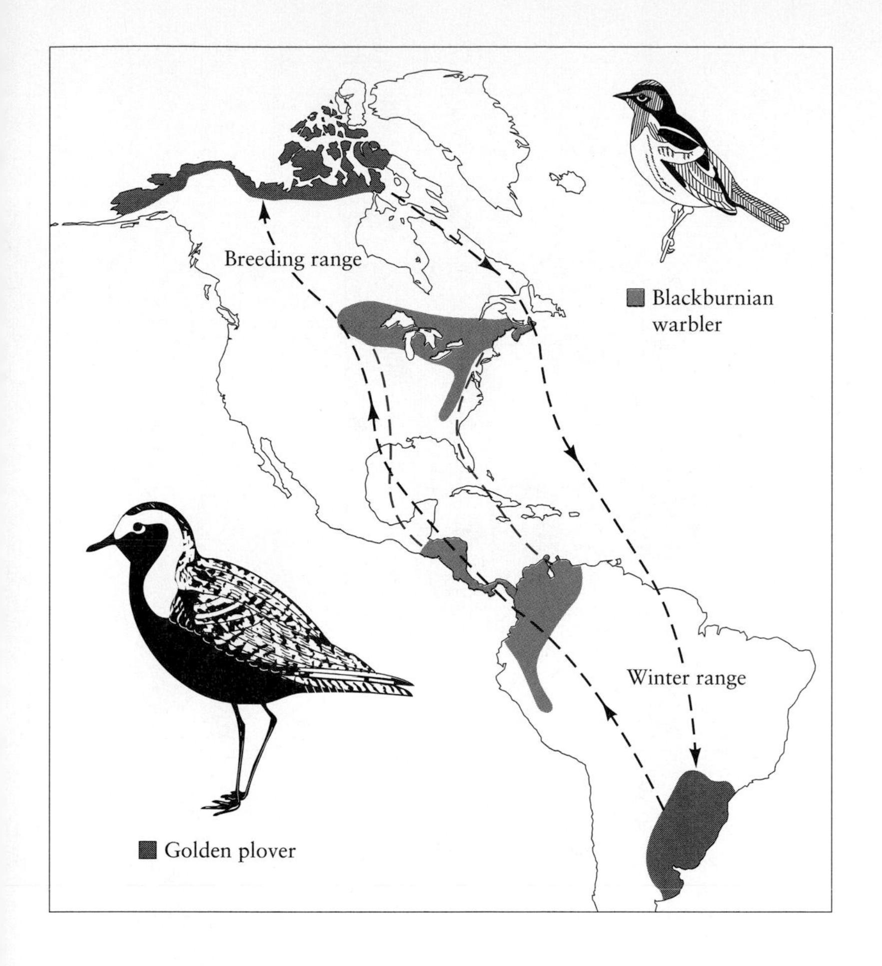

Figure 14.6 Breeding and wintering ranges of the golden plover and blackburnian warbler. Migration routes are indicated by dashed lines.

which they estimated from county reports of crop damage attributable to chinch bugs.

Consider the numbers involved. During 1873, when the bugs damaged crops severely over most of the state, ballpark estimates of the population indicated an average density of 1,000 chinch bugs per square meter over an area of 300,000 square kilometers, or a total of 3×10^{14} pests (300 trillion, more or less). By contrast, farmers reported hardly any damage in 1870 and 1875. Severe outbreaks were usually confined to small portions of the state, sometimes in the north, sometimes in the south. However, the spatial and temporal continuity of the population reveals itself in Figure 14.7 in the waxing and waning of infestations.

Populations in heterogeneous landscapes

The natural world is extremely varied. Uniform, homogeneous habitats extending over vast areas just do not exist. Satellite photographs of Amazonian rain forest in eastern Peru revealed more than 50 subtly and not so subtly different types of forest, varying on scales from a few tens of

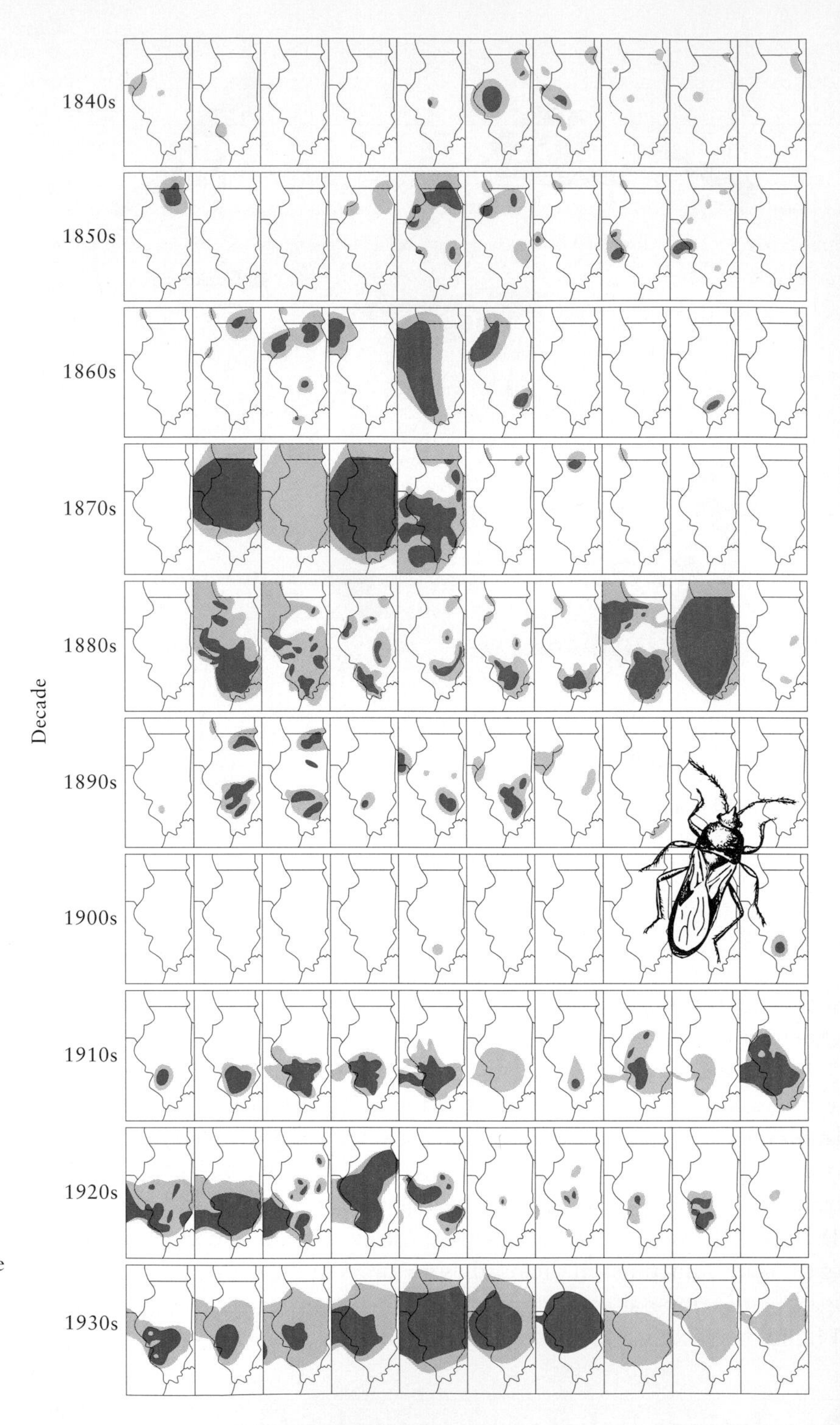

Figure 14.7 Distribution of crop damage caused by chinch bugs *(Blissus leucopterus)* in Illinois between 1840 and 1939. From V. E. Shelford and W. P. Flint, *Ecology* 24:435–455 (1943).

meters to kilometers. Ground-level studies that were coordinated with the satellite images showed that distributions of forest trees responded to small variations in drainage, soil, and recent disturbance.

The idea that the world exists as a mosaic of habitat patches is referred to as the **landscape concept.** The scale of patchiness within a landscape can exert a powerful influence on populations because it affects the distribution and quality of suitable habitat. The quality of one type of habitat can be altered by the presence of different habitats nearby. Habitat quality may be improved when other patches in the landscape provide resources such as roosting sites, nesting materials, pollinators, or water. Other kinds of neighboring habitat patches may be a serious drawback if they harbor predators and disease organisms. Throughout much of the eastern and midwestern United States, forest fragmentation has brought populations of forest birds into contact with the parasitic brown-headed cowbird, which lays its eggs in the nests of other species and greatly reduces the reproductive success of its hosts. Cowbirds prefer the open habitats of farms and fields, but do not hesitate to enter woodlots to seek out nests to parasitize. As a result, populations of many forest birds have decreased in areas of high landscape heterogeneity.

A recent experimental study of scallop populations inhabiting seagrass beds in North Carolina estuaries demonstrates the importance of patchiness to population processes in marine environments. Predation on juvenile scallops by fish, blue crabs, sea stars, and whelks was much higher in patchy areas of seagrass than in continuous seagrass habitat, even though the immediate environment of individual scallops was the same in all cases (Figure 14.8). The increased edge per unit of area of habitat in the patchy areas evidently provided increased access to the interior of the seagrass habitat by predators, just as forest edge provides access by cowbirds to birds nesting in the interior.

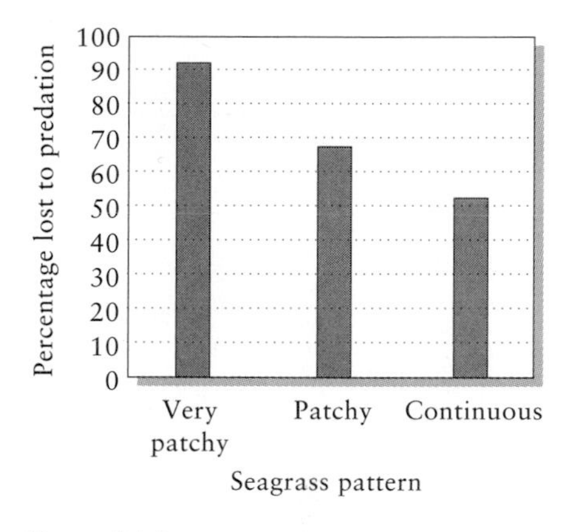

Figure 14.8 Rates of predation over a 4-week period on juvenile scallops placed in continuous seagrass habitat and in areas of seagrass with two levels of patchiness. From E. A. Irlandi, W. G. Ambrose, Jr., and B. A. Orlando, *Oikos* 72:307–313 (1995).

Ideal free distributions

When the quality of habitat patches varies, individuals tend to distribute themselves over landscapes in proportion to the suitability of environments. The suitability of a habitat patch depends not only on its intrinsic characteristics, but also on the density of other individuals living in the patch. Many organisms are capable of making behavioral decisions about where to live and where not to live, and these decisions can be based on a combination of habitat quality and occupation by other individuals in the same population. A patch becomes less attractive to newcomers as the number of individuals exploiting it increases, just as a male bird's territory becomes less attractive to a female if he already has a mate. Occupied patches are likely to have fewer remaining resources than undiscovered patches; furthermore, competing individuals may precipitate costly behavioral conflicts. When organisms have complete knowledge of patch quality and complete freedom of choice, we would expect them to occupy or exploit patches in direct proportion to their quality; that is, patches with the highest levels of resources should attract the most consumers.

Each individual chooses among patches so as to maximize its own rate of gain of resources. Imagine two patches, one containing more resources than the other. At first, individuals choose the intrinsically superior patch. But as the population builds up in that patch, the apparent quality of the patch decreases, owing to depletion of resources and antagonistic interactions, until the second patch becomes an equally good choice. At this point, individuals join the second and the first patch impartially as the quality of both continues to decline. As a result, each individual in the population exploits a patch of the same realized quality, regardless of variation in intrinsic patch quality. This outcome is called an **ideal free distribution.**

Several lines of evidence suggest the existence of, or tendencies toward, ideal free distributions in nature. Among the most compelling is the pattern of habitat selection by birds. In migratory species, individuals arrive on their breeding grounds over several weeks. Early arrivals generally fill certain habitats that confer high breeding success before newcomers begin to establish territories in poorer habitats. When individuals are removed from good breeding habitats, their places are quickly taken by others moving in from poorer habitats. When population densities are low, perhaps following a particularly severe winter, occupancy of low-quality habitats decreases more than does that of good habitats. This example is complicated by the fact that subdominant individuals are excluded from optimal habitat by antagonistic interactions with territory holders. Thus, the distribution of individuals among habitats is not completely open to free choice.

Tendencies toward an ideal free distribution have also been investigated in laboratory studies, in which the quality of patches can be controlled. One series of experiments conducted by Manfred Milinski involved sticklebacks *(Gasterosteus)* that were provided with food (water fleas) at different rates at opposite ends of an aquarium. Each end of the tank could be regarded as a patch, and the system had the following conditions conducive to establishing an ideal free distribution: (1) the two patches differed in quality, (2) quality decreased as the number of fish using a patch increased, and (3) fish were free to move between patches.

Hungry fish were placed in the aquarium about 3 hours before the start of the experiment. During trials, the number of fish in each half of the tank was recorded at the end of each 20-second interval. Before any food was added to the tank, the fish were distributed equally between the two halves. In one experiment, water fleas were added at a rate of 30 per minute to one end of the tank and at a rate of 6 per minute to the other—a ratio of 5 to 1. Within 5 minutes, the fish had distributed themselves between the two halves in the same ratio as predicted for an ideal free distribution—that is, 5 to 1. In a second experiment, water fleas were provided at rates of 30 and 15 per minute, a 2-to-1 ratio. Again, the distribution of fish followed suit. When the better and poorer patches in the tank were reversed, the fish reversed their distribution within about 5 minutes. The behavioral mechanisms they used to achieve an ideal free distribution were not determined, but cues for behavioral choices would have to incorporate both rate of food provisioning and number of competitors within a patch, both of which are probably gauged from an individual's recent experience. Such experiments demonstrate the considerable sensitivity of organisms to the conditions of their environment, as well as their behavioral flexibility in making choices.

Source and sink populations

Under an ideal free distribution, the fitness of each individual is the same regardless of the intrinsic quality of the patch it occupies. There are many reasons, however, why this ideal may rarely be attained. Among the most important of these are that individuals do not have perfect knowledge of patch quality and that territorial behavior by dominant individuals reduces free choice in subordinates. In almost every case in which it has been assessed, reproductive success varies among habitats. In habitats with abundant resources, individuals produce more offspring than required to replace themselves, and the surplus offspring usually disperse to other areas (Figure 14.9). Such populations are referred to as **source populations.** In poor habitats, populations are maintained by immigration of individuals from elsewhere, because too few offspring are produced locally to replace losses to mortality. These are referred to as **sink populations.**

In southern Europe, a small songbird called the blue tit *(Parus caeruleus)* breeds in two kinds of habitats, one dominated by the deciduous downy oak *(Quercus pubescens)* and the other by the evergreen Holm oak *(Quercus ilex).* Comparisons of population densities and reproductive success in the two habitats suggest that deciduous oak habitat is superior for tits, and supports source populations (Table 14.1). Indeed, tit populations in deciduous oak habitats produce so many young that they would grow at almost 10% annually if individuals did not disperse. The net rate of population decline in evergreen oak habitats would be about 13% per year in the absence of immigration. Even though densities of blue tits in deciduous oak habitats are six times higher than those in evergreen oak habitats, tits do not achieve an ideal free distribution. Recent genetic analyses of local populations of blue tits in southern France, which live in habitat patches separated by 10 kilometers on average, have shown that movement of young birds

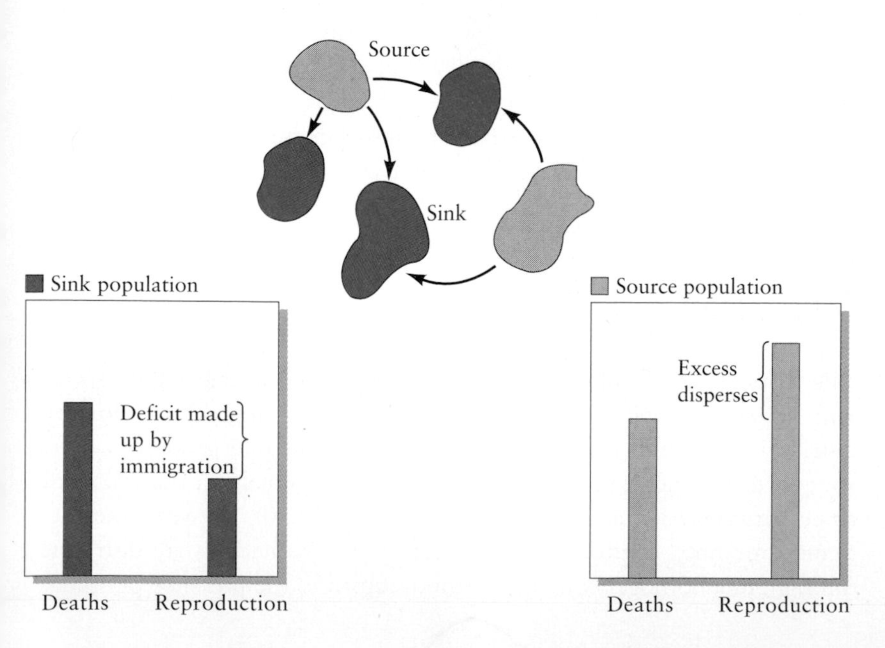

Figure 14.9 Diagram of the source-sink model of population structure. Populations in high-quality habitat patches produce excess offspring, which disperse to less suitable habitat patches, where immigration maintains less productive populations.

TABLE 14.1 Comparisons of the density and productivity of blue tit *(Parus caeruleus)* populations from deciduous and evergreen oak woodlands

	HABITAT	
	Deciduous oak	*Evergreen oak*
Breeding density (pairs/100 ha)	90	14
Laying date (average)	10 April	21 April
Clutch size	9.8	8.5
Survival to fledging	0.60	0.43
Fledglings per parent	2.9	1.8
Probable number of recruits per parent	0.59	0.37
Likelihood of death of parent	0.50	0.50
Net productivity per parent	+0.09	–0.13
Type of population	Source	Sink

Source: J. Blondel, P. C. Dias, M. Maistre, and P. Perret, *Auk* 110:511–520 (1993).

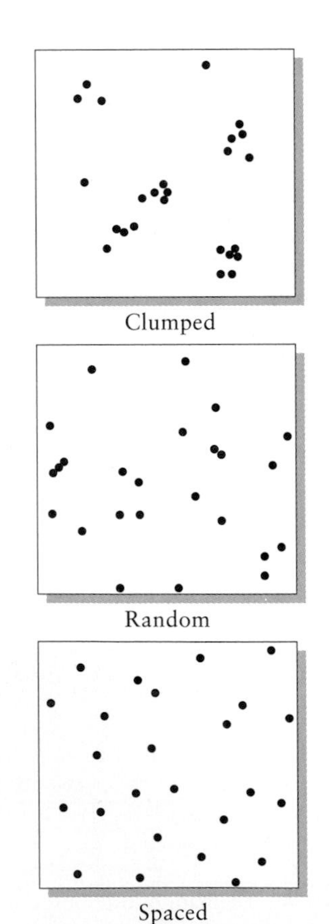

Figure 14.10 Diagram of the spatial distribution of individuals in clumped, random, and evenly spaced populations.

from deciduous oak forest to breed in evergreen oak forest (estimated at 2,000 individuals per year in the region) is about one hundred times higher than movement between patches of the same kind of habitat.

Each landscape mosaic has a mixture of patches of varying habitat quality, some of which support source populations and some of which are sinks. Dispersal is generally a one-way street from source to sink, and sink populations would rapidly dwindle to extinction were it not for immigration from more productive source populations. The source-sink model emphasizes the importance of the landscape setting to understanding the structures and dynamics of populations. Each local population within a patch also has a characteristic structure and dynamic, which reflect more closely the relationship of the individual to its immediate environment and its neighbors.

Dispersion

The **dispersion** of individuals within a population describes their spacing with respect to one another. Patterns of spacing range from **clumped** distributions, in which individuals are found in discrete groups, to evenly **spaced** distributions, in which each individual maintains a minimum distance between itself and its neighbors (Figure 14.10). Between these extremes one finds **random** dispersion, in which individuals are distributed throughout a homogeneous area without regard to the presence of others.

Spaced and clumped distribution patterns derive from different processes. Even spacing—sometimes called **hyperdispersion**—most commonly arises from direct interactions between individuals. Maintenance of a minimum distance between oneself and one's nearest neighbor results in even spacing; for example, in their crowded colonies, seabirds place their nests just beyond their neighbors' reach (Figure 14.11). Plants situated too close to larger neighbors often suffer from shading and root competition; as these individuals die, the spacing of individuals becomes more even.

Clumping, or **aggregation,** results from (1) the social predispositions of individuals that form groups, (2) clumped distributions of resources, which may be the commonest cause of aggregation in most organisms, and (3) tendencies of progeny to remain in the vicinity of their parents. Birds that travel in large flocks often aggregate to find safety in numbers. Salamanders that prefer to live under logs exhibit clumped distributions corresponding to the patterns of fallen deadwood. Trees form clumps of individuals by vegetative reproduction or when seeds disperse poorly (Figure 14.12).

Finally, in the absence of social antagonism or mutual attraction, individuals may distribute themselves at random, without regard to the positions of other individuals in the population. The three general patterns of dispersion—clumped, random, and spaced—may be distinguished by various statistical tests, one of which is illustrated in Box 14.1.

Figure 14.11 A nesting colony of Cape gannets on an island off the coast of South Africa. The densely packed birds space their nests more or less evenly, the distance being determined by behavioral interactions between individuals. Along the entire length of the South African coast, however, seabird populations are clumped during the breeding season on a few offshore islands that offer suitable nesting sites.

Dispersal and the spatial coherence of populations

Movements of individuals between populations and between localities within populations influence local population dynamics. Biologists refer to movements within populations as **dispersal** and to movements between populations—as between sources and sinks—as **emigration** (leaving) and **immigration** (entering).

Dispersal, particularly over long distances, is difficult to measure directly because detecting such movements requires marking and recapturing individuals. Most estimates of dispersal distances come from studies of population genetics, in which investigators wish to determine the genetic structure of a population and to estimate the effective population size for evolutionary processes as well as gene flow between populations. Indeed, the first attempts to measure dispersal in natural populations involved measuring movements away from a release point by fruit flies *(Drosophila)* that could be distinguished by a visible mutation.

Population biologists describe dispersal by means of a number of mathematical indices, each of which reflects different assumptions about the structure of populations. The simplest model describes dispersal as random movement through a homogenous environment, analogous to Brownian motion in physics. With respect to a fixed point of origin, some random movements take an individual farther away and some bring it closer, although on average, distance tends to increase with time. The position of any individual at any moment is the sum of many random increments of distance. The probability of finding an individual at a given distance from a release site is described by the normal frequency distribution (Figure 14.13), a bell-shaped curve whose peak coincides with the point of origin (that is, distance = 0) and whose breadth is characterized by a single variable, the **standard deviation** (s). The value of s provides a convenient index to dispersal distance: st is the dispersal distance estimated over the average life span of an individual, t (Box 14.2). **Neighborhood size** is defined as the number of individuals included within a circle of radius $2st$; it indicates the subset of other individuals with which a member of a population can potentially interact (mate, compete) over its life span.

Small songbirds of eight species marked with leg bands as nestlings and recaptured as breeding adults exhibited lifetime dispersal distances (st) between 344 and 1,681 meters, densities between 16 and 480 individuals per square kilometer, and neighborhood sizes between 151 and 7,679 individuals. For three populations of the land snail *Cepaea nemoralis,* dispersal distances (s_1) varied between 5.5 and 10 meters after 1 year, but because populations are dense and individuals have long life spans, neighborhood sizes were quite large: 1,800 to 7,600 individuals. Note that the neighborhood sizes of these slowly moving snails are on the same order as those of small birds. Mark-recapture data on the rusty lizard *(Sceloporus olivaceous)* population near Austin, Texas, revealed st to be 89 meters and neighborhood size to be between 225 and 270 individuals.

In most studies of dispersal, the principal source of information is movements of individuals away from a point of release. Other kinds of

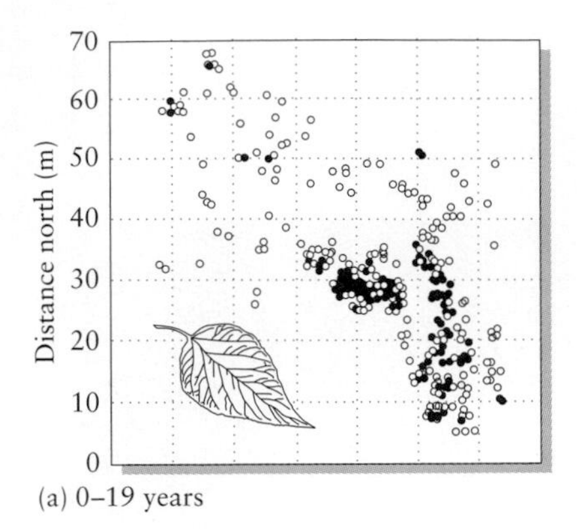

(a) 0–19 years

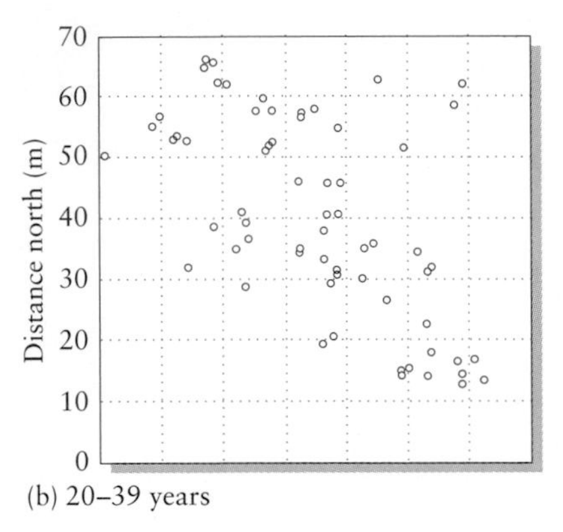

(b) 20–39 years

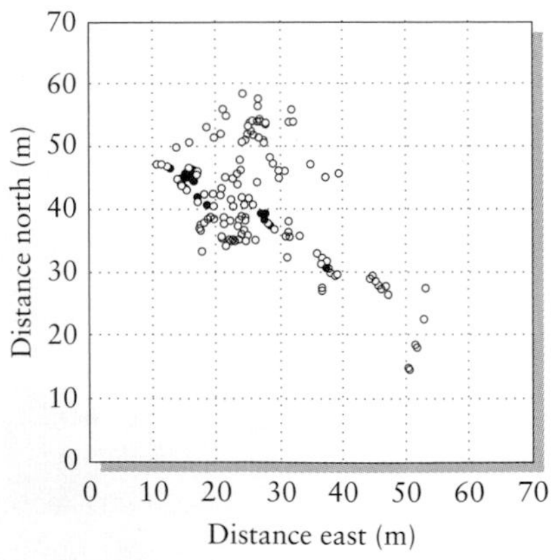

(c) 40–54 years

Figure 14.12 Maps showing the distribution of stems within three age classes in a single clone of balsam poplar *(Populus balsamifera)* in Quebec, Canada. The clone reveals a highly clumped distribution of stems as well as a shift toward the southeast. After C. Brodie, G. Houle, and M. Fortin, *J. Ecol.* 83: 309–320 (1995).

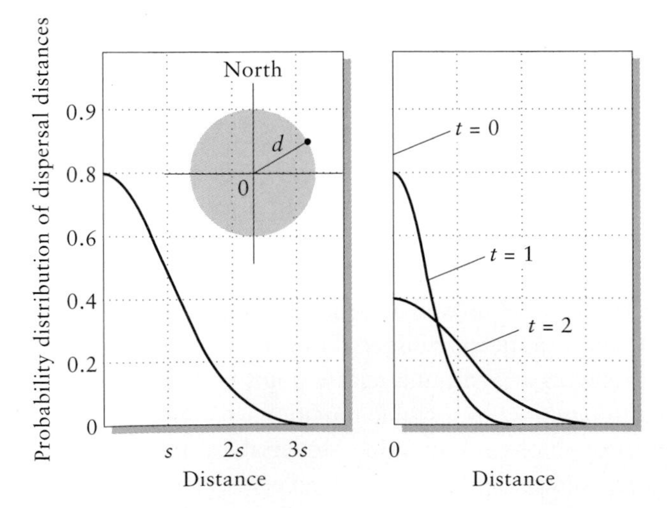

Figure 14.13 The probability density of individuals around a release point assumes a normal frequency distribution when movement occurs at random. The distribution of distances dispersed (d) may be characterized by the standard deviation of the curve (s), which increases in direct proportion to time t since release. The shaded circle indicates that dispersal may be in any direction away from the point of origin (0).

BOX 14.1 The Poisson distribution

We can distinguish patterns of dispersion found in nature by comparing them to the values expected of random spatial distributions. One of the most important of these is the Poisson distribution, which describes the expected number of individuals placed at random within a sample plot.

Suppose we divide a study area into sample plots, or other units of equal area, and count the number of individuals in each plot. Some of the plots will contain a single individual, some many, and others none, depending on the density of the population and the size of the plots. Consider, for example, the number of red mites counted on each of 150 apple leaves: 70 of the leaves had no mites, 38 had 1, 17 had 2, and so on, as shown in the following table.

MITES PER LEAF	NUMBER OF LEAVES OBSERVED	POISSON DISTRIBUTION	NUMBER OF LEAVES EXPECTED
0	70	0.3177	47.65
1	38	0.3643	54.64
2	17	0.2089	31.33
3	10	0.0798	11.98
4	9	0.0229	3.43
5	3	0.0052	0.79
6	2	0.0010	0.15
7	1	0.0002	0.02
8 or more	0	0.0000	0.00
Total	150	1.0000	149.99

Source: R.W. Poole, *An Introduction to Quantitative Ecology*, McGraw-Hill, New York (1974).

How would a random distribution of mites over the leaves appear? Suppose that each mite in the population had an equal probability of finding itself on each of the leaves (a probability of 1/150, or 0.067, per leaf) irrespective of the number of other mites present on that leaf. When mites are distributed according to a random process such as this, the expected proportion of leaves (P) with a given number of mites (x) obeys the Poisson distribution

$$P(x) = \frac{M^x e^{-M}}{x!}$$

where M is the mean number of individuals per sampling unit and $x!$ is the factorial of x (for instance, $5! = 5 \times 4 \times 3 \times 2 \times 1$). In the example, 172 mites were recorded on 150 leaves, so $M = 172/150 = 1.1467$ mites per leaf. To calculate the expected number of leaves with, say, 3 mites, we substitute $M = 1.1467$ and $x = 3$ into the equation for $P(x)$; this gives a value of 0.080, or 8% of the 150 leaves (12 leaves). As you can see, fewer leaves than expected had 1 or 2 mites, and more leaves than expected had 0 and 4 or more mites, indicating a clumped distribution.

BOX 14.2 Dispersal distance and neighborhood size

The standard deviation (s) of a normal frequency distribution is the square root of its variance (s^2). The variance is estimated by the expression

$$s^2 = \frac{1}{N} \sum_{i=1}^{N} d_i^2$$

where d_i is distance of the ith individual from the point of origin and N is the number of individuals in the sample.

In one study, investigators measured the rate of dispersal of fruit flies *(Drosophila pseudoobscura)* in grasslands in central Colorado. Flies were marked with minute fluorescent dust particles; individuals were caught in traps placed at regular intervals in eight directions of the compass at distances of up to 351 meters from the release point. After 1 day of dispersal, the estimated values of s were 139 and 171 meters at two localities.

A felicitous property of s as a measure of dispersal is that variances (s^2) of the distances add over time. Thus when fruit flies disperse distance s_1 in one day, the variance in distances after 2 days will be $2s^2{}_1$, and after t days it will be $ts^2{}_1$. Accordingly, if adult fruit flies live an average of 23 days and the average dispersal parameter (s) per day is 151 meters, s_{23} is the square root of 23×151^2, or 724 meters.

Neighborhood size within a population is the number of individuals within a circle whose radius is twice the dispersal distance (s) within the average reproductive life span of the individual. Neighborhood size provides an index to the number of individuals in a population that are potentially coupled by strong interactions. In the case of the fruit flies we just described, densities were about 0.38 flies per 100 m^2. The area of a circle of radius $2s$ is $4\pi s^2$, which would have included an estimated 26,387 individual flies.

observations are also pertinent, however, particularly the spread of introduced populations, which can occur only by movements of individuals. The European starling spread almost 4,000 kilometers across the United States in 60 years, at an average rate of about 67 kilometers per year. This figure greatly exceeds the estimates just reported for dispersal distances within populations of small songbirds. Adult starlings tend to nest in the same area year after year, so most dispersal is accomplished by young birds; established populations act as sources for population extension. Furthermore, the westward spread of the starling was characterized by frequent sightings outside the breeding season before breeders were settled in an area. It is unlikely, of course, that such long-distance movements of juveniles would be detected within an established population.

Whereas a few long-distance dispersers may expand the range of a population, they have less effect on the dynamics of widely separated but established subpopulations. Given the propensity of populations to increase, a small number of colonizing individuals can grow to a large population in a short period. But those same individuals immigrating to an established population may have a negligible effect on its size.

The genetic structure of populations

All populations contain genetic variation, although the amount of this variation differs substantially from species to species and from place to place. The genetic structure of a population describes the distribution of this variation among individuals and among localities or subpopulations, as well as the way in which organisms manage the consequences of genetic variation by means of their mating systems. Genetic variation is important to a population because it is the basis of the population's capacity to respond to environmental change through evolution. Genetic variation is also important to individuals: variation among an individual's progeny may increase the likelihood that at least some will be well adapted to particular habitat patches or to changed conditions. Most genetic variation is negative, however, in the sense that mutations usually make the individual less suited to the environment in which it lives. Genetic variation is maintained in a population primarily by mutation and by gene flow from other localities in which different genes have a selective advantage. As we shall see below, organisms have a variety of mechanisms for reducing the negative effects of such genetic variation.

Mating systems, inbreeding, and outcrossing

Everyone knows that close inbreeding is bad. **Inbreeding** is mating among close relatives. Brother-sister matings and, where it is possible (especially in plants), self-mating **(selfing)** may result in the expression of deleterious recessive genes, which can affect survival and reproductive performance. The way this happens is as follows: Suppose that an individual is heterozygous for a rare, deleterious, recessive gene (the gene pool is full of them, and most individuals have some, perhaps even inherited from their parents as new mutations). If that individual were to mate with an individual from the general population, which probably would not have the same rare deleterious allele, half of their progeny would be heterozygous for that allele, like the one parent, and half would be homozygous for the common form of the gene, like the other parent. None of the progeny would be disadvantaged by the union. If the individual selfed, however, one-quarter of its offspring would be homozygous for the deleterious allele, and would suffer loss of fitness as a result (Figure 14.14). Matings between close relatives produce the same result, only less frequently.

Most species employ mechanisms—including dispersal of progeny, recognition of close relatives, and **negative assortative mating** (choosing a mate that differs genetically from oneself)—to reduce the occurrence of inbreeding. Mammals, including humans, can and do distinguish

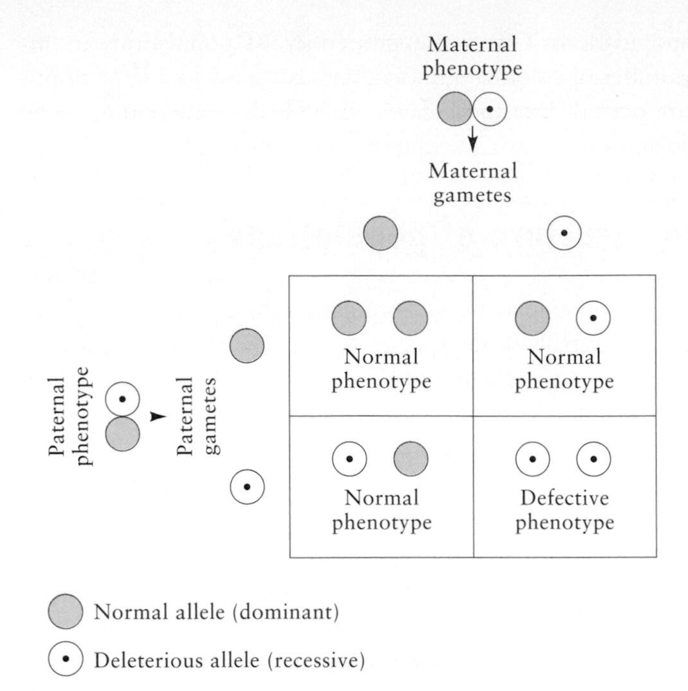

Figure 14.14 A diagram showing how inbreeding can uncover deleterious recessive genetic variation. A parent that is heterozygous for a deleterious recessive allele produces two types of gametes in equal frequency, one having a copy of the common form of the gene and the other having the deleterious recessive allele. Four combinations of parental gamete types appear with equal probability in offspring when a heterozygous parent mates with another heterozygous individual. Thus, one-quarter of the progeny of the two heterozygous parents will express the deleterious allele in homozygous form.

genetic differences in the major histocompatibility (MHC) genes of potential mates by smell. Hermaphroditic species of plants, in which individuals bear both male and female sexual organs, have additional mechanisms to prevent selfing, including self-incompatibility, temporal separation of male and female function, and elaborate flower structures designed to make self-fertilization difficult.

Although inbreeding generally creates genetic problems, it may confer important benefits in certain situations. In particular, plants that self guarantee fertilization of their flowers in habitats that lack suitable pollinators or where individuals are widely spaced. Many weedy species that colonize isolated patches of disturbed habitat (for example, dandelions) are habitual or even obligate selfers. It is assumed that most deleterious variation is weeded out of such populations a little at a time as it is exposed in homozygous individuals during the transition between outcrossing and selfing.

Outcrossing at great distances may reduce fitness when populations of plants exhibit spatially defined variation over small scales of distance, particularly in complex, heterogeneous environments. In such cases, local adaptation to particular habitat patches enhances fitness, and receiving pollen from individuals adapted to different habitat conditions may reduce the fitness of progeny that become established near their female parent. Several studies have reported an **optimal outcrossing distance** in populations of plants. Nearby individuals are likely to be close relatives, which creates the problem of inbreeding. Distant individuals are likely to be adapted to different conditions. Optimal outcrossing distance should be somewhere in between. In a study conducted in central Colorado, flowers of the larkspur *Delphinium nelsoni* were fertilized with pollen obtained from the same individual and from individuals located at distances of 1, 10, 100, and 1,000 meters. The number of seeds set per flower was greatest when the pollen came from individuals 10 meters away, and smallest for selfed pollen and for pollen obtained from individuals 1,000 meters distant. Furthermore, when these seeds were planted, survival to 1 and 2 years greatly favored matings across the intermediate distance of 10 meters.

Plants may not be able to control the distances that their pollinators travel between flowers, but they can manage genetic variability by establishing competition between pollen grains for opportunities to fertilize ovules and by selectively aborting developing ovules on the basis of the genotype of the embryo. Most plants produce many more flowers than they can mature as fruits: flowers are relatively cheap, seeds and fruits expensive. Pollen tubes must grow through the style (maternal tissue) to reach and fertilize an ovule. It has been postulated that biochemical interactions that control the rate of pollen tube growth are sensitive to the genotypes of pollen grains and female flower parts. (This is one of the ways in which self-incompatibility is achieved.) A heavy pollen load is advantageous to a female because many more pollen tubes compete to fertilize her ovules, resulting in strong selection among pollen genotypes. Excess fertilized ovules are reduced in part by predation or other extrinsic damage, but also in part by programmed abortion. Abscission of flowers and fruits can be highly selective with respect to pollen loads, number of ovules fertilized per flower, and the genotypes of developing embryos.

Banksia spinulosa is a partially self-compatible, but normally outcrossing, Australian shrub that is pollinated by small nectar-feeding birds. Each inflorescence has about 800 flowers, but produces fewer than 50 fruits. Fruit production appears to be resource-limited rather than pollen-limited because removal of one-third of the flowers from either the base or the top of an inflorescence does not significantly depress fruit set. To determine whether these plants can distinguish the quality of pollen, Australian botanists Glenda Vaughton and Susan Carthew hand-pollinated *Banksia* inflorescences with pollen obtained either from the same plant (selfed pollen) or from neighboring plants (outcrossed pollen). In other plants, they pollinated half the inflorescence with selfed pollen and half with outcrossed pollen (mixed pollination). After fruits had developed, the numbers of fruits and seeds (no more than one seed per fruit) were counted on each side of the inflorescence. Compared with cross-pollination, selfing reduced seed set by 38% (24 versus 39 seeds per half-inflorescence); and fruits with aborted seeds increased from 8% to 16% in selfed compared with outcrossed inflorescences (Figure 14.15). These results clearly indicate **inbreeding depression,** or reduction of fitness caused by inbreeding. When one-half of an inflorescence was cross-pollinated and the other half was selfed, the number of seeds produced per selfed half dropped further to 14, and 28% of the fruits aborted their seeds. This experiment shows that self-pollinated ovules, which are likely to have inferior genotypes, do not fare well in competition with cross-pollinated ovules. Thus, plants are capable of making distinctions among developing embryos on the basis of their genotypes. The extent to which this mechanism is used to manage the transmission of genetic variability to offspring in natural populations has yet to be determined.

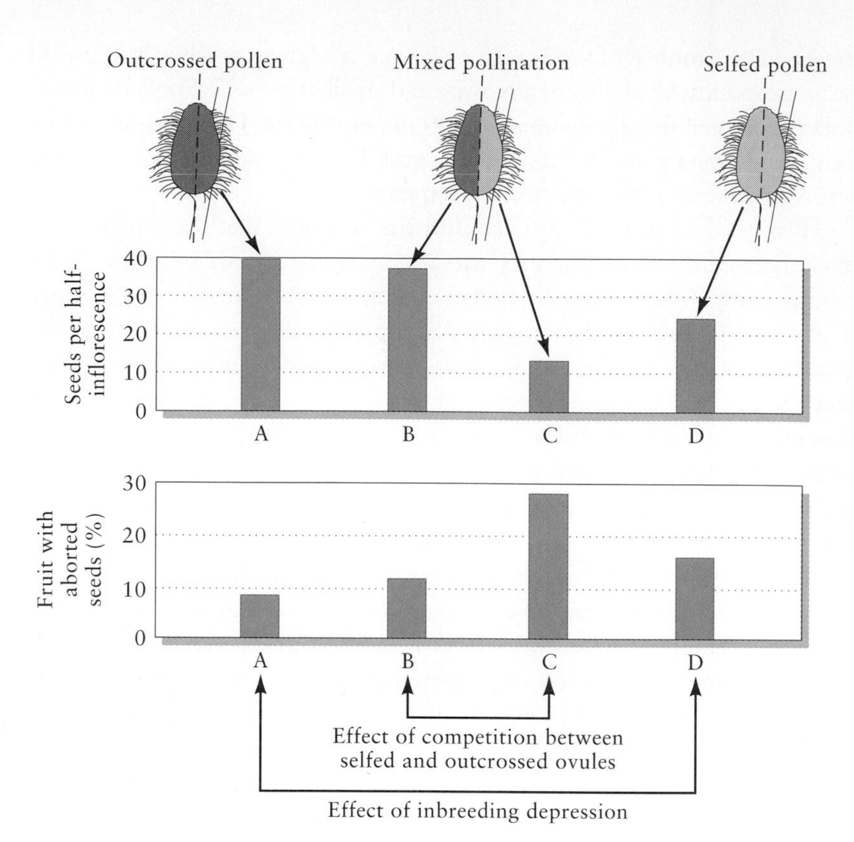

Figure 14.15 Results of a pollination experiment with the Australian shrub *Banksia spinulosa,* in which the two halves of individual inflorescences were fertilized with pollen either from the same plant or from other plants. The results show the negative effects of inbreeding (D compared with A) and the further discrimination against selfed ovules in competition with outcrossed ovules growing in the same inflorescence (C compared with B). Data from G. Vaughton and S. M. Carthew, *Biol. J. Linn. Soc.* 50:35–46 (1993).

Genetic changes in small populations

The subdivision of a landscape into patches of habitat means that populations of many species are broken up into subpopulations with little gene flow between them. This pattern has important consequences both in natural settings and in managed landscapes because several types of population processes have greater effects as the number of individuals in the population declines. In particular, small populations tend to lose genetic variation more rapidly than large populations. The result is less genetic diversity within each subpopulation and greater genetic differences between subpopulations.

Any gene may have two or more alternative forms, called **alleles,** which result from slight differences in the DNA sequence of the gene. Different alleles often cause slight differences in form or function. Because of this, they are liable to come under the influence of natural selection, which may favor an increase of one allele in a population at the expense of others. In small populations, the frequencies of alleles also may change because of random variations in birth and death rates due to chance events, even in the absence of selection and mutation. Such changes are referred to as **genetic drift.** Suppose that a population contains two alleles of the same gene. The rate of increase or decrease of each allele has a random component; just by chance, one may increase and the other decrease, resulting in a change of frequency. Even though these changes are not biased in one direction or the other, the occurrence, by chance, of a long series of changes in one direction may lead to the disappearance of one allele from the population. In this case, we say that the other allele becomes **fixed.** Further variation in allele frequency cannot take place, unless mutation introduces new alleles into the population.

The rate of fixation of alleles is inversely related to the size of a population. Thus, genetic variation decreases more rapidly over time in small populations than in large ones. Furthermore, a single episode of small population size, which might occur during the colonization of an island or a new habitat by a few individuals from a large parent population, can reduce genetic variation in the colonizing population. Such episodes are known as **founder events.** When founding populations consist of ten or fewer individuals, they typically contain a substantially reduced sample of the total genetic variation of the parent population. Continued existence at a low population size results in further loss of genetic variation due to genetic drift and close inbreeding. This situation is often referred to as a **population bottleneck.** Such a condition appears to have occurred in the recent past in the population of cheetahs in East Africa, which exhibit practically no genetic variation whatsoever. The significance of founder effects and bottlenecks for natural populations is that the fragmentation of populations into small subpopulations may eventually restrict the evolutionary responsiveness of each subpopulation to the selective pressures of changing environments, thus making these small subpopulations more vulnerable to extinction. As yet, there is no agreement as to whether the cheetah's genetic uniformity poses a serious threat to its future.

Geographic variation in the gene pool

The genotypes of individuals within a population often vary geographically. This may happen because of differences in selective factors in different parts of the population's range or because of random changes (genetic drift, founder effects) in isolated or partially isolated subpopulations. Populations do not have to be subdivided for genetic differences to arise within them. If the difference in selective factors between two localities is strong relative to the rate of gene flow between them, then differences in allele frequency can be maintained by differential selection. This often results in a gradual change, or **cline,** in allele frequency, or some phenotypic character under genetic influence, over distance.

Botanists have long recognized that individuals of a species growing in different habitats may exhibit varied forms corresponding to local ecological conditions. In many cases, these differences result from developmental responses, but experiments on some species have revealed genetic adaptations to local conditions. Almost 80 years ago, the Swedish botanist Göte Turesson collected seeds from several species of plants that lived in a variety of habitats and grew them in his garden. He found that even when grown under identical conditions, many of the plants exhibited different forms depending on their habitat of origin. Turesson called these forms **ecotypes,** a name that persists to the present, and suggested that ecotypes represent genetically differentiated lineages of a population, each restricted to a specific habitat. Because Turesson grew his plants under identical conditions, he realized that the differences between ecotypes must have had a

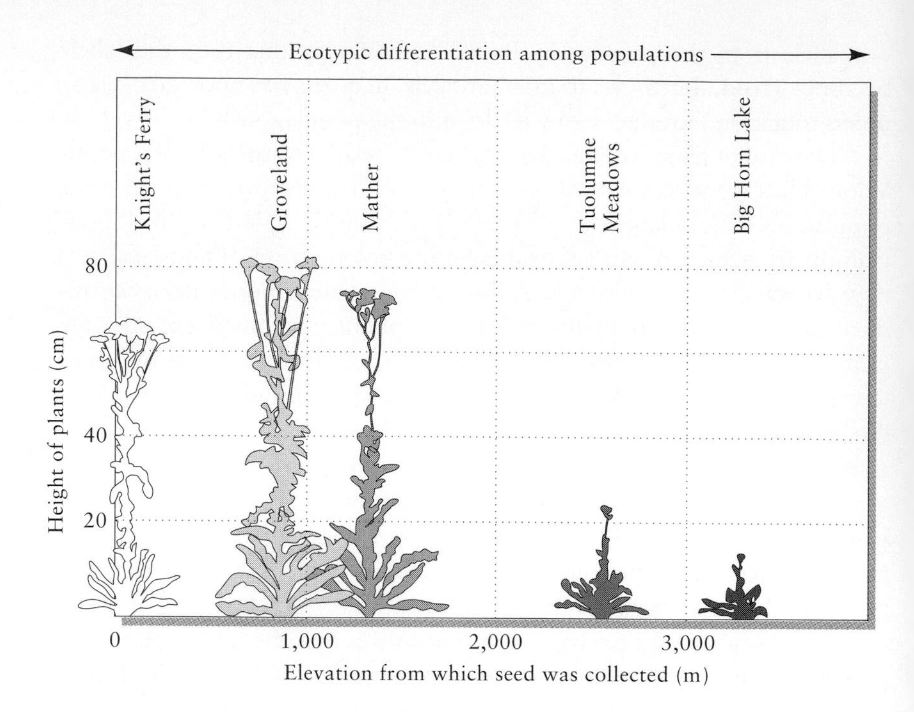

Figure 14.16 Ecotypic differentiation in populations of the yarrow, *Achillea millefolium,* was demonstrated by raising plants derived from different elevations under identical conditions in the same garden at Stanford, California, which is close to sea level. After J. Clausen, D. D. Keck, and W. M. Hiesey, *Carnegie Inst. Wash. Publ.* 581:1–129 (1948).

genetic basis and that they must have resulted from evolutionary differentiation within the species according to habitat.

Similar experiments in California on a species of yarrow, *Achillea millefolium,* also revealed ecotypic variation. *Achillea* is a member of the sunflower family (Compositae) and grows in many habitats ranging from sea level to more than 3,000 meters in elevation. Plants raised from seed collected at various points along an altitude gradient and all grown at sea level at Stanford, California, retained the distinctive size and level of seed production typical of the populations from which they came (Figure 14.16). Similar differentiation has been found over distances of only a few meters where contrasting selective pressures are strong enough to overcome the migration of individuals, seeds, or pollen between habitat patches. Such situations frequently arise on soils that develop on mine tailings, which sometimes exert strong selective pressure for tolerance to toxic metals (for example, copper, lead, zinc, and arsenic).

A trait may exhibit a gradual response (cline) to variation in the environment over a geographic gradient. In the Japanese field cricket *Teleogryllus,* for example, body size and duration of nymphal (larval) development both increase clinally from north to south. We know that these clines have a genetic basis because individuals from different localities raised under the same temperature and photoperiod regimes retain their regional differences (Figure 14.17).

In contrast, the frequencies of alleles that control color and banding patterns on shells of the snail *Bradybaena* bear no relationship to habitat or locality (Figure 14.18). This snail was introduced to Japan, probably many times over the last 200 years, with the widespread cultivation of sugar cane. Such variation in phenotypic frequencies among subpopulations could

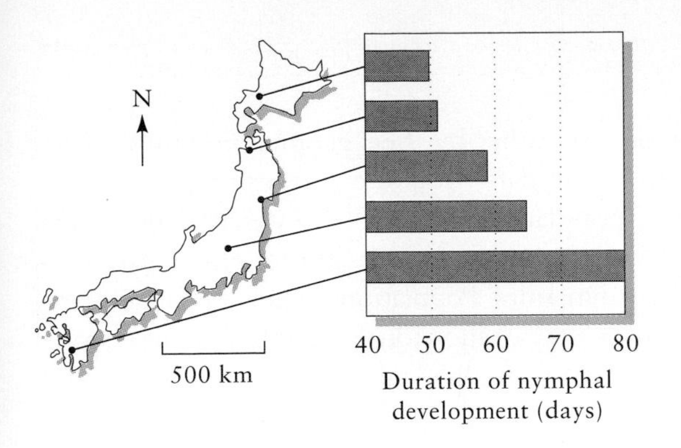

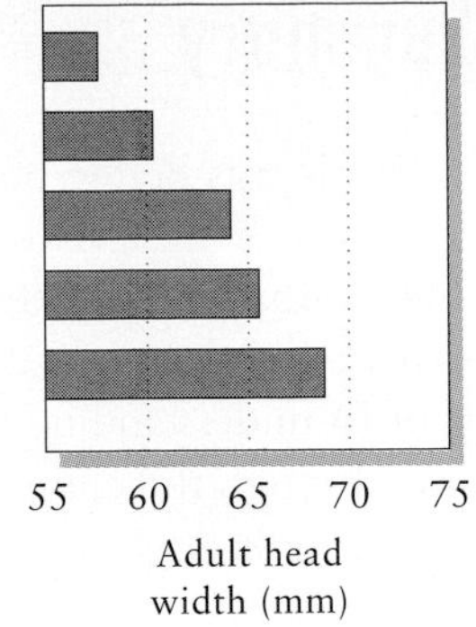

Figure 14.17 Clinal geographic variation in duration of nymphal development and width of the adult head in males from five local populations of the field cricket *Teleogryllus emma* in Japan. The nymphs were all raised at 28°C on a cycle of 16 hours of light and 8 hours of dark. Data from S. Masaki, *Evolution* 21:725–741 (1967).

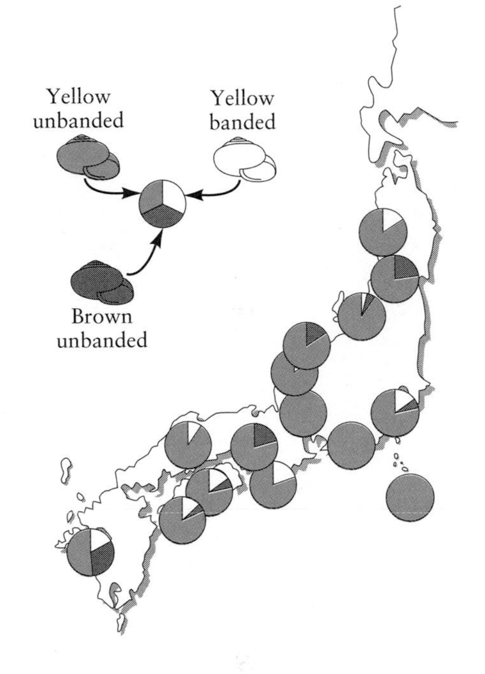

Figure 14.18 Geographic variation in frequencies of phenotypes of the land snail *Bradybaena similaris* in Japan. Data from T. Komai and S. Emura, *Evolution* 9:400–418 (1955).

occur if each local population was established by a small group of colonists that contained a random, but not necessarily representative, sample of the genetic variation of the parent population. A similarly haphazard distribution of genotypes occurs in populations of the small annual plant *Linanthus parryae.* In one area of southern California, blue flowers (the normal flower color is white) are haphazardly distributed without apparent relation to environmental factors (Figure 14.19). Evidently, gene flow occurs so infrequently between populations isolated by as little as a kilometer that frequencies of alleles for flower color diverge by genetic drift.

This chapter has illustrated the complex relationship of populations to factors, including spatial variation, in their environments. The static picture of a population usually can be mapped fairly well onto features of the environment that exert control over the population. In the next chapter, we shall begin to look at the processes responsible for changes in the sizes of populations.

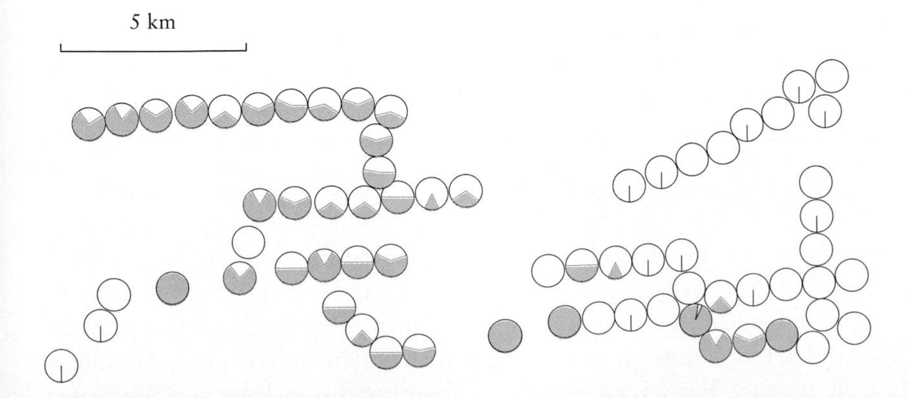

Figure 14.19 Frequencies of blue flowers (colored portion of circle) in subpopulations of *Linanthus parryae* from a small region of southern California. Genotype frequency varies abruptly over small distances in some cases, suggesting genetic drift or founder effects in small, isolated subpopulations. From C. Epling and T. Dobzhansky, *Genetics* 27:317–332 (1942).

Summary

1. Ecologists define populations either by their geographic limits of distribution or by an arbitrary boundary around a smaller area of interest. The primary objectives of population studies are to describe the spatial structure and temporal behavior of populations and to gather data relevant to understanding their dynamics. Populations achieve genetic cohesion through the common ancestry of individuals, and they achieve spatial cohesion through the movements of individuals.

2. The distribution of a population is its geographic range, which is generally limited by the extent of suitable habitat and by barriers to dispersal. Within the limits of distribution, the density of a population may vary according to differences in habitat quality.

3. Population density expresses more closely the local interactions of a species with its environment than does the total population size, which also reflects the spatial extent of suitable habitat. Changes in density over time and space provide the basic knowledge needed to understand the processes that regulate population size.

4. The landscape is a mosaic of habitat patches, and most populations have patchy distributions determined by the presence and absence of suitable habitat. Thus, populations are often divided into more or less discrete subpopulations with varied amounts of migration of individuals between them.

5. Faced with variation in habitat quality and complete freedom to choose where to live, organisms tend to distribute themselves in proportion to available resources in what is known as an ideal free distribution. Poorer habitats are eventually settled because dense populations reduce the quality of intrinsically superior habitats.

6. Ideal free distributions are rarely realized, and populations in some habitats are more productive than those in others. Populations in which reproduction exceeds mortality are referred to as source populations. Individuals disperse from source populations to sink populations, where local reproduction cannot maintain a population without immigration.

7. Dispersion describes the spacing of individuals with respect to others in a population. Clumped distributions may result from independent aggregation of individuals in suitable habitats, from spatial proximity of parents and offspring, or from tendencies to form social groups. Evenly spaced distributions may result from antagonistic interactions between individuals.

8. Movement within populations may be characterized by the variance in distances of individuals from their initial position, which may be a point of release in an experimental study. Neighborhood size is the number of individuals within a circle whose radius is twice the average dispersal distance of an individual over its reproductive life span. Neighborhood size provides an index to the number of neighbors with which an individual can potentially interact. For several species, neighborhood sizes have been estimated as on the order of 10^2 to 10^4 individuals.

9. Genetic variation may have negative effects on the ability of individuals to adapt to local variation in their environments. To some degree, individuals can minimize the negative effects of genetic variation on the genotypes of their offspring by selective mating to control level of inbreeding and outcrossing distance.

10. In small populations, random variations in reproductive success result in changes in gene frequencies and the occasional loss of alleles, causing a decrease in genetic diversity. This process is called genetic drift.

11. Different selective pressures and genetic drift in small populations result in variation in gene frequencies within the geographic range. This variation is accentuated by population subdivision and poor dispersal.

Suggested readings

Begon, M., and M. Mortimer. 1986. *Population Ecology.* 2d ed. Blackwell Scientific Publications, Oxford.

Caughley, G. 1977. *Analysis of Vertebrate Populations.* Wiley, New York and London.

Cook, R. E. 1983. Clonal plant populations. *American Scientist* 71:244–253.

Dunning, J. B., B. J. Danielson, and H. R. Pulliam. 1992. Ecological processes that affect populations in complex landscapes. *Oikos* 65:169–175.

Harper, J. L. 1977. *Population Biology of Plants.* Academic Press, New York and London.

Krebs, C. J. 1989. *Ecological Methodology.* Harper and Row, New York.

Marquet, P. A., S. A. Naverrete, and J. C. Castilla. 1995. Body size, population density, and the Energetic Equivalency Rule. *Journal of Animal Ecology* 64:325–332.

Milinski, M. 1979. An evolutionarily stable feeding strategy in sticklebacks. *Z. Tierpsychol.* 51:36–40.

Ralls, K., J. D. Ballou, and A. Templeton. 1988. Estimates of the cost of inbreeding in mammals. *Conservation Biology* 2:185–193.

Schemske, D. W. 1984. Population structure and local selection in *Impatiens pallida* (Balsaminaceae), a selfing annual. *Evolution* 38:817–832.

Silva, M., and J. A. Downing. 1995. The allometric scaling of density and body mass: A nonlinear relationship for terrestrial mammals. *American Naturalist* 145:704–727.

Southwood, T. R. E. 1978. *Ecological Methods.* 2d ed. Chapman and Hall, London; Wiley, New York.

Stephenson, A. G. 1981. Flower and fruit abortion: Proximate causes and ultimate functions. *Annual Review of Ecology and Systematics* 12:253–279.

Varley, G. C., G. R. Gradwell, and M. P. Hassell. 1975. *Insect Population Ecology.* Blackwell Scientific Publications, Oxford.

Wiens, J. A., N. C. Stenseth, B. Van Horne, and R. A. Ims. 1993. Ecological mechanisms and landscape ecology. *Oikos* 66:369–380.

CHAPTER 15

POPULATION GROWTH AND REGULATION

We cannot find a better example of the immense capacity of populations to increase than our own population, which has grown prolifically, at times doubling each quarter-century. Ever since humankind began to understand the rapid increase in its numbers, human population growth has been cause for concern. This concern led to the development of mathematical techniques to predict the growth of populations—the discipline of **demography,** or the study of populations—and to intensive study of natural and laboratory populations to learn about mechanisms of population regulation. As a result, we now have a general understanding of the causes of fluctuations in natural populations and the effects of crowding on birth and death rates.

In this chapter, we shall explore the nature of population growth and examine the factors that limit population size, showing how their effects increase with increasing population density in such a way as to bring growth under control.

Exponential growth

A population experiencing exponential growth increases in proportion to its size, just as a bank account earns interest more rapidly with a larger principal. Increase in numbers depends on reproduction by *individuals*. Therefore, a population that grows at a constant exponential rate gains individuals ever faster as the population increases. For example, a 10% annual rate of increase adds 10 individuals in 1 year to a population of 100, but the same rate of increase adds 100 individuals to a population of 1,000. Allowed to increase at this rate, the population would rapidly climb toward infinity. Charles Darwin wrote, in *On the Origin of Species*, "There is no exception to the rule that every organic being naturally increases at so high a rate, that, if not destroyed, the earth would soon be covered by the progeny of a single pair." To make his case as forcefully as possible, Darwin offered a conservative example:

> *The elephant is reckoned the slowest breeder of all known animals, and I have taken some pains to estimate its probable minimum rate of natural increase; it will be safest to assume that it begins breeding when thirty years old, and goes on breeding till ninety years old, bringing forth six young in the interval, and surviving till one hundred years old; if this be so, after a period of from 740 to 750 years there would be nearly nineteen million elephants alive, descended from the first pair.*

Because baby elephants grow up, mature, and themselves have babies, the elephant population grows exponentially.

Such a population increases according to the equation

$$N(t) = N(0)e^{rt},$$

where $N(t)$ is the number of individuals in a population after t units of time, $N(0)$ is the initial population size ($t = 0$), and r is the exponential growth rate. The constant e is the base of the natural logarithms; it has a value of approximately 2.72. Exponential growth results in a continuously accelerating curve of increase (or decelerating curve of decrease) whose slope varies directly with the size of the population (Figure 15.1).

The rate at which individuals are added to a population undergoing exponential growth is the derivative of the exponential equation; that is,

$$\frac{dN}{dt} = rN.$$

This equation encompasses two principles: First, the exponential growth rate (r) expresses population increase (or decrease) on a per individual basis. Second, the rate of increase (dN/dt) varies in direct proportion to

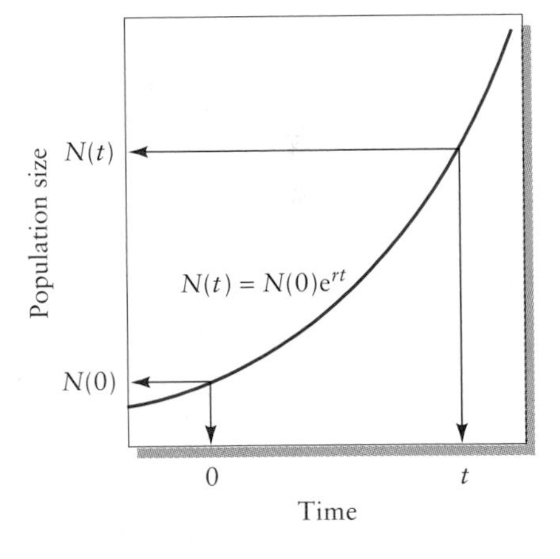

Figure 15.1 The curve of exponential growth for a population growing at rate r between time 0 and time t. During this period, number of individuals increases from $N(0)$ to $N(t)$. Notice that the slope becomes steeper as the population increases.

the size of the population (N). In words, this equation could read: (the rate of change in population size) = (the contribution of each individual to population growth) × (the number of individuals in the population).

The individual, or **per capita,** contribution to population growth represents the difference between birth rate (b) and death rate (d) calculated on a per capita basis; that is, $r = b - d$.

Rates of birth and death are abstractions having little meaning for the individual. An elephant dies only once, so it can't have a rate of death. Babies are produced in discrete litters separated by the intervals required for gestation and offspring care, not at constant rates, such as 0.05 young per day. But when births and deaths are averaged over a population as a whole, they take on meaning as rates of demographic events in the population. If 1,000 individuals produced 10,000 progeny in a year, we could reasonably assign a per capita birth rate of 10 per year, and we could assume that a population of 1 million would produce 10 million progeny—still 10 per individual—under the same conditions. If, of the groundhogs alive on their big day in one particular year, only half survived to 2 February of the next year, we would ascribe a death rate of 50% per year to the population, even though some of the groundhogs had died "completely" and others hadn't died at all.

Geometric growth

The human population grows continuously because babies are born and added to the population at all seasons of the year. This situation is unusual in natural populations, most of which restrict reproduction to a particular time of year. Accordingly, populations grow during the breeding season, then decline between one breeding season and the next (Figure 15.2). In the case of the California quail, the number of individuals doubles or triples each summer as adults produce their broods of chicks, but then dwindles by nearly the same amount during autumn, winter, and spring. Within each year, the population growth rate varies tremendously due to seasonal changes in the balance of birth and death processes. If we were interested in projecting population growth, it would be pointless to compare numbers in August, recently augmented by the chicks born that year, with numbers in May, after winter had taken its toll. One must count indi-

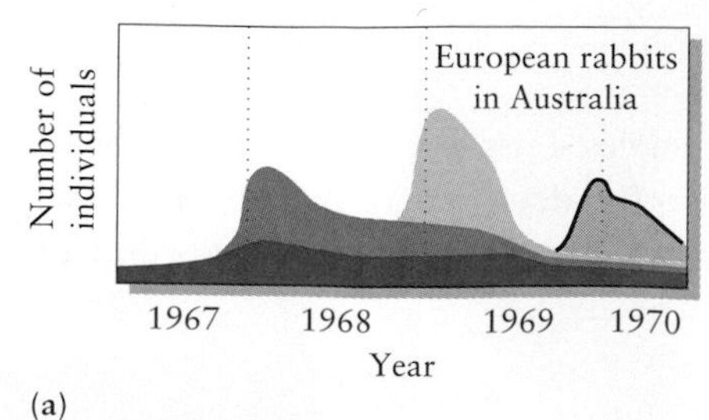

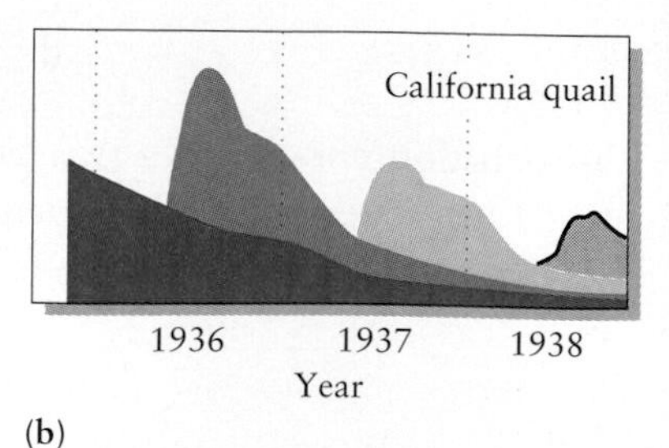

Figure 15.2 Growth of populations with discrete (seasonal) reproduction. (a) Rabbits in a subalpine population in New South Wales, Australia. (b) California quail. The curves are shaded differently to show that each curve represents the population of individuals born (or hatched) in a different year. After K. Myers, in P. J. den Boer and G. R. Gradwell (eds.), *Dynamics of Populations.* Centre Agric. Publ. Doc., Wageningen, The Netherlands (1970), pp. 478–506; J. T. Emlen Jr., *J. Wildl. Mgmt.* 4:2–99 (1940).

viduals at the same time each year, so that all counts are separated by the same cycle of birth and death processes. Such an increase (or decrease) over discrete intervals is referred to as **geometric growth.**

The rate of geometric growth is most conveniently expressed as the ratio of a population size in one year to that in the preceding year (or other time interval). Demographers have assigned the symbol λ, the lowercase Greek letter lambda, to this ratio; hence $\lambda = N(t+1)/N(t)$, where t is an arbitrary time. This definition of λ can be rearranged to provide a formula for projecting the size of the population through a single time interval:

$$N(t+1) = N(t)\lambda.$$

To project the growth of a population over many time intervals, we multiply the original population size by the geometric growth rate, once for each interval of time passed. Hence $N(1) = N(0)\lambda$, $N(2) = N(0)\lambda^2$, $N(3) = N(0)\lambda^3$, and

$$N(t) = N(0)\lambda^t.$$

Note that this equation for geometric growth is identical to the equation for exponential growth except that λ takes the place of e^r, which equals the amount of exponential growth accomplished in one time period. Because of this relationship, curves depicting the two models of growth can be superimposed (Figure 15.3), and there is a direct correspondence between the values of λ and r. For example, when a population's size remains constant, $r = 0$ and $\lambda = 1$ ($r = \log_e\lambda$, and $\log_e 1 = 0$). Decreasing populations have negative exponential growth rates and geometric growth rates of less than 1 (but greater than 0; a real population cannot have a negative number of individuals). Increasing populations have positive exponential growth rates and geometric growth rates greater than 1.

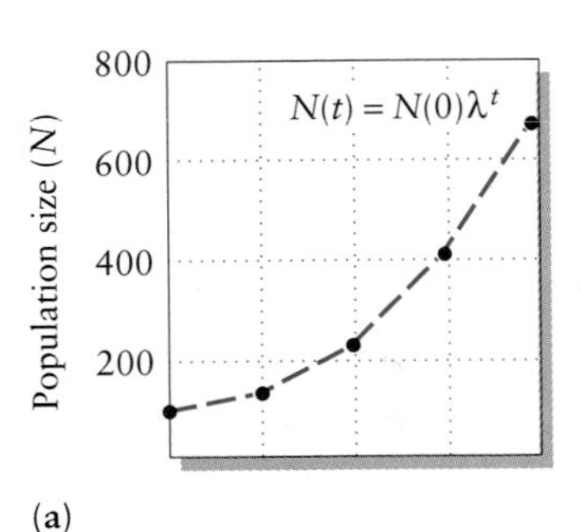

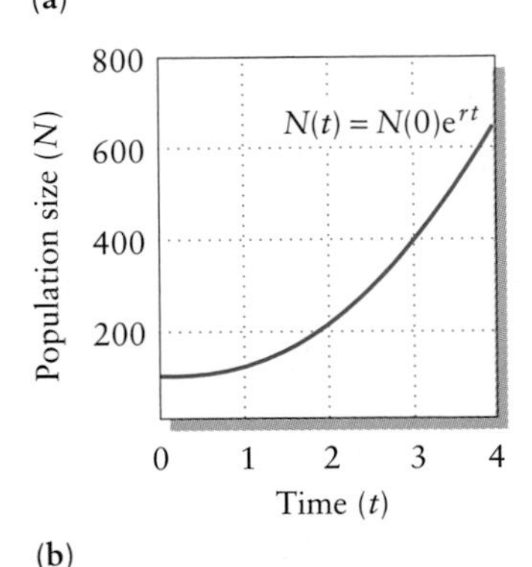

Figure 15.3 Increase in the number of individuals in populations undergoing geometric growth (a) and exponential growth (b) at equivalent rates ($\lambda = 1.6$, $r = 0.47$).

Age structure and population growth rates

When birth rates and death rates have the same values for all members of the population, the total population size (N) provides the proper basis for projecting a population increase. But when birth and death rates vary with respect to age, the contributions of younger and older individuals to population growth must be figured separately. Two populations having identical birth and death rates at corresponding ages, but different **age structures** (proportions of individuals in each age class), will grow at different rates, at least for a while. A population composed wholly of prereproductive adolescents and postreproductive oldsters, for example cannot increase until the young individuals reach reproductive age. This represents an extreme case, but smaller variations in age distribution can also profoundly influence population growth rates.

TABLE 15.1 Life table for a hypothetical population of 100 individuals

AGE (x)	SURVIVAL (s_x)	FECUNDITY (b_x)	NUMBER OF INDIVIDUALS (n_x)
0	0.5	0	20
1	0.8	1	10
2	0.5	3	40
3	0.0	2	30

When age-specific birth and survival rates remain unchanged for a long enough period, a population will assume a **stable age distribution.** Under such conditions, each age class in a population grows or declines at the same rate and, therefore, so does the total size of the population. A little pencil-and-paper figuring with a hypothetical population will demonstrate this result. Imagine a population of 100 individuals having the characteristics shown in Table 15.1. Because all 3-year-olds die, there are no 4-year-olds in the population; newborns have a fecundity of zero, as seems biologically reasonable. The population is counted at the end of the reproductive season, after young have been born (n_0). A projection of this population into the future is illustrated in Table 15.2.

In this example, the population at first grows very erratically, with λ fluctuating between 1.05 and 1.69. Eventually λ settles down to a constant value of 1.49. At this point, the population has achieved a stable age distribution; the percentage of individuals in each age class after eight time intervals appears in the rightmost column of Table 15.2. Even by the end of the fourth interval, the population has approached its stable age distribution closely, with proportions in each of the age classes of 62.7%, 21.8%, 10.0%, and 5.4%. Under stable-age conditions, each age class grows at the same rate (Figure 15.4).

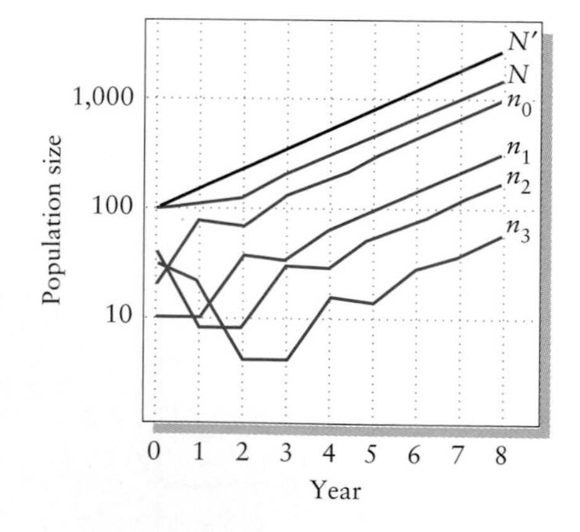

Figure 15.4 Growth of age classes as a population achieves its stable age distribution. Notice that each class eventually grows at the same exponential rate. The data used to create this graph are taken from Table 15.2. N' represents the growth of a population of 100 individuals in a stable age distribution at year 0.

The stable age distribution and growth rate of a particular population depend on birth and survival values. Any change in survival or fecundity rates alters the stable age distribution and results in a new rate of population growth. Consider the life table in Table 15.3, in which we have reduced the survival and fecundity of our hypothetical population enough to make the population decline (Table 15.4). As often happens in a decreasing population, the distribution of individuals shifts toward older age classes. In this example, the shift to a new stable age distribution involves small changes and occurs quickly, after which the population achieves a growth rate of $\lambda = 0.82$. Early on, however, some age classes briefly increase (n_2 between $t = 0$ and $t = 1$) as the age distribution of the population readjusts to new survival and fecundity values. The effect of population growth rate on age structure stands out in comparisons of

TABLE 15.2 **Projection of age distribution and total size through time for the hypothetical population described in Table 15.1**

	TIME INTERVAL									
	0	*1*	*2*	*3*	*4*	*5*	*6*	*7*	*8*	PERCENT
n_0	20	74	69	132	175	274	399	599	889	63.4
n_1	10	10	37	34	61	87	137	199	299	21.3
n_2	40	8	8	30	28	53	70	110	160	11.4
n_3	30	20	4	4	15	14	26	35	55	3.9
N	100	112	118	200	279	428	632	943	1,403	100
λ	1.12	1.05	1.69	1.40	1.53	1.48	1.49	1.49		

Note: The population was projected by multiplying the number of individuals in each age class by the survival to obtain the number in the next older age class in the next time period: $n_x(t) = n_{x-1}(t-1)s_x$. Then the number of individuals in each age class was multiplied by its fecundity to obtain the number of newborns: $n_0(t) = \Sigma n_x(t)b_x$.

human populations with stable and with growing populations (Figure 15.5). Rapid growth leads to a bottom-heavy age structure with large proportions of young individuals.

The life table

Accurate projection of change in population size requires knowledge of the number of individuals in each age class and their probabilities of survival and rates of fecundity. These statistics, which collectively are known as the **life table,** determine the addition and removal of individuals from a local population (in the absence of immigration and emigration).

TABLE 15.3 **Life table for a hypothetical population subjected to a negative growth rate**

AGE (x)	SURVIVAL (s_x)	FECUNDITY (b_x)	INITIAL NUMBER OF INDIVIDUALS (n_x)
0	0.3	0	889
1	0.6	1	299
2	0.3	2	160
3	0.0	1	55

TABLE 15.4 **Projection of age distribution and total size through time for the hypothetical population described in Table 15.3**

	TIME INTERVAL									
	0	*1*	*2*	*3*	*4*	*5*	*6*	*7*	*8*	PERCENT
n_0	889	673	576	463	383	312	257	210	172	57.7
n_1	299	267	202	173	139	115	94	77	63	21.1
n_2	160	179	160	121	104	83	69	56	46	15.4
n_3	55	48	54	48	36	31	25	21	17	5.7
N	1,403	1,167	992	805	662	541	445	364	298	
λ	0.83	0.85	0.81	0.82	0.82	0.82	0.82	0.82		

Because it is hard to ascertain paternity in many species, life tables are usually based entirely on females. For some populations with highly skewed sex ratios or unusual mating systems, this can pose difficulties, but in most cases a female-based life table provides a workable population model.

Age is designated in a life table by the symbol x, and age-specific variables are indicated by the subscript x. When reproduction occurs during a brief breeding season each year, each age class is composed of a discrete group of individuals born at approximately the same time. When reproduction is continuous, as it is in the human population, each age class x is designated arbitrarily as comprising individuals between ages $x - \frac{1}{2}$ and $x + \frac{1}{2}$.

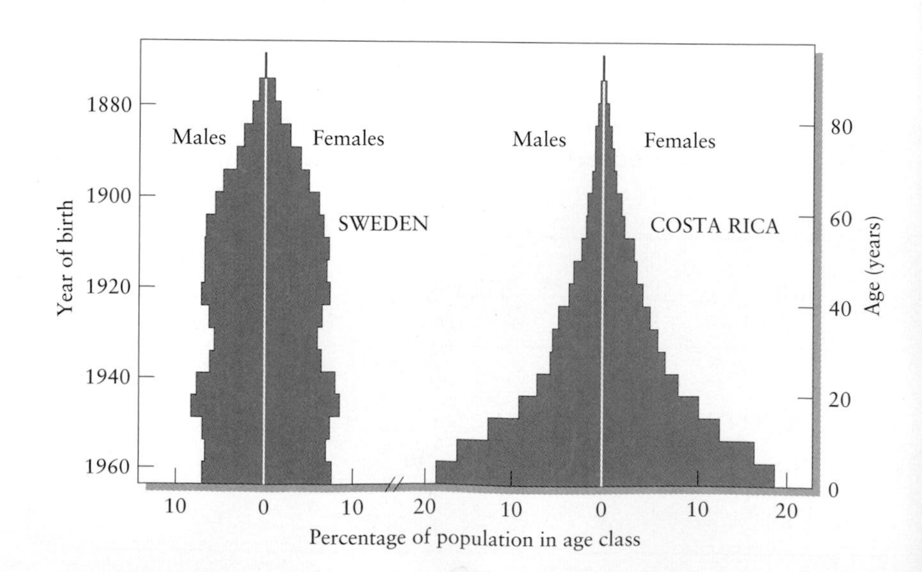

Figure 15.5 Population age structures of Sweden in 1965 and of Costa Rica in 1963. Because Sweden's population had grown slowly, its population was distributed toward older ages. Declining birth rates during the Depression and the baby boom that followed World War II were responsible for irregularities in the age structure. Costa Rica's rapid population growth, caused by a high birth rate, resulted in a bottom-heavy age structure. After data in N. Keyfitz and W. Flieger, *World Population: An Analysis of Vital Data,* University of Chicago Press, Chicago (1968).

The **fecundity** of females, which is often expressed in terms of female offspring produced per breeding season or age interval, is designated by b_x (think of b for "births"). Life tables portray the statistics of mortality in several ways. The fundamental measure is probability of survival (s_x) between ages x and $x + 1$. Probabilities of survival over many age intervals are summarized by survivorship to age x, designated by l_x (think of l for "living"), which is the probability that a newborn individual will be alive at age x. Because by definition all newborn individuals are alive at age 0, $l_0 = 1$. The proportion of newborns alive at age 1 is the probability of surviving from age 0 to age 1, hence $l_1 = s_0$. Similarly, $l_2 = s_0 s_1$ and, by extension, $l_x = s_0 s_1 s_2 s_{x-1}$. An additional measure is sometimes included in the life table: the expectation of further life (e_x) of an individual of age x.

Life table variables are summarized in Table 15.5 and illustrated in Table 15.6 by data for the annual meadow grass *Poa annua*. This life table follows the survival and fecundity of a planting of the grass under experimental conditions over 2 years, by which time the last individual had died. Age is tabulated in units of 3 months. Because *Poa* is hermaphroditic, sexes are not distinguished. Of 843 plants alive at time 0 (germination), 722, or 85.7%, were alive at 3 months ($t = 1$); hence s_0 and l_1 both equal 0.857. The life table shows that probability of dying increased with age; hence expectation of further life decreased with age. Fecundity rose to a peak of 620 seeds per 3-month period at 6 months of age and then declined.

Estimating survival in natural populations

One can estimate survival with varying degrees of reliability from four kinds of information: (1) survival of individuals to a particular age (survivorship), (2) survival of individuals in each age class from one time period to the next, (3) ages of individuals at death, and (4) the age structure of a population. Data of the first kind form the basis of a **cohort life table,** or **dynamic life table,** which follows the fate of a group of individuals born at the same time from birth to the death of the last individual. The

TABLE 15.5 Summary of life table variables

l_x	Survival of newborn individuals to age x
b_x	Fecundity at age x
m_x	Proportion of individuals of age x dying by age $x + 1$
s_x	Proportion of individuals of age x surviving to age $x + 1$
e_x	Expectation of further life of individuals of age x
k_x	$-\log_e s_x$, the exponential mortality rate between age x and $x - 1$

TABLE 15.6 **Life table of the grass *Poa annua***

AGE (x)*	NUMBER ALIVE	SURVIVORSHIP (l_x)	MORTALITY RATE (m_x)	SURVIVAL RATE (s_x)	EXPECTATION OF LIFE (e_x)	FECUNDITY (b_x)
0	843	1.000	0.143	0.857	2.114	0
1	722	0.857	0.271	0.729	1.467	300
2	527	0.625	0.400	0.600	1.011	620
3	316	0.375	0.544	0.456	0.685	430
4	144	0.171	0.626	0.374	0.503	210
5	54	0.064	0.722	0.278	0.344	60
6	15	0.018	0.800	0.200	0.222	30
7	3	0.004	1.000	0.000	0.000	10
8	0	0.000				

*Number of 3-month periods; in other words, 3 = 9 months.
Source: M. Begon and M. Mortimer, *Population Ecology,* 2d ed., Blackwell Scientific Publications, Oxford (1986). After data of R. Law.

life table for *Poa annua* (Table 15.6) is of this type. This method is readily applied to populations of plants and sessile animals in which marked individuals can be continually resampled over the course of their life spans. Herein lies one of its disadvantages, however: it can take a long time to collect the data (particularly if the subjects are redwood trees!). It is also difficult to apply to highly mobile animals.

Another method, employing a **static life table** or **time-specific life table,** sidesteps the time problem by considering the survival of individuals of known age during a single time interval. The investigator estimates each age-specific survival value independently for each age class of a population during the same period. Of course, to apply this technique it is necessary to know the ages of individuals (estimated by growth rings, tooth wear, or some other reliable index).

The distribution of ages at death was used to construct a life table for Dall mountain sheep *(Ovis dalli)* in Mount McKinley (now Denali) National Park, Alaska. The size of the horns, which grow continuously during the lifetime of an individual sheep (Figure 15.6), provided an estimate of age at death. Of 608 skeletal remains, 121 were judged to have been less than 1 year old at death, 7 between 1 and 2 years old, 8 between 2 and 3 years old, and so on, as shown in Table 15.7. The life table is constructed by the following reasoning: All 608 dead sheep must have been alive at birth; all but the 121 that died during the first year must have been alive at the age of 1 year (608 – 121 = 487), all but 128 (the 121

Figure 15.6 A group of Dall mountain sheep in Alaska. The size of the horns increases with age. Courtesy of the American Museum of Natural History.

dying during the first year and the 7 dying during the second) must have been alive at the end of the second year (608 – 128 = 480), and so on, until the oldest sheep died during their fourteenth year. Survival (l_x; the rightmost column in Table 15.7) was calculated by converting the number of sheep alive at the beginning of each interval to a decimal fraction of those alive at birth. Thus, for example, the 390 sheep alive at the beginning of the seventh year represented 64.0% (decimal fraction 0.640) of the original newborns in the sample.

Intrinsic rates of increase

The **intrinsic rate of increase** of a population, indicated by r_m (often called the Malthusian parameter), is the exponential rate of increase assumed by a population with a stable age distribution. In practice, populations rarely achieve stable age distributions and therefore rarely grow at their intrinsic rates of increase. When changing environmental conditions alter life table values, the age structure of a population continuously readjusts to the new schedule of birth and death rates. Because the actual growth performance of a population depends as much on past conditions, which determine its age structure, as on the present life table values, the intrinsic rate of increase is more useful as an indication of the effects of environmental conditions or individual attributes on population growth

TABLE 15.7 **Life table for Dall mountain sheep constructed from the age at death of 608 sheep in Denali National Park**

AGE INTERVAL (YEARS)	NUMBER DYING DURING AGE INTERVAL	NUMBER SURVIVING AT BEGINNING OF AGE INTERVAL	NUMBER SURVIVING AS A FRACTION OF NEWBORNS (l_x)
0–1	121	608	1.000
1–2	7	487	0.801
2–3	8	480	0.789
3–4	7	472	0.776
4–5	18	465	0.764
5–6	28	447	0.734
6–7	29	419	0.688
7–8	42	390	0.640
8–9	80	348	0.571
9–10	114	268	0.439
10–11	95	154	0.252
11–12	55	59	0.096
12–13	2	4	0.006
13–14	2	2	0.003
14–15	0	0	0.000

Source: Based on data of O. Murie, *The Wolves of Mt. McKinley,* U.S. Department of the Interior, National Park Service, Fauna Series No. 5, Washington, D.C. (1944); quoted by E. S. Deevey, Jr., *Quarterly Review of Biology* 22:283–314 (1947).

than as a number that can accurately project population growth. Each life table has a single intrinsic rate of increase. To find the exact value of r_m, one must solve a complicated equation, but r_m may be approximated (r_a) by the following expression, which uses the terms of the life table:

$$r_a = \frac{R_0 \log_e R_0}{\Sigma x l_x b_x},$$

where R_0—the **net reproductive rate**—is the sum of the $l_x b_x$ column. One may think of R_0 as the expected total number of offspring of an individual over the course of his or her life span. The calculation of r_a is illustrated in Table 15.8.

The growth potential of populations

We can best appreciate the capacity of a population for growth by following the rapid increase of organisms introduced into a new region with a

TABLE 15.8 **Estimation of the exponential rate of increase for the hypothetical population described in Table 15.1**

x	s_x	l_x	b_x	$l_x b_x$	$x l_x b_x$
0	0.5	1.0	0	0.0	0.0
1	0.8	0.5	1	0.5	0.5
2	0.5	0.4	3	1.2	2.4
3	0.0	0.2	2	0.4	1.2
Net reproductive rate (R_0)				2.1	
Expected number of births weighted by age					4.1

Note: The sums of the $l_x b_x$ column (net reproductive rate) and the $x l_x b_x$ column are used to estimate r_a according to the equation given in the text. In this case, we calculate r_a to be 0.38; this is equivalent to $\lambda = 1.46$, close to the observed value of about 1.48 after the population achieved a stable age distribution.

suitable environment. In 1937, 2 male and 6 female ring-necked pheasants were released on Protection Island, Washington. They increased to 1,325 adults within 5 years. This 166-fold increase represents a 178% annual rate of increase ($r = 1.02$, $\lambda = 2.78$). In other words, the population almost tripled, on average, each year. When domestic sheep were introduced to Tasmania, a large island off the coast of Australia, the population increased from less than 200,000 in 1820 to more than 2 million in 1850 (see Figure 16.1). This tenfold increase in 30 years is equivalent to an annual rate of increase of 8% ($r = 0.077$, $\lambda = 1.08$). Even such an unlikely creature as the elephant seal, whose population along the western coast of North America had been all but obliterated by hunting during the nineteenth century, increased from 20 individuals in 1890 to 30,000 in 1970 ($r = 0.091$, $\lambda = 1.096$). If you are unimpressed, consider that another century of unrestrained growth would find 27 million elephant seals crowding surfers and sunbathers off southern California beaches. Before the end of the following century, the shorelines of the Western Hemisphere would give lodging to a trillion of the beasts.

Elephant seal populations do not hold any growth records—quite the contrary. Life tables of populations maintained under optimal conditions in the laboratory have exhibited potential annual growth rates (λ) as great as 24 for the field vole, 10 billion (10^{10}) for flour beetles, and 10^{30} for the water flea *Daphnia* (Table 15.9). These growth rates are equivalent to population **doubling times** of 79, 11, and less than 4 days respectively. The potential growth rates of populations of bacteria and viruses under ideal conditions are almost unimaginable.

The intrinsic growth rate of a population varies with environmental conditions and population density. It is strictly determined by the life table values, which express the interaction between the individual and its environment. As a consequence of this interaction, the life table (and hence the

TABLE 15.9 Doubling times of populations

SPECIES	λ	$\log_e\lambda = r$	t_2 (YEARS)	t_2 (DAYS)
Elephant seal	1.096	0.091	7.5	
Ring-necked pheasant	2.78	1.02	0.67	246
Field vole	24	3.18	0.22	79
Flour beetle	10^{10}	23	0.03	11
Water flea	10^{30}	69	0.01	3.6

Note: Rapid growth rates may be expressed conveniently in terms of the time required for the population to double. The relationship between geometric growth rate (λ) and doubling time (t_2), derived from the equation for geometric growth, is $t_2 = \log_e 2/\log_e\lambda$; that is, $0.69/\log_e\lambda$. Hence for the field vole ($\lambda = 24$), $t_2 = 0.69/\log_e 24$, which is 0.22 year, or 79 days.

intrinsic growth rate) varies with the conditions of the environment. These conditions vary spatially and temporally, creating differences in population dynamics from place to place and leading to changing population dynamics over time.

Such effects of environmental conditions are best revealed in experimental studies in which groups of individuals from the same population are subjected to different conditions. For example, the intrinsic rates of increase of two species of grain beetles were found to be influenced differently by temperature and moisture (Figure 15.7). Neither *Rhizopertha* nor *Calandra* performed well at low temperatures and humidities. Moreover, the optimal conditions for growth differed between the species: *Rhizopertha* populations grew most rapidly at somewhat warmer temperatures. Of the two, *Rhizopertha* has the more tropical distribution in nature.

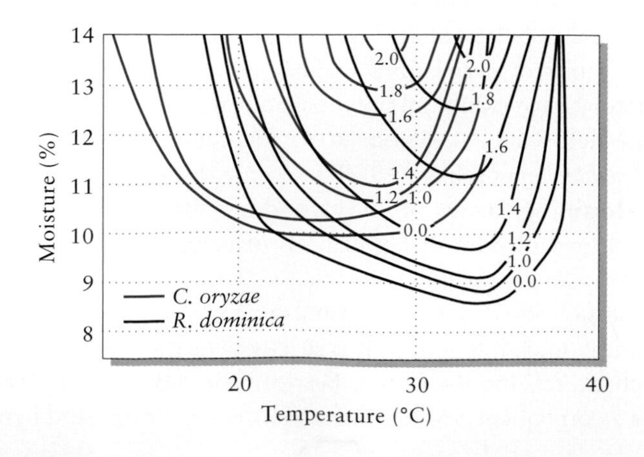

Figure 15.7 Influence of the temperature and moisture content of grain on intrinsic rates of increase in populations of the grain beetles *Calandra oryzae* and *Rhizopertha dominica* living in wheat. Rates of increase are indicated by contour lines that describe conditions with identical values of λ, which are indicated on each line. After L. C. Birch, *Ecology* 34:698–711 (1953).

TABLE 15.10 **Condensed life tables for European rabbits in arid and Mediterranean climates of Australia**

AGE (MONTHS)	PIVOTAL AGE* (MONTHS)	SURVIVAL (l_x)†	PROPORTION OF FEMALE POPULATION	FEMALES PREGNANT AT ANY ONE TIME	LITTER SIZE (EMBRYOS)	(b_x)	($l_x b_x$)	($x l_x b_x$)
Arid								
3–6	4.5	0.570	0.140	0.05	3.8	0.2	0.057	0.257
6–12	9.0	0.457	0.265	0.15	4.7	1.8	0.411	3.702
12–18	15.0	0.302	0.217	0.32	4.6	3.1	0.468	7.022
18–24	21.0	0.186	0.192	0.37	4.3	3.1	0.288	6.054
24	37.5	0.061	0.187	0.34	4.5	2.0	0.089	3.317
Total							1.313	20.352
Mediterranean								
3–6	4.5	0.222	0.177	0.23	4.4	1.9	0.198	0.891
6–12	9.0	0.260	0.359	0.43	5.6	8.7	0.696	6.264
12–18	15.0	0.075	0.303	0.53	5.9	9.4	0.357	5.358
18–24	21.0	0.028	0.107	0.45	5.9	2.8	0.039	0.823
24	37.5	0.006	0.053	0.55	6.2	1.8	0.005	0.203
Total							1.295	13.539

*Midpoint of the age interval, frequently used to calculate generation time when reproduction is more or less continuous, as it is in rabbits in Australia.
†Based on rabbits aged 0 to 3 months; hence $l_{1.5} = 1$.
Source: K. Myers, in P. J. der Boer and G. R. Gradwell (eds.), *Dynamics of Populations,* Centre Agric. Publ. Documentation, Wageningen, The Netherlands, pp. 478–506.

Climate and other environmental conditions have also been shown to be important determinants of life table values for populations in their natural settings, as one would expect. For example, the European rabbit, introduced to Australia in the nineteenth century and now widespread throughout many habitats, survives longer but produces fewer offspring per year in arid regions than in the moister Mediterranean climate regions (Table 15.10).

This particular example raises the important issue of regulation of population size. Continuous exponential growth leads, with time, to inconceivable numbers. As Darwin put it in 1872, "Even slow-breeding man has doubled in twenty-five years, and at this rate, in less than a thousand years, there would literally not be standing-room for his progeny." The growth potential of populations is driven home by the history of the rabbit population in Australia. In 1859, 12 pairs were released on a ranch in Victoria to provide sport for hunters. Within 6 years, the population had increased so rapidly that 20,000 rabbits were killed in a single hunting drive. Even by conservative estimates, the population must have increased by a factor of at least 10,000 in 6 years, an exponential growth rate (r) of about 1.5 per year (a doubling time of about 5.5 months). Yet the life tables of present-day populations (see Table 15.10) suggest growth rates (r_a) of 0.30 and 0.21. Considering the difficulty of estimating survival of the young to reproductive age and the statistical errors involved in such studies, these values probably do not differ much from $r = 0$. How can the initial rapid growth rate be reconciled with the eventual stabilization of the rabbit population? Either birth rates decreased, death rates increased, or both changed when the population became more numerous. When there are more rabbits, there is less food for each; fewer resources mean that fewer offspring can be nourished, and those offspring survive less well. Crowded populations also aggravate social strife, promote the spread of disease, and attract the attention of predators. Many such factors may act together to slow, and finally halt, population growth.

Regulation of population size

Even the most slowly reproducing species would cover the earth in a short time if its population growth were unrestrained. Nearly two centuries ago, Thomas Malthus understood that this fact "implies a strong and constantly operating check on population from the difficulty of subsistence." In *An Essay on the Principle of Population* (1798), he wrote:

> *Through the animal and vegetable kingdoms, nature has scattered the seeds of life abroad with the most profuse and liberal hand. She has been comparatively sparing in the room and the nourishment necessary to rear them. The germs of existence contained in this spot of earth, with ample food, and ample room to expand in, would fill millions of worlds in the course of a few thousand years. Necessity, that emperious all pervading law of nature, restrains them within the prescribed bounds. The race of plants, and the race of animals shrink under this great restrictive law.*

Darwin echoed this view in *On the Origin of Species:*

> *As more individuals are produced than can possibly survive, there must in every case be a struggle of existence, either one individual with another of the same species, or with the individuals of distinct species, or with the physical conditions of life. It is the doctrine of Malthus applied with manifold force to the whole animal and vegetable kingdoms; for in this case there can be no artificial increase of food, and no prudential restraint from marriage. Although some species may be now increasing, more or less rapidly, in numbers, all cannot do so, for the world would not hold them.*

This essentially modern view of the regulation of populations grew out of an awareness of the immense capacity of populations for exponential increase. In a sense, a population's growth potential and the relative constancy of its numbers cannot be logically reconciled otherwise.

The logistic equation

In 1920, Raymond Pearl and L. J. Reed, at the Institute for Biological Research of Johns Hopkins University, published a paper in the *Proceedings of the National Academy of Sciences* entitled "On the Rate of Growth of the Population of the United States since 1790 and Its Mathematical Representation." Thorough and accurate population data had been gathered even in colonial times. Indeed, the phenomenal population growth of the American colonies had greatly impressed upon Malthus how rapidly humans could multiply; this was not so evident in the more crowded European countries of his time.

Pearl and Reed wished to project the future growth of the population of the United States, which they supposed must eventually reach a limit. Data for the population to 1910, the latest census then available, had revealed a decline in the exponential rate of growth (Figure 15.8). Pearl and Reed reasoned that if this decline followed a regular pattern that could be described mathematically, it would be possible to predict the future course of the population, as long as the decline in the exponential growth rate continued. They also reasoned that changes in the exponential rate of growth must be related to the size of a population rather than to time, because any time scale is arbitrary with respect to any particular population. And so, in place of a constant value of r in the differential equation for unrestrained population growth ($dN/dt = rN$), Pearl and Reed suggested that r decreases as N increases, according to the relation

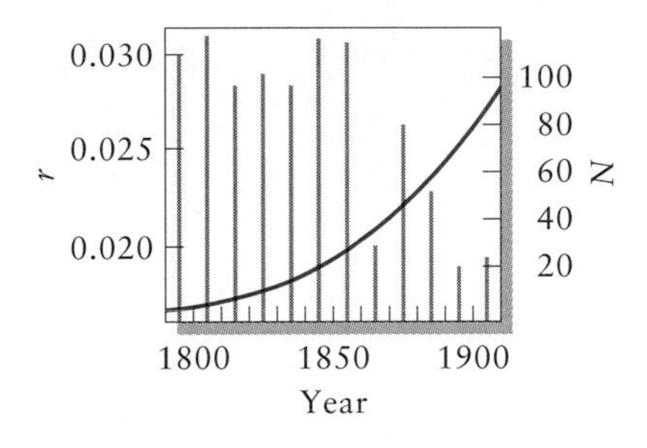

Figure 15.8 Increase in the population of the United States between 1790 and 1910 (curve) and the exponential rate of increase during each 10-year period (vertical bars). From data in R. Pearl and L. J. Reed, *Proc. Natl. Acad. Sci.* 6:275–288 (1920).

$$r = r_0 \left(1 - \frac{N}{K}\right),$$

where r_0 represents the intrinsic exponential growth rate of a population when its size is very small (that is, close to 0), and K—the **carrying capacity** of the environment—represents the number of individuals that the environment can support. Accordingly, the differential equation describing restricted population growth became

$$\frac{dN}{dt} = r_0 N \left(1 - \frac{N}{K}\right).$$

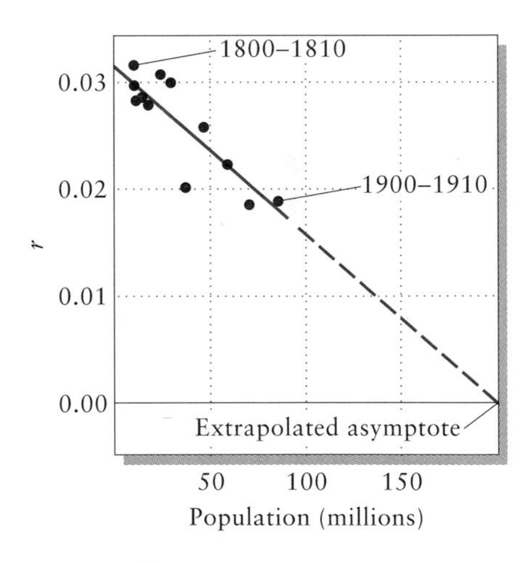

Figure 15.9 Exponential rates of population increase in the United States during each decade between 1790 and 1910, plotted as a function of population size during that decade (the geometric mean of the beginning and ending numbers) using the data shown in Figure 15.8. The dashed line is the extrapolation of a straight line (logistic equation) fitted to the data.

According to this equation, which is called the **logistic equation,** the exponential rate of increase decreases as a linear function of the size of a population. Such a decrease reasonably approximated the data for the population of the United States (Figure 15.9).

So long as population size N does not exceed the carrying capacity K—that is, N/K is less than 1—a population continues to increase, albeit at a slowing rate. When N exceeds the value of K, the ratio N/K exceeds 1, the term in parentheses ($1 - N/K$) becomes negative, and the population decreases. Because populations below K increase and those above K decrease, K is the eventual equilibrium size of a population growing according to the logistic equation. The relationships among r, dN/dt, and N are illustrated in Figure 15.10. The curve for dN/dt has a maximum at

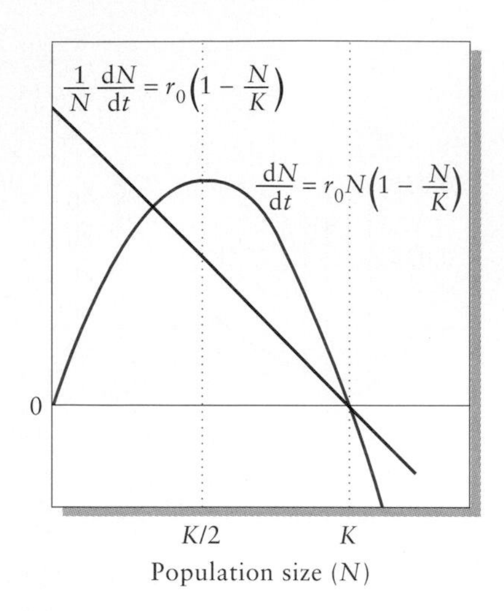

Figure 15.10 Various representations of the logistic curve of population growth. The straight black line represents the exponential growth rate of a population ($1/N\ dN/dt$) as a function of population size; r declines to 0 when $N = K$. The colored line represents the absolute rate of growth (individuals per unit of time: dN/dt), which is the product of population size (N) and the exponential growth rate, reaches a maximum when population size is one-half the carrying capacity (K).

intermediate population size, specifically when $N = K/2$, and falls to 0 as N approaches either 0 or K.

The time course of population growth according to the logistic equation can be found by integrating the differential, which yields

$$N(t) = \frac{K}{1 + be^{-r_0 t}},$$

where b is a constant equal to $[K - N(0)]/N(0)$. The value of b depends arbitrarily on the size of the population at the designated time zero—1790 in the case of Pearl and Reed's data. This equation describes a sigmoid, or S-shaped, curve (Figure 15.11): the population grows slowly at first, then more rapidly as the number of individuals increases, and finally more slowly, gradually approaching the equilibrium number K. Applied to the growth of the population of the United States from 1790 to 1910, this curve is illustrated in Figure 15.12.

Pearl and Reed obtained the best fit of their equation to the population data when the value of K was 197,273,000 and that of r_0 was 0.03134, which is equivalent to a population doubling time of 22 years. Thus, even though the population in 1910 was only 91,972,000, Pearl and Reed were able to extrapolate its future growth to twice the 1910 level from earlier growth performance. Projections often prove incorrect, however, when circumstances change. The U.S. population reached 197 million between 1960 and 1970, when it was still growing vigorously. A leveling off in the mid-200 millions can now be predicted on the basis of a much reduced birth rate in recent years, but all this could easily change.

Density-dependent factors

The logistic equation has been applied successfully to describing the growth of populations in the laboratory and in natural habitats. The equation suggests that factors limiting growth exert stronger effects on mortality and fecundity as a population grows. But what are these factors, and how do they operate? Many things influence rates of population growth, but only **density-dependent factors,** whose effect increases with crowding, can bring a population under control. Of prime importance among these factors are limitations on food supply and places to live, as well as predators, parasites, and diseases whose effects are felt more strongly in crowded than in sparse populations. Other factors, such as temperature, precipitation, and catastrophic events, alter birth and death rates largely without regard to numbers of individuals in a population. Thus, such **density-independent** factors may influence the exponential growth rate of a population, but they do not regulate its size.

Density dependence in animals

Numerous experimental studies have revealed various mechanisms of density dependence. For example, when fruit flies are confined to a bottle with a fixed supply of food, the descendants of a single pair of flies increase in number rapidly at first, but soon reach a limit. When different numbers of pairs of flies are introduced into otherwise identical culture bottles, the number of progeny raised per pair varies inversely with density of flies in the bottle (Figure 15.13). This effect results from competition among the larvae for food, which causes high mortality in dense cultures. Adult life span also declines, but only at high densities, well above the levels that affect the survival of larvae. It is often the case that juvenile stages suffer the adverse effects of density-dependent factors more than adults do.

The effects of density on the life tables of grain beetles result from fights between larvae under crowded conditions. *Rhizopertha dominica* is a tiny beetle that completes its larval development within a single grain (of wheat, for example), living off the kernel of the seed. Females lay their eggs on the surfaces of seeds. Immediately after hatching, a larva bores its way into a seed, where it commences development. Once inside, the larva enjoys security as long as it is alone; a kernel of wheat cannot support more than one beetle.

What happens when two larvae meet in the same grain? The British ecologist A. C. Crombie described it thus:

> *When two larvae in the first, second, third or fourth instars were put together into a small hole drilled in a wheat grain and watched under a binocular microscope, they were often seen to attack each other with their mandibles, and eventually either one or both left the hole. When a larva entered such a hole it always went to the bottom and turned round so as to face outward. Other larvae trying to enter the hole were fiercely attacked. Sometimes such combats resulted in the body wall of one of the antagonists becoming punctured and its bleeding to death. In their tunnels in wheat grains larvae of all instars were always found curled up with the head facing toward the way they had entered. Furthermore, in all grains dissected during the experiments to be described, whenever two larvae were found in the same tunnel at least one of them was always dead (J. Exp. Biol. 20:135–151, 1944).*

The experiments Crombie referred to were performed in the following manner. He infested a single wheat grain with from one to eight larvae, giving them 6 hours to become established in the grain. He then transferred each infested grain to a dish with five other fresh grains and left them together for 48 hours. At the end of the experiment, he dissected all the grains to determine how many larvae had been killed and how many had migrated to a new grain. The results showed conclusively the strong effects of density on the survival and movement of larvae, even within the first 2 days of life (Figure 15.14).

Crombie's beetles contested each grain of wheat through direct confrontation. After all, sole possession of a single grain is a larva's ticket to success. In this population, density dependence came into play abruptly as the number of larvae approached the number of grains of wheat. Above that level, regardless of the number of larvae relative to wheat grains, the end result was always the same: one grain, one larva. Territorial behavior in many kinds of animals regulates population size in a density-dependent

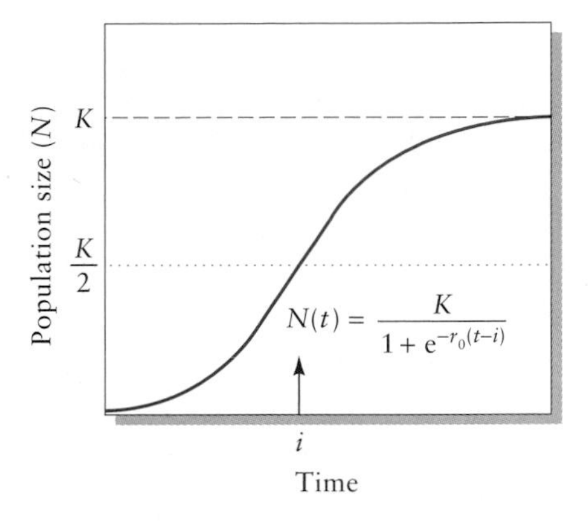

Figure 15.11 According to the logistic growth equation, increase in numbers over time follows an S-shaped curve that is symmetrical about the inflection point ($K/2$). That is, accelerating and decelerating phases of population growth have the same shape. In the form of the equation shown on the graph, i is the time at which the population size reaches the inflection point.

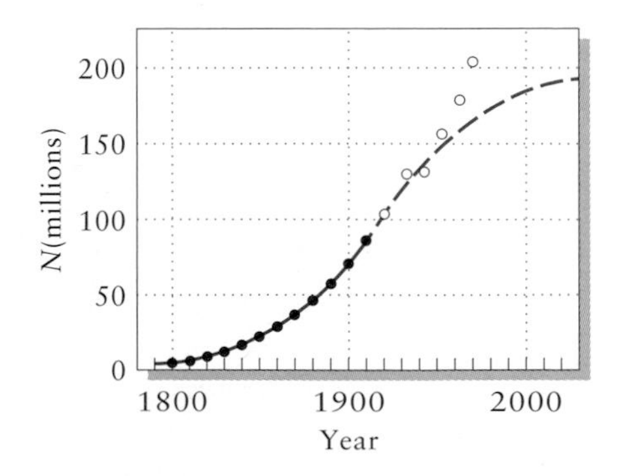

Figure 15.12 A logistic curve fitted to the population of the United States between 1790 and 1910 (solid dots). Subsequent censuses (open dots) have shown numbers above the projected population curve.

fashion, forcing young or socially subordinate individuals to leave the local population and seek space elsewhere.

Water fleas *(Daphnia pulex)* influence one another less directly by eating the same resources. Each water flea consumes millions of prey: single-celled green algae and diatoms of the plankton. As prey are eaten, the availability of food to other water fleas decreases gradually, leading to a graded response of birth and death rates to prey density. When water fleas were maintained in small beakers at densities between 1 and 32 individuals per cubic centimeter and were fed identical cultures of green algae, fecundity decreased markedly with increasing population density (Figure 15.15). Somewhat unexpectedly, however, survival increased at densities up to 8 individuals per cubic centimeter before decreasing at higher densities. At densities of 8 individuals per cubic centimeter and above, stunted body growth suggested that depletion of food resources between periodic feedings limited birth rates and survival; the effect was clearly density-dependent.

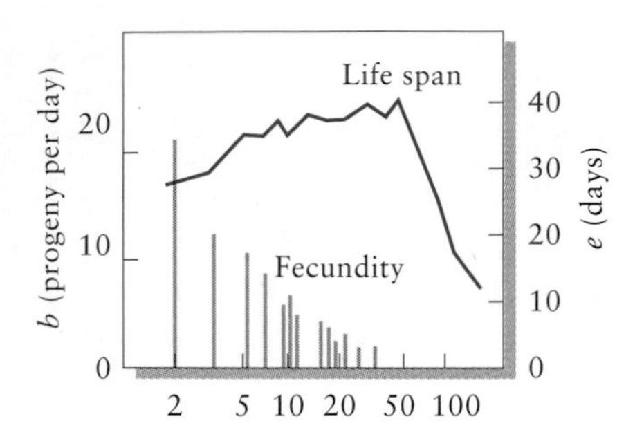

Figure 15.13 The continuous line shows the influence of density on life span (e, days) and the vertical bars its influence on fecundity (b, progeny per day) in laboratory populations of the fruit fly *Drosophila melanogaster.* After R. Pearl, *Q. Rev. Biol.* 2:532–548 (1927).

The geometric rate of population growth (λ) calculated from the water flea life tables decreased linearly with increasing density and fell below 1.0 at a density of about 20 individuals per cubic centimeter (Figure 15.16). Therefore, under the conditions of temperature, light, water quality, and food availability provided in the laboratory, *Daphnia* populations would attain a stable size of about 20 individuals per cubic centimeter, regardless of the initial density of a culture.

Most studies of density dependence have focused on laboratory populations, in which factors affecting populations can be controlled experimentally. The simplicity of such systems leaves some doubt about the relevance of laboratory findings to populations in more complex natural surroundings, where physical conditions change continually and food and predation are not controlled by the experimenter. For example, winter weather and other factors have caused the population of song sparrows on Mandarte Island—a 6-hectare speck of land off the coast of British Columbia—to fluctuate widely between 4 and 72 breeding females and between 9 and 100 breeding males during recent years. In response to this environmentally induced variation in population size, density-dependent factors clearly limit the productivity of the population during years of high density by restricting the number of breeding males (territoriality), reducing the number of offspring per female (decreases in breeding season food supply), and reducing survival of juveniles in autumn and winter (Figure 15.17).

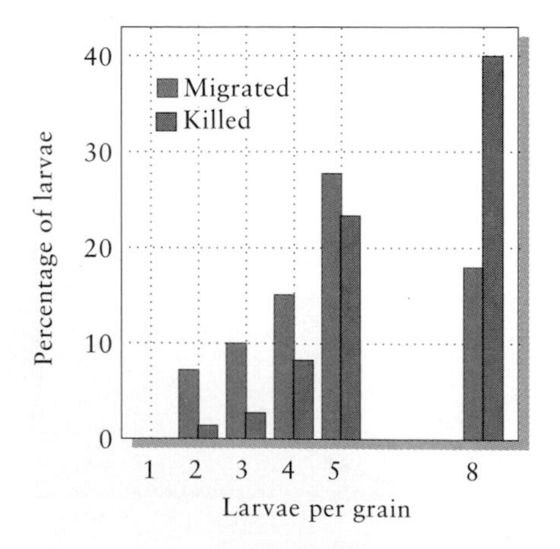

Figure 15.14 Effects of density (number of larvae per grain) on migration and mortality in the grain beetle *Rhizopertha dominica*. From data in A. C. Crombie, *J. Exp. Biol.* 20:135–151 (1944).

Although natural variation in population size provides a method for visualizing density dependence, ideally we would like to conduct the same experiment in nature as in the laboratory—that is, to alter the density of individuals in the population while keeping everything else constant. In practice, this difficult experiment can be accomplished only with populations managed intensively for some other purpose. Game animals are sometimes maintained at altered levels by management practices, and ecologists have taken advantage of such situations to study population processes. A survey of harvested white-tailed deer in New York State in the 1940s provides an example of this method.

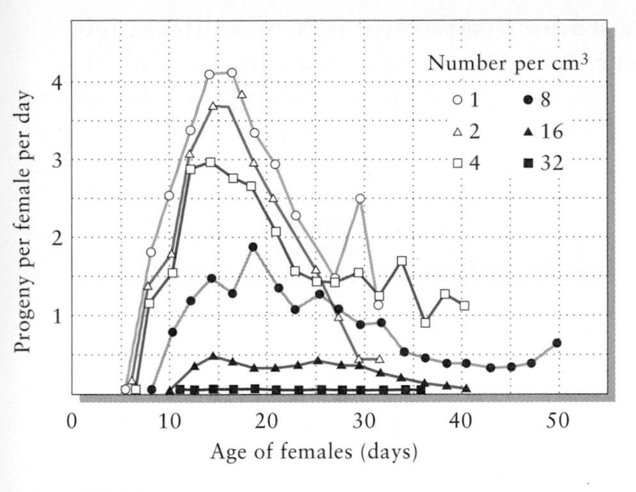

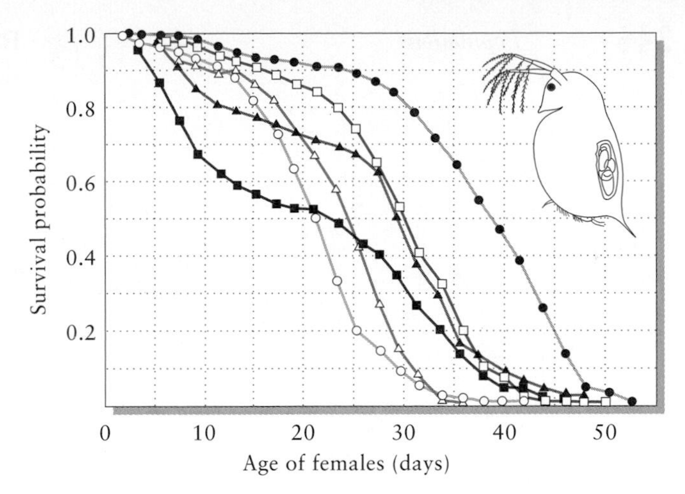

Figure 15.15 Fecundity and survival in laboratory populations of *Daphnia pulex* as a function of age at different densities. From P. W. Frank, C. D. Boll, and R. W. Kelly, *Physiol. Zool.* 30:287–305 (1957).

The reproduction and survival of deer depend directly on the quality of their food. Deer browse leaves, and they require large quantities of new growth with high nutritional content to maintain high growth rates and normal reproduction. In white-tailed deer in New York State, the proportion of females pregnant and the average number of embryos per pregnant female were found to be directly related to range conditions (Table 15.11). The number of corpora lutea in each ovary indicates the number of eggs ovulated and hence the reproductive potential of a female. A difference between number of corpora lutea and number of embryos shows that

Figure 15.16 Values of λ calculated from the life table data for *Daphnia pulex* portrayed in Figure 15.15. The population growth rate, which can be calculated by the method outlined in Table 15.8, decreases as a function of density. After R. Laughlin, *J. Anim. Ecol.* 34:77–91 (1965).

TABLE 15.11 Reproductive parameters of white-tailed deer *(Odocoileus virginianus)* in five regions of New York State, 1939–1949

REGION*	PERCENT OF FEMALES PREGNANT	EMBRYOS PER FEMALE	*CORPORA LUTEA* PER OVARY
Western (best range)	94	1.71	1.97
Catskill periphery	92	1.48	1.72
Catskill central	87	1.37	1.72
Adirondack periphery	86	1.29	1.71
Adirondack center (worst range)	79	1.06	1.11

*Arranged by decreasing suitability of range.

Source: E. L. Chaetum and C. W. Severinghaus, *Trans. N. Am. Wildl. Conf.* 15:170–189 (1950).

TABLE 15.12 Reproductive parameters of white-tailed deer *(Odocoileus virginianus)* in the DeBar Mountain area of the Adirondack Mountains of New York State before and after hunting

REGION	PERCENT OF FEMALES PREGNANT	EMBRYOS PER FEMALE	*CORPORA LUTEA* PER OVARY
1939–1943 (prehunting)	57	0.71	0.60
1947 (after heavy hunting)	100	1.78	1.86

Source: E. L. Chaetum and C.W. Severinghaus, *Trans. N. Am. Wildl. Conf.* 15:170–189 (1950).

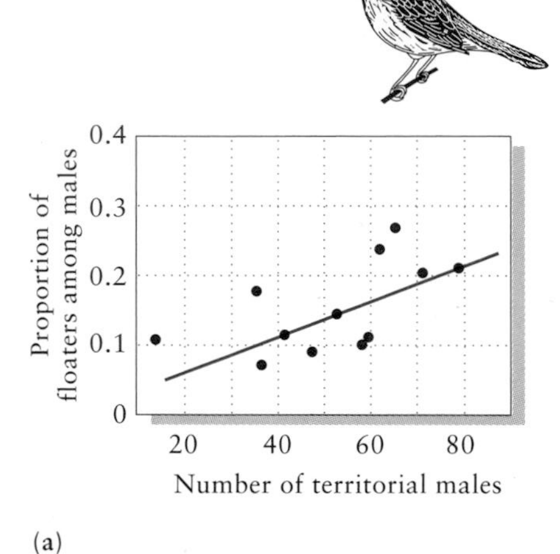

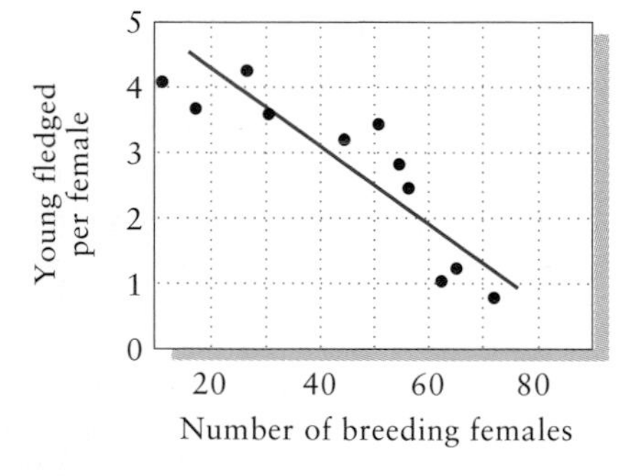

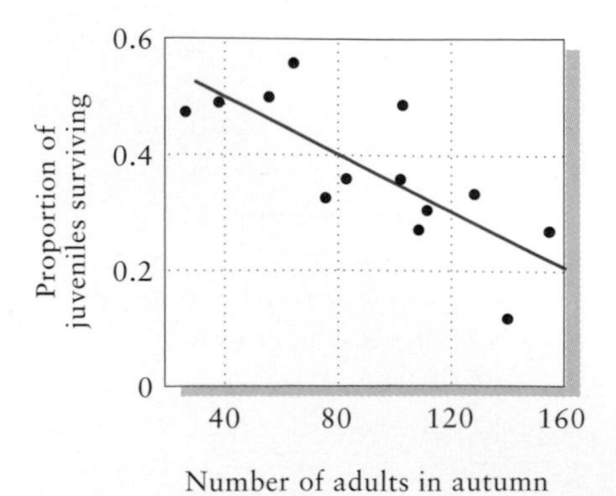

poor range conditions, resulting in poor nutrition of pregnant females, have caused embryo death and resorption. In the central Adirondack area, where habitat for deer was very poor, even ovulation was greatly reduced. Range deterioration caused by overgrazing can often be reversed by selective hunting to thin dense populations. When DeBar Mountain, an area of very poor range, was opened to hunting, the population of white-tailed deer decreased, range quality recovered, and reproduction improved dramatically (Table 15.12).

Density dependence in plants

Like animals, plants experience increased mortality and reduced fecundity at high densities. A common response of plants to intense competition for resources is slowed growth, which has consequences for fecundity and, to a lesser extent, survival. The sizes of flax *(Linum)* plants grown to maturity at different densities reveal this flexibility (Figure 15.18). When seeds were sown sparsely at a density of 60 per square meter, the modal dry weight of individuals fell between 0.5 and 1 g, and many plants attained weights exceeding 1.5 g. When seeds were sown at densities of 1,440 and 3,600 per square meter, most of the individuals weighed less than 0.5 g, and few

Figure 15.17 Density dependence in the Mandarte Island song sparrow *(Melospiza melodia)* population. With increased crowding on the small island, the proportion of males prevented from acquiring territories ("floaters") increases, and the number of fledglings produced per female, as well as the survival of those offspring through autumn and winter, decreases. After P. Arcese and J. N. M. Smith, *J. Anim. Ecol.* 57:119–136 (1988), and J. N. M. Smith, P. Arcese, and W. M. Hochachka, in C. M. Perrins, J.-D. Lebreton, and G. J. M. Hirons (eds.), *Bird Population Studies: Relevance to Conservation and Management,* Oxford University Press, Oxford, pp. 148–167 (1991).

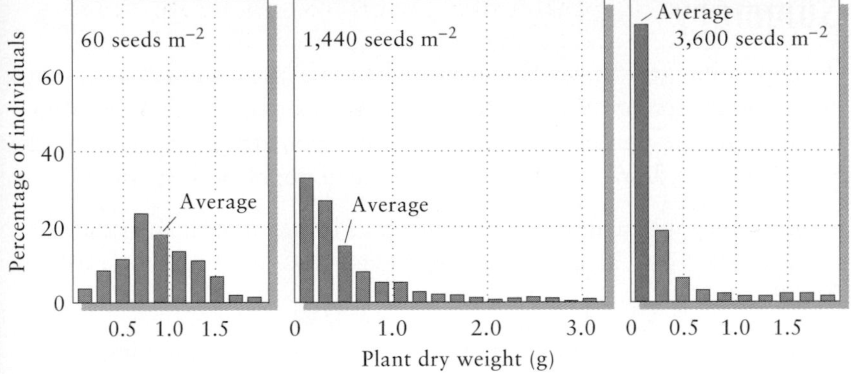

Figure 15.18 Distribution of dry weights of individuals in populations of flax plants sown at different densities. After J. L. Harper, *J. Ecol.* 55:247–270 (1967).

grew to large size. Variation in size within a planting results from chance factors early in the seedling stage, particularly date of germination and quality of the site in which the seedling grows. Early germination in a favorable spot gives a plant an initial growth advantage over others, which increases as larger plants grow and crowd their smaller neighbors.

The flexibility of plant growth does not preclude mortality in crowded situations. When horseweed *(Erigeron canadensis)* seed was sown at a density of 100,000 per square meter (equivalent to about 10 seeds in the area of your thumbnail), young plants competed vigorously. As the seedlings grew, many died, and the density of the surviving seedlings decreased (Figure 15.19). At the same time, however, the growth rates of surviving individual plants exceeded the rate of decline of the population, and the total weight of the planting increased. Over the entire growing season, a thousandfold increase in the average weight of each plant more than balanced the hundredfold decrease in population density.

When the logarithm of average plant weight is plotted as a function of the logarithm of density, data points recorded during the growing season fall on a line with a slope of approximately −3/2 (Figure 15.20). Plant ecologists call this relationship between average plant weight and density a **self-thinning curve.** Such is the regularity of this relationship that many have referred to it as the **−3/2 power law.**

Density-dependent factors tend to bring populations under control and maintain their size at close to the carrying capacity set by the availability of resources and conditions of the environment. Changes in these conditions and resources continually establish new equilibrium values toward which populations grow or decline. Furthermore, catastrophic changes in the environment brought about by a sudden freeze, a violent storm, or a shift in an ocean current often reduce populations far below their carrying capacities and initiate periods of population recovery. Thus, although density-dependent factors regulate all populations, variations in the environment also cause all populations to fluctuate, to a greater or lesser extent, about their equilibrium sizes. We shall explore population variation in more detail in the next chapter.

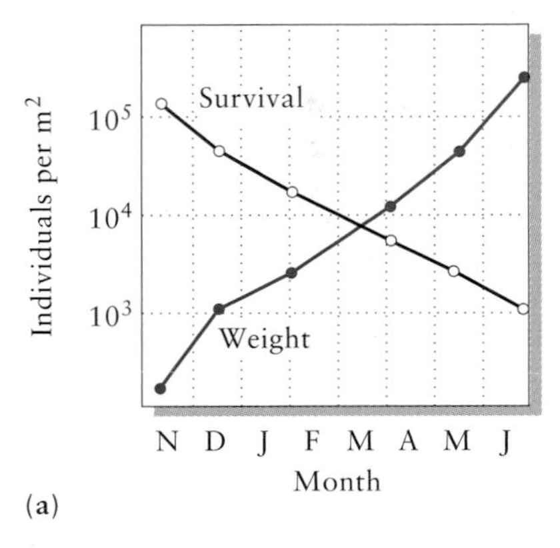

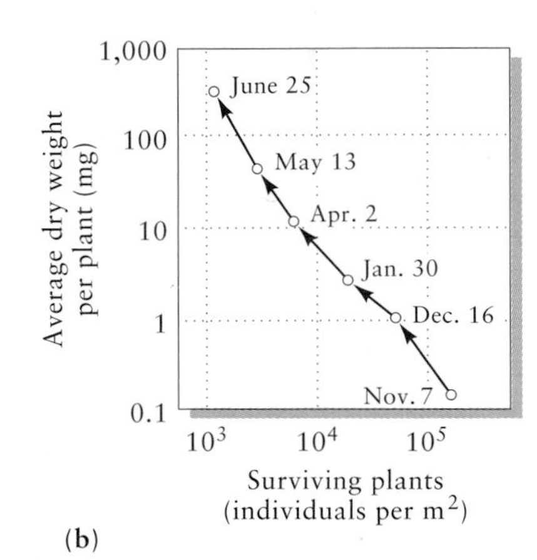

Figure 15.19 (a) Progressive change in plant weight and population density in an experimental planting of horseweed *(Erigeron canadensis)* sown at a density of 100,000 seeds per square meter. (b) Relationship between plant density and plant weight as the season progressed. After J. L. Harper, *J. Ecol.* 55:247–270 (1967).

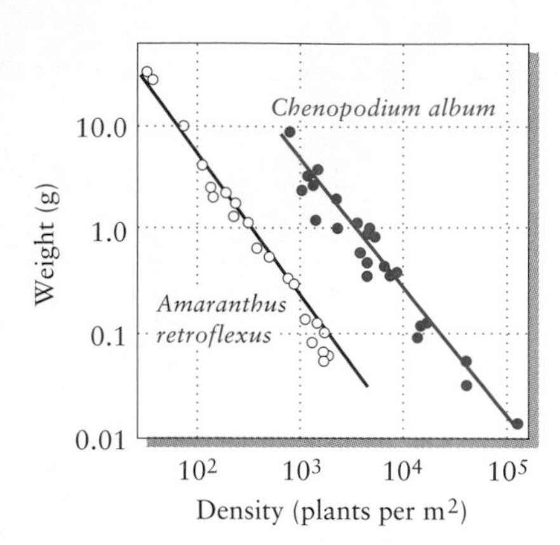

Figure 15.20 Changes in plant density and mean plant weight with time for plantings of *Amaranthus retroflexus* and *Chenopodium album*. The two variables are related by a slope of −3/2 (−1.5), which has been extremely consistent in experiments of this type. After J. L. Harper, *Population Biology of Plants,* Academic Press, New York (1977).

Summary

1. Population growth can be described by the exponential rate of increase (r) in the expression $N(t) = N(0)e^{rt}$. The factor by which a population increases in one unit of time (e^r) is also the geometric growth rate of the population (λ). λ and e^r are interchangeable in population equations.

2. Exponential growth rate is the difference between birth and death rates averaged over individuals (that is, per capita rates) in a population (that is, $r = b - d$).

3. The instantaneous rate of increase of an exponentially growing population is $dN/dt = rN$, which says that the growth rate of a population depends both on its size and on per capita birth and death rates.

4. Populations with discrete breeding seasons increase geometrically by periodic increments according to the relation $N(t + 1) = N(t)\lambda$.

5. When birth rates and death rates vary according to the age of individuals, we must also know the proportion of individuals in each age class in order to project population growth.

6. The life table of a population displays fecundities (b_x) and probabilities of survival (s_x) of individuals by age class (x). These are the principal variables used in models of the dynamics of populations. Survival rates may be estimated from fates of individuals born at the same time (a cohort or dynamic life table), survival of individuals of known age during a single time period (a static or time-specific life table), the age structure of a population at a particular time, or the age distribution of deaths.

7. A population with fixed life table values assumes a stable age distribution in which numbers in each age class, as well as the population as a whole, increase at the same exponential or geometric rate, known as the intrinsic rate of increase of the population (r_m).

8. The life table of a population varies with the conditions of the environment and with the density of the population.

9. The incongruity between the potential of all populations for rapid growth and the relative constancy of populations over long periods led naturally to the idea that environmental factors exert density-dependent effects on population processes. Accordingly, population growth can be slowed only if birth rates and survival decrease as populations grow. Dwindling supplies of food and increasing pressure from predators and disease exert their effects on population processes in such a density-dependent manner.

10. Density-dependent population growth can be described by the logistic equation, which has the differential form

$$\frac{dN}{dt} = r_0 N \left(1 - \frac{N}{K}\right)$$

and the integral form

$$N(t) = \frac{K}{1 + be^{-r_0 t}}.$$

11. Laboratory and field studies of animal and plant populations have shown how density-dependent factors are expressed in population processes. During the 1930s, experimental laboratory systems were developed using organisms such as fruit flies, grain beetles, and water fleas. Similar studies of plants and of animals in natural habitats, based largely on such economically important species as crop plants, weeds, insect pests, and game animals, came later.

Suggested readings

Andrewartha, H. G., and L. C. Birch. 1954. *The Distribution and Abundance of Animals.* University of Chicago Press, Chicago.

Andrewartha, H. G., and L. C. Birch. 1984. *The Ecological Web: More on the Distribution and Abundance of Animals.* University of Chicago Press, Chicago.

Clutton-Brock, T. H., M. Major, and F. E. Guinness. 1985. Population regulation in male and female red deer. *Journal of Animal Ecology* 54:831–846.

Gotelli, N. J. 1995. *A Primer of Ecology.* Sinauer Associates, Sunderland, Mass.

Kingsland, S. E. 1985. *Modeling Nature: Episodes in the History of Population Ecology.* University of Chicago Press, Chicago.

Lack, D. 1954. *The Natural Regulation of Animal Numbers.* Oxford University Press, London.

Murdoch, W. W. 1994. Population regulation in theory and practice. *Ecology* 75:271–287.

Myers, R. A., N. J. Barrowman, J. A. Hutchings, and A. A. Rosenberg. 1995. Population dynamics of exploited fish stocks at low populations levels. *Science* 269:1106–1108.

Skogland, T. 1985. The effects of density-dependent resource limitations on the demography of wild reindeer. *Journal of Animal Ecology* 54:359–374.

Weiner, J. 1988. Variation in the performance of individuals in plant populations. In A. J. Davy, M. J. Hutchings, and A. R. Watkinson (eds.), *Plant Population Ecology,* pp. 59–81. Blackwell Scientific Publications, Oxford.

Weller, D. E. 1987. A reevaluation of the –3/2 power rule of plant self-thinning. *Ecological Monographs* 57:23–43.

Wynne-Edwards, V. C. 1986. *Evolution through Group Selection.* Blackwell Scientific Publications, Oxford and Boston, Mass.

CHAPTER 16

TEMPORAL AND SPATIAL DYNAMICS OF POPULATIONS

Under the influence of density-dependent factors, populations tend to increase or decrease toward steady-state numbers determined by the carrying capacities of their environments. However, environments vary over time, and therefore, so do populations. Moreover, patterns of variation derive not only from changing environments, but also from the intrinsic dynamics of population responses. Thus, the study of variation in population size has its roots in two distinct kinds of phenomena. The first includes responses of populations to perceptible change in their environments; the second includes regular cycles or irregular fluctuations in numbers that are unrelated to obvious periodic variation in environmental factors.

Ecological conditions also vary from place to place, which may cause the dynamics of individual subpopulations to differ. Individuals migrating between two localities can couple the dynamics of

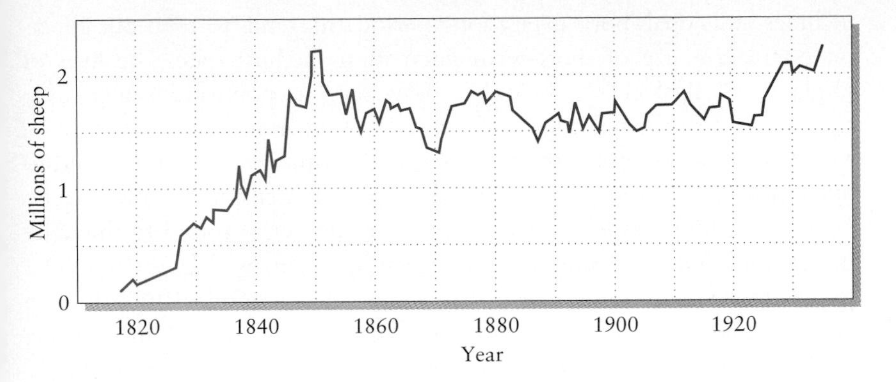

Figure 16.1 Numbers of domestic sheep on the island of Tasmania since their introduction in the early 1800s. After J. Davidson, *Trans. R. Soc. S. Aust.* 62: 342–346 (1938).

subpopulations and make them behave as a single population. More often, distance isolates subpopulations, and they behave at least partly independently. Changes in an entire population are the sum of changes in all of its subpopulations, but because the dynamics of large and small populations differ, subdivided populations possess unique properties.

In this chapter, we shall discuss the causes of variation in population size, explore how this variation affects small and large populations differently, and examine the consequences of dispersal among subpopulations. The dynamics of small populations have become increasingly relevant as species dwindle toward extinction and landscape use by humans fragments habitats into smaller, more isolated parcels.

Fluctuation in natural populations

Variation in the density of a population depends on the magnitude of fluctuation in its environment and on the inherent stability of the population. After domestic sheep became established on the island of Tasmania, their population varied irregularly between 1,230,000 and 2,250,000—less than a factor of two—over nearly a century (Figure 16.1). Much of this variation was related to changes in grazing practices, markets for wool and meat, and pasture management, which could be considered factors in the environment of the sheep industry, if not in that of the sheep themselves.

In sharp contrast, populations of small, short-lived organisms may fluctuate wildly over many orders of magnitude within short periods. Populations of the green algae and diatoms that make up the phytoplankton may soar and crash over periods of a few days or weeks (Figure 16.2); these rapid fluctuations overlay changes with longer periods that occur on, for example, a seasonal basis. Sheep and algae differ in their degree of sensitivity to environmental change and in the response times of their populations. Because sheep are larger, they have a greater capacity for homeostasis and better resist the physiological effects of environmental change. Furthermore, because sheep live for several years, the population at any one time

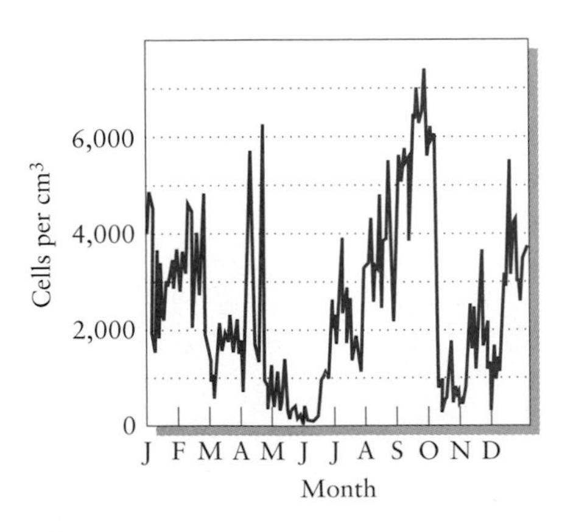

Figure 16.2 Variation in the density of phytoplankton in samples of water taken from Lake Erie during 1962. After C. C. Davis, *Limnol. Oceanogr.* 9:275–283 (1964).

includes individuals born over a long period; this tends to even out effects on population size of short-term fluctuations in birth rate. The lives of single-celled algal cells span only a few days, so populations turn over rapidly and bear the full impact of a capricious environment.

Populations of similar species living in the same place often respond to different environmental factors. For example, the densities of four species of moths, whose larvae all feed on pine needles, were found to fluctuate more or less independently in a pine forest in Germany (Figure 16.3). The populations varied over three to five orders of magnitude (1,000-fold to 100,000-fold change) with irregular periods of a few years. Furthermore, the highs and lows of the four populations did not coincide closely, suggesting that even though each species fed on the same trees, their populations were governed independently by different factors.

Among the most striking population phenomena found in nature are the regular cycles of abundance of certain mammals and birds at high latitudes. For example, regular trapping of small mammals over a 26-year period in northern Finland revealed six peaks of abundance separated by intervals of 4 or 5 years (Figure 16.4). This regularity is distinctly nonrandom, and the peaks and troughs exhibited by the most abundant species, the vole *Clethrionomys rufocanus,* are roughly paralleled by the less common species.

Population variation and age structure

Temporal variation in population dynamics often leaves its mark on the age structure of a population—that is, the relative frequencies of individuals of each age. As we have seen in Chapter 15, a changing age structure can affect the rate of population growth. The sizes of age classes also pro-

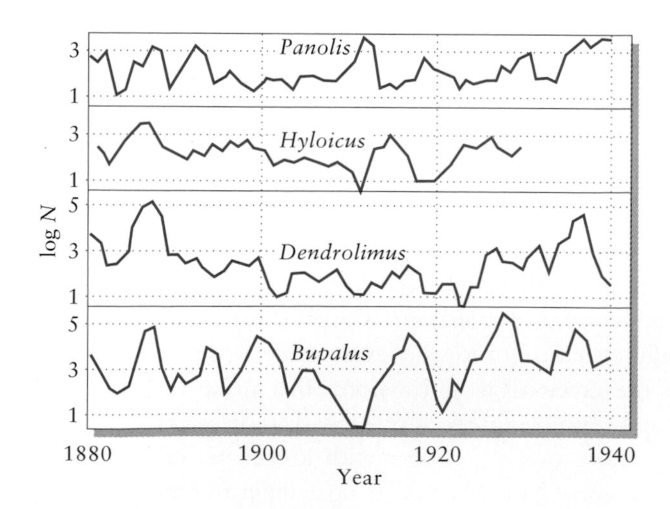

Figure 16.3 Fluctuations in numbers of pupae of four species of moths (hibernating larvae in the case of *Dendrolimus*) in a managed pine forest in Germany over 60 consecutive midwinter counts. After G. C. Varley, *J. Anim. Ecol.* 18:117–122 (1949).

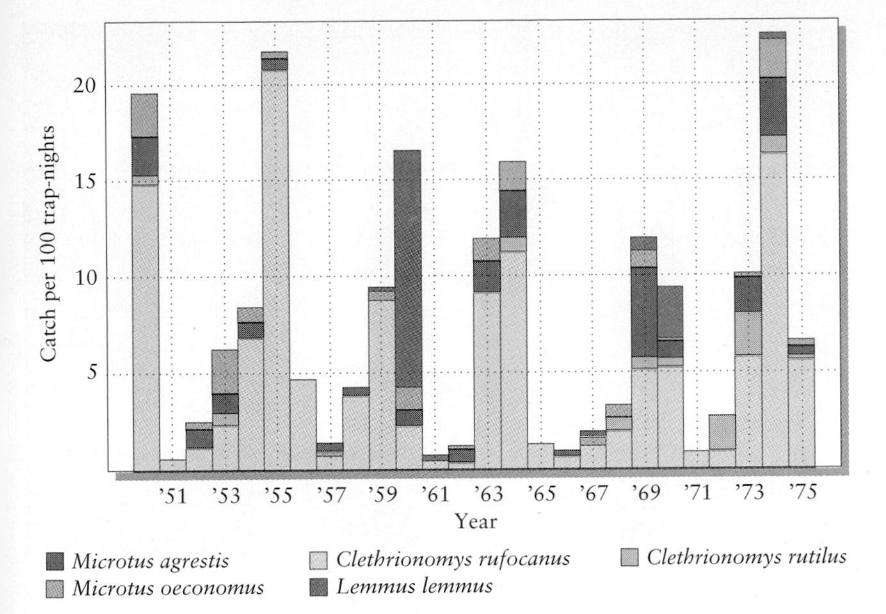

Figure 16.4 Numbers of small mammals trapped in an area of northern Finland from 1950 to 1975. The population cycles are dominated by variations in the numbers of the most common species, *Clethrionomys rufocanus,* but other species tended to reach peak abundances more or less in synchrony with it. From R. Brewer, *The Science of Ecology,* 2d ed., Saunders, New York (1994), after S. Lahti, J. Tast, and H. Uotila, *Luonnon Tutkija* 80:97–107 (1976).

vide a history of population change in the past. For example, the age composition of samples from the Lake Erie commercial whitefish catch for the years 1945–1951 shows that during 1947, 1948, and 1949, most of the individuals caught belonged to the 1944 year class (Figure 16.5). Biologists estimated the ages of fish from growth rings on their scales; their data showed that 1944 was an excellent year for spawning and recruitment, particularly compared with several years that followed.

Variation in annual recruitment of individuals into a population is also evident in the age structure of stands of trees. Age may be estimated by counting the growth rings in the woody tissue of the trunk of a tree, one ring being added each year under normal circumstances (Figure 16.6). The pattern in a virgin stand of timber surveyed near Hearts Content, Pennsylvania, in 1928 shows that individuals of most species were recruited sporadically over the nearly 400-year span of the record (Figure 16.7). Many white pines became established between 1650 and 1710, undoubtedly following a major disturbance, possibly associated with the serious drought and fire year of 1644. Fire can open a forest enough to allow the establishment of white pine seedlings, which do not tolerate deep shade. In contrast, beech—a species whose seedlings can grow under the canopy of a closed forest—exhibited a relatively even age distribution.

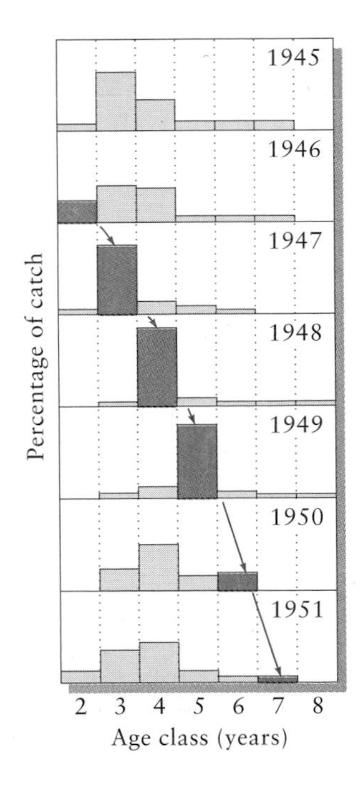

Figure 16.5 Age composition of samples from the commercial whitefish catch from Lake Erie between 1945 and 1951. Fish spawned in 1944 are highlighted by colored bars. From G. H. Lawler, *J. Fish. Res. Bd. Can.* 22:1197–1227 (1965).

Key-factor analysis

Recognizing that population fluctuations result from the impact of environmental factors on birth and death rates, biologists began to examine the effects of each factor separately to determine which exert the greatest influence on populations—that is, which are **key factors**—and whether they act in a density-dependent fashion. Forest entomologist R. F. Morris explained the approach as follows:

Figure 16.6 Cross section of the trunk of a Monterey pine, showing annual rings formed by winter (dark) and summer (light) growth.

> *A preliminary examination of rather extensive life-table data for the spruce budworm . . . suggested that the factors affecting this species in any one place are of two types—those that cause a relatively constant mortality from year to year and contribute little to population variation, and those that cause a variable, though perhaps much smaller, mortality and appear to be largely responsible for the observed changes in population. . . . A factor of the latter type will here be called a "key factor," meaning simply that changes in population density from generation to generation are closely related to the degree of mortality caused by this factor, which therefore has predictive value. (Ecology 40:580–588, 1959)*

Morris reasoned that change in a population from time t to time $t + 1$ (λ) depends on the fecundity and survival of individuals alive at time t. Furthermore, overall survival includes the separate probabilities of survival through each stage of the life cycle (egg, various larval stages, pupa,

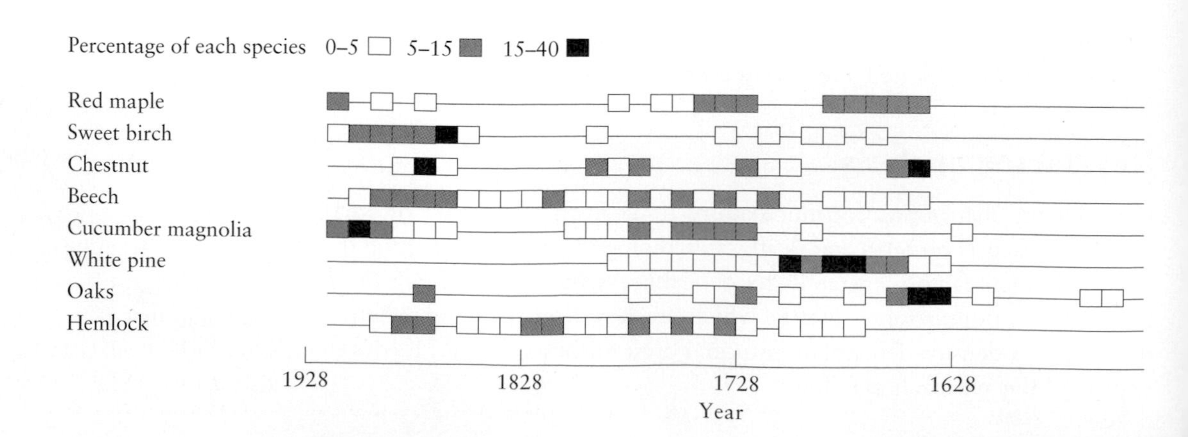

Figure 16.7 Age distribution of forest trees near Hearts Content, Pennsylvania, in 1928. After A. F. Hough and R. D. Forbes, *Ecol. Monogr.* 13:299–320 (1943).

and adult). Each of these stages suffers from its own mortality factors. Accordingly, Morris represented population changes by the simple model

$$\lambda = FS_1S_2S_3 \ldots$$

where F represents fecundity and S_i survival through each stage i. In logarithmic form,

$$r = \log F + \log S_1 + \log S_2 + \log S_3 + \ldots$$

(remember, $r = \log_e\lambda$). Morris then determined statistically whether variation in r followed upon variation in one or a few of the values of F and S_i.

By way of example, Morris applied his analysis to data for the black-headed budworm *(Acleris variana)* (Figure 16.8). The budworm belongs to the moth family Tortricidae (leaf-rollers), is native to eastern Canada, and is a major defoliator of fir trees in New Brunswick. One generation of adults appears each year; individuals overwinter in the egg stage. Larvae are at risk of being parasitized by wasps and flies. Twelve years of observations on densities of budworm populations and proportions of larvae parasitized revealed a fluctuation in numbers of over two orders of magnitude with a period of about 9 years. Survival of larvae (S_1) followed a similar course, being high during population buildup and low during decline. In this example, the exponential growth rate of the population ($r = \log_e\lambda$) was strongly correlated with $\log_e S_1$. From this analysis, Morris concluded that larval parasitism is a key factor in population processes of the black-headed budworm.

In other studies of Canadian agricultural and forest insect pests, the critical stage at which key factors acted, as well as the nature of the key factors themselves, varied from species to species. In only one, the Colorado potato beetle, was food supply a key factor. Weather was the most important factor in three of the species, and disease, parasitism, and emigration from local populations ranked highly in the rest. In nine of twelve studies, the key factor exerted a density-dependent influence; that is, it caused survival to decrease as the size of the population increased.

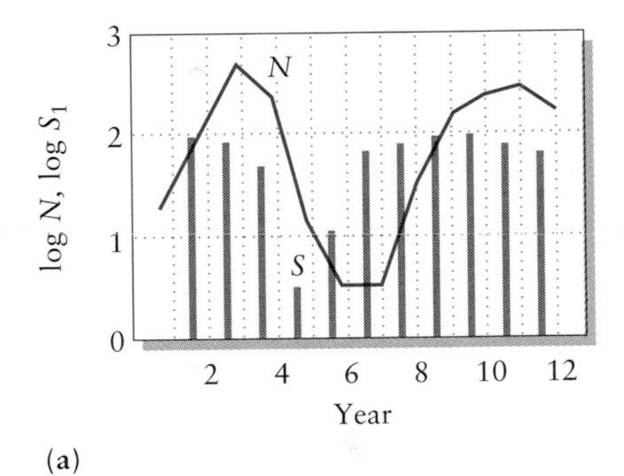

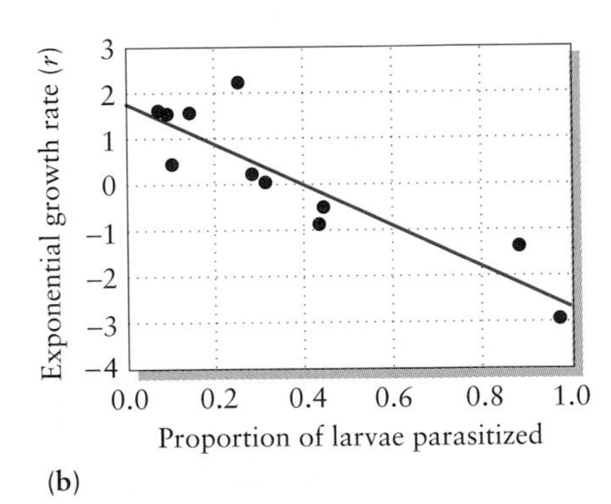

Figure 16.8 (a) Population size (N, colored line) and larval survival (S_1, vertical bars) for the black-headed budworm. (b) Relationship between exponential growth rate (r) and number of larvae surviving parasitism (S_1). After data in R. F. Morris, *Ecology* 40:580–588 (1959).

Tracking of environmental change

Fluctuations in conditions and resources continually increase and decrease the carrying capacity of the environment of each population. How a population responds to such changes through density-dependent effects depends on the intrinsic capacity of the population to increase in size, or r_m. The faster the potential rate of growth or decline of the population—that is, the higher the fecundity and the shorter the life span of individuals—the greater is its capacity to track change in its environment. Theoretical studies indicate that, as a general rule, populations with r much greater than 1 will track their environments closely, quickly responding to changes in carrying capacity. Populations with r much less than 1, such as the Tasmanian sheep, tend to be sluggish and unresponsive, their numbers

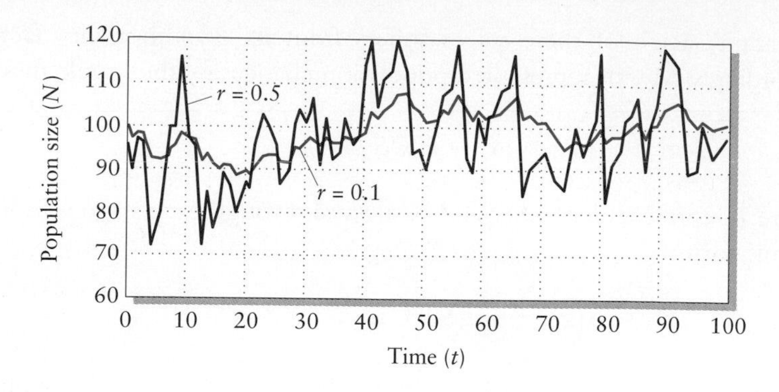

Figure 16.9 Population growth according to a logistic equation with random variation in carrying capacity. A population with a higher growth rate ($r = 0.5$) tracks fluctuations in carrying capacity more closely than a population with a lower growth rate ($r = 0.1$). From N. J. Gotelli, *A Primer of Ecology,* Sinauer Associates, Sunderland, Mass. (1995).

ironing out short-term variations in their environments. Differences in the responses of high-r and low-r populations can be demonstrated by simulations using logistic population growth equations with randomly varying carrying capacities (Figure 16.9).

Population cycles and intrinsic demographic processes

Except for factors associated with daily, lunar (tidal), and seasonal cycles, environmental fluctuations tend to be irregular rather than periodic. Historical records reveal that years of abundant rain or drought, extreme heat or cold, and natural disasters such as fires and hurricanes occur irregularly, perhaps even at random. Biological responses to these factors are similarly aperiodic. For example, widths of growth rings of trees vary in direct proportion to temperature and rainfall; the frequency distribution of numbers of years between peaks in ring width cannot be distinguished from a random series (Figure 16.10).

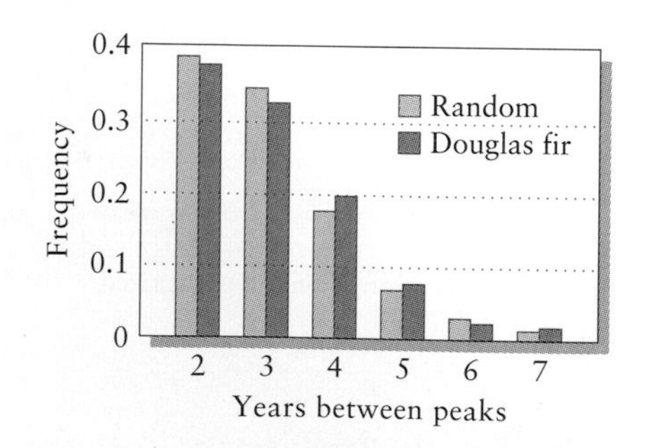

Figure 16.10 Frequency distributions of intervals between peaks in widths of growth rings of Douglas fir and a series of random numbers. After L. C. Cole, *J. Wildl. Mgmt.* 15:233–252 (1951).

The sizes of many populations do, however, change with periodic frequency (for example, see *Clethrionomys* in Figure 16.4). For many years, ecologists believed that such **cycles** must be caused by environmental factors that exhibit similar periodic variation. A regular 11-year cycle in sunspot numbers was frequently mentioned, but the sunspot cycle never matched population cycles well, and no one could see a direct connection between the two.

With the development of population models in the 1920s and 1930s, it became evident that because of their inherent dynamic properties, populations subjected to even minor, random environmental fluctuations could be caused to oscillate. Such cycling can result from **time delays** in the response of birth and death rates to changes in the environment, as we shall see below. Just as momentum imparted to a pendulum by the acceleration of gravity carries it past the equilibrium point and causes it to swing back and forth periodically, "momentum" imparted to a population by high birth rates at low density or high death rates at high density carries the population past its equilibrium when demographic responses have a time lag.

Time delays that cause populations to oscillate are inherent in models based on discrete generations. The discrete-time logistic equation, for example, has the form

$$N_{t+1} = N_t + rN_t\left[1 - \frac{N_t}{K}\right].$$

According to this model, populations respond by discrete increments from one time (t) to the next ($t + 1$) and therefore cannot continuously readjust their growth rates as population size approaches equilibrium. This can cause a population to overshoot its equilibrium, first in one direction and then in the other, as population size N draws closer to carrying capacity K. When the exponential growth rate r is less than 1, the increase in a population between t and $t + 1$ will be less than the difference between population size and equilibrium level K. Therefore, the population will approach K directly, as shown in Figure 16.11. When r exceeds 1 but is less than 2, a population will overshoot its equilibrium but will nonetheless end up closer to the equilibrium than before. Thus the population will **oscillate** back and forth across the equilibrium value, getting closer with each generation. This behavior is called **damped oscillation.** When r exceeds 2, the population ends up farther from the equilibrium each generation, and oscillations increase. With increasing r, these oscillations take on very complex, eventually unpredictable forms referred to as **chaos.** Short of this point, however, populations may settle into stable oscillations called **limit cycles,** in which numbers bounce back and forth between high and low values.

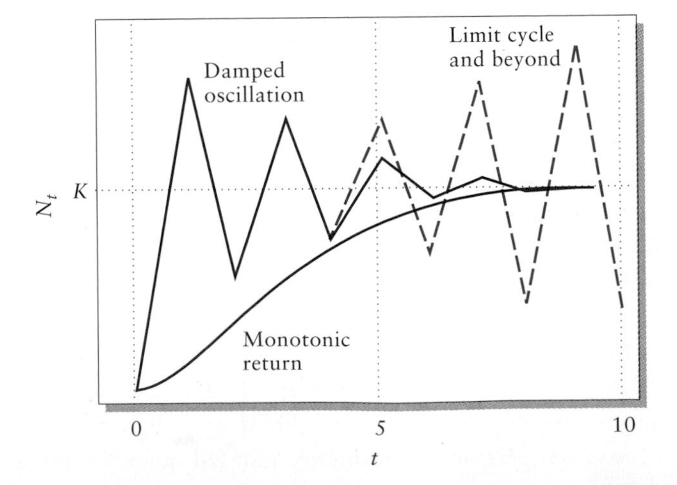

Figure 16.11 Approach to equilibrium according to a discrete logistic process when $r < 1$ (monotonic return), $1 < r < 2$ (damped oscillation), and $r > 2$ (limit cycle).

In reality, few populations have such a discrete structure as to make these models realistic, and such complicating factors as age structure and variations in r and K alter the behavior of such models considerably. Time delays nonetheless contribute importantly to the dynamics of populations. Discrete-time models have an intrinsic time delay of one time unit in population response. In continuous-time models, which have no intrinsic time delay, time lags in population processes, resulting from the developmental period that separates reproductive episodes between generations, can create cyclic population behavior.

Time delays and oscillations in continuous-time models

Oscillations are produced in continuous-time models when the response of population growth to density is time-delayed; in other words, oscillations occur when the effect of density dependence reflects the density of a population τ time units in the past (τ is the lowercase Greek letter tau). Modified accordingly, the logistic equation becomes

$$\frac{dN}{dt} = rN(t)\left[1 - \frac{N(t-\tau)}{K}\right].$$

This model produces damped oscillations in N as long as the product $r\tau$ is less than $\pi/2$ (about 1.6). Below $r\tau = e^{-1}$ (0.37), the population increases or decreases **monotonically**—without oscillation—to the equilibrium point. For $r\tau$ greater than $\pi/2$, the oscillations increase to form limit cycles, with the maximum population size reaching $N/K = e^{r\tau}$. For example, for $r\tau = 2$, oscillations increase in amplitude until the maximum value of N is e^2 (= 7.4) times K. The periods of these limit cycles, measured from peak to peak, increase from about 4τ to more than 5τ with increasing $r\tau$.

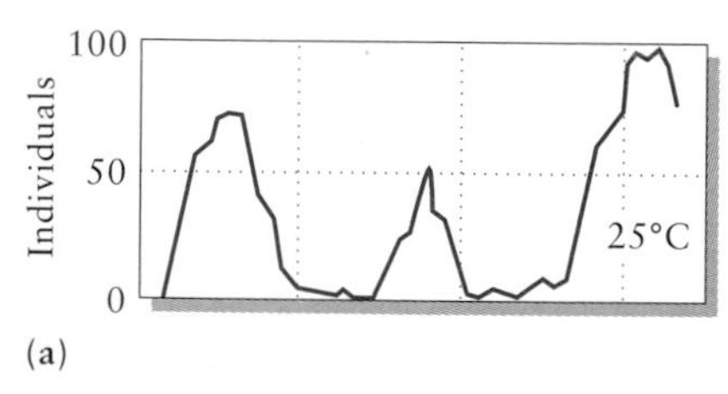

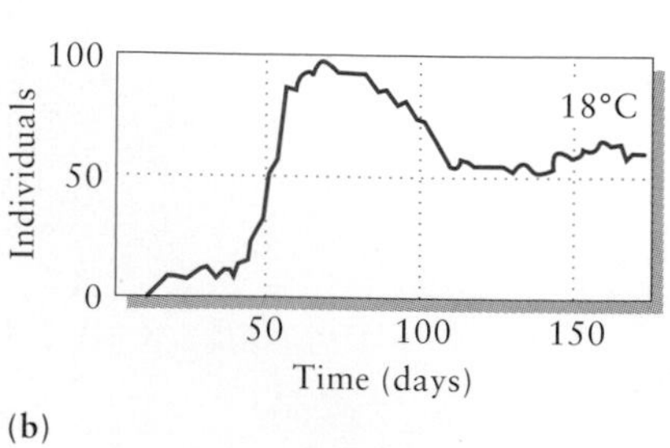

Figure 16.12 Growth of *Daphnia magna* populations (a) at 25°C and (b) at 18°C, showing the development of population cycles at the warmer temperature. After D. M. Pratt, *Biol. Bull.* 85:116–140 (1944).

Cycles have been observed in many laboratory cultures of single species. In one such study, populations of the water flea *Daphnia magna* exhibited marked oscillations when cultured at 25°C, but these oscillations disappeared at 18°C (Figure 16.12). The period at 25°C appeared to be just over 40 days for two cycles, suggesting a time delay in the density-dependent response of about 10 days. This is about the average age at which water fleas give birth at 25°C. The time lag arose in the following manner: As population density increased, reproduction decreased, falling nearly to zero when the population exceeded 50 individuals. Survival was less sensitive to density even at the highest densities, and adults lived at least 10 days. Crowding at the peak of the cycle prevented births. Then, when the population fell to densities low enough to permit reproduction, it contained only senescent, nonreproducing individuals, and thus the population continued to decline. The beginning of a new cycle awaited the accumulation of young, fecund individuals. The length of the time delay was approximately the average adult life span at high densities.

At the lower temperature, reproductive rate fell quickly with increasing density, and life span increased greatly over that seen at 25°C at all densities. Populations at the colder temperature apparently lacked a time delay because deaths were more evenly distributed over all ages, and some individuals gave birth even at high population densities. Consequently, generations overlapped more broadly. At the higher temperature, water fleas behaved according to a discrete-generation model with its built-in time delay of one generation. At the lower temperature, they behaved according to a continuous-generation model with little or no time delay.

Storage of lipid reserves by some species of water fleas reduces the sensitivity of mortality to crowding and therefore introduces a time delay into population processes. *Daphnia galeata,* a large species, stores energy in the form of lipid droplets (Figure 16.13) during periods of high food abundance (that is, low density). It can then live on these stored reserves when food supplies dwindle as a result of overgrazing at high population densities. Females also transfer lipids to each offspring through oil droplets in their eggs, thereby increasing the survival of young, prereproductive water fleas under poor feeding conditions. The smaller *Bosmina longirostris* stores a smaller amount of lipids, so starvation increases quickly in response to increases in population density. The consequences of this difference for population growth are predictable and reveal a connection between life history adaptations and population dynamics: in one study, *Daphnia* exhibited pronounced limit cycles with a period of 15 to 20 days, whereas *Bosmina* populations grew quickly to an equilibrium with perhaps a single strongly damped overshoot (Figure 16.14). The r value of *Daphnia* populations was about 0.3 days. With a cycle period of 15 to 20 days, τ must have been about 4 to 5 days, and therefore $r\tau$ was about 1.2–1.5. Because the value of $r\tau$ was somewhat less than $\pi/2$, the cycles in the *Daphnia* population should have damped out eventually.

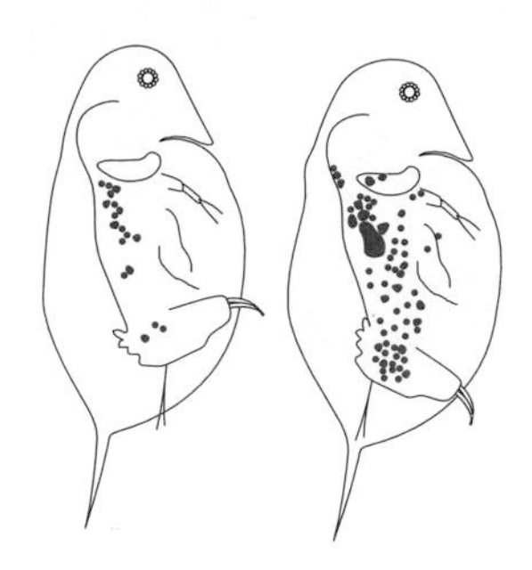

Figure 16.13 Energy stores in the form of oil droplets in individuals of *Daphnia galeata* grown under conditions of (left) low and (right) high food abundance. From C. E. Goulden and L. L. Hornig, *Proc. Natl. Acad. Sci. (USA)* 77:1716–1720 (1980).

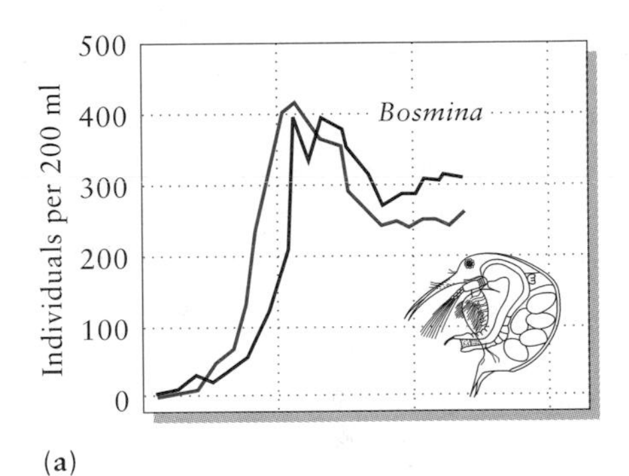

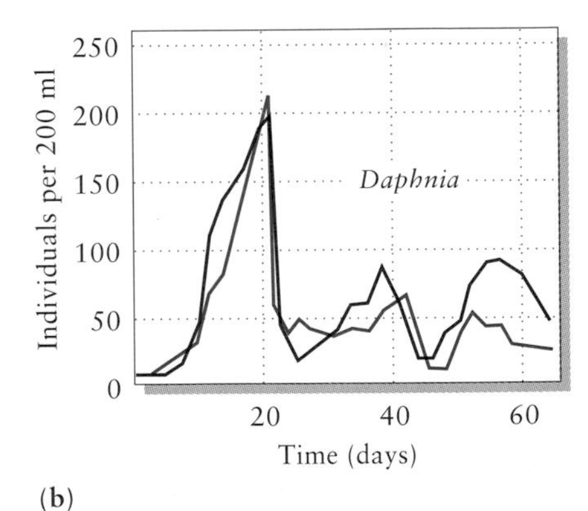

Figure 16.14 Densities of (a) two populations of the cladoceran *Bosmina longirostris* and (b) two populations of its larger relative *Daphnia galeata*. From C. E. Goulden, L. L. Henry, and A. J. Tessier, *Ecology* 63:1780–1789 (1982).

Time lags and oscillations in blowfly populations

The behavior of a population with respect to its equilibrium is sensitive to many aspects of life history that govern time delays in responses to density. Slight differences in culture conditions or intrinsic properties of species can tip the balance between a monotonic approach to equilibrium and a limit cycle. Australian entomologist A. J. Nicholson's experimental manipulations of time delay in laboratory cultures of the sheep blowfly *Lucilia cuprina* provide a dramatic demonstration of the relationship of time delays to population cycles. Under one set of culture conditions, Nicholson provided larvae with 50 g of ground liver per day, while giving adults unlimited food. The number of adults in the population cycled through a maximum of about 4,000 to a minimum of 0 (at which point all the individuals were either eggs or larvae) with a period of between 30 and 40 days (Figure 16.15).

In this experiment, the regular fluctuations of blowfly populations were caused by a time delay in the responses of fecundity and mortality to the density of adults in the cages. At high population densities, adults laid many eggs, resulting in strong larval competition for the limited food supply. None of the larvae that hatched from eggs laid during adult

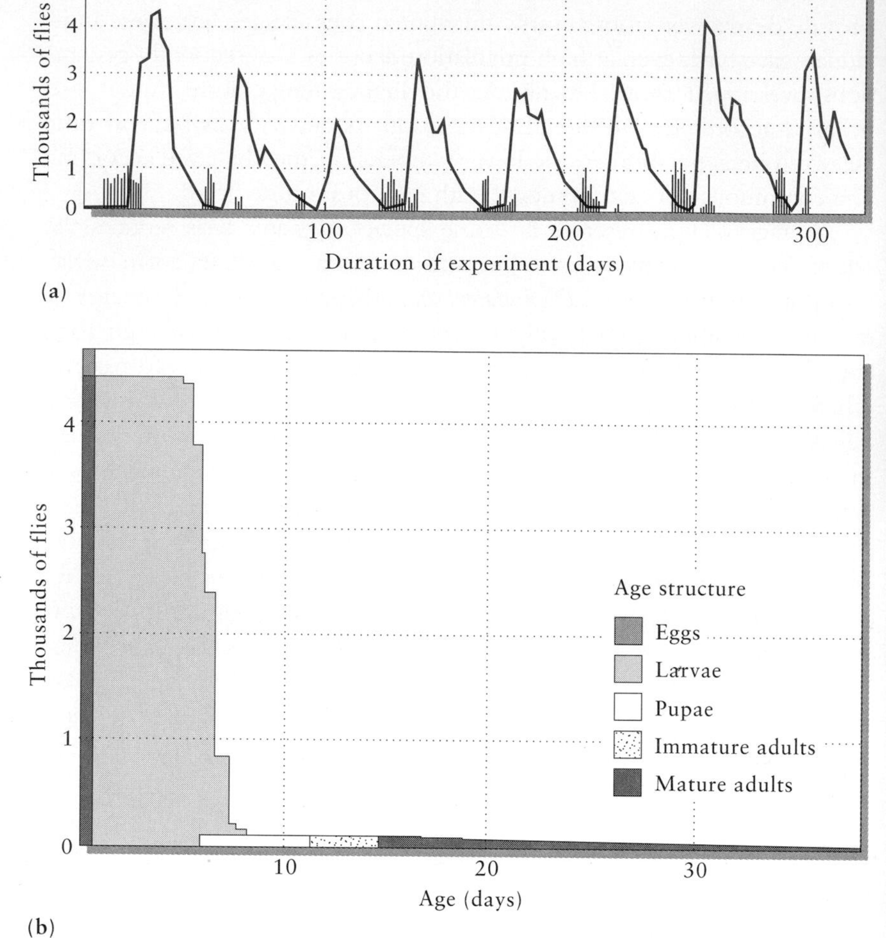

Figure 16.15 (a) Fluctuations in laboratory populations of the sheep blowfly *Lucilia cuprina.* Larvae were provided with 50 g of liver per day; adults were given unlimited supplies of liver and water. The continuous line represents the number of adult blowflies in the population cage. Vertical lines represent numbers of adults that eventually emerged from eggs laid on the days indicated by the lines. (b) Average age structure of the population. After A. J. Nicholson, *Cold Spring Harbor Symp. Quant. Biol.* 22:153–173 (1958).

population peaks survived, primarily because they did not grow large enough to pupate. Therefore, large adult populations gave rise to few adult progeny, and because adults lived less than 4 weeks, the population soon began to decline. Eventually, so few eggs were laid on any particular day that most of the larvae survived, and the size of the adult population began to increase again.

We may interpret the behavior of this population in terms of a time-delayed logistic process, which provides a good fit to the observed oscillations, with $r\tau = 2.1$. This value predicts the ratio of the maximum to the minimum population to be 84 and the cycle period to be 4.54τ. The experiment clearly reveals that density-dependent factors did not immediately affect the mortality rates of adults as the population increased, but were felt a week or so later when their progeny were larvae. Larval mortality did not express itself in the size of the adult population until those larvae emerged as adults about 2 weeks after eggs were laid. As in the *Daphnia* population maintained at a high temperature, crowding in the blowfly population created discrete, nonoverlapping generations with an inherent time delay equal to the larval development period, about 10 days.

The hypothesis that time delays cause population cycles can be tested directly by eliminating time delays in density-dependent responses—that is, by making the deleterious effects of resource depletion at high density felt immediately. Nicholson did this by adjusting the amount of food provided to his flies so that food availability limited adults as severely as it did larvae. Adult flies require protein to produce eggs. By restricting the amount of liver available to adults to 1 g per day, Nicholson curtailed egg production to a level determined by the availability of liver rather than by the number of adults in the population. Under these conditions, recruitment of new individuals into the population was determined at the egg-laying stage by the influence of food supply on per capita fecundity, and most of the larvae survived. As a result, fluctuations in the population subsided (Figure 16.16).

We have seen that responses of populations to density can be delayed by development time and by storage of nutrients, both of which put off deaths to a later point in the life cycle or to a later time. Density-dependent effects on fecundity can act with little delay when adults produce eggs quickly from resources accumulated over a short period. Populations controlled primarily by such factors should not exhibit marked oscillations.

Regardless of any time delay in its density-dependent response, a population at its equilibrium point will remain there until perturbed by some outside influence, whether a change in the equilibrium level (K) or a catastrophic change in population size (N). Once displaced from the equilibrium, some populations will move toward stable limit cycles, depending on the nature of the time delay and the response time. Others will return to the equilibrium directly or through damped oscillations. Cycles may be reinforced through interactions with other species—prey, predators, parasites, perhaps even competitors—with similar rates of response to population change, as we shall see in later chapters.

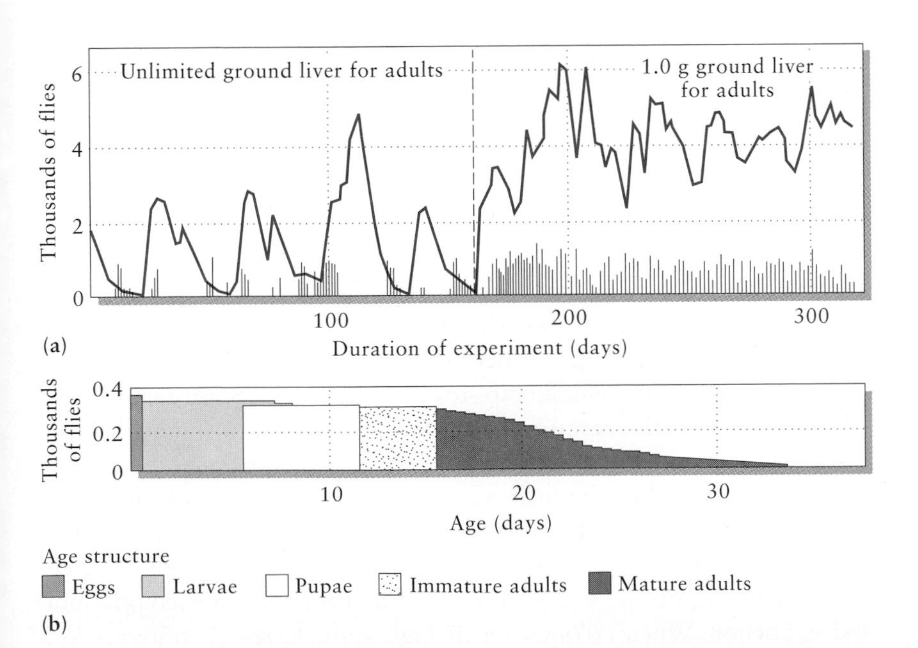

Figure 16.16 (a) Effect on the fluctuations of a sheep blowfly population of limiting the food supply available to adults. This experiment was similar to that depicted in Figure 16.15 in all other respects. (b) Average age structure of the population under the limited food conditions. After A. J. Nicholson, *Cold Spring Harbor Symp. Quant. Biol.* 22:153–173 (1958).

Metapopulations

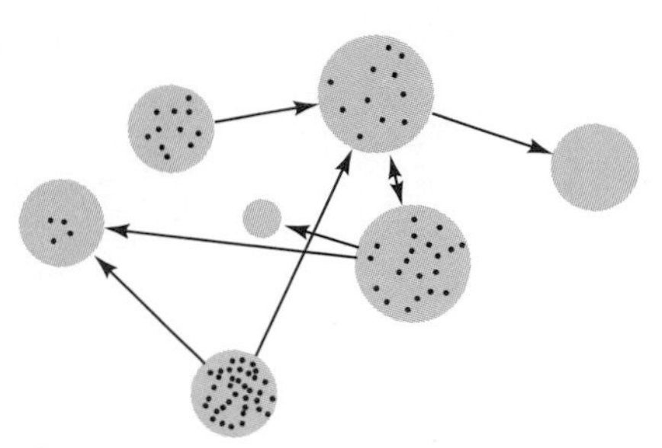

Figure 16.17 A metapopulation portrayed as a set of discrete subpopulations with partially independent local population dynamics. At any one time, some patches of suitable habitat may be occupied while others are not. Each dot represents an individual; arrows depict migration between patches.

Populations that are divided into discrete subpopulations linked by emigration and immigration are called **metapopulations** (Figure 16.17). One may think of metapopulations as "populations of populations," with each subpopulation having a probability of birth (colonization) and death (extinction). Two types of processes contribute to the dynamics of metapopulations. The first includes growth and regulation of populations within patches—processes that we have already discussed in detail. The second includes **migration** of individuals between patches, or **colonization** in the case of individuals that emigrate to an empty patch. Because subpopulations are typically much smaller than the metapopulation as a whole, local catastrophes and chance fluctuations in numbers of individuals have greater effects on their population dynamics. Indeed, the smaller the patch, the higher the probability of extinction of a subpopulation during a particular time interval. As a result, the dynamics of the total population depend critically on the rate of migration between patches, which results in recolonization of empty patches.

When individuals move frequently between subpopulations, local population fluctuations damp out, and changes in local population size mirror those of the larger population. Thus, a high rate of migration transforms metapopulation dynamics into the dynamics of a single large population. At the other extreme, when no individuals move between patches, the subpopulations in each patch behave independently. When these subpopulations are small, they have high probabilities of extinction, and the total population gradually goes extinct as individuals die out in each of the patches the population occupies. Intermediate levels of migration result in the colonization of some patches left unoccupied by extinction. Under such circumstances, the total population exists as a shifting mosaic of occupied and unoccupied patches. This mosaic has its own dynamics and equilibrium properties, which are illustrated simply in Box 16.1. The equilibrium fraction of occupied patches is sensitive to the relative rates of colonization (c) and extinction (e) of a patch. When the rate of colonization exceeds the rate of extinction, the fraction of occupied patches reaches an equilibrium between 0 and 1. When extinction exceeds colonization, the fraction of occupied patches declines to zero, and the entire metapopulation goes extinct. This pattern makes clear the importance of keeping habitat patches from becoming too isolated or, alternatively, maintaining dispersal corridors between patches in a managed landscape.

Immigration from large or increasing subpopulations can keep declining populations from dwindling to small numbers and eventual extinction. This is known as the **rescue effect,** and it can be incorporated into a metapopulation model by making the rate of extinction decrease as the fraction of occupied patches increases (that is, with more numerous sources of migrants, or rescuers). In one version of a metapopulation model with the rescue effect built in, patch occupancy either increases to 1 or decreases to 0, depending on the relative values of the parameters for colonization and extinction. When colonization is high enough, rescue amounts to a

BOX 16.1 **A simple model of metapopulation dynamics**

Consider a population subdivided into discrete patches. Without going into the details of the population dynamics within each patch, we assume that within a given time interval, each subpopulation has a probability of going extinct, which we shall refer to as e. Therefore, if p is the fraction of patches occupied by subpopulations, then subpopulations go extinct at the rate ep. The rate of colonization of empty patches depends on the fraction of patches that are empty $(1 - p)$ and the fraction of patches sending out potential colonists (p). Thus we may express the rate of colonization as a single constant c times the product $p(1 - p)$. Putting the extinction and colonization terms together yields the metapopulation dynamic

$$\frac{dp}{dt} = cp(1 - p) - ep$$

From this equation, it follows that the metapopulation attains an equilibrium ($dp/dt = 0$) when $c(1 - p) = e$. This equation may be rearranged to give

$$\hat{p} = 1 - \frac{e}{c}$$

the equilibrium proportion of unoccupied patches (indicated by the little hat over the p). The equilibrium is stable, because when p is below the equilibrium point, colonization exceeds extinction, and vice versa.

This simple model shows the critical importance of the relative rates of extinction and colonization (e/c). When $e = 0$, $p = 1$, and all patches are occupied. (This does not mean that the patches cease to have independent dynamics, only that they are large enough or otherwise stable enough not to suffer extinction.) When $e = c$, $p = 0$, and the metapopulation heads toward extinction. Intermediate values of e result in a shifting mosaic of occupied and unoccupied patches.

positive density dependence, in which the survival of subpopulations increases in the presence of more numerous neighboring subpopulations.

Stochastic processes

Small populations respond strongly to chance events, which can cause random fluctuations in population size. So far, we have considered models that assume large population sizes and no variation in the average values of birth and death rates due to chance. Such models, whose outcomes we can predict with certainty, are called **deterministic models.** In the real world, however, random variations can influence the course of population growth. The death of an individual is a chance event that has a certain probability of occurring during some interval. The number of deaths within a population, however, has a probability distribution. The average value of this distribution is equal to the number of individuals in the population times the probability of death. But the actual number of deaths observed in a particular population will vary above or below this value just by chance. Processes that are influenced by chance events in this manner are called **stochastic processes.**

Coin flipping provides a useful analogy. Suppose you repeatedly flip a set of 10 pennies. Although the probability of a head turning up on each toss is one-half, any one set of trials might turn up 6 heads or 3 heads. When the test is repeated frequently enough, the average of the outcomes settles down to 5 heads, but many trials turn up 4 or 6 heads, somewhat fewer yield 3 or 7 heads, and runs with all heads occur once in 1,024 trials, on average.

Turning to population models, let us suppose that adults successfully rear offspring with a probability of 0.5 per year. We would therefore expect a population of 10 individuals to produce 5 offspring on average, but the actual number is likely to vary from that value. What effect will this have on population growth? Consider a simple birth process (no deaths) in which a population grows exponentially according to $N(t) = N(0)e^{bt}$. Now suppose that the product bt equals 0.5. A deterministic model predicts a population at time t equal to 1.65 times the initial population ($e^{0.5}$). Thus, if the initial population were 500 individuals, the number after interval t would be 824. Over many small populations—say, $N(0) = 5$ individuals—the realized number of individuals at time t would average 8.24, but it would vary from as few as 5 (no births occurred) to as many as 20, just by chance (Figure 16.18).

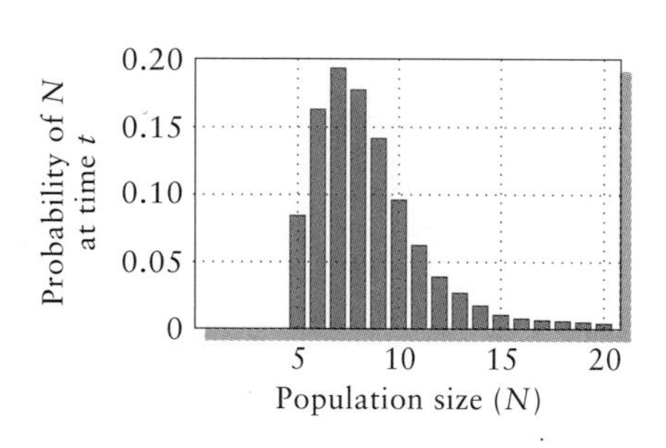

Figure 16.18 Probability distribution of numbers of individuals (N) at time t in a population undergoing a pure birth process with initial size $N(0) = 5$ and $bt = 0.5$. After E. C. Pielou, *Mathematical Ecology,* Wiley, New York (1977).

Stochastic extinction of small populations

Population change follows upon births and deaths. Birth rates and death rates depend on a variety of ecological factors, but whether a particular individual dies or successfully rears one or more progeny during an interval depends largely on chance. With an annual probability of death of one-half, for example, some individuals live and, on average, an equal number die. There exists a finite probability, however, that all the individuals in a population will die, just as 10 out of 10 coin tosses will come up tails with a small but finite probability.

Chance events exert their influence more forcefully in small populations than in large ones. This becomes clear when we consider that the probability of obtaining 5 tails in a row with 5 successive tosses of a coin is 1 in 32, compared with the smaller chance of 1 in 1,024 of obtaining twice as many tails in a row. If we visualize each individual in the population as a coin, and equate turning up tails with death, it is clear that a population of 5 individuals has a higher probability of extinction, just by chance, than a population of 10.

TABLE 16.1 Probability of extinction when birth rate = death rate = 0.5 per year, for populations of initial size *i* within period *t*

POPULATION SIZE	TIME			
	1	*10*	*100*	*1000*
1	0.33	0.83	0.98	0.998
10	10^{-4}	0.16	0.82	0.980
100	10^{-48}	10^{-7}	0.14	0.819
1,000	10^{-99}	10^{-79}	10^{-8}	0.135

Theorists have devoted considerable attention to the probability of extinction of populations. They have derived mathematical expressions relating probability of extinction at time t $[p_0(t)]$ to birth rate b, death rate d, and population size N. The simplest of these expressions represents the case in which b and d are equal—that is, births balance deaths, and the average change in population size is zero—for which

$$p_0(t) = \left[\frac{bt}{1 + bt}\right]^N.$$

Because the term within the brackets is always less than 1, the probability of extinction decreases with increasing population size and increases with larger b and d, indicating more rapid population turnover.

The relationship of the probability of extinction within time period t to population size N is shown in Table 16.1 for a population in which $b = d = 0.5$. These are reasonable values for adult death and recruitment in a population of terrestrial vertebrates. We see, for example, that for a population with 10 individuals, the probability of extinction is 0.16 within a 10-year period, 0.82 within a 100-year period, and virtually certain (0.98) within 1,000 years, provided that b and d do not vary with population size. Even for such a population with an initial size of 1,000, the probability of extinction is more than 10% within a millennium and becomes virtually certain (0.999) within a million years.

Most stochastic extinction models do not include density-dependent changes in birth and death rates; in those that do, extinction becomes exceedingly rare except in the smallest populations. Accordingly, we should consider whether density-independent stochastic models are relevant to natural populations. The answer is that they are, for several reasons. First, land use patterns and habitat fragmentation are such that many species now exist as collections of exceedingly small subpopulations, often so isolated that their eventual demise cannot be prevented by immigration from other populations. Second, changing environmental conditions are likely to reduce the productivity of populations trapped in isolated patches of habitat and bring them closer to the abyss of extinction. Third, when endangered species compete for resources with other species, the advantages that they would gain by reason of their low density (perhaps more food per individual) may be usurped by their competitors. In this case, small populations of one species may be influenced by high population densities in other species.

When populations dwindle to a small size, they become more and more susceptible to extinction, particularly on small islands where populations are restricted geographically and are rarely augmented by immigration. In fact, extinction occurs so often in some island groups that we can determine its probability from historical records. These data confirm theoretical predictions from models of stochastic extinction. For example, species lists compiled in 1917 and 1968 for birds on the Channel Islands, off the coast of southern California, reveal that during the 51-year interval between censuses, 7 of 10 species disappeared from Santa Barbara Island (3 km^2 in area), but only 6 of 36 species disappeared from the larger Santa Cruz Island (249 km^2). (Some of the species extinct on each island were replaced by new colonists of different species.) On an annual basis, these figures can be expressed as 0.1% and 1.7% of the avifauna per year, respectively, with extinction rate and island size inversely related. Comparable rates have been determined for two tropical islands: 0.2% per year on Karkar, an island of 368 km^2 located 16 km off the coast of New Guinea, and 0.23% per year on Mona, 26 km^2, located between Puerto Rico and Hispaniola in the Greater Antilles.

Disappearances of populations from isolated islands dramatize the role of extinction of subpopulations in local patches in the dynamics of metapopulations. The extinction rate, which influences the equilibrium number of patches occupied, depends on the number of individuals in a subpopulation and hence on the size of the patch it occupies. These considerations emphasize the interaction of spatial and temporal dynamics in population processes and the fact that we must understand the spatial structure of populations if we are to manage them intelligently.

Summary

1. Most populations fluctuate, either in response to variations in the environment or because they have oscillatory properties intrinsic to their dynamics. Owing to their well-developed mechanisms for homeostasis, species with larger body sizes and longer life spans tend to respond less rapidly to changes in their environments. A conspicuous exception occurs in the regular cycling of populations of small mammals in northern latitudes.

2. The age structure of a population often indicates temporal heterogeneity in the recruitment of individuals. For example, seedlings of certain species tend to become established in forests primarily following a major disturbance such as fire, drought, or storm. Thus, population processes may be sporadic rather than uniform over time.

3. Key-factor analysis is a method of identifying factors that cause fluctuations in population size based on the relationship of changes in population

size (r) to logarithms of survival rates (S) of individuals at risk from various factors. Field studies have revealed that key factors are often those that cause high and variable mortality from generation to generation. Many of these factors act in a density-dependent fashion.

4. Discrete-time models of populations with density dependence show that populations tend to oscillate when perturbed. For r between 0 and 1, population size (N) approaches equilibrium (K) monotonically. For r between 1 and 2, N undergoes damped oscillations and eventually settles down to K. When r exceeds 2, oscillations in N increase in amplitude until either a stable limit cycle is achieved or the population fluctuates irregularly (chaos).

5. Continuous-time models can produce cyclic population changes when density-dependent responses are time-delayed. Defining the time delay as τ, we find that such models exhibit monotonic damping when the product $r\tau$ lies between 0 and e^{-1} (0.37), damped oscillations when $r\tau$ lies between e^{-1} and $\pi/2$ (1.6), and limit cycles with a period of 4τ or more when $r\tau$ exceeds $\pi/2$.

6. Many laboratory populations of animals exhibit oscillations that arise from time delays in the responses of individuals to density. These time delays are related to the period of development from egg to adult and may be enhanced by storage of nutrients. In laboratory populations of sheep blowflies, A. J. Nicholson experimentally circumvented a time delay and was thereby able to eliminate cycles in numbers.

7. When populations are subdivided into discrete subpopulations occupying patches of suitable habitat, ecologists refer to them as metapopulations. The dynamics of metapopulations depend not only on birth and death processes within patches but also on migration of individuals between patches. When the rate of extinction of subpopulations is small compared with the rate of colonization of unoccupied patches, a metapopulation exists as a changing mosaic of an equilibrium number of occupied patches.

8. The dynamics of small populations, such as a subpopulation within an individual patch, depend to a large degree on chance events. Stochastic models demonstrate that the probability of extinction due to random fluctuation in population size is greater in smaller populations.

Suggested readings

Belovsky, G. 1987. Extinction models and mammalian persistence. In M. Soulé (ed.), *Viable Populations for Conservation,* pp. 35–57. Cambridge University Press, Cambridge and New York.

Berryman, A. A. 1996. What causes population cycles of forest lepidoptera? *Trends in Ecology and Evolution* 11:28–32.

Burkey, T. V. 1995. Extinction rates in archipelagoes: Implications for populations in fragmented habitats. *Conservation Biology* 9:527–541.

Daniel, C. J., and J. H. Myers. 1995. Climate and outbreaks of the forest tent caterpillar. *Ecography* 18:353–362.

Elliott, J. K. 1985. Population regulation for different life-stages of migratory trout *Salmo trutta* in a Lake District stream, 1966–1983. *Journal of Animal Ecology* 54:617–638.

Gilpin, M. E., and I. Hanski. 1991. *Metapopulation Dynamics: Empirical and Theoretical Investigations.* Cambridge University Press, Cambridge and New York.

Goulden, C. E., and L. L. Hornig. 1980. Population oscillations and energy reserves in planktonic Cladocera and their consequences to competition. *Proceedings of the National Academy of Sciences USA* 77:1716–1720.

Hassell, M. P., J. H. Lawton, and R. M. May. 1976. Patterns of dynamical behaviour in single-species populations. *Journal of Animal Ecology* 45:471–486.

Keith, L. B. 1990. Dynamics of snowshoe hare populations. *Current Mammalogy* 2:119–195.

Lindström, J., E. Ranta, V. Kaitala, and H. Lindén. 1995. The clockwork of Finnish tetraonid population dynamics. *Oikos* 74:185–194.

Myers, J. H. 1993. Population outbreaks in forest lepidoptera. *American Scientist* 81:240–251.

Taylor, A. D. 1990. Metapopulations, dispersal, and predator-prey dynamics: An overview. *Ecology* 71:429–433.

Villard, M.-A., G. Merriam, and B. A. Maurer. 1995. Dynamics in subdivided populations of Neotropical migratory birds in a fragmented temperate forest. *Ecology* 76:27–40.

CHAPTER 17

POPULATION GENETICS AND EVOLUTION

Each individual in a population is endowed with a unique genetic constitution, made up of a combination of genetic factors from its mother and its father. Genetic variability within a population has many consequences, the most important of which for the study of ecology is evolution by natural selection. The term **evolution** pertains to any change in the genetic makeup of a population, including the consequences of gene flow between populations and genetic drift. When genetic factors cause differences in fecundity and survival among individuals, evolutionary change comes about through **natural selection.** Individuals whose attributes enable them to achieve higher rates of reproduction leave more descendants, and therefore the genes responsible for these good attributes increase in a population. Consider the following example of evolutionary change in a California citrus pest.

Early in the twentieth century, certain species of scale insects were serious pests in citrus orchards in southern California. An effective means of controlling scale populations was to fumigate orchards with cyanide gas. After a few years of such treatment, however, fewer of the insects were killed by the gas, and before long the scale regained its pest status. Researchers determined that scale insects had evolved a genetically based resistance to cyanide poisoning. Furthermore, when they surveyed orchards in areas that had never been fumigated, they found that small numbers of individuals possessed an innate resistance to cyanide. Thus, despite their initial successes, fumigation programs in the end had favored reproduction by cyanide-resistant individuals, whose progeny then increased to epidemic proportions (Figure 17.1).

The citrus scale story illustrates the three main ingredients of evolution by natural selection: (1) variation among individuals, (2) inheritance of that variation (the genetic basis of evolution), and (3) differences in reproductive success, or **fitness,** related to that variation. Evolution by natural selection results from heritable variation in fitness within populations. In this chapter, we shall discuss the origin of genetic variation and its consequences for evolution.

Most evolutionary biologists believe that the diversification of living beings over the long history of life has been guided primarily by natural selection. It is important to understand, however, that natural selection is not an external force that urges organisms toward some preplanned goal, in the sense that humans "artificially" select cows to achieve a higher rate of milk production in their herds. Quite the opposite. Selection occurs—it is generated—as a consequence of differences in reproductive success between different phenotypes in a particular environment. The process that creates selection is ecological—namely, the interaction of individuals with their environment, including its physical conditions, food resources, predators, and so on. The cold winter wind doesn't care whether a bird is well insulated by its plumage. Whether a rabbit runs fast or not is irrelevant to evolution. All that matters is whether fast rabbits leave more offspring, perhaps because they are more likely to escape foxes. One presumes that a fox would prefer to chase slow rabbits, but, alas, by catching slow ones, it ends up encouraging faster ones to reproduce.

Three kinds of natural selection

Natural selection may be stabilizing, directional, or disruptive (Figure 17.2). In the case of **stabilizing selection,** individuals with intermediate, or average, phenotypes have higher reproductive success than those with

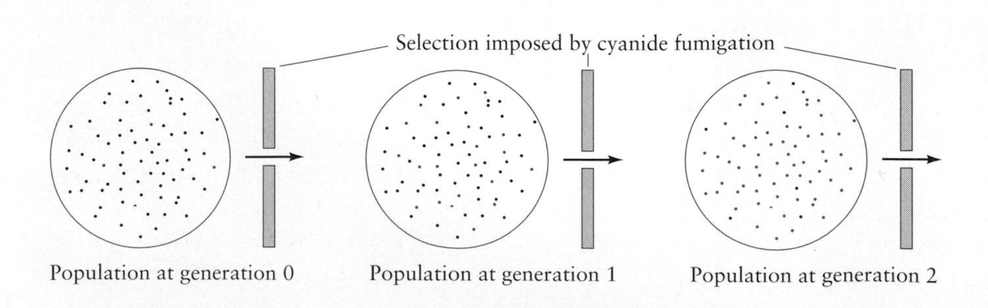

Figure 17.1 Diagrammatic representation of evolutionary change in a population in response to a change in a selective factor in the environment. Genetic factors that confer resistance to cyanide fumigation are present at low frequency in populations of scale insects that have never been exposed to cyanide, simply because of recurrent mutations. In the absence of cyanide, the trait may actually be mildly harmful. When populations are fumigated on a regular basis, however, the gene for cyanide resistance confers high fitness, and its frequency in populations rapidly rises until it becomes nearly fixed.

extreme phenotypes. Stabilizing selection tends to draw the distribution of phenotypes within a population toward an intermediate, optimum point, and counteracts the tendency of phenotypic variation to increase through mutation and gene flow between populations. Stabilizing selection performs housekeeping for a population, sweeping away harmful genetic variation. When the environment of a population is relatively unchanging, we would expect stabilizing selection to be the dominant mode, and little evolutionary change to take place.

Directional selection occurs when the most fit individuals have a more extreme phenotype than the average of the population. In this case, individuals whose phenotypes are to one side of the population average produce the most progeny, and the distribution of phenotypes in succeeding generations shifts toward a new optimum. When that new optimum is reached, selection becomes stabilizing. Evolutionary biologists have likened the high and low points of fitness that relate to different phenotypes to peaks and valleys of an **adaptive landscape.** Because selection favors individual phenotypes with higher fitness, evolution leads populations to "climb" adaptive peaks. If the world were static, all populations would eventually arrive upon one peak or another, and directional evolution would cease. Because the environment changes, however, adaptive peaks constantly shift their positions, like waves on the surface of a stormy sea, and generate new directional selection.

A population may occasionally find itself straddling an adaptive valley, across which extreme phenotypes have a higher fitness than do individuals having intermediate phenotypes. This situation leads to **disruptive selection,** which tends to increase genetic and phenotypic variation within a population and, in the extreme, to create a bimodal distribution of phenotypes. Disruptive selection is thought to be uncommon. It might occur, for example, when individuals can specialize on one of a small number of food sources that differ according to size or some other attribute. Disruptive selection might also ensue when interactions among individuals create alternatives to the prevalent life history; for example, territorial behavior by large males might place a premium on small males that sneak copulations with females. Few such cases have come to light, however.

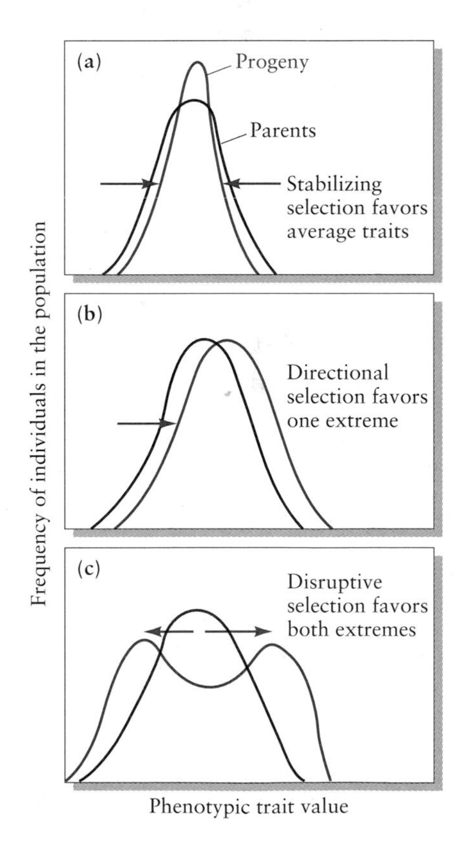

Figure 17.2 Diagrammatic representation of three forms of selection: (a) stabilizing, (b) directional, and (c) disruptive. Arrows indicate the phenotypic responses of the populations to each type of selection when phenotypic variation has a genetic basis.

The origin of genetic variation

Regardless of the form selection takes, it cannot cause evolutionary change unless there is genetic variation within a population. Genetic information is contained in the molecule **deoxyribonucleic acid,** or **DNA** for short, and genetic variation is caused by changes in the DNA molecule. The structure of DNA is an unbranched chain of subunits held together by sugar-phosphate links. DNA has four kinds of subunits, which are called **nucleotides**: adenine, thymine, cytosine, and guanine, whose acronyms are their first letters: A, T, C, and G. Thus, a stretch of DNA may be represented by the order of its subunits; for example, ATGGCATTAACGT. Genetic information is encoded in the particular order of the different nucleotides, just as the order of letters in a word conveys information.

A DNA strand serves as a template from which a cell manufactures proteins. Proteins themselves also are unbranched chains of subunits, composed of up to 20 different amino acids. Each amino acid is encoded by one or more unique sequences of three subunits of the DNA molecule. These coding triplets are referred to as **codons.** For example, the DNA sequence AAA (adenine-adenine-adenine) specifies the amino acid phenylalanine; GAG specifies leucine, CTT specifies glutamic acid, and so on. Because four different subunits taken three at a time yield 64 (4^3) different sequences, the **genetic code** contains considerable redundancy. In other words, because DNA must encode only 20 amino acids, and 64 different codons are possible, several codons may specify a single amino acid. For example, at the extreme, leucine is encoded by the sequences AAT, AAC, GAA, GAG, GAT, and GAC.

A change (substitution) in one of the subunits in a DNA codon may change the amino acid that it specifies. For example, consider the following sequence of subunits in a DNA molecule and the corresponding amino acids:

Position:	1	4	7	10	13	16
DNA:	GAA	TGG	CGA	GAA	ATA	GGG
Amino acid:	Leucine	Serine	Alanine	Leucine	Tyrosine	Proline

If the guanine subunit occupying the eighth position were changed to thymine, the third codon would be altered from CGA to CTA, and it would now encode the amino acid aspartine instead of alanine. Such changes do occur, and geneticists call them **mutations.** Because of the redundancy in the genetic code, changes in some nucleotides do not change the amino acid specified by a codon and therefore are not expressed in the phenotype. Such changes are often referred to as **silent mutations,** because they are not "heard from," or **neutral mutations,** because they have no consequence for fitness. Other mutations may result in the substitution of one amino acid for another in a protein chain. The new protein produced by a mutant gene may or may not have properties different from those of the original protein, and any altered properties may be beneficial or, more likely, harmful to an individual. Processes that cause mutations are blind to the adaptive landscape within which the population wanders.

Sequences of DNA subunits are read in groups of three from specially encoded starting points, so the deletion of a single nucleotide, which sometimes results from unrepaired damage to the DNA molecule, offsets the triplets by one place from their original position and changes the nucleotide sequences of all the ensuing codons. For example, if guanine were deleted from the eighth position of the sequence presented above, the third triplet of the code would now be read as the seventh, ninth, and tenth nucleotides, resulting in the following amino acid sequence:

Position:	1	4	7	10	13	16
DNA:	GAA	TGG	CAG	AAA	TAG	GGT
Amino acid:	Leucine	Serine	Valine	Phenyl-alanine	Isoleu-cine	Proline

In general, deletions and additions of subunits have far more pervasive and harmful effects on proteins than do substitutions.

Many proteins and the genes from which they are produced have been fully sequenced so that we know how the order of amino acid subunits in a protein relates to the order of nucleotides in its gene coding region. In some cases it has been possible to see the full connection between genetic material, protein, organism function, and reproductive success. The hemoglobin molecule is one of these. Hemoglobin consists of four protein chains; the predominant type of hemoglobin in adult humans has two kinds of chains, referred to as alpha and beta, each encoded by a different gene. The disease sickle-cell anemia is caused by a mutation in the gene for the beta chain—the seemingly trifling substitution of adenine for thymine in the seventeenth nucleotide position of the gene. This particular change has drastic consequences, however. The amino acid glutamic acid (codon CTT) is replaced by valine (CAT) as the sixth amino acid in the sequence. This alters the structure of hemoglobin such that when the molecules release oxygen from the red cells in the bloodstream, they become stacked close together in long helices, which impairs their function and causes severe and debilitating anemia. The aberrant hemoglobin molecules give red blood cells a peculiar sicklelike shape—hence the name of the disease. Of course, other mutations have been responsible for the subtle changes in the oxygen-binding properties of hemoglobin that have accompanied the adaptation of animals to different oxygen environments and to more or less active lifestyles.

Mutations are caused by various kinds of errors in the sequences of nucleotides in DNA: substitutions, deletions, additions, and rearrangements of nucleotides. These mistakes are caused by random copying errors when genetic material replicates during cell division, by certain highly reactive chemical agents, or by ionizing radiation. For any particular nucleotide in a DNA sequence, the rate of mutation is extremely low, on the order of one in 100 million per generation. However, this low rate multiplied by the hundreds or thousands of nucleotides in a gene, and by the trillion or so nucleotides in such complex organisms as vertebrates, means that each individual is likely to sustain one or more mutations in some part of its genome. Many of these mutations do not express themselves because they occur in portions of the DNA molecule that do not encode proteins, because they create redundant codons that do not change the specified amino acids, or because the amino acid changes caused by the mutation have no physiological effect. Nonetheless, measured rates of expressed mutation average about 1 in 100,000 to one in a million per generation for each gene, with rates for particular genes being much higher.

Most mutations are harmful to organisms. The reason for this is simple: Over millions of years of evolution, natural selection weeds out most deleterious genes, leaving behind only those genes that suit organisms to their environments (stabilizing selection). Any new variant is more likely to disrupt the well-tuned interaction between an organism and its surroundings—push it off an adaptive peak, so to speak—than to improve this interaction. Mutations may be beneficial, however, when the environment changes and organisms are no longer so well adapted. Then, small changes in the genetic material inherited from one parent may sometimes shift the structure and functioning of the individual to better match it to its new environment (directional selection). Such changes in environments come about through variations in climate, introductions of new organisms, or genetic changes in predators, prey, and disease organisms that are already present.

Genotype and phenotype

Before we consider the details of how natural selection produces evolutionary change, we should quickly review some useful definitions. A **genotype** includes all the genetic characteristics that influence the structure and functioning of an organism, which themselves make up the **phenotype.** Thus, a genotype is a set of genetic instructions, and a phenotype is the rendering, or expression, of a genotype in the form of an organism. Of course, this rendering is also influenced by the environment. To put it another way, the genotype is to the phenotype as blueprints are to the structure of a building. In this analogy, the effects of environmental influences are like details in a blueprint that are left up to the discretion of the building contractor, which may hinge, for example, on unpredictable changes in the availability of certain construction materials.

Each gene encodes a particular protein, which may be used as part of an organism's structure or may function as an enzyme or hormone. Different nucleotide sequences of a particular gene are referred to as **alleles.** In many cases, alleles create perceptible and measurable differences in an organism's phenotype. For example, blue-eyed and brown-eyed humans have different alleles of a single gene. Many genetic disorders, such as sickle-cell anemia, Tay-Sachs disease, cystic fibrosis, and albinism, are caused by defective alleles of individual genes.

Every individual has two copies of each gene, one inherited from its mother and one from its father (some exceptions include sex-linked genes and organisms that reproduce without the sexual union of gametes). An individual that has two different forms (alleles) of a particular gene is said to be **heterozygous** for that gene. When both copies of a gene are the same, that individual is **homozygous.** When an individual is heterozygous, either the two different alleles may produce an intermediate phenotype, or one may mask the expression of the other. In the latter case, one allele is said to be **dominant** and the other **recessive.** When heterozygotes have an intermediate phenotype, the alleles are said to be **codominant.** Most harmful alleles are recessive because the defective function of their gene products may be masked by the normal gene product of the dominant allele.

Hardy–Weinberg equilibrium

Members of each generation of a population produce gametes from which the next generation is constituted following fertilization and formation of zygotes (Figure 17.3). It is an important principle of genetics that, in a large population with random mating, no selection, no mutation, and no migration between populations, the frequencies of alleles and genotypes remain constant from generation to generation. In other words, no evolutionary change occurs through the sexual process of reproduction itself. This principle is called the **Hardy–Weinberg law** after the two geneticists who independently described it in 1908. Hardy's and Weinberg's important discovery showed that changes in allele and genotype frequencies can result only from the action of additional forces on the gene pool of a population. Understanding the nature of these forces has occupied evolutionary biologists ever since.

When a population exists in Hardy–Weinberg equilibrium, the proportions of homozygotes and heterozygotes take on equilibrium values, which we can calculate from the proportions of each allele in a population. Suppose that a particular genetic locus A has two alleles, A_1 and A_2, which occur in proportions p and q ($p + q = 1$, and therefore $q = 1 - p$). At Hardy–Weinberg equilibrium, the three genotypes that can result will occur in the following proportions:

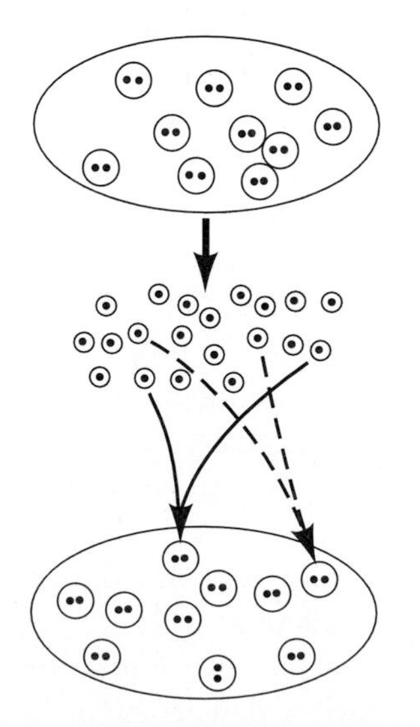

Figure 17.3 Diagram of the formation of each generation by sexual reproduction, showing how genotypes are reconstituted in descendant populations from gametes produced by the parental population. This diagram is particular to an organism, such as a clam or mussel, that sheds its gametes into the water, in which the gametes fuse at random to form zygotes. When fertilization is internal, each combination of genotypes in a mating pair will produce a unique combination of, or unique proportions of, offspring genotypes. Hardy–Weinberg frequencies nonetheless pertain.

Genotype:	A_1A_1	A_1A_2	A_2A_2
Frequency:	p^2	$2pq$	q^2

Notice that $p^2 + 2pq + q^2 = 1$. These proportions come from the probabilities that each type of zygote will be formed from random combination of any two gametes. To form an A_1A_1 homozygote, both gametes must have an A_1 allele. When the probability of one result (A_1) in one trial is p, the probability of two such results in two trials is simply the probability of each one multiplied together, or p^2. The proportion of heterozygotes is $2pq$ because a heterozygote will result from an A_1 egg and an A_2 sperm, with probability pq, and from an A_2 egg and an A_1 sperm, also with probability pq.

We can calculate the numeric values of proportions of genotypes under Hardy–Weinberg equilibrium as in the following example. Suppose one allele (A_1) occurs with a frequency of 0.7, and the other (A_2) with a frequency of 0.3 ($0.7 + 0.3 = 1$). Accordingly, 49% ($0.7^2 = 0.49$) of the genotypes in the population will be A_1 homozygotes, 42% ($2 \times 0.7 \times 0.3 = 0.42$) will be heterozygotes, and 9% ($0.3^2 = 0.09$) will be A_2 homozygotes. Any deviations from these genotype frequencies indicate the presence of selection, nonrandom mating, or other factors that influence the genetic makeup of the population.

Mutation, assortative mating, selection, migration, and genetic drift all cause deviations from Hardy–Weinberg equilibrium. Mutation is a weak force compared with selection, but it is the ultimate source of all genetic variability within populations. Genetic drift is most important in small populations, in which stochastic variation in birth and death rates can cause substantial changes in allele frequencies between generations just by chance. **Assortative mating** simply means that mates are chosen nonrandomly with respect to their genotypes: like mating with like is referred to as **positive assortative mating;** a tendency for mates to differ genetically is referred to as **negative assortative mating.** It should be obvious that negative assortment—for example, an A_1A_1 homozygote mating with an A_2A_2 homozygote—increases the proportion of heterozygotes in a population at the expense of homozygotes. Positive assortment, including inbreeding, has the opposite effect of reducing the proportion of heterozygotes. One consequence of positive assortative mating is an increase in the expression of recessive alleles in progeny, including rare harmful genes that are usually masked by dominant alleles in heterozygous form. When two A_1A_2 heterozygotes mate, or an A_1A_2 heterozygote selfs, for example, one-quarter of the progeny will have A_2A_2 genotypes.

Migration results in the transfer of genes from one population to another. When two populations have different allele frequencies, which may happen if they experience different selective forces, any mixture of the two populations will generally have a deficiency of heterozygotes. This is called the **Wahlund effect.** To create an extreme example, suppose we mix equal numbers of individuals from two populations that are each in Hardy–Weinberg equilibrium, but have different allele frequencies, say, $p = 0.2$ and $p = 0.8$. The genotype frequencies in the two populations before mixing are 0.04, 0.32, 0.64 and 0.64, 0.32, 0.04. Immediately after the populations are combined, the frequency of the A_1 allele is 0.5, and the proportions of the three genotypes are 0.34, 0.32, and 0.34, which clearly differ from Hardy–Weinberg frequencies. In general, a deficiency of heterozygotes in a population strongly indicates either positive assortative mating or a Wahlund effect due to population mixing. A closer look at the biology of the population is likely to help us distinguish between these possibilities. For example, one would not expect mussels, which shed their gametes into the ocean, to exercise strong assortative mating, but it would not be surprising to find that ocean currents had caused some mixing between populations.

Both stabilizing and directional selection tend to increase the proportion of heterozygotes in a population, but a more interesting consequence of directional selection is the change it produces in the frequencies of alleles, because this is the basis of long-term evolutionary change.

Natural selection and changes in allele frequencies

The reproductive success of a phenotype is a measure of its **evolutionary fitness**—that is to say, the rate at which it leaves descendants. In principle, fitness may be calculated from the life table of individuals sharing a phenotype and expressed as the exponential (r) or geometric rate of increase (λ). Models of evolutionary change by natural selection take a simpler approach that uses a relative measure of fitness. For example, suppose we say that the phenotype produced by genotype A_1A_1 has a fitness of 1. The phenotype produced by genotype A_2A_2 might have a higher fitness, by fraction s, giving a fitness of $1 + s$, or it might have a lower fitness, by fraction t, giving a fitness of $1 - t$. The same applies to the heterozygote phenotype. When we have specified the fitness of each genotype in a population, we can ask how much the frequency of each allele will change in the course of one generation of selection.

For each generation, the frequencies of the genotypes are multiplied by the fitnesses of the phenotypes they produce to obtain the relative numbers of their descendants in the next generation. Suppose that genotypes A_1A_1, A_1A_2, and A_2A_2 have the Hardy–Weinberg frequencies p^2, $2pq$, and q^2, and the A_1 allele is dominant, meaning that genotype A_1A_2 produces a phenotype that is indistinguishable from, and has the same fitness as, the A_1A_1 phenotype. Furthermore, let's say that the fitnesses of the A_1A_1 and A_2A_2 phenotypes are 1 and $1 - s$. The relative numbers of progeny of each of the genotypes are p_2, $2pq$, and $(1 - s)q_2$. We can now count up the relative numbers of A_1 and A_2 alleles in the progeny and calculate the change in frequency of the A_2 allele caused by selection against homozygous genotypes (Box 17.1). The result is a formula that gives the

BOX 17.1 Rate of change of allele frequency under selection

To illustrate the development and application of models in population genetics, we shall derive an equation to predict the change in frequency of a harmful, recessive allele. A population has two alleles at a particular genetic locus, A_1 and A_2. The initial frequencies of these alleles are p and q. A_1 is dominant over A_2, and the fitness of the homozygous genotype A_2A_2 is less, by factor s, than the fitness of the A_1A_1 genotype. We set up the following table to follow the reproductive success of each genotype and its contribution to the gene pool of the next generation.

	GENOTYPE		
	A_1A_1	A_1A_2	A_2A_2
Initial genotype frequency	p^2	$2pq$	q^2
Reproductive success (fitness)	1	1	$(1-s)$
Relative proportion of descendants	p^2	$2pq$	$(1-s)q^2$
Relative proportion of A_1 alleles in the descendant population	p^2	pq	
Relative proportion of A_2 alleles in the descendant population		pq	$(1-s)q^2$

Now, the proportion of A_2 alleles in the descendant population (q') is the ratio of A_2 alleles to the total, or

$$q' = \frac{(pq + [1-s]\, q^2)}{(p^2 + 2pq + [1-s]\, q^2)}$$

which may be simplified to

$$q' = \frac{q\,(1-sq)}{(1-sq^2)}$$

The change in allele frequency from one generation to the next, Δq, is $q' - q$, which, with a little algebra, can be rearranged to give

$$\Delta q = \frac{-sq^2(1-q)}{(1-sq^2)}$$

change in frequency of the A_2 allele as a function of its frequency in the parental generation (q) and the strength of selection acting upon it in the homozygous state (s), namely,

$$\Delta q = \frac{-sq^2(1 - q)}{(1 - sq^2)}.$$

When the recessive homozygote is lethal, that is, when $s = 1$, this equation can be simplified to

$$\Delta q = \frac{-q^2}{(1 + q)}.$$

When selection is very weak (s is small, perhaps less than 0.01, or 1%), the equation simplifies approximately to

$$\Delta q = -sq^2(1 - q).$$

These equations tell us several important things about the course of evolution. First, selection against the A_2A_2 genotype always causes a decrease in the frequency of the A_2 allele (Δq is always less than 0). Second, the rate of change in q depends on both the selective pressure on a population (fitness differential) and the frequency of the A_2 allele. For example, change in q is fastest when q is relatively large because a larger proportion of the A_2 alleles are exposed in homozygous form (Figure 17.4). Third, evolution stops ($\Delta q = 0$) only when q is equal to either 0 or 1, in which cases either the A_1 or the A_2 allele is fixed in the population and there is no longer any genetic variation for selection to act upon.

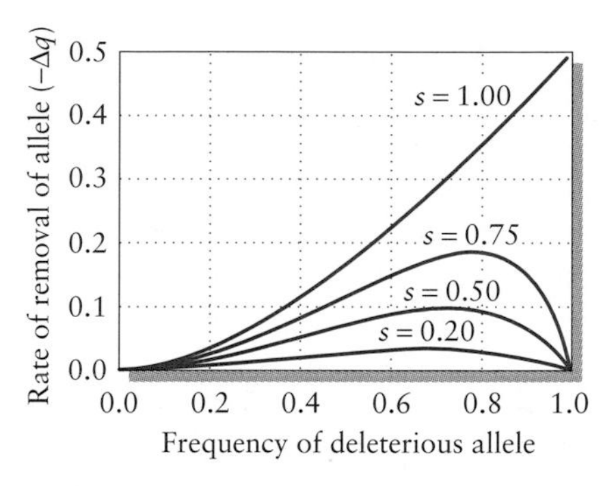

Figure 17.4 Rate of change in the frequency of a recessive, harmful allele as a function of allele frequency and strength of selection. Values were calculated from the equation in Box 17.1, and show that the rate of evolution is greatest when a recessive allele is common and therefore frequently exposed in homozygous phenotypes.

Maintenance of genetic variation in populations

Natural selection cannot produce evolutionary change without genetic variation, yet one consequence of most kinds of selection is the reduction of genetic variation. How does evolution continue under this circumstance? Does the availability of genetic variation ever limit rate of evolutionary change?

Every population is supplied with new genetic variation by mutation and migration. Mutation rates are difficult to estimate, and they vary widely among species and among genes. Rates of 1 in 10,000 to 1 in 100,000 gametes per generation have been observed in many genes in which mutations have readily apparent visible effects that can be scored easily, such as the color of kernels on an ear of corn or defects in the wing structure of fruit flies. Plants with chlorophyll deficiencies, which are lethal, appear by mutation at rates of almost 1 in 100 to 1 in 10,000, but any of a number of genes involved in producing chlorophyll may be responsible for such "phenotypic" mutations. In laboratory strains of the fruit fly *Drosophila,* mutations anywhere in the genome having lethal effects arise at a rate of about 2% per generation, and mutations with mildly detrimental effects arise at a rate of about one per fly. There is no question that the gene pool is actively churning out genetic variation.

Spatial and temporal variation in the environment tends to maintain this genetic variation within a population by favoring different alleles at different times and places. It is perhaps remarkable that populations contain so much genetic variability. About a third of the genes that encode enzymes involved in cellular metabolism show variation in most species surveyed, and fully 10% of these genes may be heterozygous in any given individual. At any particular time, most of this genetic variation either has no consequences for individuals—that is, it is neutral—or has bad effects when expressed. Thus, most genetic variation either produces no variation in fitness among individuals, and therefore no evolutionary change, or creates stabilizing selection that performs the housekeeping function of weeding harmful genetic variation out of populations. In the event that the environment changes, some of this genetic variation may take on positive survival value, and it then fuels the fires of evolution. But this is purely a hit or miss consequence of the randomness of mutation. It is almost certainly true that evolution could proceed faster if mutations were designer-made for positive change. Nonetheless, there seems to be enough genetic variation in most populations so that evolutionary change is a constant presence.

Fitness and evolution in natural populations

The story of cyanide resistance in scale insects mentioned above was one of the first documented cases of genetic change in a population in response to a change in selective factors in its environment. Similar cases of pesticide and herbicide resistance among agricultural pests and disease vectors, as well as the increasing resistance of bacteria to antibiotics, are further examples of how rapidly the gene pools of populations can respond to changes wrought by humankind in their environments. In each case, genetic variation that was present in the gene pool before the environmental change allowed the population to respond to the changed conditions. In most of these cases, the populations have very high rates of increase and pass through many generations each year. Thus, these populations are able to withstand powerful selective pressures that can bring about rapid evolutionary change.

From the standpoint of population growth, selection is equivalent to death. Indeed, one often refers to the relative decrease in number of descendants caused by selection as **selective deaths** or, because these "deaths" result from genetic factors, as **genetic deaths.** Because these deaths take away from the growth potential of a population, they are also referred to as the **genetic load** that a population must bear. In the case of selection against a deleterious recessive allele, the relative number of selective deaths is sq^2. Thus, the genetic load on the population increases in direct proportion to the fitness differential and the frequency of the selected genotype. When selection is weak and the frequency of a harmful allele is low, the genetic load on the population is not so high as to cause a population decline. Most populations produce an excess of progeny, and their numbers are trimmed by density-dependent factors. Thus, the primary effect of selection is to change allele frequencies, not population size.

There are two circumstances in which genetic load can be a major problem for a population. One is the situation in which the environment changes so drastically that most of the population suffers reduced fitness, so much so that the population growth rate becomes negative. If there is no genetic variation in the population, it will soon decline to extinction. If a small number of genetic variants resist the effects of the change (the small proportion of cyanide-resistant scale insects, for example), then the population may recover eventually owing to increases in these superior genotypes. The second situation can occur when population size is drastically reduced—by habitat fragmentation or when an island is colonized by a small number of immigrants, for example—and the level of inbreeding goes up. Inbreeding (positive assortative mating) tends to increase the proportion of homozygous genotypes in a population, which exposes harmful alleles in phenotypes. Because practically all individuals in a population have harmful recessive alleles of one or more genes, close inbreeding can create a very heavy genetic load, potentially heavy enough, along with stochastic variation, to drive a small population to extinction.

For the most part, slow changes in the environment do not create excessive genetic loads on populations, and if there is sufficient genetic variation, populations will respond through evolutionary change. One of the most striking cases of such evolution in action is that of industrial melanism in the peppered moth in England. Early in the nineteenth century, occasional dark (or **melanistic**) specimens of the common peppered moth *(Biston betularia)* were collected. Over the next 100 years, this dark form, referred to as *carbonaria,* became increasingly common in forests near heavily industrialized regions of England, which is why the phenomenon is often referred to as **industrial melanism.** In the absence of factories and other heavy industry, the light form of the moth still prevailed. This phenomenon aroused considerable interest among geneticists, who showed by cross-mating light and dark forms that melanism is an inherited trait determined by a single dominant gene. Because the melanistic trait is an inherited characteristic, its spread reflected genetic changes (evolution) in the population. Melanism is not unique to the peppered moth; dark forms have appeared in many other moths and in other insects.

Peppered moths inhabit dense woods and rest on tree trunks during the day. Where melanistic individuals had become common, the environment must somehow have been altered so as to give dark forms a survival advantage over light forms. It seemed reasonable to suppose that natural selection had led to the replacement of typical light individuals with *carbonaria* individuals. To test this hypothesis, the English biologist H. B. D. Kettlewell measured relative fitnesses of the two forms independently of the fact that the frequency of one had increased over that of the other.

To determine whether the *carbonaria* form had greater fitness than the typical peppered moth in areas where melanism occurred, Kettlewell chose the mark-recapture method. He marked adult moths of both forms with a dot of cellulose paint and then released them. The mark was placed on the underside of the wing so that it would not attract the attention of predators to a moth resting on a tree trunk. Kettlewell recaptured moths by attracting them to a mercury vapor lamp in the center of the woods or to caged virgin females at the edge of the woods. (Only males could be used in the study because females are attracted neither to lights nor to virgin females.)

In one experiment, Kettlewell marked and released 201 typicals and 601 melanics in a wooded area near industrial Birmingham. The results were as follows:

	TYPICALS	MELANICS
Number of moths released	201	601
Number of moths recaptured	34	205
Percentage recaptured	16	34

These figures indicated that more of the dark form survived over the course of the experiment. A similar experiment in a nonindustrial area revealed higher survival by the typical salt-and-pepper form of the moth.

The specific agent of selection was easily identified. Kettlewell reasoned that in industrial areas, pollution had darkened the trunks of trees so much that typical moths stood out against them and were readily found by predators. Any aberrant dark forms were better camouflaged against darkened tree trunks, and their coloration conferred survival value (Figure 17.5). Eventually, differential survival of dark and light forms would lead to changes in their relative frequency in a population. To test this idea, Kettlewell placed equal numbers of light and dark forms on tree trunks in polluted and unpolluted woods and watched them carefully at some distance from behind a blind. (A blind is a tentlike structure intended to conceal observers from their subjects; it is more often called a hide in England.) He quickly discovered that several species of birds regularly searched tree trunks for moths and other insects, and that these birds more readily found a moth that contrasted with its background than one that resembled the bark it clung to. Kettlewell tabulated the following instances of predation:

	INDIVIDUALS TAKEN BY BIRDS	
	Typicals	*Melanics*
Unpolluted woods	26	164
Polluted woods	43	15

These data were consistent with the results of the mark-recapture experiments. Together they clearly demonstrated the operation of natural selec-

Figure 17.5 Typical and melanistic forms of the peppered moth at rest on (a) a lichen-covered tree trunk in an unpolluted countryside and (b) a soot-covered tree trunk near Birmingham, England. From the experiments of H. B. D. Kettlewell.

tion, which over a long period resulted in genetic changes in populations of the peppered moth in polluted areas.

One of the most gratifying aspects of the peppered moth story is that, with the advent of smoke control programs and the return of forests to a cleaner state, frequencies of melanistic moths have decreased. In the area around the industrial center of Kirby in northwestern England, for example, the *carbonaria* form decreased from more than 90% of the population to about 30% over a period of 20 years (Figure 17.6).

Population genetics and rates of evolution

The evolutionary mechanics of selection and genetic responses are part of the science of **population genetics.** A primary task of population geneticists since the late 1920s has been the development of quantitative methods for predicting changes in gene frequencies in response to selection. Such predictions are possible through the application of models such as the one presented in Box 17.1 for selection upon a single gene with one allele dominant over the other. Equations that predict the change in allele frequency resulting from one generation of selection can be used to show how a population evolves over many generations of continued selection and to predict how rapidly a population can respond genetically to a change in its environment.

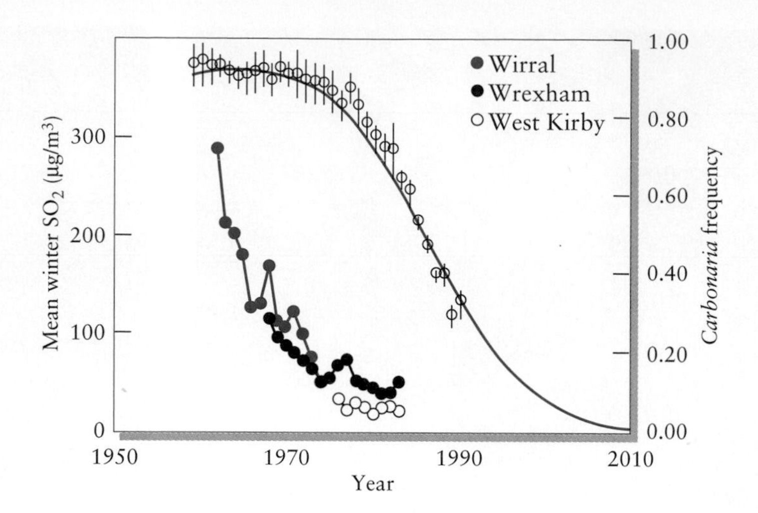

Figure 17.6 Changes in frequency of the melanistic *carbonaria* form of the peppered moth since the beginning of pollution control programs in England in the 1950s. The index to pollution is the winter level of sulfur dioxide, where 1 = 300 micrograms per cubic meter of air. Sulfur dioxide directly affects lichens growing on tree trunks, against which moths rest by day. The lag in evolutionary response to changes in air pollution levels reflects the time required for forests to return to a more natural (unpolluted) state, as well as a low initial frequency of the recessive allele for typical coloration. After C. A. Clarke, G. S. Mani, and G. Wynne, *Biol. J. Linn. Soc.* 26:189–199 (1985); G. S. Mani and M. E. N. Majerus, *Biol. J. Linn. Soc.* 48:157–165 (1993).

The time required for a dominant allele to replace a recessive allele depends on its initial frequency and, particularly, on the strength of selection. In the case of the replacement of the typical peppered moth by the *carbonaria* form in polluted woods of England, this substitution is known to have taken about a century. From the results of Kettlewell's experiments on the peppered moth, we can estimate that the fitness of the recessive homozygous genotype for typical coloration was only 47% that of the **carbonaria** genotype in polluted woods; hence the fitness differential, or strength of selection against the typical form, was $s = 0.53$. We can use this value in the equation

$$\Delta q = \frac{-sq^2(1-q)}{(1-sq^2)},$$

where q is the frequency of the allele for typical coloration, to follow the time course of increase in the *carbonaria* allele (Figure 17.7). Note that because $p = 1 - q$, $\Delta p = -\Delta q$. Because *carbonaria* is a dominant allele, it is exposed to selection even at low frequency, and the initial frequency of the allele has little effect on the rate of evolutionary change. According to a simulation of this process, the transition to a population that consists mostly of melanistic forms takes about 50 generations. When selection is weaker, however, the transition takes much longer. The peppered moth has 1 generation each year, so the prediction from population genetics appears to be consistent with the observed time course of the gene substitution, assuming an initial frequency before the Industrial Revolution of 0.1–1.0% and selection on the order of $s = 0.1–0.5$.

The equations of population genetics can be turned around and used to estimate fitness differentials from changes in gene frequency. For example, the rate of decrease in the frequency of the *carbonaria* allele since the introduction of pollution control programs in England is consistent with a 12% selective disadvantage ($s = 0.12$). Another example comes from records of trapped foxes kept by Moravian mission posts in Labrador for a hundred years (1834–1933). These records show that the proportion of silver foxes in the catch declined from about 0.15 to 0.05 during that time. Because silver coat is a recessive phenotype, the frequency of the genotype is equal to q^2, and the frequency of the silver allele (q) must have decreased from 0.39 to 0.22 over a period of 100 years. This much change would have required a fitness differential (s) of about 0.035, or 3.5% per year. It does not seem unreasonable that a greater demand for silver fox furs might have led trappers to cause an annual mortality of silver foxes 3% higher than that of the typical red form.

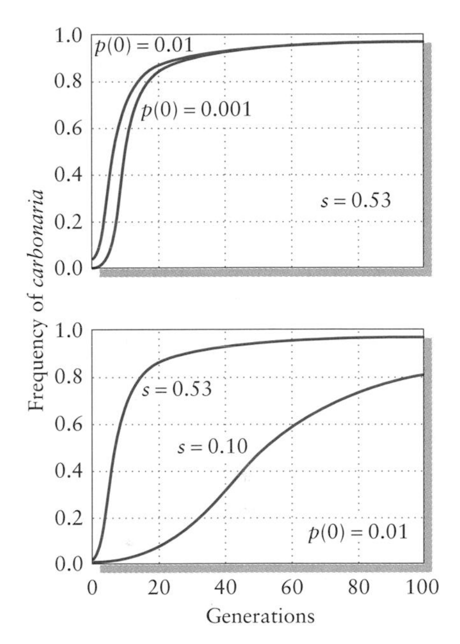

Figure 17.7 Simulation of the change in frequency of the *carbonaria* allele in a population of peppered moths in a region of industrial pollution, assuming selection differentials against homozygous genotypes giving typical coloration of $s = 0.53$, as measured in Kettlewell's experiments, and $s = 0.1$; initial allele frequencies, $p(0)$, of *carbonaria* for the simulations were 0.01 and 0.001. The curves were simulated by repeatedly iterating the equation in Box 17.1.

Variation in quantitative traits

Although many attributes of organisms are controlled by individual genes, as in the case of melanism in the peppered moth, much evolutionary change involves modifications of continuously varying traits, such as length of appendages, body size and shape, thickness of the hair or cuticle, and continuous gradations of behavior. These traits are often called **quantitative characters.** Variation in such traits depends on the contributions of many genes, and their expression in the phenotype cannot be analyzed in the same manner as genotype frequencies for a single gene. Animal and plant breeders interested in such attributes as milk production, oil content of seeds, and rate of egg production have developed a mathematical treatment of continuously varying traits, known as **quantitative genetics.** The theory of quantitative genetics assumes that variation within a population results from additive contributions of many genes with similar effects. Thus, the length of an appendage may come under the influence of a dozen genetic loci, each of which may cause a small increase or decrease in length relative to the population average. Individuals with a net excess of length-increasing alleles at these twelve loci would have appendages longer than the average.

The heart of quantitative genetics is the **variance** of a trait within a population. Variance (V) is a statistical measure of variation; specifically, it is the average of the squared deviations of individuals from the mean of the population (Box 17.2). Continuously variable traits often exhibit a bell-shaped distribution of values among individuals: most individuals are clustered near the mean, and frequency diminishes toward both extremes (Figure 17.8). Each individual's value of a particular trait (the **phenotypic value**) is determined by deviations from the population mean caused by

BOX 17.2 Calculation of the variance within a small population

The variance in a trait X (a measurement) in a population of n individuals is

$$V = \frac{1}{n} \sum_{i=1}^{n} (X_i - \bar{X})^2$$

where X_i is the value for each individual i (i = 1 to n), and $\bar{X}$ is the mean value of X in the population. The calculation of a variance is shown by example in the following tabulation.

Individual (i)	Height, in inches (X_i)	Deviation from mean ($X_i - \bar{X}$)	Squared deviation $(X_i - \bar{X})^2$
1	75	5	25
2	73	3	9
3	72	2	4
4	72	2	4
5	71	1	1
6	70	0	0
7	68	–2	4
8	67	–3	9
9	67	–3	9
10	65	–5	25
	Total = 700		Sum of squared deviations = 90
	Mean ($\bar{X}$) = 70		Variance = 9.0

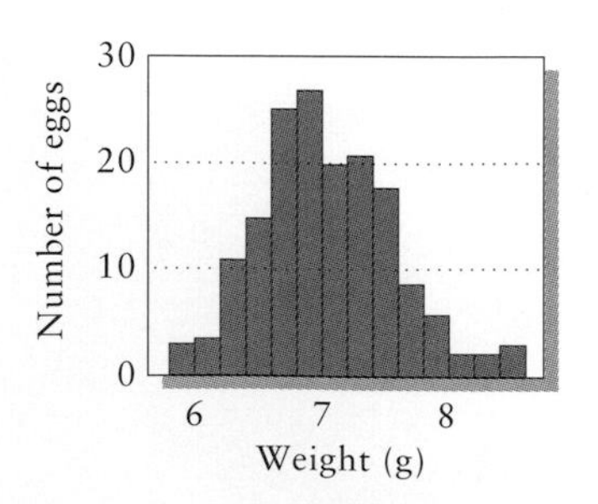

Figure 17.8 Frequency distribution of weights of eggs of European starlings (*Sturnus vulgaris*) near Philadelphia, Pennsylvania, showing a bell-shaped distribution of phenotypic variation.

genetic and environmental influences. Because both sources of deviation enter into the calculation of variance for all values in a population, we may speak of phenotypic variance (V_P) as having two components, one attributable to genetic constitution (V_G) and one resulting from environmental factors (V_E). These two components added together equal the total phenotypic variance; that is, $V_P = V_G + V_E$.

Genotypic variance can be subdivided further: V_A is the **additive variance** determined by the different expressions of alleles in homozygous form; V_D is the **dominance variance** produced by interactions between alleles in heterozygous form; and V_I is the **interaction variance,** comprising the influences of different genes on the expression of alleles at a particular locus. The principal task of quantitative genetics has been to estimate the magnitudes of the several components of phenotypic variance. This is made necessary by the fact that response to selection derives only from the additive genetic component of variance: only V_A reflects the genetic diversity of a population—that is, the different alleles that replace, and are replaced by, others during evolutionary change. Phenotypic variance can be partitioned into its several components by statistical analyses of the results of breeding programs designed for this purpose. These analyses are based on correlations of phenotypic values between close relatives—usually between parents and their offspring or between siblings (see Box 17.3).

The following example drawn from the human population illustrates the principle of correlation among relatives, as well as some pitfalls in using correlations to estimate the genetic basis of a trait. Eighteen male

BOX 17.3 Correlation

The strength of the relationship between two variables in a sample of observations or individuals is indicated by the correlation between pairs of measurements. These pairs of values may be two different traits measured on the same individual (height and weight, for example), or the same trait measured on two different individuals (father and son, or two brothers, for example). The strength of the relationship is measured by the correlation coefficient (r), which is the square root of the ratio of the covariance between measurements X and Y to the product of the standard deviations of X and Y. Thus,

$$r = \left(\frac{\mathrm{COV}_{XY}}{\sqrt{V_X}\ \sqrt{V_Y}} \right)^{1/2}$$

The covariance of X and Y over n observations is calculated by

$$\mathrm{COV}_{XY} = \frac{1}{n} \sum_{i=1}^{n} (X_i - \bar{X})\ (Y_i - \bar{Y})$$

Variances of X and Y are calculated as in Box 17.2. The value of r can range between +1 and −1, which represent perfect positive and perfect negative relationships between pairs of values. Diagrams of representative distributions and their correlations are shown in the figures below.

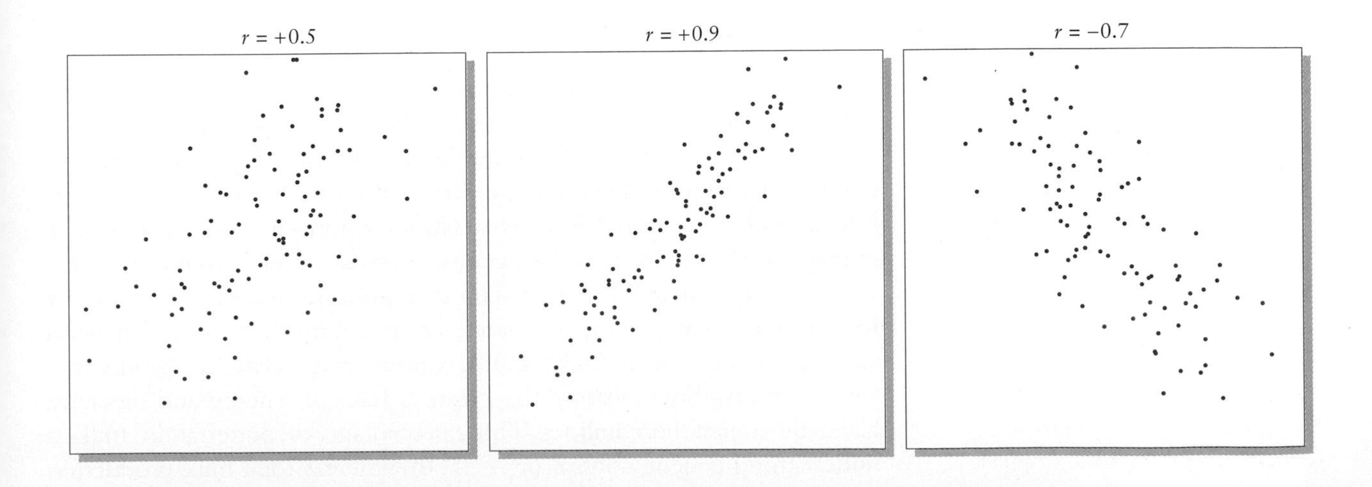

students were asked the heights of their mothers and fathers. The data showed, not surprisingly, that heights of sons were positively correlated with the average height of their parents (correlation coefficient, r = .62; Figure 17.9). Much of this correlation was undoubtedly due to genetic factors inherited by the sons from their parents, but part may also have reflected shared environmental factors, such as quality of diet and other factors that influence height, and maternal effects caused by variations in the environment of the fetus and nursing infant. That a maternal effect might have been important is shown by the fact that heights of sons were more strongly correlated with heights of their mothers (r = .67) than with heights of their fathers (r = .51). One more complication in this study is that of positive assortative mating with respect to height. There was a correlation of r = .60 between the height of the mother and the father; tall men marry tall women.

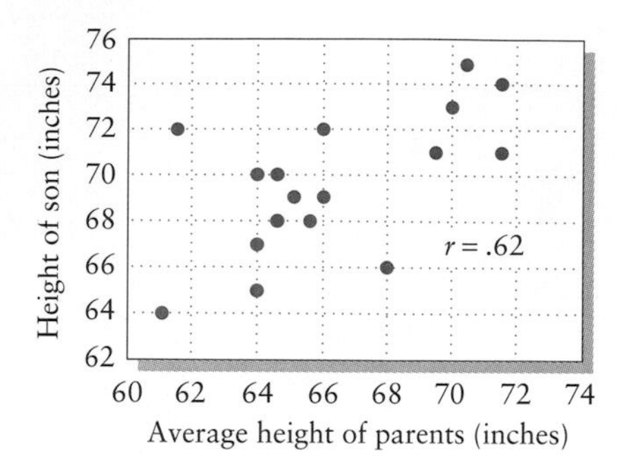

Figure 17.9 Relationship of heights of a small sample of male college students to the average of the heights of their mothers and fathers. The correlation coefficient is .62.

Human geneticists have gotten around some of the complications raised by common environments and other factors that get in the way of estimating genetic components of variation by comparing pairs of identical and fraternal twins. Identical (monozygotic, or MZ) twins have the same genotypes (G), but correlations between their phenotypic values also consist in part of shared environmental effects (E). The same goes for fraternal (dizygotic, DZ) twins, except that DZ twins share only about half of their genes (½G). To see how this fact can be used to advantage, consider the following somewhat simplified reasoning. Think of the phenotypic correlation (P) between MZ twins as $P_{MZ} = G + E$, and that between DZ twins as $P_{DZ} = \frac{1}{2}G + E$. These expressions can be combined and rearranged to give $G = 2\ (P_{MZ} - P_{DZ})$. Thus, one may estimate the genetic contribution to variation in a trait by the difference between the correlations between monozygotic and dizygotic pairs of twins, multiplied by 2. The results of such twin comparisons for several genetic diseases of humans are shown in Figure 17.10. As you can see, alcoholism, especially in women, has a relatively large basis in shared environment, but there appears to be little genetic variation for the trait (P_{MZ} and P_{DZ} are similar). In contrast, schizophrenia, autism, and rheumatoid arthritis evidently have strong genetic components (P_{MZ} exceeds P_{DZ}).

Heritability

The proportion of phenotypic variance that is due to additive genetic factors is often expressed as their ratio and is called **heritability** ($h^2 = V_A / V_P$). Historically, most studies of heritability have focused on traits of commercial value in livestock, poultry, and crops, for which heritabilities have been estimated from correlations among relatives produced in highly structured breeding programs. Representative values of h^2 for such traits appear in Table 17.1. The data indicate that sizes have higher heritabilities (0.50–0.70), and hence are less sensitive to environmental variation, than weights (0.20–0.35). Among traits related to production and fecundity, those creating the greatest drain on energy and nutrients have the lowest heritabilities. Thus, percentage of butterfat in milk is under strong genetic control (h^2 = 0.60), whereas total milk production

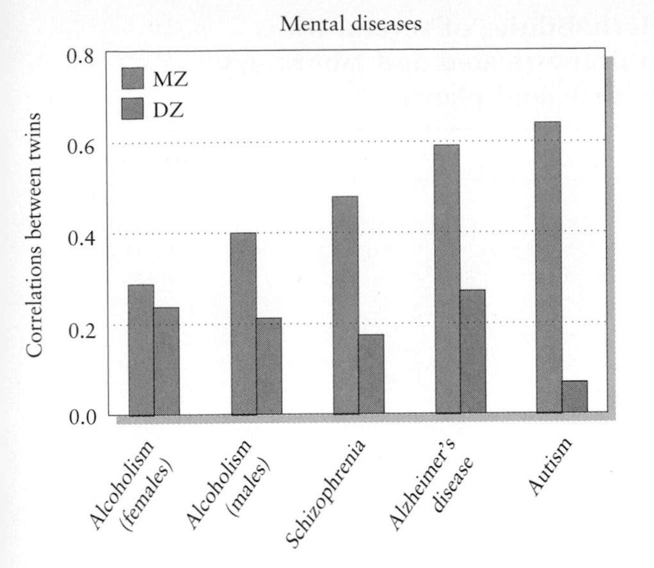

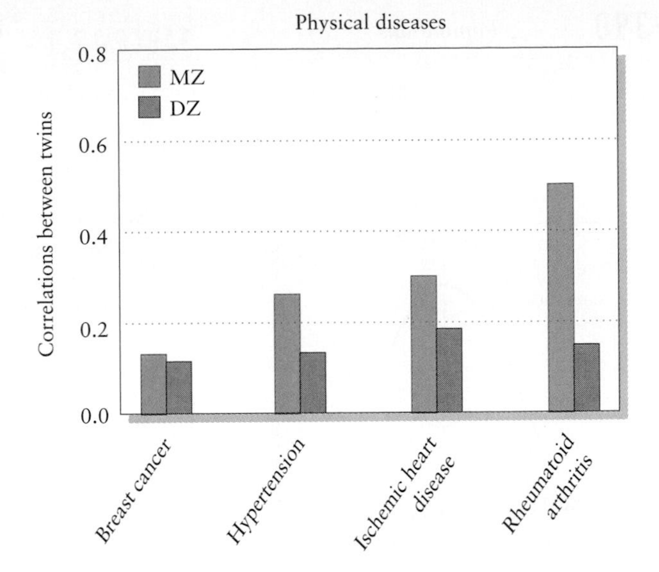

Figure 17.10 Correlations between pairs of identical (monozygotic, MZ) and fraternal (dizygotic, DZ) twins for expression of certain mental and physical diseases in the human population. A high degree of genetic influence is indicated when the MZ correlation (identical genotypes) greatly exceeds the DZ correlation (50% of genes shared). From R. Plomin, M. J. Owen, and P. McGuffin, *Science* 264:1733–1739 (1994).

has a low heritability (0.30); variation in egg size in chickens has a large additive genetic component ($h^2 = 0.60$), whereas rate of egg production has a lower heritability (0.30). Any trait that requires a large commitment of resources is sensitive to environmental variation in those resources. Heritabilities of fecundity and life history characteristics generally are low (0.05–0.50). Population biologists are now finding that heritabilities of traits in wild populations resemble those of domestic animals and crops.

Selection and the evolutionary response of quantitative traits

The change in a quantitative trait resulting from a single generation of selection (R, for **selection response**) depends on the deviation of the selected individuals from the mean value of the population (S, for **selection differential**) and on the heritability of the trait, according to the relationship

$$R = h^2 S.$$

For example, if h^2 were 0.5 and if males and females that were 10 size units larger than the population average were bred together, their progeny would be 5 size units, overall, larger than the average of the unselected population. The greater the heritability of a trait, the more rapidly it can respond to selection.

Values of R and S are conveniently expressed as multiples of the standard deviation of measurements within a population, the **standard deviation** (SD) being the square root of the variance. Each value expressed in standard deviation units corresponds to a particular percentile rank in the population. Zero SD units represents the population mean; when values are symmetrically distributed about the mean, half the individuals lie above that value and half below. When values have a normal distribution, as in Figure 17.11, 31% of individuals lie above +0.5 SD and 31% lie below

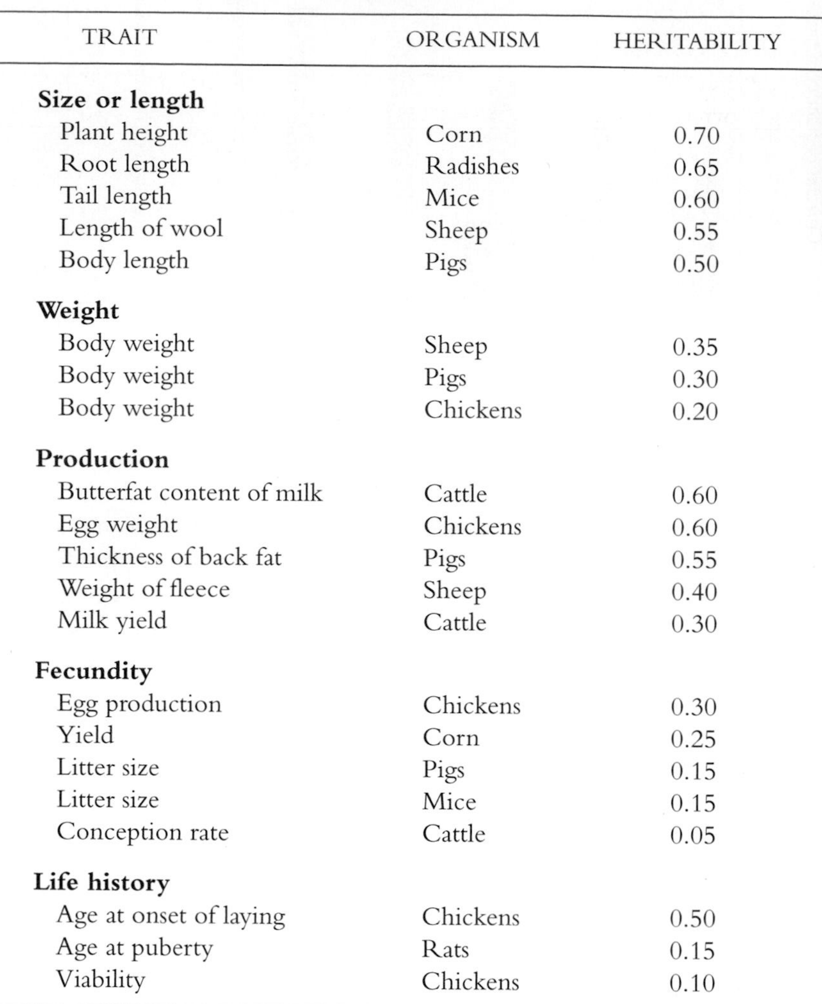

TABLE 17.1 **Heritabilities of several traits in domesticated and laboratory animals and plants**

TRAIT	ORGANISM	HERITABILITY
Size or length		
Plant height	Corn	0.70
Root length	Radishes	0.65
Tail length	Mice	0.60
Length of wool	Sheep	0.55
Body length	Pigs	0.50
Weight		
Body weight	Sheep	0.35
Body weight	Pigs	0.30
Body weight	Chickens	0.20
Production		
Butterfat content of milk	Cattle	0.60
Egg weight	Chickens	0.60
Thickness of back fat	Pigs	0.55
Weight of fleece	Sheep	0.40
Milk yield	Cattle	0.30
Fecundity		
Egg production	Chickens	0.30
Yield	Corn	0.25
Litter size	Pigs	0.15
Litter size	Mice	0.15
Conception rate	Cattle	0.05
Life history		
Age at onset of laying	Chickens	0.50
Age at puberty	Rats	0.15
Viability	Chickens	0.10

Source: D. S. Falconer, *Introduction to Quantitative Genetics,* 2d ed., Ronald Press, New York (1981).

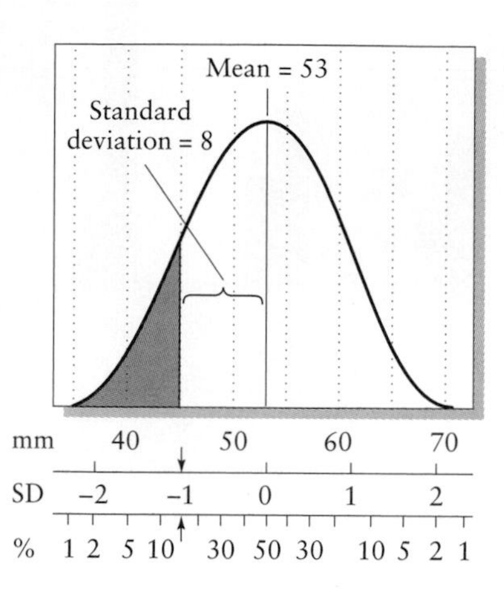

Figure 17.11 Schematic diagram of variation in a normally distributed hypothetical trait, showing the relationships among measurement scale, standard deviation, and the proportion of a population with phenotypic values more extreme than a particular value. For example, 16% of the members of a population have phenotypic values greater than 1 standard deviation below the mean (shaded portion); the same is true of phenotypic values greater than 1 SD above the mean.

−0.5 SD from the mean; 16% have values more extreme than either +1.0 SD or −1.0 SD; 7% exceed 1.5 SD; and only 2.3% exceed 2.0 SD in each direction from the mean. As a result, the more intense selection is (the larger the value of S), the smaller the number of individuals selected (conversely, the larger the number of selective deaths) and the smaller the number of resulting progeny for the next generation of selection to act upon. When selection is too strong, a population dwindles, eventually to extinction. Even in artificial selection programs, the strength of selection is limited by the size of the stock population and by the reproductive rate of the selected individuals, which must be at least as large as the number of individuals eliminated by selection each generation.

The relationship among selection intensity (percentage of individuals selected), phenotypic selection differential (S), and selection response (R)

can be illustrated by a program of selection for rate of egg laying in the chestnut flour beetle, *Tribolium castaneum*. In the stock population, the number of eggs laid by an adult female from 7 to 11 days after she emerged from the pupa (the phenotypic trait investigated) had a mean of 19.0, a standard deviation of 11.8, and a heritability of 0.30. Investigators established one unselected line and five lines with different levels of selection corresponding to between 50% and 95% selective deaths (Table 17.2). Knowing the variability, heritability, and selection differential, we can estimate the initial response of the population to selection. For example, in the C line, a selection intensity of 80% selective deaths corresponds to a selection differential (S) of 1.4 SD, or 16.5 eggs (1.4×11.8). With a heritability of 0.30, response to selection should be about 5 eggs per generation ($R = h^2S = 0.30 \times 16.5$). The observed response fell somewhat short of this prediction (about 3 eggs per generation), probably because the estimate of heritability included maternal and dominance effects as well as additive genetic variation. But the beetle population did behave as predicted in that rate of response varied in direct proportion to intensity of selection (Figure 17.12).

With continued selective pressure on experimental populations, response to selection eventually stops, as Figure 17.12 shows. The slowed response results from several factors, one of which is erosion of genetic variation by selection. Selection works only when individuals vary genetically within a population. When all unfit alleles are removed by selection, evolution pauses until new mutations or gene combinations appear.

Laboratory studies show that the large gains predicted by quantitative genetics models can, in fact, be realized. Because quantitative variation is multigenic, a selection response can be pushed several standard deviations beyond the mean phenotypic value of an unselected population. For example, mature body weights of unselected Japanese quail average about 91 g with a standard deviation of 8 g. After 40 generations of selection for high body weight, the population average in one study increased to 200 g, or almost 14 SD units above the mean of the unselected population. As predicted, the heritability of body weight in the selected population decreased with time as selection removed genetic factors contributing to smaller size from the population.

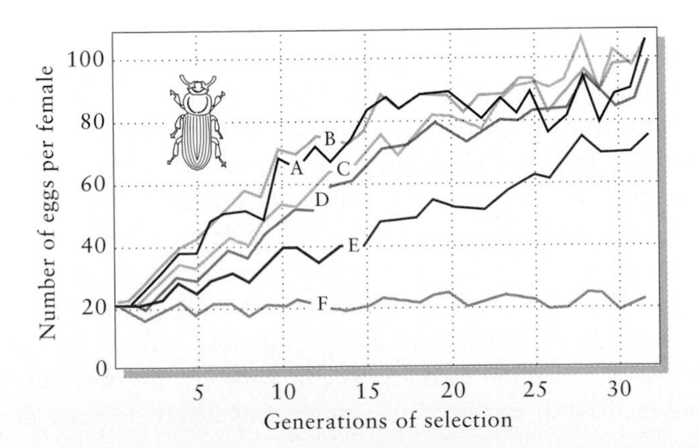

Figure 17.12 Change in rate of egg laying in *Tribolium castaneum* lines exposed to different levels of selection, increasing from F (no selection) to A. From R. G. Ruano, F. Orozco, and C. Lopez-Fanjul, *Genet. Res.* 25:17–27 (1975).

TABLE 17.2 **Selection procedure in six lines of *Tribolium castaneum* under selection for fecundity**

LINE	NUMBER OF FAMILIES SCORED PER GENERATION	NUMBER OF FEMALES SCORED PER FAMILY	TOTAL SCORED	TOTAL SELECTED	SELECTION INTENSITY (PERCENT REMOVED)	SELECTION DIFFERENTIAL (*S*)*
A	10	20	200	10	95	2.0
B	20	10	200	20	90	1.8
C	40	5	200	40	80	1.4
D	66	3	198	66	67	1.1
E	100	2	200	100	50	0.8
F	200	1	200	200	0	0.0

*Standard deviation units.
Source: R. G. Ruano, F. Orozco, and C. Lopez-Fanjul, *Genet. Res.* 25:17–27 (1975).

What is often more difficult to explain than the response of a trait to selection is a characteristic leveling off of the response, often after only modest progress. This pattern usually does not result from exhaustion of genetic variation for the trait: reverse selection (back toward the mean of the unselected population) typically produces an immediate response, which can happen only when genetic variation remains in a population. Furthermore, when selection is relaxed and all phenotypes are bred with equal frequency, a selected trait sometimes returns toward the preselection measurement, apparently by itself. The most reasonable explanation of these results is that selection applied to one trait causes changes in other traits that affect the fitness of the organism. An increase in the rate of egg laying, for example, may cause physiological or morphological changes that reduce viability and thus oppose the artificial selection regime.

Correlated responses to selection

Evolutionary responses often include traits other than the one selected. The various parts of organisms are integrated through their development and functioning, bringing about an interdependence of phenotypic traits, particularly those involving size or rate of growth and production. As a result, selection on one trait often produces a **correlated response to selection** in another trait. Correlated responses reflect genetic correlations between traits; that is to say, many genetic factors are expressed in more than one feature of a phenotype. Animal and plant breeders have estimated genetic correlations between traits in many domestic species. For example, in poultry, the genetic correlation between body weight and egg weight is +0.50; between body weight and egg production, it is –0.16. Therefore, one cannot apply selection to body weight without also obtaining a relatively rapid increase in egg size and a slower but steady decrease in rate of laying. Simultaneous selection for large egg size and small body size goes against the grain of genetic correlation, and it is usually unsuccessful.

Many traits, such as body weight and egg size, form developmentally, genetically, or functionally related groups that tend to respond to selection in concert. For example, genetic correlations among skeletal measurements of mice reveal four clusters of traits that are highly integrated genetically among themselves but relatively independent of one another: (1) skull length; (2) skull width, body weight, and tail length; (3) skull width (providing a link to group 2), scapula length, and other measurements associated with the pectoral girdle; and (4) limb bone length and total body length. In mice, therefore, selection for body weight produces responses in tail length and the proportions of the skull as well as in body weight itself.

In some cases, a plant or animal breeder selects for a particular trait, which produces negative correlated responses in other traits, thereby reducing the effectiveness of, and ultimately halting, the selection program. Flour beetles *(Tribolium)* selected for fast or slow larval development exhibit a number of correlated responses that decrease fitness regardless of the direction of selection. Selection for rapid development results in decreased size and larval survival and an increased incidence of adult abnormalities. Selection for slow development results in increased adult weight, an increased incidence of adult abnormalities, and decreased fertility in females. Such correlated responses tend to work against a selection program because they produce counteracting selective pressures. Similar correlated responses in reproductive fitness have appeared in selected lines of chickens. Selection for both increases and decreases in body weight and egg weight causes declines in reproductive fitness, as indicated by rate of egg production, hatch rate, and survival of chicks. Regardless of the character or direction of selection, fitness in the selected lines varied from 54% to 85% of that in the nonselected line. These results of artificial selection programs suggest that populations are balanced genetically and that their adaptations are both well tuned to the environment and finely adjusted to one another.

Conclusions for ecologists

The field of population genetics has a number of important messages for ecologists. First, every population harbors some genetic variation that influences fitness. This means that evolution is a continuing process in all populations. It also means that individual organisms should be expected to have adaptations that help them to reduce the harmful effects of deleterious mutations on themselves and their offspring. Adaptations to ensure outcrossing are one kind of mechanism by which organisms manage the ubiquitous genetic variability in populations.

Second, changes in selective factors in the environment of a population will almost always be met by evolutionary responses that lead to shifts in phenotypes within the population. The response itself is not always predictable and depends on the particular genetic variation present in the population at a given time. Most quantitative traits have enough genetic variation to respond to selection, but the range and extent of a response may be limited by correlated responses of other traits that have negative fitness consequences. Given enough time, populations may reach some sort of evolutionary optimum (adaptive peak) and become stabilized, but we have little idea of how much time is required.

Third, rapid environmental changes brought about by human-caused changes in the environment, the introduction of predatory or disease organisms, or the appearance of genetic novelties in those enemies will often exceed the capacity of a population to respond by evolution. In these circumstances, the decline of populations toward extinction is a distinct possibility.

Summary

1. Population genetics demonstrates how evolution can proceed by means of changes in the frequencies of alleles according to their relative fitnesses. Genetic changes in populations also occur through mutation, immigration and emigration, assortative mating, and, in small populations, through random processes.

2. Selection may be either stabilizing, in which case intermediate phenotypes in a population are most fit; directional, in which case one or another extreme phenotype is favored over the most common ones; or disruptive, in which case several extreme phenotypes are favored simultaneously.

3. Mutations result from changes in the nucleotide subunits that make up DNA molecules. These occur at a very low rate, but they are the ultimate source of all genetic variability. Most mutations are detrimental to the well-being of their carriers.

4. The genotype includes all the genetic factors that determine the structure and functioning (which together constitute the phenotype) of an individual. Many genetic factors have unique, measurable effects on the phenotype.

5. The frequencies of homozygous and heterozygous genotypes in the absence of selection, mutation, migration, and nonrandom mating can be estimated by the Hardy–Weinberg law, which states that alleles with frequencies p and q will form homozygous genotypes with frequencies p^2 and q^2, and heterozygotes with frequency $2pq$.

6. Deviations from Hardy–Weinberg equilibrium may be caused by mutation, migration, nonrandom mating, genetic drift, and selection. The frequency of heterozygotes in a population tends to be increased by negative assortative mating and decreased by mixing between populations.

7. Even though selection tends to remove genetic variation from a population, variation is maintained at a high level by mutation and gene flow from other populations and by varying selective pressures within populations.

8. Evolution by natural selection occurs when genetic factors influence survival and fecundity. Those individuals that achieve the highest reproductive rate are said to be selected, and their proportion increases with time.

Simple equations show that the rate of change in allele frequency varies with respect to the strength of selection and allele frequency itself. The example of melanism in the peppered moth illustrates these principles.

9. Many adaptations of ecological interest involve modifications of continuously varying traits. Variation in a trait within a population is described by its variance, which has environmental and genetic components. The science of quantitative genetics has developed statistical analyses to tease apart these components from the results of certain breeding programs.

10. Heritability (h^2) is the ratio of additive genetic variance to phenotypic variance; its value, which may be estimated from correlations of phenotypic values among relatives, ranges between 0 and 1. Heritabilities in natural populations are on the order of 0.5–0.7 for many size traits, but are often lower for production-related traits.

11. The response of a trait (R) to selection is equal to its heritability times the selection differential (S). In animal and plant breeding, the stronger the selection, the faster the response, as long as enough offspring are produced to replace the individuals that are selectively removed.

12. Response to selection levels off when genetic variation is exhausted or, more frequently, when correlated responses in other traits reduce the fitness of selected individuals.

13. The phenotype often consists of groups of genetically intercorrelated traits that tend to respond to selection in concert. Such correlations may inhibit the independent evolutionary response of two traits to opposing selective pressures. The response μ of one trait to selection applied on a second trait depends on the genetic correlation between them.

Suggested readings

Cook, L. M., C. S. Mani, and M. E. Varley. 1986. Postindustrial melanism in the peppered moth. *Science* 231:611–613.

Endler, J. A. 1986. *Natural Selection in the Wild*. Princeton University Press, Princeton, N.J.

Falconer, D. S. 1989. *Introduction to Quantitative Genetics.* 3d ed. Longman, Harlow, England.

Ford, E. B. 1975. *Ecological Genetics.* 4th ed. Chapman and Hall, London; Wiley, New York.

Gould, F. 1991. The evolutionary potential of crop pests. *American Scientist* 79:496–507.

Grant, P. R. 1991. Natural selection and Darwin's finches. *Scientific American* 265:82–87.

Hartl, D. L. 1988. *A Primer of Population Genetics.* 2d ed. Sinauer Associates, Sunderland, Mass.

Kettlewell, H. B. D. 1959. Darwin's missing evidence. *Scientific American* 200:48–53.

Maynard Smith, J. 1989. *Evolutionary Genetics.* Oxford University Press, Oxford.

PART V

SPECIES INTERACTIONS

CHAPTER 18

RELATIONSHIPS AMONG SPECIES

Survival rates and reproductive success determine the growth rates of populations and the evolutionary fitness of individuals within populations. These measures depend on how well individuals cope with biological as well as physical factors in their environments. Successful reproduction requires resources sufficient to defend territories, attract mates, make eggs, and provision offspring. To survive, organisms must not only tolerate physical stresses but also avoid detection and capture by predators and infection by disease organisms.

Plants and animals exhibit a variety of structures and behaviors used to obtain food and avoid being eaten or parasitized. Indeed, this variety is one of the most remarkable features of life. Each type of organism has its own unique place in nature, with a corresponding array of adaptations to its own particular environment. Much of this diversity has resulted from natural selection acting on the ways in which plants and animals procure resources and escape predation.

Wing markings that blend artfully into their resting backgrounds by day enable moths to escape the notice of most predators. Flowers, by their insistent colors and fragrances, call attention to themselves and attract the notice of insects and birds that carry

TABLE 18.1 Categories of relationships between species

TYPE OF INTERACTION	EFFECTS OF INTERACTION ON	
	Species 1	*Species 2*
Competition	Negative (–)	Negative (–)
Consumer-resource	Positive (+) for consumer	Negative (–) for resource
Detritivore-detritus	Positive (+)	Indifferent (0)
Mutualism	Positive (+)	Positive (+)

pollen from one flower to the next. The agents whose influence has shaped such adaptations are biological, and their effects differ from those of physical factors in two ways. First, biological factors stimulate mutual evolutionary responses in the traits of interacting populations, referred to as **coevolution**: a predator shapes its prey's adaptations for escape, but its own adaptations for pursuit and capture are just as surely shaped by the attributes of its prey. Second, biological factors foster **diversity** of adaptations rather than promote similarity. In response to similar physical stresses in the environment, many kinds of organisms evolve similar solutions. This phenomenon is called **convergence.** Most desert plants, for example, have reduced or finely divided leaves that minimize heat stress and water loss. In response to biological factors, however, organisms tend to specialize, pursuing different assortments of prey, striving to avoid different combinations of predators and disease organisms, and engaging in cooperative arrangements with unique sets of pollinators, seed dispersers, or gut microorganisms.

Types of interactions between species

Most relationships among species fall conveniently into four broad categories defined by the effect of the interaction on each of the parties. These categories are summarized in Table 18.1.

Consumer-resource interactions

All life forms are both **consumers** and victims of consumers. Predation, herbivory, parasitism, and other kinds of consumption are the most fundamental interactions in nature because everything must eat, and most organisms risk being eaten. Predator-prey, herbivore-plant, and parasite-host relationships are all examples of **consumer-resource interactions,** which organize biological communities into series of **consumer chains.** We have seen these above in the guise of food chains. It is typical of consumer-resource interactions that consumers benefit, and their population sizes may increase, while resources suffer, individual and population alike. Thus, although energy and nutrients move up a consumer chain, populations are controlled both from below by resources and from above by consumers. Similarly, natural selection exerts its influence from both directions.

Detritivore links in food chains are special because detritus eaters do not affect the rate of supply of their food, and they do not select for traits in populations of organisms that are sources of their food. For example, earthworms in soil do not directly influence leaf fall from the trees above them, even though they may indirectly increase plant production by speeding the return of nutrients to forms in the soil that plants can reuse. Burying beetles have little effect on the populations of small rodents that supply the carcasses they feed upon. Because detritivores are not engaged in reciprocal interactions, we will have little to say about them in this part of the book.

Mutualism

Mutualism refers to a wide range of interactions between species that benefit both participants. Flowers provide bees with a supply of nectar, and bees carry pollen between plants and effect fertilization; mycorrhizal fungi extract from the soil inorganic nutrients that plants can use, and plants supply their fungus partners with carbohydrates. In most cases, each party to a mutualism is specialized to perform a complementary function for the other. In lichens, photosynthetic algae team up with fungi that can obtain nutrients from difficult substrates, such as bark and rock surfaces. Such intimate associations, in which the members together form a distinctive entity, are referred to as **symbioses**—literally, a "living together."

Competition between species

Competition results when many species seek the same resources, and the depressing effect that each one has on the availability of the shared resources adversely affects the others. This relationship arises whenever two or more consumer chains join together (Figure 18.1). Usually, individuals compete indirectly through their mutual effects on shared resources. Less

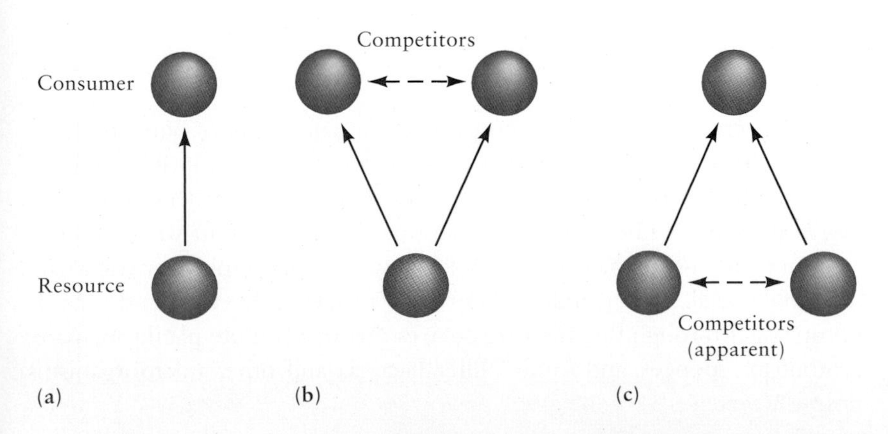

Figure 18.1 A schematic diagram of the relationships between species involved in (a) resource-consumer interactions, (b) competition, and (c) indirect competition. Solid arrows indicate the direction of consumption; dashed arrows indicate competitors.

frequently, when consumers can profitably defend resources, competitors may interact directly through various antagonistic behaviors. Hummingbirds chase other hummingbirds, not to mention bees and moths, from flowering bushes. Encrusting sponges use poisonous chemicals to overcome other species of sponges as they expand to fill open space on rock surfaces. Many shrubs release toxic chemicals into the soil that depress the growth of competitors. Even bacteria wage chemical warfare with each other to tip the balance of their competitive interactions.

Competitors may also interact indirectly through their shared consumers, as well as through shared resources. For example, when two competitors have different levels of vulnerability to a certain predator, an increase in the population of the less vulnerable species may result in an increase in the population of predators, which, in turn, will depress the population of the more vulnerable species. This kind of interaction is referred to as **apparent competition,** "apparent" because the interaction need not involve the shared resources of the competitors. Disease organisms are potent agents of apparent competition. Songbirds from all over the world have been introduced to the Hawaiian Islands and have caused the decline and extinction of most of the native species. In this case, the successful invaders brought diseases, such as avian malaria and pox virus, to which the native Hawaiian birds had no innate resistance. Thus, the outcomes of interactions between native birds and the outlanders were controlled by their mutual consumers.

Interactions of all kinds involve adaptations of each participant species that reduce the negative effects of the interaction for itself and increase its positive effects. In the case of mutualism, the participants cooperate, and we can expect a fair degree of mutual coordination of their structure and functioning. In the case of predator and prey, each vies to gain an advantage over the other, and new adaptations continually arise that shift the balance in favor of one or the other. This evolutionary arms race has spawned an array of fascinating mechanisms for eating and for avoiding being eaten. We shall examine some of those mechanisms in this chapter, wherein we explore a variety of predator-prey, parasite-host, and herbivore-plant interactions and conclude with a look at several examples of mutualism. In subsequent chapters, we shall consider the implications of these relationships for the regulation of populations and for evolution.

Predator proficiency

When we think of **predator** and **prey,** we usually think of lynx and hare, or bird and beetle—predators that pursue, capture, and eat individual prey. Though smaller than their predators, such prey are large enough to be worth pursuing. Other organisms consume minute prey in vast numbers, and they are also predators. Blue whales weigh many tons but eat small, shrimplike krill, fish fry, and the like. On a smaller scale, clams and mussels pump water through tiny filtering devices that trap minute plankton; many protozoans, sponges, and rotifers filter bacteria and other microorganisms from the water.

Figure 18.2 African lions taking a midday break in Amboseli National Park, Kenya. With their powerful legs and jaws, lions can subdue prey somewhat larger than themselves. But because they cannot maintain speed over long distances, successful hunting relies on stealth and surprise.

As the size of prey increases in relation to that of the predator, prey become more difficult to capture, and predators become specialized for pursuing and subduing their prey (Figure 18.2). Beyond a certain size ratio, however, predators lack sufficient strength and swiftness to capture potential prey items. An individual lion will attack an animal its own size or a little larger, but it is no match for a fully grown elephant. A few species, including lions, wolves, hyenas, and army ants, hunt cooperatively and thus can run down and subdue prey substantially larger than themselves.

At the other end of the relative size spectrum we find **parasites**—the myriad viruses, bacteria, protozoans, worms, and others that attach themselves to, or invade, the bodies of their hosts and feed on their tissues or blood or on partially digested food in their intestines. Parasitism differs from predation and filter feeding because the survival of many parasites depends on their hosts' surviving rather than dying; a parasite must not bite the hand that feeds it, unless it or its offspring can quickly invade a new host.

Between these two ends of the spectrum are the **parasitoids,** which are insects whose larvae consume the tissues of living hosts, but inevitably kill their hosts by the time they mature. Most parasitoids are wasps or flies, and their hosts are often the eggs, larvae, or pupae of other kinds of insects, especially moths and butterflies. Because they reside within and eat the tissues of a living host, parasitoids are sometimes referred to as parasites. However, because they inevitably kill their hosts, they are more properly considered predators.

Depending on what parts of a plant they eat, herbivores act as either predators or parasites. Parasites live on the productivity of a host organism without killing it, so a deer, browsing on a few leaves and stems, functions as a parasite. A sheep that consumes an entire plant, pulling it up by the roots and macerating it into lifeless shreds, behaves as a predator. Beetle larvae that develop within individual seeds, thereby destroying the embryonic plants they contain, are also predators.

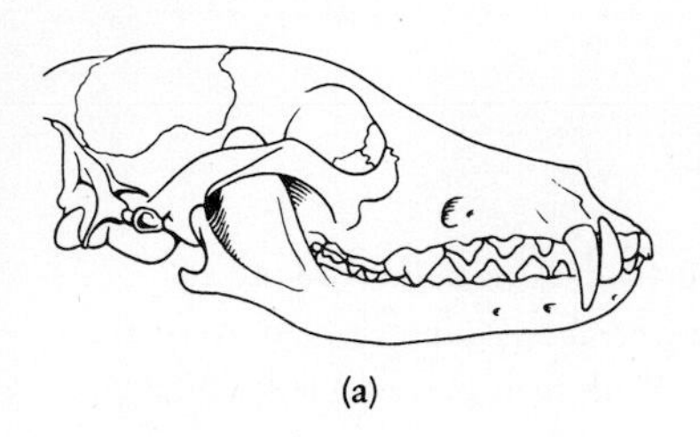

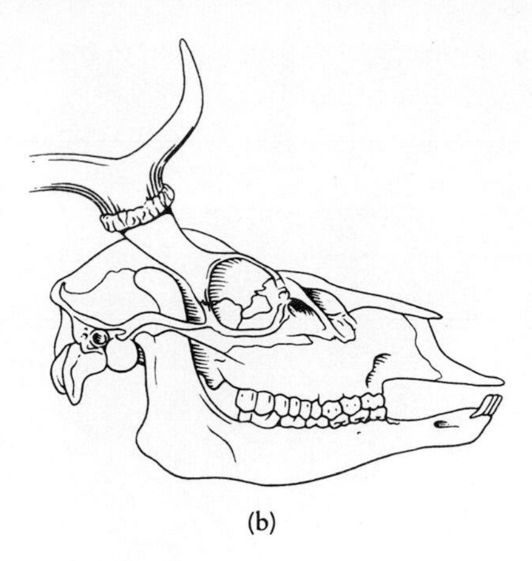

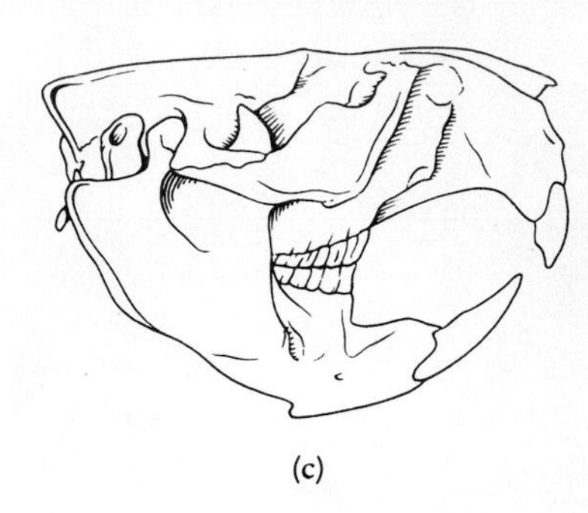

Figure 18.3 Skulls of three mammals, illustrating adaptations of the jaws and teeth to different diets. (a) Coyote, with daggerlike canine teeth and knifelike premolars for securing prey and tearing flesh. (b) Fallow deer, with well-developed, flat-surfaced molars and premolars for grinding plant materials; note the absence of canines and upper incisors. The lower incisors are used to secure vegetation against the upper jaw; the deer then rips the leaves from the plant. (c) Beaver, with greatly enlarged chisel-like incisors that are used to gnaw on wood; canines and premolars are not present. After T. A. Vaughan, *Mammalogy,* 3d ed., Saunders, Philadelphia (1986).

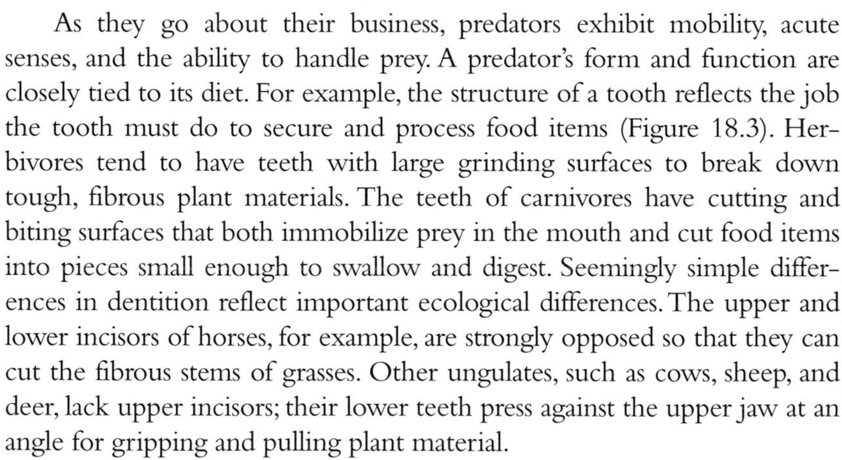

As they go about their business, predators exhibit mobility, acute senses, and the ability to handle prey. A predator's form and function are closely tied to its diet. For example, the structure of a tooth reflects the job the tooth must do to secure and process food items (Figure 18.3). Herbivores tend to have teeth with large grinding surfaces to break down tough, fibrous plant materials. The teeth of carnivores have cutting and biting surfaces that both immobilize prey in the mouth and cut food items into pieces small enough to swallow and digest. Seemingly simple differences in dentition reflect important ecological differences. The upper and lower incisors of horses, for example, are strongly opposed so that they can cut the fibrous stems of grasses. Other ungulates, such as cows, sheep, and deer, lack upper incisors; their lower teeth press against the upper jaw at an angle for gripping and pulling plant material.

Many predators use their forelegs to help them tear their food into small morsels. Birds such as hawks, eagles, owls, and parrots use their powerful sharp-clawed feet and hooked beaks for this purpose. Diving birds often eat large fish, but they must swallow them whole because their hind legs are specialized for swimming and diving rather than for grasping and dismantling prey. Some species of snakes compensate for their lack of grasping appendages with distensible jaws that enable them to swallow large prey whole (Figure 18.4).

Figure 18.4 Some species of snakes have enlarged their gape by as much as 20% by shifting the articulation of the jaw with the skull from the quadrate bone (black) to the supratemporal (color). The figure shows the position of the jaw elements when the mouth is closed and open. After C. Gans, *Biomechanics,* Lippincott, Philadelphia (1974).

The quality of its diet influences the adaptations of a predator's digestive and excretory systems as well as structures directly related to procuring food. Plants contain long, fibrous molecules, such as **cellulose** and **lignin,** that form supportive structures in stems and leaves. These components make vegetation more difficult to digest than the high-protein diets of carnivores; consequently, the digestive tracts of herbivorous animals are often greatly elongated and may have powerful gizzards for grinding and shredding plant materials. In addition, the guts of many herbivores have saclike offshoots—the caeca (singular: caecum) of rabbits and the rumens of cows, for example—which, like fermentation vats, house bacteria and protozoans that aid digestion. With a larger volume of intestines, an herbivore can keep meals in the digestive tract longer and digest them more thoroughly. However, such herbivores must carry quantities of undigested food in their bellies, adding weight and reducing their mobility. This is one reason that few species of birds make use of fermentative digestion.

Prey escape

The ways in which prey organisms avoid being eaten are as diverse as the hunting tactics of their predators. Hiding, escape, and an active defense can all be effective, depending on the particular circumstances of a predator-prey relationship. Grasslands offer few hiding places for deer, antelope, and other grazers, so escape depends on early detection of predators and swift movement. Plants cannot flee like animals, but many produce thorns and defensive chemicals that dissuade herbivores.

Protective defenses rarely involve physical combat because few prey can match their predators, and predators carefully avoid those that can. Instead, many seemingly defenseless organisms produce foul-smelling or stinging chemical secretions to dissuade predators. Whip scorpions and bombardier beetles direct sprays of noxious liquids at threatening animals. Many plants and animals contain chemical substances that make them inedible or poisonous. Slow-moving animals, such as porcupines and armadillos, protect themselves with spines or armored body coverings.

Crypsis and warning coloration

The evolution of the camouflaged appearances and resting positions by which various prey avoid detection by predators fascinates observers of the natural world and testifies to the force and pervasiveness of natural selection.

Figure 18.5 This Central American mantis of the genus *Acanthops* resembles a dead, curled-up leaf and thus escapes the notice of most predators.

Many organisms achieve **crypsis**—which means that they avoid predation by blending in with their backgrounds—by matching the color and pattern of bark, twigs, or leaves. Elaborate concealment of the head, antennae, and legs underscores the importance of these cues to predators. Various animals resemble sticks, leaves, flower parts, or even bird droppings. These organisms are not so much concealed as they are mistaken for inedible objects and passed over. The stick-mimicking phasmids (stick insects) and leaf-mimicking katydids often conceal their legs in resting positions either by folding them back upon their bodies or by protruding them in a stiff, unnatural fashion. The dead-leaf-mimicking mantis *Acanthops* partially conceals its head under its folded front legs (Figure 18.5). Asymmetry is also a good cover for animals, but it is difficult to achieve. The leaf-mimicking moth *Hyperchiria nausica* produces the appearance of an asymmetrical midvein by folding one forewing over the other (Figure 18.6). A moth may sometimes rest with a leg protruding to one side but not to the other, or with its abdomen twisted to one side to break its symmetry.

Figure 18.6 The moth *Hyperchiria nausica* partly disguises its symmetry by folding one wing over the other.

When discovered, many cryptic organisms confront their would-be predators with second-line defenses, including startle displays and various attack-and-escape mechanisms. The green caterpillar of the hawkmoth *Leucorampha omatus* normally assumes a cryptic position. When disturbed, however, it puffs up its head and thorax, looking for all the world like the head of a small poisonous snake, complete with a false pair of large, shiny eyes; the caterpillar consummates this display by weaving back and forth while hissing like a serpent (Figure 18.7). The eyespots that many moths and other insects display when disturbed (Figure 18.8) resemble the eyes of large birds of prey and may frighten or startle predators enough to open a window of escape.

Crypsis is a strategy of palatable, or edible, animals. Others take a bolder approach to antipredator defense: they produce noxious chemicals or accumulate them from food plants, and they advertise the fact with *conspicuous* color patterns in the form of **warning coloration,** or **aposematism.** Predators learn quickly to avoid markings such as the black and orange stripes of the monarch butterfly, which is so foul-tasting that a single experience with this prey is remembered for a long time. It is not a coincidence that many noxious forms adopt similar patterns: black and either red or yellow stripes adorn such diverse animals as yellow-jacket wasps and coral snakes. These color combinations so consistently advertise noxiousness that some predators have evolved innate aversions to such patterns and need not learn to avoid such prey by experience.

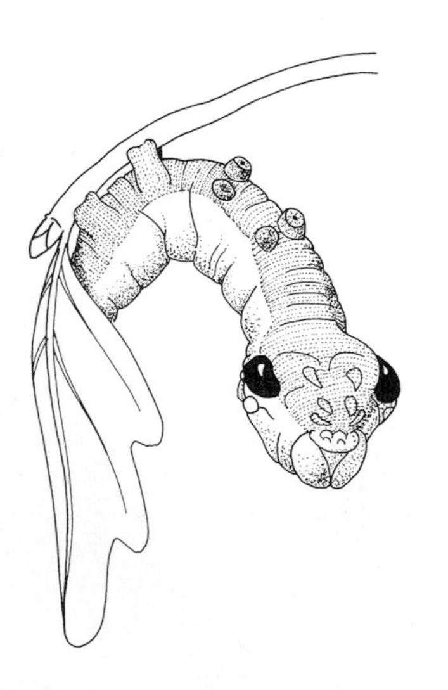

Figure 18.7 The snake display of the caterpillar of the sphingid moth *Leucorampha omatus.* From M. H. Robinson, *Evol. Biol.* 3:225–259 (1969).

Why aren't all potential prey species noxious or unpalatable? Part of the answer is that chemical defenses can be costly to manufacture and maintain. In many cases, defensive compounds use up a large portion of an individual's energy or nutrients that might otherwise be allocated to growth or reproduction. Furthermore, many prey organisms rely on food plants to supply toxic organic compounds that they cannot manufacture themselves. Of course, the consumers must themselves avoid the toxic effects of such chemicals in order to use them effectively against their potential predators.

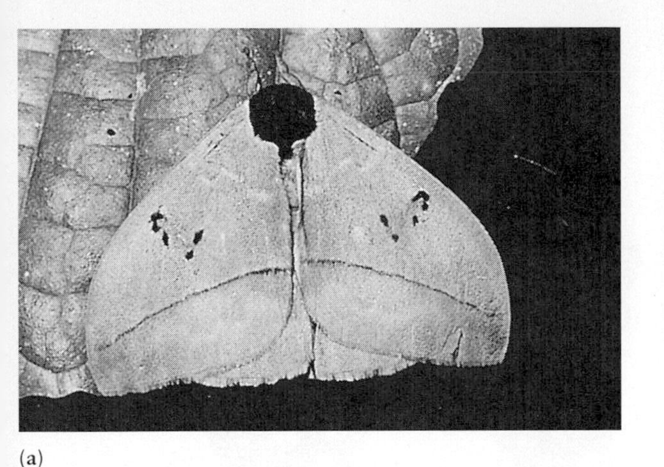

(a)

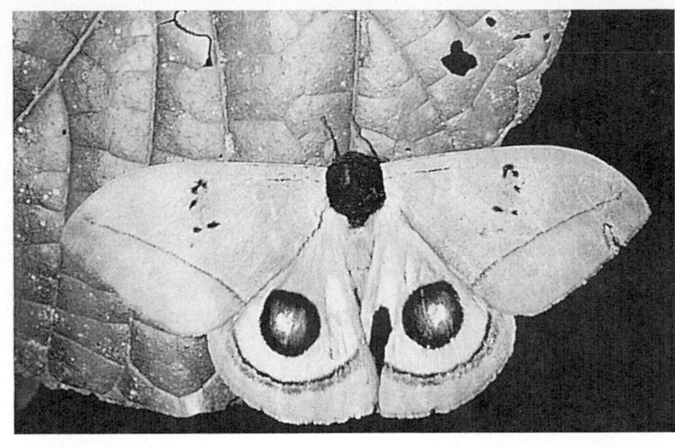

(b)

Figure 18.8 The eyespot display of an automerid moth from Panama. (a) Normal resting position. (b) Reaction when touched.

Batesian mimicry

Unpalatable animals and plants that display warning coloration often serve as **models** for mimicry by palatable forms, which evolve to resemble the noxious organisms. Some potential prey even contrive to resemble their predators to throw them off guard (Figure 18.9). These relationships are collectively referred to as **Batesian mimicry,** which was named after its discoverer, the nineteenth-century English naturalist Henry Bates. In his journeys to the Amazon region of South America, Bates found numerous cases of palatable insects that had forsaken the cryptic patterns of close relatives and had come to resemble brightly colored, unpalatable species.

Experimental studies have demonstrated convincingly that mimicry does confer an advantage on mimics. For example, toads that were fed live bees, and were stung on the tongue, thereafter avoided palatable drone flies, which mimic bees. But when naive toads were fed only dead bees from which the stings had been removed, they relished the drone fly mimics. Thus, toads learned to associate the conspicuous and distinctive color patterns of bees with an unpleasant experience. Similar results were obtained with blue jays as predators: unpalatable monarch butterflies were the models, and their (palatable) viceroy butterfly mimics were the experimental subjects.

In some cases, mimicry relationships involve several different models. For example, in an African swallowtail butterfly, *Papilio dardanus,* females are polymorphic, each individual resembling one of a variety of different models (Figure 18.10). (The genes for mimetic appearance are not expressed in males, presumably because females choose males on the basis of their coloration and prefer the typical swallowtail color pattern. Moreover, because males do not produce or carry eggs, or spend time searching for suitable oviposition sites, they may be less vulnerable to predators.) Why do such mimicry polymorphisms evolve? When mimics become common relative to models, predators do not learn to avoid either one as quickly because they often sample palatable mimics rather than the noxious models. In this case, a rare mimetic form has an advantage: predators that are learning to associate warning coloration with unpalatability are not confused by

Figure 18.9 The wing markings of the tephritid fly *Rhagoletis pomonella* closely resemble the forelegs and pedipalps of jumping spiders (Salticidae). Courtesy of the U.S. Department of Agriculture.

Figure 18.10 Mimicry polymorphism in some East African butterflies. The three butterflies in the left-hand column are unpalatable species that serve as models for several mimetic species. At right, *Papilio dardanus,* showing the nonmimetic male (top) and three mimetic forms of females. After V. C. Wynne-Edwards, *Animal Dispersion in Relation to Social Behavior,* Oliver and Boyd, Edinburgh (1962).

frequent encounters with common mimetic forms. Thus natural selection, by favoring odd types of mimics, acts to diversify a population.

Müllerian mimicry

Another type of mimicry, called **Müllerian mimicry** after its discoverer, occurs among unpalatable species that come to resemble one another. Many species form Müllerian mimicry complexes in which each participant is both model and mimic. When a single pattern of warning coloration is adopted by several unpalatable species, avoidance learning by predators is made more efficient because a predator's bad experience with one species confers protection on all the other members of the mimicry complex. For example, most of the bumblebees and wasps that co-occur in Rocky Mountain meadows share a pattern of black and yellow stripes. In the Tropics, dozens of species of unpalatable butterflies, many of them distantly related, share patterns of black and orange "tiger stripes" or black, red, and yellow coloration patterns in Müllerian mimicry complexes.

Parasites

Parasites are usually much smaller than their prey, or hosts, and live either on their surfaces **(ectoparasites)** or inside their bodies **(endoparasites).** Both types demonstrate characteristic adaptations to their way of life. Parasites that live inside of or in close association with a larger organism enjoy a benign physical environment regulated by their host. Parasites living in the gut (tapeworms, for example) are bathed in a predigested food supply and retain for themselves little more than a highly developed capacity to produce eggs.

In spite of the advantages of a comfortable environment and a ready supply of nourishment close at hand, the life of a parasite is not easy. Host organisms have a variety of mechanisms to recognize invaders and destroy them. The immune systems of vertebrates produce antibodies in response to the presence of foreign proteins. These antibodies attach themselves to the surfaces of parasites, disabling them and allowing macrophage cells to attack them and the spleen to clear them from the body. Parasites counter these defenses by coating themselves with proteins from their hosts; by hiding in tissues or within cells where they are inaccessible to the lymphocytes (white blood cells) that produce immune responses; or by changing the protein molecules on their outer surfaces so often that their hosts cannot produce new antibodies fast enough to keep up.

As if life in the host were not difficult enough, parasites must also disperse through a hostile environment to get from one host to another. Many accomplish this via complicated life cycles, one or more stages of which can cope with the external environment. *Ascaris,* an intestinal roundworm that parasitizes humans, has a relatively simple life cycle. A female *Ascaris* may lay tens of thousands of eggs per day, which pass out of the host's body in its feces. Where sanitation is poor or where human excrement is applied to farmland as fertilizer, the eggs may be inadvertently ingested. The egg is the only life cycle stage of *Ascaris* that occurs outside the host, and it is well protected by a sturdy, impermeable outer covering. These parasites rely on their hosts to consume their eggs and complete their life cycles.

Schistosoma, a trematode worm (blood fluke) that commonly infects humans and other mammals in tropical regions, has a more complicated life cycle that involves a freshwater snail as an intermediate host. Male-female pairs of adult worms live in the blood vessels that line the human intestine or bladder, depending on the species of *Schistosoma.* The eggs pass out of a host's body in feces or urine. When the eggs are deposited in water, they develop into a free-swimming larval form (the miracidium), which locates and burrows into a snail within 24 hours. In the snail, a miracidium produces cells that eventually develop into free-swimming cercariae, which leave the snail and can penetrate the skin of the next host in the cycle. Once inside the body of a human or other host, the cercariae travel a circuitous route through the blood vessels until they become lodged in an appropriate place, where they metamorphose into adult worms. A snail infected with one miracidium may liberate from 500 to 2,000 cercariae per day over a period of a month. During their lifetimes, which average between 4 and 5 years in human hosts, adults may produce as many as several hundred eggs per day. Of course, few of these eggs and larvae survive to become sexually mature adults. Nonetheless, *Schistosoma,* which causes the incurable disease schistosomiasis, is known to affect hundreds of millions of humans and countless domestic animals.

The life cycle of the protozoan parasite *Plasmodium,* which causes malaria, resembles that of *Schistosoma* in that it involves two hosts, a mosquito and a human or some other mammal, bird, or reptile. But whereas

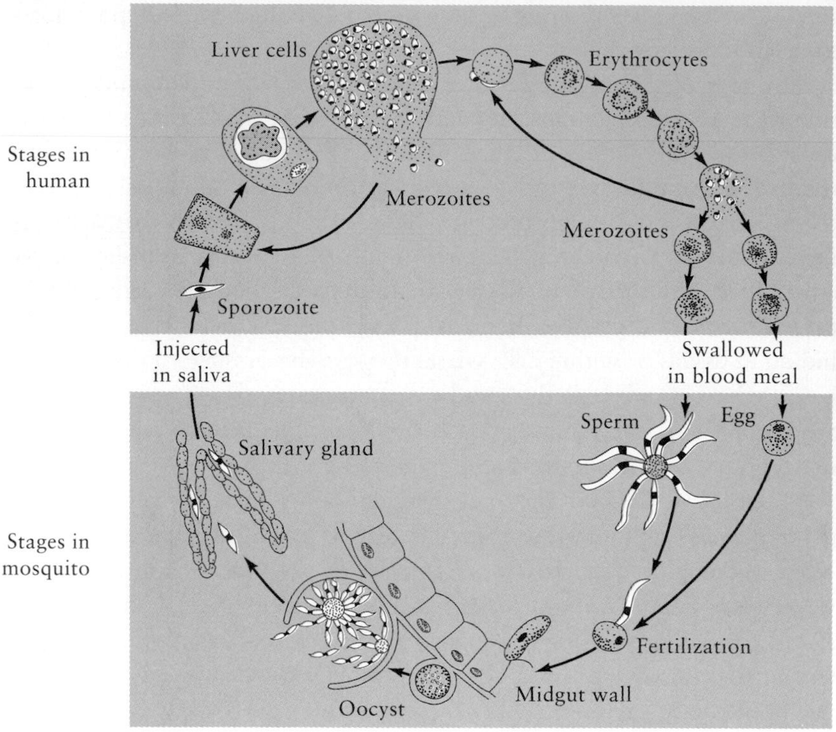

Figure 18.11 Stages in the life cycle of the malaria parasite *Plasmodium*. After R. Buchsbaum, *Animals without Backbones,* 2d ed., University of Chicago Press, Chicago (1948); M. Sleigh, *The Biology of Protozoa,* American Elsevier, New York (1973).

the sexual phase of the schistosome life cycle occurs in a human host, that phase of the *Plasmodium* life cycle takes place within a mosquito (Figure 18.11). When an infected mosquito bites a human, mobile cells called sporozoites are injected into the bloodstream with the mosquito's saliva. The sporozoites at first proliferate by mitosis in liver cells, after which they enter red blood cells (erythrocytes) as merozoites, where they feed on hemoglobin and grow. When a merozoite becomes large enough, it undergoes a series of divisions (asexual reproduction), and the daughter merozoites break out of the red blood cell. Each merozoite can enter a new red blood cell, grow, and repeat the cycle, which takes about 48 hours. (When the infection has built up to a high level, the emergence of daughter cells corresponds to periods of high fever.) After several of these cycles, some of the merozoites that enter red blood cells change into sexual forms. If these are swallowed by a mosquito along with a meal of blood, the sexual cells are transformed into eggs and sperm, and fertilization (sexual reproduction) takes place. Fertilized eggs penetrate the mosquito's gut wall and then undergo a series of divisions to produce sporozoites. These work their way into the salivary glands of the mosquito, from which they may enter a new intermediate host, thereby completing the life cycle.

Herbivory and plant defenses

The conflict between herbivore and plant resembles that between parasite and host in that both are waged primarily on biochemical battlegrounds.

Plant defenses against herbivores include the inherently low nutritional value of most plant tissues and the toxic properties of so-called secondary compounds produced and sequestered for defense. Sessile marine organisms, including algae, plants, and animals, also employ a variety of chemical defenses. Structural defenses, such as spines, hairs, tough seed coats, and sticky gums and resins, are important as well (Figure 18.12).

The nutritional quality and digestibility of algal and plant foods is critical to herbivores. Because young animals require a lot of protein for growth, the reproductive success of grazing and browsing mammals depends on the protein content of their food. Herbivores usually select plant food according to its nutrient content. Young leaves and flowers are often preferred over mature leaves because of their low cellulose content; fruits and seeds are particularly nutritious compared with leaves, stems, and buds because of their higher nitrogen, fat, and sugar contents.

Many plants use chemicals to reduce the availability of their proteins to herbivores. For example, **tannins** sequestered in vacuoles in the leaves of oaks and other plants combine with leaf proteins and digestive enzymes in an herbivore's gut, thereby inhibiting protein digestion. As a consequence, tannins considerably slow the growth of caterpillars and other herbivores, an effect that reduces the quality of tannin-laden hosts as food plants. With the buildup of tannins in some oak leaves over the summer, fewer and fewer leaves are attacked by herbivores. Insects that feed on tannin-rich plants can reduce the inhibitory effects of tannins by producing detergent-like surfactants in their gut fluids, which tend to disperse tannin-protein complexes.

Whereas tannins exhibit a generalized reaction with proteins of all types, many **secondary compounds** of plants (that is, compounds used

Figure 18.12 Spines protect the stems and leaves of many plants. (a) A cholla cactus *(Opuntia)* from Arizona. (b) An agave (century plant) from Baja California.

(a)

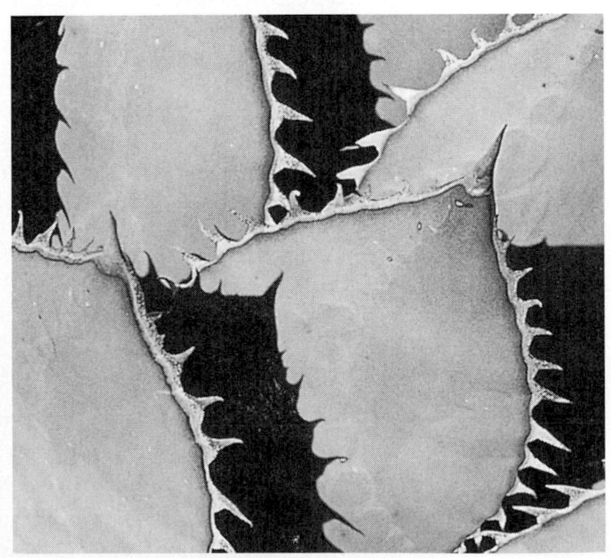

(b)

not for metabolism but for other purposes, chiefly defense) interfere with specific metabolic pathways or physiological processes of herbivores. However, because the sites of action of such compounds are localized biochemically, herbivores may counter their toxic effects by modifying their own physiology and biochemistry. Such **detoxification** may involve one or several biochemical steps, including oxidation, reduction, or hydrolysis of the toxic substance or its conjugation with another compound.

Consider the chemical give-and-take between larvae of bruchid beetles and the seeds of legumes (pea family) that they consume. Adult bruchids lay their eggs on developing seed pods. The larvae then hatch and burrow into the seeds, which they consume as they grow. To counter this attack, legumes have mounted a variety of defenses. One approach has been the evolution of tiny seeds. Each larva feeds on only one seed. To pupate successfully and metamorphose into an adult, a larva must attain a certain size, which is ultimately limited by the amount of food in the seed. The small seeds of some species of legumes contain too little food to support the growth of a single bruchid larva. Of course, small seed size imposes costs on the plant as well, because small seedlings with few resources often survive less well than large ones.

Most legume seeds also contain substances that inhibit the proteolytic enzymes produced in an herbivore's digestive organs. Although these toxins provide an effective biochemical defense against most insects, many bruchid beetles have metabolic pathways that either bypass them or are insensitive to them. Among legume species, however, soybeans stand out as being resistant to attack even by most bruchid species. When bruchids lay their eggs on soybeans, the first instar larvae die soon after burrowing beneath the seed coat; chemicals isolated from soybeans have been shown to inhibit the development of bruchid larvae in experimental situations.

Seeds of the tropical leguminous tree *Dioclea megacarpa* contain 13% L-canavanine by dry weight. This nonprotein amino acid is toxic to most insects because it interferes with the incorporation into proteins of arginine, which it closely resembles. One species of bruchid, *Caryedes brasiliensis,* possesses enzymes that discriminate between L-canavanine and arginine during protein formation as well as enzymes that degrade L-canavanine to forms that can be used as a source of nitrogen. For every defense, a new counterattack can be devised.

Tobacco hornworms (larvae of the hawkmoth *Manduca sexta*) can tolerate nicotine concentrations in their food far in excess of those that kill other insects. Nicotine disrupts the normal functioning of the nervous system by preventing transmission of impulses from nerve to nerve. Hornworms have circumvented this defense by excluding nicotine from their nerves at the cell membranes (in other species of moths, nicotine readily diffuses into nerve cells). Resistance to nicotine enables hornworms to feed on tobacco *(Nicotiana tabacum),* a member of the tomato family (Solanaceae), but some other species of *Nicotiana* produce other alkaloid toxins that tobacco hornworms cannot tolerate. When tobacco hornworms were grown on 44 species of *Nicotiana* in greenhouse experiments, the larvae grew normally on 25 species, but their growth was retarded or stopped completely on the others. In addition, 15 of the species

TABLE 18.2 **Secondary plant compounds involved in plant-herbivore interactions**

CLASS	APPROXIMATE NUMBER OF CHEMICAL STRUCTURES	DISTRIBUTION	PHYSIOLOGICAL ACTIVITY
Nitrogen compounds			
Alkaloids	5,500	Widely in angiosperms, especially in roots, leaves, and fruits	Many toxic and bitter-tasting
Amines	100	Widely in angiosperms, often in flowers	Many repellent-smelling, some hallucinogenic
Amino acids (nonprotein)	400	Especially in seeds of legumes, but relatively widespread	Many toxic
Cyanogenic glycosides	30	Sporadic, especially in fruits and leaves	Poisonous (as HCN)
Glucosinolates	75	Cruciferae and 10 other families	Acrid and bitter
Terpenoids			
Monoterpenes	1,000	Widely, in essential oils	Pleasant-smelling
Sesquiterpene lactones	600	Mainly in Compositae, but increasingly found in other angiosperms	Some bitter and toxic, also allergenic
Diterpenoids	1,000	Widely, especially in latex and plant resins	Some toxic
Saponins	500	In over 70 plant species	Hemolyze blood cells
Limonoids	100	Mainly in Rutaceae, Meliaceae, and Simaroubaceae	Bitter-tasting
Curcurbitacins	50	Mainly in Cucurbitaceae	Bitter-tasting and toxic
Cardenolides	150	Especially common in Apocynaceae, Asclepiadaceae, and Scrophularaceae	Toxic and bitter
Phenolics			
Simple phenols	200	Universal in leaves, often in other tissues as well	Antimicrobial
Other			
Polyacetylenes	650	Mainly in Compositae and Umbelliferae	Some toxic

Source: J. B. Harborne, *Introduction to Ecological Biochemistry,* 2d ed., Academic Press, New York (1982).

caused moderate to severe mortality. These results emphasize the degree of specialization that can develop in consumer-resource interactions.

Most plants produce toxic defensive compounds. Many of these, like pyrethrin, are important sources of pesticides (which is how plants use them, of course); others, like digitalis, have found use as drugs (some of their pharmacological effects are beneficial in small doses). Secondary plant compounds can be divided into three major classes based on their chemical structure: nitrogen compounds ultimately derived from amino acids, terpenoids, and phenolics (Table 18.2). Among the nitrogen-based substances are lignin, a highly condensed polymer that resists digestion; **alkaloids,** such as morphine (derived from poppies), atropine, and nicotine (from various members of the tomato family); nonprotein amino acids, such as L-canavanine; and cyanogenic glycosides, which produce cyanide (HCN). **Terpenoids** include essential oils, latex, and plant resins; among the **phenolics,** many simple phenols have antimicrobial properties.

Where herbivory is most intense, plants have the most varied and concentrated toxins. And where plant defenses are strong, adaptations of herbivores to detoxify poisonous substances proliferate. This coevolutionary arms race between plants and herbivores promotes the biochemical specialization of herbivores on certain restricted groups of plants with similar toxins. Associations of plants and herbivores in groups based on plant chemistry and structure have been referred to as **plant defense guilds.**

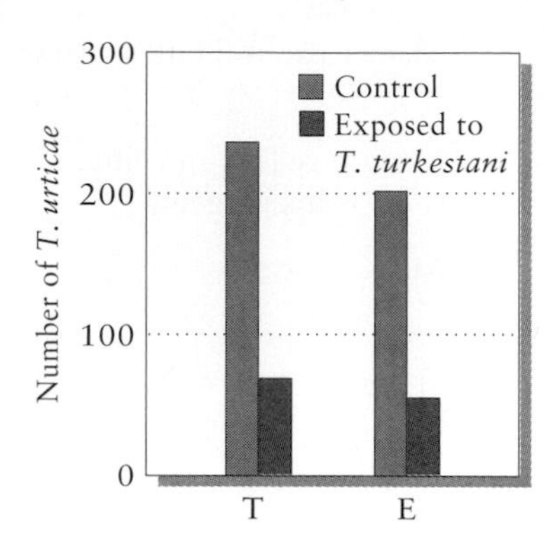

Figure 18.13 Mean numbers of the mite *Tetranychus urticae* on cotton plants previously exposed to a closely related mite species, *T. turkestani,* and on control plants with no previous mite exposure. T = total population; E = eggs. From R. Karban and J. R. Carey, *Science* 225:53–54 (1984).

Plant defenses may be **induced** by herbivore damage in a manner analogous to the way that foreign proteins induce an immune response in vertebrate animals. Alkaloids, phenolics, N-oxidases, and proanthocyanins, all of which are linked to antiherbivore defenses, increased dramatically in many plants following defoliation by herbivores (or the clipping of leaves by investigators). Other studies have shown that plant responses to herbivory can substantially reduce subsequent herbivory (Figure 18.13). This inducibility suggests that some chemical defenses are too costly to maintain economically under light grazing pressure. Several studies have shown trade-offs between production of defensive chemicals and growth. In addition, where soils are low in the nutrients required for the production of some defensive chemicals, the costs of defense are relatively higher. Undoubtedly, the offensive biochemical tactics of herbivores are also expensive.

Mutualism

A relationship between two species in which both benefit is referred to as a **mutualism.** In very general terms, mutualisms fall into three categories: trophic, defensive, and dispersive.

Trophic mutualisms

Trophic mutualisms usually involve partners specialized in complementary ways to obtain energy and nutrients; hence the term *trophic,* which pertains to feeding relationships. We have seen trophic mutualisms in the symbiotic associations of algae and fungi to form lichens, of fungi and plant roots to form mycorrhizae, and of *Rhizobium* bacteria and plant roots to form nitrogen-fixing root nodules. In these cases, each of the partners supplies a limiting nutrient or energy source that the other cannot obtain. *Rhizobium* can assimilate molecular nitrogen (N_2) from the soil—a useful feature in nitrogen-poor soils—but requires carbohydrates supplied by a plant for the energy needed to do this. Bacteria in the rumens of cows and other ungulates can digest the cellulose in plant fibers, which a cow's own digestive enzymes cannot do. The cows benefit because they assimilate some of the by-products of bacterial digestion and metabolism for their own use (they also digest some of the bacteria themselves). The bacteria benefit by having a steady supply of food in a warm, chemically regulated environment that is optimal for their own growth.

Ants belonging to the tropical group Attinae harvest leaves and bring them to their underground nests, where they use them to cultivate highly specialized species of fungus. These leaf-cutting ants consume the fungus; in fact, it is their only source of food. They also provide a living environment for the fungus, which can live nowhere else in nature. Thus, both organisms are totally dependent on each other. Such mutualistic relationships are extremely stable, especially compared with consumer-resource interactions, because both partners cooperate and are mutually evolved to each other's benefit as well as to their own. Genetic studies indicate that some of these relationships go back more than 20 million years.

Defensive mutualisms

Defensive mutualisms involve species that receive food or shelter from their mutualistic partners in return for defending those partners against herbivores, predators, or parasites. For example, in marine systems, specialized fishes and shrimps clean parasites from the skin and gills of other species of fish. These cleaners benefit from the food value of parasites they remove, and the groomed fish are unburdened of some of their parasites. Such relationships, often referred to as **cleaning symbioses,** are most highly developed in clear, warm tropical waters, where many cleaners display their striking colors at locations, called cleaning stations, to which other fish come to be groomed. As might be expected, a few species of predatory fish mimic the cleaners: when other fish come and expose their gills to be groomed, they get a bite taken out of them instead.

Figure 18.14 The thorns of *Acacia hindsii,* like those of *A. cornigera,* are greatly enlarged and are filled with a soft pith that ants excavate for nests. Thorns from a non-ant acacia are shown at left for comparison. Courtesy of D. H. Janzen; from D. H. Janzen, *Evolution* 20:249–275 (1966).

Facultative and obligate mutualisms

Cleaners and the fish they groom do not live in close association, and each can survive without the other. Their mutualism is **facultative,** meaning that the partners can do with it or without it. When two species are inextricably bound through mutual dependence, they form what is known as an **obligate mutualism.** The algae and fungi in lichens are for the most part obligate trophic mutualists. Among defensive mutualisms, the interdependence between certain kinds of ants and swollen-thorn acacias in Central America provides a fine example of obligate mutualism.

Acacia plants provide food and nesting sites for ants in return for the protection from insect pests and competing plants that the ants provide. The bull's-horn acacia *(Acacia cornigera)* has large hornlike thorns with a tough woody covering and a soft pithy interior (Figure 18.14). To start a colony in an acacia, a queen ant of the species *Pseudomyrmex ferruginea* bores a hole in the base of one of the enlarged thorns and clears out some of the soft material inside to make room for her brood. In addition to

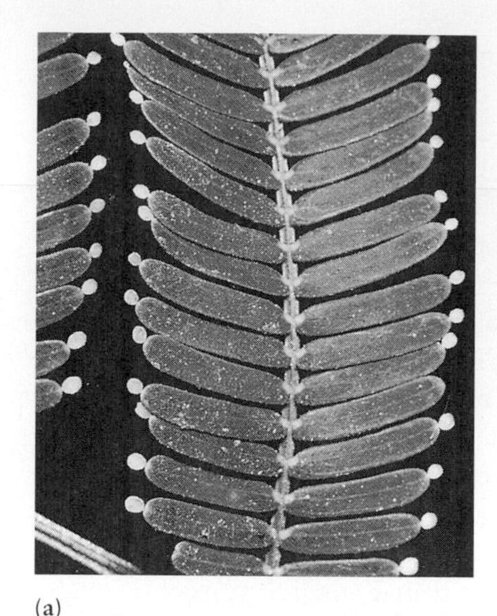

(a)

(b)

Figure 18.15 The leaves of *Acacia collinsii,* like those of *A. cornigera,* provide ants with food in the form of Beltian bodies at the tips of leaflets (a) and nectaries at the leaf base (b). Courtesy of D. H. Janzen; from D. H. Janzen, *Evolution* 20:249–275 (1966).

housing ants, acacias provide carbohydrate-rich food for them in nectaries at the bases of the leaves as well as fats and proteins in the form of nodules, called Beltian bodies, at the tips of some leaves (Figure 18.15). In return, the ants protect their host plant from insect pests. As a colony grows, more and more thorns on the plant are filled. A colony may grow to more than a thousand workers within a year and, eventually, may include tens of thousands of workers. At any one time, about a quarter of the ants are outside their nests actively gathering food and defending the acacia against herbivorous insects. The relationship between *Pseudomyrmex* and *Acacia* is obligatory: neither the ant nor the acacia can survive without the other. Other ant-acacia associations are facultative; that is, the ant and the acacia can co-occur to mutual benefit, but both can exist independently as well. Species of acacia that altogether lack the protection of ants frequently produce toxic compounds to defend their leaves against herbivores.

To test for the influence of ants on the growth and survival of acacia plants, one can keep ants off new acacia shoots and compare their growth with that of shoots that house ants. After 10 months of one such experiment conducted in southern Mexico, shoots lacking ants weighed less than one-tenth as much as those with intact ant colonies, and they produced fewer than half the leaves and one-third the number of swollen thorns.

The mutualism between ants and acacias has been accompanied by adaptations of both species to increase the effectiveness of their association. For example, *Pseudomyrmex* is active both night and day, an unusual trait for ants, and thereby provides protection for the acacia at all times. Also, these ants have a true sting like their wasp relatives and will swarm vertebrate herbivores that attempt to feed on their host plants. The ants also clear away potential plant competitors by attacking seedlings near their host plant's base as well as any vines or overhanging branches of other plants. In a similar adaptive gesture, the acacia retains its leaves throughout the year, and thereby provides a year-round source of food for the ants. Most related species lose their leaves during the dry season.

Dispersive mutualisms

Dispersive mutualisms generally involve animals that transport pollen between flowers in return for rewards such as nectar, or that disperse seeds to suitable habitats and eat the nutritional fruits that contain the seeds. Dispersive mutualisms rarely involve close living arrangements between members of the mutualistic association. Seed dispersal mutualisms are not usually highly specialized: a single bird species may eat many kinds of fruit, for example, and each kind of fruit can be eaten by many kinds of birds. Plant-pollinator relationships tend to be more restrictive because it is in a plant's interest that a flower visitor carry pollen to another plant of the same species. We'll see below how this specialization is achieved.

Pollination

Some plants are wind-pollinated, and pollen grains land on the receptive flowers of other individuals just by chance. Where many species live together within an area and where great distances separate individuals, wind pollination is relatively inefficient. Various sorts of animals visit flowers to feed on highly nutritious pollen and nectar, to gather other substances such as oils or fragrances, to lay their eggs, or to mate. In doing so, they also transport pollen from flower to flower with a relatively high efficiency. Plant-pollinator relationships may have originated as purely consumer-resource interactions: pollen is an excellent food, and the ovaries of flowers, where seeds develop, are excellent brood sites for insect larvae. Even pure acts of consumption result in some pollen being transferred fortuitously between plants.

Since this pattern began, floral structures have been modified through evolution to increase the efficiency of pollen transfer. Many of these modifications involve offering accessible rewards, such as nectar, which are relatively economical for a plant to produce, and arranging flower parts in such a way that pollen is transferred to the bodies of particular animal visitors (Figure 18.16). As flower structure becomes more highly specialized, fewer and fewer types of animals fit a flower in such a way that they contact the anthers and transfer pollen efficiently to the stigmas of other flowers. Thus, flower morphology can exclude certain types of flower visitors and increase the efficiency of pollen transfer.

Plant-pollinator relationships are highly developed in the orchid family, with its variety of flower shapes, colors, and smells. The intricate tie between flower and pollinator is exemplified by the orchid *Stanhopea grandiflora* and the tropical bee *Eulaema meriana*. This obligate mutualism is unusual in that *Stanhopea* flowers produce no nectar and only male *Eulaema* bees visit them. The flowers are extremely fragrant, and each species of *Stanhopea* orchid has its own unique combination of odors so that its specialist pollinator can find it without confusion. Each type of orchid tends to attract a single type of bee.

When a male *Eulaema* bee visits an orchid, it brushes part of the flower with specially modified forelegs and then transfers collected substances to the tibia of its hind leg, which is enlarged and has a storage cavity. This

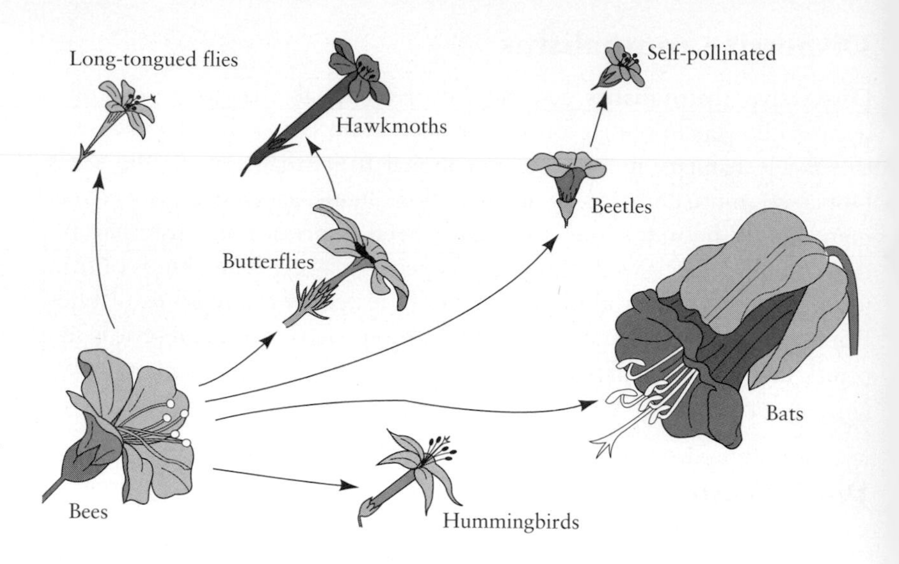

Figure 18.16 Diversity of flower types in the plant family Polemoniaceae, with their major pollinators indicated. The arrows suggest possible pathways in the evolution of the various pollination syndromes. After P. R. Ehrlich and R. W. Holm, *The Process of Evolution,* McGraw-Hill, New York (1963); based on Verne Grant.

behavior provides the bee with a perfume that males use to attract female bees. Each bee species uses a slightly different scent to attract mates. In *Stanhopea,* a bee enters a flower from the side and brushes at a saclike structure on the lip of the flower (Figure 18.17). The surface of the lip is very smooth, and bees often slip when they withdraw from a flower. (The orchid fragrances may also intoxicate bees and cause them to lose their footing.) When a bee slips, it may brush against the column of the orchid flower, where the pollinaria (saclike structures filled with pollen) are precisely placed so as to stick to the hindmost part of the thorax of the bee. If a bee with an attached pollinarium slips and falls out of another flower, the pollinarium catches on the stigma and pollinates the flower. Thus flower structure and bee behavior are mutually adapted to increase the efficiency of pollen transfer.

Taking advantage of mutualisms

Volumes have been written about pollination, but the wonderful complexity of species interactions revolving around pollination systems can be illustrated by a few additional phenomena. One of these is that many flowers offer no rewards, yet are still visited by pollinators. Plants with these "empty" flowers are true cheaters and succeed by luring pollinators with the promise of a reward (most flowers actually do offer them) without actually delivering. Cheating works because most pollinators visit many different kinds of flowers and are attracted to general features of flowers, which cheaters can mimic. In addition, some flowers that normally offer rewards are empty because they have been visited recently and emptied of their nectar. Thus, pollinators are accustomed to encountering empty flowers from time to time.

Many organisms other than plants that produce flowers take advantage of flower visitors for transportation. Mites living in *Heliconius* (banana family) flowers in tropical America hitch rides from plant to plant in the nostrils of hummingbirds. Many pathogenic fungi, such as anther smuts, are transmitted between flowers by pollinators, much like a venereal disease. The fungi are often colored to mimic the flowers in which they reside, or have even evolved to resemble the structures and fragrances of flowers and attract pollinators on their own.

Cheating based on defensive and other kinds of mutualisms is more widespread in nature than one might suppose. Virtually any mutualism or cooperative arrangement can be exploited by deceptive behavior. For example, certain predatory fireflies (which are beetles, in fact) produce light signals that mimic the mating signals of males of other species. When a female is attracted to the display of such an impostor, she is eaten rather than fertilized. A salticid spider of New Zealand employs the same trick, mimicking the mating displays of other species to trick unsuspecting amorous prey. This type of mimicry is referred to as **aggressive mimicry;** it is a tactic that enables predators to avoid being detected by their prey, or even to attract prey.

A final point to round out this discussion is that flowers are dangerous places for many visitors because predators are also attracted to them to feed on pollinators. For example, crab spiders live in inflorescences and match the color of the flowers to conceal themselves; they sit and wait for flower visitors and grab them when they can. One group of hawkmoths (Acherontiinae) has evolved some very clever defenses against this type of attack. The primary innovation is an exceptionally long tongue, which enables a moth to sip nectar from a flower without coming close enough to be grabbed by a lurking predator (Figure 18.18). This strategy, however, makes hawkmoths more vulnerable to bats, which are presented with a line of attack free of obscuring or obstructing vegetation. Faced with this danger, a feeding hawkmoth swings rapidly back and forth in front of a flower, like a pendulum with its fulcrum at the point where the moth's long tongue is inserted into the flower. Bats have a harder time picking off such a moving target.

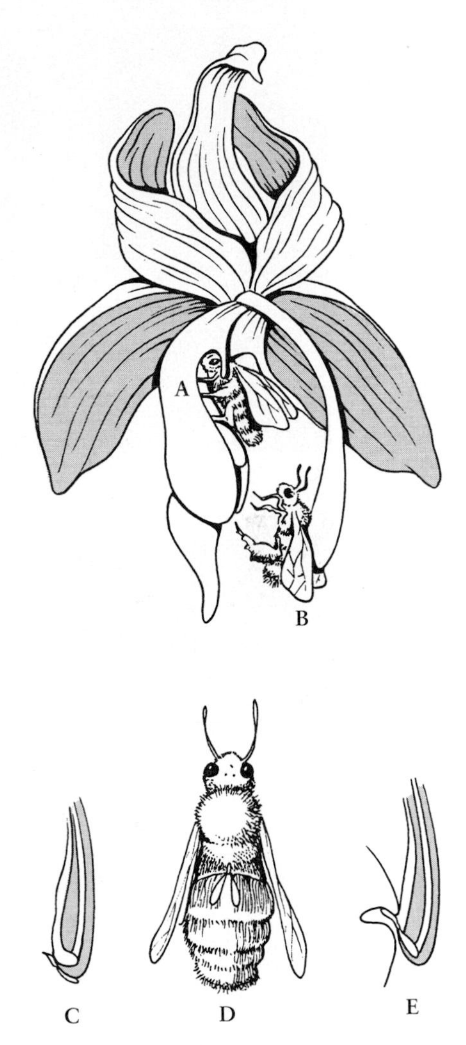

Figure 18.17 Pollination of the orchid *Stanhopea grandiflora* by *Eulaema meriana.* A bee enters from the side and brushes at the base of the orchid flower's lip (A). If it slips (B), the bee may fall against the pollinarium, which is placed on the end of a column (C; longitudinal section through column), in which case the pollinarium becomes stuck to the hind end of the bee's thorax (D). If a bee with an attached pollinarium later falls out of another flower, the pollinarium may catch in the stigma (E), which is so placed on the column that the flower cannot be self-fertilized. After R. L. Dressler, *Evolution* 22:202–210 (1968).

Seed dispersal

Long-distance dispersal of seeds can be advantageous to many plants that germinate and grow best in patches of disturbed habitat, or whose offspring are particularly susceptible to pathogens and herbivores that are abundant in the vicinity of the parent plant. Often, seeds that germinate close to a parent fare poorly because they must compete with another established individual. Of course, seed dispersal requires a mechanism to disperse the seeds. Many plants rely on wind to carry their seeds away, but such seeds must be tiny to be carried effectively on currents of air, and they land more or less at random.

Other plants, particularly in the Tropics, entice various animals—usually birds and mammals—to disperse their seeds for them. In return, such plants offer a reward in the form of a fruit or other edible structure attached to the seed. Attributes of the fruit itself, including its size, color, and structure and its position on the plant, partly determine which species are likely to be effective dispersers. Large fruits can be eaten only by large birds, for whom small fruits often aren't worth the effort. Fruits suspended

(a)

(b)

(c)

Figure 18.18 (a) A long-tongued hawk moth sipping nectar from an inflorescence, out of reach of the predatory crab spider sitting in wait. (b) Multiple exposure showing the pendulum motion of a hawk moth in front of the flower from which it is taking nectar. (c) Occasionally a hawk moth comes too close to a flower and is caught. Photographs courtesy of L. T. Wasserthal.

from the ends of branches may be accessible only to birds that can pick them off while hovering. Most fruits tend to be edible by most consumers and are there for the picking. It is no surprise, then, that seed dispersal does not promote a high degree of specialization. Nevertheless, a general picture of dispersal syndromes has emerged, in which large species of birds that feed on large fruits appear to form a set of mutualists with a greater degree

TABLE 18.3 **Characteristics of trees and birds involved in specialized and generalized dispersal systems**

TREES	BIRDS
Specialized dispersal systems	
Often large seeds (e.g., > 1 g), not mechanically tough	Often large (e.g., > 250 g), with reduced gizzard, intestine
Fruits nutritious and energy-rich, with high lipid or protein content	Depend on fruits for critical portion of diet; reduced use of animal food
Low annual fecundity (e.g., median crop 100–5,000)	Small populations of a few (e.g., 5–10) species/community
Consistently high fruit removal	Efficient daily use of crops
Extended fruiting seasons make use of a limited disperser assemblage	Specialists habituated to consistently use few fruit species per day
Seed dispersal away from parents critical for recruitment	A few reliable specialists most effective for recruitment
Generalized dispersal systems	
Often small seeds (e.g., < 0.5 g), mechanically tough	Often small (e.g., < 50 g), with generalized gut
Fruits less energy rich, with high carbohydrate and water content	Fruits complement primarily nonfruit diet
High annual fecundity (e.g., median crop 20,000–1,000,000+)	Large populations of many (e.g., 20–100) species/community
Low and variable seed removal	Some species may fail to visit for days or weeks at a time
Sharply peaked fruiting season, drawing numerous opportunists to a superabundant display	May use many fruiting species, switch opportunistically, or complement mostly insect diet
Seed dispersal may or may not be critical; many seeds dormant	Many frugivores play roughly similar roles

Source: H. F. Howe, *Vegetatio* 107/108:3–13 (1993).

of interdependence with the plant species whose fruits they eat than do the set of birds that consume smaller fruits (Table 18.3).

When a bird removes a fruit from a plant, it may consume the entire structure, digest and assimilate the fruit pulp, and pass the undigested seeds in the feces. Birds often perch over suitable germination sites, such as at the edges of clearings in forests. A seed dispersed in this manner may find itself in a good habitat with a modest supply of highly nutritive manure. Indeed, the seeds of some species of plants will germinate only after passing through the digestive tract of an animal.

Many mammals pass large numbers of viable seeds through their digestive tracts, and carry others that stick to their hair. Some mammals, however, are not good seed dispersers because they can gnaw through tough seed coats and get at the seed itself. Thus, they become seed predators. But even mammals, like some birds, may bury or otherwise cache

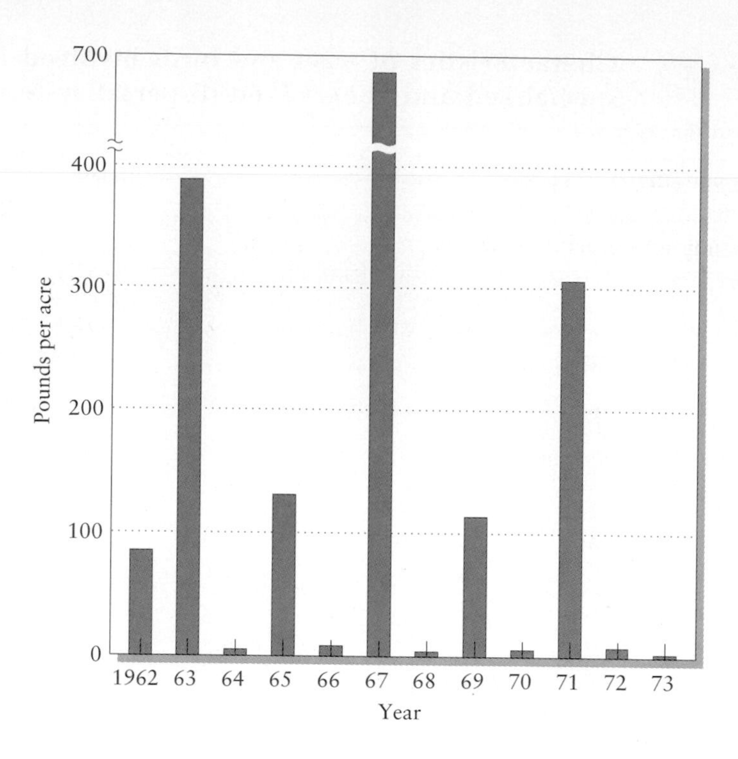

Figure 18.19 Production of white oak acorns in a North Carolina forest, showing alternating years of large and small crops. From R. Brewer, *The Science of Ecology,* 2d. ed., Saunders, Philadelphia (1994); after data in D. E. Beck, *USDA Forest Serv. Res. Note* SE-244:1–8 (1977).

excess seeds that they cannot eat at a single sitting, putting them aside as insurance against poorer times ahead. (Blue jays are notorious hoarders, as anyone who has tried to keep a bird feeder east of the Rockies stocked with sunflower seed can attest.) Many of these seeds are never recovered, in which case they may germinate and grow, the mammal or bird having effectively planted them. Some plants take advantage of this **hoarding** behavior by producing large crops of seeds all at once, overwhelming the capacity of seed predators to consume them. As a result, many are buried, never to be retrieved. This tactic is known as **mast seeding** (mast is an Old English word referring to acorns and beechnuts accumulated on the forest floor that are gathered to feed hogs). In eastern North America, for example, white oaks produce large acorn crops every other year, alternating with a virtual absence of acorns (Figure 18.19). The lean years prevent the buildup of populations of seed parasites and predators and are thought to result in higher germination success overall.

Summary

1. The environments of organisms include biological factors—prey, predators, and diseases—that interact with the physical environment in molding adaptations. Unlike physical factors, biological factors themselves evolve and diversify. The relationship between predators and their prey illustrates the mutual influences of organisms and the biological factors in their environment.

2. Interactions between species can be classified as consumer-resource (including predator-prey, herbivore-plant, and parasite-host) interactions, mutualisms, or competition.

3. Depending on the relative sizes of predators and their prey, predatory behavior may range from the active pursuit of individual prey to the filtering of tiny organisms from large quantities of water.

4. Predators are well adapted to pursue, capture, and eat particular types of prey. Carnivorous and herbivorous mammals, for example, differ in the conformation of their teeth and in the size and design of their digestive systems.

5. Organisms escape predation by avoiding detection; by means of chemical, structural, and behavioral defenses; and by escape. Crypsis and warning coloration are examples of the defenses of edible and unpalatable organisms respectively.

6. In Batesian mimicry, edible organisms evolve to resemble unpalatable organisms that are rejected by predators, thus gaining some protection by deceiving potential attackers. Müllerian mimicry complexes comprise noxious species that use similar displays to advertise that they are unpalatable.

7. Host-parasite relations are specialized interactions characterized by complex life cycles of the parasite that include stages specialized to make the difficult journey from one host to another. Parasite and host often evolve a delicate balance when the parasite's well-being depends on the survival of its host.

8. Plants have evolved numerous structural and chemical defenses to deter herbivores. These defenses include factors that influence the nutritional quality and digestibility of plant parts as well as specialized chemicals—secondary compounds—that have toxic effects on animals and microorganisms. Most of these chemicals are nitrogen-containing compounds, terpenoids, or phenolics derived by modification and elaboration of normal metabolic pathways.

9. Many herbivores have evolved ways to detoxify secondary chemicals of plants, enabling them to specialize on plant hosts that are poisonous to most other species.

10. Mutualisms are relationships that benefit both parties. They may be classified as trophic, defensive, or dispersive. In trophic mutualisms, each partner is specialized to provide a different limiting nutrient, as in the cases of lichens, ruminant mammals, and root nodules. In defensive mutualisms, such as the ant-acacia interaction, one partner provides protection, usually in return for food.

11. Dispersive mutualisms are plant-animal interactions in which the animal disperses pollen or seeds in the course of harvesting or processing food that the plant supplies. The structures of flowers and fruits limit the variety of animals that perform this function for a particular species of plant,

thereby increasing the efficiency of pollen transfer and the likelihood that seeds will reach suitable sites for germination and growth.

Suggested readings

Abrahamson, W. G. 1989. *Plant-Animal Interactions.* McGraw-Hill, New York.

Allen, M. F. 1991. *The Ecology of Mycorrhizae.* Cambridge University Press, New York.

Armbruster, W. S. 1992. Phylogeny and the evolution of plant-animal interactions. *BioScience* 42:12–20.

Barbosa, P., P. Gross, and J. Kemper. 1991. Influence of plant allelochemicals on the tobacco hornworm and its parasitoid, *Cotesia congregata. Ecology* 72:1567–1575.

Fleming, T. H. 1993. Plant-visiting bats. *American Scientist* 81:460–467.

Fritz, R. S., and E. L. Simms (eds.). 1992. *Plant Resistance to Herbivores and Pathogens: Ecology, Evolution, and Genetics.* University of Chicago Press, Chicago.

Hay, M. E. 1991. Marine-terrestrial contrasts in the ecology of plant chemical defenses against herbivores. *Trends in Ecology and Evolution* 6:362–365.

Jackson, R. R., and R. S. Wilcox. 1990. Aggressive mimicry, prey-specific predatory behaviour and predator-recognition in the predator-prey interactions of *Portia fimbriata* and *Euryattus* sp., jumping spiders from Queensland. *Behavioral Ecology and Sociobiology* 26:111–119.

May, M. 1991. Aerial defense tactics of flying insects. *American Scientist* 79: 316–329.

Real, L. (ed.). 1983. *Pollination Biology.* Academic Press, Orlando, Fla.

Rice, E. L. 1984. *Allelopathy.* 2d ed. Academic Press, Orlando, Fla.

Robinson, M. H. 1969. Defenses against visually hunting predators. *Evolutionary Biology* 3:225–259.

Sagers, C. L., and P. D. Coley. 1995. Benefits and costs of defense in a Neotropical shrub. *Ecology* 76:1835–1843.

Stein, B. A. 1992. Sicklebill hummingbirds, ants, and flowers. *BioScience* 42:27–33.

Trager, W. 1986. *Living Together: The Biology of Animal Parasitism.* Plenum, New York.

Whittaker, R. H., and P. P. Feeny. 1971. Allelochemics: Chemical interactions between species. *Science* 171:757–770.

Wickler, W. 1968. *Mimicry in Plants and Animals.* World University Library, London.

CHAPTER 19

COMPETITION

Competition is any use or defense of a resource by one individual that reduces the availability of that resource to other individuals. Competition is one of the most important ways in which the activities of individuals affect the well-being of others, whether they belong to the same species **(intraspecific competition)** or to different species **(interspecific competition).** As we saw earlier, competition within populations reduces resource levels in a density-dependent manner and thereby affects fecundity and survival. The more crowded a population, the stronger is competition between individuals. Thus, intraspecific competition underlies the regulation of population size. Furthermore, when genetic factors cause individuals to differ in the efficiency with which they exploit resources, more efficient individuals may leave more descendants than less efficient ones, and the proportion of their genes may increase in populations over time. In this way, intraspecific competition is intimately related to evolutionary change.

Competition between individuals of different species causes a mutually depressing effect on the populations of both; each species contributes to the regulation of the other as well as its own population. Under some conditions, particularly when interspecific competition is intense, it may lead to the elimination of one species by the other. Because of this potential, competition is an important factor in determining which species can coexist within a habitat.

The outcome of competition between two populations depends on the relative efficiencies with which they exploit their shared resources. Every population consumes resources. When resources are scarce relative to demand for them, each act of consumption by one individual makes a resource less available to others, as well as to itself. As consumption continues, resources decline to levels that no longer support the growth of the consuming population, and the population may reach an equilibrium size. When one population can continue to grow at a resource level that curtails the growth of a second population, the first will eventually replace the second. Thus, competition and its various outcomes depend on the relationship of consumers to their resources.

In this chapter, we shall consider some of the general principles of competition between species, illustrate the potential effects of competition by examining the results of laboratory experiments, and demonstrate the importance of competition in natural systems.

Resources

Ecologist David Tilman of the University of Minnesota defines a **resource** as any substance or factor that is consumed by an organism and that can lead to increased population growth rates as its availability in the environment is increased. Two qualifications are key to this definition. First, a resource is consumed and its amount is thus reduced. Second, a resource is used by a consumer for its own maintenance and growth. Thus food is always a resource, and water is a resource for terrestrial plants and animals. Water is consumed, and it is critical to maintenance and growth; furthermore, when its availability is reduced, biological processes are affected in such a way as to reduce population growth.

Consumption connotes more than the act of eating. For sessile animals, space (open, available sites) is considered a resource. Among barnacles growing on rocks within the intertidal zone, individuals require space to grow, and larvae require space to settle and take up adult life (Figure 19.1). Crowding increases adult mortality and reduces fecundity by limiting the growth of adults and the recruitment (settling) of larvae. Open space fosters reproduction and recruitment, and individuals "consume" open sites as they colonize and grow on them. Hiding places and other safe sites constitute another kind of resource. Each area of habitat has a limited number of holes, crevices, or patches of dense cover in which an organism may escape predation or seek refuge from inclement weather. As some individuals occupy the best sites, others must settle for less favorable places; they may suffer higher mortality as a consequence.

What factors are not resources? Temperature is not a resource. Higher temperatures may raise reproductive rates, but individuals do not consume temperature. Temperature and other nonconsumable physical and biological factors are important, of course, but they must be considered differently from resources. Temperature, humidity, salinity, hydrogen ion concentration (pH), buoyancy, and viscosity are all **conditions** that influence the rates of processes and therefore the individual's ability to consume

Figure 19.1 Competition for space among barnacles on the Maine coast. (a) Above their optimal range in the intertidal zone, the barnacles are sparse, and young can settle in the bare patches. (b) Lower in the intertidal zone, dense crowding of barnacles precludes further population growth; young barnacles can settle only on older individuals. Courtesy of the American Museum of Natural History.

resources, but they are not themselves used and thereby transformed by the activities of organisms.

Resources can be classified according to how their consumers affect them. **Nonrenewable resources,** such as space, are not altered by use. Once occupied, space becomes unavailable; it is "replenished" only when the consumer leaves. In contrast, **renewable resources** are constantly regenerated, or renewed. Births in a prey population continually supply food items for predators. The continuous decomposition of organic detritus in the soil provides a fresh supply of nitrate for plant roots.

Among renewable resources, we recognize three types. The first type includes resources that have a source external to the system, beyond the influence of consumers. Sunlight strikes the surface of the earth regardless of whether plants "consume" it; local precipitation is largely independent of the consumption of water by plants; for all practical purposes, detritus rains down from the sunlit surface to the abyssal depths of the oceans uninfluenced by the consumers groping there in everlasting darkness.

Renewable resources of the second type are generated within the ecosystem and are directly affected by the activities of consumers. Most predator-prey, plant-herbivore, and parasite-host interactions involve resources of this type.

Renewable resources of the third type issue from within a system, but resource and consumer are linked indirectly, either through other resource-consumer steps or through abiotic processes. For example, in the nitrogen cycle of a forest, plants assimilate nitrate from the soil. Herbivores and detritivores consume plant biomass, returning large quantities of organic nitrogen compounds to the soil. These are attacked by microorganisms, which release the nitrogen in a form the plants can use. The uptake of nitrate by plants has little direct effect on its release by detritivores. Similarly, consumption of detritus cannot immediately influence plant production. Clearly, however, detritivores and microorganisms do influence plant production indirectly through the rate at which they release nutrients into the soil.

Limiting resources

Consumption reduces the availability of a resource. What is used by one organism cannot be used by another. By diminishing their resources, consumers limit their own population growth. As a population grows, its overall resource requirement grows as well. When the requirement increases so much that the decreasing supply of resources can no longer fulfill the need, population size levels off or even begins to decrease. However, whereas all resources, by definition, are reduced by their consumers, not all resources limit consumer populations in this way. All terrestrial animals require oxygen, for example, but they do not depress its level in the atmosphere even noticeably before some other resource, such as food supply, limits population growth.

The potential of a resource to limit population growth depends on its availability relative to demand. At one time, ecologists believed that populations were limited by the single resource that had the greatest relative scarcity. This principle has been called **Liebig's law of the minimum,** after Justus von Liebig, a German chemist who set forth the idea in 1840. According to this law, each population increases until the supply of some resource (the **limiting resource**) no longer satisfies the population's requirement for it. The growth of a population under a given set of conditions responds uniquely to the level of each of its resources. When the diatom *Cyclotella meneghiniana* was grown under silicate and phosphate limitation in a laboratory culture, population growth and resource depletion ceased when phosphate levels were reduced to 0.2 μM or silicate levels were reduced to 0.6 μM. According to Liebig's law of the minimum, whichever of these resources is reduced to this limiting value first regulates the growth of the *Cyclotella* population.

Liebig's law applies strictly only to resources having an independent influence on the consumer. In many cases, two or more resources interact to determine the growth rate of a consumer population; that is, the growth rate of a population at a particular level of one resource depends on the level of one or more other resources.

When two resources together enhance the growth of a consumer population more than the sum of both individually, the resources are said to be synergistic (from the classical roots *syn,* "together," and *ergon,* "work"). This principle can be illustrated by a study of the small herbaceous plant *Impatiens parviflora,* which is common in woodlands of England. In one experiment, fertilized (with nitrate and phosphate) and nonfertilized (control) *Impatiens* were exposed to different levels of light from the time of seed germination until the end of the experiment at 5 weeks. Added light enhanced the growth of fertilized plants more than that of controls (Figure 19.2); hence the ability of *Impatiens* to use light depends on the presence of other resources. Plant growth requires both the carbon assimilated by photosynthesis, as a source of energy and for structural carbohydrates, and nitrogen and phosphorus for the synthesis of proteins and amino acids. At the highest light intensities used in the experiment, nitrogen and phosphorus were also shown to interact in their effect on plant growth (Figure 19.3), demonstrating that both are required for normal growth.

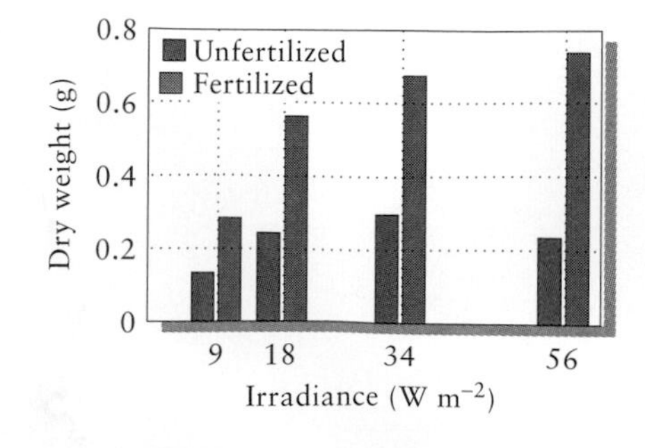

Figure 19.2 Joint influence of light levels and fertilizer on the growth of *Impatiens*. After W. J. H. Peace and P. J. Grubb, *New Phytol.* 90:127–150 (1982).

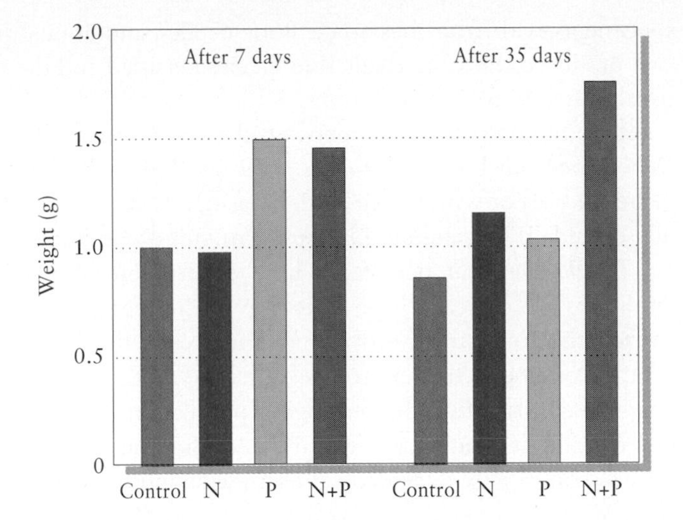

Figure 19.3 Joint influence of nitrogen (N) and phosphorus (P) fertilization on growth of *Impatiens.* Note that in the first week of growth, adding nitrogen alone had little effect (there was so much in the soil that it was not a limiting factor for young plants). After 35 days, however, neither nutrient alone enhanced growth as much as the addition of both. Clearly, nitrogen and phosphorus were synergistic in promoting plant growth. After W. J. H. Peace and P. J. Grubb, *New Phytol.* 90:127–150 (1982).

The experimental demonstration of competition

When ecologists began to consider the dynamics of population growth in the late 1920s, many investigators initiated experiments to determine the effects of one species on the population growth of another. In these experiments, two species were first grown separately under controlled conditions and resource levels to determine the equilibrium sizes of their populations (carrying capacities) in the absence of interspecific competition. The two species were then grown together under the same conditions to determine the effect of each on the other. The difference between the population growth of one species in the presence and in the absence of the other was taken as a measure of competition between them. Experiments by the Russian biologist G. F. Gause on protozoans were among the first and most influential on subsequent work in population biology. When the protozoans *Paramecium aurelia* and *P. caudatum* were established separately on the same type of nutritive medium, both populations grew rapidly to limits imposed by resources. When the two species were grown together, however, only *P. aurelia* persisted (Figure 19.4).

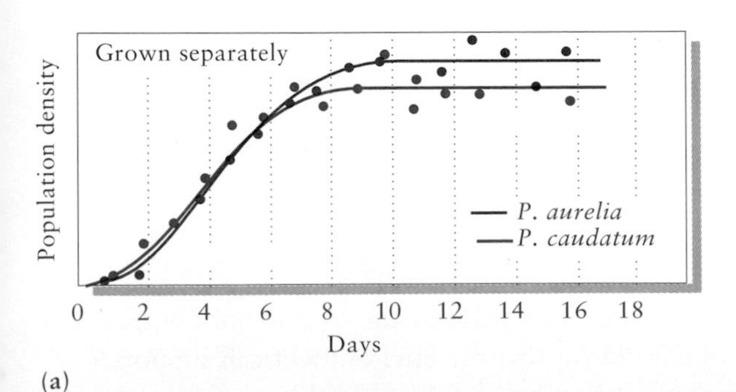

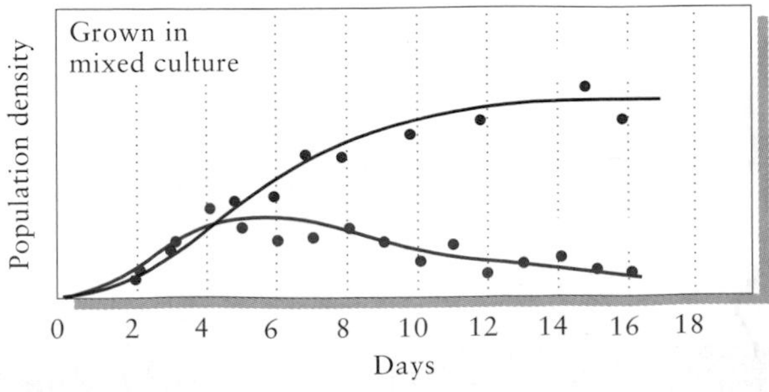

Figure 19.4 Population densities of two species of *Paramecium* when grown in separate cultures (a) and when grown together (b). Although both species thrive when grown separately, *P. caudatum* cannot survive together with *P. aurelia*. After G. F. Gause, *The Struggle for Existence,* Williams & Wilkins, Baltimore (1934).

Similar experiments with fruit flies, mice, flour beetles, and annual plants almost always produced the same result: one species persisted and the other died out, usually after 30 to 70 generations.

The outcome of competition depends on the conditions of the environment. When the flour beetles *Tribolium castaneum* and *T. confusum* were grown together in vials of wheat flour under warm, dry conditions, *T. confusum* usually excluded *T. castaneum*. Under warm, moist conditions, however, it was *T. castaneum* that persisted. These experiments showed that physical conditions are critical to the outcome of competition, but they also emphasized that only the relative performance of the two species under the same conditions matters to the outcome of their interaction. *Tribolium confusum* consistently achieved higher population densities under warm, moist conditions than under warm, dry conditions when grown alone, but it was consistently excluded under moist conditions by *T. castaneum,* whose population growth was favored even more (Figure 19.5).

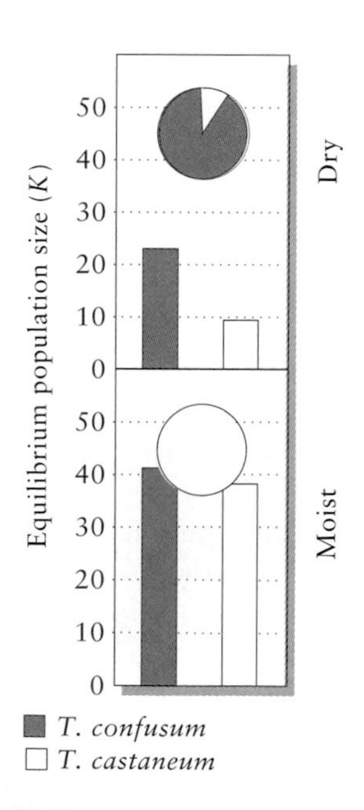

Figure 19.5 The outcome of competition between two species of flour beetles of the genus *Tribolium* at warm temperatures under dry and moist conditions. The vertical bars illustrate the equilibrium populations (total number of adults, larvae, and pupae per gram of flour) when each species is grown alone; the pie charts present the percentage of 20 to 30 contests won by each species. After data in T. Park, *Physiol. Zool.* 27:177–238 (1954); T. Park, *Science* 138:1369–1375 (1962).

The competitive exclusion principle

The accumulating results of laboratory experiments on competition eventually appeared so general as to warrant their elevation to the status of a general rule, the **competitive exclusion principle.** This principle can be summarized as follows: Two species cannot coexist indefinitely on the same limiting resource. The qualification *limiting* is required in the definition of the principle because competitive exclusion expresses itself only when consumption depresses resources and thereby limits population growth.

The competitive exclusion principle states that species having identical resource requirements cannot coexist. Similar species do, of course, coexist in nature. But as we shall see in later chapters, detailed observations always reveal ecological differences between such species, often based on subtle differences in habitat or diet preference. These observations prompt us to ask how much ecological segregation is sufficient to allow coexistence. Although this question has been very difficult to answer, theoretical analyses of competition have suggested some of the general conditions under which species may coexist.

The theory of competition and coexistence

Most competition theory springs from the formulations of A. J. Lotka and G. F. Gause, who used the logistic equation for population growth as their starting point. Remember that according to the logistic equation, the rate of increase of population i is expressed by

$$\frac{dN_i}{dt} = r_i N_i \left(\frac{K_i - N_i}{K_i}\right),$$

where r is the exponential rate of increase in the absence of competition and K is the number of individuals that the environment can support (carrying capacity). Intraspecific competition appears as the term $(K - N)/K$;

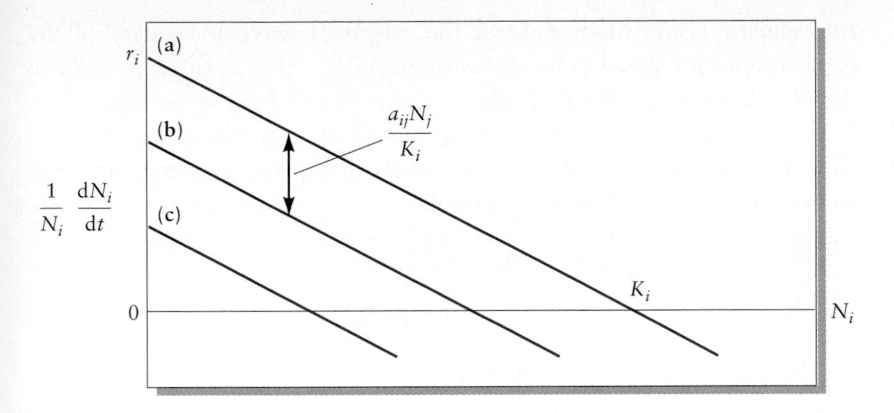

Figure 19.6 The per capita population growth rate of species i as a function of the population density (N_i) under purely intraspecific competition (a) and under low (b) and high (c) levels of interspecific competition. Interspecific competition reduces the equilibrium level of the population below the carrying capacity (K_i) in the absence of competition.

as N approaches K—that is, as population size approaches the carrying capacity—$K - N$ approaches 0. As we have seen before, a stable equilibrium is reached when $N = K$ and population size has reached the carrying capacity (Figure 19.6).

The Italian ecologist Vito Volterra incorporated interspecific competition into the logistic equation by adding the term $-a_{ij}N_j/K_i$ to the quantity within the parentheses. Hence

$$\frac{dN_i}{dt} = r_iN_i\left[\frac{K_i - N_i - a_{ij}N_j}{K_i}\right],$$

where N_j is the number of individuals of a second species (j), and a_{ij} is the coefficient of competition—that is, the effect of an individual of species j on the exponential growth rate of the population of species i. We may think of the competition coefficient a_{ij} as the degree to which individuals of species j use the resources of individuals of species i. That is why $a_{ij}N_j$ is divided by K_i, the carrying capacity of species i. How much individuals of species j usurp the resources of species i determines the effect of j on i's rate of population growth and the equilibrium size of population i under interspecific competition (see Figure 19.6).

Strictly speaking, the term N_i/K_i should have coefficient a_{ii}, but this is assumed to be 1 and is left out. Although it need not be, the value of a_{ij} is usually less than 1; individuals of the same species are likely to compete more intensely than individuals of different species. Because each species of a pair exerts an effect on the other, the mutual relationship between them requires two equations: one, presented above, for the effect of species j on species i, and a second, similar equation for the effect of species i on species j. If two species are to coexist, the populations of both must reach a stable size greater than 0. That is, both dN_i/dt and dN_j/dt must equal 0 at some combination of positive values of N_i and N_j. From the previous equation, we see that $dN_i/dt = 0$ when

$$\hat{N}_i = K_i - a_{ij}N_j$$

The little "hat" (^) over the N indicates that it is an equilibrium value. In the absence of interspecific competition ($a_{ij} = 0$), the equilibrium population size $\hat{N}_i$ is equal to K_i, a measure of the resources available to species i.

Interspecific competition reduces the effective carrying capacity of the environment for species i by the amount $a_{ij}\hat{N}_j$ —that is, in proportion to the population size and coefficient of competition of the second species.

By making some algebraic substitutions, we can express the equilibrium population size of species i in terms of just the carrying capacities and competition coefficients, specifically

$$\hat{N}_i = \frac{K_i - a_{ij}K_j}{1 - a_{ij}\,a_{ji}}$$

Because competition coefficients generally are less than 1, the denominators of these equations normally assume positive values. Therefore, the equilibrium value $\hat{N}_i$ will be positive only when the numerator of the equation is positive and hence when a_{ij} is less than the ratio K_i/K_j. In words, coexistence is more likely when coefficients of competition are relatively weak and the carrying capacities of both competitors are similar. Any number of species may coexist as long as these criteria are met for all pairs of them. In the most general terms, the coexistence of two species requires that

$$a_{ij}a_{ji} < 1.$$

Competition in nature

Competitive exclusion in nature is a transient phenomenon. The evidence of exclusion having taken place is lost when a poor competitor disappears. We can observe the process in the laboratory when we mix populations according to our whim and follow the course of their interaction. The closest natural analogy to a laboratory experiment is the accidental or intended introduction of species by humans. For example, when several species of parasites are introduced to an area simultaneously to control a weed or insect pest, the control species are brought together in the same locality to exploit the same resource. It is not surprising that competitive exclusion occurs under these conditions.

Between 1947 and 1952, the Hawaii Agriculture Department released thirty-two parasitoid species to combat several species of fruit pests, including the Oriental fruit fly. Of these species, thirteen became established, but only three kinds of braconid wasps proved to be important parasitoids of fruit flies. Populations of these three wasps, all closely related members of the genus *Opius,* successively replaced each other from early 1949 to 1951, after which only *Opius oophilus* was commonly found to parasitize fruit flies (Figure 19.7). As each parasite population was replaced by a more successful species, the level of parasitism on fruit flies by wasps also increased, suggesting superior competitive ability.

A similar pattern of replacement, in this case involving wasps that parasitize scale insects, has been more thoroughly documented in southern California. Scale insects are pests of citrus groves and can cause extensive damage to trees. As the evolution of resistance by pests reduced the effectiveness of chemical pesticides, agricultural biologists turned to the im-

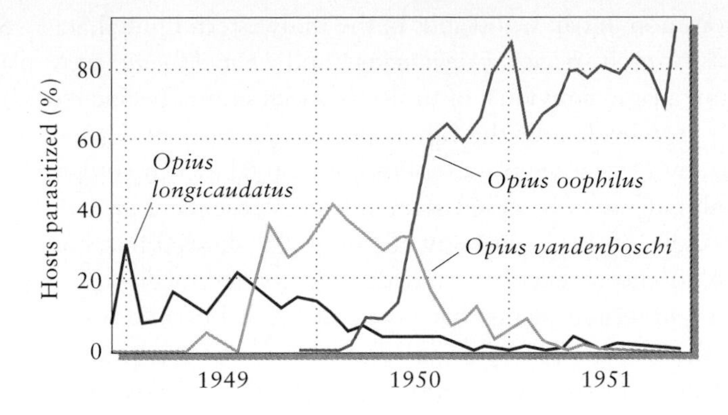

Figure 19.7 Successive changes in the predominance of three species of wasps of the genus *Opius* that are parasitic on the Oriental fruit fly. After H. A. Bess, R. van den Bosch, and F. A. Haramoto, *Proc. Hawaiian Entomol. Soc.* 17:367–378 (1961).

portation of insect parasitoids and predators. Yellow scales have infested California citrus groves since oranges and lemons were first planted there. In the late 1800s, the red scale was accidentally introduced and replaced the yellow scale throughout most of its range, perhaps itself a case of competitive exclusion.

Of the many species introduced in an effort to control citrus scale, tiny parasitoid wasps of the genus *Aphytis* (from the Greek *aphyo,* "to suck") have been most successful. One species, *A. chrysomphali,* was accidentally introduced from the Mediterranean region and became established by 1900. Despite its tremendous population growth potential, *A. chrysomphali* did not effectively control scale insects, particularly not in the dry interior valleys. In 1948 a close relative from southern China, *A. lingnanensis,* was introduced as a control agent. This species increased rapidly and widely replaced *A. chrysomphali* within a decade. When both species were grown in the laboratory, *A. lingnanensis* was found to have the higher net reproductive rate, whether the two species were placed separately or together in population cages.

Although *A. lingnanensis* had excluded *A. chrysomphali* throughout most of southern California, it still did not provide effective biological control of scale insects in the interior valleys because cold winter temperatures greatly reduced parasitoid populations. Its larval development slows to a standstill at temperatures below 16°C (60°F), and adults cannot tolerate temperatures below 10°C (50°F). In 1957 a third species of wasp, *A. melinus,* was introduced from areas in northern India and Pakistan where temperatures range from below freezing in winter to above 40°C in summer. As was hoped, *A. melinus* spread rapidly throughout the interior valleys of southern California, where temperatures resemble those of the wasp's native habitat, but it did not become established in the milder coastal areas.

Experimental studies of competition in plants

The depressing effect of interspecific competition on the growth of plants has been demonstrated in many experimental studies. One such study used two species of *Desmodium,* small herbaceous legumes (members of the pea

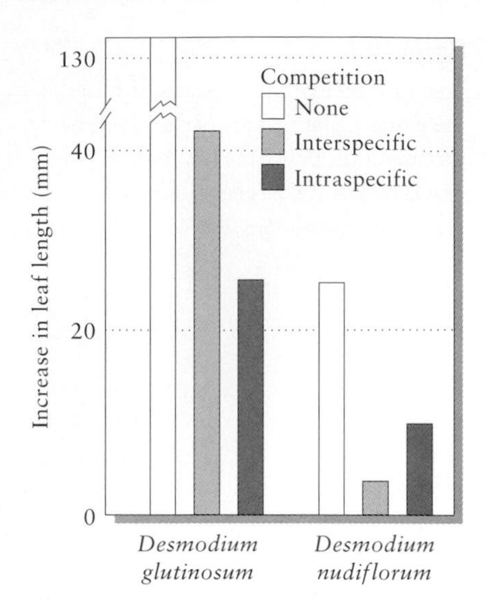

Figure 19.8 The growth responses of two species of *Desmodium* when planted near individuals of the same species, near individuals of the other species, and at a distance from individuals of either species. After W. G. Smith, *Am. Midl. Nat.* 94:99–107 (1975).

family) common in oak woodlands in the midwestern United States. Small individuals of each species, *D. glutinosum* and *D. nudiflorum,* were planted 10 cm from a large individual of the same species (intraspecific test), 10 cm from a large individual of the other species (interspecific test), or at least 3 m from any *Desmodium* plant (control). The total increase in length of all leaves, both old and new, served as an index to subsequent growth.

The results of the experiment (Figure 19.8) showed that both species grew best in the absence of individuals of either species (though surrounded by unrelated plants that occur in the habitat). It was also clear, however, that the growth of *D. nudiflorum* was depressed more by interspecific than by intraspecific competition, and that interspecific competition was asymmetrical, with *D. glutinosum* exerting the stronger effect and *D. nudiflorum* the weaker.

Experimental studies have also shown that the presence of one species may enhance the growth or survival of another that potentially competes for the same resources. In one study in a tundra habitat in northern Finland, removal of individuals of several dwarf shrub species revealed that the presence of *Vaccinium uliginosum* had a negative effect on *Vaccinium vitis-idaea,* as one would expect of competition between closely related species, but a positive effect on *Empetrum nigrum.* Evidently, the larger *Vaccinium* provides protective shelter for the smaller *Empetrum* against the harsh arctic climate, which counterbalances the effects of competition for soil nutrients.

Plant competition in nutrient-rich and nutrient-poor habitats

Ecologists are divided over the role of competition in nutrient-rich versus nutrient-poor habitats. One might think that where nutrient levels are high, they are less likely to be limiting to plant populations, and therefore that interspecific competition should be weaker. Accordingly, plant ecologists P. J. Grubb and David Tilman have suggested that the intensity of competition is greater where resources are less abundant. The opposite point of view, espoused by J. P. Grime and Paul Keddy, suggests that on poor soils, abiotic factors, such as water and mineral nutrients, limit plant populations so much that the intensity of biotic interactions is low. Where individual plants are spaced far apart and aboveground biomass is low, there is probably relatively little competition for light. Under these same circumstances, however, the root systems of plants may compete intensely for limiting soil water and nutrients.

One way to distinguish between these hypotheses is to conduct competition experiments in high-productivity and low-productivity environments. The results often depend on the particular system used and the way in which the experiments are designed. For example, a competition experiment in Israel with the desert annual *Stipa capensis* showed that competition intensity increased with increasing plant productivity, measured by aboveground biomass (Figure 19.9). Variation in productivity was related to variation in soil water, which differed between the 2 years of the study and which was also manipulated during each year. With little water, plants achieved small stature and did not interfere with each other; apparently, most of the competition in this system occurs aboveground. Furthermore, the experiment involved an annual plant, which grows up from seed each year and then dies. Low water availability could prevent plants from growing large enough root systems to compete for limited water or nutrients in the soil.

In another study conducted in prairie habitat in Minnesota, low-, medium-, and high-productivity plots were established by adding ammonium nitrate fertilizer. Aboveground biomass varied among the plots by factors of only two to three, not several orders of magnitude as in the study in Israel. Competition intensity for three species of prairie grasses did not vary significantly over the nutrient gradient (Figure 19.10). The investigators suggested that competition was intense belowground on low-nutrient plots and aboveground on high-nutrient plots, resulting in strong competition across the gradient. What are we to make of such studies? Competition appears to be pervasive, but the manifestation of competition depends very strongly on the characteristics of the species and habitats in which competitive effects are investigated.

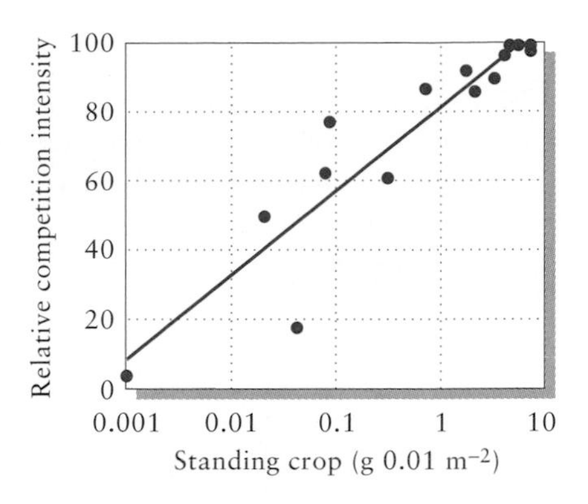

Figure 19.9 The relative intensity of interspecific competition between desert annual plants *(Stipa capensis)* increases as a function of the productivity of the habitat, measured by standing crop. Differences in productivity were produced by water supplements. After R. Kadmon, *J. Ecol.* 83:253–262 (1995).

Experimental studies of competition in animals

Plants occupy space. In dense plantings, roots and leaves crowd together so closely that individuals constantly vie for sunlight, water, and soil nutrients. The animals that most closely resemble plant systems in this regard are sessile invertebrates of rocky shores. Among the most prominent of these are barnacles, which may form dense, continuous populations. Just as plants rely on light as a source of energy, barnacles gather food in the form of plankton from the water that washes over them.

One of the first experimental demonstrations of competition in nature resulted from the work of Joseph Connell on two species of barnacles within the intertidal zone of the rocky coast of Scotland. Adults of *Chthamalus stellatus* normally occur higher in the intertidal zone than those of *Balanus balanoides,* the more northerly of the two species. Although the vertical distributions of newly settled larvae of the two species overlap broadly within the intertidal zone, the line between the vertical distributions of adults is sharply drawn.

Connell demonstrated that adult *Chthamalus* live only in that portion of the intertidal zone above *Balanus* not because of physiological tolerance limits, but because of interspecific competition. When Connell removed *Balanus* from rock surfaces, *Chthamalus* thrived in the lower portions of the intertidal zone where they normally did not occur.

The two species compete directly for space. *Balanus* have heavier shells and grow more rapidly than *Chthamalus;* as individuals expand, the shells of *Balanus* edge underneath those of *Chthamalus* and literally pry them off the rock! *Chthamalus* can occur in the upper parts of the intertidal zone because they are more resistant to desiccation than *Balanus;* even when

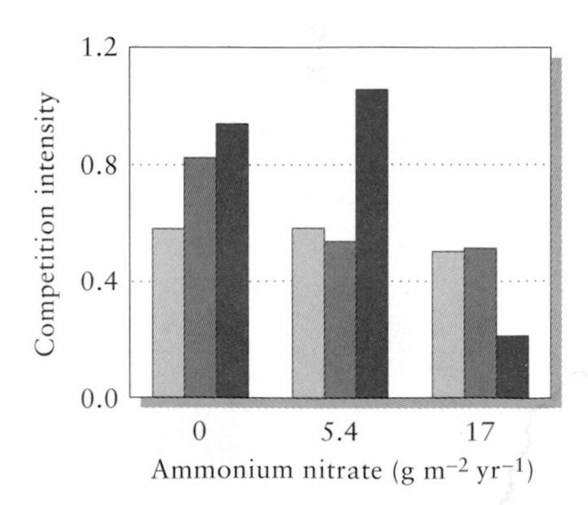

Figure 19.10 The relative intensity of interspecific competition in three species of prairie grasses under conditions of low, medium, and high nutrient availability. None of the three species showed a significant trend in competition intensity across the nutrient gradient. After S. D. Wilson and D. Tilman, *Ecology* 72:1050–1065 (1991).

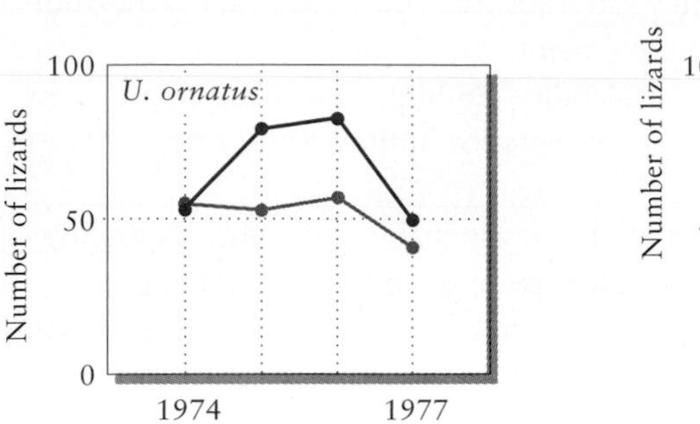

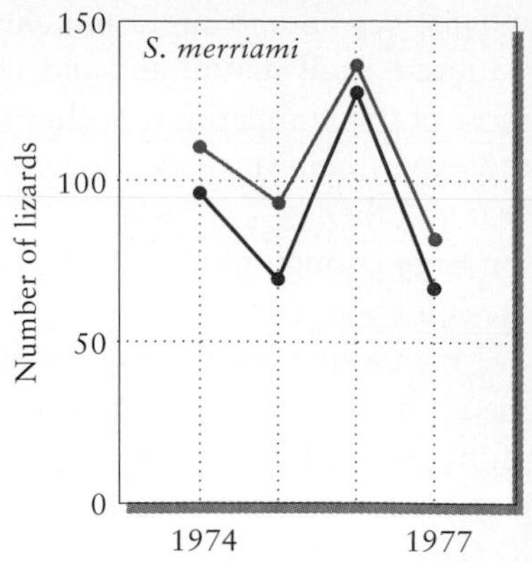

Figure 19.11 Populations of two species of lizards (individuals per hectare) on plots from which the other species was removed and on control plots where both species remained. From A. E. Dunham, *Ecol. Monogr.* 50:309–330 (1980).

surfaces in the upper levels are kept free of *Chthamalus, Balanus* do not invade.

Competition between barnacles results from physical interference rather than from depression of shared food or other resources. Mobile animals may exhibit similar interference competition through occasional aggressive encounters. For example, two species of voles (small mouselike rodents of the genus *Microtus*) co-occur in some areas of the Rocky Mountain states. In western Montana, the meadow vole *(M. pennsylvanicus)* normally lives in wet habitats surrounding ponds and watercourses, whereas the mountain vole *(M. montanus)* is restricted to dry habitats. When meadow voles were trapped and removed from an area of wet habitat, mountain voles began to move in from surrounding dry habitats. And when mountain voles were trapped and removed from a dry habitat, which they occupied exclusively, meadow voles began to show up there. Each species excludes the other from one habitat by aggressive behavior.

Although territorial defense and social aggression occur frequently within species, they are not as common between species. Interspecific competition occurs more regularly through exploitation of resources. Because exploitative competition expresses its effects indirectly, through differential survival and reproduction of individuals of different species, it may be difficult to detect.

In Big Bend National Park, Texas, the canyon lizard *Sceloporus merriami* and the tree lizard *Urosaurus ornatus* appear to compete for a shared food resource. Both species search for insect prey on exposed surfaces of large rocks. When *Sceloporus* were removed from experimental areas, the numbers of *Urosaurus* increased over those in control areas during 2 years of a 4-year study (Figure 19.11). In contrast, removal of *Urosaurus* did not result in an increase in *Sceloporus* populations. Fluctuations in control and experimental populations of *Sceloporus* were closely related to rainfall (1975 and 1977 were dry years), suggesting that physical factors may limit *Sceloporus* more severely, whereas competition influences *Urosaurus* more strongly, a situation recalling the relative ecological positions of *Balanus* and *Chthamalus* in the intertidal zone.

Mechanisms of competition

Earlier in this chapter, **exploitation competition,** in which individuals deprive others of the benefits of resources, was distinguished from **interference competition,** in which individuals directly affect others by physical (chasing and fighting, for example) or chemical means (toxins). Thomas Schoener, an ecologist at the University of California at Davis, subdivided mechanisms of competition into six categories in order to determine whether they varied consistently among species and habitats. His categories were as follows:

1. Consumptive competition, based on the use of some renewable resource (this is equivalent to exploitation competition)
2. Preemptive competition, based on the occupation of open space
3. Overgrowth competition, which occurs when one individual grows upon or over another, thereby depriving the second of light, nutrient-laden water, or some other resource
4. Chemical competition, by production of a toxin that acts at a distance after diffusing through the environment
5. Territorial competition (defense of space)
6. Encounter competition, which involves transient interaction over a resource that may result in physical harm, loss of time or energy, or theft of food

Categories 4–6 are different manifestations of interference competition.

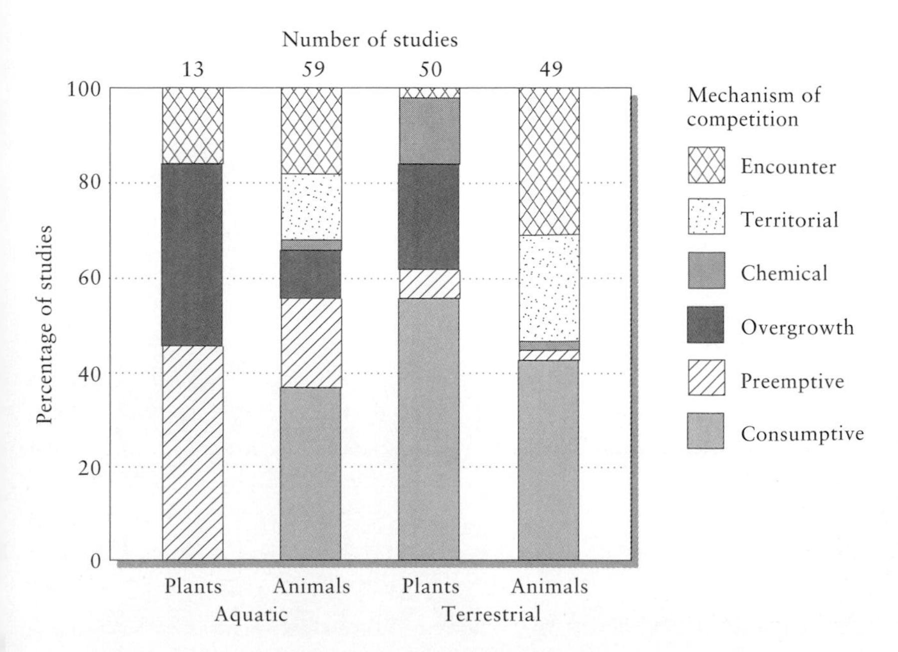

Figure 19.12 Frequencies of various mechanisms of competition among plants and animals in aquatic and terrestrial habitats, as revealed by experimental field studies. From data in T. W. Schoener, *Am. Nat.* 103:277–313 (1983).

The occurrence of each of these mechanisms of competition depends on the capabilities of organisms and the habitats in which they occur (Figure 19.12). Preemptive and overgrowth competition appear among sessile space users, primarily terrestrial plants and marine organisms living on hard substrates. Territorial and encounter competition appear among actively moving animals. Chemical competition appears among terrestrial plants (toxins are diluted too readily in aquatic systems). By Schoener's count, consumptive (exploitation) competition is the commonest, especially in terrestrial environments. Preemptive and overgrowth competition predominate among plants and algae in marine habitats, where, particularly on hard substrates, space and light are the most limiting resources; in terrestrial habitats, plants additionally compete by exploitation (consumption) of nutrients in the soil.

Of Schoener's categories, we have not yet discussed chemical competition, or **allelopathy.** Although the causing of injury *(-pathy)* to other individuals *(allelo-)* by chemical means has been reported most frequently in terrestrial plants, such interactions may take on a variety of forms. Some parasites appear to exclude other, potentially competing parasites from a host by stimulating the host's immune system against them. It has also been suggested that the abundant oils in the eucalyptus trees of Australia promote frequent fires in the leaf litter, which kill the seedlings of competitors. More frequently, it is the direct effect of a toxic substance that does the damage.

In shrub habitats in southern California, several species of sage (genus *Salvia)* use chemicals to inhibit the growth of other vegetation. Clumps of *Salvia* are usually surrounded by bare areas separating the sage from neighboring grassy areas (Figure 19.13). When observed over long periods, *Salvia* may be seen to expand into the grassy areas. But because sage roots extend only to the edge of the bare strip and not beyond, it is unlikely that a toxin is extruded into the soil directly by the roots. The leaves of *Salvia* produce volatile terpenes (a class of organic compounds that includes cam-

(a)

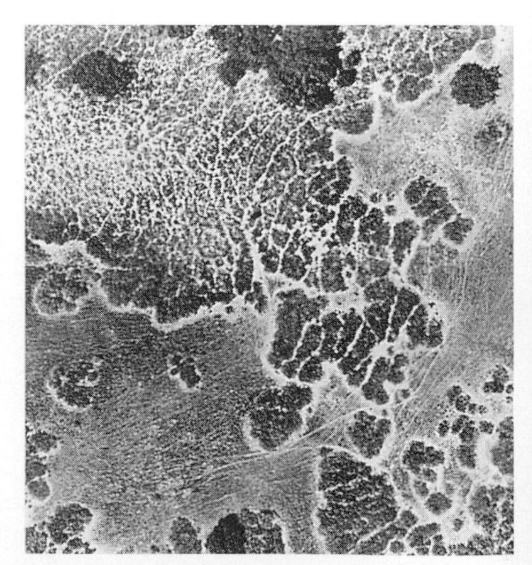
(b)

Figure 19.13 (a) Bare patch at the edge of a clump of sage (at left) includes a 2 m wide strip with no plants (A to B) and a wider area of inhibited grassland (B to C) lacking wild oat and bromegrass, which are found with other species to the right of C in unaffected grassland. (b) An aerial view shows sage and California sagebrush invading annual grassland in the Santa Inez Valley of California. Courtesy of C. H. Muller. From C. H. Muller, *Bull. Torrey Bot. Club* 93:332–351 (1966).

THE TEMPERATE ZONE

Temperate forests superficially resemble tropical rain forests in their structure but differ in being less diverse and less complex and in having a cold season of dormancy. The canopy may reach 40 meters in this red oak forest in Michigan. Most of the production of a forest enters detritus food chains in which fungi play a special role of attacking lignin in wood. Although wind plays a more prominent role in pollination and seed dispersal in temperate zones, many species have showy flowers (dogwood) and conspicuous fruits (cherry) to attract animals (temperate deciduous forest biome).

Photo by R. E. Ricklefs.

Photo by Marcos A. Guerra, courtesy of Smithsonian Tropical Research Institute.

Photo by R. E. Ricklefs.

Photo by R. E. Ricklefs.

Compared to the rather uniform leaves of tropical forest trees (right, above) which tend to be simple in shape and have entire (smooth) margins, the leaves of temperate forest species (right, below) have greater variety and tend to have serrated (toothed) margins. This fact had been used to infer ancient climates from samples of fossil leaves. Autumn brings dormancy to a broad-leaved forest in upstate New York. Winter comes to a birch-maple forest in Massachusetts. Alpine environments, such as this scene in the Rocky Mountains of Colorado, have harsh conditions throughout much of the year (tundra biome). The uneven snow melt creates patchiness in the alpine meadow habit. Lichens colonize bare rock surfaces in the alpine zone, while some plants establish themselves in the meager soil.

Photo by R.E. Ricklefs.

Photo by R. E. Ricklefs.

Photo by R. E. Ricklefs.

Photo by R. E. Ricklefs.

Photo by R. E. Ricklefs.

Photo by R. E. Ricklefs.

In arid parts of the subtropics, such as the Sonoran Desert in southern Arizona, desert environments are dominated by drought-adapted shrubs and succulent cacti (subtropical desert biome). Even though production is low, such communities may be diverse. The abundant nectar and pollen produced by flowers of the tall columnar saguaro cactus are a major source of food for many animals. As drought stress increases, plants are spaced more widely, leaving more exposed soil surface, and diversity decreases; Mohave Desert of southern California. Freezing winter temperatures in southern Utah exclude cacti but support grassland vegetation (temperate grassland biome).

Photo by R. E. Ricklefs.

Photo by R. E. Ricklefs.

Photo by R. E. Ricklefs.

Photo by R. E. Ricklefs.

The forests of temperate parts of Australia (below) and New Zealand (right, temperate rain forest biome) superficially resemble those of other temperate regions. In Australia, dominance of forests by eucalyptus trees gives them unique characteristics, including vertically oriented evergreen leaves (compare with the more familiar leaves of maple), prolific nectar production in the abundant flowers, and a more prominent role for fire, which frequently sweeps through the oil-rich leaf litter. The temperate rain forests of New Zealand are dominated by broad-leaved podocarp trees, which are closely related to needle-leaved conifers, and also contain many typically tropical elements such as tree ferns.

Photo by R. E. Ricklefs.

Photo by R. E. Ricklefs.

Photo by R. E. Ricklefs.

Photo by R. E. Ricklefs.

Photo by R. E. Ricklefs.

THE MARINE ENVIRONMENT

Even in the cold waters surrounding Antarctica, the oceans teem with marine life and support immense populations of whales, seals, and penguins. Under the surface, one finds an abundance of organisms both within the water column and at the bottom. In this scene, starfish, crinoids, sponges, and various worms are conspicuous.

Photo by R. E. Ricklefs.

Photo by Paul Dayton.

A cliff in the Red Sea off Israel reveals the abundant and often colorful life that abounds in tropical oceans.

Photo by Eric Hanauer.

The coral reefs of the Caribbean Sea off the northern coast of Panama reveal a seemingly endless variety of animals. The fanlike gills and feeding appendages of polychaete worms contrast with the coral in which they burrow. The green color of the coral comes from the symbiotic algae that live within the coral organism. Encrusting sponges cover bits of dead coral. A nudibranch—a snail without a shell—grazes algae and small animals on the surface of coral rubble. Anemones, which are sessile relatives of jellyfish, trap small prey from the water with their stinging tentacles; parrotfish chew up the living coral with their powerful jaws. The abundant food and complex structure of the coral reef provide habitat for hundreds of species of fish, many of which travel in schools to reduce risk of predation.

Photo by Carl C. Hansen, courtesy of Smithsonian Tropical Research Institute.

Photo by Carl C. Hansen, courtesy of Smithsonian Tropical Research Institute.

Photo by Carl C. Hansen, courtesy of Smithsonian Tropical Research Institute.

Photo by Carl C. Hansen, courtesy of Smithsonian Tropical Research Institute.

Photo by Carl C. Hansen, courtesy of Smithsonian Tropical Research Institute.

Satellite image of the California coast on June 15, 1981, showing concentration of phytoplankton (red and yellow) near the coast and in large eddy currents that extend out from the coast. This image emphasizes the spatial heterogeneity of the marine system resulting from water currents. Satellite image of the North Atlantic Ocean during the first week of June, 1984, in which warm water is indicated by red and cold water by green or blue. The image shows the Gulf Stream along the coast of Florida breaking up into large eddies as it crosses the Atlantic. The Gulf Stream not only transports considerable heat to the vicinity of northern Europe but also conveys a tropical community of marine organisms far to the north, where the warm-water environment becomes increasingly heterogeneous in time and space.

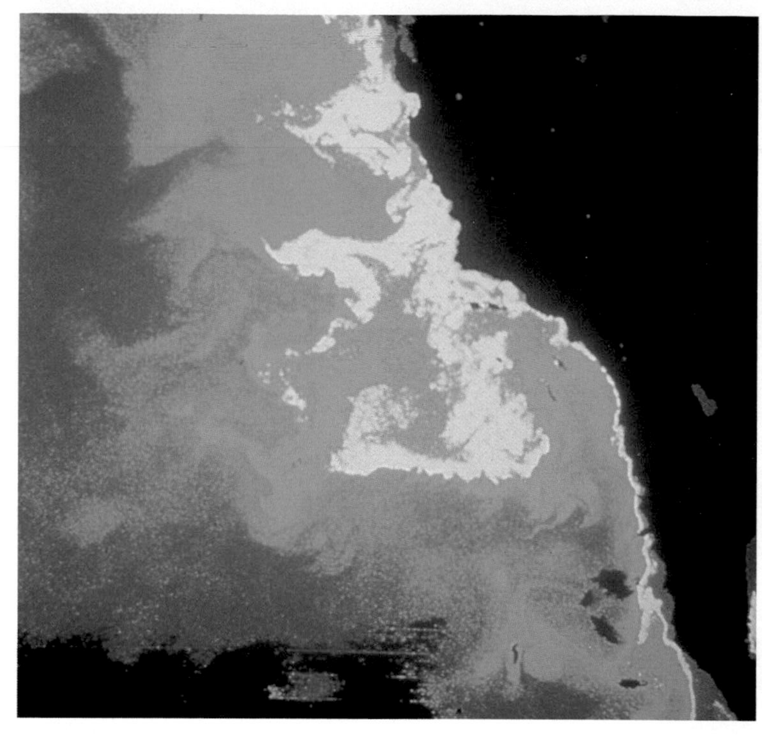

Coastal Zone Color Scanner image courtesy of J. A. McGowan from J. Palaéz and J. A. McGowan, *Limnol. Oceanog.* 31:927–950 (1986).

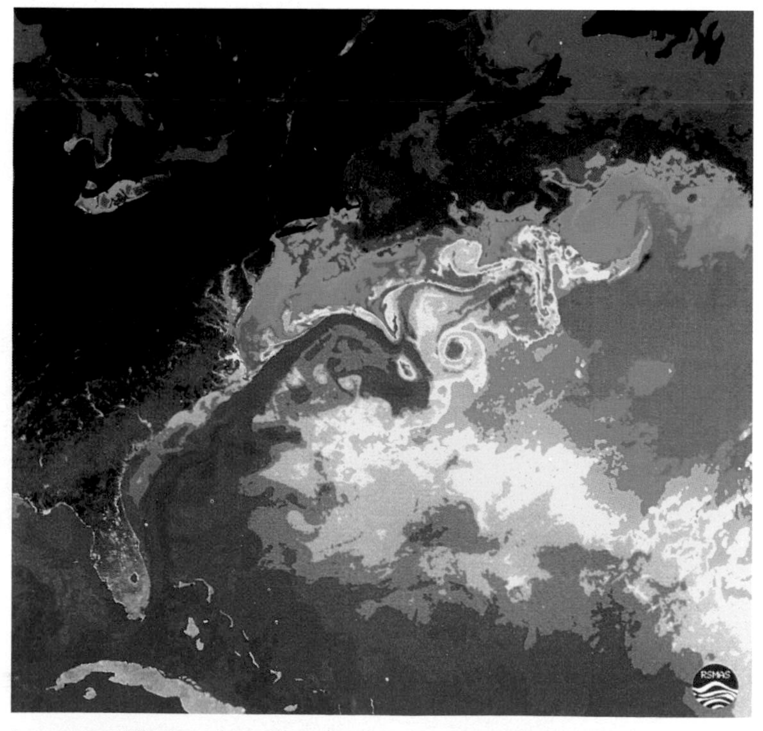

Image courtesy of Otis Brown, Robert Evans, and Mark Carle, University of Miami Rosentiel School of Marine and Atmospheric Science.

phor and gives foods spiced with sage part of their distinctive taste), which apparently harm nearby plants directly through the atmosphere.

Asymmetry of competition

Field experiments have revealed many cases in which one member of a species pair responded to the addition or removal of the other species, but not vice versa. According to one compilation, among 98 reciprocal tests of competition between species reported in the literature, no response was reported for either member of 44 pairs, reciprocal negative effects were reported for 21 pairs, and response by only one species was reported for 33 pairs. The high percentage of cases in the last category suggests that asymmetry is more the rule than the exception in competitive interactions.

Asymmetry in competition must derive from asymmetry in ecology. The superior competitor is almost always more strongly limited by some other factor—such as environmental tolerances or predators—that is external to the competitive interaction. In the case of the barnacles mentioned earlier, *Balanus* could effectively exclude *Chthamalus* from lower levels of the intertidal zone, but it lacked the physiological tolerance that allowed *Chthamalus* to exist higher up. Therefore, when *Balanus* was removed from the lower portion of the tidal zone, where it dominated, *Chthamalus* could become established, showing a strong effect of *Balanus* on *Chthamalus.* However, when *Chthamalus* was removed from rocks at high tide levels, where it dominated, *Balanus* failed to colonize them.

Predators often balance strong asymmetry between competitors by their preference for the competitively dominant species, which is the most abundant. On marine hard substrates (such as rocky shores and reef rubble), rapidly growing leafy species of algae often outcompete slow-growing encrusting forms, which they overgrow and shade out of existence. But predators such as sea urchins and herbivorous fish readily graze the leafy forms of algae while leaving the less accessible, encrusting ones untouched. In this situation, competition appears to be asymmetrical: removal of leafy algae results in the establishment of encrusting forms. Removal of encrusting forms (whose presence depends on herbivores) does not lead to the establishment of leafy species, however, because these are eaten as soon as they colonize the rock surface.

Along the shores of the Caribbean Sea, patch reefs have three ecological zones: (1) the reef flat, the topmost part of the reef that is covered by shallow water and is often exposed at low tide; (2) the reef slope, the outer edge of the reef as it descends to deep water; and (3) the sand plain, the area of sandy bottom between patches of coral reef. Because few fish can inhabit the shallow water on the reef flat, herbivory there is slight. On the sand plain, predatory fish drive away herbivorous species, which cannot find hiding places in the exposed habitat. Therefore, only the structurally complex reef slope experiences strong herbivory by fish. As a result, species of algae typical of the reef slope must effectively resist herbivory.

This resistance was discovered by an experiment in which species of algae from each of the three habitats were transplanted to new locations within each habitat. As we might expect, algae from the reef slope were

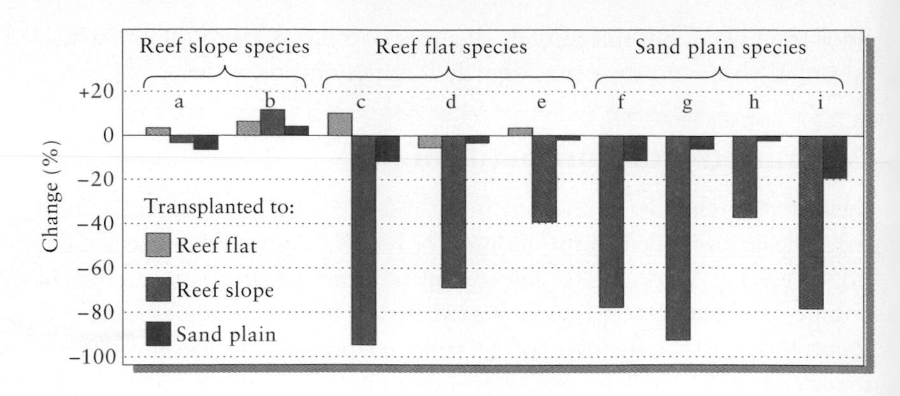

Figure 19.14 Change in biomass of species of algae from three coral reef habitats after transplantation to each of the other habitats. Reef slope species resist grazing by fish; other species are heavily grazed when transplanted to the reef slope. After M. E. Hay, T. Colburn, and D. Downing, *Oecologia* 65:591–598 (1983).

virtually untouched in all the habitats, but algae from the reef flat and sand plain were heavily grazed (40–90% of their biomass was consumed in 48 hours) when transplanted to the reef slope (Figure 19.14). In another experiment, algae from all three habitats were transplanted to the reef slope, where some individuals were enclosed in cages and others were exposed to herbivory in partial cages. The sand plain species were grazed heavily in the partially exposed cages, but grew vigorously when protected from herbivores. The reef flat species did not grow well even when protected, possibly because of low levels of light on the reef slope.

Because sand plain algae species grow more rapidly than those from the slope and flat, they would probably exclude those species from those habitats through overgrowth competition were it not for their vulnerability to herbivory on the reef slope and their inability to tolerate the physically stressful conditions of desiccation and temperature fluctuation on the reef flat. Reef flat species form a dense turf of algae that is resistant to both herbivory and desiccation, but the price of this adaptation is slow growth. These examples show that asymmetrical competition results when species that share one resource (such as space or food) are differently specialized with respect to other aspects of the environment (such as physical conditions or consumers).

Competition between distantly related species

Charles Darwin emphasized that competition should be most intense between closely related species or organisms. In *On the Origin of Species,* he remarked, "As species of the same genus have usually, though by no means invariably, some similarity in habits and constitution, and always in structure, the struggle will generally be more severe between species of the same genus, when they come into competition with each other, than between species of distinct genera." Darwin reasoned that similar structure indicates similar ecology, especially with respect to resources consumed. Although this must generally be the case, many of the same resources are used by distantly related organisms. Barnacles and mussels, as well as algae, sponges, bryozoans, tunicates, and others, occupy space in the intertidal zone and actively compete by preemption and overgrowth. Both fish and aquatic birds prey on aquatic invertebrates. Krill *(Euphausia superba),*

(a)

(b)

(c)

Figure 19.15 (a) A congregation of sea stars *(Pisaster)* at low tide on the coast of the Olympic Peninsula, Washington. This sea star (b) is an important predator on mussels (c).

shrimplike crustaceans that abound in subantarctic waters, are fed upon by virtually every type of large animal, including fish, squid, diving birds, seals, and whales. Recent increases in seal and penguin populations in the Southern Ocean have been related to decreased competition from whales, whose populations have been decimated by commercial exploitation. In terrestrial habitats, invertebrates in forest litter are consumed by spiders, ground beetles, salamanders, and birds. In desert ecosystems, many of the same insect species are eaten by birds and lizards, and the seeds of many of the same plants species are consumed by ants, rodents, and birds. These examples illustrate the strong potential for competition between very distantly related as well as unrelated organisms.

Predation and the outcome of competition

Darwin also noted that grazing can maintain a high diversity of plants in grasslands. In the absence of grazers, dominant competitors grow rapidly and exclude others. Similar results have been obtained from experiments on marine algal communities under the pressure of grazing by limpets, snails, and urchins. These studies indicate that predation has a strong hand in shaping the structure of biological communities by influencing the outcome of competitive interactions between prey species.

University of Washington ecologist Robert Paine was one of the first to demonstrate this point experimentally. On the exposed rocky coast of the state of Washington, the intertidal zone harbors several species of barnacles, gooseneck barnacles, mussels, limpets, and chitons (a kind of grazing mollusk); these are preyed upon by the sea star *Pisaster* (Figure 19.15). In one experiment, Paine removed sea stars from a study area 8 m in length and 2 m in vertical extent; an adjacent area of similar size was left undisturbed. Following the removal of sea stars, the number of prey species in the experimental plot decreased rapidly, from fifteen at the beginning of the study to eight at the end. Diversity declined in the experimental area because populations of barnacles and mussels increased and crowded out many of the other species. Paine concluded that sea stars maintain the diversity of the area by limiting populations of barnacles and mussels, which are superior competitors for space in the absence of predators.

Studies conducted in artificial ponds have shown that predatory salamanders can reverse the outcome of competition among frog and toad tadpoles. In one experiment, ponds were supplied with 200 hatchlings of the spadefoot toad *(Scaphiopus holbrooki),* 300 of the spring peeper *(Hyla crucifer),* and 300 of the southern toad *(Bufo terrestris).* Each of the ponds, which were identical in all other respects, also received 0, 2, 4, or 8 individuals of the predatory broken-striped newt *(Notophthalmus viridescens).* In the absence of newt predation, *Scaphiopus* tadpoles grew rapidly, survived well, and dominated the ponds along with smaller numbers of *Bufo; Hyla* tadpoles were all but eliminated (Figure 19.16). *Notophthalmus* apparently prefer toad tadpoles, and at higher numbers of predators, survival of both *Scaphiopus* and *Bufo* decreased markedly. With fewer toads per pond, levels of food increased, and survival and growth of *Hyla* tadpoles improved immensely, as did the growth of surviving *Scaphiopus* and *Bufo* tadpoles.

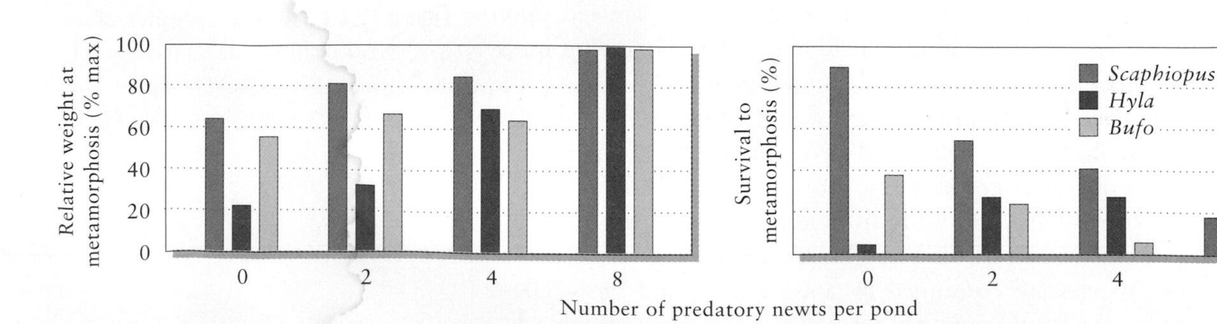

Figure 19.16 Effect of predators on survival and on weight at metamorphosis in three species of anurans (frogs and toads) raised in artificial ponds. From P. J. Morin, *Science* 212:1284–1286 (1981).

Experimental studies illustrate the pervasiveness of interspecific competition in ecological communities. The nature of competitive interactions depends on the species being compared, limiting factors in the physical environment, and the role of predators in tipping the balance between competing species. As we shall see in the next chapter, predation—indeed, all kinds of consumption—also exerts powerful effects on the regulation of population size and the coexistence of species within ecological communities.

Summary

1. Competition is the use or contesting of a resource by more than one individual consumer. When the individuals belong to the same species, their interaction is called intraspecific competition; when they belong to different species, it is called interspecific competition.

2. A resource may be defined as any factor that is consumed and whose increase promotes population growth. Thus, light, food, water, mineral nutrients, and space are resources. Temperature, salinity, and other such conditions are not.

3. Resources may be classified as nonrenewable (space) or renewable (light and food), and the latter may be further distinguished according to the influence of the consumer on the provisioning of the resource: no influence, direct influence, or indirect influence through other consumers.

4. Of all the resources consumed, only one or a few limit the population growth of the consumer; these are normally those resources whose supply relative to demand is least. This principle is known as Liebig's law of the minimum.

5. Competition may be demonstrated in the laboratory and in the field by a change in the population size of one species following the addition or removal of another. When two species compete strongly, the population of the first is sensitive to changes in numbers of the second, and vice versa.

6. Theoretical investigations and laboratory studies have led to the generalization that no two species of competitors can coexist on the same limiting resource. This has come to be known as the competitive exclusion principle.

7. Some mathematical treatments of competition are based on the logistic equation of population growth, to which a term is added for the effect of interspecific competition. The strength of this effect in the model is specified by the coefficient of competition.

8. The equilibrium population sizes of two competing species can be described by an equation including the carrying capacities and competition coefficients for each of the species. In the most general terms,

coexistence requires that the product of the competition coefficients of the first and second species be less than 1.

9. Laboratory experiments present clear evidence of competition among species, but natural populations are limited by physical conditions and consumers as well as by shared resources. Ecologists have also conducted numerous field studies designed to reveal the influence of competition on the sizes of natural populations.

10. Transplant experiments with plants creating varying conditions of intraspecific and interspecific competition illustrate differences in the mechanisms of competition between habitats of high and low productivity.

11. Removal experiments involving intertidal invertebrates have demonstrated strong competition among space-filling animals such as barnacles, mussels, and encrusting sponges. Competitive exclusion is accomplished by direct physical interaction.

12. Rapid invasion of a habitat by a species of small mammal following the removal of another, related species demonstrates direct interference competition through aggressive behavior.

13. Exploitation competition is most convincingly demonstrated in studies that show appropriate changes in resource levels accompanying the demographic response of one species after removal of a competitor.

14. Field studies have revealed many mechanisms of competition: consumptive, preemptive, overgrowth, chemical, territorial, and encounter. The first is usually classified as exploitation competition, and the last three as interference competition. Preemptive and overgrowth competition are based on the use of space and renewable resources respectively, but involve close contact of competing individuals.

15. Experiments have demonstrated the prevalence of asymmetry in competition, in which the effect of one species is clearly greater than that of the other. Coexistence is possible, however, when the asymmetry is balanced by physical conditions or by consumers whose actions favor the poorer competitor.

Suggested readings

Connell, J. H. 1961. The influence of interspecific competition and other factors on the distribution of the barnacle *Chthamalus stellatus. Ecology* 42: 710–723.

Connor, E. F., and D. Simberloff. 1986. Competition, scientific method, and null models in ecology. *American Scientist* 74:155–162.

Goldberg, D. E., and A. M. Barton. 1992. Patterns and consequences of interspecific competition in natural communities: A review of field experiments with plants. *American Naturalist* 139:771–801.

Grace, J. B., and D. Tilman (eds.). 1990. *Perspectives on Plant Competition.* Academic Press, San Diego, Calif.

Hardin, G. 1960. The competitive exclusion principle. *Science* 131:1292–1297.

Inchausti, P. 1995. Competition between perennial grasses in a Neotropical savanna: The effects of fire and of hydric-nutritional stress. *Journal of Ecology* 83:231–243.

Keddy, P. 1989. *Competition.* Chapman and Hall, London.

Paine, R. T. 1974. Intertidal community structure: Experimental studies on the relationship between a dominant competitor and its principal predator. *Oecologia* 15:93–120.

Schoener, T. W. 1982. The controversy over interspecific competition. *American Scientist* 70:586–595.

Schoener, T. W. 1983. Field experiments on interspecific competition. *American Naturalist* 122:240–285.

Tilman, D. 1982. *Resource Competition and Community Structure.* Princeton University Press, Princeton, N.J.

Wilson, S. D., and D. Tilman. 1993. Plant competition and resource availability in response to disturbance and fertilization. *Ecology* 74:599–611.

CHAPTER 20

PREDATION

The basic question of population biology is this: What factors influence the size and stability of populations? In previous chapters, we saw how density-dependent factors and time delays modify the responses of birth and death rates to population density. As we expanded our perspective on populations to include interactions among species, the basic question of population biology was elaborated: How do species affect each other's populations? We have seen that exploitation competition and interference competition from other species can depress the growth rate of a population. Because most species are both consumers and resources for other consumers, it also becomes necessary to ask whether populations are limited primarily by what they eat or by what eats them.

The study of predator-prey interactions attempts to answer at least two important questions: First, do predators reduce the size of their prey populations to below their carrying capacities? Second, do the dynamics of predator-prey interactions cause populations to oscillate? The first question is of great practical concern to those interested in the management of crop pests, game populations, and endangered species. It also has far-reaching implications for our understanding of the interactions among species that share resources and, therefore, for our understanding of the structure of biological communities. The second question is motivated by observations of predator-prey cycles in nature and directly addresses the issue of stability in natural systems. Ecologists have tried to answer these questions with a combination of observation, theory, and experimentation.

In this chapter, we shall explore the effects of consumers—predators, parasites, herbivores, and others—on resource populations. Consumer-resource interactions have been investigated theoretically and experimentally in both laboratory and natural populations. The insights and practical knowledge gained through these studies have also contributed importantly to our understanding of the structure and dynamics of natural systems.

A few distinctions

Consumers go by many names, the most familiar of which are predators, parasites, parasitoids, herbivores, and detritivores. From the standpoint of population interactions, some of these distinctions are useful, whereas others are confusing. Let's start with **predators.** The images of an owl eating a mouse and a spider eating a fly capture the essentials of predation. Predators catch individuals and consume them, thereby removing them from the prey population. In contrast, a **parasite** consumes a living host. Although it may increase the probability of a host's dying from other causes or reduce its fecundity, a parasite does not by itself remove an individual from a resource population.

As we have seen, the term **parasitoid** is applied to species of wasps and flies whose larvae consume the tissues of living hosts, usually the eggs, larvae, and pupae of other insects, inevitably leading to the host's death (Figure 20.1). Parasitoids are like parasites in that they consume living tissue; in most other ways, however, they resemble predators.

Herbivores eat whole plants or parts of plants. From the standpoint of consumer-resource relations, herbivores function as predators when they consume whole plants, and as parasites when they consume living tissues but do not kill their victims. When a sparrow eats a seed, it acts as a predator because it kills the entire living embryo of a plant contained in that seed. When deer consume some leaves on a shrub, they affect their resources much as mosquitoes and vampire bats do when they take their meals of blood. In the case of herbivory, consumption of a portion of a plant's tissues is referred to as **grazing** (generally applied to grasses and other herbaceous vegetation, and also to algae) or **browsing** (applied to woody vegetation).

Detritivores consume dead organic material—leaf litter, feces, carcasses—and have no direct effect on the populations that produce those resources. Detritivores live off the waste of other species. As a consequence, they do not affect the abundance of their food supplies, and their activities consequently do not influence the evolution of the living sources of their food. Because detritivore populations are not dynamically coupled to the resource populations that produce their food, they will not be considered further in this chapter.

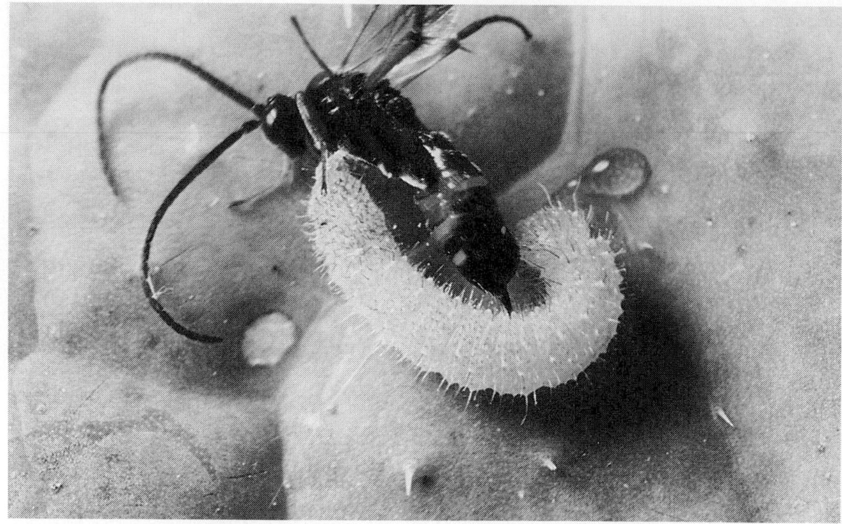

(a)

(b)

Figure 20.1 Two parasitoid wasps, *Apanteles rubecula* (a) and *Ptermalus puparum* (b), laying eggs in larvae of the cabbage worm moth at different stages of development. Courtesy of the U.S. Department of Agriculture.

Limitation of resource populations by consumers

The cyclamen mite is a pest of strawberry crops in California. Populations of the mites are usually kept under control by a species of predatory mite of the genus *Typhlodromus.* Cyclamen mites typically invade a strawberry crop shortly after it is planted, but their populations usually do not reach damaging levels until the second year. *Typhlodromus* mites usually invade fields during the second year, and because they are such efficient predators, these mites rapidly subdue cyclamen mite populations and prevent further outbreaks.

Greenhouse experiments have demonstrated the role of predation in keeping cyclamen mites in check. One group of strawberry plants was stocked with both predator and prey mites; a second group was kept predator-free by regular application of parathion, an insecticide that kills the predatory species but does not affect the cyclamen mite. Throughout the study, populations of cyclamen mites remained low in plots they shared with *Typhlodromus,* but their infestation attained damaging proportions on predator-free plants (Figure 20.2). In field plantings of strawberries, cyclamen mites also reached damaging levels where predators were eliminated by parathion (a good example of an insecticide having an undesired effect), but they were effectively controlled in untreated plots. When a cyclamen mite population began to increase in an untreated planting, the predator population quickly mushroomed and reduced the outbreak. On average, cyclamen mites were about 25 times more abundant in the absence of predators than in their presence.

Typhlodromus owes its effectiveness as a predator to several factors in addition to its voracious appetite. Its population can increase as rapidly as that of its prey. Both species reproduce by **parthenogenesis;** that is, females lay fertile eggs produced without sexual reproduction, all of which produce female offspring. Consequently, mite populations can grow very rapidly, because each individual in a population produces eggs. Cyclamen mites lay three eggs per day over the 4 or 5 days of their reproductive life span; *Typhlodromus* lay two or three eggs per day for 8–10 days. The seasonal synchrony of *Typhlodromus* reproduction with the growth of prey populations, their ability to survive at low prey densities, and their strong dispersal powers also contribute to the predatory efficiency of *Typhlodromus.* During winter, when cyclamen mite populations dwindle to a few individuals hidden in crevices and folds of leaves in the crowns of strawberry plants, the predatory mites subsist on honeydew produced by aphids and whiteflies. They do not reproduce except when they feed on other mites. Whenever predators appear to control prey populations—and *Typhlodromus* is no exception—the predators usually exhibit a high reproductive capacity compared with that of the prey, combined with strong dispersal powers and an ability to switch to alternative food resources when the primary prey are unavailable.

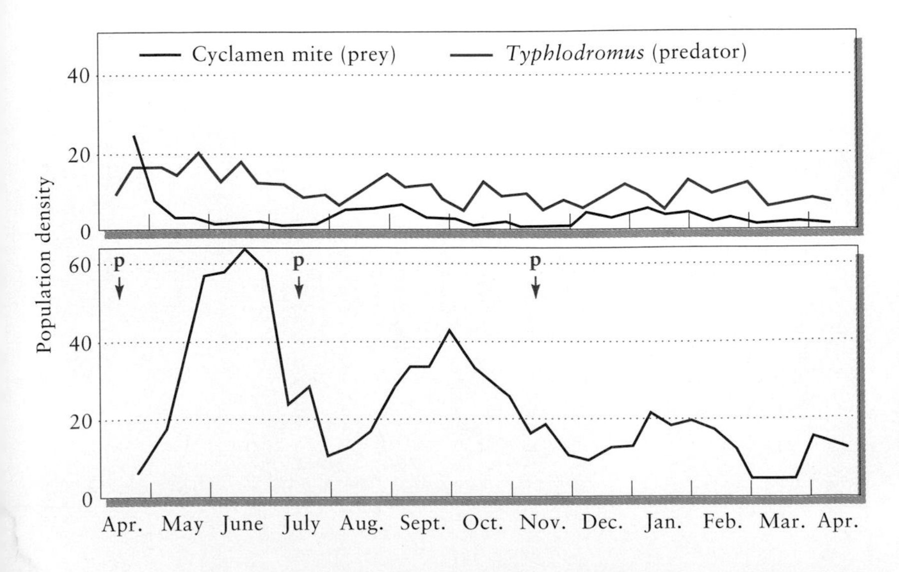

Figure 20.2 Infestation of strawberry plots by cyclamen mites *(Tarsonemus pallidus)* in the presence of the predatory mite *Typhlodromus* (above) and in its absence (below). Prey populations are expressed as numbers of mites per leaf; predator levels are the numbers of leaflets out of 36 on which one or more *Typhlodromus* were found. Parathion treatments are indicated by "p." After C. B. Huffaker and C. E. Kennett, *Hilgardia* 26:191–222 (1956).

Inadequate dispersal is perhaps the only factor that keeps the cactus moth from exterminating its principal food source, the prickly pear cactus. When prickly pear *(Opuntia)* was introduced into Australia, it spread rapidly over the island continent, covering thousands of acres of valuable pasture and rangeland. After several unsuccessful attempts to eradicate the plant, the cactus moth *(Cactoblastis cactorum)* was introduced from South America. The caterpillar of the moth feeds on growing shoots of the prickly pear and quickly destroys the plant—literally by nipping it in the bud and inoculating it with various pathogens and rot-causing organisms. After they became established in Australia, cactus moths exerted such effective control that within a few years, prickly pear became a pest of the past (Figure 20.3).

The cactus moth has not eradicated the prickly pear, however, because the cactus manages to disperse to predator-free areas, thereby keeping one jump ahead of the moth and maintaining a low-level equilibrium in a continually shifting mosaic of isolated patches. Indeed, one would probably not guess that the cactus moth keeps the prickly pear at its present low population levels; the moths are scarce in the remaining stands of cactus in Australia today. (The same moth probably controls prickly pear populations in some areas of its native home in Central and South America, but its decisive role might have gone unnoticed if the appropriate experiment had not been performed in Australia.)

Consumer control is not unique to terrestrial ecosystems—quite the opposite. Experiments on the effect of sea urchins on populations of algae that live on hard substrates have demonstrated consumer control in some rocky shore communities. The simplest experiments consisted of removing sea urchins and following the subsequent growth of their algal prey. When urchins are kept out of tide pools and off subtidal rock surfaces, the biomass of algae quickly increases, indicating that herbivory reduces algal populations below the level that the environment can support. Different kinds of algae also appear after herbivore removal. Large brown algae flourish and begin to replace both coralline algae (whose hard, shell-like structure deters grazers) and small green algae (whose short life cycles and high

(a)

(b)

Figure 20.3 Photographs of a pasture in Queensland, Australia, (a) 2 months before and (b) 3 years after the introduction of the cactus moth to control the prickly pear cactus. From A. P. Dodd, in A. Keast, R. L. Crocker, and C. S. Christian (eds.), *Biogeography and Ecology in Australia,* W. Junk, The Hague (1959). Courtesy of W. H. Haseler, Department of Lands, Queensland, Australia.

reproductive rates enable algal population growth and reestablishment to keep ahead of grazing pressure by sea urchins). In subtidal plots kept free of predators, brown kelps become established in thick stands that shade out most small species. We shall consider additional examples of consumer control in the following pages.

Parasite-host systems

Parasitism resembles predation in that a living resource is consumed, but it differs from predation in that the resource is not killed, at least not immediately. Parasites can nonetheless have profound physiological and behavioral consequences for their hosts, adversely affect their populations, and limit their geographic distributions. The symptoms in a host caused by a parasite infection are usually referred to as a **disease;** conversely, **disease organisms,** or **pathogens,** are those parasites that produce recognizable symptoms in their hosts. Parasite-host interactions also differ from predator-prey interactions in that many parasites pass through complex life cycles, alternating between different hosts and free-living stages (see Figure 18.11), and in that hosts can mount immune responses to prevent, control, or eliminate parasite infections.

The complex life histories of parasites involve a variety of interactions with hosts, and different sets of factors affect each stage, both parasitic and free-living, of the life cycle. Problems of locating hosts (**contagion** or dispersal) are crucial in parasite and, particularly, disease populations. In the case of endoparasites—those living within the bodies of their hosts—population growth may occur within a single host, resulting in the subdivision of the parasite population and fostering population interactions among the descendants of infecting individuals.

The balance between parasite and host populations is influenced by the **immune response** and other defenses of the host. Immunity depends in part on the production of **antibodies** that recognize and bind to foreign proteins **(antigens),** such as protein molecules on the outer surfaces of parasites and disease organisms. An immune response takes time to develop, which gives the parasite a chance to develop and multiply within a host. However, antibodies persist long after an infection has been controlled, thereby reducing the probability of subsequent infection.

Parasites have their own means of circumventing immune mechanisms. Some microscopic and submicroscopic disease organisms produce chemical factors that suppress the immune systems of their hosts; this is the most troublesome feature of the AIDS virus. Others have surface proteins that mimic host antigens and thus escape notice by the immune system. Some schistosomes are known to excite an immune response when they enter the host, but do not succumb to antibody attack because they coat themselves with proteins of the host before antibodies become numerous. As a consequence, parasites that subsequently infect a host face a barrage of antibodies stimulated by the earlier entrance of the now entrenched parasite individuals. When this response affects closely related species of parasites, it is known as **cross-resistance.** For example, most of the predominantly human forms of schistosomiasis are extremely virulent. But when a person has been infected previously by other schistosome organisms, some of which have little effect on humans, the effect of the parasite is moderated considerably.

On a population level, a serious outbreak of a viral or bacterial disease is often followed by a period during which most of the individuals in the host population have achieved some degree of immunity to reinfection. Until this immunity is lost, or until susceptible individuals are recruited into the population, disease organisms may be unable to spread.

Herbivores and plant populations

We have seen the role cactus moths play in controlling populations of the prickly pear cactus in Australia. Herbivorous insects have been employed in many other situations to control imported weeds. Consider the example of Klamath weed, a European species toxic to livestock, which accidentally became established in northern California in the early 1900s. By 1944 the weed had spread over 2 million acres of rangeland in 30 counties. Biological control specialists borrowed an herbivorous beetle of the genus *Chrysolina* from an Australian control program. In a success of similar proportions to that of the war against the prickly pear, within 10 years after the first beetles were released, Klamath weed was all but obliterated as a range pest. Its abundance is estimated to have been reduced by more than 99%.

The proportion of net primary production consumed by herbivores is least in forests, intermediate in grasslands, and greatest in aquatic environments. In the littoral zones of aquatic communities, herbivores consume most of the plant and algal production; in terrestrial habitats, the bulk of plant production moves up food chains through detritus pathways. The differences in the functional roles of herbivores in these two habitats are related to the greater digestibility of aquatic vegetation. Because aquatic herbivores consume algae so efficiently, there is greater potential for production to be regulated by herbivores in freshwater and marine ecosystems than in terrestrial ecosystems.

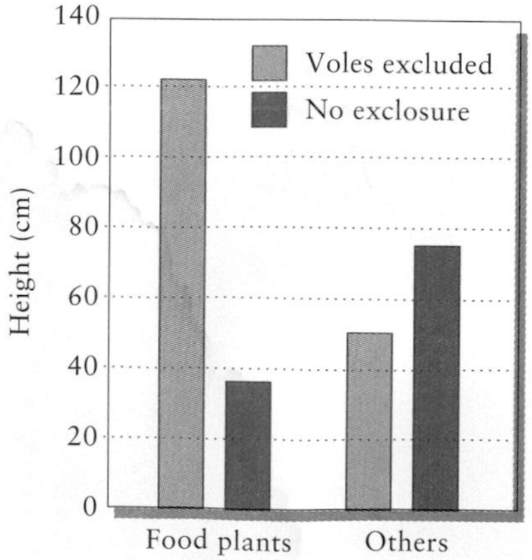

In grasslands, herbivores (mostly insects and grazing mammals) consume 30–60% of the aboveground vegetation. Their influence on plant production is revealed by exclosure experiments. In one study in California, wire fences were constructed to keep voles out of small areas of grassland. At the end of the 2-year study, the food plants of voles (mostly annual grasses) had grown more and produced more seeds within the fenced plots than outside the exclosures, where voles continued to graze. Perennial grasses and herbs not included in the voles' diet were not directly affected by the exclosures (Figure 20.4).

Figure 20.4 Relative biomass (summed height per 100 cm^2) of food plants and nonfood plants in grassland plots fenced to exclude voles and in unfenced control plots after 2 years. Food plants are mostly annual grasses; nonfood plants include perennial grasses and herbs. After G. O. Batzli and F. A. Pitelka, *Ecology* 51:1027–1039 (1970).

Although herbivores rarely consume more than 10% of forest vegetation, occasional outbreaks of tent caterpillars, gypsy moths, and other insects can completely defoliate or otherwise eradicate entire forests. In addition, long-term studies of growth and survival of trees after defoliation by insects demonstrate that there may be a considerable lag between an infestation and the expression of its effects.

Grazing resembles infection by parasites and pathogens in that, like the infected host, the individual grazed-upon plant usually is not killed. Thus plants have time to respond to grazing and browsing by producing defensive chemical compounds in a manner recalling the immune mechanisms of animals. Depending on how long a plant takes to produce these defenses, time lags may be incorporated into grazer-plant systems, thereby promoting oscillations.

Plants have many inducible defenses against herbivory, most of them chemical. Wounding may cause the production of toxic, noxious, or nutrition-reducing compounds—in the area of a wound or systemically throughout the plant—that reduce subsequent herbivory. In some cases, these responses may take only minutes or hours; in others, they require a new season of growth. When shoots of aspen, poplar, birch, and alder are heavily browsed by snowshoe hares, shoots produced during the following year have exceptionally high concentrations of terpenes and phenolic resins, which are extremely unpalatable to hares. When resins were applied in varying concentrations to shoots of unbrowsed trees, hares were found to avoid shoots containing 80 milligrams or more of resin per gram of dry weight, regardless of the amount of other food available.

Predator-prey cycles

Population cycles have contributed to the lore of population ecology since Charles Elton's paper "Periodic Fluctuations in the Numbers of Animals: Their Causes and Effects" was published in the *British Journal of Experimental Biology* in 1924. Most of Elton's data concerned fur-bearing mammals in the Canadian boreal forests and tundra, where the Hudson's Bay Company had kept detailed records of the numbers of furs brought in by trappers each year. Data for the lynx and its principal prey, the snowshoe hare, revealed regular fluctuations of great magnitude (Figure 20.5). Each cycle lasted approximately 10 years, and cycles of the two species were highly synchronized, with peaks in lynx abundance tending to trail those in hare abundance by a year or two.

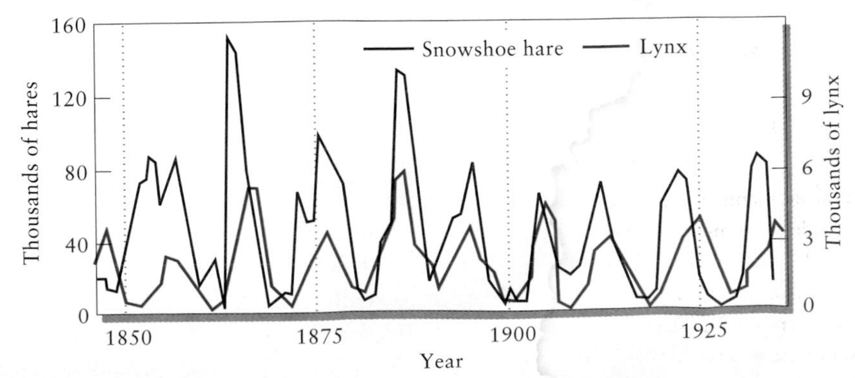

Figure 20.5 Population cycles of the lynx and the snowshoe hare (see photograph) in the Hudson Bay region of Canada, as indicated by fur returns to the Hudson's Bay Company. Photograph courtesy of the U.S. Fish and Wildlife Service. After D. A. MacLulich, *University of Toronto Studies, Biol. Ser.* no. 43 (1937).

The periods of population cycles vary from species to species, and even within a species. In Canada, most cycles have periods of either 9–10 years or 4 years. Although colored foxes have a 10-year cycle over most of their range, they exhibit a pronounced 4-year cycle in Labrador and on the Ungava Peninsula, where they prey mostly on lemmings, which also have 4-year cycles. In general, small herbivores such as voles and lemmings have 4-year cycles; large herbivores—snowshoe hares, muskrat, ruffed grouse, and ptarmigan—have 9- or 10-year cycles. Predators that feed on short-cycle herbivores (arctic foxes, rough-legged hawks, snowy owls) themselves have short population cycles. Predators of larger herbivores (red foxes, lynx, marten, mink, goshawks, horned owls) have longer cycles. The length of the cycle also appears to be related to habitat: longer cycles are observed in forest-dwelling species and shorter ones in tundra-dwelling species.

The closely linked population cycles of some predators and prey suggest that these oscillations could result from the way in which predator and prey interact with each other. We saw in an earlier chapter that delays in the responses of birth rates and death rates to changes in the environment can cause population cycles. Most predator-prey interactions also have response lags because of the time required to produce offspring or to mount an immune response. Theoretical considerations of population dynamics have shown that the period of a population cycle is about four to five times the response lag. Thus, the 4-year and 9–10-year population cycles of mammals inhabiting boreal forest and tundra are consistent with time delays of 1 or 2 years. Such time delays could result from the typical periods between birth and sexual maturity in these mammals. In other words, the influence of conditions in a particular year may not be felt in a predator population until young born in that year are themselves old enough to breed.

Long-term observations have confirmed that population fluctuations continue more or less unchanged over many cycles, so this dynamic behavior appears to represent stable interactions between predators and prey. One of the earliest goals of population biologists was to establish such cycles in experimental populations, for which one could work out the dynamics of the relationship and study potential causes of the cycles.

Laboratory experiments on predator-prey cycles

When azuki bean weevils *(Callosobruchus chinensis)* are maintained in cultures with parasitoid braconid wasps *(Heterospilus),* the populations of predators and prey fluctuate out of phase with each other in regular cycles (Figure 20.6). Introduced to a population of weevils that is ultimately limited by a constant ration of seeds, wasps rapidly increase in number. As the wasp population grows, parasitism becomes a major source of mortality for the weevils. When parasitism exceeds the reproductive capacity of the weevil population, the number of weevils begins to decline, but because *Heterospilus* is an efficient parasitoid, it continues to prey heavily upon the weevils even as the population is reduced. Eventually the weevils are

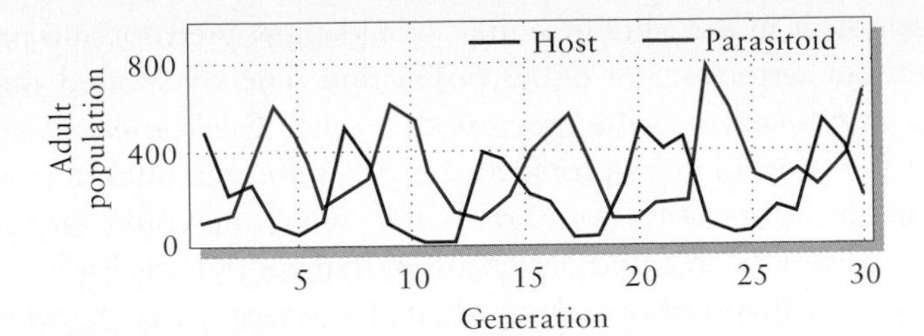

Figure 20.6 Population fluctuations of the azuki bean weevil (the host) and its braconid wasp parasitoid. After S. Utida, *Cold Spring Harbor Symp. Quant. Biol.* 22:139–151 (1957).

nearly exterminated, and the predator population, lacking adequate food, decreases rapidly. The wasp is not so efficient that all weevil larvae are attacked; hence a small but persistent reserve of weevils always remains to initiate a new cycle of prey population growth after predators have become scarce.

Extremely efficient predators often eat their prey populations to extinction and then become extinct themselves. This hopeless situation can be stabilized, however, if some of the prey can find refuges in which to escape. G. F. Gause demonstrated this principle in one of the earliest experimental studies on predator-prey systems. Gause employed *Paramecium* as the prey and another ciliated protozoan, *Didinium,* as the predator. In one experiment, predator and prey individuals were introduced to a nutritive medium in a plain test tube. By creating such a simple environment, Gause had "stacked the deck" against the prey; the predators readily found all of them, and when the last *Paramecium* had been consumed, the predators starved. In a second experiment, Gause added some structure to the environment by placing glass wool, in which the *Paramecium* could escape predation, at the bottom of the test tube. The tables thus having been turned, the *Didinium* population starved after consuming all readily available prey, but the *Paramecium* population was restored by individuals concealed in the glass wool.

Gause finally achieved recurring oscillations in predator and prey populations by periodically adding small numbers of predators—restocking the pond, so to speak. Repeated addition of individuals to an experimental culture corresponds to natural repopulation by colonists from other areas in a locality where extinction of either predator or prey has occurred. This pattern is reminiscent of the interaction between the cactus moth and the prickly pear, in which the cactus escapes complete annihilation by dispersing to predator-free areas.

Huffaker's experiments on mite populations

C. B. Huffaker, a biologist at the University of California at Berkeley who pioneered the biological control of crop pests, attempted to produce a mosaic environment in the laboratory that would allow predator and prey to persist without restocking of either population. The six-spotted mite, *Eotetranychus sexmaculatus,* was the prey; another mite, *Typhlodromus occidentalis,* was the predator; oranges provided the prey's food. Huffaker established experimental populations on trays within which he could vary the number, exposed surface area, and dispersion of oranges (Figure 20.7).

Each tray had 40 positions arranged in 4 rows of 10 each; where Huffaker did not place oranges, he substituted rubber balls of about the same size. The exposed surface area of the oranges was varied by covering them with different amounts of paper, the edges of which were sealed in wax to keep mites from crawling underneath. In most experiments, Huffaker first established a prey population with 20 females per tray, then introduced 2 female predators 11 days later. (Both species reproduce parthenogenetically, so males were not required.)

When six-spotted mites were introduced to the trays alone, their populations leveled off at between 5,500 and 8,000 mites per tray. When predators were added, their numbers increased rapidly, and they soon wiped out the prey population. Their own extinction followed shortly.

Although predators always eliminated the six-spotted mites, the positions of the exposed areas of the oranges influenced the course of extinction. When the exposed areas were in adjacent positions, minimizing

Figure 20.7 (a) One of Huffaker's experimental trays in which four oranges, half exposed, are distributed at random among the 40 positions in the tray. Other positions are occupied by rubber balls. (b) Each orange is wrapped with paper and its edges sealed with wax. The exposed area has been divided into numbered sections to facilitate counting the mites. Courtesy of C. B. Huffaker. From C. B. Huffaker, *Hilgardia* 27:343–383 (1958).

(a)

(b)

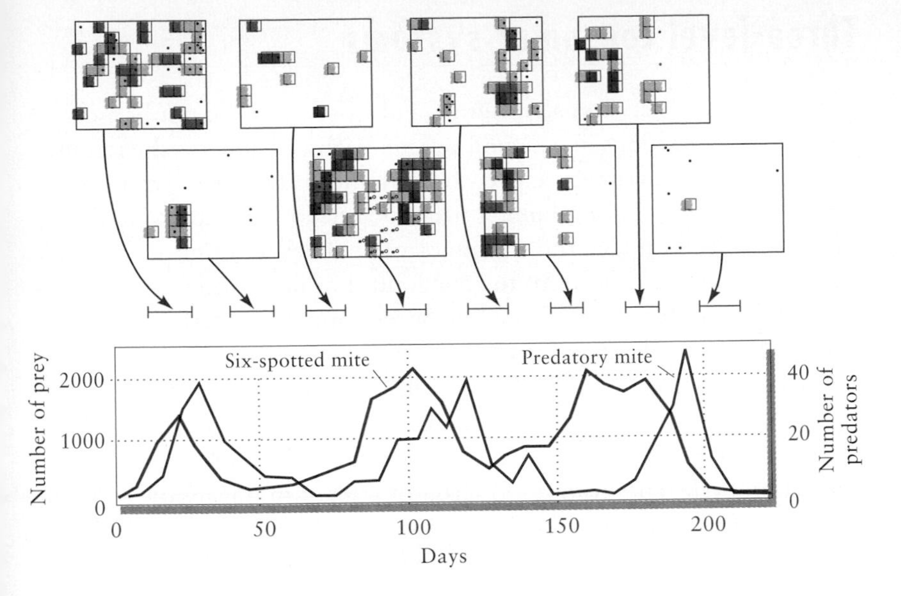

Figure 20.8 Population cycles of the six-spotted mite and the predatory mite *Typhlodromus* in a laboratory situation. The shaded boxes show the positions and relative densities of mites in the trays at the eight times indicated. The shading records the relative density of six-spotted mites; the black dots indicate the presence of predatory mites. After C. B. Huffaker, *Hilgardia* 27:343–383 (1958).

dispersal distance between food sources, prey populations reached maxima of only 113–650 individuals and were driven to extinction within 23–32 days after the beginning of the experiment. The same area of exposed oranges dispersed randomly throughout the 40-position tray supported prey populations that reached maxima of 2,000 to 4,000 individuals and persisted for 36 days. Thus, survival of the prey population could be prolonged by providing remote areas of suitable habitat to which predators dispersed slowly.

Huffaker reasoned that if predator dispersal could be further retarded, the two species might coexist. To accomplish this, he increased the spatial complexity of the environment and introduced barriers to dispersal. The number of possible food positions was increased to 120, and a feeding area equivalent to six oranges was dispersed over all 120 positions. A mazelike pattern of Vaseline barriers was placed among the food positions to slow dispersal of the predators. *Typhlodromus* must walk to get where it is going, but six-spotted mites spin a silk line that they can use like a parachute to float on wind currents. To take advantage of this behavior, Huffaker placed vertical wooden pegs throughout the trays, which the mites used as jumping-off points in their wanderings. This arrangement finally produced a series of three population cycles over the 8 months of the experiment (Figure 20.8). The distribution of predators and prey throughout the trays continually shifted as the prey, on the way to extermination in one feeding area, recolonized the next a jump ahead of their predators.

Despite the tenuousness of the predator-prey cycle that was achieved, this experiment demonstrated that a spatial mosaic of suitable habitats enables predator and prey populations to coexist through time. But, as we saw in Gause's experiments with protozoans, predator and prey may also coexist locally if some prey can take refuge in hiding places. And when the environment is so complex that predators cannot easily find scarce prey, stability again can be achieved.

Three-level consumer systems

The relationship between a consumer and its resource cannot be isolated from other interactions in a natural system. Each consumer is the resource for some other consumer, and each consumer-resource interaction is a single link in a food chain linking many populations. Recently, ecologists have begun to consider the dynamics of multilevel consumer-resource systems and their implications for the control of populations and the stability of food webs in natural systems. Several ambitious studies are beginning to shed some light on these issues.

In one experiment in the Yukon Territory of Canada, Charles J. Krebs, of the University of British Columbia, and his colleagues studied the effects of food supply and predation on densities of snowshoe hares over a population cycle. The major food chain of interest in this relatively simple system is willow-hare-lynx. The investigators established large tracts of land measuring 1 km on a side as experimental plots for the following treatments: predator exclosure to keep out lynx (–P), food addition to provide extra nourishment to the hares (+F), and both treatments. Compared with control plots (no treatments), –P led to a twofold increase in the hare population during its peak and decline phases, +F to a threefold increase, and both –P and +F together to an elevenfold increase. These differences in population densities between plots could be accounted for by differences in both survival and reproductive rates, particularly during the decline phase of the population cycle. Predators caused most of the known deaths (83% of radio-collared hares). Although only 9% of the hares starved, the investigators assumed that undernourishment increased the vulnerability of hares to predators. The investigators concluded that both food and predation influenced peak numbers of hares and that both factors together had a greater influence than the sum of the two acting alone. In this case, the density of an herbivore population was regulated both from above by its predators and from below by its resources.

Robert Marquis, at the University of Missouri-St. Louis, has studied the effects of herbivores on plant growth in a three-level consumer system consisting of oaks, leaf-chewing insects, and insectivorous birds. The experiments were simple: some branches of oak saplings were enclosed in coarse nets, which kept birds out but let insects pass; others were treated with insecticide. Leaf damage after 2 years was 34% inside the bird exclosures, which contained high densities of leaf-chewing insects, 24% on control branches without any treatment, and only 9% on branches treated with insecticide (Figure 20.9). The extreme herbivory within bird exclosures also depressed leaf production during the following year and adversely affected the subsequent growth of oak saplings. In a similar study conducted on tropical trees, Carlos Fonseca, of the University of Campinas in Brazil, showed that ants effectively controlled levels of herbivores. *Tachigalia* trees harbor colonies of the stinging ant *Pseudomyrmex concolor* in the hollow petioles of their leaves. When ants were experimentally removed from seedlings and saplings, the number of leaf-chewing herbivores increased fourfold, removal of leaf biomass increased tenfold,

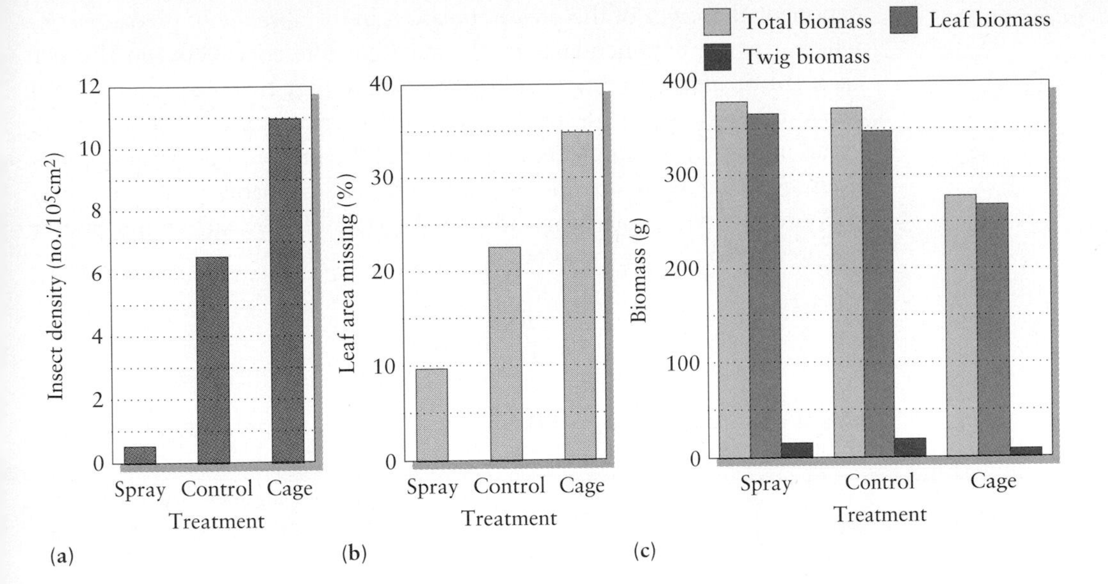

Figure 20.9 The influence of herbivore removal (spray) and predator exclosure (cage) on damage to leaves of white oak *(Quercus alba)* saplings in a Missouri forest: (a) density of insect herbivores; (b) percentage of leaf area removed by chewing insects; (c) biomass of leaves produced during the year following the treatments. From R. J. Marquis and C. J. Whelan, *Ecology* 75:2007–2014 (1994).

individual leaf longevity was cut by half, and apical growth was reduced by one-third. These experiments show clearly that the influence of a consumer population extends downward through at least two links in the food chain. This inherent complexity in nature makes it difficult to understand how natural systems are regulated. It is not surprising that the first attempts to come to grips with species interactions involved simple, two-species models.

Models of predator-prey interactions

In an attempt to understand the origin of population cycles, Alfred J. Lotka and the Italian biologist Vito Volterra independently provided the first mathematical descriptions of predator-prey interactions during the 1920s. As we shall see, their models predict oscillations in the abundances of predator and prey populations, with predator numbers lagging behind those of their prey. The success of these early modeling efforts led to the development of ever more complex quantitative depictions of species interactions, which have made great contributions to our understanding of the dynamics of natural systems. In this discussion, however, the basic principles and possibilities of modeling can be illustrated by the earlier, simpler treatments.

The Lotka-Volterra predator-prey model

Following a common convention, we shall designate the number of predator individuals by P and the number of prey individuals by R (think of R for resource; previous editions of this book used the symbol H for herbivore, but organisms other than herbivores can be prey resources). The growth rate of the prey population has two components: (1) unrestricted exponential growth of the prey population in the absence of predators, rR, where r is the exponential growth rate (the difference between the per capita birth and death rates); and (2) removal of prey by predators, over and above other causes of death. Lotka and Volterra assumed that predation varies in direct proportion to the product of the prey and predator populations, RP, and therefore in proportion to the probability of a random encounter between predator and prey. Accordingly, the rate of increase of the prey population is given by

$$\frac{dR}{dt} = rR - cRP,$$

where c is a coefficient expressing the efficiency of predation, or capture efficiency.

The growth rate of the predator population balances the birth rate, which depends on number of prey captured, against a constant death rate imposed from outside the system (for example, by weather):

$$\frac{dP}{dt} = acRP - dP,$$

The birth term is the number of prey captured (cRP) times a coefficient (a) for the efficiency with which food is converted to population growth. The death rate is a constant (d) times the number of predator individuals.

When both predator and prey populations achieve equilibrium ($dR/dt = 0$ and $dP/dt = 0$), $rR = cRP$ and $acRP = dP$. We can rearrange these equations to give

$$\hat{P} = \frac{r}{c} \quad \text{and} \quad \hat{R} = \frac{d}{ac},$$

where $\hat{P}$ and $\hat{R}$ are the equilibrium sizes of the predator and prey populations, respectively. Note that both $\hat{P}$ and $\hat{R}$ are constant values, each independent of the abundance of the other population. The point at which the lines representing $\hat{P}$ and $\hat{R}$ cross is called the **joint equilibrium,** which is the only combination of population sizes P and R that is stable.

According to the Lotka-Volterra model, when the populations stray from their joint equilibrium rather than return to the equilibrium point, they oscillate around it in a continuous cycle. The period of the oscillation (T) is approximately $2\pi/\sqrt{rd}$. Hence the higher the population growth potential of the prey or the death rate of the predator—that is, the higher the rate of population turnover—the faster the system oscillates.

The relationship between predator and prey can be portrayed as a graph with axes representing the sizes of the populations (Figure 20.10). By convention, predator numbers increase along the vertical axis and prey numbers along the horizontal axis. The equilibrium population values of predator ($\hat{P}$) and prey ($\hat{R}$) partition the graph into four regions. The horizontal line $\hat{P} = r/c$, representing the condition $dR/dt = 0$, is called the **equilibrium isocline** (or **zero growth isocline**) of the prey. For any

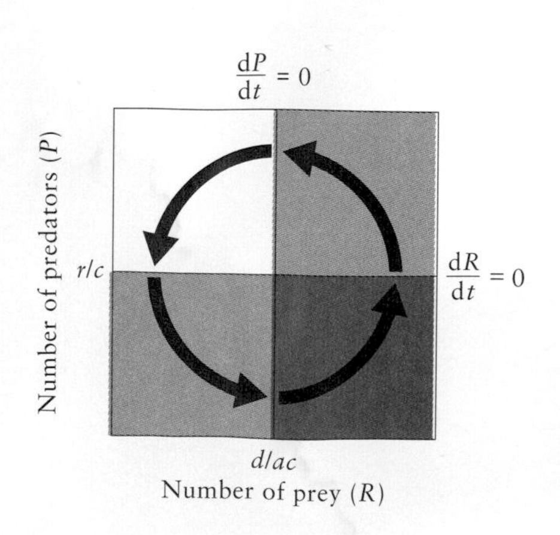

Figure 20.10 Representation of the Lotka-Volterra predator-prey model on a population graph. Trajectories of the populations show that the predator and prey will continually oscillate out of phase with each other.

combination of predator and prey numbers that lies in the region below this line, the prey increase because there are few predators to eat them. In the region above the prey isocline, prey populations decrease because of overwhelming predator pressure. For the predators, population increases are limited to the region to the right of the predator equilibrium isocline ($\hat{R} = d/ac$), where prey are abundant enough to sustain population growth. To the left of the line, predator populations decrease as a result of insufficient prey.

The change in predator and prey populations together follows a path, shown in each of the four sections of the graph, that combines the individual changes of the predator and prey populations. This path is called a **population trajectory.** In the lower right-hand section of the graph, both predators and prey increase, and their joint population trajectory moves up and to the right. The trajectories in the four regions together define a counterclockwise cycling of the predator and prey populations one-quarter cycle out of phase, with the prey population increasing and decreasing just ahead of its predator population (Figure 20.11).

The equilibrium isocline for the predator ($dP/dt = 0$) defines the minimum level of prey ($\hat{R} = d/ac$) that can sustain the growth of the predator population. The equilibrium isocline for the prey ($dR/dt = 0$) defines the greatest number of predators ($\hat{P} = r/c$) that the prey population can sustain. If the reproductive rate of the prey (r) increased or the capture efficiency of the predators (c) decreased, or both, the prey isocline (r/c) would increase—that is, the prey population would be able to bear the burden of a larger predator population. If the death rate of the predators (d) increased and either the predation efficiency (c) or the reproductive efficiency of predators (a) decreased, the predator isocline (d/ac) would move to the right, and more prey would be required to support the predator population. Increased predator hunting efficiency (c) would simultaneously reduce both isoclines: fewer prey would be needed to sustain a given capture rate (the predator isocline would decrease), and the prey population would be less able to support the more efficient predators (the prey isocline would decrease).

Lotka pointed out that adding terms representing density-dependent limitation of predators and prey would tend to create an inward spiraling of the population trajectory toward the joint equilibrium. For example, the expression $dR/dt = rR - cRP - bR^2$ is a logistic equation ($dR/dt = rR - bR^2$; b is a coefficient with an arbitrary value) with an added predation term ($-cRP$). The logistic equation may look more familiar to you in the form $dR/dt = rR\ (1 - b/r\ R)$, where b/r is equivalent to $1/K$, the inverse of the carrying capacity. Regardless of the form of the equation, the density-dependent term ($-bR^2$) leads to a damped oscillation and a stable equilibrium. Density dependence in the predator population has the same effect.

Finnish ecologists Ilkka Hanski and E. Korpimäki developed a number of models of predator-prey interaction, which they fit to data on the cycling of voles (*Microtus*) and their primary predator, the weasel *Mustela nivalis.* The models that most closely mimicked the 3–5-year population

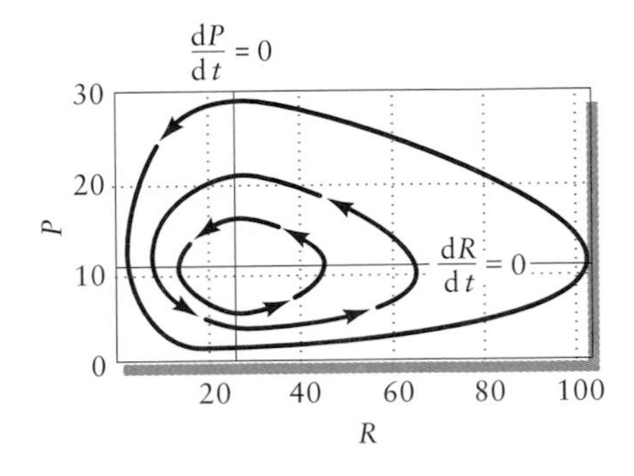

Figure 20.11 Simulated population trajectories of predators and prey according to a Lotka-Volterra model. The degree of oscillation about the equilibrium point reflects the initial population size in each simulation. From G. D. Elseth and K. D. Baumgardner, *Population Biology,* Van Nostrand, New York (1981).

cycles found in nature included a predator on the weasel as well as a refuge for the weasel at low weasel density (to keep the weasel's predator from driving it to extinction and to allow for rapid increase of weasels from the bottom of their population cycle). This more complicated model contains three levels of consumers, or two consumer-resource interactions, and a refuge for the intermediate consumer at low density. Refuges generally have a stabilizing effect on resource-consumer interactions, as we shall see.

According to the Lotka-Volterra model, when either the predator or the prey population is displaced from its equilibrium, the system will oscillate in a closed cycle. Any further perturbation of the system will give the population fluctuations a new amplitude and duration until some other outside influence acts upon them. This state of oscillation is said to be a **neutral equilibrium** because no internal forces act to restore the populations to the intersection of the predator and prey isoclines. Therefore, random perturbations will eventually increase fluctuations to the point at which the trajectory reaches one of the axes of the predator-prey graph, and one or both populations die out. This property in itself suggests that the Lotka-Volterra equations greatly oversimplify nature.

Other concerns about the adequacy of the Lotka-Volterra model focus on the predation term (cRP). For a given density of predators, the rate of consumption increases in direct proportion to the density of prey (R). Accordingly, predators cannot be satiated; they just keep on eating, no matter how many prey they catch. Clearly this aspect of the model is unrealistic. How would adding a bit more reality here affect the behavior of the model?

The functional response

The relationship of an individual predator's rate of food consumption to the density of its prey has been labeled the **functional response** by Canadian entomologist C. S. Holling. This relationship can take several forms, each having different effects on the dynamics of the predator-prey relationship. In the particular case of the Lotka-Volterra model, the rate of predation is independent of the density of the prey population. This type of relationship is called a **type I functional response,** and it gives the Lotka-Volterra model its distinctive property of having a neutral equilibrium. This lack of dependence of the proportion of prey consumed on the density of the prey is related to the Lotka-Volterra model in the following way. According to the model, the total rate of prey consumption by predators is cRP, where P is the number of predators, R is the number of prey, and c is the capture efficiency. Dividing this term by the number of predators, P, we see that the rate of consumption per predator is simply cR. Thus, prey are consumed in direct proportion to their abundance or density at rate c, as illustrated in Figure 20.12. The fact that predators can continue to consume prey at a high rate even when prey are very abundant means that the fecundity of individual predators, which in the model is proportional to the number of prey consumed, increases without limit in direct proportion to prey availability. Thus, when there are few predators

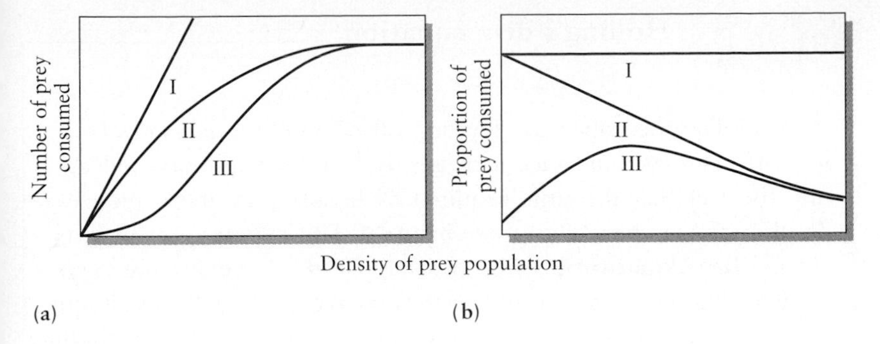

Figure 20.12 Three types of functional responses of predators to increasing prey density: (a) the functional response in terms of the number of prey consumed; (b) the functional response in terms of the proportion of prey consumed. Type I: The predator consumes a constant proportion of the prey population regardless of its density. Type II: Predation rate decreases as predator satiation sets an upper limit on food consumption. Type III: Predator response is depressed at low prey density because of low hunting efficiency or absence of a search image.

but many prey, the birth rate of the predators is very high, the predator population grows rapidly, and the prey population can be brought under control. This particular circumstance causes the neutral equilibrium that is characteristic of the Lotka-Volterra model.

According to the type I functional response, predators cannot be satiated. As the density of the prey increases, each individual predator just eats more of them. An obvious modification of the type I functional response is the **type II functional response,** in which the number of prey consumed per predator initially rises quickly as the density of prey increases, but then levels off with further increases in prey density. A **type III functional response** resembles type II in having an upper limit to prey consumption, but it differs in that the response of predators to prey is depressed at low prey density.

Two factors dictate that a functional response should reach a plateau: First, predators may become satiated—constantly full—at which point their rate of feeding is limited by the rate at which they can digest and assimilate food. Second, as a predator captures more prey, the time it spends handling and eating prey cuts into its searching time. Eventually, these two factors reach a balance, and prey capture rate levels off (Box 20.1).

At high prey densities, type II and type III response curves differ little: they are both inversely density-dependent. In other words, as the density of prey increases, the proportion of prey consumed decreases. Over the lower range of prey densities, however, type III responses differ from type II responses in that the proportion of the prey consumed increases with density of prey. Several factors may cause the predator response to decrease at lower prey densities, thereby resulting in a type III functional response: (1) a heterogeneous habitat may afford a limited number of safe hiding places, which protect a larger proportion of the prey at lower densities than at higher densities; (2) lack of reinforcement of learned searching behavior

BOX 20.1 Holling's disc equation

C. S. Holling described the leveling off of consumption rates by a predator as prey abundance increases by a simple expression reflecting the fact that the time required to handle prey items reduces hunting time as prey abundance increases. This expression is known as the **disc equation** because of its application to experiments in which blindfolded human subjects were required to discover and pick up small discs of paper on a flat surface. Any such task, including the capture and eating of prey, requires a certain handling time, T_h. The total handling time is therefore the handling time per item times the number of encounters, E. The time left over for searching (T_s) is the total time minus the total handling time; that is, $T_s = T - T_h E$. The number of encounters can itself be defined as the product of search time, prey density (P), and a constant (a) for the efficiency of searching: $E = a(T - T_h E)P$. Combining the expressions for T_s and E and solving for E yields

$$E = \frac{aPT}{1 + aPT_h}.$$

When prey are scarce, the denominator term aPT_h is much less than 1, and the number of prey encountered approaches aPT. Hence encounters are directly proportional to prey density. When prey are dense, aPT_h is much greater than 1, and E approaches the ratio T/T_h, which is a constant value defining the maximum number of prey that can be captured in time T. Thus at high prey density, search time drops to near zero, and the number of prey captured is limited only by how long the predator requires to handle each one; the shorter the handling time, the more prey can be captured.

owing to a low rate of prey encounters may reduce hunting efficiency at low prey densities; and (3) switching to alternative sources of food when prey are scarce may reduce hunting pressure. The relationship of prey consumption to prey density depends on a change in average prey vulnerability in the case of factor 1, on search and capture efficiency in the case of factor 2, and on motivation to hunt or searching time in the case of factor 3.

Learning influences the searching behavior of many predators, especially vertebrates with complex nervous systems. An object is always easier to find if one has a preconception of what it looks like and where it might be. Such **search images** are acquired by experience. The denser a prey population, the more frequently predators encounter prey; search efficiency increases and handling time decreases as a result of experience. At low prey densities, neither experience nor efficiency is so well developed, and capture rate is therefore depressed.

Regardless of the mechanisms underlying a type III response, predators often switch to a second prey species as the density of the first is reduced. For example, when the predatory water bug *Notonecta glauca* was presented with two types of prey—isopods and mayfly larvae—the predators consumed the more abundant prey species, whichever it was, in a proportion greater than its percentage of occurrence (Figure 20.13). The predators' **switching** depended to some degree on variation in the success of attacks on prey as a function of their relative density. When water bugs encountered mayfly larvae infrequently, fewer than 10% of attacks were successful. At higher densities, and therefore higher encounter rates, attack success rose to almost 30%.

The numerical response

Individual predators can increase their consumption of prey only to the point of satiation. Continued predator response to increasing prey density above the level that results in satiation can be achieved only through an increase in the *number* of predators, either by immigration or by population growth, which together constitute a **numerical response.** Populations of most predators grow slowly, especially when the reproductive potential of a predator is much lower than that of its prey and the predator's life span is longer. Immigration from surrounding areas contributes to the numerical responses of mobile predators, which may opportunistically congregate where resources become abundant. The bay-breasted warbler, a small insectivorous bird of eastern North America, exhibits such behavior during periodic outbreaks of the spruce budworm. During years of outbreak in a particular area, the density of warblers may reach 120 breeding pairs per 100 acres, compared with about 10 pairs per 100 acres during nonoutbreak years. This population behavior shows how a predator can take advantage of a shifting mosaic of prey abundance.

Consider the different responses of three predatory birds, the pomarine jaeger, snowy owl, and short-eared owl, to varying densities of lemmings on the arctic tundra (Table 20.1). Lemming populations exhibit great fluctuations: high and low points in a population cycle may differ by

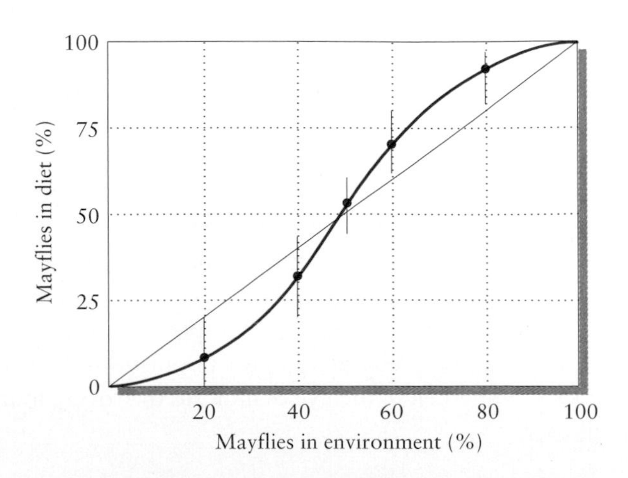

Figure 20.13 The percentage of mayfly larvae in the diet of the predaceous water bug *Notonecta* is lower than expected by chance when mayfly larvae are uncommon, but higher than expected when larvae are abundant. The straight line is the hypothetical point at which a predator would exhibit no preference for mayflies over other prey. Data are presented as means (dots) and ranges (indicated by vertical lines) of several separate trials. From M. Begon and M. Mortimer, *Population Ecology,* 2d ed., Blackwell Scientific Publications, Oxford (1981); after J. H. Lawton, J. R. Beddington, and R. Bonser, pp. 141–158, in M. B. Usher and M. H. Williamson (eds.), *Ecological Stability,* Chapman & Hall, London (1974).

TABLE 20.1 Responses of predatory birds to varying densities of the brown lemming near Barrow, Alaska

	1951	1952	1953
Brown lemming (individuals per acre)	1–5	15–20	70–80
Pomarine jaeger	Uncommon, no breeding	4 breeding pairs per square mile	18 breeding pairs per square mile
Snowy owl	Scarce, no breeding	0.2–0.5 breeding pairs per square mile; many non-breeders	0.2–0.5 breeding pairs per square mile; few non-breeders
Short-eared owl	Absent	One record	3–4 breeding pairs per square mile

Source: F. A. Pitelka, P. O. Tomich, and G. W. Treichel, *Ecol. Monogr.* 25:85–117 (1955).

a factor of 100. At Barrow, Alaska, during the summer of 1951, when lemmings were scarce, none of the predatory birds bred; short-eared owls did not even appear in the area. During the following summer, one of moderate lemming density, both the jaeger and the snowy owl bred, but short-eared owls again were absent. In 1953, a peak year for lemmings, all three species of avian predators bred. Jaegers were four times more abundant in 1953 than in 1952, showing a strong numerical response. In contrast, the density of snowy owls did not increase, but each pair of birds reared more young. Whereas most snowy owl nests contained 2–4 eggs during the year of moderate lemming abundance, clutches of up to 12 eggs were laid during the peak year of 1953.

Stability in predator-prey systems

Studies of the functional and numerical responses of predators, as well as of the population dynamics of prey populations, have revealed the biological limitations of simple predator-prey models such as the Lotka-Volterra equations. Many complications of biological reality have been incorporated into models and their contributions to the stability of predator-prey interactions studied. The findings can be summarized by a number of general rules of predator-prey stability. Stability in predator-prey systems may be thought of as the tendency of populations of predators and prey to achieve nonvarying equilibrium sizes. This is a special, restrictive meaning of the word *stability* because predator and prey populations may also oscillate in stable cycles. Nonetheless, the cycles indicate the strong influence of destabilizing factors.

Five factors tend to damp predator-prey cycles and thus promote the long-term persistence of the relationship: (1) predator inefficiency (or enhanced prey escape or defense); (2) density-dependent limitation of the population of either the predator or the prey by factors external to their relationship; (3) alternative food sources for the predator; (4) refuges from predation at low prey densities; and (5) reduced time lags in predator population response to changes in prey abundance.

Predator inefficiency (lower *c* in the Lotka-Volterra model) results in higher equilibrium levels for both prey and predator populations (more predators can be supported by the larger prey populations) and in lower turnover rates for both at equilibrium. Both these consequences would seem to enhance the stability of a predator-prey system. Alternative food sources stabilize predator populations because individuals may switch between food types in response to changing prey abundance. Similarly, safe refuges from predation allow prey populations to maintain themselves at higher levels in the face of intense predation, thereby facilitating the recovery phase of the population cycle. Indeed, so many factors tend to stabilize predator-prey relationships that the periodic cyclic behavior of some systems seems to require special explanations. Most probably, population cycles reflect a balancing of the stabilizing effects of density dependence, alternative prey, and refuges against the destabilizing effects of time lags in predator-prey interactions. Time lags are ubiquitous in nature, arising from the developmental periods of animals and plants, the time required for numerical responses by predator populations, and the time course of immune responses and of induced defenses in plants. In some circumstances, perhaps in simple ecological systems, these factors outweigh stabilizing influences and result in population cycles.

Multiple stable states in predator-prey systems

The size of any population is influenced by the abundance of its resources and by its own consumers. Extremely efficient predators may depress a prey population to levels far below its carrying capacity. Alternatively, a prey population may be limited primarily by its own food supplies while predators remove an inconsequential number of prey. As we have already seen, equilibrium population size in many situations reflects a balance between the limiting influences of food supply and predators. Under some circumstances, however, a population may have two or more stable equilibrium points, only one of which may be occupied at a given time. This situation is referred to as **multiple stable states.** How do multiple stable states arise? What factors determine which of the stable points is occupied?

A simple model based on a type III functional response curve illustrates how multiple stable states can develop (Figure 20.14). This model of a hypothetical system describes changes in recruitment and predation rates in prey populations as a consequence of increasing prey density. The recruitment curve represents the net contribution of births and deaths in the absence of predators. Hence per capita recruitment is high when the prey population is small and decreases to zero as the population approaches its carrying capacity. The predation curve is the sum of the functional and numerical responses of predator populations. The predation rate may be low at low prey densities because of switching to alternative prey or difficulty in locating scarce prey; it also tails off at high prey densities because of predator satiation and extrinsic limits to predator populations. The predation curve in the graph thus represents a type III functional response (see Figure 20.12b).

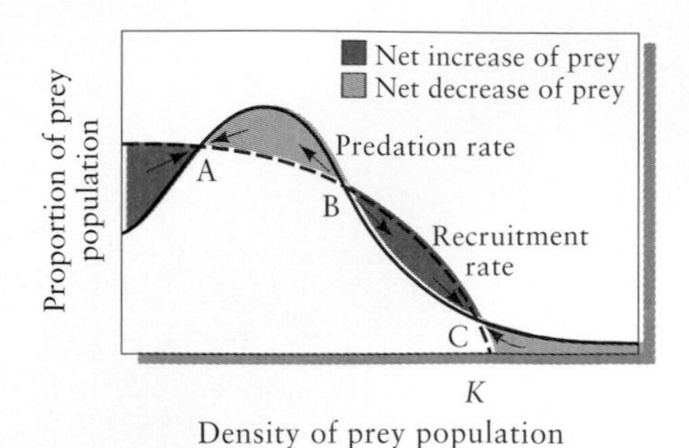

Figure 20.14 Predation and recruitment rates in a hypothetical predator-prey system. The predation curve is based on a type II functional response (see Figure 20.12). When predation exceeds recruitment, prey populations decrease, and vice versa (as shown by the arrows). Points A and C are stable equilibria for the prey population; the lower point (A—lower in terms of prey density) represents population control by predators; the higher point (C) represents population control by food and other resources.

The recruitment and predation curves shown in Figure 20.14 produce three equilibrium points. The highest and lowest of these three points represent stable equilibria around which populations are regulated; the middle equilibrium is unstable. The lower (in terms of prey density) equilibrium point (A) corresponds to the situation in which predators regulate a prey population at substantially below its carrying capacity (K). Below point A—that is, when prey are very sparse—either they are too difficult to find, or predators develop preferences for other, more abundant prey. As the density of prey increases above point A, however, predators are able to capture them more efficiently, and they tend to eat the prey population back down to point A. Above point B and below point C, predation efficiency is not high enough to regulate the population of prey. However, above point C, the prey become limited by their own resources, and the predator-prey system reaches a stable equilibrium close to the carrying capacity of the environment for the prey population. Point B represents a changeover from strong predator control to strong resource control of a prey population. Below point B, prey populations are further reduced to stable point A; above point B, the prey escape control and increase to stable point C. The transition occurs because, as the prey become more abundant, the predators become less efficient at capturing them: individual predators become satiated more quickly and do not need to eat more prey, and predator populations become limited by factors other than their food supply, perhaps by disease or social interactions. The upper equilibrium (C) corresponds to the situation in which a prey population is regulated at close to K primarily by the availability of food and other resources; here predation exerts a minor depressing influence on population size.

The implications of Figure 20.14 for practical concerns such as the control of crop pests are clear. Predators maintain a shaky hold on prey populations at point A. If a heavy frost or an introduced disease reduces the predator population long enough to allow the prey population to slip above point B, the prey may continue to increase until they reach the higher (in terms of prey density) stable equilibrium point C, regardless of how quickly the predator population recovers. To the farmer, this means that a crop pest population that is normally controlled at harmless levels by predators and parasites suddenly becomes a menacing outbreak. After such a change, predators exert little control over the pest population until some quirk of the environment brings its numbers below point B, back within the realm of predator control.

A pertinent experiment in Australia that investigated the ability of foxes and feral cats to regulate a population of European rabbits showed that predator control may be tenuous. Predators were removed from an area by continual shooting, and the rabbit population began to increase rapidly, as one would expect following the removal of predator control of the rabbit population. The predator removal program was stopped after a few months and predators were allowed to reinvade the area. At this point, however, the predators were unable to control the rabbit population, which continued to grow at a rapid rate. Thus, it appeared that the prey population had escaped control by its predators and may have been growing toward a higher stable size, regulated perhaps by its own food supply.

Using the predation-recruitment diagram in Figure 20.14, we can examine the consequences of different levels of predation for control of a prey population (Figure 20.15). Inefficient predators cannot regulate prey populations at low densities; they depress prey numbers slightly, but the prey population remains near the equilibrium level set by resources (Fig. 20.15a, point C). Increased predation efficiency at low prey density, however, can result in predator control at point A (Fig. 20.15b). When functional and numerical responses are sufficient to maintain high densities of predators, predation may effectively limit prey growth under all circumstances, and equilibrium point C disappears (Fig. 20.15c). Finally, predation may be so intense at all prey densities that the prey are eaten to extinction (Fig. 20.15d, no equilibrium point).

In all likelihood, predators are able to drive prey populations to extinction only in simple laboratory systems or when predator populations are maintained at high densities by the availability of some alternative but less preferred prey (hence, there is no switching). Indeed, many ecologists have advocated providing parasites and predators of pest species with innocuous alternative prey to enhance biological control. At the very least, the curves in Figure 20.14 suggest that the position of a predator-prey equilibrium, whether at very low prey levels or close to their carrying capacity, may shift between extremes with small changes in closely

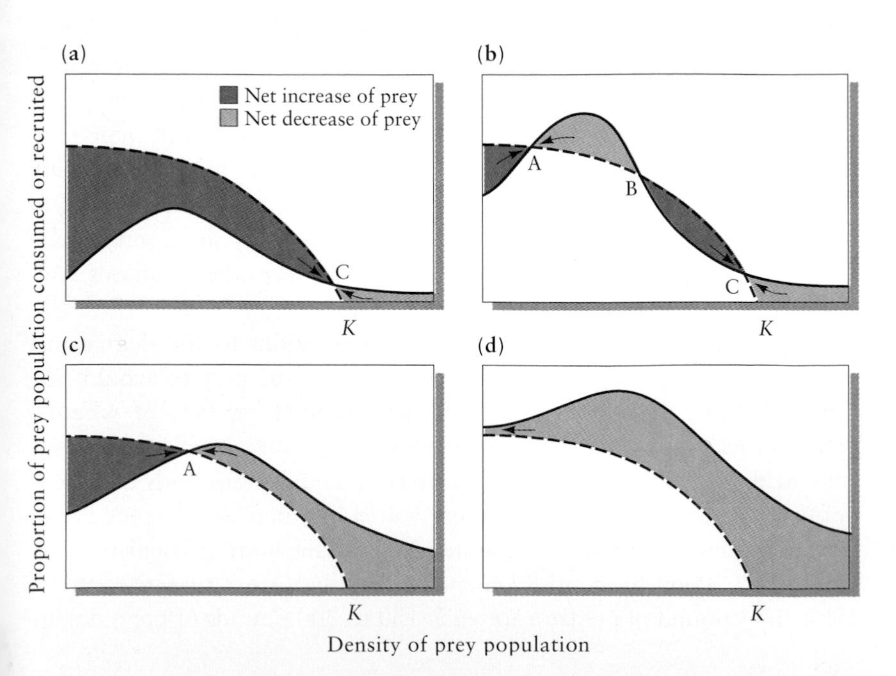

Figure 20.15 Predation (solid lines) and recruitment curves (dashed lines) at different intensities of predation, showing the effect of predation intensity on the number of equilibria.

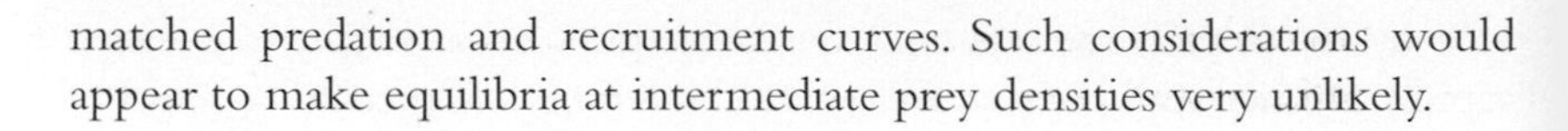
matched predation and recruitment curves. Such considerations would appear to make equilibria at intermediate prey densities very unlikely.

Predator-prey population ratios and maximum sustainable yield

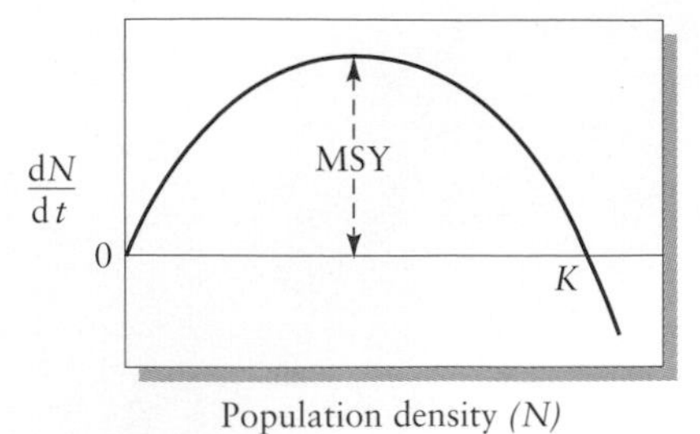

Figure 20.16 The absolute rate of growth of a population (dN/dt) as a function of population density. Predators can remove prey individuals at rate dN/dt and maintain the prey population at a constant (sustainable) level. The peak of dN/dt, which is the maximum sustainable yield (MSY), occurs at intermediate prey densities.

The ability of a prey population to support a predator population varies with its density (Figure 20.16). A prey population that is very small relative to its carrying capacity can support few predators because, although each prey individual's reproductive potential may be high, the total recruitment rate of a small population is low. A prey population near its carrying capacity is also unproductive because, though members are numerous, each individual's reproductive potential is severely limited by the effects of crowding.

At some intermediate density, the overall recruitment rate of a prey population reaches a maximum. Because predators can remove a number of individual prey equal to the recruitment rate without reducing the size of the prey population, the prey population density that yields the maximum recruitment generally will support the greatest number of predators. The rate of recruitment at this point is known as the **maximum sustainable yield** because it is the maximum rate at which predators can remove individuals without depressing the prey population.

Ranchers and game managers strive to maintain populations of beef cattle, deer, and geese at their most productive levels to maximize the harvest of those species without reducing their populations. We may ask whether predators also prudently manage their prey populations in such a way as to maximize the productivity of their own populations. If so, how can such behavior evolve?

Territorial animals, which exclude competitors from their feeding areas, could indeed space themselves with respect to their prey to achieve maximum yields. When the feeding areas of predators overlap, however, intraspecific competition dictates that each predator must maximize its immediate harvest at the expense of long-term yields. Our own species behaves no differently. Intelligently managed ranches, with fences to exclude competing livestock, can achieve maximum sustainable yields. Alas, in highly competitive situations—fishing in international waters, to name one—we have shown ourselves to be pathetically short-sighted and imprudent, and stocks of many species of fish and other seafoods have declined dramatically.

In most cases, predators exploit prey populations to the degree that their ability to capture prey exceeds the ability of the prey to avoid being captured. Both skills are evolved characteristics. Regardless of whether predators act prudently to manage prey populations for maximum sustainable yield, they often achieve a characteristic equilibrium with their prey populations. The relationship between wolves and their various prey populations in several areas demonstrates this equilibrium particularly well (Table 20.2). Population ratios and, particularly, biomass ratios are relatively constant (1 pound of predator for each 150 to 300 pounds of prey) despite

TABLE 20.2 **Relationships between populations of predators and their prey in several localities**

LOCALITY	PREDATOR	PRINCIPAL PREY	DENSITY OF PREDATORS (INDIVIDUALS PER 100 SQUARE MILES)	PREDATOR: PREY RATIO	
				Numbers	*Biomass*
Jasper National Park	Wolf	Elk, mule deer	1	1:100	1:250
Wisconsin	Wolf	White-tailed deer	3	1:300	1:300
Isle Royale	Wolf	Moose	10	1:30	1:175
Algonquin Park	Wolf	White-tailed deer	10	1:150	1:150
Canadian Arctic	Wolf	Caribou	1.7	1:84	1:186
Utah	Coyote	Jackrabbits	28	1:1,000	1:100
California	Mountain lion	Deer	—	1:550	1:900
Idaho (primitive area)	Mountain lion	Elk, mule deer	7.5	1:116	1:524
Ngorongoro Crater, Tanzania	Hyena	Ungulates	440	1:135	1:46
Nairobi Park, Kenya	Felids	Ungulates	96	1:97	1:140
Alaska	Pomarine jaeger	Lemmings	—	1:1,263	1:90

Source: R. E. Ricklefs, *Ecology,* 3d ed., W. H. Freeman, New York (1990).

the fact that the species and density of the wolves' principal prey vary considerably with locality.

Different predator-prey systems appear to achieve different equilibria. The population ratio of mountain lions to deer in California is 1:500 to 1:600, which is equivalent to a biomass ratio of about 1:900, and the exploitation rate is only 6%, compared with values of 18% and 37% for wolves in two areas. A mountain lion-elk-mule deer system in Idaho has a biomass ratio of 1:524 and an exploitation rate of 5% for the elk population and 3% for the mule deer population. Evidently, wolves exploit their prey more efficiently than mountain lions, perhaps because of their social hunting habits. Where predators feed on more abundant prey populations, as in savanna, grassland, and tundra habitats, predators not only are more numerous but also achieve higher biomass ratios (1:50 to 1:150: Table 20.2). Large cats—lions and cheetahs—remove 16% of prey biomass in Nairobi Park, Kenya, where their biomass ratio is 1:140.

Summary

1. Ecologists distinguish three basic kinds of consumers: predators, which remove a prey individual from the prey population as they consume it; parasitoids (mostly small flies and wasps), which kill their hosts, but only

after the parasitoid larvae have pupated; and parasites and grazers, both of which consume portions of the living animal or plant organism, but do not kill it.

2. One source of time delays in parasite and grazing systems is the acquisition of immunity or other resistance. Plants exhibit chemical responses to browsing and grazing that discourage further consumption. Time delays in these prey responses may be responsible for the cycles observed in disease outbreaks and in herbivore populations (and those of the carnivores that prey on them).

3. Experimental studies of pest species and their natural predators have demonstrated that, in many cases, consumers can reduce resource populations far below their carrying capacities.

4. Experimental studies have demonstrated that predator and prey populations can be made to oscillate in the laboratory. Maintenance of population cycles usually requires a complex environment in which prey are able to establish themselves in refuges.

5. Experimental studies in natural systems have demonstrated strong predator control of prey populations in many cases, and have also shown that the influences of consumers extend downward through several links of the food chain.

6. Alfred J. Lotka and Vito Volterra, in the 1920s, devised simple models of predator and prey dynamics that predicted population cycles. The models used differential equations in which the rate of prey removal was directly proportional to the product of the predator and prey populations.

7. The functional response describes the relationship between prey removal rate per predator and prey density. Whereas the Lotka-Volterra models, which employ a type I functional response curve, are inherently unstable, type III functional response curves can result in stable regulation of prey populations at low density.

8. The numerical response describes the response of a predator population to increasing prey density by population growth and immigration.

9. Stability in predator-prey interactions is promoted by density dependence in either predator or prey, by refuges or hiding places in which prey can escape predation, by low predator efficiency, and under some circumstances by the availability of alternative prey. Stable population cycles in nature apparently express the balance between these stabilizing factors and the destabilizing influence of time delays in population responses.

10. Models of consumer-resource systems suggest that such systems can have two stably regulated points (multiple stable states) between which populations may shift, depending on environmental conditions. The lower equilibrium is determined by the strong depressing influence of predators on prey populations; the upper equilibrium lies close to the carrying capacity of the prey in the absence of predation. Sudden climatic or biotic stresses may shift a system from one to the other of these points, resulting in successive controlled and outbreak conditions.

11. Each prey population has a density at which the number of individuals that can be harvested without causing a population decrease is greatest. This is called the point of maximum sustainable yield. When an individual predator can control its prey, as humans can do with many domestic and game species, maximum sustainable yields can be achieved. When predators compete for the same resources, maximization of short-term yields generally precludes the achievement of a maximum sustainable yield. Nevertheless, the balance between consumer efficiency and resource escape leads to consistent ratios in biomass of predators to prey.

Suggested readings

Anderson, R. M., and R. M. May. 1980. Infectious diseases and population cycles of forest insects. *Science* 210:658–661.

Brooks, J. L., and S. I. Dodson. 1965. Predation, body size and composition of the plankton. *Science* 150:28–35.

Crawley, M. J. 1983. *Herbivory: The Dynamics of Animal-Plant Interactions.* University of California Press, Berkeley.

DeBach, P., and D. Rosen. 1991. *Biological Control by Natural Enemies.* 2d ed. Cambridge University Press, New York.

Errington, P. L. 1963. The phenomenon of predation. *American Scientist* 51: 180–192.

Godfray, H. C. J. 1994. *Parasitoids: Behavioral and Evolutionary Ecology.* Princeton University Press, Princeton, N.J.

Jefferies, R. L., D. R. Klein, and G. R. Shaver. 1994. Vertebrate herbivores and northern plant communities: Reciprocal influences and responses. *Oikos* 71: 193–206.

Lindroth, R. L., and G. O. Batzli. 1986. Inducible plant chemical defences: A cause of vole population cycles? *Journal of Animal Ecology* 55:431–449.

Marquis, R. J., and C. J. Whelan. 1994. Insectivorous birds increase growth of white oak through consumption of leaf-chewing insects. *Ecology* 75:2007–2014.

May, R. M. 1983. Parasite infections as regulators of animal populations. *American Scientist* 71:36–45.

May, R. M., J. R. Beddington, C. W. Clark, S. J. Holt, and R. M. Laws. 1979. Management of multispecies fisheries. *Science* 205:267–277.

McNaughton, S. J. 1985. Ecology of a grazing ecosystem: The Serengeti. *Ecological Monographs* 55:259–294.

Myers, J. H. 1993. Population outbreaks in forest lepidoptera. *American Scientist* 81:240–281.

Price, P. W. 1980. *Evolutionary Biology of Parasites.* Princeton University Press, Princeton, N.J.

Shrag, S. J., and P. Wiener. 1995. Emerging infectious disease: What are the relative roles of ecology and evolution? *Trends in Ecology and Evolution* 10:319–324.

CHAPTER 21

EVOLUTIONARY RESPONSES AND COEVOLUTION

Populations of consumers, resources, and competitors fashion the environment of every species. Each selects traits in the others that tend to alter their interactions. For example, by capturing those prey that it finds most readily or can catch most easily, a predator leaves behind more cryptic or swifter prey. These prey reproduce and pass on to future generations their protective coloring and speed. As a result, the efficiency with which the population of predators exploits that prey population as a whole tends to diminish over evolutionary time, all other things being equal. Of course, predators also evolve to capture their prey more easily.

When two populations interact, each may evolve in response to characteristics of the other that affect its well-being—that is to say, its evolutionary fitness. This process in a very broad sense can be referred to as **coevolution,** a term that recognizes the effects of each species on those populations with which it interacts. In many usages, however, the term *coevolution* is restricted narrowly to the situation in which one species evolves an adaptation specifically in response to an adaptation in another species that affects their interaction. When this restrictive definition is applied, ecologists have difficulty identifying unambiguous cases of coevolution. The following example is a bit of a stretch, but makes the point. Hyenas have jaws and associated muscles that are strong enough to crack the bones of their prey. These modifications clearly are adaptations for eating selected by attributes of the prey. They cannot, however, be considered an example of coevolution because the properties of the bones of gazelles did not evolve to resist being eaten by hyenas, or any other predator. By the time a hyena has reached that part of its meal, bone structure has no consequence for prey survival. In contrast, when an herbivore evolves the ability to detoxify substances produced by a plant specifically to deter herbivory, the requirements of the strict definition of coevolution are more likely to be met, and this may be considered an example of coevolution.

When the coevolutionary relationship between two species is antagonistic, as it is between predator and prey, the species can become locked in an evolutionary battle to increase their own fitnesses, each at the other's expense. Such a struggle may lead to an evolutionary stalemate in which both antagonists continuously evolve in response to each other, or they may simply run out of the genetic variability needed to fuel further evolutionary change. In either event, the net outcome of their interaction may be a steady state. Alternatively, when one of the antagonists cannot evolve fast enough, it may be driven to extinction. Coevolution between mutualists may lead to stable arrangements of complementary adaptations that promote their interaction.

Evolutionary relationships between antagonists

In this chapter, we shall explore some of the consequences of evolutionary responses for interactions between predators and their prey, between competitors, and within mutualistic associations. Interactions between antagonists have been characterized most fully because of the importance of such relationships in agriculture and game management. The following example was one of the first to be analyzed in the context of coevolution, and it still provides a particularly vivid case study of evolution in action within a pathogen-host system.

The myxoma virus and the rabbit

Shortly after the release of a few pairs on a ranch in Victoria in 1859, the European rabbit became a major pest in Australia. To give an idea of how fast the rabbit populations increased, within a few years local ranchers were erecting rabbit fences and organizing rabbit brigades—shooting parties—in vain attempts to keep their numbers under control. Eventually, hundreds of millions of rabbits were distributed throughout most of the continent, where they destroyed range and pasture lands and threatened wool production. The Australian government tried poisons, predators, and other potential controls, all without success. After much investigation, the answer to the rabbit problem seemed to be a myxoma virus (a relative of smallpox) discovered in populations of a related South American rabbit. The myxoma virus produced a small, localized fibroma (a fibrous cancer of the skin). Its effect on South American rabbits was not severe, but European rabbits infected by the virus died quickly of myxomatosis.

In 1950, the myxoma virus was introduced locally in Victoria. An epidemic of myxomatosis broke out among the rabbits and spread rapidly. The virus was transmitted primarily by mosquitoes, which bite infected areas of the skin and carry the virus on their mouthparts. The first epidemic killed 99.8% of the infected rabbits, reducing their populations to very low levels. But during the following myxomatosis season (which coincides with the mosquito season), only 90% of the remaining population was killed. During the third outbreak, only 40–60% of infected rabbits succumbed, and their population began to grow again.

The decline in the lethality of myxomatosis in the Australian rabbits resulted from evolutionary responses in both the rabbit and the virus populations. Before the introduction of the myxoma virus, some rabbits had genetic factors that conferred resistance to the disease. Although nothing had spurred an increase in these factors before, they were strongly selected by the myxoma epidemic, until most of the surviving rabbit population consisted of resistant animals (Figure 21.1). At the same time, virus strains with less virulence increased because reduced virulence lengthened the survival time of infected rabbits and thus increased the mosquito-borne dispersal of the virus (mosquitoes bite only living rabbits). A virus organism that kills its host quickly has less chance of being carried by mosquitoes to other hosts.

Left on its own, the Australian rabbit-virus system would probably evolve to an equilibrial state of benign, endemic disease, as it had in the population of South American rabbits from which the myxoma virus was isolated. Currently, pest management specialists keep the system out of equilibrium and maintain the effectiveness of myxoma as a control agent by finding new strains of the virus to which the rabbits have yet to evolve immunity.

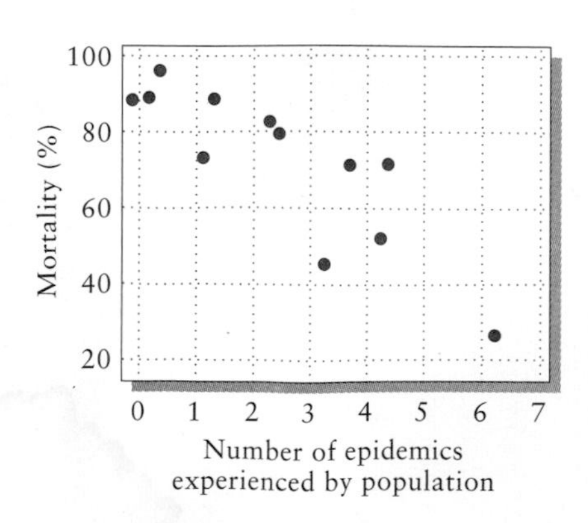

Figure 21.1 The decrease in susceptibility of wild rabbits to a myxoma virus with grade III virulence (that is, causing 90% mortality in genetically unselected wild rabbits). From F. Fenner and F. N. Ratcliffe, *Myxomatosis,* Cambridge University Press, London (1981).

The decreasing virulence of the myxoma virus results from the fact that mosquitoes disperse the virus to new hosts more efficiently when the infected host has a longer life span. Thus, less virulent strains of myxoma have a higher rate of growth in populations of hosts as a whole, if not within individual hosts. Highly contagious diseases that are spread directly through the atmosphere or water have no such constraint on dispersal and often exhibit high levels of virulence, with debilitating or even fatal consequences for their hosts. Similarly, most predators do not rely on a third party to find prey, and rather than evolving toward a benign equilibrium of restraint and tolerance, predator and prey tend to become locked in an evolutionary battle of persistent intensity. The outcome of the battle depends on which population gets the evolutionary upper hand.

Pimentel's studies on evolution in parasitoid-host systems

Several years ago, David Pimentel and his colleagues at Cornell University explored the evolution of host-parasitoid relationships using the housefly and a wasp parasitoid of the pupal stage of the fly, *Nasonia vitripennis* (Figure 21.2). In one population cage, *Nasonia* was allowed to parasitize a fly population that was kept at a constant level by replenishment from a stock that had not been exposed to the wasp. Any flies that escaped attack by wasp parasitoids were removed from the population cage, so the wasps were provided only with evolutionarily "naive" hosts.

In a second population cage, fly hosts were kept at the same constant number, but because emerging flies were allowed to remain, the population could evolve resistance to the wasps. The population cages were maintained for about 3 years, long enough for evolutionary change to occur. Over the course of this experiment, the reproductive rate of wasps in the cage that permitted evolution dropped from 135 to 39 progeny per female, and their longevity decreased from 7 to 4 days. The average level of the parasitoid population also decreased (1,900 adult wasps versus 3,700 in the nonevolving system), and population size was more constant than in the nonevolving cage. These results suggest that the flies evolved additional defenses when subjected to intense parasitism.

Experiments were then established in thirty-compartment population cages in which the numbers of flies were allowed to vary freely. One such experiment was started with flies and wasps that had had no previous contact with each other, and a second was established with animals from the evolving population described above. In the first cage, the wasps were efficient parasitoids, and the system underwent severe oscillations. In the second cage, however, the wasp population remained low, and the flies attained a high and relatively constant population level (Figure 21.3). This result strongly reinforced the conclusion, drawn from the earlier experiments, that the flies had evolved resistance to the wasp parasitoids.

Figure 21.2 The wasp *Nasonia* parasitizing a pupa of the housefly. Courtesy of D. Pimentel. From D. Pimentel, *Science* 159:1432–1437 (1968).

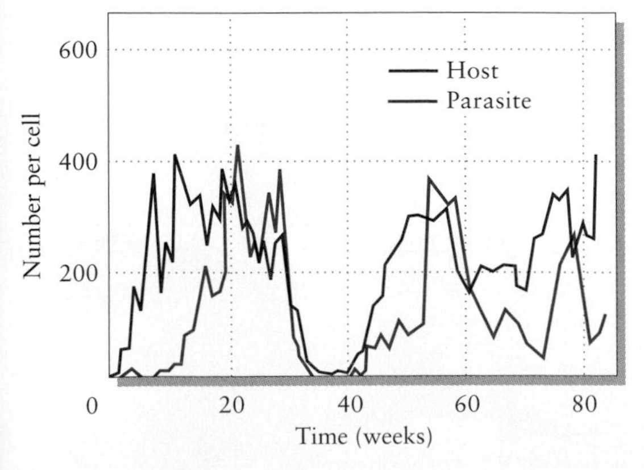

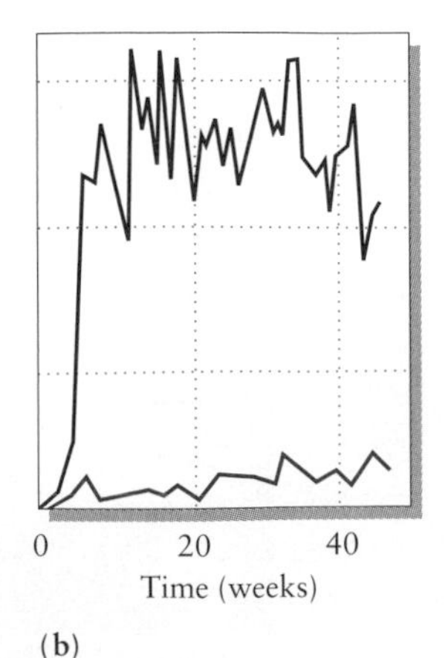

Figure 21.3 Populations of houseflies and the wasp parasitoid *Nasonia vitripennis* in thirty-cell laboratory cages. (a) Control: flies had no previous experience with the wasp. (b) Experimental: flies had been exposed to wasp parasitism for more than 1,000 days. After D. Pimentel, *Science* 159:1432–1437 (1968).

The genetics of coevolution in plant-pathogen systems

The suggestion that consumer and resource populations evolve in response to each other presupposes that each contains genetic variability for traits that influence their interactions. In the case of the wasp-fly interaction, it was clear that evolution had occurred, but the genetic basis of the evolutionary change could not be determined. Such determination has been less of a problem in the case of studies on diseases of plants, in which the difference between virulence and benignness may depend on a single gene and thus is amenable to simple Mendelian genetic analysis.

Plant geneticists have developed strains of domestic crops, such as flax and wheat, that are resistant to particular genetic strains of various pathogens, such as rusts (teliomycetid fungi). The crop strains differ from one another by a few (perhaps even by single) genetic factors that make them either susceptible or resistant to infection by particular strains of rust. Over the course of crop improvement programs, when new strains of rust appear, crop geneticists select new resistant strains of the crop by exposing experimental populations to the pathogen. New strains of the pathogen appear in an area either by migration or by mutation, creating continual evolutionary flux in such a system.

A survey of genetic strains of wheat rust *(Puccinia graminis)* in Canada revealed that new virulence genes appear from time to time and sweep through a population (Figure 21.4). Genetic races of wheat rust are distinguished both by their physiological characteristics and by their virulence when tested on lines of wheat containing different resistance alleles. Most of the virulence strains within a single physiological race of a rust differ by only one gene, and it is possible to trace evolutionary relationships among races of wheat rust by examining their virulence on wheat with different resistance genes. For example, in 1969, race 15 B-1L of the wheat rust was virulent on strains of wheat with resistance genes (Sr) 8, 10, and 11, and it was avirulent on strains with genes 15 and 17. In 1970 a strain of the rust was isolated that had become avirulent on Sr 11. In 1971, a new strain appeared that was virulent on Sr 15. That same strain lost its virulence on Sr 8 the following year; another lost its virulence on Sr 11. In 1973, a new strain virulent on Sr 17 appeared.

The rust-wheat system contains the essential element of coevolution: an interaction between the fitnesses of the genotypes of the host and those of the pathogen. The system is kept in flux by the introduction of new vir-

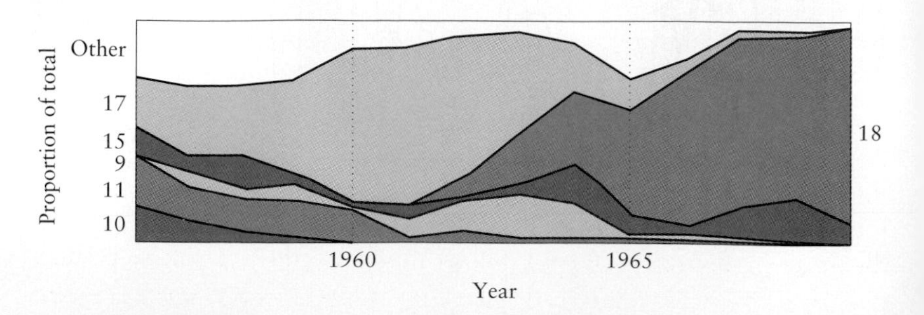

Figure 21.4 Relative proportions of different virulence genes in the rust *Puccinia graminis* infecting Canadian wheat. From G. J. Green, *Can. J. Bot.* 53: 1377–1386 (1975).

(a)

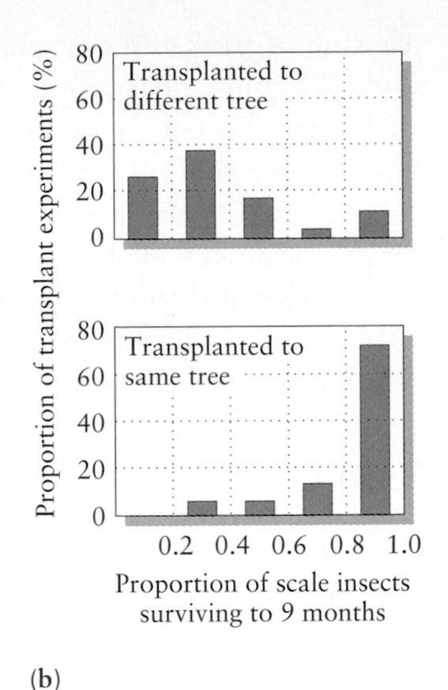

(b)

Figure 21.5 (a) Black pineleaf scale on needles of the ponderosa pine, illustrating the damage caused by feeding. (b) Survival of individual scale insects, which depends on the adaptation of localized populations to the genotypes of particular trees, decreases markedly when scales are transplanted to different trees. Courtesy of D. N. Alstad and G. F. Edmunds, Jr. From G. F. Edmunds and D. N. Alstad, *Science* 199:941–945 (1978).

ulence genes in the rust—and perhaps by new resistance genes in the wheat, although the latter are pretty much controlled by plant geneticists nowadays.

Genotype-genotype interactions have been found in several natural systems and may turn out to be the rule in populations of plants and herbivores and of hosts and pathogens. The genetics of most plant defenses are difficult to work out in as much detail as has been done for wheat resistance genes, but genetic effects can nonetheless be detected in experiments on natural populations. For example, studies have shown that variation among individual trees in the defenses of ponderosa pines are paralleled by variation in the genotypes of the scale insects that infest them (Figure 21.5). The scales are extremely sedentary, exhibiting so little migration from tree to tree that local populations on individual trees have apparently evolved independently of those on other trees. This is shown by the differing success of scale insects experimentally transferred between trees and between branches on the same tree: specifically, the survival of scales transplanted between trees is much lower than that of controls transferred within the same tree. It is reasonable to assume that differences between trees and between local populations of scales are genetic, so this finding represents a case of genotype-genotype interaction.

Evolutionary equilibria of consumer and resource populations

The evolutionary responses of consumer and resource populations can be depicted by a simple graphical model that relates the rates of evolution of the two populations to the efficiency of consumption (Figure 21.6). For a prey population, for example, the rate at which new adaptations useful in the escape or avoidance of predators are selected should vary in direct proportion to the predation rate. In the absence of predation, there can be no selection of adaptations for predator avoidance. But as predation increases, so do selection and evolutionary response, at least up to limits set by the availability of genetic variation.

The selection of new adaptations useful to the predator in exploiting the prey should vary in the opposite fashion. When a particular prey species is not heavily exploited, adaptations of predators that enable them to use that resource are selected, and predation on that prey population increases. As exploitation of the prey increases, however, intraspecific competition among predators reduces the selective value of further increases in predation. Very high rates of predation conceivably could select individuals that shifted their diets toward other prey species. Hence evolution by a predator population could result in decreased efficiency in its use of a particular prey species, as indicated in that portion of Figure 21.6 where the "predator" curve sinks below the horizontal axis.

In this simple model, the balancing influences of predator and prey adaptation achieve an evolutionary steady state where the two curves cross. When predator adaptations are relatively effective and the prey are exploited at a high rate, selection on the prey population tends to improve its escape mechanisms faster than selection on the predator population improves its ability to exploit the prey. Conversely, when the exploitation rate is low, prey evolve more slowly than predators. This steady state between the adaptations of predators and prey should result in a relatively constant rate of exploitation regardless of the specific predator and prey adaptations. As in any steady state, both antagonists continually evolve to maintain this balance, just as nations continually develop new weapons and defenses to maintain a stalemated arms race.

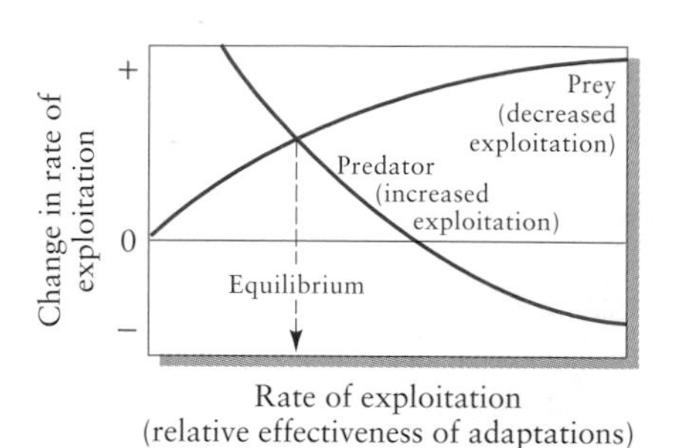

Figure 21.6 A graphical model of the evolutionary equilibrium between a consumer and its resource, based on the way in which the rate of exploitation influences the rates of evolutionary change in the adaptations of consumer and resource. Consumer adaptations tend to increase the rate of resource exploitation, but as the exploitation rate increases, selective pressure to increase it further diminishes because resource populations are driven to lower and lower levels and become less suitable resources. When exploitation is very high, consumers may be selected to switch to alternative resources, and the rate of change in exploitation levels resulting from new adaptations may become negative. Adaptations of resource populations tend to decrease the rate of exploitation, and selective pressure on resource populations increases as the exploitation rate increases. Exploitation is brought into equilibrium when the population consequences of consumer and resource adaptations balance.

Pimentel's experiments on host-parasite interactions that were discussed above illustrate the dynamics of the predator-prey equilibrium. The housefly (host) and the parasitoid wasp *Nasonia* undoubtedly had achieved an evolutionary equilibrium in their natural habitat. When brought into a simple laboratory habitat, *Nasonia* wasps were able to exploit housefly populations at a greatly increased rate because they required little time to search out hosts. (Setting up these experimental conditions was equivalent to shifting the exploitation rate of *Nasonia* on houseflies far above the equilibrium level in Figure 21.6.) This shift increased the selective pressure on the housefly to escape parasitism much more than the selective pressure on the predator to further increase its exploitation rate. As a result, the ability of houseflies to escape parasitoids increased, and the level of exploitation by *Nasonia* decreased toward a new steady state.

Genetic variation in competitive ability

When two species are closely matched in competition, the different competitive abilities of genotypes within each population may result in different outcomes of competition. Sometimes genes that influence competitive ability express themselves in the phenotype so subtly as to avoid detection by direct examination of individuals. Instead, they must be inferred from the outcome of competition. In this sense, competitive ability summarizes the interaction of a phenotype with its environment. The experiments that follow demonstrate genetic variation in competitive ability. Clearly, such genetic variation also makes it possible for competitive ability to evolve.

Competition experiments involving the flour beetles *Tribolium confusum* and *T. castaneum* sometimes have uncertain outcomes; that is, at certain combinations of temperature and humidity, the two species are so closely matched that small differences in the individuals used in the experiments might tip the balance of competition in favor of one species or the other. In one series of experiments, for example, at 29°C and 70% relative humidity, *T. castaneum* "won" in about five-sixths of the experiments and *T. confusum* in one-sixth. These experiments were begun with two pairs of individuals of each species. Each individual carries only a small sample of the genetic diversity of the population, and the genetic makeup of any two males and two females is likely to vary widely between samples. When the experiments were repeated under identical conditions, but were begun with ten pairs of each species, *T. castaneum* won twenty out of twenty times. These results suggest that in the first set of experiments, genetic variation in the parent individuals might have altered the outcome of competition. The larger parental populations used in the second set of experiments probably resembled the average genetic makeup of the species more closely, thereby producing more consistent results.

Several strains of each species of *Tribolium* were inbred to reduce their genetic variability, and then the strains were tested against each other in competition experiments. When strains were inbred for more than twelve generations, certain strains of *T. confusum* consistently outcompeted certain strains of *T. castaneum*—a reversal of their normal competitive relationship. These experiments indicated that changes, even minor ones, in the genetic constitution of a population can greatly influence its competitive ability.

When laboratory populations of the fruit flies *Drosophila serrata* and *D. nebulosa* were grown together, the competitive ability of *D. serrata* appeared to increase over time. The two species were established in population cages at 19°C, and they quickly achieved a pattern of stable coexistence with 20–30% *D. serrata* and 70–80% *D. nebulosa.* In one cage, however, the frequency of *D. serrata* began to increase after the 20th week and attained about 80% by the 30th week, a reversal of the initial predominance of *D. nebulosa*. When individuals of both species were removed from the competing populations after the 30th week and tested against stocks maintained in single-species cultures, the competitive ability of each species was found to have increased after exposure to the other in the competition experiment. When the competitive ability of *D. serrata* individuals from the one cage in which that species predominated was tested against that of unselected stocks of *D. nebulosa, D. serrata* again showed superior competitive ability.

It is quite clear from these experiments that competitive ability has a genetic basis and can evolve in laboratory populations. The particular adaptations responsible for changes in competitive ability were not determined in these experiments; they could conceivably include any increase in the efficiency of use of a food resource, in the rate of offspring production per unit of food consumed, or in survival at any stage of the life cycle. One of the few generalizations to come from this body of work is that sparse populations can evolve interspecific competitive ability more rapidly than dense populations. Why? Possibly because if different, somewhat conflicting adaptations determine the outcomes of intraspecific and interspecific competition, then selection for increased interspecific competitive ability will be stronger on the rarer of two competitors.

We return to the work of David Pimentel for laboratory evidence that a poor competitor can evolve a competitive advantage (judged by relative population density) over a formerly superior adversary. Pimentel and his colleagues conducted laboratory experiments with flies to determine whether two species could coexist on one food resource by means of frequency-dependent evolutionary changes in their competitive ability. In other words, can one species, as it is being excluded by the second and becoming rare, evolve increased interspecific competitive ability rapidly enough to regain the upper hand?

The housefly *(Musca domestica)* and the blowfly *(Phaenicia sericata),* which have similar ecological requirements and comparable life cycles (about 2 weeks), were chosen for the experiments. Both species feed on dung and carrion in nature, and they are often found together on the same food resources. The flies were raised in small population cages at 27°C, with a mixture of agar and liver provided as food for the larvae and sugar for the adults. The outcomes of an initial series of four competition experiments begun with individuals from wild populations of the housefly and the blowfly were split, with each species winning twice. The mean extinction time for the blowfly, when the housefly won, was 92 days; it was 86 days for the housefly when the blowfly won. These results showed that the two species were close competitors, but the small cages used did not allow enough time for evolutionary change before one of the populations was excluded.

To prolong the housefly-blowfly interaction, Pimentel started a population in a sixteen-cell cage, which consisted of single cages in four rows of four with connections between them (Figure 21.7). Under these conditions, populations of houseflies and blowflies coexisted for almost 70 weeks

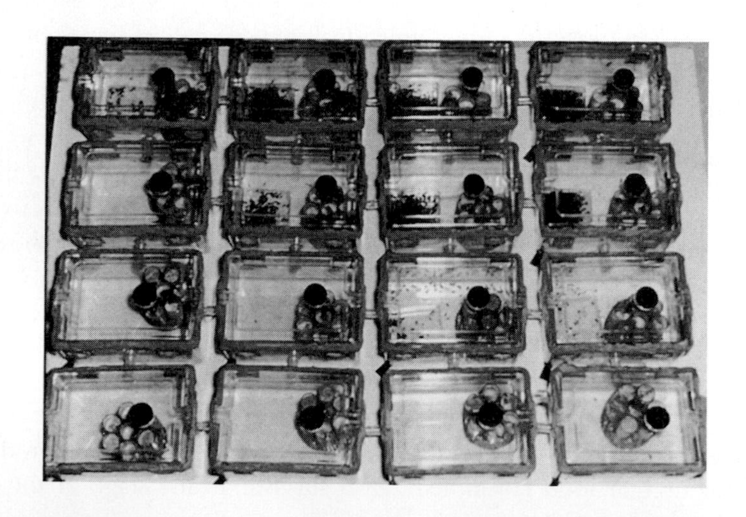

Figure 21.7 The sixteen-cell cage used by Pimentel to study competition between populations of flies. Note the vials with larval food in each cage and the passageways connecting the cells. The dark objects concentrated in the upper right-hand cells are fly pupae. Courtesy of D. Pimentel. From D. Pimentel, E. H. Feinberg, P. W. Wood, and J. T. Hayes, *Am. Nat.* 99:97–109 (1965).

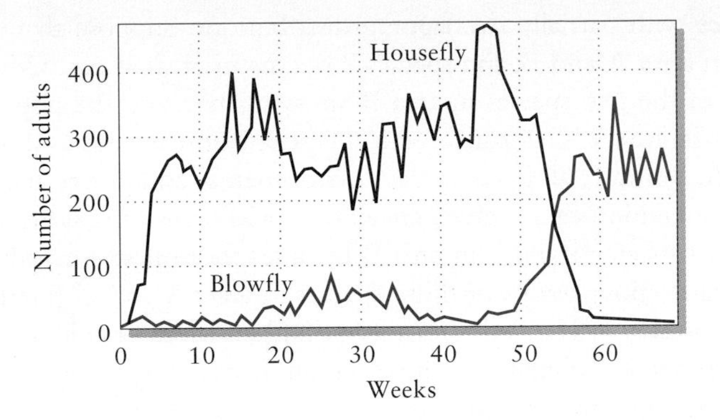

Figure 21.8 Changes in competing populations of houseflies and blowflies in a sixteen-cell cage. After D. Pimentel, E. H. Feinberg, P. W. Wood, and J. T. Hayes, *Am. Nat.* 99:97–109 (1965).

and showed a striking reversal of numbers between the two species at about 50 weeks (Figure 21.8). After 38 weeks, when the blowfly population was still low, and just a few weeks prior to its sudden increase, individuals of both species were removed from the population cage and tested in competition with each other and with wild strains of the housefly and blowfly. Captured wild blowflies turned out to be inferior competitors to wild and experimental strains of the housefly. But blowflies that had been removed from the population cage at 38 weeks consistently outcompeted both wild and experimental populations of the housefly. Apparently, the experimental blowfly population had evolved superior competitive ability while it was rare and on the verge of extermination.

Character displacement and evolutionary divergence of competitors

Theory suggests that if resources are sufficiently varied, competitors should diverge and specialize. Laboratory experiments such as those described above have shown that competitive ability may have a genetic component and therefore may be under the influence of selection, although particular adaptations have not been identified. But if competition exerts a potent evolutionary force in nature, we should find evidence there that competitors have partly molded each other's adaptations.

Although related species that live together in the same place differ in the way they use the environment (using different food resources, for example), these differences have not necessarily evolved as a result of their interaction. An alternative explanation for such differences is that each of the species became adapted to different resources in different places, and when their populations subsequently overlapped as a result of range extensions, these ecological differences remained.

We may get around this objection by comparing the ecology of a species in an area where it co-occurs with a competitor with its ecology in another area where the competitor is absent. When two species coexist within the same geographic area, they are said to be **sympatric;** when their distributions do not overlap, they are said to be **allopatric.** The terms sympatry and allopatry also can be applied to different parts of the ranges of species with partially overlapping distributions. Suppose that species 1 occurs in areas A and B, and species 2 occurs in areas B and C. The populations of the two species in area B are sympatric, and the population of species 1 in area A is allopatric with the population of species 2 in area C. If areas A, B, and C all have similar environmental conditions and habitats, and if competition causes divergence, we would expect the sympatric populations of species 1 and 2 in area B to differ more from each other than the allopatric populations of those species in areas A and C (Figure 21.9). This phenomenon is called **character displacement.** Ecologists disagree on the prevalence of character displacement in nature.

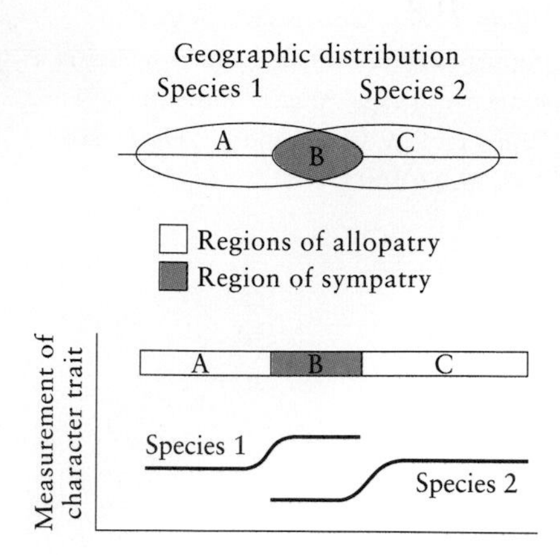

Figure 21.9 The phenomenon of character displacement, in which character traits of two closely related species differ more where they occur in sympatry than in allopatric portions of their geographic ranges.

A large number of examples do seem to fit the pattern of character displacement, one of which involves ground finches *(Geospiza)* of the Galápagos Islands. On islands with more than one finch species, the finches usually have beaks of different sizes, indicating different ranges of preferred food size. For example, on Marchena Island and Pinta Island, the beak size ranges of the three resident species of ground finches do not overlap (Figure 21.10). On Floreana and San Cristobal, the two resident species, *G. fuliginosa* and *G. fortis,* have beaks of different sizes. On Daphne Island, however, where *G. fortis* occurs alone, its beak is intermediate in size between those of the two species on Floreana and San Cristobal. On Los Hermanos Island, *G. fuliginosa* occurs alone, and its beak is intermediate in size.

The Galápagos ground finches illustrate the diversifying influence of competition because of the chance distribution of these species on small islands within the archipelago: some islands have two or three species and

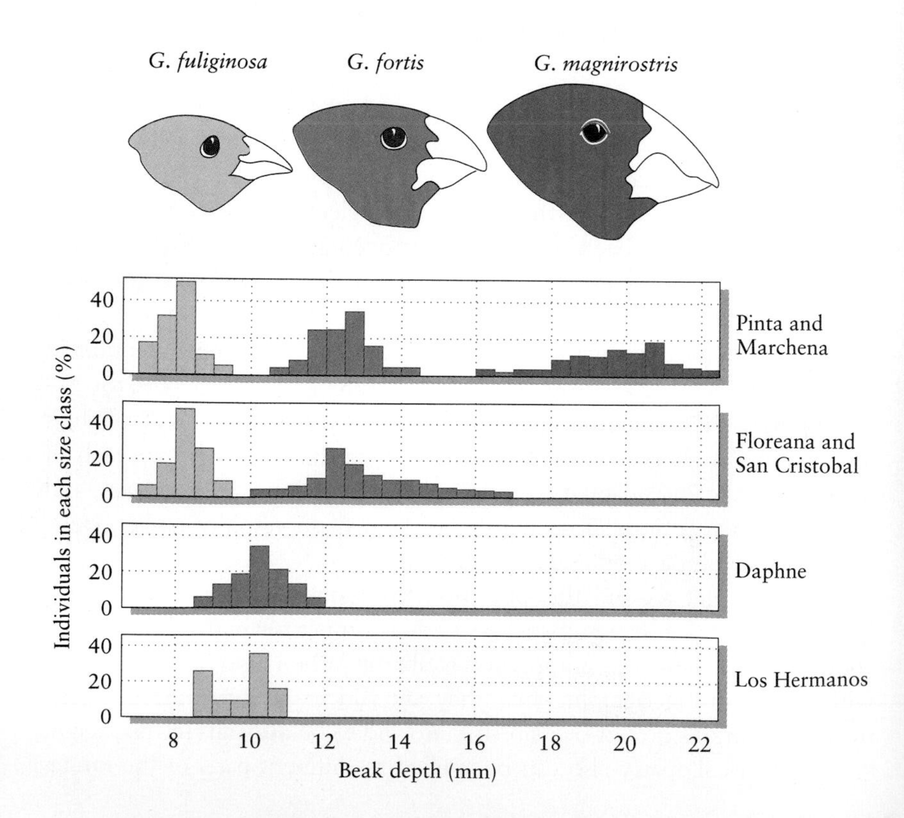

Figure 21.10 Proportions of individuals with beaks of different sizes in populations of ground finches *(Geospiza)* on several of the Galápagos Islands. After D. Lack, *Darwin's Finches,* Cambridge University Press, Cambridge (1947).

some only one. In many other cases, it is difficult to know whether differences between two species arose because of competition between them or because they evolved in response to selection by other environmental factors in different places, and then retained their differences when their populations reestablished contact. In most cases, certainly, genetic differences that lead to speciation occur in allopatry. So why not differences that allow two species to avoid strong competition? In either case, coexistence depends on some degree of ecological difference between species, whether it is achieved in allopatry or as an evolutionary consequence of competition in sympatry.

Coevolution

Reciprocal evolutionary responses between populations are referred to as coevolution. The phenomenon of coevolution raises several questions. First, do pairs of populations undergo reciprocal evolution, or do "coevolved" traits arise from the responses of populations to selective pressures exerted by a variety of environmental factors, followed by an ecological sorting out of species with compatible features? Second, are species organized into interacting sets based on their evolved adaptations, whether "coevolved" or not? And third, do these adaptations enhance such system properties as the productivity of the biological community and its resistance to perturbation?

Complementary adaptations among pairs, or small groups, of species have often been attributed to coevolution without evidence having been presented for the evolutionary history of the relationship itself. A close match between the adaptations of different species does not prove coevolution. An extreme example will make this point: the adaptations of organisms clearly match many physical aspects of the environment, yet the physical environment does not respond adaptively to its living inhabitants. Thus the match between organism and physical environment cannot be called coevolution.

Coevolution in ants and aphids?

The difficulty we have in recognizing true cases of coevolution is illustrated by the case of a mutualism involving ants that protect aphids and leafhoppers from predators and, in return, harvest the nutritious honeydew that they excrete. In one such system, aphids, which are small, sedentary homopteran insects, form dense colonies on inflorescences of ironweed *(Veronia)* in New York State. Also occurring on ironweed is the larger membracid bug (leafhopper) *Publilia,* another homopteran that sucks plant juices from leaves. These insects are tended by three species of ants. One, in the genus *Tapinoma,* is tiny (2–3 mm) but abundant. The other two (both species of *Myrmica*) are larger (4–6 mm) and more aggressive, but less common. The two genera of ants rarely co-occur on the same plant.

The presence of *Tapinoma* greatly enhances the survival of aphid colonies but has less effect on the survival of leafhoppers. The larger

Myrmica offers substantial protection to leafhoppers but is less effective in warding off predators of aphids. On plants from which both species of ants were excluded, predators were more numerous (recalling experiments on acacias in Central America, discussed in Chapter 18); when the predatory larvae of ladybird beetles were added to the system, *Myrmica* ants and, to a lesser extent, *Tapinoma* ants effectively reduced their predation on leafhoppers.

The ant-aphid-leafhopper system has all the elements expected of coevolution, but how can we be sure that the adaptations of the ant and homopteran participants evolved in response to each other? Most insects that suck plant juices produce large volumes of excreta from which they either do not or cannot extract all the nutrients. Therefore, honeydew production may simply reflect diet rather than having evolved to encourage protection by ants. For their part, many ants are voracious generalists that are likely to attack any insect they encounter; they may need no special adaptations to confer benefits on the aphids and leafhoppers upon whose excreta they also feed. The fact that the different genera of ants more effectively protect different honeydew sources may simply reflect their different sizes and levels of aggression, which may have evolved in response to unrelated environmental factors.

Why don't the ants eat the aphids and leafhoppers they tend? Perhaps this restraint is an evolved trait of ants that facilitates the ant-homopteran mutualism. It may even have arisen as an extension of the common ant behavior of defending plant structures that produce nectar—flowers or specialized nectaries (see, for example, Figure 18.15). Although this system has apparent specificity of interactions, that is not sufficient to prove coevolution.

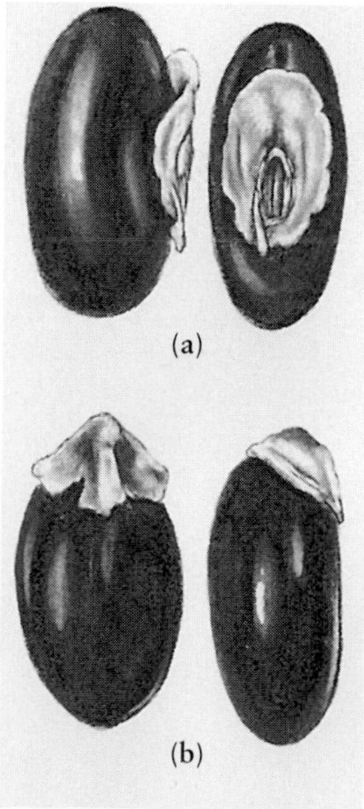

(a)

(b)

Figure 21.11 Ant-dispersed seeds of two Australian plants, (a) *Kennedia rubicunda* and (b) *Beyeria viscosa,* showing the edible, light-colored appendage (elaiosome) that attracts ants. After R. Y. Berg, *Aust. J. Bot.* 23:475–508 (1975).

Ants and seed dispersal

An analogous situation in the Cape region of South Africa highlights the importance of particular adaptations of ants, whether coevolved or not, in maintaining an ant-plant mutualism. There, many species of plants in the family Proteaceae have seeds with fleshy, edible attached structures called elaiosomes (illustrated for some Australian plants in Figure 21.11). Foraging ants pick up seeds and transport them to their underground nests, where the elaiosomes are eaten. The seeds themselves, which the ants cannot eat, are then discarded, either in underground chambers or in refuse heaps on the surface. In many regions these disposal sites are suitable for seed germination and seedling establishment. But in fynbos (the South African term for brushy, chaparral-like habitat), germination of many species of plants occurs only after fires have swept an area, and the only seeds that germinate are those stored by ants in underground nest chambers.

Recently, the introduced Argentine ant *Iridomyrmex humilis* has invaded areas of fynbos shrublands and displaced many of the less aggressive native ants. The Argentine ant differs from native ants in that it does not store seeds within its nests; instead, it removes the elaiosomes on the surface and drops the seeds there. As a result, germination of one plant

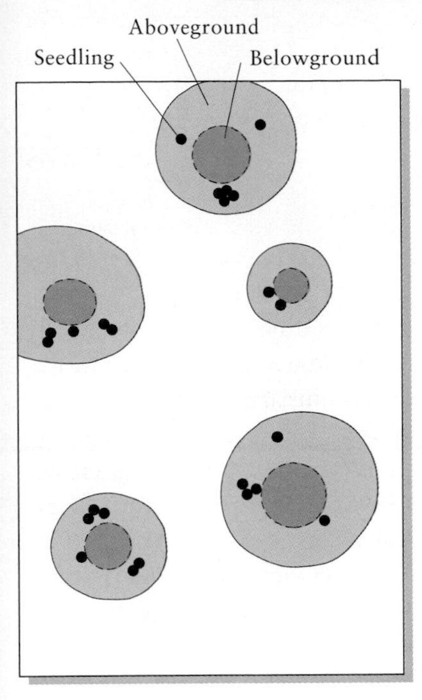

(a) *Iridomyrmex* present

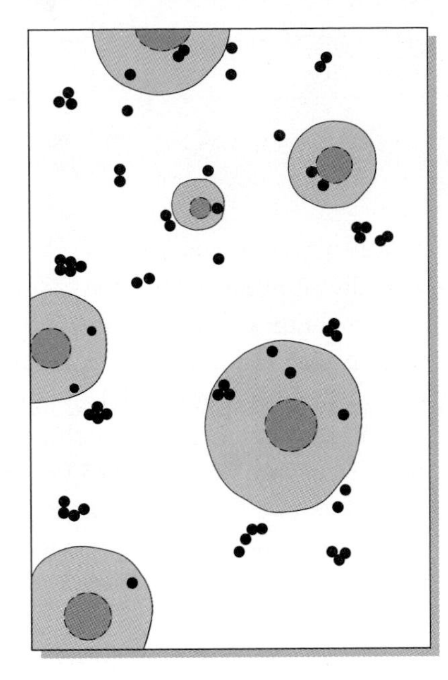

(b) *Iridomyrmex* absent

Figure 21.12 Seedling dispersion in *Mimetes cucullatus* populations after a burn (a) in the presence of *Iridomyrmex* and (b) in its absence. The extent of aboveground and belowground parts of mature plants are indicated by shading. In the absence of *Iridomyrmex,* native ants distribute the seeds of *Mimetes* widely throughout the habitat. From W. Bond and P. Slingsby, *Ecology* 65:1031–1037 (1984).

species *(Mimetes)* following fire was drastically reduced in areas invaded by *Iridomyrmex* (Figure 21.12). With the continued persistence of the Argentine ant, it is likely that much of the native Cape flora will disappear as underground seed reserves are depleted.

Clearly, particular adaptations of ants influence their effectiveness as seed dispersers. But these adaptations may be evolutionarily independent of the plants themselves. The South African ants that disperse the seeds of *Mimetes* are diet generalists, and their food-caching behavior probably evolved for reasons unrelated to their role as seed dispersers. The plants may have evolved merely to take advantage of this fortuitous element in their environment without any reciprocal evolution on the part of the ant. Whereas some mutualisms, such as that between ant and acacia, clearly involve specialized adaptations of both parties and seem to represent cases of coevolution, many mutualisms may involve more serendipitous arrangements.

Herbivores and the chemical defenses of plants

University of Illinois biologist May Berenbaum has placed elements of the relationship between certain butterflies and their umbelliferous host plants in the context of coevolution. Umbellifers (parsley family) produce many noxious chemicals, among the most prominent of which are the furanocoumarins. The biosynthetic pathway of these chemicals leads from para-coumaric acid (which, being a precursor of lignin, is found in

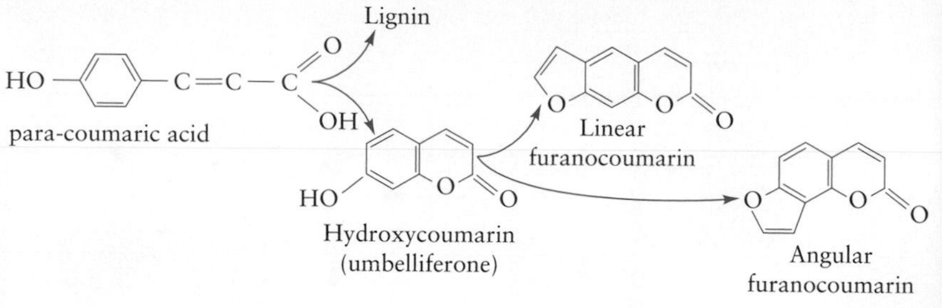

Figure 21.13 Biosynthetic pathways in the synthesis of furanocoumarins. Each step is controlled by a different enzyme.

virtually all plants) to hydroxycoumarins such as umbelliferone, and finally to the furanocoumarins. These last include linear and angular forms, which are produced directly from hydroxycoumarins by different enzyme reactions (Figure 21.13).

As one proceeds down this biosynthetic pathway, toxicity increases and occurrence among plant families decreases. Hydroxycoumarins have some biocidal properties; linear furanocoumarins (LFCs) interfere with DNA replication in the presence of ultraviolet light; and angular furanocoumarins (AFCs) interfere with growth and reproduction quite generally. Para-coumaric acid is widespread among plants, occurring in at least a hundred families; only thirty-one families possess hydroxycoumarins. LFCs are restricted to eight plant families and are widely distributed in only two: Umbelliferae and Rutaceae (the citrus family). AFCs are known only from two genera of Leguminosae (pea family) and ten genera of Umbelliferae.

Among species of herbaceous umbellifers in New York State, some (especially those growing in woodland sites with low levels of ultraviolet light) lack furanocoumarins, others contain LFCs only, and some contain both LFCs and AFCs. Surveys of herbivorous insects collected from these plant species showed that (1) host plants containing both AFCs and LFCs were, somewhat surprisingly, attacked by more species of insects than were plants with only LFCs or with no furanocoumarins; (2) insect herbivores on AFC/LFC plants tended to be extreme diet specialists, most having been found on no more than three genera of plants; and (3) these specialists tended to be abundant compared with the few generalists found on AFC/LFC plants and compared with all herbivores found either on LFC plants or on umbellifers lacking furanocoumarins.

Although LFCs and (especially) AFCs are extremely effective deterrents to most species of herbivorous insects, some genera that have evolved to tolerate these chemicals have become successful specialists. One can make a strong case for coevolution here. The taxonomic distribution of hydroxycoumarins, LFCs, and AFCs across host plants suggests that plants containing LFCs are a subset of those containing hydroxycoumarins and that those containing AFCs are an even smaller subset of those containing LFCs. This pattern is consistent with an evolutionary sequence of plant defenses progressing from hydroxycoumarins to LFCs and AFCs. Furthermore, insects that specialize on plants containing LFCs belong to groups that characteristically feed on plants containing hydroxycoumarins, and those that specialize on plants containing AFCs have close relatives that feed on plants containing LFCs. These patterns of taxonomic distribution of insects across host plants are consistent with coevolution within the system.

The story of the evolution of chemical resistance in plants and the breaking of that resistance by certain groups of insects is conjecture, based on the logic of the evolutionary relationships of the taxa involved. We have no means of directly watching such evolutionary interactions unfold; evolution occurs too slowly in natural systems. Berenbaum's inferences about evolution build on the idea that ancestral and derived characters (such as absence and presence of AFCs) should be found among close relatives if they are linked by evolution. This logic has been elaborated into a branch of evolutionary biology known as **phylogenetic reconstruction,** which uses similarities and differences among species to determine their evolutionary relationships—that is, the history of the derivation of descendant taxa from common ancestors. The techniques of phylogenetic reconstruction are introduced in Box 21.1, and their application to the problem of coevolution is illustrated in the following example of a pollination mutualism.

The yucca moth and the yucca

The curious pollination relationship between yucca plants (*Yucca,* a member of the lily family) and moths of the genus *Tegeticula* (Figure 21.14) was

Figure 21.14 The mohave yucca *(Yucca shidigera)* and a yucca moth of the genus *Tegeticula.* After J. A. Powell and R. A. Mackie, *Univ. Calif. Publ. Entomol.* 42:1–59 (1966).

BOX 21.1 Inferring phylogenetic history

In the analysis of coevolutionary relationships between species, it has been very informative to relate the coevolved traits of each species to its evolutionary history. Of course, we can't actually go back in time and watch evolution occur. We can, however, infer the evolutionary history of a lineage, particularly the order in which various traits evolved, by examining the occurrence of particular traits among related species. To do so, we must first work out the evolutionary relationships among species that have descended from a common ancestor. Such a set of species is called a **monophyletic group,** referring to the fact that all of the species can be traced back to a single common ancestor. We can characterize each species by the character states that it exhibits for a set of traits. In the case of a monophyletic group of plants, for example, we might describe each species by such traits, or characters, as red versus yellow flowers, fleshy versus dry fruits, hairy versus smooth stems, woody versus herbaceous growth form, and presence versus absence of a certain defensive alkaloid. Each of these five characters has two states, which we can designate A or a, B or b, and so on. Therefore, each taxon can be described by its five character states, for example, ABcdE.

We don't know what the common ancestor of a particular monophyletic group looked like, but we can nonetheless make inferences about the evolution of the group from the distribution of traits among its species. We can represent the speciation events that produced the contemporary species in the group as a series of branch points on a phylogenetic tree. Such a tree is called a **cladogram,** from the word *clade,* which is a term for an evolutionary lineage. Once a particular evolutionary lineage has split into two independent evolutionary lineages, then any changes in character state that occur within the newly formed lineages must be restricted to them. These new character states are shared by all members of the new lineage, and these characters are called shared derived traits, or **synapomorphies** (a fancy way of saying the same thing with Latin roots). Once we understand that characters that arise progressively later in evolutionary time are restricted to progressively smaller subsets of the monophyletic group, then we can use character traits to reconstruct phylogenetic history, or, at least, our best guess at it. This process of phylogenetic reconstruction is called **cladistics.**

Suppose we have a group of five closely related species that exhibit character states aBCDE, ABCDe, aBcdE, AbCDe, and aBcDE. We can now construct a simple hypothesis for the evolutionary relationships among the species by assuming that species that share traits not found in the others are closely related. For example, species aBcdE and aBcDE differ only by trait (D,d); they share trait c, found nowhere else in the clade. These two species also share traits a, B, and E with species aBCDE. In turn, these three species form a single lineage characterized by trait a, which is a synapomorphy for that group. Similarly, species ABCDe and AbCDe share trait e; the latter additionally has trait b instead of B. Now we see that the five species can be linked to one another by a branching tree of the form shown below (the evolutionary steps required to produce this pattern are shown by the dark bars). This tree represents the simplest way in which the observed character states can link the species together.

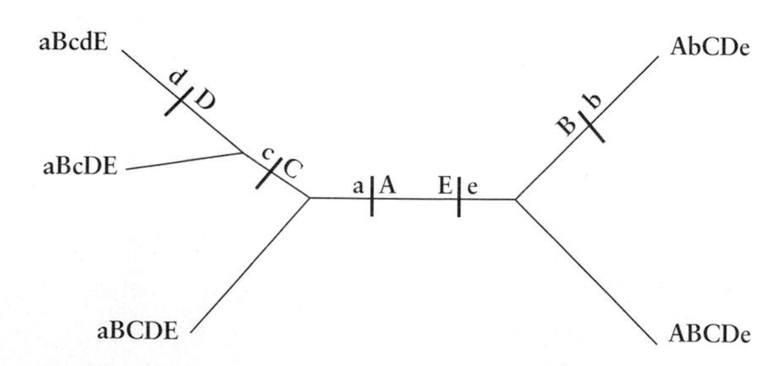

It is also possible, of course, to imagine a more complicated scenario in which the same traits arise in two different lineages, or in which traits are gained and lost within the same lineage. For example, in the tree portrayed below, the species are arranged more or less at random, and the evolutionary steps required to produce the observed character traits are indicated. Note that in this random tree, more evolutionary steps (8) are needed than in the first (5), which was constructed by grouping species on the basis of shared characters. As it is shown, the random tree exhibits independent evolution of the same trait in different lineages—a phenomenon called **homoplasy**—in two cases (c and A), and a trait reversal (to e and back to E) in another lineage. Which of these is more likely to be the correct tree? Biologists usually apply the principle of parsimony, which says that the simplest is the best, and choose the tree with the fewest steps. Although homoplasies and reversals happen, such a tree makes the fewest assumptions about evolutionary changes within the clade. Needless to say, when many species and many traits are considered, phylogenetic trees can be very complex, and often several different phylogenetic hypotheses are equally parsimonious—that is, there is no single best tree.

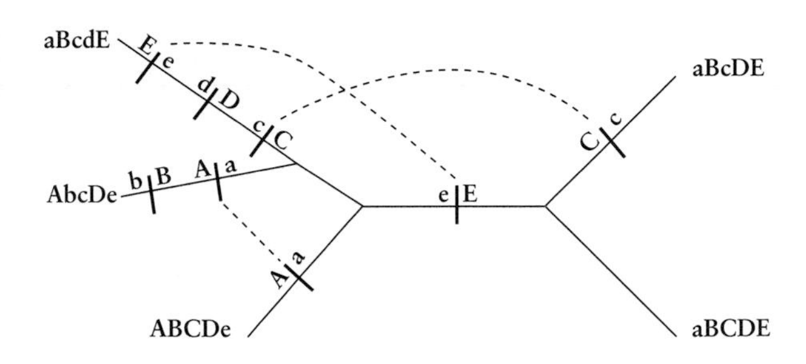

Notice that the trees portrayed above have no base; that is to say, they are not **rooted.** We cannot tell what combination of traits was possessed by the earliest ancestor of these modern taxa. One way of getting around this problem is to look at the characters of related lineages of organisms, which are called **outgroups.** Suppose that the nearest relatives of our monophyletic group all have traits AβcδE. β and δ are character states that appear only in the outgroup. We can infer then that A, C, and E are primitive characters and that a, c, and e are derived within our monophyletic group. Accordingly, the ancestor of the group must have had traits ABCDE. We can now root our phylogenetic tree with the inferred ancestor at the lowest branchpoint of the tree, as shown in the figure below.

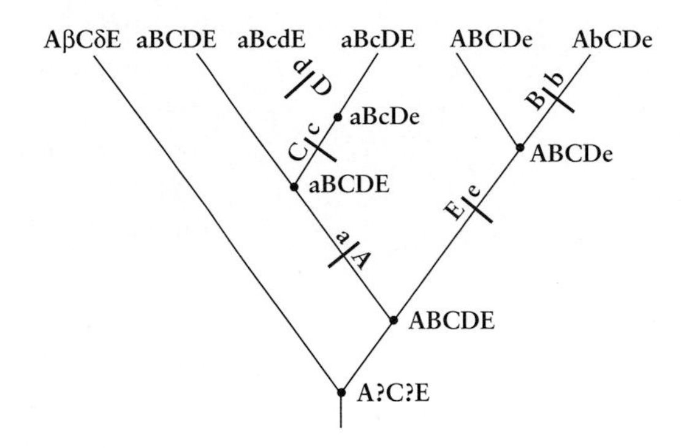

Early cladograms were based primarily on morphological characters, with biochemical or behavioral traits sometimes included when these were suitable. Such characters are often difficult to work with, especially among distantly related taxa, because it is sometimes hard to know whether the characters exhibited by two taxa are homologous—that is, whether they share a common evolutionary history. It would be pointless to compare the characteristics of structures—such as horns and antlers—that are not homologous. More recently, phylogenetic trees have been based on sequences of nucleotides in specific regions of DNA, the genetic material. Each position in the DNA sequence can be treated as a character that may have one of four states: the nucleotides adenine, guanine, cytosine, and tyrosine. At each position in the sequence, mutations can cause changes in the bases, which are passed on to progeny down through many generations. Estimated mutation rates at individual positions in the DNA sequence range between one in a million and one in a billion per generation. Thus, the sequence is highly conserved, and mutations that do occur are likely to remain as shared derived characters for long periods of the evolutionary history of a lineage. Reconstructions of phylogenetic trees from DNA sequence data are often based on 500–1,000 nucleotides. Depending on the degree of relationship among the taxa considered and the particular portion of the DNA analyzed, such a segment of DNA may have dozens of variable sites that are easily compared across species.

first described nearly a century ago, but the details of this mutually beneficial and obligatory relationship have been worked out only during the past few years. Adult female yucca moths carry balls of pollen from flower to flower by means of specialized mouthparts. During the act of pollination, a female moth enters a yucca flower and deposits 1 to 15 eggs in the ovary through cuts it makes with its ovipositor. After each egg is laid, the moth crawls to the top of the pistil of the flower and deposits a bit of pollen on the stigma. This guarantees that the flower is fertilized and that the moth's offspring will have developing seeds to feed on. After the moth has laid her eggs, she may scrape some pollen off the anthers and add it to the ball she carries in her mouthparts before flying to another flower. Males also come to the flowers to mate with the females, but only the females carry the pollen.

The relationship between the moth and the yucca is **obligatory.** *Tegeticula* larvae can grow nowhere else; *Yucca* has no other pollinator. In return for the pollination of its flowers, the yucca seemingly tolerates the moth larvae feeding on its seeds, but the extent of this loss of potential reproduction is small, rarely exceeding 30% of a seed crop. The moth's restraint concerning the number of eggs laid per flower is a puzzling aspect of the yucca-moth relationship. Over the short term, it would seem that moths laying larger numbers of eggs per flower might have higher individual reproductive success and evolutionary fitness, even though such behavior over the long term might lead to extinction of the yucca. In fact, it is the yucca that regulates the number of eggs laid per flower. When too many eggs are laid in the ovary of a particular flower—too many being enough to eat a majority of the developing seeds—the flower is aborted and the moth larvae die. While this would also seem to reduce the seed production of the yucca, resources that would have supported the production of seeds in the now aborted flower are diverted to other flowers. Yuccas, like many plants, produce far more flowers than they need. Flowers are relatively cheap to produce, and seeds are expensive, so it is almost always the case among plants that some flowers, even those that are pollinated, are aborted before setting seed. Selective abortion of insect-damaged fruit occurs widely among plants, and yuccas use this mechanism to keep their moth pollinators in line.

The moth and the yucca plant have many adaptations that support their mutualistic interaction. On the yucca's part, its pollen is sticky and can easily be formed into a ball that the moth can carry; the stigma is specially modified as a receptacle to receive pollen. On the moth's part, individuals visit flowers of only one species of yucca, mate within the flowers, lay their eggs in the ovary within the flower, exhibit restraint in the number of eggs laid per flower, and have specially modified mouthparts and behaviors to obtain and carry pollen. Because the mutualism of *Tegeticula* and *Yucca* is so tight, one might expect all these characteristics to have evolved as a result of coevolution between the two.

In fact, however, many aspects of the mutualism are present in the larger lineage of nonmutualistic moths (Prodoxidae) within which *Tegeticula* evolved. Examination of a phylogenetic tree of the moths (Figure

(a)

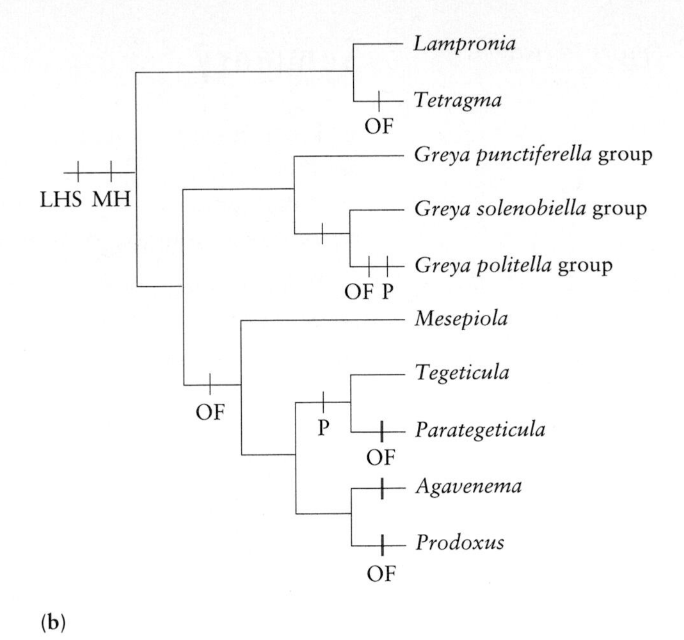

(b)

Figure 21.15 Phylogenetic (evolutionary) tree of the moth family Prodoxidae, showing the sites of the evolution of traits critical to the yucca moth-yucca mutualism in moths of the genus *Tegeticula*. The characters are: LHS, local host specificity; MH, mating on host; OF, oviposition in flower; P, pollinator. Black bars indicate the gain of a character shared by descendants of the lineage; colored bars indicate loss of a character. From O. Pellmyr and J. N. Thompson, *Proc. Natl. Acad. Sci. USA* 89:2927–2929 (1992).

21.15) shows that several of the highly specialized characters of *Tegeticula* are found in other members of the family. Indeed, host specialization and mating on the host plant are basal (primitive) features of the family found in all its members. The trait of ovipositing in flowers has evolved independently at least three times in the family and has reversed (reverted to the ancestral state) at least twice, in *Parategeticula* and *Agavenema*. Of the species that oviposit in flowers, only *Tegeticula* and one species of *Greya* actually function as pollinators; the others are strictly parasites of the plants in which their larvae grow. It should be mentioned that *Greya politella* pollinates *Lithophragma parviflorum* in the saxifrage family, which is not even closely related to the yuccas. We see from this example that many of the adaptations that occur in the yucca-moth mutualism appear to have been present in the moth lineage before the establishment of the mutualism itself. Such traits are often referred to as **preadaptations.**

Where does this leave us with regard to coevolution? The general consensus among ecologists is that species interactions strongly affect evolution and shape the adaptations of consumers and resource populations alike. This may be thought of as coevolution in a broad sense—sometimes called diffuse coevolution—in that populations simultaneously respond to an array of complex interactions with many other species. Coevolution in the narrow sense, in which changes in one evolving lineage stimulate evolutionary responses in the other, and vice versa, may be limited to very tight mutualisms in which strong interactions are limited to a pair of species. Even in such cases as that of the yucca and its moth pollinator, however, what appear to be coevolved traits may have been preadaptations that were critical to the establishment of the obligate mutualism in the first place. However, no subtlety of definition can detract from the reality that interactions among species are major sources of selection and evolutionary response.

Summary

1. Populations of consumers, resources, and competitors respond to one another through evolutionary changes in the characteristics that determine consumer efficiency and the outcome of competition. Hence their interactions have evolutionary as well as population dynamics.

2. Evidence of evolutionary changes in consumer-resource systems has been obtained in laboratory studies on host-parasitoid interactions. After periods of co-occurrence, rates of parasitoid attack decreased and host populations increased, apparently following selection for improved host defenses against parasitoids.

3. Studies on pathogens of plant crops—wheat rust, for example—have revealed a simple genetic basis for virulence and resistance, which determine the outcomes of parasite-host interactions.

4. Because selection for prey defenses increases in proportion to predation rate, and selection for predatory efficiency decreases as predation rate increases, predator and prey reach an evolutionary steady state at some intermediate level of predation.

5. The outcome of competition in many experimental systems has been shown to depend on the genotypes of the competitors, which indicates that competitive abilities, or carrying capacities, or both, have a genetic basis.

6. Experiments on competition between species of flies have revealed reversals of competitive ability over the course of tens of generations. By testing populations against unselected controls, investigators have confirmed genetic changes in competing populations.

7. One may test whether competition can result in evolutionary divergence in nature by comparing ecological (or related morphological) traits of a population in the presence and in the absence of a competitor. When the two differ, the pattern is referred to as character displacement.

8. Coevolution is the interdependent evolution of species that interact ecologically. The interactions may be antagonistic (consumer-resource, competitive) or cooperative (mutualism). Because each species in a coevolved pair is an important component of the environment of the other, changes in one select adaptive responses in the other, and vice versa.

9. The most convincing demonstrations of coevolution involve cases of mutualism, such as the obligate interdependence of *Pseudomyrmex* ants and *Acacia*. The ant keeps the plant free of herbivores while the plant provides the ant with food and housing; both have adaptations of morphology and behavior that promote their interrelationship.

10. Analysis of biosynthetic pathways has shown how increasingly toxic chemical defenses may evolve in response to herbivore pressure. When variations in these pathways (and in the abilities of insects to detoxify the chemicals) are overlaid upon taxonomic relationships within each group, one can infer the evolutionary history of a plant-insect interaction.

11. The interaction between yucca moths and yuccas is an obligate mutualism in which the moth pollinates the plant, but its larvae consume developing seeds. Both the moth and the yucca have specializations that promote this relationship, but phylogenetic analysis shows that some of the adaptations of the moth are present in close relatives that are not mutualists of yuccas. Such traits are called preadaptations.

Suggested readings

Armbruster, W. S. 1992. Phylogeny and the evolution of plant-animal interactions. *BioScience* 42:12–20.

Berenbaum, M. R. 1983. Coumarins and caterpillars: A case for coevolution. *Evolution* 37:163–179.

Bogler, D. J., J. L. Neff, and B. B. Simpson. 1995. Multiple origins of the yucca-yucca moth association. *Proceedings of the National Academy of Sciences USA* 92: 6864–6867.

Boucher, D. H. (ed.). 1985. *The Biology of Mutualism.* Croom Helm, London.

Brooks, D. R., and D. A. McLennan. 1991. *Phylogeny, Ecology, and Behavior.* University of Chicago Press, Chicago.

Brower, L. P. 1969. Ecological chemistry. *Scientific American* 220:22–29.

Davies, N. B., and M. Brooke. 1991. Coevolution of the cuckoo and its hosts. *Scientific American* 264:92–98.

Ewald, P. W. 1994. *Evolution of Infectious Disease.* Oxford University Press, Oxford.

Futuyma, D. J., and M. Slatkin (eds.). 1983. *Coevolution.* Sinauer Associates, Sunderland, Mass.

Handel, S. N., and A. J. Beattie. 1990. Seed dispersal by ants. *Scientific American* 263:76–83.

Janzen, D. H. 1966. Coevolution of mutualism between ants and acacias in Central America. *Evolution* 20:249–275.

Janzen, D. H. 1985. The natural history of mutualisms. Pp. 40–99 in D. H. Boucher (ed.), *The Biology of Mutualism.* Croom Helm, London.

Nitecki, M. H. (ed.). 1983. *Coevolution.* University of Chicago Press, Chicago.

Pellmyr, O., and C. J. Huth. 1994. Evolutionary stability of mutualism between yuccas and yucca moths. *Nature* 372:257–260.

Pellmyr, O., J. Leebens-Mack, and C. J. Huth. 1996. Non-mutualistic yucca moths and their evolutionary consequences. *Nature* 380:155–156.

Pellmyr, O., and J. N. Thompson. 1992. Multiple occurrences of mutualism in the yucca moth lineage. *Proceedings of the National Academy of Sciences USA* 89: 2927–2929.

Price, P. W. 1977. General concepts on the evolutionary biology of parasites. *Evolution* 31:405–420.

Real, L. (ed.). 1983. *Pollination Biology.* Academic Press, Orlando, Fla.

Thompson, J. N. 1994. *The Coevolutionary Process.* University of Chicago Press, Chicago.

PART VI

COMMUNITIES

CHAPTER 22

COMMUNITY STRUCTURE

Every place on earth—each meadow, each pond, each rock at the edge of the sea—is shared by many coexisting organisms. These plants, animals, and microbes are linked to one another by their feeding relationships and other interactions, forming a complex whole often referred to as the **biological community.** Interrelationships within communities govern the flow of energy and cycling of elements within the ecosystem. They also influence population processes, and in doing so determine the relative abundances of species. Finally, interrelationships within communities exert natural selection on the gene pools of populations and therefore stimulate evolution among coexisting species.

We have discussed ecosystem processes, population dynamics, and evolution at length in earlier chapters. Observations and experiments have clarified the mechanisms underlying these processes, which are now understood in broad outline. However, several related issues have perplexed and polarized ecologists for decades and have not yet been clearly resolved. A common element underlying these issues is a century-old debate about the nature of communities as ecological entities: whether they are integrated wholes or merely assemblages of independently evolved species. In this chapter, we shall consider the various concepts of community, and then examine the ways in which ecologists have characterized the structure of biological communities so that they can compare community properties in different environments and regions of the earth.

Concepts of the community

One of the great ecological debates of the twentieth century has contrasted polar viewpoints of the organization of ecological communities. One extreme—the **holistic concept**—champions the view that a community is a superorganism whose functioning and organization can be appreciated only when it is considered as an entire entity. Common sense tells us that we cannot ponder the significance of a kidney's functioning apart from the organism to which it belongs. Many ecologists have argued that it is equally pointless to consider soil bacteria without reference to the detritus they feed on, their own predators, and the plants nourished by their wastes. Accordingly, they argue, one can understand each species only in terms of its contribution to the dynamics of the whole system. Most importantly, ecological and evolutionary relationships among species make a community much more than the sum of its individual parts.

The other extreme viewpoint—the **individualistic concept**—holds that community structure and functioning simply express the interactions of individual species that make up local associations and do not reflect any organization, purposeful or otherwise, above the species level. Because natural selection acts on the reproductive output of individuals, each population in a community fosters only its own self-interest. The flow of energy and cycling of nutrients within ecosystems result from the predatory endeavors of individuals that make up communities.

There is also an intermediate, or mixed, point of view, which accepts the individualistic premises that species evolve out of self-interest and that communities may be assembled haphazardly, but admits to the holistic premise that attributes of community structure and function arise from unique interactions among species that happen to live in a particular place. Furthermore, these interactions are often reinforced by coevolution, reflecting the strong reciprocal forces of selection that occur among interacting species.

Independently of the debate over the nature of the community, ecologists have devoted considerable effort to describing community-level properties. The simplest measure of a community's structure is the number of species it includes, which is often referred to as **species richness** or **species diversity.** Early naturalists knew that more species live in tropical localities than in temperate and boreal zones. For example, Barro Colorado Island, a 16-km^2 island in Gatun Lake, Panama, supports 211 species of trees that grow to be taller than 10 m, more species than are found in all of Canada. Plots of 0.1 ha in some regions of Amazonian Peru and Ecuador contain more than 300 species; every other individual tree in such plots belongs to a different species! With the exception of taxa especially adapted to harsher conditions unique to higher latitudes, most types of organisms exhibit their highest diversity in the Tropics.

Biologists have hardly catalogued all the species of plants and animals, let alone microbes. About a million and a half species have been described and named worldwide; estimates of the total run well into the tens of millions. Because many of these taxa are succumbing to human encroachment before they become known to science, ecologists feel an urgent need to understand the causes of variation in diversity among biological communities, and to find ways to preserve as much of this natural heritage as possible.

Because even the simplest biological communities contain overwhelming numbers of species, ecologists have often partitioned diversity into numbers of species at each trophic level (that is, primary producers, herbivores, carnivores) and, within trophic levels, among different **guilds** distinguished by method or location of foraging (for example, herbivores include leaf eaters, stem borers, root chewers, nectar sippers, and bud nippers). It sometimes proves more practical to make comparisons based on these smaller subdivisions of communities.

Whenever ecologists have attempted to tabulate the diversity of a community or part of a community by identifying all individuals encountered within a given area, they have found that a few species are abundant and many more are rare. Regardless of the particular species included in a sample, numbers of individuals per species often assume regular patterns of distribution. These patterns of **relative abundance** are another way in which ecologists have quantified the structure of communities, as we shall see below.

Regular patterns of community structure do not argue for or against a holistic (superorganismal) interpretation of the community. Organization can arise either according to some design imposed on an entire system or by the independent activities of, and interactions between, a system's components. In the former case, which is a holistic point of view, community structure reflects attributes of species selected to enhance the functioning of the community as a whole. In the latter case, which is an individualistic concept, the structure of a community is a collective property of its individual components, each of which endeavors to function in its own right within the community. Ecologists have come to understand that neither extreme viewpoint is tenable, and they now strive to determine the extent of community integration and its biological mechanisms.

Definition of the community

Ecologists have given **community** a variety of meanings. Usually, the term is applied to a group of populations that occur together, but any similarity among definitions ends there. Throughout the development of ecology as a science, the term has often denoted **associations** of plants and animals occurring in a particular locality and dominated by one or more prominent species or by some physical characteristic. We speak of an oak community, a sagebrush community, and a pond community, meaning all the plants and animals found in a particular place dominated by the community's namesake. Used in this way, the term is unambiguous: a community is spatially defined and includes all the populations within its boundaries.

Ecologists also define communities on the basis of interactions among associated populations. This implies a functional rather than a descriptive use of the term. Ecologists sometimes use the term *association* when they describe populations that occur in the same area without regard to their interactions and use *community* when they study interactions among populations in an association.

Communities defy delineation when populations extend beyond arbitrary spatial boundaries. Migrations of birds between temperate and tropical regions link communities in each area; within some tropical localities, as many as half the birds present during the northern winter are migrants. Salamanders, which complete their larval development in streams and ponds but pursue their adult existence in the surrounding woods, tie together aquatic and terrestrial communities, just as trees do when they shed their leaves into streams and thereby support aquatic detritus-based food chains.

Community structure and functioning blend a complex array of interactions, directly or indirectly tying together all members of a community into an intricate web. The influence of each population extends to ecologically distant parts of the community. Insectivorous birds, for example, do not eat trees, but they do prey on many of the insects that feed on foliage or pollinate flowers. By eating pollinators, birds may indirectly affect seed production, the availability of food to animals that feed on fruits and seedlings, and the predators and parasites of those animals. The ecological and evolutionary effects of a population extend in all directions throughout the trophic structure of a community by way of its influence on predators, competitors, and prey, but this influence dissipates as it passes through each successive link in the chain of interaction. Its effects similarly spread through space and forward through time by way of the movements of individuals and the momentum of population processes. Because of this momentum, the present-day community bears an imprint of the past.

The community as a natural unit of ecological organization

Ecologists have described functional relationships within associations of species just as physiologists have connected the functions of various parts of the body. The analogy between community and organism is obvious, as we have seen. As Victor E. Shelford, in a 1931 paper entitled "Some concepts of bioecology," wrote:

It is an old practice to liken organisms to cosmic systems, and cosmic systems to organisms. Again in this case, it is convenient to liken the biome (plant-animal formation) to an amoeboid organism, a unit of parts, growing, moving, and manifesting internal processes which may be likened to metabolism, locomotion, etc., in an organism. (Ecology *12:455–467, 1931)*

Shelford even drew parallels between certain processes of community change and wound healing.

Certainly the most influential advocate of the organismal viewpoint was the American plant ecologist Frederick E. Clements, who, early in the twentieth century, perceived communities as discrete units with sharp boundaries and a unique organization. The conspicuousness of many dominant vegetation types reinforced Clements's view. A forest of ponderosa pine, for example, is distinct from the fir forests that grow in moister habitats and from the shrubs and grasses typical of drier sites. Boundaries between these community types are often so sharp that we may cross them within a few meters along a gradient of climatic conditions. Some community boundaries, such as those between deciduous forest and prairie in the midwestern United States and between broad-leaved forest and needle-leaved forest in southern Canada, are respected by most species of plants and animals.

An opposite view of community organization was held at about the same time by H. A. Gleason, who suggested that a community, far from being a distinct unit like an organism, is merely a fortuitous association of organisms whose adaptations enable them to live together under the particular physical and biological conditions that characterize a particular place. A plant association, he said, is "not an organism, scarcely even a vegetational unit, but merely a coincidence."

Clements's and Gleason's concepts of community organization predict different patterns of species distribution over ecological and geographic gradients. On one hand, Clements believed that species belonging to a community are closely associated with one another, which implies that the ecological limits of distribution of each species will coincide with the distribution of the community as a whole. Ecologists call this type of community organization a **closed community.** On the other hand, Gleason believed that each species is distributed independently of others that co-occur in a particular association. Such an **open community** has no natural boundaries; therefore, its limits are arbitrary with respect to the geographic and ecological distributions of its member species, which may extend their ranges independently into other associations.

Ecotones

The structures of closed and open communities are shown schematically in Figure 22.1. In the diagram depicting closed communities, the distributions of the species in each community coincide closely along a gradient of environmental conditions—for example, from dry to moist. Closed

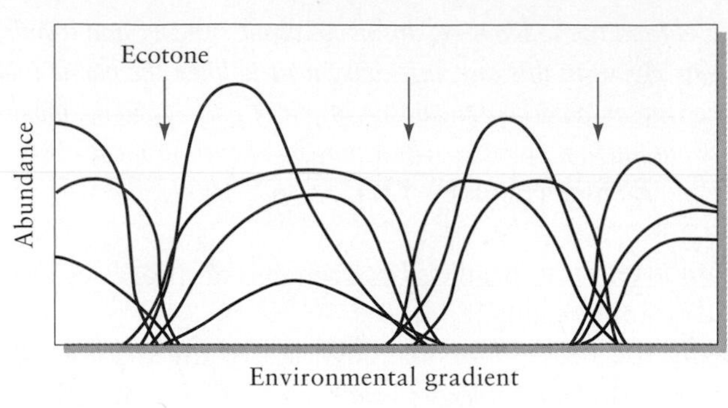

(a) Closed communities

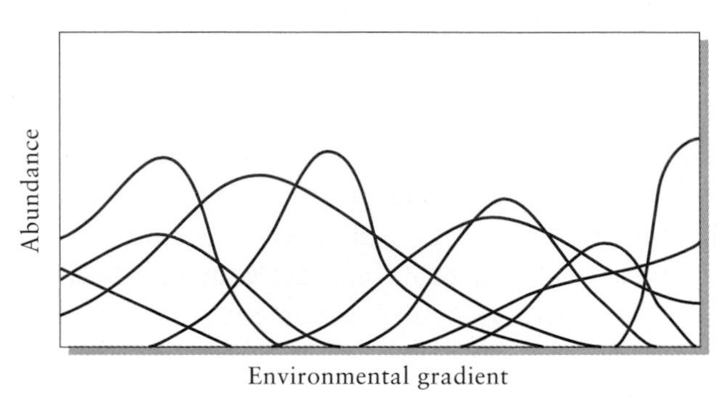

(b) Open communities

Figure 22.1 Hypothetical distributions of species organized into distinct assemblages (closed communities, a) and distributed at random along a gradient of environmental conditions (open communities, b). Ecotones between closed communities are indicated by arrows. Each curve represents the abundance of a different individual species along the ecological gradient.

communities are natural ecological units with distinct boundaries. The boundaries of such communities, called **ecotones,** are regions of rapid replacement of species along the gradient. In the diagram depicting open communities, species are distributed at random with respect to one another, giving an open structure. We might arbitrarily delimit a "community" at some point, perhaps a dry forest community near the left-hand end of a moisture gradient, while recognizing that some of the species included would be more characteristic of drier portions of the gradient and that others would reach their greatest abundance in wetter sites.

The concepts of open and closed communities both have some validity in nature. We observe distinct ecotones between associations under two circumstances: first, when the physical environment changes abruptly—for example, at the transition between aquatic and terrestrial communities, between distinct soil types, and between north-facing and south-facing slopes of mountains; and second, when one species or life form so dominates the environment that the edge of its range determines the distributional limits of many other species.

Sharp physical boundaries create well-defined ecotones. These occur at the interface between most terrestrial and aquatic (especially marine) communities (Figure 22.2) and where underlying geologic formations cause the mineral content of soil to change abruptly. An ecotone between plant associations on serpentine and on nonserpentine soils in southwestern Oregon is represented in detail in Figure 22.3. Levels of nickel, chromium, iron, and magnesium increase across the boundary into serpen-

Figure 22.2 A sharp community boundary (ecotone) in the Bay of Fundy, New Brunswick, associated with an abrupt change in the physical properties of adjacent habitats. Seaweeds extend only to the high tide mark. Between the high tide mark and the spruce forest, waves wash soil from rocks and salt spray kills pioneering land plants, leaving the area devoid of vegetation.

tine soils; copper and calcium levels of the soil drop off. The edge of the serpentine soils marks the boundaries of many species that are either excluded from, or restricted to, serpentine outcrops. A few species exist only within the narrow zone of transition; others, seemingly unresponsive to variations in these soil minerals, extend across the ecotone.

A change in soil acidity often accompanies a transition between broad-leaved and coniferous needle-leaved forests, an example of the second type of ecotone. The decomposition of conifer needles produces organic acids more abundantly than the breakdown of leaves of flowering plants; furthermore, because needles tend to decompose slowly, a thick layer of partly decayed organic material accumulates at the soil surface. This dramatic shift in environment between broad-leaved and needle-leaved forests marks the edges of distributions of many understory species within

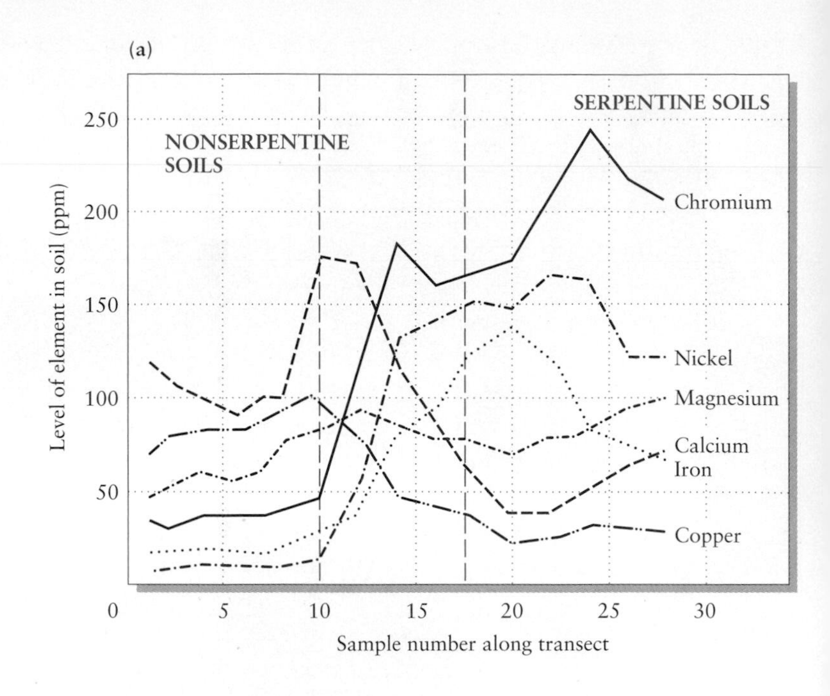

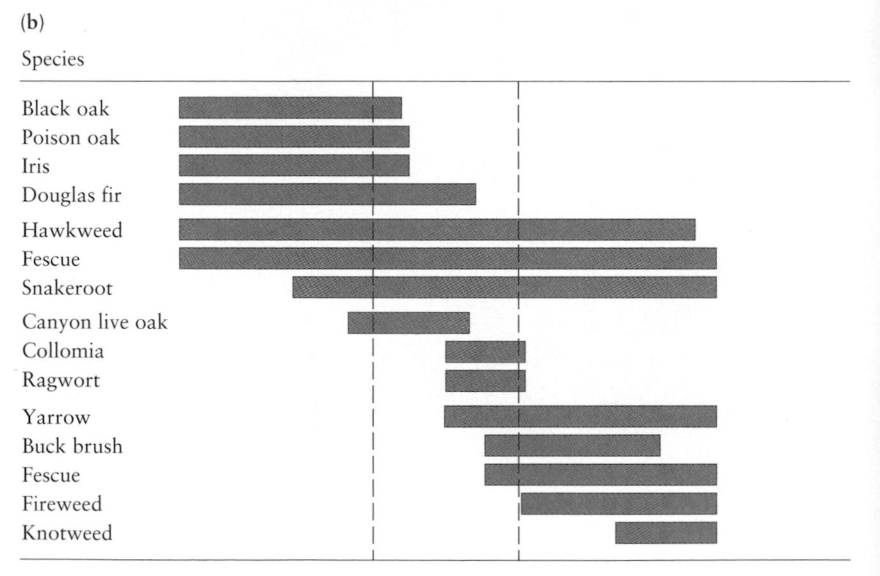

Figure 22.3 Changes in the concentration of elements in soil (a) and replacement of plant species (b) across the boundary between nonserpentine soils (samples 1–10) and serpentine soils (samples 18–28) in southwestern Oregon. The transect diagrammed here is somewhat atypical in that magnesium does not increase abruptly across the serpentine ecotone. After C. D. White, *Vegetation-Soil Chemistry Correlations in Serpentine Ecosystems,* Ph.D. dissertation, University of Oregon, Eugene (1971).

each forest type. Similarly, at boundaries between grassland and shrubland or between grassland and forest, sharp changes in surface temperature, soil moisture, light intensity, and fire frequency result in many species replacements. Boundaries between grasslands and shrublands are often sharp because when one or the other vegetation type holds a slight competitive edge, it dominates the community. Grasses prevent the growth of shrub seedlings by reducing the moisture content of surface layers of soil; shrubs depress the growth of grass seedlings by shading them. Fire evidently maintains a sharp edge between prairies and forests in the midwestern

United States. Perennial grasses resist fire damage that kills tree seedlings outright, but fires do not penetrate deeply into the moister forest habitats.

Plant ecologists have long recognized the influence of climate on plant associations. For example, in 1936, Forrest Shreve described a chaparral-desert transition in Baja California in relation to moisture and freezing temperatures. Desert species, particularly cacti, do not tolerate prolonged frost and dwindle in diversity and abundance north of the frost line; chaparral species drop out more gradually to the south within the transition zone as a consequence of water stress. As Shreve concluded:

> *Plants of the desert are more sharply confined to their own formation than are the species and genera of the chaparral and other northern types of vegetation. This appears to be due to the fact that the only requirement for the long southward extension of a chaparral plant is the occurrence in the desert region of relatively moist habitats, however restricted in area, while the northward extension of a desert plant requires a well-drained soil, a high percentage of sunshine and freedom from freezing temperatures of more than a few hours' duration. These more exacting requirements are met only in close proximity to the edge of the desert or else in light soils or on steep south slopes near the sea.* (Madroño *3:257–264, 1936)*

The continuum concept

The broad-leaved deciduous forests of eastern North America are bounded to the north by cold-tolerant evergreen needle-leaved forests, to the west by drought- and fire-resistant grasslands, and to the southeast by fire-resistant pine forests. Within the region of their distribution, one broad-leaved forest looks pretty much like another at first glance. As a result of early botanical explorations, ecologists were aware that different species of trees and other plants occurred in different areas within the deciduous forest biome. According to Clements's closed-community viewpoint, the distinctive vegetation of each area represented a distinct community separated by sharp vegetational transitions from other communities. But as ecologists described plant distributions in more detail, they found that plant associations fit the closed-community concept less and less well; classifications of plant communities became more and more finely split until absurd levels of distinction were reached.

Out of this mounting chaos there arose a new concept of community organization, referred to as the **continuum concept,** which is an open-community point of view. Within broadly defined habitats, such as forest, grassland, or estuary, populations of plants and animals gradually replace one another along gradients of physical conditions. Environments of the eastern United States form a continuum, with a north-south temperature gradient and an east-west rainfall gradient. The species of trees found in any one region (for example, those native to eastern Kentucky) have different geographic ranges, suggesting a variety of evolutionary backgrounds and ecological relationships. Some species reach their northern limits in

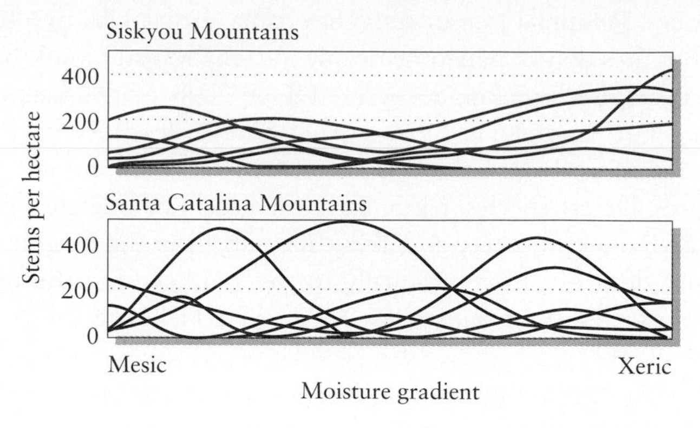

Figure 22.4 Distributions of species along moisture gradients at 460–470 m elevation in the Siskyou Mountains of Oregon and at 1,830–2,140 m elevation in the Santa Catalina Mountains of southeastern Arizona. Fewer species of plants reside in the mountains of Oregon, but each species has a wider ecological distribution, on average. After R. H. Whittaker, *Ecol. Monogr.* 30:279–338 (1960); R. H. Whittaker and W. A. Niering, *Ecology* 46:429–452 (1965).

Kentucky, some their southern limits. Because few species have broadly overlapping geographic ranges, the association of plant species found in any given spot does not represent a closed community. Each species has a unique evolutionary history and present-day ecological position, with a variable degree of association with other species in the local community.

A more detailed view of the forests of Kentucky would reveal that many tree species segregate along local gradients of conditions. Some grow along ridge tops, others along moist river bottoms; some on poorly developed, rocky soils, others on rich, organic soils. The species represented in each of these more narrowly defined associations might exhibit correspondingly closer ecological distributions, but the open-community concept would still better describe these associations of plants.

Gradient analysis

The validity of the open-community/continuum concept depends on the way in which species are distributed along ecological gradients. Therefore, a compelling resolution of the open versus closed community debate was finally reached when ecologists plotted abundances of species over ranges of conditions in the physical environment. In such a **gradient analysis,** closed community organization would reveal itself by the presence of sharp ecotones in species distributions, as suggested in Figure 22.1. A gradient analysis is usually undertaken by measuring both abundances of species and physical conditions at a number of localities and then plotting the abundances of each species as a function of the value of the physical condition. The range of conditions might embrace any number of physical variables, such as moisture, temperature, salinity, exposure, or light level. Ecologists might pick sampling localities at regular intervals along a known physical gradient, such as that of temperature as it decreases up an elevation gradient.

Cornell University ecologist Robert Whittaker pioneered gradient analysis in North America, and his work was influential in putting to rest the extreme Clementsian view of closed communities. Whittaker conducted most of his work in mountainous areas where moisture and temperature vary over short distances according to elevation, slope, and exposure. These variables in turn determine the light, temperature, and moisture levels at a particular site. When Whittaker plotted abundances of species found at sites at the same elevation distributed over a range of soil moisture, he found that the species occupied unique ecological distributions, with their peaks of abundance scattered along the environmental gradient (Figure 22.4). These findings were consistent with an open community structure.

In the Great Smoky Mountains of Tennessee, dominant species of trees occur widely outside the plant associations that bear their names (Figure 22.5). For example, red oak grows most abundantly in relatively dry sites at high elevations, but its distribution extends into forests dominated by beech, white oak, chestnut, and even hemlock (an evergreen conifer) and reaches throughout the entire range of elevation in the Smoky Mountains. Beech prefers moister situations than red oak, and white oak reaches its greatest abundance in drier situations, but all three species occur together in many areas. Whittaker's studies, and many others on plant and animal distributions since then, have failed to reveal evidence of distinct ecotones between associations of species. Even Forrest Shreve, who emphasized the

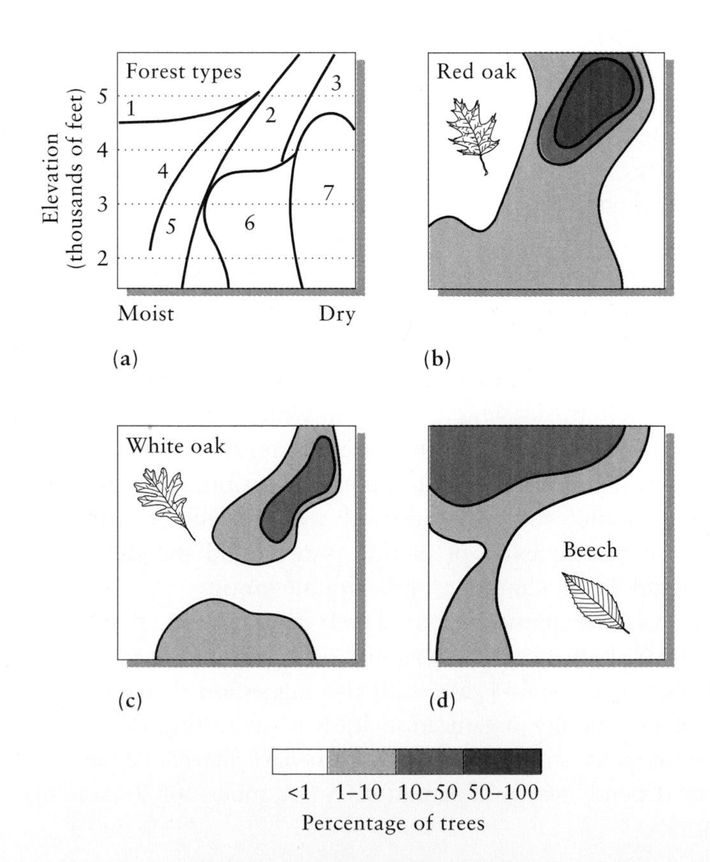

Figure 22.5 Distributions of (b) red oak, (c) white oak, and (d) beech with respect to altitude and soil moisture in the Great Smoky Mountains of Tennessee. The approximate boundaries of major forest associations are shown in (a): forest types are 1, beech; 2, red oak-chestnut; 3, white oak-chestnut; 4, cove; 5, hemlock; 6, chestnut oak-chestnut; 7, pine. Relative abundance, represented by the degree of shading, is measured as the percentage of tree stems more than 1 cm in diameter in samples of approximately 1,000 stems. After R. H. Whittaker, *Ecol. Monogr.* 26:1–80 (1956).

distinct boundary between desert and chaparral plant associations, recognized that

> *As is true of the meeting ground between any two great plant formations, the dominant plants of each formation are found to vary in the distance to which they extend into the other. This indicates that their habitat requirements are not so nearly identical as their close association in the midst of their respective formations would suggest. (*Madroño *3: 257–264, 1936)*

By the 1960s, the field studies of Whittaker and other observers had put to rest Clements's view of the closed community by demonstrating that plant species are distributed more or less independently over ranges of ecological conditions. Few cases of consistent association between species were apparent, and these were overwhelmed by the predominately open structure of ecological communities.

Trophic structure and food webs

When the community is viewed from an ecosystem perspective, one sees that species occur in functional groups whose members occupy similar trophic positions. Thus, plants can be lumped together as producers, all herbivores (from ant to zebra) share the herbivore label, and so on. When we describe trophic structure in this way, we emphasize the functional similarities of species. Placing species together in a small number of functional categories obscures the distinctiveness of communities that arises from their differences in numbers of species or in their evolutionary histories.

When we apply a food web perspective to the community, we tend to emphasize diversity. Although food webs are based on functional relationships, they stress connections between populations and recognize, for example, that not all herbivores consume all producers. Because food web analysis includes species-level information about a community, it has greater power than ecosystem analysis to differentiate structure.

Food web analysis has progressed through two historical periods, which we shall refer to as its descriptive and analytical phases. The descriptive phase began early in the twentieth century. Initially, ecologists portrayed food webs by drawing diagrams in which arrows connected species in a community according to their feeding relationships. These diagrams were often complex and difficult to compare quantitatively between communities. To some ecologists, food web diagrams merely emphasized the overwhelming complexity of natural systems and the need to simplify their structure by dividing them into trophic groups.

The analytical approach to food web study asks whether the structure of a food web influences the dynamics of its constituent populations. This phase began in the mid-1950s with the suggestion that increasing complexity of community organization leads to increasing dynamic stability. The reasoning was simple: When predators have alternative prey, their own numbers depend less on fluctuations in numbers of a particular prey species. Where energy can take many routes through a system, disruption of one pathway merely shunts more energy through another, and the overall flow continues uninterrupted.

This idea linked community stability directly to species diversity and food web complexity, and it stimulated a flurry of theoretical, comparative, and experimental work. These studies have yet to produce a consensus, partly because both structure and stability elude definition and measurement and partly because different theories lead to different predictions about stability. For example, an alternative to the idea that diversity generates stability is that as communities become more diverse, species exert greater influence on one another through their various interactions; these biological links in turn may create pervasive time lags in population processes, which tend to destabilize diverse systems.

An influential early experimental study on food webs was conducted by Robert T. Paine, of the University of Washington, on how consumers determined the structure of rocky shore intertidal communities. Paine emphasized the role of predators. He compared food webs in the Gulf of California and on the coast of Washington, both of which are dominated by imposing predators, the sea stars *Pisaster* and *Heliaster* (Figure 22.6). By removing sea stars from experimental areas on the coast of Washington, Paine demonstrated the crucial role of these predators in maintaining the structure of the community. Released from predation by this manipulation, mussels *(Mytilus)* spread very rapidly, crowding other organisms out of the experimental plots and reducing the diversity and complexity of local food webs. Removal of the urchin *Strongylocentrotus,* an herbivore, similarly allowed a small number of competitively superior algae to dominate an area, crowding out many ephemeral or grazing-resistant species. Paine showed that predators and herbivores can manipulate competitive relationships among species at lower trophic levels and thereby control the structure of a community. Such species are called **keystone predators** because when they are removed, the edifice of the community tumbles.

Whereas many ecologists had emphasized interactions among competitors within a particular trophic level, Paine and others who followed his example stressed the additional importance of consumer-resource relationships as a key to understanding community organization. Paine also distinguished different concepts of food webs. His concepts of connectedness webs, energy flow webs, and functional webs describe different ways in which populations influence one another within communities. **Connectedness webs** emphasize feeding relationships among organisms, portrayed as links in a food web. **Energy flow webs** represent an ecosystem viewpoint, in which connections between species are quantified by flux of energy between a resource and its consumer. In **functional webs,** the importance of each population in maintaining the integrity of a community is reflected in its influence on the growth rates of other populations. This controlling role, which only experiments can reveal, need not correspond to the amount of energy flowing through a particular link in the food web of an intact community, as shown dramatically for an intertidal zone food web in Figure 22.7.

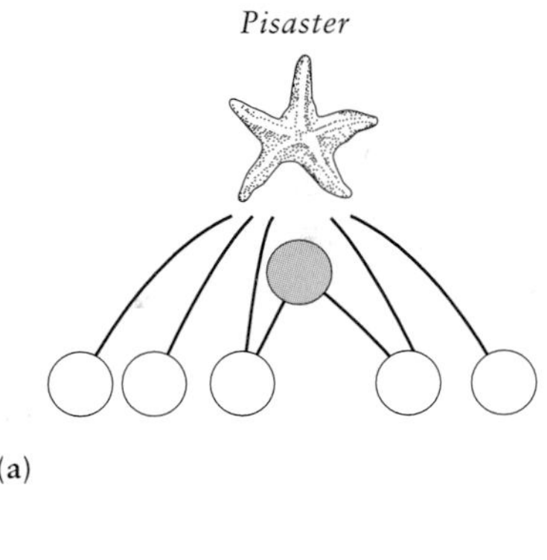

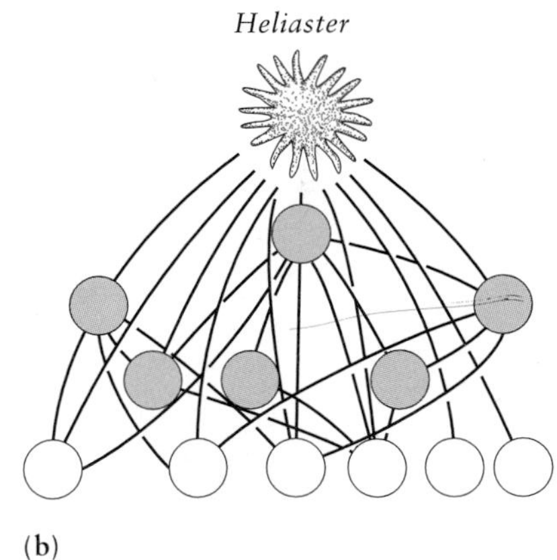

Figure 22.6 Intertidal food webs dominated by keystone predators: the sea stars (a) *Pisaster,* on the coast of Washington, and (b) *Heliaster,* in the northern Gulf of California. The lowest trophic levels of these food webs include such herbivores as chitons, limpets, herbivorous gastropods, and barnacles (unshaded circles). After R. T. Paine, *Am. Nat.* 100:65–75 (1966).

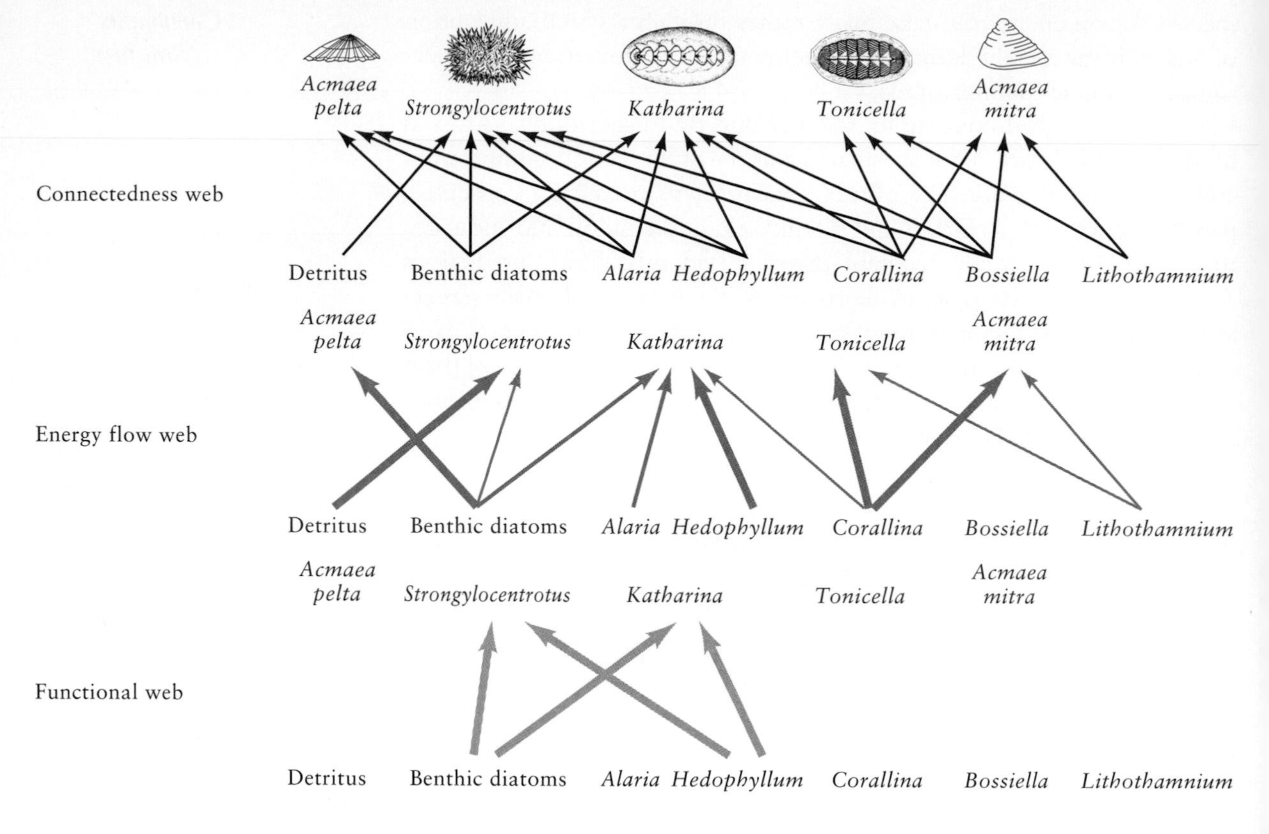

Figure 22.7 Three approaches to depicting trophic relationships, illustrated for species in a rocky intertidal habitat on the coast of Washington. From R. T. Paine, *J. Anim. Ecol.* 49:667–685 (1980).

As Paine and others employed food web diagrams to depict the structure of biological communities, a few ecologists questioned whether differences in the structure of food webs could affect the dynamics, stability, and persistence of communities. This issue is crucial to ecology because it bears upon the fundamental contrast between holistic and individualistic philosophies. We raise the issue here by posing such questions as: Is one particular arrangement of feeding relationships among species intrinsically more stable than a different arrangement among the same number of species? How important is the consideration of food web stability to the structure of natural communities?

In the two food webs illustrated in Figure 22.8, similar numbers of species are organized in strikingly different structures. The mudflat community is relatively simple, having seven links among the seven species portrayed in the diagram, and with only one species preying on more than one trophic level. By contrast, the plant-insect-parasitoid system is complex; it exhibits twelve links among eight species and several cases of omnivory (feeding on more than one trophic level). Theoretical work indicates that, all other things being equal, increasing the degree of feeding on more than one trophic level reduces a food web's stability—that is, its resistance to perturbation and its ability to return to equilibrium.

With respect to the food web structure of communities, theory and observation are deeply divided. On one hand, some attributes of food web design are consistent with qualities that enhance the intrinsic dynamic stability of food webs—that is, the tendency of food webs to regain their structure following disturbance or displacement from an equilibrium state. On the other hand, we observe radically different food webs in nature, such as those shown in Figure 22.8. Does this variation mean that the rules of food web stability vary depending on which organisms and ecological circumstances are involved? Or is it the case that feeding relationships depend on the attributes of species, and the organization of feeding relationships within a community has little influence on its stability?

These are tough questions. An important first step toward answering them is to characterize community organization in ways that match theory. Five attributes of communities seem important: diversity, **connectance** (complexity of links between species), food chain length, omnivory, and **compartmentalization** (subdivision of the food web). A highly compartmentalized community is one in which subsets of species interact regularly among themselves but only infrequently with species in other subsets. At one extreme, each compartment could be considered a separate community. Compartments might coincide with patches of habitat or distinctive species within habitats. For example, in southern Canada, broad-leaved trees, pines, firs, and hemlocks all have distinctive associations of lepidopteran herbivores; few species or genera of moths and sawflies feed on plants in more than one of these groups of trees. In mixed forests, therefore, one might consider the insect faunas of broad-leaved and needle-leaved trees as separate communities. In contrast, moths feed widely among different species of broad-leaved trees, and although most specialize to a greater or lesser degree, they seem not to distinguish discrete subsets. Indeed, analysis of many food webs turns up little evidence for compartmentalization despite considerable feeding specialization within associations of species occupying more or less uniform habitats.

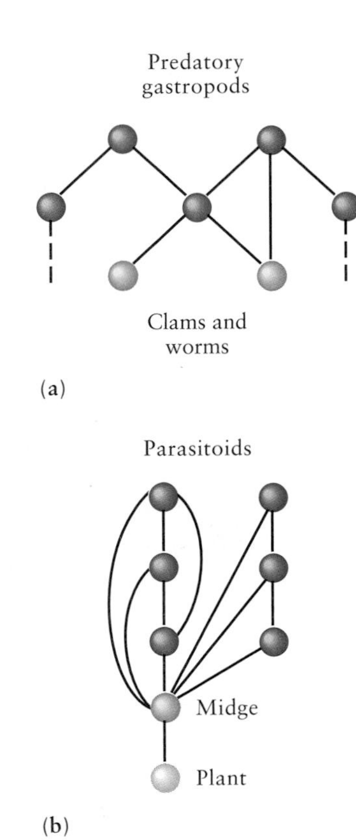

Figure 22.8 Examples of food webs with (a) little and (b) frequent omnivory). Part (a) is based on a mudflat community containing intertidal gastropods, bivalves, and their prey; Part (b) is based on the plant *Bacharis,* its insect herbivores, and their parasitoids. In food web diagrams such as these, lines connect the resource below to the consumer above. Not all prey species are depicted. After S. L. Pimm, *Food Webs,* Chapman & Hall, London and New York (1982).

Relative abundance

The Danish botanist Christen Raunkiaer noted early in the twentieth century that abundances of species within local assemblages assumed regular distributions. When he plotted the number of species in each of several abundance classes, the points followed a reversed J-shaped curve, like that shown in Figure 22.9. This pattern suggests that within a particular community, a few species attain high abundance—they are the **dominants** in the community—whereas most others are represented by relatively few individuals. Raunkiaer did not use a mathematical expression to describe his "law" of frequency, but its generality and pervasiveness among different assemblages have tempted others to mathematical description ever since.

Mathematics can serve two purposes here. We can describe many data (species abundances, in this case) with a simple equation and use values of variables in the equation to make comparisons among different samples of species. Or we can use the logic of a mathematical model to investigate processes that might produce the observed distributions. Without going into detail, let us simply say that models of relative abundance have served better as descriptive tools than they have as a way to elucidate the processes that determine relative abundance. Within every community, some species are common and others are rare. Species abundance appears to reflect the variety and abundance of resources available to each population, as well as the influences of competitors, predators, and diseases. Why some species fare well in this arena and others do not continues to puzzle ecologists.

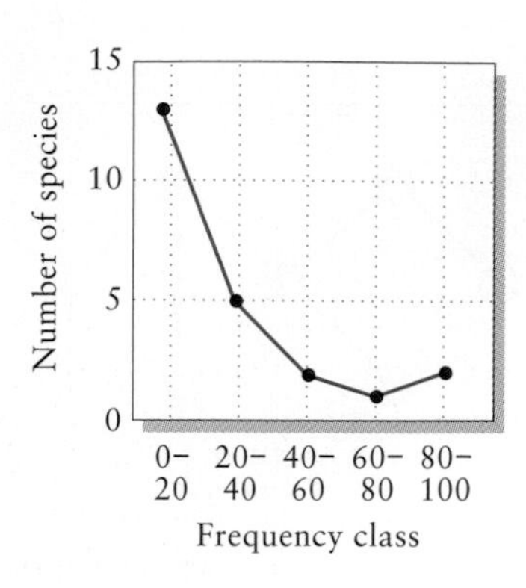

Figure 22.9 Number of species of plants in a peat bog near Kalamazoo, Michigan, in each of five frequency classes, based on percentages of twenty-five 0.1-m^2 sampling areas occupied. From data in L. A. Kenoyer, *Ecology* 8:341–349 (1927).

The lognormal distribution

The abundance of a particular species reflects a balance among a large number of factors and processes, variations in each of which result in small increases or decreases in abundance. Statisticians have shown that sums of many independent factors with small effects tend to assume a **normal distribution,** the familiar bell-shaped curve. Because factors that affect population size tend to multiply together in exerting their influence, one might expect a sample of population sizes to assume a **lognormal distribution,** in which species are distributed along a bell-shaped curve with respect to the logarithm of species abundances. Typically, such a distribution has many species with intermediate levels of abundance and relatively few rare or common species.

In 1948, Frank Preston published a seminal paper, entitled "The commonness, and rarity, of species," in which he characterized distributions of species abundances by lognormal curves. Preston assigned species to arbitrary classes of abundance on the basis of a logarithmic scale of numbers of individuals per species: 1–2 individuals, 2–4 individuals, 4–8 individuals, 8–16 individuals, and so on. Preston called these classes **octaves** because each was twice as large as the preceding class. (In the musical scale, the vibration frequency of each note is twice that of the note one octave lower.) In a large sample of individuals, species often distribute themselves normally over logarithmic abundance categories, as shown in Figure 22.10.

The normal distribution is described by the equation

$$n_R = n_0 e^{\frac{1}{2}\left(\frac{R}{\sigma}\right)^2},$$

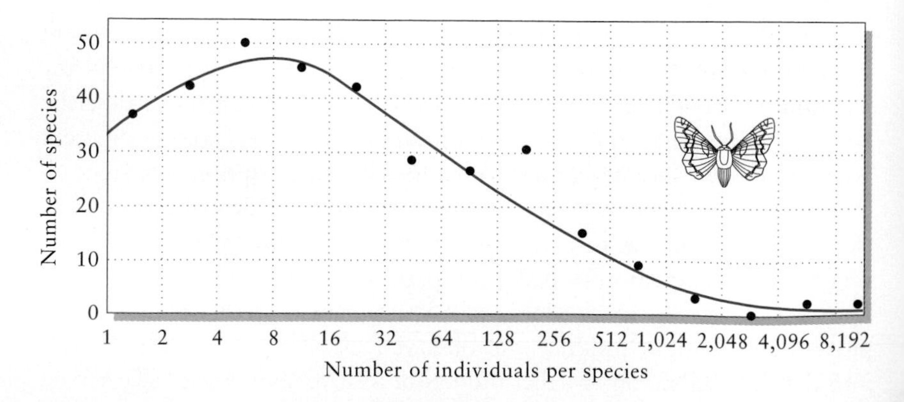

Figure 22.10 Relative abundances of species of moths attracted to light traps near Orono, Maine. Size classes (octaves), which increase by a factor of 2 from one class to the next, are shown on the horizontal axis; the number of species in each size class is plotted on the vertical axis. The distribution of abundances is hump-shaped, with a mode of 48 species in the "4–8 individuals" size class. After F. W. Preston, *Ecology* 29:254–283 (1948).

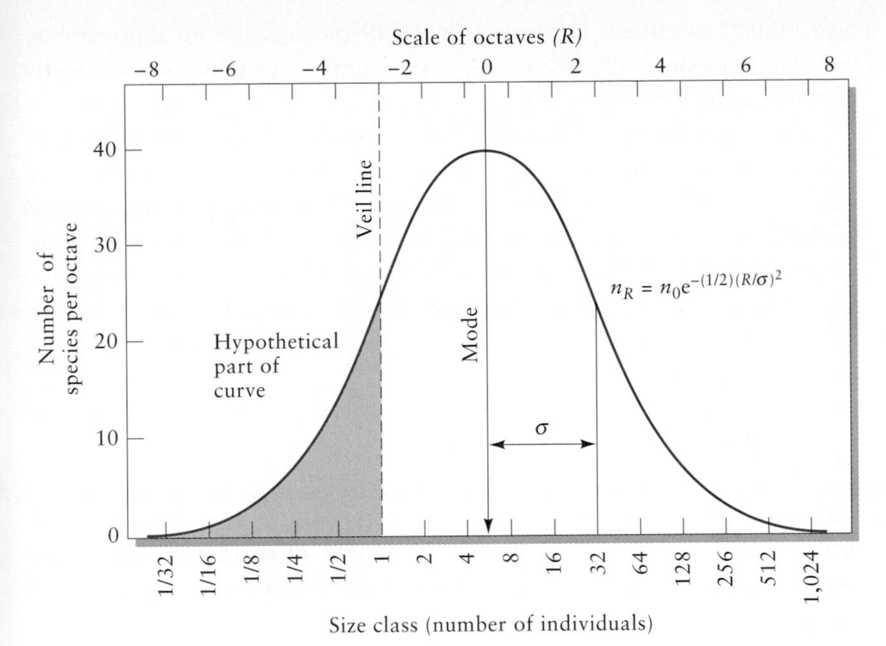

Figure 22.11 Lognormal distribution of species abundances. The part of the curve to the left of the veil line (shaded) corresponds to species with less than one individual in the sample, and thus not represented. The scale of octaves (R) begins at the modal octave (0). One standard deviation (σ) on both sides of the mode includes about two-thirds of all the species in the sample. After F. W. Preston, *Ecology* 29:254–283 (1948).

in which n_R represents the number of species whose abundance is R octaves greater or less than the modal abundance of species within a community, n_0 represents the number of species in the modal abundance class, and σ (the standard deviation) measures dispersion (the breadth of the normal curve). We can see these relationships in a graph of the lognormal distribution (Figure 22.11). The dispersion of the curve—whether it is narrow or broad—is proportional to σ.

In theory at least, the entire lognormal distribution of species abundances in a community can never be sampled. Some species are too rare to be represented by one or more individuals in a sample. These species fall below the **veil line** of the distribution, and their presence can be revealed only by increasing the total number of individuals examined. When the size of a sample doubles, the modal abundance of species moves one octave to the right (the abundance of all species doubles, on average), and additional species, each represented by one individual, appear in the distribution at the veil line.

A useful feature of Preston's lognormal curve is that it takes sample sizes into account. We can predict the total number of species (N) in a community, including those not represented in the sample, if we know only the number of species in the modal abundance class (n_0) and the dispersion of the lognormal distribution (σ). The appropriate equation is

$$N = n_0 \sqrt{2\pi\sigma^2} = 2.5\sigma n_0.$$

For the sample of moths graphed in Figure 22.10, n_0 is 48 and σ is 3.4 octaves; therefore, $N = 2.5 \times 3.4 \times 48 = 408$ species. The actual sample of over 50,000 specimens contained only 349 species, 86% of the number that would theoretically be present in the sample area if abundances were distributed lognormally.

The dispersion (σ) of lognormal curves fitted to large samples of associations varies somewhat among groups of organisms. Preston obtained values of 2.3 for birds but between 3.1 and 4.7 for moths; among samples of diatoms, values ranged between 2.8 and 4.7. Censuses of forest birds revealed values of 0.98 for lowland tropical localities, 1.36 for temperate localities, and 1.97 for islands; these data indicate a greater range of species abundances on islands than in temperate areas or, especially, in lowland tropical areas.

Diversity indices

Differences in abundances of species in communities pose two practical problems for ecologists. First, the total number of species included in a sample varies with sample size because as more individuals are sampled, the probability of encountering rare species increases. Thus, we cannot compare diversity among areas sampled at different intensities merely by counting species. Second, not all species should contribute equally to our estimate of total diversity because their functional roles in a community vary, to some degree, in proportion to their overall abundance.

Ecologists have tackled the second problem by formulating **diversity indices** in which the contribution of each species is weighted by its relative abundance. Two such indices are widely used in ecology: Simpson's index and the Shannon-Weaver index. In both cases, we calculate the indices from proportions (p_i) of each species (i) in the total sample of individuals. Simpson's index is

$$D = \frac{1}{\sum p_i^2}$$

where D is a measure of diversity. For any particular number of species in a sample (S), the value of D can vary from 1 to S, depending on the **evenness** of species abundances. When five species have equal abundance, each p_i is 0.20. Therefore, each $p_i^2 = 0.04$ and the sum of the p_i^2s is 0.20. D is equal to the reciprocal of the sum, or 5, which is the number of species in the sample. Similar calculations for some hypothetical communities are presented in Table 22.1, where it is apparent that rarer species contribute less to the value of the diversity index than do common species.

The Shannon-Weaver index, developed from information theory, is calculated by the equation

$$H = -\sum p_i \log_e p_i,$$

where H is a logarithmic measure of diversity. As in the case of Simpson's index, higher values of H represent greater diversity. Also like Simpson's

TABLE 22.1 **Comparison of diversity indices D, H, and e^H for hypothetical communities of five species having different relative abundances**

PROPORTION OF SAMPLE REPRESENTED BY SPECIES					DIVERSITY INDEX		
A	B	C	D	E	D	H	e^H
0.28	0.25	0.25	0.25	0.00	4.00	1.386	4.00
0.20	0.20	0.20	0.20	0.20	5.00	1.609	5.00
0.24	0.24	0.24	0.24	0.04	4.30	1.499	4.48
0.25	0.25	0.25	0.25	0.001	4.02	1.393	4.03
0.50	0.30	0.10	0.07	0.03	2.81	1.229	3.42

index, the Shannon-Weaver index gives less weight to rare species than to common ones. Because H is roughly proportional to the logarithm of number of species, it is sometimes preferable to express the index as e^H, which is proportional to the actual number of species. Table 22.1 presents values of e^H, which we may compare directly to Simpson's index.

Sample size and species richness

As mentioned above, a serious problem in estimating the number of species in an association arises from the fact, readily apparent from the log-normal distribution, that number of species increases in direct proportion to number of individuals sampled. If we wish to standardize measurements of diversity for comparison, we must base them on comparable samples. When samples include different numbers of individuals, comparability can be achieved by a statistical procedure known as **rarefaction,** in which equal-sized subsamples of individuals are drawn at random from the total.

Rarefaction can be thought of as a means of portraying the relationship between number of species and sample size. Howard Sanders of the Woods Hole Oceanographic Institution applied this technique to samples of benthic marine organisms dredged from soft sediments at various localities in several marine habitats. The number of specimens varied between samples because of differences in densities of organisms in the substrate and unavoidable variation in sampling procedures. These samples revealed a general relationship between sample size and diversity, but it was not possible to tell whether that relationship appeared as an artifact of sampling or reflected consistent differences between habitats unless one rarefied the samples to make them comparable (Figure 22.12). As the computer randomly removed specimens from samples, diversity decreased. But rarefaction curves clearly distinguished the localities, showing that for comparable samples, diversity varied considerably among different habitats.

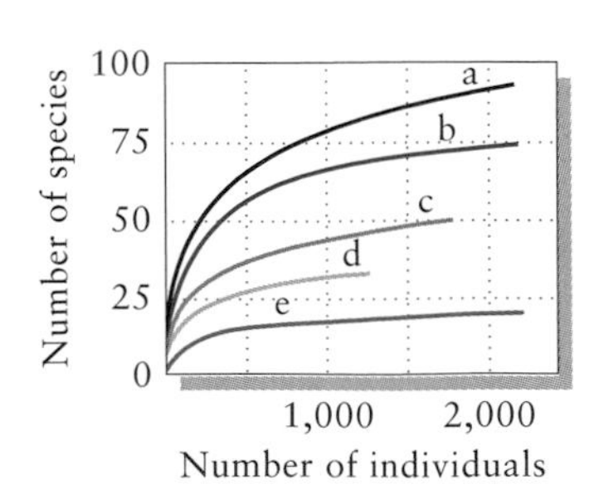

Figure 22.12 Number of species of bivalves and polychaete worms as a function of sample size for different marine environments. Tropical and deep-sea faunas tend to be more diverse than faunas in less constant environments. Habitats are: a, tropical shallow water; b, slope (deep sea); c, outer continental shelf; d, tropical shallow water; e, boreal shallow water. After H. L. Sanders, *Brookhaven Symp. Biol.* 22:71–81 (1969).

Species and area

As a rule, more species occur within large areas than within small areas. The botanist Olaf Arrhenius first formalized this **species-area relationship** in 1921. Since then, it has been common practice to characterize relationships between species (S) and area (A) with power functions of the form

$$S = cA^z,$$

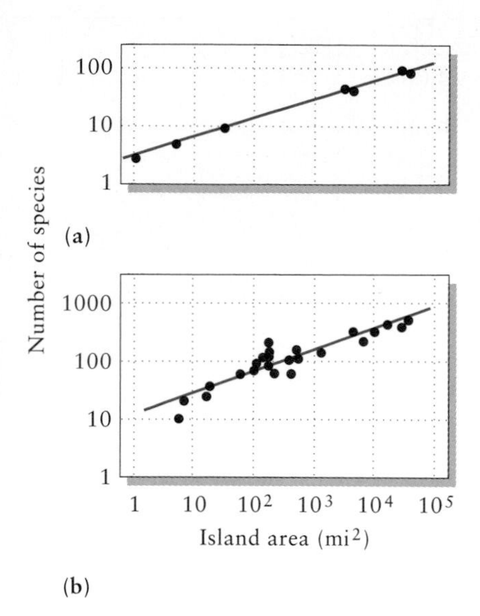

Figure 22.13 Species-area curves for (a) amphibians and reptiles in the West Indies and (b) birds in the Sunda Islands, Malaysia. From R. H. MacArthur and E. O. Wilson, *Evolution* 17:373–387 (1963), and *The Theory of Island Biogeography,* Princeton University Press, Princeton, N.J. (1967).

where c and z are constants fitted to the data. Graphical portrayals of species-area relationships plot the logarithm of species number against the logarithm of area, as shown in Figure 22.13. After log-transformation, the species-area relationship becomes

$$\log S = \log c + z \log A,$$

which is the equation for a straight line.

Analysis of species-area relationships among many groups of organisms has revealed that most values of z fall within the range 0.20–0.35—that is, number of species increases in proportion to the one-fifth to one-third power (fifth root to cube root) of area. This consistency of z suggests several possibilities. Observed z values might be a simple statistical consequence of the lognormal distribution of species abundances. As we increase the area of a sample, we usually increase the number of individuals included within it, and as sample size increases, the veil line moves to the left, exposing more and more species. However, empirical studies have demonstrated variation in z associated with differences in attributes of samples, suggesting that biological processes, in addition to statistical factors, influence the species-area relationship. For example, z values obtained for continental areas of a certain size tend to be lower than those obtained for islands within a comparable size range.

Where a flora or fauna is perfectly known (that is, all species have been sampled), the sampling properties of the lognormal distribution cannot be held accountable for any relationship between species and area; no species hide behind the veil line. For example, we possess near-perfect knowledge of the land bird fauna of the West Indies, among which species increase with area with a slope of about $z = 0.24$. Here, differences in diversity between large and small islands must signify differences in the islands' intrinsic qualities. Likely candidates include habitat heterogeneity, which undoubtedly increases with the size (and resulting topographic heterogeneity) of an island; and size per se, as larger islands make better targets for potential immigrants from the mainland. In addition, larger populations on larger islands probably persist longer, being endowed with greater genetic diversity, broader distributions over area and habitat, and numbers large enough to prevent chance extinction.

Summary

1. A biological community is an association of interacting populations. Questions about communities address the evolutionary origins of community properties, relationships between community organization and stability, and the regulation of species diversity.

2. Generally speaking, communities do not form discrete units separated by abrupt transitions in species composition. Species tend to distribute themselves over ecological gradients of conditions independently of the distributions of other species. Ecologists refer to this pattern as an open-community structure.

3. Discontinuities between associations of plants and animals, called ecotones, sometimes occur at sharp physical boundaries or accompany changes in the growth forms that dominate a habitat. The aquatic-terrestrial transition provides an example of the first kind of ecotone, the prairie-forest transition an example of the second.

4. To analyze distributions of species with respect to environmental conditions and with respect to distributions of other species, ecologists have devised various types of gradient analysis, in which they position sample localities on scales of physical conditions. The distributions of species along these environmental gradients emphasize the open structure of communities.

5. Within local areas, ecologists have characterized communities in terms of the number of species present, their relative abundances, their organization into guilds of species with similar feeding habitats, and food webs portraying feeding relationships among species.

6. The lognormal distribution of species abundances, which has proved to be a useful empirical device, characterizes frequency distributions by a modal abundance class (n_0) and dispersion of abundances about the mode (σ). The term n_0 depends on sample size, but the value of σ is an intrinsic property of the community. The lognormal concept stresses the dependence of diversity estimates on sample size.

7. Various indices of diversity, most notably Simpson's index and the Shannon-Weaver (information) index, have been devised to account for variations in abundance when making comparisons between samples. Because the number of species increases as sample size increases, as predicted by the lognormal distribution of abundances, ecologists have also devised rarefaction procedures and other statistical techniques to make samples of different sizes comparable.

8. The number of species in a sample increases in direct proportion to the area sampled. This results in part from larger areas giving rise to larger total samples. However, studies of well-known faunas and floras also indicate that larger areas are more heterogeneous ecologically, providing

opportunities to sample more kinds of habitats, and that larger islands have more species because they are better targets for colonization and because larger populations better resist extinction.

Suggested readings

Brown, J. H. 1995. *Macroecology.* University of Chicago Press, Chicago.

Brown, J. H., and E. J. Heske. 1990. Control of a desert-grassland transition by a keystone rodent guild. *Science* 250:1705–1707.

Diamond, J., and T. J. Case (eds.). 1986. *Community Ecology.* Harper & Row, New York.

Gee, J. H. R., and P. S. Giller. 1987. *Organization of Communities Past and Present.* Blackwell Scientific Publications, Oxford.

Gleason, H. A. 1926. The individualistic concept of the plant association. *Torrey Botanical Club Bulletin* 53:7–26.

Jackson, J. B. C. 1994. Community unity? *Science* 264:1412–1413.

MacArthur, R. H., and E. O. Wilson. 1967. *The Theory of Island Biogeography.* Princeton University Press, Princeton, N.J.

Magurran, A. E. 1988. *Ecological Diversity and Its Measurement.* Princeton University Press, Princeton, N.J.

Paine, R. T. 1980. Food webs: Linkage, interaction strength and community infrastructure. *Journal of Animal Ecology* 49:667–685.

Palmer, M. W., and P. S. White. 1994. Scale dependence and the species-area relationship. *American Naturalist* 144:717–740.

Pimm, S. L. 1982. *Food Webs.* Chapman & Hall, London and New York.

Pimm, S. L. 1991. *The Balance of Nature?* University of Chicago Press, Chicago.

Reice, S. R. 1994. Nonequilibrium determinants of biological community structure. *American Scientist* 82:424–435.

Ricklefs, R. E., and D. Schluter (eds.). 1993. *Species Diversity in Ecological Communities: Historical and Geographical Perspectives.* University of Chicago Press, Chicago.

Risser, P. G. 1995. The status of the science of examining ecotones. *BioScience* 45:318–325.

Terborgh, J. 1985. The role of ecotones in the distribution of Andean birds. *Ecology* 66:1237–1246.

Whittaker, R. H. 1953. A consideration of climax theory: The climax as a population and pattern. *Ecological Monographs* 23:41–78.

Whittaker, R. H. 1967. Gradient analysis of vegetation. *Biological Reviews* 42: 207–264.

CHAPTER 23

COMMUNITY DEVELOPMENT

Communities exist in a state of continuous flux. Organisms die and others are born to take their places; energy and nutrients pass through the community. Yet the appearance and composition of most communities do not change appreciably over time. Oaks replace oaks, squirrels replace squirrels, and so on, in continual self-perpetuation. But when a habitat is disturbed—a forest cleared, a prairie burned, a coral reef obliterated by a hurricane—the community slowly rebuilds. Pioneering species adapted to disturbed habitats are successively replaced by others until a community often attains its former structure and composition (Figure 23.1).

The sequence of changes initiated by disturbance is called **succession,** and the ultimate association of species achieved is called a **climax** community. These terms describe natural processes that caught the attention of early ecologists, including Frederick Clements. By 1916, Clements had outlined the basic features of succession, supporting his conclusions with detailed studies of change in plant communities in a variety of environments. Since

Figure 23.1 Stages of succession in an oak-hornbeam forest in southern Poland. From (a) to (f), time since clear-cutting progresses from immediately after clearing (a) to 7, 15, 30, 95, and 150 years. Photographs by Z. Glowacinski, courtesy of O. Jarvinen. From Z. Glowacinski and O. Jarvinen, *Ornis Scand.* 6:33–40 (1975).

then, the study of community development has grown to include the processes that underlie successional change, adaptations of organisms to the different conditions of early and late succession, and interactions between colonists and the species that replace them. Ecologists have come to realize that succession and community perpetuation differently express the same processes and, further, that so-called climax communities actually consist of patchwork quilts of successional stages following upon localized disturbances. In many cases, alternative "climax communities" exist; successional sequences may lead to different climaxes, depending on conditions during the development of the sere and on which species, by chance, happen to become established at the outset. Thus, communities may exhibit multiple stable states.

Succession and the concept of the sere

The creation of any new habitat—a plowed field, a sand dune at the edge of a lake, an elephant's dung, a temporary pond left by a heavy rain—attracts a host of species particularly adapted as good invaders. These first colonists are followed by others that are slower to take advantage of a new habitat but are eventually more successful than the pioneering species. In this way, the character of a community changes with time. Successional species themselves change the environment. For example, plants shade the earth's surface, contribute detritus to the soil, and alter soil moisture levels. These changes often inhibit the continued success of the species that cause them and make the environment more suitable for other species, which then exclude those responsible for the change.

The opportunity to observe succession presents itself conveniently in abandoned agricultural fields of various ages. On the Piedmont of North Carolina, bare fields are quickly covered by a variety of annual plants (see Figure 23.10). Within a few years, most of these annuals are replaced by herbaceous perennials and shrubs. Shrubs are followed by pines, which eventually crowd out earlier successional species; pine forests are in turn invaded and then replaced by a variety of hardwood species that constitute the last stage of the successional sequence. Change comes rapidly at first. Crabgrass quickly enters an abandoned field, hardly allowing time for the plow's furrows to smooth over. Horseweed and ragweed dominate the field in the first summer after abandonment, aster in the second, and broomsedge in the third. The pace of succession falls off as slower-growing plants appear: the transition to pine forest requires 25 years, and another century must pass before the developing hardwood forest begins to resemble the natural climax vegetation of the area.

The transition from abandoned field to mature forest is only one of several successional sequences that may lead to the same climax within a given region. In various regions of the eastern United States and Canada, a particular kind of forest is the end point of several different successional series, or **seres,** each of which has a different beginning. For example, the sequence of species on newly formed sand dunes at the southern end of Lake Michigan in Indiana differs from the sere that develops on abandoned fields a few miles away. Sand dunes are first invaded by marram and bluestem grasses. Individuals of these species established in soils at the edge of a dune send out rhizomes (runners) under the surface of the sand, from which new shoots sprout. These grasses stabilize a dune's surface and add organic detritus to the sand. Numerous annuals follow perennial grasses onto the dunes, further enriching and stabilizing them and gradually creating conditions suitable for shrub species. Sand cherry, dune willow, bearberry, and juniper form shrub layers before pines take hold. As in abandoned fields in North Carolina, pines do not reseed well after initial

(a)

(b)

Figure 23.2 Initial stages of plant succession on sand dunes along the coast of Maryland. (a) Beach grass on the frontal side of a dune. This grass is used widely to stabilize dune surfaces. (b) Invasion of back dune areas by bayberry and beach plum. Courtesy of the U.S. Soil Conservation Service.

establishment and persist for only one or two generations. In northern Indiana, pines give way in the end to the forests of beech, oak, maple, and hemlock that are characteristic of the region.

Succession follows a similarly predictable course on Atlantic coastal dunes, where the beach grass that initially stabilizes dune surfaces is followed by bayberry, beach plum, and other shrubs. Shrubs act like the snow fencing often used to keep dunes from blowing out; they are called dune-builders because they intercept blowing sand and cause it to pile up around their bases (Figure 23.2). Succession in nearby estuaries leading to the establishment of terrestrial communities begins with salt-tolerating plants and progresses as sediments and detritus build the soil surface above the water level. Whether succession begins on sand or marsh, the vegetation eventually resembles that of upland sites in the vicinity. Many different seres tend toward the same climax within a region.

Primary succession

Beginning with Clements's classic work on succession, published in 1916, ecologists have classified seres into two types according to their origin. The establishment and development of plant communities in newly formed habitats previously lacking plants—sand dunes, lava flows, rock bared by erosion or landslides or exposed by receding glaciers—is called **primary succession.** The return of vegetation to its former state following a disturbance is called **secondary succession.** The distinction between the two blurs because disturbances vary in the degree to which they destroy the fabric of a community and its physical support systems. A tornado that levels a large area of forest usually leaves intact the soil's bank of nutrients, seeds, and sproutable roots. In contrast, a severe fire may burn through organic layers of the soil, destroying hundreds or thousands of years of biologically mediated development.

Species colonizing thin deposits of clay left by receding glaciers in the Glacier Bay region of southern Alaska (Figure 23.3) must cope with

Figure 23.3 A valley exposed by a receding glacier, visible at top center, in North Tongass National Forest, Alaska. The recently bared rock surfaces at the bottom of the valley just below the glacier have not yet been recolonized by shrubby thickets. Courtesy of the U.S. Forest Service.

Figure 23.4 Stages of bog succession illustrated by a bog formed behind a beaver dam in Algonquin Provincial Park, Ontario, Canada. The open water in the center is stagnant, poor in minerals, and low in oxygen. These conditions result in accumulation of detritus and lead to a gradual filling in of the bog, which passes through stages dominated by shrubs and, later, black spruce.

deficiencies of nutrients, particularly nitrogen, and with stressful wind and cold. Here the sere begins with mat-forming mosses and sedges (grasslike plants) and then progresses through prostrate willows, shrubby willows, alder thicket, and sitka spruce to spruce-hemlock forest. Succession moves ahead rapidly, reaching an alder thicket stage within 10 to 20 years and achieving tall spruce forest within 100 years.

Succession also provides a means by which dry land is reclaimed from certain aquatic habitats, such as **bogs** that form in beaver ponds and deep kettleholes in cool north temperate and subarctic regions. Bog succession begins when rooted aquatic plants become established at the edge of a pond (Figure 23.4). Some species of sedges form mats on the water surface extending out from the shoreline. Occasionally these mats grow completely over a pond before it fills in with sediments, producing a more or less firm layer of vegetation over the water surface, a so-called "quaking bog." Detritus produced by the sedge mat accumulates as layers of organic sediments on the bottom of the pond, where the stagnant water contains little or no oxygen to sustain microbial decomposition. Eventually these sediments become peat, which is used by humans as a soil conditioner and sometimes as a fuel for heating (Figure 23.5). As a bog accumulates sediments and detritus, *Sphagnum* moss and shrubs, such as Labrador tea and cranberry, become established along the edges, themselves adding to the development of a soil with progressively more terrestrial qualities. In northern peatlands, *Sphagnum* is the major contributor to peat accumulation. At the edges of peatlands, shrubs may be followed by black spruce and larch, which eventually give way to climax species of forest trees, including birch, maple, and fir, depending on the locality.

Disturbance and secondary succession

Breaks in the canopy of a forest tend to close over as surrounding individuals take advantage of their new opportunities. A small gap, such as that left by a falling limb, is quickly filled by growth of branches from surrounding trees. A big gap left by a fallen tree may provide saplings in the understory with a chance to reach the canopy and claim a permanent place in the sun. A large area cleared by intense fire may have to be colonized anew by seed blown or carried in from surrounding intact forest. Even when reseeding initiates a successional sequence, the size and type of disturbance influence which species become established first. Some plants require abundant sunlight for germination and establishment, and their seedlings are intolerant of competition from other species. These species usually have strong powers of dispersal; they often have small seeds that are easily blown about and can reach centers of large disturbances inaccessible to members of the climax community.

The influence of gap size on succession has been investigated in several marine habitats, where disturbance and recovery frequently follow upon each other. In southern Australia, Michael Keough investigated the colonization of artificially created patches, ranging in size from 25 to 2,500 cm^2 (5–50 cm on a side), by various subtidal encrusting invertebrates that grow on hard surfaces. The major epifaunal (surface-growing) taxa of the region vary considerably in their colonizing ability and competitive ability, which are generally inversely related (Table 23.1). When Keough created bare patches of different sizes within larger areas of rock occupied by encrusting invertebrates, the exposed areas were quickly filled in by growth from surrounding areas of such highly successful competitors as tunicates and sponges. In this case, patch size had little influence on community development, because the distances from the edges to the centers of the patches (less than 25 cm) were easily spanned by growth. The many bryozoan and polychaete larvae that attempted to colonize these patches were quickly overgrown.

Among isolated patches—hard substrates placed in sand to mimic the shells of *Pinna* clams—patch size had a much greater effect on the pattern of colonization. Just by chance, a few of the small patches failed to be

Figure 23.5 A 1-m vertical section through a peat bed in a filled-in bog in Quebec, Canada. The layers represent accumulations of organic detritus from plants that successively colonized the bog as it was filled. The peat beds are several meters thick.

TABLE 23.1 Life history attributes of the major epifaunal (surface-growing) marine invertebrates at Edithburgh, southern Australia

TAXON	GROWTH FORM	COLONIZING ABILITY	COMPETITIVE ABILITY	CAPACITY FOR VEGETATIVE GROWTH
Tunicates	Colonial	Poor	Very good	Very extensive, up to 1 m^2
Sponges	Colonial	Very poor	Good	Very extensive, up to 1 m^2
Bryozoans	Colonial	Good	Poor	Poor, up to 50 cm^2
Serpulid polychaetes	Solitary	Very good	Very poor	Very poor, up to 0.1 cm^2

Source: M. J. Keough, *Ecology* 65:423–437 (1984).

colonized by tunicates and sponges, which do not disperse well, thereby giving bryozoans and polychaetes a chance to obtain a foothold. Because larger patches make bigger targets, many of these were settled by small numbers of tunicates and sponges, which then spread rapidly and eliminated other species that had colonized along with them. As a result, tunicates and sponges predominated on the larger isolated patches, but bryozoans and polychaetes—which, once established, can obstruct the colonization of tunicate and sponge larvae—dominated many of the smaller patches. In this system, bryozoans and polychaetes are disturbance-adapted species—what botanists call **weeds.** They get into open patches quickly, mature and produce offspring at an early age, and then are often eliminated by more slowly colonizing but superior competitors. Such weedy species require frequent disturbances to stay in the system.

The size of a patch partly determines whether predators and herbivores will be active there; these consumers can affect the course of succession. Some consumers may select larger patches for feeding because they are easy to find and require less travel time between patches. Other consumers that are themselves vulnerable to predators may require the cover of intact habitat, from whose edges they venture to feed in newly exposed areas. In this case, small patches are likely to be grazed more intensively than the centers of large patches.

Rabbits rarely feed far from the cover of brush or trees, so as to avoid being discovered by predators far from safety. Limpets (grazing mollusks) similarly do not venture far from the safety of mussel beds to feed on algae. In an intertidal rocky shore habitat in central California, University of California ecologist Wayne Sousa cleared patches of either 625 cm^2 or 2,500 cm^2 in mussel beds and excluded limpets from half the patches in each of these sets by applying a barrier of copper paint along their edges. He then monitored the colonization of the cleared patches by several species of algae during the following 3 years (Figure 23.6). Limpets live in the crevices between mussels when they are not feeding; doing so enables them to avoid predation. Because this behavior limits their foraging range, densities of limpets in the small patches (surrounded by more edge compared with area) exceeded those in the larger patches. Also, and not surprisingly, throughout the course of the experiment, algae grew more densely in the larger patches. Where limpet grazing was prevented, total cover by all species of algae was high and did not differ between patches of different size.

As we would expect, limpet grazing depressed the establishment and growth of most species of algae but favored three species of rare, presumably inferior, competitors: the brown alga *Analipus,* the green alga *Cladophora,* and the red alga *Endocladia.* These algae have low-lying, crustose growth forms that make them vulnerable to shading and overgrowth by other species, but help them to resist grazing. In addition, establishment of *Endocladia* was sensitive to patch size, generally being more common in larger patches regardless of limpet grazing. Where limpets grazed freely, colonizing mussels, which eventually crowd out all other space-occupying species, reached greater abundance in large patches than in small ones. This difference resulted from the interaction of patch size and limpet grazing, rather than from patch size per se, because mussels colonized areas protected from grazing independently of patch size.

The result of succession: The climax community

Ecologists have traditionally viewed succession as leading inexorably to an ultimate expression of community development, a climax community. Early studies of succession demonstrated that the many seres found within

Figure 23.6 A natural cleared patch in a bed of mussels *(Mytilus californianus)* on the central coast of California. The patch is about 1 m across and has been colonized by a heavy growth of the green alga *Ulva.* Note the distinct browse zone around the perimeter of the patch. It is created by limpets, which feed only short distances away from refuge in the mussel bed. Courtesy of W. P. Sousa. From W. P. Sousa, *Ecology* 65:1918–1935 (1984).

a region, each developing under a particular set of local environmental circumstances, often progress toward the same climax. These observations led to a concept of mature communities as natural units—even as closed systems—which was clearly stated by Frederick Clements in 1916:

> *The developmental study of vegetation necessarily rests upon the assumption that the unit or climax formation is an organic entity. As an organism the formation arises, grows, matures, and dies. Its response to the habitat is shown in processes or functions and in structures which are the record as well as the result of these functions. Furthermore, each climax formation is able to reproduce itself, repeating with essential fidelity the stages of its development. The life history of a formation is a complex but definite process, comparable in its chief features with the life history of an individual plant.* (Carnegie Inst. Wash. Publ. *242:1–512)*

Clements recognized fourteen climaxes in the terrestrial vegetation of North America, including two types of grassland (prairie and tundra), three types of scrub (sagebrush, desert scrub, and chaparral), and nine types of forest ranging from pine-juniper woodland to beech-oak forest. He believed that climate alone determined the nature of the local climax. Aberrations in community composition caused by soils, topography, fire, or animals (especially grazing) represented interrupted stages in the transition toward the local climax—immature communities.

In recent years, the concept of the climax as an organism or a discrete unit has been greatly modified—to the point of outright rejection by many ecologists—because it has become clear that communities are open systems whose composition varies continuously over environmental gradients. In addition, various factors, including the size of a disturbance and physical conditions during early succession, may result in different "climax" communities. Whereas in 1930 plant ecologists described the climax vegetation of much of Wisconsin as a sugar maple–basswood forest, by 1950 ecologists placed this forest type on an open continuum of climax communities extending over both broad, climatically defined regions and local, topographically defined areas. To the south, beech increased in prominence; to the north, birch, spruce, and hemlock were added to the climax community; in drier regions bordering prairies to the west, oaks became prominent. Locally, quaking aspen, black oak, and shagbark hickory, long recognized as successional species on moist, well-drained soils, came to be accepted as climax species on drier upland sites.

Mature stands of forest in Wisconsin, representing the end points of local seres, were ordered by J. T. Curtis and R. P. McIntosh along a **continuum index** based on forest stand composition, ranging from dry sites dominated by oak and aspen to moist sites dominated by sugar maple, ironwood, and basswood. A continuum index for Wisconsin forests was calculated from the species composition of each forest type, and its value varied between arbitrary extremes of 300 for a pure stand of bur oak to 3,000 for a pure stand of sugar maple. Although increasing values of the index correspond to seral stages leading to the sugar maple climax, they

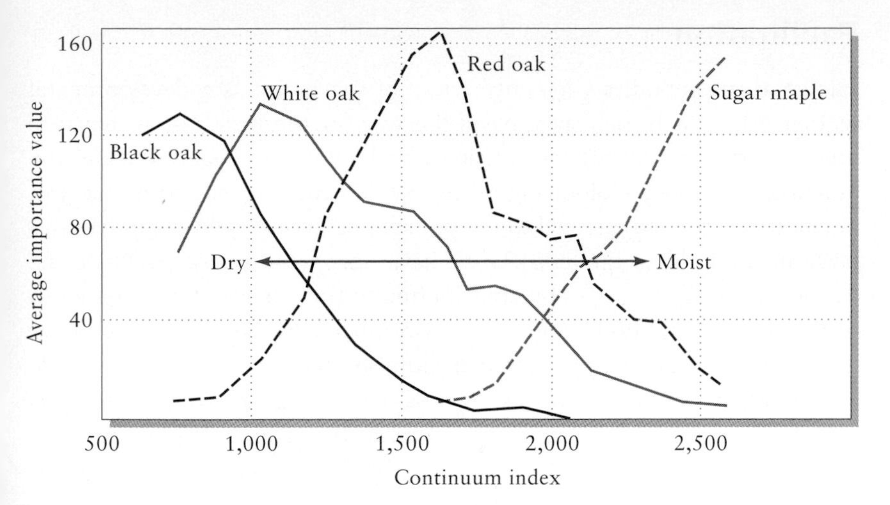

Figure 23.7 Relative importance (a measure of abundance) of several species of trees in forest communities of southwestern Wisconsin, arranged along a continuum index. Soil moisture, exchangeable calcium, and pH increase toward the right on the continuum index. After J. T. Curtis and R. P. McIntosh, *Ecology* 32:476–496 (1951).

may also represent local climax communities determined by topographic or soil conditions. Thus, the so-called climax vegetation of southern Wisconsin actually represents a continuum of forest (and, in some areas, prairie) types (Figure 23.7).

The causes of succession

Two factors determine the position of a species in a sere: the rate at which it invades a newly formed or disturbed habitat, and its response to changes that occur in the environment over the course of succession. Some species disperse slowly, or grow slowly once established, and therefore become dominant late in the sequence of associations in a sere. Rapidly growing plants that produce many small seeds, carried long distances by wind or by animals, have an initial advantage over species that disperse slowly, and they dominate the early stages of a sere. In a habitat that burns frequently, many species have fire-resistant seeds or root crowns that germinate or sprout soon after a fire and quickly reestablish their populations.

Early successional species sometimes modify environments in such a way as to allow later-stage species to become established. The growth of herbs on a cleared field shades the soil surface and helps the soil to retain moisture, providing conditions more congenial to the establishment of less tolerant plants. Conversely, colonization by some species may inhibit the entrance of others into a sere, either by superior competition for limiting resources or by direct interference.

This diverse array of processes governing the course of succession was classified by American ecologist Joseph Connell and Australian ecologist R. O. Slatyer into three categories of mechanisms—facilitation, inhibition, and tolerance—which relate to the consequences of early stages of succession for the development of later stages. The terms facilitation, inhibition, and tolerance describe the effect of the presence of one species on the probability of establishment of a second, and whether that effect is positive, negative, or neutral.

Facilitation

Facilitation embodies Clements's view of succession as a developmental sequence in which each stage paves the way for the next, just as structure follows structure during an organism's development—and during the building of a house. Colonizing plants enable climax species to invade, just as wooden forms are essential to the pouring of a concrete wall but have no place in the finished building. As we have noted, early stages facilitate the development of later stages by contributing to the nutrient and water levels of the soil and by modifying microenvironments at the soil surface. Alder trees *(Alnus),* which harbor nitrogen-fixing bacteria in their roots, provide an important source of nitrogen to soils developing on sandbars in rivers and in areas exposed by retreating glaciers. Black locust plays the same role in early succession in the southern Appalachian region of the United States, as does the introduced tree *Myrica faya* on lava flows in Hawaii.

Soils do not develop in marine systems, but facilitation often occurs when one species enhances the quality of a site for the settling and establishment of another. Working with experimental panels placed subtidally in Delaware Bay, T. A. Dean and L. E. Hurd found that for some species combinations, the presence of one species inhibited establishment of a second, but that hydroids enhanced settlement of tunicates, and both hydroids and tunicates facilitated settlement of mussels. In southern California, early-arriving, fast-growing algal stands provide dense protective cover for re-establishment of kelp plants following disturbance by winter storms. In areas kept clear of early successional species of algae, grazing fish quickly removed settling kelp sporophytes. In a parallel manner, establishment of the surfgrass *Phyllospadix scouleri* in rocky intertidal communities depends on the presence of certain early successional algae to which its seeds cling before germinating (Figure 23.8). In the absence of these algae, the seagrass cannot invade the community.

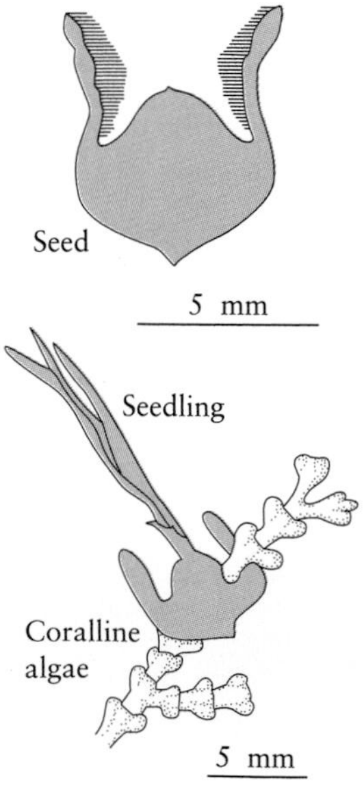

Seed
5 mm
Seedling
Coralline algae
5 mm

Figure 23.8 Seeds of the surfgrass *Phyllospadix* have barbs that enable them to become attached to certain types of erect algae. After T. Turner, *Am. Nat.* 121: 729–738 (1983).

Inhibition

Inhibition of one species by the presence of another is a common phenomenon that we have discussed in detail in the chapters on competition and predation. One species may inhibit another by eating it, by reducing resources to a level the second species cannot subsist on, or by confronting it with noxious chemicals or antagonistic behavior. With respect to succession, climax species by definition inhibit species characteristic of earlier stages: the latter cannot invade a climax community except following disturbance.

Because inhibition is so intimately related to species replacement, it forms an integral part of the orderly succession from the early stages of a sere through the climax. Inhibition can give rise to an interesting situation when the outcome of an interaction between two species depends on which becomes established first. Colonizing propagules are often the most sensitive stage of the life history, and sometimes neither species of a pair can become established in the presence of competitively superior adults of the other. In this case, the course of succession depends on precedence. Precedence, in turn, may be strictly random, depending on which species reaches a disturbed site first, or it may depend on certain properties of a disturbed site—its size, location, the season, and so on. We have seen such a case in subtidal habitats of southern Australia, where bryozoans, when they become established first, can prevent the establishment of tunicates and sponges. Because of their stronger powers of dispersal, this is more likely to happen on small, isolated substrates than elsewhere.

According to Connell and Slatyer's inhibition model, succession follows upon the establishment of one species or another only through the death and replacement of established individuals. Thus, just by chance, successional change moves toward predominance of longer-lived species.

Tolerance

The **tolerance** model of Connell and Slatyer holds that some species invade newly exposed habitat and become established independently of the presence or absence of other species, depending only upon their dispersal abilities and the physical conditions of the environment. The ensuing sere is then determined by the process of competitive exclusion—that is, by the life spans and competitive abilities of the colonists as they grow. Early stages will be dominated by poor competitors that have short life cycles but become established quickly; superior competitors will constitute the climax species, but they may grow more slowly and may not express their dominance in the sere until the others have grown up and reproduced.

An example: Old-field succession on the Piedmont of North Carolina

Clearly, all three of Connell and Slatyer's mechanisms—facilitation of establishment, inhibition of establishment, and competitive exclusion (replacement of established populations)—together with the life history characteristics of successional species, are important factors in every sere; none operates exclusively of the others. The early stages of plant succession on old fields in the Piedmont region of North Carolina (Figure 23.9)

Figure 23.9 An old field on the Piedmont of North Carolina. Such habitats develop after abandonment of agricultural land.

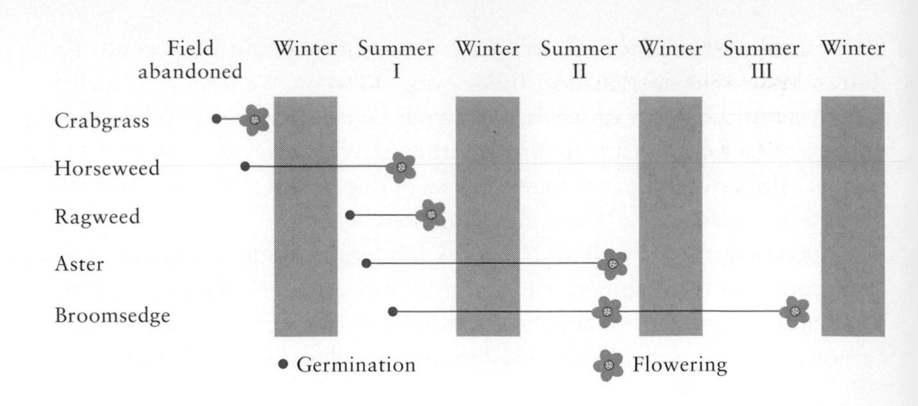

Figure 23.10 A schematic summary of the life histories of five early successional species of plants that colonize abandoned fields in North Carolina.

demonstrate how these factors combine in a particular sere. The first 3 to 4 years of old-field succession are dominated by a small number of species that replace one another in rapid sequence: crabgrass, horseweed, ragweed, aster, and broomsedge. The life history cycle of each species partly determines its place in the succession (Figure 23.10). Crabgrass, a rapidly growing annual, is usually the most conspicuous plant in a cleared field during the year in which the field is abandoned. Horseweed is a winter annual whose seeds germinate in autumn. Through winter, the plant exists as a small rosette of leaves; it blooms by the following midsummer. Because horseweed disperses well and develops rapidly, it usually dominates 1-year-old fields. But because its seedings require full sunlight, horseweed is quickly replaced by shade-tolerant species. Thus, early succession is dominated by tolerance—colonizing species disperse readily and tolerate the harsh conditions of newly exposed ground—but rapidly shifts to inhibition.

Ragweed is a summer annual; its seeds germinate early in spring and the plants flower by late summer. Ragweed dominates the first summer of succession in fields that are plowed under in late autumn, after horseweed normally germinates. Aster and broomsedge are biennials that germinate in spring and early summer, exist through winter as small plants, and bloom for the first time in their second autumn. Broomsedge persists and flowers during the following autumn as well.

Horseweed and ragweed both disperse their seeds efficiently and, as young plants, tolerate desiccation. These abilities enable them to invade cleared fields rapidly and produce seed before competitors become established. Decaying horseweed roots stunt the growth of horseweed seedlings, so the species is self-limiting in the successional sequence. Such growth inhibitors presumably are by-products of other adaptations that increase the fitness of horseweed during the first year of succession; or, perhaps, because horseweed plants have little chance of persisting during the second year as a result of invasion of the sere by superior competitors, self-inhibition has little negative selection value. Regardless, this phenomenon is fairly common in early stages of succession.

Aster successfully colonizes recently cleared fields, but grows slowly and does not dominate old-field habitats until the second year. The first aster plants to colonize a field thrive in full sunlight; their seedlings, however, are not shade-tolerant, and the adult plants shade their progeny out of existence. Furthermore, aster does not compete effectively with broomsedge for soil moisture. Catherine Keever observed this when she cleared a circular area, 1 m in radius, around several broomsedge plants and planted aster seedlings at various distances from them (Figure 23.11). Those closest to the broomsedge plants grew poorly because of reduced soil water availability.

Approaching the climax

Succession continues until the addition of new species to the sere and the exclusion of established species no longer change the environment of the developing community. The progression of different growth forms modifies conditions of light, temperature, moisture, and soil nutrients. Replacement of grasses by shrubs and then by trees on abandoned fields brings a corresponding modification of the physical environment that facilitates the establishment of species whose seedlings cannot tolerate heat and dry soil. Conditions change more slowly, however, when the vegetation achieves the tallest growth form that the environment can support. The final biomass dimensions of a climax community are limited by climate independently of events during succession.

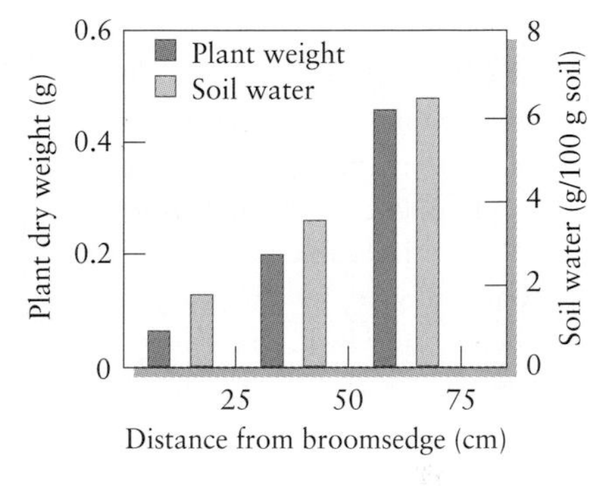

Figure 23.11 Growth response of aster (dry weight) and soil water as a function of distance from broomsedge plants in an old field. From C. Keever, *Ecol. Monogr.* 20:230–250 (1950).

Once forest vegetation establishes itself, patterns of light intensity and soil moisture do not change, except in the smallest details, with the introduction of new species of trees. For example, beech and maple replace oak and hickory in northern hardwood forests because their seedlings are better competitors in the shade of the forest floor environment, but beech and maple seedlings probably develop as well under their own parents as they do under the oak and hickory trees they replace. At this point, succession reaches a climax; the community has come into equilibrium with its physical environment.

To be sure, changes in species composition may follow attainment of a climax growth form by a sere. For example, a site near Washington, D.C., that was left undisturbed for nearly 70 years developed a tall forest community dominated by oak and beech. The community had not, however, reached an equilibrium at the time it was studied, because the youngest individuals—the saplings in the forest understory that will eventually replace existing trees—included neither white nor black oak. In another century, this forest will probably be dominated by species with the most vigorous reproduction: red maple, sugar maple, and beech (Figure 23.12).

The time required for succession to proceed from a disturbed habitat to a climax community varies with the nature of the climax and the initial quality of the soil. Obviously, succession is slower to gain momentum when it starts on bare rock than when it starts on recently exposed soil. A mature oak-hickory forest climax will develop within 150 years on a cleared field in North Carolina. Climax stages of western grasslands are reached in 20 to 40 years of secondary succession. In the humid Tropics, forest communities regain most of their climax elements within 100 years after clear-cutting, provided that the soil is not abused by farming or prolonged exposure to sun and rain. Primary succession usually proceeds much more slowly. Radiocarbon dating methods suggest that beech-maple

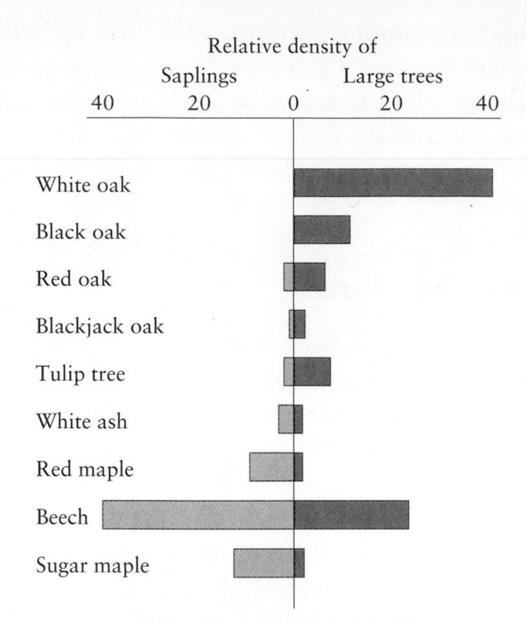

Figure 23.12 Composition of a forest undisturbed for 67 years near Washington, D.C. The relative predominance of beech and maple saplings in the understory foretells a gradual successional change in the community beyond the present oak-beech stage. After R. L. Dix, *Ecology* 30:663–665 (1957).

climax forest requires up to 1,000 years to develop on Lake Michigan sand dunes.

Ecologists generally agree that communities are more diverse and complex at intermediate stages of succession. We do not know whether this increase in the diversity of a community during its early stages of succession is related to increased production, greater constancy of physical characteristics of the environment, or greater structural heterogeneity of the habitat. Furthermore, we have no reason to suspect that gradients of diversity along a successional continuum respond to the same factors that determine diversity along a structurally analogous gradient of mature communities.

The biological properties of a developing community change as species enter and leave a sere. As a community matures, the ratio of biomass to productivity increases. The maintenance requirements of the community also increase until production can no longer meet demand, at which point net accumulation of biomass in the community stops. The end of biomass accumulation does not necessarily signal the attainment of climax; species may continue to invade a community and replace others whether the biomass of the community increases or not. Attainment of steady-state biomass does mark the end of major structural change in the community, and further changes are usually limited to the adjustment of details.

As plant size increases with succession, a greater proportion of the nutrients available to a community come to reside in organic materials. Furthermore, because the vegetation of mature communities has more supportive tissue, which is less readily digestible than photosynthetic tissue, a larger proportion of their productivity enters detritus food chains rather than consumer food chains. Other aspects of the community change as well. Forest soils hold nutrients more tightly because tree roots protect soils from erosion. The well-developed root systems of trees take up minerals more rapidly and store them to a greater degree than do the root systems of early successional plants. The forest canopy protects the environment near the ground from extremes of heat and humidity, and conditions in the litter are more favorable to detritus-feeding organisms.

Species characteristics through the sere

Succession in terrestrial habitats entails a regular progression of plant forms. Plants characteristic of early stages of succession and those typical of late stages employ different strategies of growth and reproduction. Early-stage species capitalize on their high dispersal ability to rapidly colonize newly created or disturbed habitats. Climax species disperse and grow more slowly, but their shade tolerance as seedlings and their large size as mature plants give them a competitive edge over early successional species. Plants of climax communities are adapted to grow and prosper in the environment they create, whereas early successional species are adapted to colonize unexploited environments. The progression of successional species is therefore accompanied by a shift in the balance between adaptations promoting dispersal and adaptations enhancing competitive ability.

Some characteristics of early and late successional plants are compared in Table 23.2. To enhance their colonizing ability, early seral species produce many small seeds that are usually wind-dispersed (dandelion and milkweed, for example). Their seeds can remain dormant in the soils of forest and shrub habitats for years until fires or treefalls create the bare-soil conditions required for their germination and growth. The seeds of most climax species, being relatively large, provide their seedlings with ample nutrients to get started in the highly competitive environment of the forest floor.

TABLE 23.2 **General characteristics of early and late successional plants**

CHARACTERISTIC	EARLY	LATE
Number of seeds	Many	Few
Seed size	Small	Large
Dispersal	Wind, stuck to animals	Gravity, eaten by animals
Seed viability	Long, latent in soil	Short
Root:shoot ratio	Low	High
Growth rate	Rapid	Slow
Mature size	Small	Large
Shade tolerance	Low	High

The survival of seedlings in shade is directly related to seed weight (Figure 23.13). The ability of seedlings to survive the shady conditions of climax habitats is inversely related to their ability to grow rapidly in the direct sunlight of early successional habitats. When placed in full sunlight, early successional herbaceous species grew ten times more rapidly than shade-tolerant trees. Shade-intolerant trees, such as birch and red maple, had intermediate growth rates. Thus, plants must balance shade tolerance and growth rate against each other; each species must reach a compromise between those adaptations that best suits it for survival in a sere.

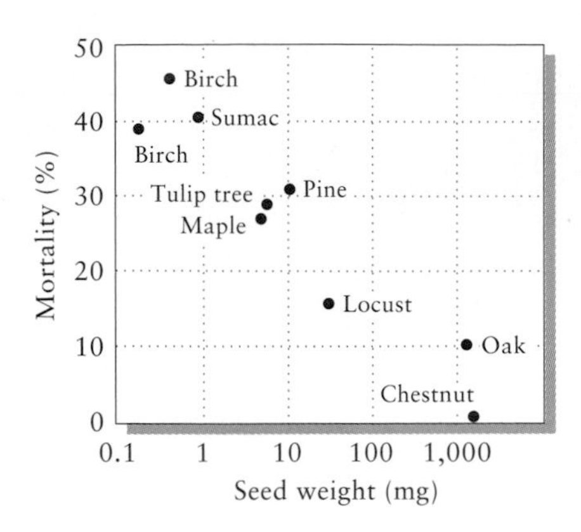

Figure 23.13 Relationship between seed weight and mortality of seedlings after 3 months under shaded conditions. After J. P. Grime and D. W. Jeffrey, *J. Ecol.* 53:621–642 (1965).

The rapid growth of early successional species results in part from their allocation of a relatively large proportion of seedling biomass to stems and leaves. A plant's allocation of tissue between roots and aboveground parts (shoots) influences its growth rate. Leaves sustain photosynthesis, and their productivity determines the net accumulation of plant tissue during growth. Thus the allocation of a large proportion of production to shoot biomass by early successional plants leads to rapid growth and production of large crops of seeds. Because annual plants must produce seeds quickly and abundantly, they never attain large size. Climax species allocate a larger proportion of their production to root and stem tissue to increase their competitive ability; thus they grow more slowly. In the seedlings of annual herbaceous species, the shoot typically accounts for 80–90% of the entire plant; in biennials, 70–80%; in herbaceous perennials, 60–70%; and in woody perennials, 20–60%.

The character of the climax community

Clements's idea that a region has only one true climax (the **monoclimax theory**) forced botanists to recognize a hierarchy of interrupted or modified seres by attaching such names as subclimax, preclimax, and postclimax. This terminology naturally gave way to a polyclimax viewpoint, which recognized the validity of many different types of vegetation as climaxes, depending on local conditions. More recently, the development of the continuum index and gradient analysis fostered the broader **pattern-climax theory** of Robert Whittaker, which recognizes a regional pattern of open climax communities whose composition at any one locality depends on particular environmental conditions at that point.

Many factors determine the composition of a climax community, among them soil nutrients, moisture, slope, and exposure. Fire is an important feature of many climax communities, favoring fire-resistant species and excluding species that otherwise would dominate. The vast southern pine forests along the Gulf Coast and southern Atlantic Coast of the United States are maintained by periodic fires. Pines have become adapted to withstand scorching under conditions that destroy oaks and other broad-leaved species (Figure 23.14). Some species of pines do not even shed their seeds unless triggered by the heat of a fire passing through the understory below. After a fire, pine seedlings grow rapidly in the absence of competition from other understory species.

(a)

(b)

(c)

Figure 23.14 (a) A stand of longleaf pine in North Carolina shortly after a fire. Although the seedlings are badly burned (b), the growing shoot is protected by the dense, long needles (shown on an unburned individual, c) and often survives. In addition, the slow-growing seedlings have extensive roots that store nutrients to support the plant following fire damage.

Any habitat that is occasionally dry enough to create a fire hazard but is normally wet enough to produce and accumulate a thick layer of plant detritus is likely to be influenced by fire. Chaparral vegetation in seasonally dry habitats in California is a fire-maintained climax that gives way to oak woodland in many areas when fire is prevented. The forest-prairie edge in the midwestern United States separates "climatic climax" and "fire climax" communities—terms that refer to the dominating effects of climate and fire, respectively, in determining their species composition. Frequent burning kills seedlings of hardwood trees, but perennial prairie grasses sprout from their roots after a fire. The forest-prairie edge occasionally shifts back and forth across the countryside, depending on the intensity of recent drought and the extent of recent fires. After prolonged wet periods, the

Figure 23.15 Zebras and Thompson's gazelles feed side by side in the Serengeti ecosystem of East Africa, but eat different plants. The gazelles prefer to feed in areas previously grazed by wildebeests and other large herbivores.

forest edge advances out onto the prairie as tree seedlings grow up and begin to shade out grasses. Prolonged drought followed by intense fire can destroy tall trees and permit rapidly spreading prairie grasses to gain a foothold. Once prairie vegetation establishes itself, fires become more frequent because of the rapid buildup of flammable litter. Reinvasion by forest species then becomes more difficult. By the same token, mature forests resist fire and are rarely damaged enough to allow encroachment of prairie grasses. Hence the stability of the forest-prairie boundary.

Grazing pressure also can modify a climax. Grassland can be turned into shrubland by intense grazing. Herbivores may kill or severely damage perennial grasses and allow shrubs and cacti that are unsuitable for forage to establish themselves. Most herbivores graze selectively, suppressing favored species of plants and bolstering competitors that are less desirable as food. On African plains, grazing ungulates make up a regular succession of species through an area, each using different types of forage. When wildebeests, the first of the successional species, were experimentally excluded from some areas, the subsequent wave of Thompson's gazelles preferred to feed in areas previously used by wildebeests or other large herbivores (Figure 23.15). Apparently, heavy grazing by wildebeests stimulates the growth of food plants that gazelles prefer and reduces cover within which predators of the smaller herbivores could conceal themselves. In western North America, grazing allows invasion of the alien cheatgrass *(Bromus tectorum),* which promotes fire and may lead succession to an alternative stable state.

Transient and cyclic climaxes

We usually view succession as a series of changes leading to a climax, whose character is determined by, and that exists in equilibrium with, the local environment. Once established, beech-maple forest perpetuates itself, and its general appearance does not change despite constant replacement of individuals within the community. Yet not all climaxes persist. Simple cases of **transient climaxes** include the development of animal and plant communities in seasonal ponds—small bodies of water that either dry up in summer or freeze solid in winter and thereby regularly destroy the communities that become established in them each year. Each spring the ponds are restocked either from larger, permanent bodies of water or from resting stages left by plants, animals, and microorganisms before the habitat disappeared in the previous year.

Succession recurs whenever a new environmental opportunity appears. For example, excreta and dead organisms provide resources for a variety of scavengers and detritus feeders. On African savannas, carcasses of large mammals are devoured by a succession of vultures (Figure 23.16), beginning with large, aggressive species that gorge themselves on the largest masses of flesh, followed by smaller species that glean smaller bits of meat from the bones, and ending with a kind of vulture that cracks open bones to feed on marrow. Scavenging mammals, maggots, and microorganisms enter the sequence at different points and ensure that nothing edible remains. This succession has no climax because all the scavengers disperse when the feast concludes. We may, however, consider all the scavengers a part of a climax: the entire savanna community.

In simple communities, the particular life history characteristics of a few dominant species can create a **cyclic climax.** Suppose, for example, that species A can germinate only under species B, B only under C, and C only under A. This situation creates a regular cycle of species dominance in the order A, C, B, A, C, B, A, . . . , the length of each stage being determined by the life span of the dominant species. Stable cyclic climaxes usually follow such a scheme, often with one stage being bare substrate. Wind or frost heaving sometimes drives such a cycle. When heaths and other types of vegetation suffer extreme wind damage, shredded foliage and broken twigs create openings for further damage, and the process becomes self-accelerating. Soon a wide swath is opened in the vegetation; regeneration occurs on the protected side of the damaged area while wind damage further encroaches upon exposed vegetation. As a result, waves of damage and regeneration move through a community in the direction of the wind

Figure 23.16 Vultures feeding on a wildebeest carcass in Masai Mara Park, Kenya.

(Figure 23.17). If we watched the sequence of events at any one location, we would witness a healthy heath being reduced to bare earth by wind damage and then regenerating in repeated cycles (Figure 23.18). Similar cycles occur in windy regions where hummocks, or small mounds of earth, form around the bases of clumps of grasses. As these hummocks grow, the soil becomes more exposed and better drained. As the soil dries out, shrubby lichens take over a hummock and exclude the grasses around which the hummock formed. However, the shrubby lichens are worn down by wind erosion, eventually giving way to prostrate lichens, which resist wind erosion but, lacking roots, cannot hold soil. Eventually the hummocks wear away completely, and grasses once more become established and renew the cycle.

Mosaic patterns of vegetation types typify any climax community where deaths of individuals alter the environment. Treefalls open a forest canopy and create patches of habitat that are dry, hot, and sunlit compared with the forest floor under unbroken canopy. These openings are often invaded by early seral forms, which persist until the canopy closes. Thus, treefalls create a shifting mosaic of successional stages within an otherwise uniform community. Indeed, adaptation by different species to growing in particular conditions created by different-sized openings in the canopy could enhance the overall diversity of a climax community. Similar ideas have developed about intertidal regions of rocky coasts, where wave damage and intense predation continuously open new patches of habitat.

Figure 23.17 Waves of wind damage and regeneration in balsam fir forests on the slopes of Mt. Katahdin, Maine. Courtesy of D. G. Sprugel. From D. G. Sprugel and F. H. Bormann, *Science* 211:390–393 (1981).

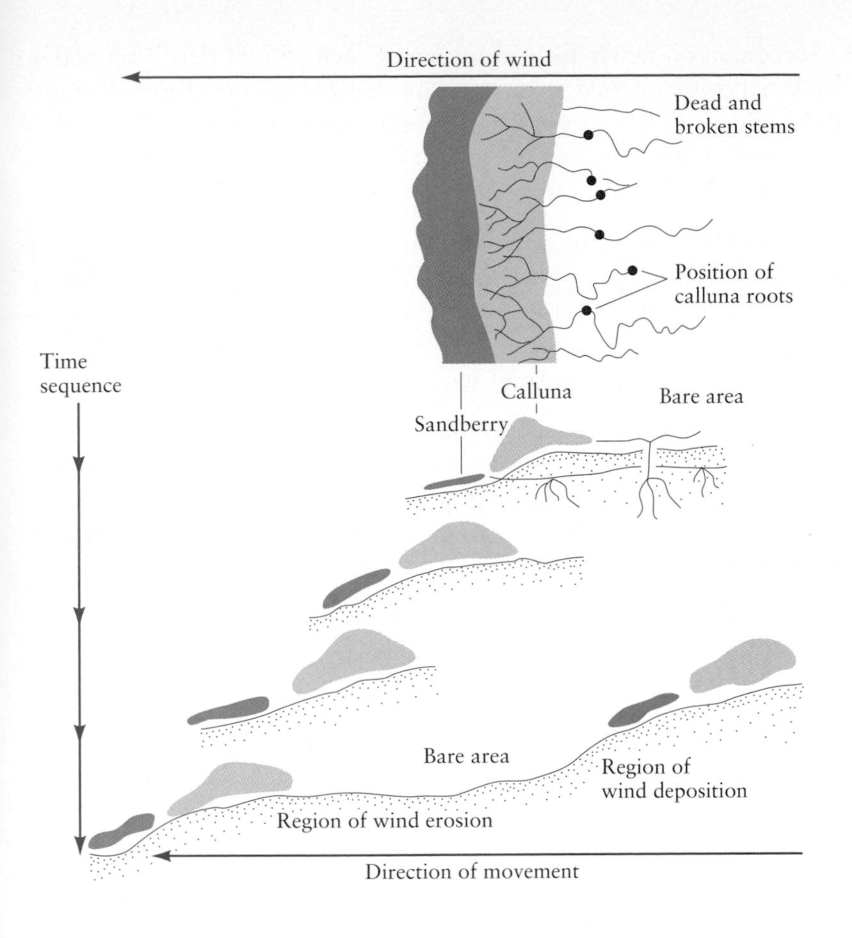

Figure 23.18 Sequence of wind damage and regeneration in dwarf heaths of northern Scotland. At the top is a view of a heath from above, showing a band of sandberry growing in the protected lee of the calluna heath and the wind damage to the heath on its upwind side (to the right). Below is a side view of the band of sandberry and heath as it appears to "migrate" downwind over time. After A. S. Watt, *J. Ecol.* 35:1–22 (1947).

The concept of the community climax must include cyclic patterns of changes, mosaic patterns of distribution, and alternative stable states. The climax is a dynamic state, self-perpetuating in composition, even if by regular cycles of change. Persistence is the key to the climax, and a persistent cycle defines a climax as well as an unchanging steady state does.

Succession emphasizes the dynamic nature of biological communities. By upsetting their natural balance, disturbance reveals to us the forces that determine the presence or absence of species within a community and the processes responsible for regulating community structure. Succession also emphasizes the idea that communities often comprise patchwork mosaics of successional stages and that community studies must consider disturbance cycles on many scales of time and space.

Summary

1. Succession is community change at a locality following either habitat disturbance or the exposure of new substrate. The particular sequence of communities at a given location is referred to as a sere, and the ultimate stable association of plants and animals that is achieved is called a climax.

2. Succession on newly formed substrates, such as sand dunes, landslides, and lava flows—referred to as primary succession—involves substantial modification of the environment by early colonists. Moderate disturbances, which leave much of the physical structure of the ecosystem intact, are followed by secondary succession.

3. The initial stages of the sere depend on the intensity and extent of the disturbance, but its end point reflects climate and topography—that is, within a region, seres tend to converge on a single climax. However, variations in the area of a disturbance and in conditions during early stages of succession may lead to alternative stable states.

4. Especially in secondary succession, a species' entrance into and persistence in a sere depend on its colonizing and competitive abilities. Members of early stages tend to disperse well and grow rapidly; those of later stages tend to tolerate low resource levels or to dominate direct interactions with other species.

5. Joseph Connell and R. O. Slatyer categorized the processes that govern succession as facilitation, inhibition, and tolerance. All three involve the effect of one established species on the probability of colonization by a second, potential invader.

6. Facilitation predominates in early stages of primary succession. Inhibition is a more common feature of secondary succession. It may be expressed in precedence effects, conferring competitive dominance on, for example, the first arrival.

7. The tolerance concept emphasizes the differing abilities of species to tolerate conditions of the environment as they change through succession, and downplays the effects of other species on their establishment.

8. Succession continues until a community is dominated by species capable of becoming established in their own and one another's presence. At this point the community becomes self-perpetuating.

9. In general, biomass increases during succession, whereas net production and diversity tend to be greatest in the middle stages.

10. Characteristics of species vary according to their place in a sere, so the overall structure and function of the community changes accordingly. Pioneering species tend to have many small seeds that disperse easily, have shade-intolerant seedlings that grow rapidly, and reach maturity quickly; late-stage species have the opposite features.

11. The character of the climax may be influenced profoundly by local conditions, such as fire and grazing, that alter interactions among seral species.

12. Transient climaxes develop on ephemeral resources and habitats, such as vernal pools and the carcasses of individual animals. In such cases, we may think of a regional climax as including transient successional sequences.

13. Cyclic local climaxes may develop where each species can become established only in association with some other one. Cyclic climaxes are often driven by harsh physical conditions, such as frost and strong winds. In this case the regional climax includes any local cyclic seres.

Suggested readings

Callaway, R. M., and F. W. Davis. 1993. Vegetation dynamics, fire, and the physical environment in coastal central California. *Ecology* 74:1567–1578.

Christensen, N. L., and R. K. Peet. 1984. Convergence during secondary forest succession. *Journal of Ecology* 72:25–36.

Connell, J. H., and R. O. Slatyer. 1977. Mechanisms of succession in natural communities and their role in community stability and organization. *American Naturalist* 111:1119–1144.

Grubb, P. J. 1977. The maintenance of species diversity in plant communities: The importance of the regeneration niche. *Biological Reviews* 52:107–145.

Keever, C. 1950. Causes of succession on old fields of the Piedmont, North Carolina. *Ecological Monographs* 20:230–250.

Keough, M. J. 1984. Effects of patch size on the abundance of sessile marine invertebrates. *Ecology* 65:423–437.

McIntosh, R. P. 1985. *The Background of Ecology: Concept and Theory.* Cambridge University Press, Cambridge and New York.

Pickett, S. T. A., and P. S. White (eds.). 1985. *The Ecology of Natural Disturbance and Patch Dynamics.* Academic Press, Orlando, Fla.

Riggan, P. J., S. Goode, P. M. Jacks, and R. N. Lockwood. 1988. Interaction of fire and community development in chaparral of southern California. *Ecological Monographs* 58:155–176.

Sousa, W. P. 1984. Intertidal mosaics: Patch size, propagule availability, and spatially variable patterns of succession. *Ecology* 65:1918–1935.

Vitousek, P. M., and L. P. Walker. 1989. Biological invasion by *Myrica faya* in Hawaii, USA: Demography, nitrogen fixation, ecosystem effects. *Ecological Monographs* 59:247–266.

Watt, A. S. 1947. Pattern and process in the plant community. *Journal of Ecology* 35:1–22.

West, D. C., H. H. Shugart, and D. B. Botkin (eds.). 1981. *Forest Succession: Concepts and Application.* Springer-Verlag, New York.

CHAPTER 24

BIODIVERSITY

Comparisons of communities of plants and animals have revealed certain patterns that suggest that many community properties are regulated by physical and ecological processes. One example of such a pattern is the trophic organization of communities, in which the laws of thermodynamics dictate that the energy available to each trophic level decreases at the next higher level in the food chain. This organizing principle produces certain regularities in the distribution of numbers of individuals and biomass among trophic levels within communities: predators generally are less numerous than herbivores, for example.

Ecologists have also noted patterns in communities that appear to be indifferent to energetic constraints. The most important of these include certain regularities in numbers of species within communities, commonly referred to as species diversity. As we noted earlier, large islands tend to support more species than small islands, suggesting that diversity is somehow regulated with respect to area or to some ecological factor correlated with area. To cite another example, biologists have found more kinds of organisms in the Tropics than at higher latitudes, even when comparisons are made between communities with similar levels of biological productivity.

The great naturalist explorers of the nineteenth century—Charles Darwin, Henry W. Bates, Alfred Russel Wallace, and others—recognized that the Tropics held a great store of undescribed species, many having bizarre forms and habits. This remains true to the present. Taxonomists so far have cataloged fewer than 2 million species. But by extrapolating rates of discovery of new insects and other life forms, some biologists have estimated that as many as 30 million species of animals and plants may inhabit the earth, most of them small insects in tropical forests.

Why are there so many different kinds of organisms in the Tropics (and why are there so few toward the poles)? The factors that regulate the diversity of natural communities, and presumably provide answers to such questions, are the subject of this chapter. Biologists hold two views on the issue of diversity. One maintains that diversity increases without limit over time; thus, tropical habitats, being much older than temperate and arctic habitats, have had time to accumulate more species. The second view holds that diversity reaches an equilibrium at which factors that remove species from a system balance those that add species. Accordingly, factors that add species would have to weigh more heavily in the balance, or factors that remove species would have to weigh less heavily, as one moves toward the Tropics.

Throughout the first half of the twentieth century, the first viewpoint enjoyed the broadest favor. Tropical habitats were thought to have persisted since the beginning of life, whereas vicissitudes of climate (particularly during the last Ice Age) had occasionally destroyed most temperate and arctic habitats, resetting the diversity clock, so to speak. More recently, however, with the integration of population ecology into community theory, ecologists have come to consider diversity as an equilibrium at which opposing diversity-dependent processes balance each other, just as equilibrium population size represents a balance between opposing density-dependent birth and death processes. This viewpoint challenges ecologists to identify the processes responsible for adding species to, and removing species from, communities and to discover why the balance between these processes differs systematically from place to place.

Geographic patterns of species diversity

Within most large taxonomic groups of organisms—plant, animal, and perhaps microbial—numbers of species increase markedly, with a few exceptions, toward the equator (Figure 24.1). For example, within a small region at 60° north latitude, we might find 10 species of ants; at 40°, between 50 and 100 species; and in a similar sampling area within 20° of the equator, between 100 and 200 species. By one count, Greenland is home to 56 species of breeding birds, New York to 105, Guatemala to 469, and Colombia to 1,395. Diversity in marine environments follows a similar trend: Arctic waters harbor 100 species of tunicates, but over 400 species are known from temperate regions and more than 600 from tropical seas. Latitudinal trends in diversity extend even to the greatest depths of the oceans, where conditions were once thought to be unvarying over the entire globe.

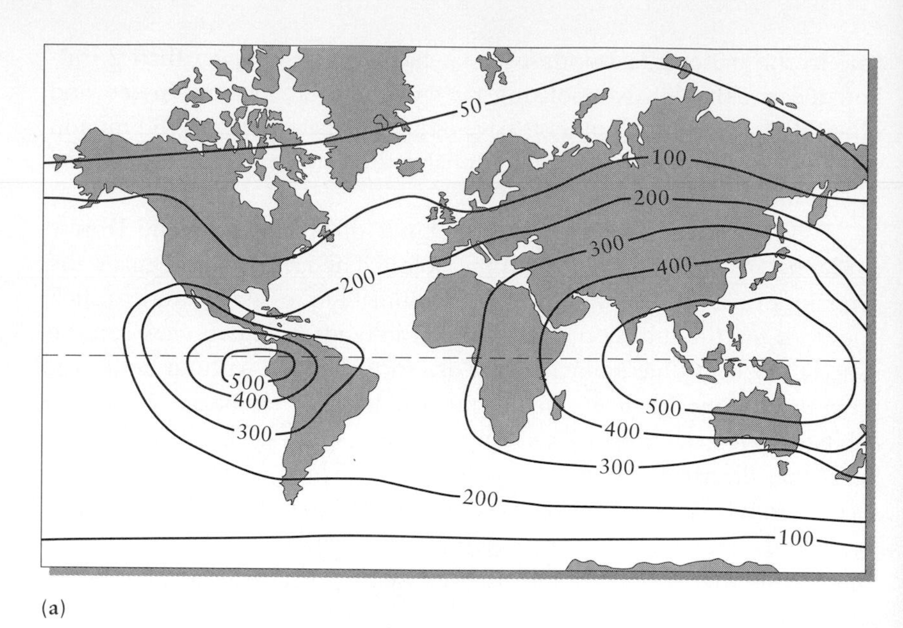

(a)

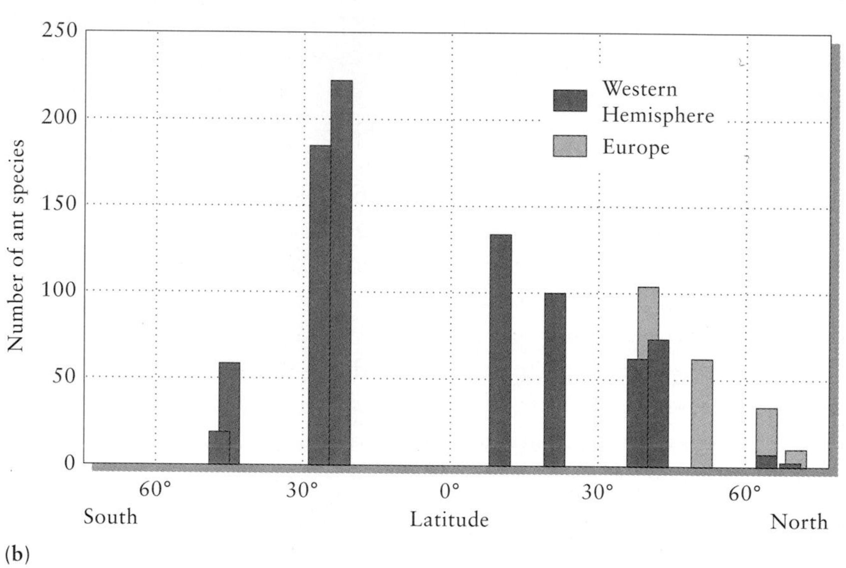

(b)

Figure 24.1 Two examples of global patterns of diversity. (a) Contour lines on the map indicate the number of species of nearshore and continental-shelf bivalves (clams, mussels, scallops, and their relatives) found at locations within the contour intervals. Maximum diversity occurs in the Tropics, particularly within Australasia and the eastern Pacific Ocean. From F. G. Stehli, A. L. McAlester, and C. E. Helsley, *Geol. Soc. Am. Bull.* 78:455 (1967). (b) Number of ant species found within small sampling areas as a function of latitude. Peak diversity appears to be in subtropical South America, rather than at the equator. European localities support greater diversity than is found at similar latitudes in North America because of their generally warmer temperatures. After data in N. Kusnezov, *Evolution* 11:298 (1957).

Within a given belt of latitude, numbers of species may vary widely among habitats according to productivity, degree of structural heterogeneity, and suitability of physical conditions. For example, censuses of birds in small areas (usually 5–20 ha) in the temperate zone reveal an average of about 6 species of breeding birds in grasslands, 14 in shrublands, and 24 in floodplain deciduous forests (Table 24.1). Habitat structure apparently overrides habitat productivity in determining species diversity. For example, marshes are very productive but are structurally uniform, and have relatively few species. Desert vegetation is less productive, but its greater variety of structure apparently makes room for more kinds of inhabitants (Figure 24.2).

Structure and diversity have always gone together in the minds of bird-watchers and other naturalists, but Robert MacArthur and John

TABLE 24.1 Plant productivity and the average number of species of birds in representative temperate zone habitats

HABITAT	APPROXIMATE PRODUCTIVITY ($g\ m^{-2}\ yr^{-1}$)	AVERAGE NUMBER OF BIRD SPECIES
Marsh	2,000	6
Grassland	500	6
Shrubland	600	14
Desert	70	14
Coniferous forest	800	17
Upland deciduous forest	1,000	21
Floodplain deciduous forest	2,000	24

Source: E. J. Tramer, *Ecology* 50:927–929 (1969); productivity data from R. H. Whittaker, *Communities and Ecosystems,* 2d ed., Macmillan, New York (1975).

MacArthur were the first to place this relationship in a quantitative framework that made it accessible to analysis. They did this simply by plotting the diversity of birds observed in different habitats according to diversity in foliage height, a measure of the structural complexity of vegetation (Figure 24.3). The trick was to quantify diversity in habitat structure, which they did by applying the Shannon-Weaver diversity index described in Chapter 22. Others were quick to demonstrate similar

(a)

(b)

Figure 24.2 Desert and marsh vegetation illustrate extremes of productivity with an inverse relationship between productivity and diversity. The desert is the Sonoran Desert of Baja California; the marsh is in the Malheur Refuge, Oregon. Courtesy of the U.S. Department of Interior.

diversity relationships. Among web-building spiders, for example, species diversity varies in direct relation to heterogeneity in heights of the tips of vegetation to which spiders attach their webs. Lizard species diversity closely parallels total volume of vegetation per unit of area in desert habitats of the southwestern United States.

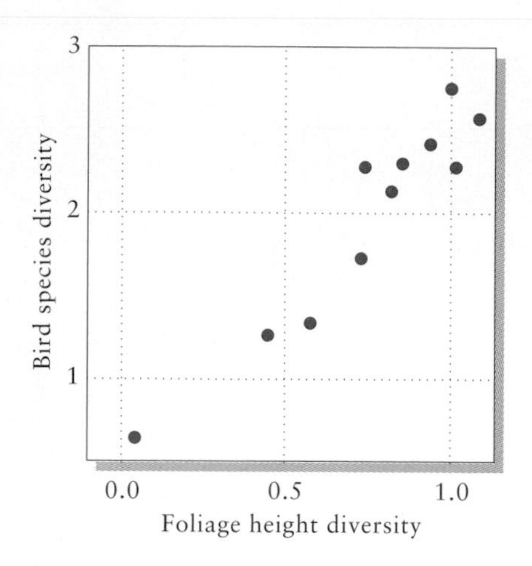

Figure 24.3 Relationship between bird species diversity and foliage height diversity, determined for areas of deciduous forest in eastern North America. From R. H. MacArthur and J. MacArthur, *Ecology* 42:594–598 (1961).

On a regional basis, number of species varies according to suitability of physical conditions, heterogeneity of habitats, and isolation from centers of dispersal. In North America, the number of species in most groups of animals and plants increases from north to south, but the influence of geographic heterogeneity and the isolation of peninsulas also is apparent. Within the area of North America to the Isthmus of Panama, the number of mammal species occurring in square blocks 150 miles on a side increases from 15 in northern Canada to more than 150 in Central America (Figure 24.4). Across the same latitude in the middle of the United States, more species of mammals live in the topographically heterogeneous western mountains (90–120 species per block) than in the more uniform environments of the East (50–75 species per block). Notice also that diversity *decreases* toward the south in Baja California, as one moves along the peninsula away from the center of diversity in the southwestern United States. The number of species of breeding land birds follows a similar pattern, but reptile and amphibian faunas do not. Reptiles are more diverse in the eastern half of the United States than in the mountainous western regions; amphibians are strikingly underrepresented in the deserts of the Southwest because most species require abundant water.

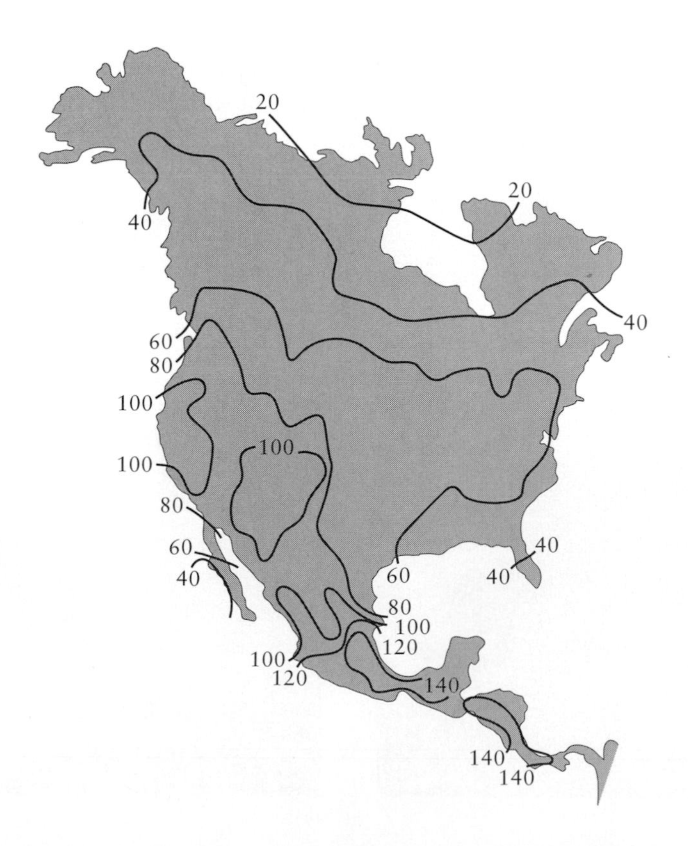

Figure 24.4 Species density contours for mammals in grid squares 150 miles on a side in continental North America. From G. G. Simpson, *Syst. Zool.* 13:57–73 (1964).

Diversity and niche relationships

Ecologists use the term **niche** to express the relationships of individuals or populations to all aspects of their environments—and hence the ecological roles of species within communities. A niche represents the range of conditions and resource qualities within which an individual or species can survive and reproduce. Thus, for example, the boundaries of a particular species' niche might extend between temperatures of 10° and 30°C, prey sizes of 4 and 12 mm, and the hours between dawn and dusk. Of course, the niche of any species would include many more variables than these three, and ecologists often cite the multidimensional nature of the niche to acknowledge the complexity of species-environment relationships. The degree to which the niches of two species overlap determines how strongly they might compete with each other. Thus the niche relationships of species provide an informative measure of the structural organization of biological communities.

We may think of a community as a group of species occupying niches within a space defined by axes of resource qualities and physical conditions of the environment. Within this niche space, adding or removing species has certain geometric consequences. The possibilities are illustrated in Figure 24.5, which portrays distributions of species niches along a single continuous axis. This axis might represent average size of prey items, perhaps, or height distribution within the intertidal zone. The implications of this figure can be understood when niche relationships are characterized by niche breadth and overlap between the niches of neighboring species in niche space. We shall assume for simplicity that resources are uniformly available throughout the niche space. Now, we can ask how species might be added to the community represented by condition A in Figure 24.5. This could be accomplished by any one, or a combination, of three types of adjustments.

1. Without any change in niche relationships (breadth and overlap), the total niche space of a community would have to increase in direct proportion to the number of species (condition B in the figure). (Note that niche size refers to variety of resources, not their amounts.)

2. Without a change in niche breadth, increased diversity could be accommodated by increased niche overlap (condition C). In this case, the average productivity of each species would decline as a consequence of increased sharing of resources, all other things being equal.

3. Without an increase in niche overlap, increased specialization could accommodate additional species within a community's niche space (condition D). Here, too, average productivity would decline because each species would have access to a narrower range of resources.

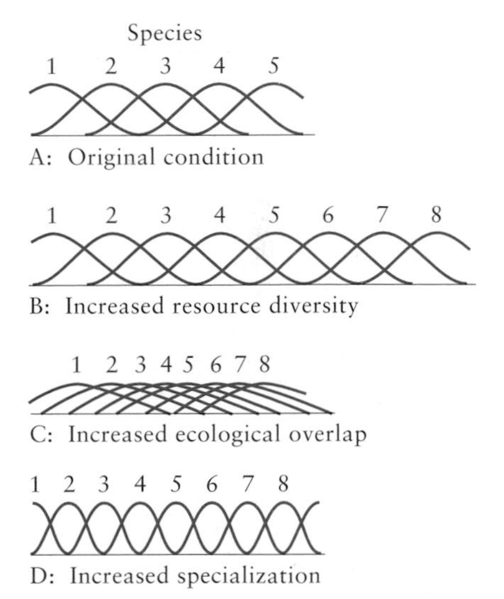

Figure 24.5 A schematic diagram showing how resource use along a continuum can be altered to accommodate more species. The horizontal axes represent the variety of ecological resources and conditions that define the total niche space of the community. The height of each curve (vertical axis) represents the intensity of utilization of resources by each species.

Most ecologists agree that the high diversity in the Tropics results at least in part from there being a greater variety of ecological roles there. That is, the total community niche occupies greater volume near the equator, where there are many species, than it does toward the poles, where there are few. This situation is illustrated by condition B in Figure 24.5. For example, part of the increase in the number of bird species toward the Tropics is related to an increase in fruit- and nectar-feeding species and in insectivorous species that hunt by searching for their prey while quietly sitting on perches—a type of behavior uncommon among birds in temperate regions. Among mammals, the Tropics are species-rich primarily because of the many species of bats in tropical communities. Nonflying mammals are no more diverse at the equator than they are in the United States and other temperate regions at a similar latitude, although their variety does decrease as one goes farther north.

One way to assess niche diversity within a fauna is to use the morphology of species as an indicator of their ecological roles—that is, to assume that differences in morphology among related species reveal different ways of life. For example, size of prey varies in relation to body size of the consumer; different shapes of appendages can be related to locomotion in ways that distinguish different methods of hunting, different habitat structures, and different ways of escaping predators. Morphological analyses of communities have consistently revealed a relatively constant density of species packing in morphologically defined niche space; therefore, as diversity increases, so does the total morphological volume occupied. This finding suggests that added species increase the variety of ecological roles played by members of communities.

To illustrate this principle, let us consider a comparison of bat communities in temperate and tropical localities. A morphological space having two dimensions captures many of the important ecological properties of bat species. The ratio of ear length to forearm length—a measure of ear size relative to body size—is related to the bat's sonar system and thus to the type and location of its prey. The ratio of the lengths of the third and fifth digits of the hand bones in the wing describes the shape of the wing—whether it is long and thin or short and broad. Therefore, this axis is an index to the flight characteristics of bats and, in turn, to the types of prey they can pursue and the habitats within which they can capture prey efficiently.

When we plot each bat species in a community on a graph whose axes are these two morphological dimensions, we can visualize niche relationships among species (Figure 24.6). In the less diverse community in Ontario, in southern Canada, the bat species all have similar morphology: all are small insectivores. The more diverse community in Cameroon, in tropical West Africa, occupies a much larger volume of morphological space, corresponding to a greater variety of ecological roles played by bats there. In addition to small insectivorous species, fruit eaters, nectar eaters, fish eaters, and large, predatory bat eaters make up the bat community.

In streams and rivers, species diversity in most taxonomic groups increases from headwaters to mouths of rivers. One presumes that as a river increases in size, it presents a greater variety of ecological opportunities, more abundant resources, and more stable and therefore more reliable physical conditions. Local communities reflect these changes. For example, a headwater spring in the Rio Tamesi drainage of east central Mexico was found to support only one species of fish, a detritus-feeding platyfish

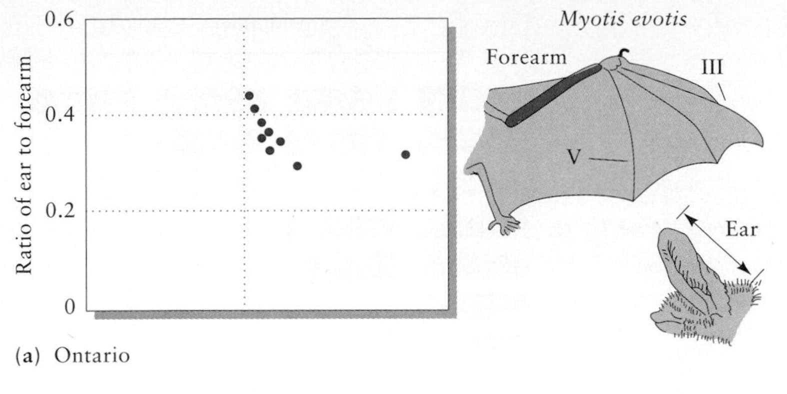

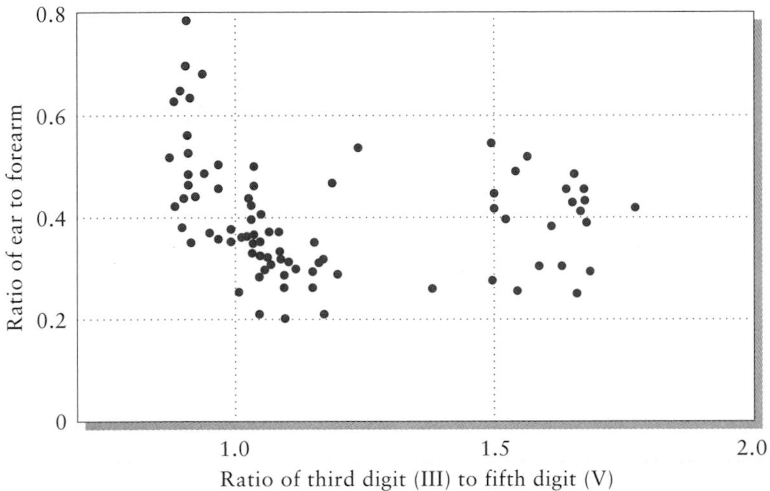

Figure 24.6 Distribution in morphological space of species in the aerial-feeding bat faunas of (a) southeastern Ontario and (b) Cameroon, West Africa. The horizontal axis is the ratio of the lengths of the third and fifth digits of the hand, and the vertical axis is the ratio of ear length to forearm length. After M. B. Fenton, *Can. J. Zool.* 50:287–296 (1972).

(Xiphophorus) (Figure 24.7). Farther downstream, three species occurred: the platyfish, a detritus-feeding molly *(Poecilia)* that prefers slightly deeper water, and a mosquito fish *(Gambusia)* that eats mostly insect larvae and small crustaceans. Fish communities even farther downstream included additional carnivores—among them, fish eaters—and other fish that feed primarily on filamentous algae and vascular plants. Downstream communities had all the species of upstream communities plus additional ones restricted to downstream localities. Thus diversity increases as a stream becomes larger and presents more kinds of habitats and a greater variety and abundance of food items. A general rule evident in these examples is that species diversity is correlated with ecological diversity.

Escape space

Niche dimensions are not limited to food supplies and physical conditions. Avoidance of predation is equally important to population processes, and avenues of predator escape make up dimensions of the niche space along which species may diversify. That part of the niche space that is defined by adaptations (including behaviors) of prey organisms that help them avoid predators is referred to as **escape space.** We

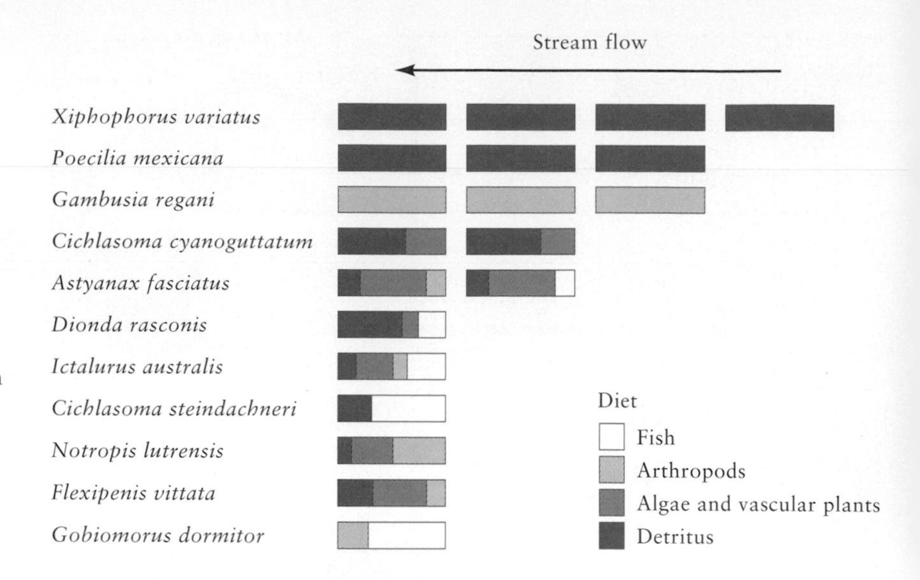

Figure 24.7 Fish species and their diets in four communities (vertical columns), from a headwater spring containing one species (rightmost column) to downstream communities with up to eleven species (leftmost column). The communities sampled were in the Rio Tamesi drainage of east central Mexico. From R. M. Darnell, *Am. Zool.* 10:9–15 (1970).

would expect predators to be most efficient when they focus their attention on portions of niche space most densely occupied by prey species. Where many prey species use the same mechanisms to escape predation, predators with adaptations or learned behaviors that enable them to exploit those prey will prosper; thus, those prey populations will suffer increased mortality. Conversely, prey having unusual adaptations for predator escape should be strongly favored by natural selection. As a result, predation pressure should diversify prey with respect to escape mechanisms, and prey species should therefore tend to become uniformly distributed within available niche space. The quality of a particular position within the niche space depends on predator attributes (hunting methods, body size, visual acuity, and color perception) and the attributes of prey that help them to escape (color and pattern of resting background for cryptic prey; availability of hiding places and other refuges; structure of the vegetation for those that rely on fleeing to escape).

One study used morphology to estimate the packing of moth species in escape space in different habitats. In this case, escape space corresponded to the variety of backgrounds against which day-resting moths conceal themselves to avoid detection by diurnal visual predators. Among cryptic moth species, appearances have evolved to match the background against which a species rests. Hence we assume that a moth's morphological appearance, or **aspect,** reflects characteristics of its resting places and the searching techniques of predators to be avoided (Figure 24.8). The variety of cryptic patterns has been referred to as **aspect diversity.**

Appearances of moths were described by several characteristics, including morphology and position of the legs, coloration, and reaction to disturbance (Figure 24.9). Samples of moths were analyzed from three areas: a spruce-aspen forest in Colorado (43 species), a Sonoran Desert habitat in Arizona (51 species), and a lowland rain forest in Panama (203 species). The total volume of escape space used, as revealed by the variety of morphological characteristics represented in each sample, was greater

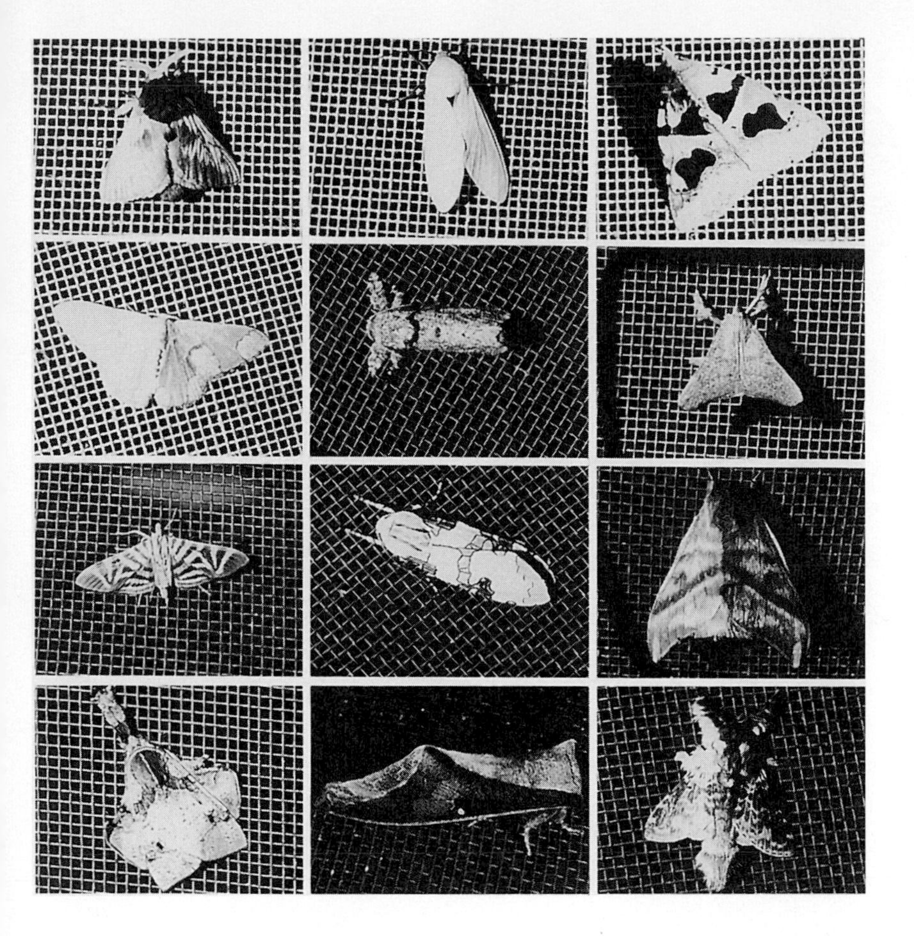

Figure 24.8 Representative species of moths from Panama, photographed against window screens to which they were attracted by ultraviolet lights. These moths show the variety of appearances and shapes exhibited by members of the community. From R. E. Ricklefs and K. O'Rourke, *Evolution* 29:313–324 (1975).

in Panama than in Colorado or Arizona. Again we see that species are added to a community by expansion of the niche space used.

Although these and other examples suggest an increase in total niche space with increasing diversity, two considerations must be kept in mind. First, changes in niche breadth (degree of specialization) and overlap may also accompany variations in diversity, although these are more difficult to measure. Second, increased community niche space may be in part a consequence of diversity rather than an underlying cause or permissive factor. The mere presence of more species makes more roles—a greater variety of interactions—possible. For example, the larger proportions of specialized parasitic species found in collections of insects from sites with higher diversity suggests enhancement of diversity through biotic interaction.

Habitat specialization and diversity

Diversity has several components, two of which are **alpha** (or **local**) **diversity** and **gamma** (or **regional**) **diversity.** Local diversity is the number of species in a small area of more or less uniform habitat. Clearly, local diversity is sensitive to definition of habitat, area, and intensity of sampling effort. Regional diversity is the total number of species observed in all habitats within a region. By **region,** ecologists generally mean a

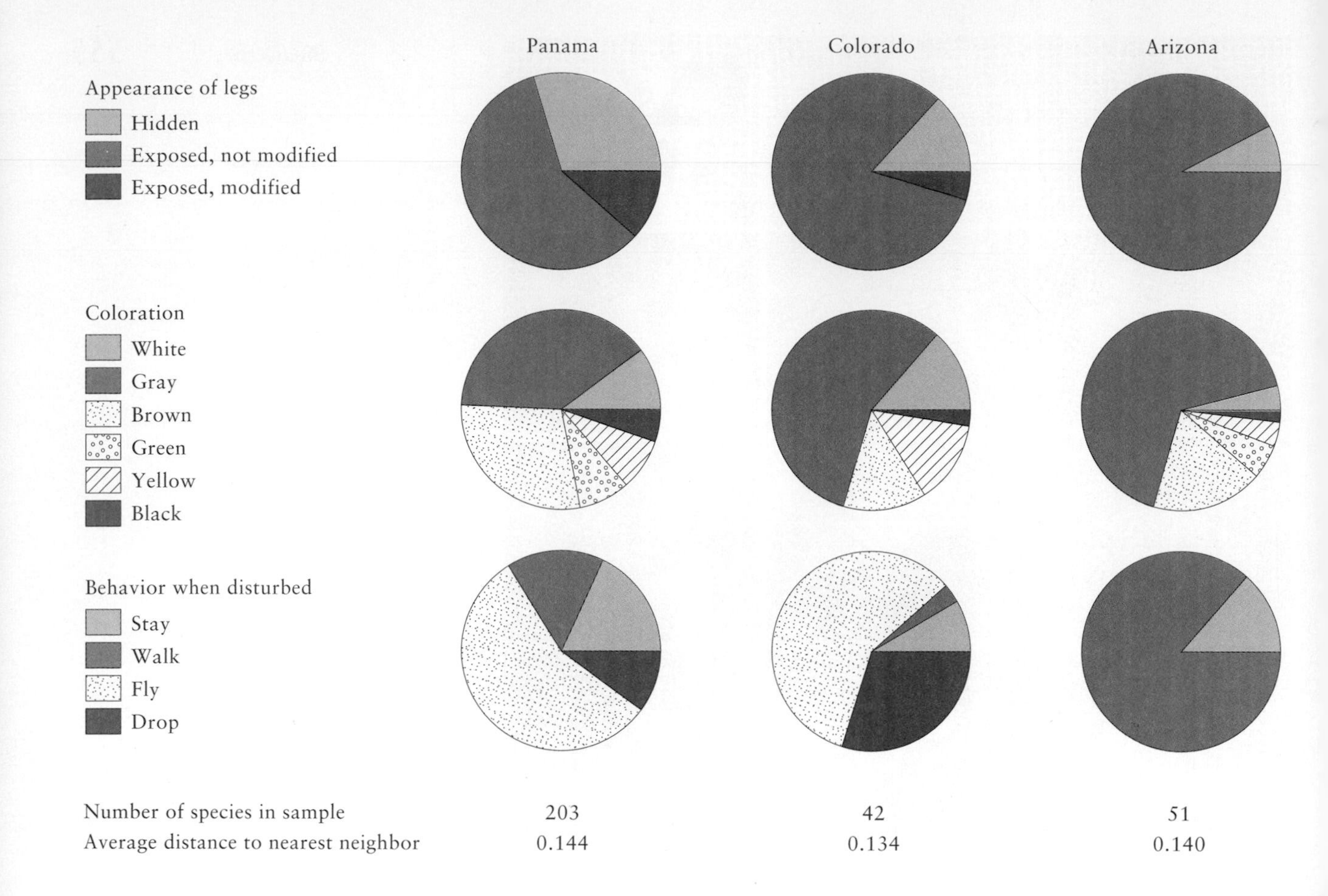

Figure 24.9 Diversity in the appearance and behavior of moths from three localities, exemplified by three characteristics related to their antipredator adaptations. Variety is greatest in the most diverse sample, that from Panama. Average distance to the nearest neighbor within escape space indicates density of species packing. From data in R. E. Ricklefs and K. O'Rourke, *Evolution* 29:313–324 (1975).

geographic area that includes no significant barriers to dispersal of organisms. Thus, the boundaries of a region depend on which organisms we consider. The important point is that within a region, distributions of species should reflect selection of suitable habitats rather than inability to disperse to a particular locality.

When each species occurs in all habitats within a region, local and regional diversities are the same. When each habitat has a unique flora and fauna, regional diversity equals the average local diversity times the number of habitats in the region. Ecologists refer to the difference in species from one habitat to the next as **beta diversity.** The greater the difference, or turnover, of species between habitats, the greater is beta diversity. There are many different ways of quantifying beta diversity, but a useful one is the equivalent number of unique habitats recognized by species within a region. When all species are habitat generalists, there is effectively only a single habitat within the region, and beta diversity is equal to 1. As habitat specialization increases, more habitats are recognized. Accordingly, gamma diversity equals alpha diversity times beta diversity. It is not practical to measure beta diversity directly because habitat distributions of species overlap. But we can calculate the equivalent number of unique habitats recognized by species within a region from the following relationship: beta diversity equals gamma diversity divided by alpha diversity. We shall use this relationship shortly to see how different components of diversity vary as species are added to a region.

Where many species coexist within a region, each occurs in relatively few kinds of habitat. Changes in gamma diversity generally result from parallel changes in both alpha and beta diversity. This relationship has been most carefully noted in studies of islands and neighboring continental regions, where one can compare different levels of diversity (resulting from different degrees of geographic isolation) among regions having similar climate and life forms. Islands usually have fewer species than comparable mainland areas, but island species often attain greater densities than their mainland counterparts, a phenomenon called **density compensation.** Also, they expand into habitats that would normally be filled by other species on the mainland, a phenomenon called **habitat expansion.** Collectively, these phenomena are referred to as **ecological release.** On the island of Puerto Rico, many bird species occupy most habitats on the island. In Panama, which has a similar variety of tropical habitats, species occupy fewer habitats, often a single type.

Ecological release has been demonstrated in surveys of bird communities in seven tropical regions and islands within the Caribbean Basin: these range in size from mainland Panama to St. Kitts, a small island in the Lesser Antilles. These surveys show that where fewer species occur, each is likely to be more abundant and to live in more habitats (Table 24.2).

TABLE 24.2 **Relative abundance and habitat distribution of resident land birds in seven tropical localities within the Caribbean Basin***

LOCALITY	NUMBER OF SPECIES OBSERVED (REGIONAL DIVERSITY)	AVERAGE NUMBER OF SPECIES PER HABITAT (LOCAL DIVERSITY)	HABITATS PER SPECIES	RELATIVE ABUNDANCE PER SPECIES PER HABITAT (DENSITY)	RELATIVE ABUNDANCE PER SPECIES	RELATIVE ABUNDANCE OF ALL SPECIES
Panama	135	30.2	2.01	2.95	5.93	800
Trinidad	106	28.2	2.35	3.31	7.78	840
Jamaica	56	21.4	3.43	4.97	17.05	955
Tobago	53	21.4	3.63	4.71	17.10	906
St. Lucia	33	15.2	4.15	5.77	23.95	790
Grenada	30	15.5	4.63	5.36	24.82	745
St. Kitts	20	11.9	5.35	5.88	31.45	629

*Based on ten counting periods in each of nine habitats in each locality. The relative abundance of each species in each habitat is the number of counting periods in which the species was seen (maximum 10); this times the number of habitats gives the relative abundance per species; this times the number of species gives the relative abundance of all species together.

Source: G. W. Cox and R. E. Ricklefs, *Oikos* 29:60–66 (1977); J. M. Wunderle, *Wilson Bull.* 97:356–365 (1985).

Similar numbers of individuals of all species added together were seen in each of the seven localities, although the numbers of species (regional diversity) differed by a factor of almost 7 between Panama and St. Kitts. In each habitat in Panama (mainland), about three times as many species (alpha diversity) were recorded, and populations of each species were about half as dense, as in corresponding habitats on St. Kitts (the smallest island). Beta diversity (the equivalent number of habitats recognized by species) increased by a factor of almost 3 between St. Kitts and Panama.

This survey of diversity patterns suggests several general conclusions. At a global scale, our perception of biodiversity is dominated by a pronounced increase in diversity as one travels from high latitudes toward the equator. Within latitudinal belts, diversity appears to be correlated with topographic heterogeneity within a region and the complexity of local habitats. Islands exhibit species impoverishment. Everywhere, higher diversity is associated with greater community niche volume.

How do we explain these patterns of diversity? Some biologists have claimed that diversity increases with time and depends on age, but the strong correlation between habitat structure and diversity would seem to cast doubt on that hypothesis. Alternatively, most ecologists now believe that diversity achieves an equilibrium value at which processes that add species and those that subtract species balance each other. Migration of species between habitats and regions, as well as production of new species within regions, adds to the number of species in local habitats. Within a local community, species are removed by competitive exclusion, elimination by efficient predators, or the plain bad luck of succumbing to a regional disaster such as a major volcanic eruption. Distinguishing among the various factors responsible for generating and maintaining diversity at local and regional scales has proved to be exceedingly difficult. In truth, many factors undoubtedly influence species diversity, and the challenge will be to determine the mechanisms and relative strengths of each of these influences.

The time hypothesis

The fossil record ought to tell us whether or not diversity has increased over time. However, this isn't as simple as it sounds because of some severe problems inherent in sampling the history of life: the abundance and quality of fossils increases toward the present; paleontologists have more difficulty sampling local diversity than regional diversity; most fossils cannot be distinguished as finely as the species level; and the record is more complete for some habitats that favor preservation, such as shallow embayments, than for others, such as arid or montane habitats. Minimizing these considerations for the moment, the data currently available suggest that during the past 65 million years, diversity has remained constant within some groups at some times and has increased in others, notably flowering plants, fishes, birds, and mammals (see, for example, Figure 24.10). The fossil record does not tell us unambiguously whether other groups declined at the same time, nor does it reveal whether diversity achieved local equilibria while it

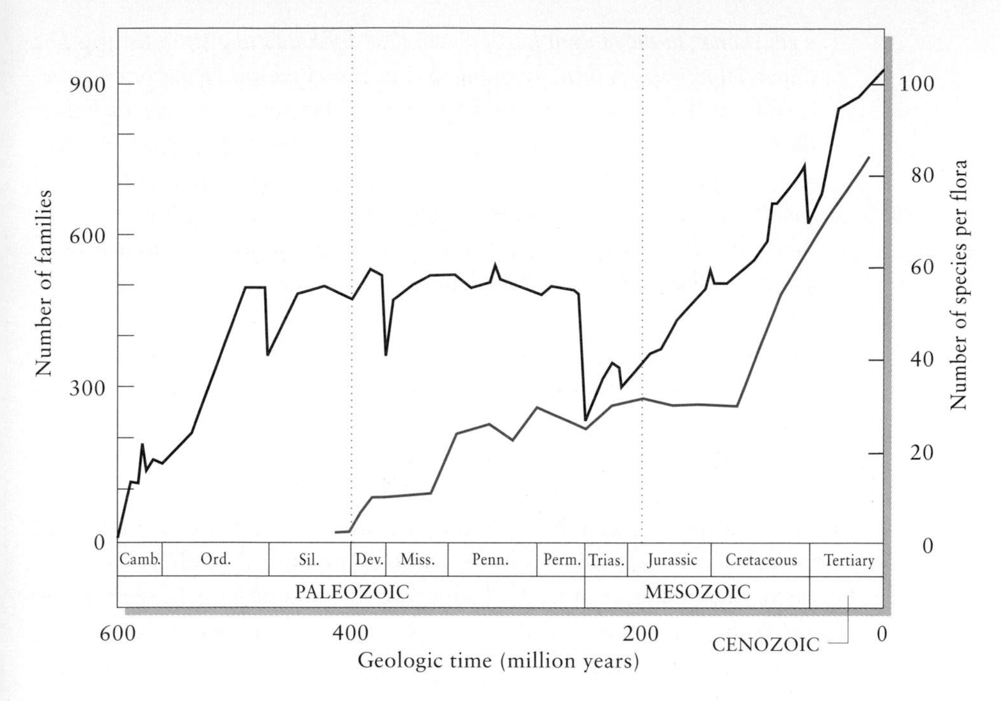

Figure 24.10 Changes in the global total number of families of marine animals since the beginning of the Paleozoic era (black line), and changes in the average number of species represented in local fossil floras of terrestrial plants since the mid-Paleozoic (colored line; scale 10×). Note the more or less constant diversity of marine animal families through the latter part of the Paleozoic era, the mass extinction at the end of the Permian period, and the continuous increase in diversity since, with the proliferation of modern marine taxa (primarily fishes, mollusks, and crustaceans). Note also the rather constant local diversity of plants throughout much of the Paleozoic and Mesozoic eras, the absence of a mass extinction at the end of the Permian period, and the rapid rise in local diversity beginning with the appearance of flowering plants in the late Jurassic period. From J. J. Sepkoski Jr., *Paleobiology* 10:246–267 (1984); A. J. Knoll, in J. Diamond and T. J. Case (eds.), *Community Ecology,* Harper & Row, New York (1986), pp. 126–141.

increased globally. It is also clear that numbers of taxa have remained remarkably stable over long stretches of the earth's history, even though taxonomic composition has changed all along.

One explanation for high diversity in the Tropics is that tropical conditions appeared on the earth's surface earlier than colder environments, allowing time for the evolution of a greater variety of plants and animals. Although this time hypothesis has been voiced recently, it is hardly new. It was fully stated in 1878 by the English naturalist Alfred Russel Wallace:

> *The equatorial zone, in short, exhibits to us the result of a comparatively continuous and unchecked development of organic forms; while in the temperate regions there have been a series of periodical checks and extinctions of a more or less disastrous nature, necessitating the commencement of the work of development in certain lines over and over again. In the one, evolution has had*

a fair chance; in the other, it has had countless difficulties thrown in its way. The equatorial regions are then, as regards their past and present life history, a more ancient world than that represented by the temperate zones, a world in which the laws which have governed the progressive development of life have operated with comparatively little check for countless ages, and have resulted in those wonderful eccentricities of structure, of function, and of instinct—that rich variety of colour, and that nicely balanced harmony of relations which delight and astonish us in the animal productions of all tropical countries.

Because the tropical zone girdles the earth about its equator—the earth's widest point—tropical latitudes include more area, both land and sea, than temperate and arctic regions. For this reason alone, it is not surprising that the Tropics should harbor more species than temperate or arctic zones. The earth's climate has undergone several cycles of warming and cooling, which have been discovered by records, in sediments and fossils, of their influence on vegetation and ocean temperatures. As the climate of the earth warmed, as it last did during the Oligocene epoch, perhaps 30 million years ago, the area of the Tropics and Subtropics expanded, reaching what is now the United States and southern Canada, as well as most of Europe, and temperate and arctic zones were squeezed into smaller areas closer to the poles. During the last 25 million years, the climate of the earth has become cooler and drier, and the Tropics have contracted.

Both high and low latitudes experienced drastic fluctuations in climate during the Ice Ages of the last 2 million years. Temperate and arctic areas witnessed the expansion and retreat of glaciers, which caused major habitat zones to be displaced geographically and, possibly, to disappear. Periods of glacial expansion were coupled with low rainfall and reduced temperature in the Tropics. The Amazonian rain forest, which today covers vast regions of the Amazon River's drainage basin, is thought by some biologists to have been repeatedly restricted to small, isolated refuges during dry periods correlated with glacial expansion in the north. Restriction and fragmentation of rain forest habitat could have driven many species to extinction; conversely, isolation of populations in patches of rain forest could have facilitated the formation of new species. It is reasonably well understood that alternating expansion and contraction of humid habitats accounts in part for the diversity of the present Australian avifauna. Other animals and plants were similarly affected: for example, about 500 species in the genus *Eucalyptus,* a type of tree or shrub, are indigenous to Australia (but now widely planted around the world).

Equilibrium theories of diversity

The integration of "population thinking" into community studies during the 1950s and 1960s led ecologists to postulate that diversity might be regulated by local interactions among species. If competitive exclusion set limits on the ecological similarity of species (and thus on the intensity of competition between them), then communities could become saturated with species, and additions would be balanced by local extinctions. Accordingly, diversity would reach an equilibrium: new species added to the local community by regional diversification and migration would be compensated by local exclusion of close competitors. The equilibrium point itself would be affected by physical conditions, variety of resources, predators, environmental variability, and perhaps other factors. Thus, conditions in the Tropics might allow greater numbers of species to coexist locally by reducing the intensity or consequences of competition.

Ecologists were attracted to this view because it placed at least part of the problem of species diversity within their conceptual domain of present-day processes taking place within small areas. The earlier idea that diversity reflected history left ecologists with little to say about the matter. The alternative, what we might call local determinism, became so attractive, however, that ecologists embraced it nearly to the point of denying that regional and historical factors influence local diversity. Instead of viewing diversity as a balance between regional production of species and local extinction, ecologists began to feel that local diversity was determined largely by local conditions. Hence their investigations concentrated almost exclusively on local interactions and niche relationships among populations. As we shall see below, the results of these studies did not provide completely satisfying explanations for diversity patterns, and the pendulum of ecological thinking is now swinging back toward a balance of local and historical factors.

Diversity on islands

During the 1960s, Robert MacArthur, then at the University of Pennsylvania, and E. O. Wilson at Harvard University developed their famous **equilibrium theory of island biogeography,** which states that the number of species on an island balances regional processes governing immigration against local processes governing extinction. On islands too small to support speciation through geographic isolation of populations, diversity increases solely through immigration from other islands or from the mainland. Whereas we know little about rates of speciation within continents, we may reasonably assume that greater distances separating an island from continental sources of species result in fewer species immigrating to that island.

Consider an offshore island. The flora and fauna of the adjacent mainland make up the **species pool** of potential colonists of the island. The rate of immigration of new species to the island decreases as the number of species on the island increases; that is, as more and more potential mainland colonists establish themselves on the island, fewer and fewer immigrating individuals belong to new species. When all mainland species occur on the island, the immigration rate of new species must be zero. If species disappear at random, the number of extinctions per unit of time increases with the number of species present on the island. Where the immigration and extinction curves cross, the corresponding number of species on the island attains an equilibrium ($\hat{S}$) (Figure 24.11).

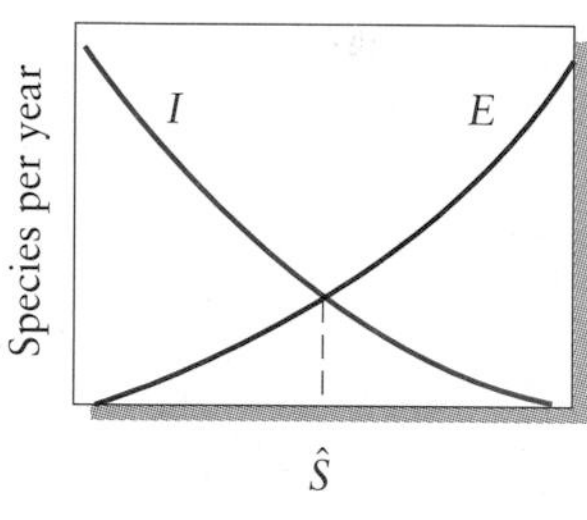

Figure 24.11 Equilibrium model of the number of species on islands. The equilibrium number of species ($\hat{S}$) is determined by the intersection of the immigration (I) and extinction (E) curves. After R. H. MacArthur and E. O. Wilson, *Evolution* 17:373–387 (1963); MacArthur and Wilson, *The Theory of Island Biogeography,* Princeton University Press, Princeton, N.J. (1967).

Immigration and extinction rates probably do not vary in strict proportion to the number of potential colonists and the number of species established on an island. Some species undoubtedly colonize more easily than others, and they reach the island first. Thus, the rate of immigration to an island initially decreases more rapidly than it would if all mainland species had equal potential for dispersal; as a result, immigration rate follows a curved line with respect to increasing island diversity, as in Figure 24.11. Competition between species on islands probably abets extinction, so the extinction curve rises progressively more rapidly as species diversity increases.

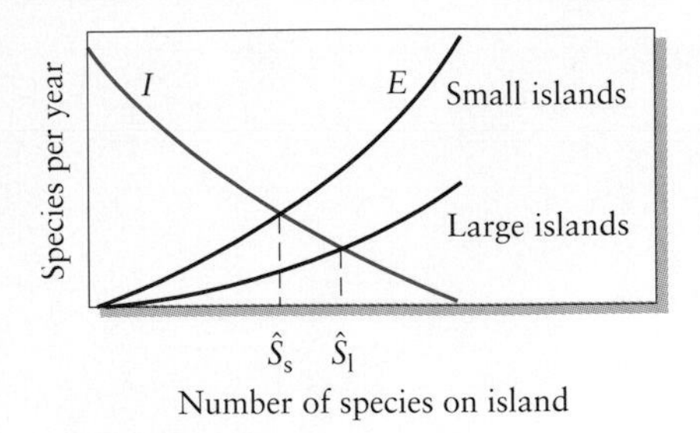

Figure 24.12 According to the MacArthur-Wilson equilibrium model, small islands support fewer species because of higher extinction rates. $\hat{S}_s$ = equilibrium number of species for small islands; $\hat{S}_l$ = equilibrium number of species for large islands.

If probability of extinction increases as absolute population size decreases, extinction curves will be higher for species on small islands than for those on large islands. Therefore, small islands will support fewer species than large islands (Figure 24.12), which is, of course, the pattern we see. If the rate of immigration to islands decreases with increasing distance from mainland sources of colonists, the immigration curve will be lower for far islands than for near islands and the equilibrium number of species for distant islands will lie to the left of that for islands close to the mainland (Figure 24.13). These predictions have been verified for islands throughout the world.

As a corollary to its predictions concerning equilibrium diversity, the MacArthur-Wilson theory also predicts that if some disaster reduces the diversity of a particular island, new colonists will, over time, restore diversity to its predisturbance equilibrium. A natural test of this prediction began quite spectacularly in 1883 when Krakatau, an island located between Sumatra and Java in the East Indies, blew up after a long period of repeated volcanic eruptions. At least half the island disappeared beneath the sea, and hot pumice and ash covered its remaining area. The entire flora and fauna of the island were certainly obliterated, as was apparent to the first visitors after the explosion. During the years that followed, plants and animals recolonized Krakatau at a surprisingly high rate: within 25 years, explorers found more than 100 species of plants and 13 species of land and freshwater birds. During the next 13 years, 2 species of birds disappeared but 16 were gained, bringing the total to 27. During the next 14-year period between exploring expeditions, bird diversity on Krakatau did not change, but 5 species disappeared and 5 new ones arrived, suggesting that the number of species had reached an equilibrium, at a level that would be expected for an island the size of Krakatau located in the Tropics. (No one knew how many species were present before the explosion.) Experimental studies of colonization of mangrove islands by arthropods have reinforced the pattern of colonization and attainment of an equilibrium seen on Krakatau (Figure 24.14).

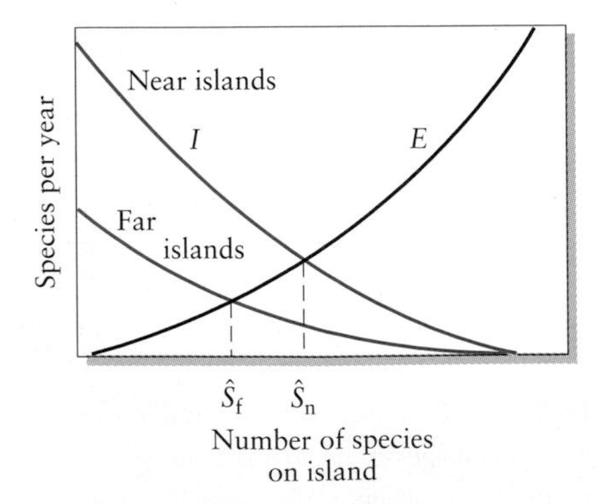

Figure 24.13 According to the MacArthur-Wilson equilibrium model, islands close to the mainland support more species because of higher immigration rates. $\hat{S}_f$ = equilibrium number of species on far islands; $\hat{S}_n$ = equilibrium number of species on near islands.

Equilibrium theory in continental communities

We can apply an equilibrium view of diversity to mainland assemblages of species as well as to those on oceanic islands. The major difference is that

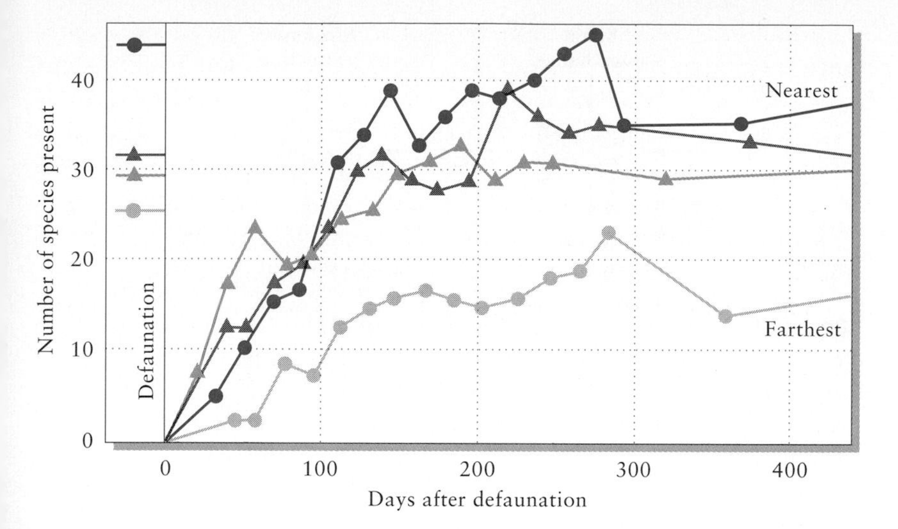

Figure 24.14 Recolonization of four small mangrove islands in the lower Florida Keys whose entire faunas, consisting almost solely of arthropods, were exterminated by methyl bromide fumigation. Estimated numbers of species present before defaunation are indicated at left. Species accumulated more slowly and achieved lower equilibrium numbers on more distant islands. From D. S. Simberloff and E. O. Wilson, *Ecology* 50:278–296 (1970).

regional production of new species, which is called **speciation,** augments the addition of species from outside the system by immigration. In a large region isolated from others by barriers to dispersal (an island continent, for example), new species originate primarily by speciation events within the region. Curves relating rates of species production and extinction to regional diversity might look like those drawn in Figure 24.15. The curvature of the lines would vary depending on the processes that produced them. Probability of extinction per species could increase if competitive exclusion increased with diversity, whereas it could decrease if mutualisms and alternative pathways of energy flow buffered diverse communities from external perturbations. The rate of speciation per species could level off if opportunities for further diversification were restricted by increasing diversity, whereas it could increase if diversity led to greater specialization and a higher probability of reproductive isolation of subpopulations.

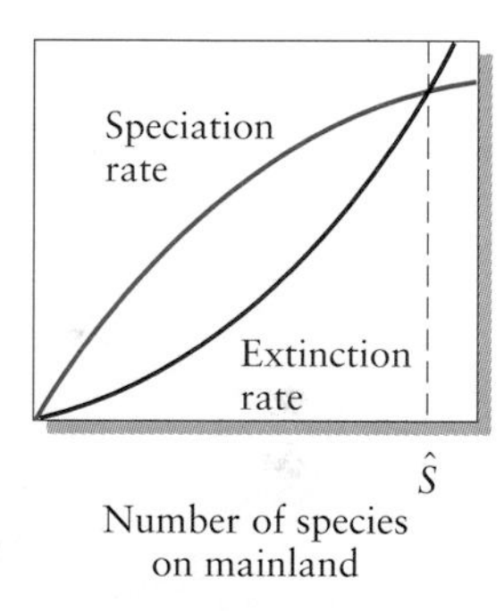

Figure 24.15 Equilibrium model of the number of species in a mainland region with a large area; in this case, new species are generated by the evolutionary process of speciation as well as by immigration from elsewhere. After R. H. MacArthur, *Biol. J. Linn. Soc.* 1:19–30 (1969).

Regardless of the particular shape of the immigration, speciation, and extinction curves, many biologically reasonable models can define an equilibrium level of diversity. Therefore, although such models provide a valuable perspective, they do not necessarily explain variation in diversity: most causal factors can be incorporated into an equilibrium model. For example, the time hypothesis would apply when present diversity remained far below the equilibrium. Alternatively, local communities might be saturated with species while diversity continued to increase regionally, the growing difference being made up by increasing beta diversity. If diversity were in equilibrium, the positions of species production and extinction curves would be affected by a variety of factors, each of which could shift the equilibrium (Figure 24.16).

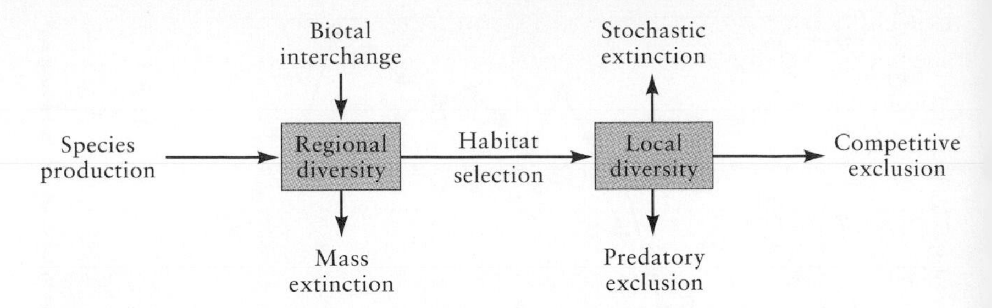

Figure 24.16 Diagram of factors affecting regional and local species diversity. Numbers of species are increased at the regional level by speciation and immigration. Ecological interactions influence diversity at local levels. Each local community contains a small sample of the total regional diversity because species are habitat specialists. Thus habitat selection connects regional and local diversity.

Competition and diversity

Both theory and experiment have shown that intense competition promotes exclusion of species from a community. Many ecologists have argued that less severe competition for shared resources allows more species to coexist. How might interspecific competition be reduced? Greater ecological specialization, greater resource availability, reduced resource demand, and intensified predation have all been cited as possibilities. In every case, however, competition between individuals of the same species, as well as between individuals of different species, influences how populations adjust to these factors. It is hard to imagine reduced competitive ability as an adaptive property in any population. If anything, natural selection should tend to increase the competitive ability of individuals within populations as well as between individuals of different species. Therefore, if variation in competition were to explain variation in diversity, the relative strengths of intraspecific and interspecific competition in different regions would have to differ in terms of their influence on a population's gene pool.

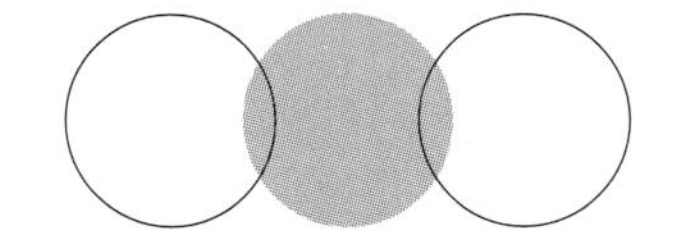

(a) One dimension (asymmetrical)

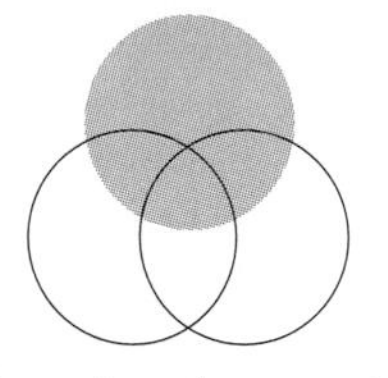

(b) Two dimensions (symmetrical)

Figure 24.17 Schematic diagram of ecological relationships between three species, organized along (a) one dimension and (b) two dimensions. Relationships between three species positioned along one dimension are asymmetrical because the middle species faces two competitors and the outer two species interact with only one. Relationships are symmetrical on two dimensions, making it more difficult for two species to exclude a third, or vice versa.

Perhaps differences in these relative strengths might be envisioned in terms of the geometry of niche relationships. In a niche space having few dimensional axes, each population has relatively few neighbors. As the number of dimensions increases, the same number of species may pack themselves differently into the niche space, increasing the number of neighbors but reducing the asymmetry of competition (Figure 24.17). This variety may prevent each individual population from adapting to compete effectively with its neighbors in communities of high niche dimensionality, preventing the emergence of competitive dominants. High dimensionality may occur in physically undemanding environments, such as the wet Tropics or the abyssal depths of the oceans. In harsher environments, a few physical factors may dominate the character of niche space by establishing a small number of critically important dimensions. Furthermore, because biological factors generate many niche dimensions, the dimensionality of niche space may increase as biological communities build over time, further enhancing their capacity to support many species against the evolutionary tendency of a small number of species to dominate.

Explanations for variation in plant species diversity

Because the diversity of plant resources rather straightforwardly influences the potential diversity of animals, the most rigorous tests of general explanations for diversity lie in their application to plant communities. The question, "Why are there so many different kinds of trees in the Tropics?" has many plausible answers. These answers may be grouped into five categories: those based on (1) environmental heterogeneity in space, (2) environmental variability in time, (3) environmental heterogeneity produced by disturbance, (4) consumer pressure, and (5) rates of production (by speciation or immigration) of competitively equivalent species.

Many ecologists have argued that diversity of trees varies in proportion to the heterogeneity of the environment. Abundant evidence suggests that tropical forest trees may be specialized to certain soil and climate conditions. But could variability in the physical environment in the Tropics account for a tenfold (or more) greater diversity of plants in tropical than in temperate forests? It seems unlikely, unless plants recognize much finer differences in tropical environments than they do in temperate regions, especially since temperate regions have greater heterogeneity in some climate factors than do tropical regions.

Unpredictable temporal heterogeneity generally is not thought to promote diversity. But theoretical considerations suggest that under some circumstances, year-to-year variation in reproductive rates, such that each species predominates in some years, can lead to coexistence. This mechanism relies on a kind of frequency dependence that favors rare species in particular years and so prevents them from being excluded by competitors. However, if there is any difference between tropical and temperate regions in temporal heterogeneity, it is likely to be greater in temperate latitudes.

This model of rareness advantage was inspired by coral reef fishes, which exhibit a diversity comparable to that of tropical forest trees and exhibit little niche diversification. For this system, there has been considerable controversy over the issue of competition among species. Some observers have suggested that juvenile fish colonize patches of coral at random, with opportunities to take the place of adults that die or otherwise leave their territories spread evenly among individuals of all species. This idea became known as the **lottery hypothesis.** Colonization by lottery, which produces random variation in time, reduces competitive exclusion and may contribute to the coexistence of large numbers of fish species in tropical reef communities. But the lottery model cannot explain the high diversity of larval fish in the plankton from which coral reef residents come. Nor can it explain the difference in fish diversity between tropical and temperate oceans.

Other observers have related the high diversity of tropical rain forests and coral reefs to intermediate levels of disturbance. We have already touched upon the **intermediate disturbance hypothesis**: disturbances to communities caused by physical conditions, predators, or other factors open space for colonization and initiate a cycle of succession by species adapted to colonizing disturbed sites. With a moderate level of disturbance, a community becomes a mosaic of patches of habitat at different stages of regeneration; together, these patches contain the full variety of species characteristic of a successional sere. For this hypothesis to account satisfactorily for differences in diversity between regions, especially of the magnitude of latitudinal differences in tree species diversity, there must be comparable differences in level of disturbance. Rates of turnover of individual forest trees (that is, the inverse of average life span) do not differ appreciably between temperate and tropical areas (Table 24.3). Nor is it likely that major disturbances such as storms and fires are more frequent in the Tropics. Thus, although disturbance may promote diversity, it seems unlikely to account for much of the observed latitudinal variation in diversity among forests or, indeed, other types of communities.

Consumers and diversity of resource species

When predators reduce prey populations below their carrying capacities, they may reduce competition among them and promote the coexistence of many prey species. Moreover, selective predation on superior competitors may allow competitively inferior species to persist in a system. Both accidental and intentional experiments have revealed an immense capacity on the part of predators to reduce populations of their prey under some conditions. Their community effects have been particularly well documented in aquatic systems, where the introduction of a predatory sea star, salamander, or fish can utterly transform a community of primary consumers and producers.

From Darwin's time at least, naturalists have believed that both selective and nonselective herbivory can influence diversity of plant species. In particular, several authors have suggested that herbivory could promote high diversity in tropical forests. Perhaps herbivores feed on buds, seeds, and seedlings of abundant species so efficiently as to reduce their densities. This would allow other, less common species to grow in their place. The key to this idea is that abundance per se, rather than any intrinsic quality of individuals as food items, makes a species vulnerable to consumers. Consumers locate abundant species easily, and therefore their own populations grow to high levels.

Several lines of evidence support this **pest pressure hypothesis.** For example, attempts to establish plants in monoculture frequently fail because of infestations of herbivores. Dense plantations of rubber trees in their native habitats in the Amazon Basin, where many species of herbivores have evolved to exploit them, have met with singular lack of success. But rubber tree plantations thrive in Malaya, where specialist herbivores

TABLE 24.3 Turnover of canopy trees in primary forests in tropical and temperate localities

LOCALITY	TURNOVER TIME (YEARS)*	TURNOVER RATE (% PER YEAR)
Tropical		
Panama	62–114	0.9–1.6
Costa Rica	80–135	0.7–1.3
Venezuela	104	1.0
Gabon	60	1.7
Malaysia	32–101	1.0–3.1
Temperate		
Great Smoky Mountains	49–211	0.5–2.0
Tionesta, Pennsylvania	107	0.9
Hueston Woods, Ohio	78	1.3

*Turnover time does not include the time required to grow into the canopy, which is estimated to be 54–185 years for various temperate zone species.

Source: Data from J. R. Runkle, in S. T. A. Pickett and P. S. White (eds.), *The Ecology of Natural Disturbance and Patch Dynamics,* Academic Press, Orlando, Fla. (1985), pp. 17–33; F. E. Putz and K. Milton, in E. G. Leigh, Jr., A. S. Rand, and D. M. Windsor (eds.), *The Ecology of a Tropical Forest,* Smithsonian Institution Press, Washington, D.C. (1982), pp. 95–100; N. V. L. Brokaw, in *The Ecology of Natural Disturbance and Patch Dynamics,* pp. 53–69.

are not (yet) present. Attempts to grow many other commercially valuable crops in single-species stands in the Tropics have met the same disastrous end that befell the rubber plantations.

The difficulties of monoculture extend beyond the Tropics. Temperate zone trees also have their herbivores; few acorns escape predation by squirrels and weevils, and seedlings are attacked by herbivores and pathogens just as they are in the Tropics. If pest pressure does promote greater diversity in the Tropics, it must operate differently in different latitude belts. In particular, either tropical herbivores and plant pathogens must be more specialized with respect to species of host plant, or their populations must be more sensitive to the density and dispersion of host populations.

The pest pressure hypothesis predicts that seedlings will be less likely to establish themselves close to adults of the same species than at a distance from them (Figure 24.18). Adult individuals may harbor populations of specialized herbivores and pathogens that could readily infest nearby progeny; furthermore, because most seeds fall close to their parent, herbivores may be attracted to the abundance of seedlings there while overlooking the few that disperse to a more distant location. The prediction that success in germination and establishment should increase with distance from the parent has been tested in a number of studies, which have yielded varied but generally supportive results.

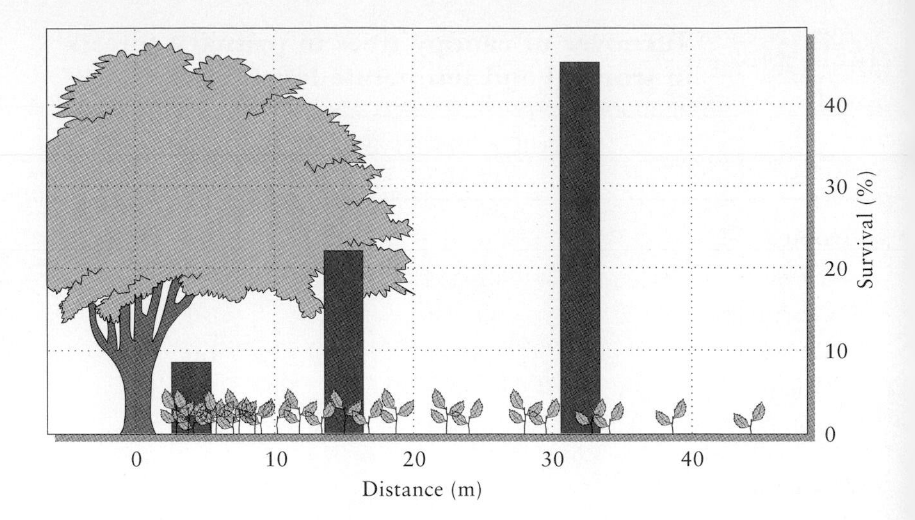

Figure 24.18 A schematic diagram of the pest pressure hypothesis. Seedlings are most dense close to the parent tree, but survival of seedlings is highest at a distance, where seedlings escape the detrimental effects of crowding close to the parent on survival. Data on survival of seedlings to 18 months of age are for the Neotropical tree *Dipteryx panamensis.* From D. H. Janzen, *Am. Nat.* 104: 501–528 (1970); D. A. Clark and D. B. Clark, *Am. Nat.* 124:769–788 (1984).

If seedlings establish themselves more readily at greater distances from their parent or other adults of the same species, then conspecific individuals should be widely dispersed within a forest rather than randomly associated or clumped. This prediction can be tested by mapping the locations of individuals of each species and applying one of several statistics used to test degree of dispersion. Most studies have shown that many species of trees, especially the more common ones, have significantly clumped distributions. This result weighs against a major role for pest pressure in forest community structure, although degree of clumping is relative, and herbivores and pathogens may cause greater dispersion of mature individuals than would occur otherwise.

Rates of species production

As we have seen, environmental heterogeneity and consumer activities can influence local population processes in such a way as to affect species diversity. Regional processes are more difficult to contemplate, much less observe directly, and correspondingly little has been written about their role in determining local and regional diversity. In nonequilibrium systems, rates of species production and extinction, as well as the age of a region or habitat type, directly determine diversity. Where diversity attains an equilibrium, the rate of species production will influence regional diversity; it will also influence local diversity unless local communities become saturated with species at levels determined solely by local ecological conditions. Many systematists and biogeographers have suggested that differing rates of species production, along with the temporal history of environments, have contributed to latitudinal differences in diversity. But because thousands of generations may be needed for isolated populations to develop barriers to reproduction and become distinct species, ecologists can neither observe the process nor experiment with it.

A few theoretical papers have addressed the problem of tree species diversity in the context of species production. For example, one lottery-type model of replacement of forest trees suggested that the probability that the individual that fills a gap in the forest will belong to a particular species varies in direct proportion to the frequency of that species in the forest. Hence gaps are more likely to be filled by abundant species than by rare ones. Without the appearance of new species, such a system will eventually become dominated by a single species, just by chance. Thus, in such a system, diversity must be maintained, and its level determined, by the rate of species production (in the model, this is represented by random appearances of unique individuals).

The fragmentation of tropical forests during dry periods of the recent Ice Age (Figure 24.19) provided opportunities for allopatric speciation in the Tropics at a time when harsh conditions and restriction of habitats in temperate and arctic zones may have caused an increase in extinction rates there. However, if differences in diversity between temperate and tropical forests express recent high rates of species production in the Tropics, we would expect to find more species per genus in tropical forests than in temperate zone forests. In fact, tropical forests manifest their tremendous diversity as much at the family and genus levels as at the species level (Figure 24.20), and indeed appear to be relatively lacking in closely related (congeneric) species compared with temperate forests. Their great number of higher taxa (genera and families) provides evidence of the ancient roots of diversity there. It seems reasonable to conclude that if differences in speciation rate are responsible for differences in the diversity of forests, they have been persistent differences acting over very long periods.

Explaining global patterns of biodiversity remains one of the greatest challenges to ecologists today. The answers we seek are likely to involve both local ecological processes and processes acting over large scales of space and time. The latter have not traditionally been a part of ecological investigation, and they should give ecologists an impetus to expand their concepts of ecological systems. The next chapter presents an introduction to some historical and biogeographic processes and events that may have important implications for the structure and functioning of contemporary ecological systems.

(a) Glacial

(b) Pluvial

Figure 24.19 One reconstruction of the distribution of lowland rain forest in South America (a) during the height of glacial periods in the Northern Hemisphere and (b) at present.

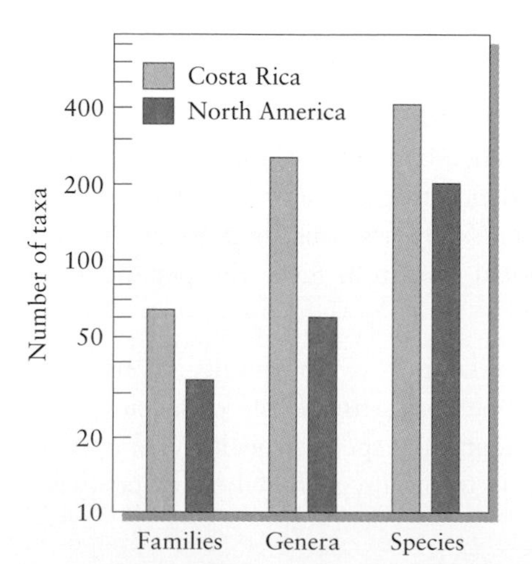

Figure 24.20 Diversity of trees at three taxonomic levels in a rain forest in eastern Costa Rica and in broad-leaved deciduous forests of eastern North America. From data in R. E. Ricklefs, in D. Otte and J. Endler (eds.), *Speciation and Its Consequences,* Sinauer Associates, Sunderland, Mass. (1989), pp. 599–622.

Summary

1. A conspicuous pattern revealed by studies of biological communities is the tendency of species diversity in tropical regions to greatly exceed that at higher latitudes.

2. Diverse tropical communities contain a greater variety of ecological niches, as well as a greater number of species, than temperate communities.

3. In regions of high species diversity, individual species are more likely to be habitat specialists than are their counterparts in regions of lower diversity. Hence, although more species occupy each habitat (alpha diversity), species also distinguish differences between habitats more finely (beta diversity). The total diversity of species within a large region containing many habitats is that region's gamma diversity.

4. One explanation for latitudinal differences in diversity—the time hypothesis—suggests that tropical habitats are older than temperate habitats and consequently have had more time to accumulate species. According to this hypothesis, diversity is nonequilibrial, and the number of species within a region continually increases.

5. Recently, thinking about diversity has been dominated by equilibrium theories, which state that habitats become saturated by species, which reach numbers determined by local ecological conditions. According to these theories, latitudinal differences in diversity result from differences in variety of resources and in how species interact to partition those resources that facilitate the coexistence of species.

6. Differences in the numbers of species on islands emphasize the importance of regional processes—immigration from a continent or from other islands—to the maintenance of species diversity. On continents, immigration of species to local areas reflects, in part, the rate of production of new species, which is also a regional process.

7. Several explanations for high plant diversity in the Tropics focus on the role of disturbance in creating mosaics of successional stages and in creating heterogeneous conditions for seedling establishment within gaps in the forest canopy.

8. Predators may enhance diversity among their prey by reducing prey populations (and hence competition for resources), thereby making coexistence easier. Evidence that predators and diseases may act in a density-dependent manner supports this hypothesis. Density-dependent and frequency-dependent predation favor the persistence of rare species and enhance diversity.

9. Several issues pertaining to community diversity have not been adequately resolved. One concerns the rate of species production, particularly whether environmental and species-specific characteristics promote higher rates of speciation in tropical regions, where diversity is high.

Suggested readings

Bush, M. B., R. J. Whittaker, and T. Partomihardjo. 1995. Colonization and succession on Krakatoa: An analysis of the guild of vining plants. *Biotropica* 27:355–372.

Case, T. J., and M. L. Cody. 1987. Testing island biogeographic theories. *American Scientist* 75:402–411.

Connell, J. H. 1978. Diversity in tropical rain forests and coral reefs. *Science* 199:1302–1310.

Cornell, H. V., and J. H. Lawton. 1992. Species interactions, local and regional processes, and limits to the richness of ecological communities: A theoretical perspective. *Journal of Animal Ecology* 61:1–12.

Currie, D. J. 1991. Energy and large-scale patterns of animal- and plant-species richness. *American Naturalist* 137:27–49.

Heywood, V. H. (ed.). 1996. *Global Biodiversity Assessment.* Cambridge University Press, Cambridge.

Janzen, D. H. 1970. Herbivores and the number of tree species in tropical forests. *American Naturalist* 104:501–528.

MacArthur, R. H. 1965. Patterns of species diversity. *Biological Reviews* 40:510–533.

MacArthur, R. H. 1972. *Geographical Ecology: Patterns in the Distribution of Species.* Harper & Row, New York.

Morton, S. R., and C. D. James. 1988. The diversity and abundance of lizards in arid Australia: A new hypothesis. *American Naturalist* 132:237–256.

Ricklefs, R. E. 1987. Community diversity: Relative roles of local and regional processes. *Science* 235:167–171.

Ricklefs, R. E., and D. Schluter (eds.). 1993. *Species Diversity in Ecological Communities: Historical and Geographical Perspectives.* University of Chicago Press, Chicago.

Rosenzweig, M. 1995. *Species Diversity in Space and Time.* Cambridge University Press, Cambridge.

Roughgarden, J. 1989. The structure and assembly of communities. In J. Roughgarden, R. M. May, and S. A. Levin (eds.), *Perspectives in Ecological Theory,* pp. 203–226. Princeton University Press, Princeton, N.J.

Terborgh, J. 1992. *Diversity and the Tropical Rain Forest.* Scientific American Library, New York.

Westoby, M. 1988. Comparing Australian ecosystems to those elsewhere. *BioScience* 38:549–556.

CHAPTER 25

HISTORY AND BIOGEOGRAPHY

The earth provides an ever-changing setting for the development of biological systems. Over millions of years of earth history, animals and plants have witnessed changes in climate and other physical conditions, rearrangements of geographic positions of continents and ocean basins, growth and wearing down of mountain ranges, evolution of biological novelties such as new tactics of predators and disease organisms, and catastrophic impacts with extraterrestrial bodies. These changes have helped to direct the course of evolution and diversification of organisms and have influenced the development of biological communities. The history of life reveals itself to us in the geochemical record of past environments, in fossil traces left by long-extinct taxa, and in the geographic distributions and evolutionary relationships of living species. All these lines of evidence show that life has taken tortuous and unpredictable paths and has suffered occasional setbacks.

The most obvious consequence of this history is a nonuniform distribution of animal and plant forms on the surface of the earth. Australia, for example, has many unique forms—koalas, kangaroos, and eucalyptus trees—because of its long isolation as an island continent surrounded by ocean barriers to the dispersal of terrestrial creatures. Every part of the earth has its own distinctive fauna and flora. Even the major ocean basins, interconnected as they are by continuous corridors of water, have somewhat differentiated biotas, isolated by ecological barriers of temperature and salinity.

For ecologists, biological history raises two potential problems. First, the structure and functioning of organisms may be influenced as much by ancestry as by local environment. The truth of this proposition diminishes our confidence in matching the morphology, physiology, and behavior of organisms to the conditions and resources of their environments. For example, the marsupial mode of reproduction (involving, among other characteristics, early birth and subsequent development of young in a pouch) is uniquely a property of the marsupial line of mammalian evolution, not a result of unique ecological properties of the continent of Australia, where marsupials are now most diverse. Biologists refer to characteristics shared by a lineage irrespective of environmental factors as **phylogenetic effects.** These effects reflect the inertia of evolution—the lack of change of some attributes in the face of change in the environment. Ecologists recognize that such effects may influence structure and functioning in ecological systems, although this is difficult to demonstrate experimentally. Imagine that the plants and animals of Australia were replaced by a similar number of taxa from other regions with a similar climate. Would these new ecosystems function in the same manner, with similar levels of biological productivity and responses to environmental perturbation?

The second problem raised by biological history is that history and geography also affect the diversification of species. To the extent that the histories of each region of the earth have differed, we might also expect biological diversity and the development of biological communities to differ. Because of this, it may be difficult to interpret patterns of diversity solely in terms of local environmental conditions.

A simple way to test for historical and phylogenetic effects is to compare systems that occur under similar environmental conditions in different geographic regions. According to the principle of convergence, which we shall discuss in greater detail later in this chapter, inhabitants of similar environments with disparate historical origins often resemble one another because they adapt to similar ecological factors. If, on one hand, the characteristics of two systems closely parallel environmental conditions and do not differ by region, we may conclude that the influence of ecology predominates over that of history, and we may safely ignore historical factors. If, on the other hand, similar habitats support different numbers of species in different regions, then ecologists have to accept the impact of history on the present and incorporate historical factors into their analyses. Thus, consideration of history and biogeography is important to ecologists not only because of the intrinsic interest of these topics, but also because of their potential for explaining the character of present-day ecological systems.

In this chapter, we shall first briefly examine some historical processes that have shaped the distribution and development of ecological systems. Then we shall examine the principle of convergence, paying particular attention to the diversity of biological communities. We shall see that history and biogeography have indeed influenced the character of local communities and have played an important role in the development of patterns of diversity.

The geologic time scale

The earth formed about 4.5 billion years ago, and life arose within its first billion years. For most of the history of the earth, life forms remained primitive. Physical conditions at the earth's surface and the ecological systems that developed at that time were strikingly different from those of the present. The atmosphere had extremely little oxygen, and early microbes used strictly anaerobic metabolism. The oxygen in our present-day atmosphere was largely produced by photosynthetic microorganisms during this Archeozoic time of earth history. At some point, however, the oxygen concentration of the atmosphere became high enough to sustain oxidative metabolism and made it possible for more complex life forms to evolve. The eukaryotic cell, which is the basic building block of all modern complex organisms, is a product of the last billion years of evolution. Little record of this development exists because most very ancient life forms did not have hard skeletons or shells that form fossils. Much evidence of early complex life forms consists of tracks and burrows in the mud in which they lived.

All of this changed about 600 million years ago with the appearance in the fossil record of most of the modern phyla of invertebrate organisms. Echinoderms, arthropods, mollusks, and brachiopods rose to prominence in oceans of that period, as did other life forms—evolutionary experiments, so to speak—that are no longer with us. No one knows for sure why animals began to protect themselves with hard shells or outer skeletons at that moment in history, but paleontologists regard the occasion as the beginning of life in its modern form. The interval between that critical moment and the present, occupying about one-eighth of the total history of the earth, has been divided into a series of eras, periods, and epochs (Table 25.1). The first of these divisions is the Paleozoic era—Paleozoic means "old animals"—and the Cambrian is the first period within the Paleozoic era.

The divisions of geologic time coincide with noticeable changes in the fauna and flora of the earth, changes easily perceived in the fossil record. Thus the end of the Cambrian period marks the disappearance of several prominent groups from the fossil record and their replacement by others not seen before. The major divisions at the end of the Paleozoic era and between the Mesozoic ("middle animals") and Cenozoic ("recent animals") eras also coincide with major extinctions of animal taxa: trilobites, among others, in the first case and dinosaurs in the second. Thus, boundaries between various periods of paleontological time signal either major disruptions or less pervasive changes in the course of development of life forms. In some cases, discontinuities in the rock strata at these boundaries reflect changes in the earth's crust. In at least one case, at the end of the Cretaceous period (Mesozoic era), disruption was caused by the explosive impact of an extraterrestrial object on the surface of the earth.

TABLE 25.1 **The geologic time scale**

ERA	PERIOD	EPOCH	DISTINCTIVE FEATURES	YEARS BEFORE PRESENT
Cenozoic	Quaternary	Recent	Modern humans	11,000
		Pleistocene	Early humans	1,700,000
	Tertiary	Pliocene	Large carnivores	5,000,000
		Miocene	First abundant grazing animals	23,000,000
		Oligocene	Large running mammals	38,000,000
		Eocene	Many modern types of mammals	54,000,000
		Paleocene	First placental mammals	65,000,000
Mesozoic	Cretaceous		First flowering plants; extinction of dinosaurs and ammonites at end of period	135,000,000
	Jurassic		First birds and mammals; dinosaurs and ammonites abundant	192,000,000
	Triassic		First dinosaurs; abundant cycads and conifers	223,000,000
Paleozoic	Permian		Extinction of many kinds of marine animals, including trilobites	280,000,000
	Carboniferous	Pennsylvanian	Great coal-forming forests; conifers; first reptiles	321,000,000
		Mississippian	Sharks and amphibians abundant; large primitive trees and ferns	345,000,000
	Devonian		First amphibians and ammonites; fishes abundant	405,000,000
	Silurian		First terrestrial plants and animals	438,000,000
	Ordovician		First fishes; invertebrates dominant	495,000,000
	Cambrian		First abundant record of marine life; trilobites dominant, followed by massive extinction at end of period	570,000,000
	Precambrian		Fossils extremely rare, consisting of primitive aquatic plants	

Source: J. H. Brown and A. C. Gibson, *Biogeography,* Mosby, St. Louis (1983).

Continental drift

The earth's surface has been very restless over its history. Continents are islands of low-density rock floating on the denser material of the earth's interior. Giant convection currents in the semimolten material of the mantle carry continents along like gigantic logs on the surface of water. At times in the past continents have coalesced, and at other times they have drifted apart. This movement of landmasses on the surface of the earth is

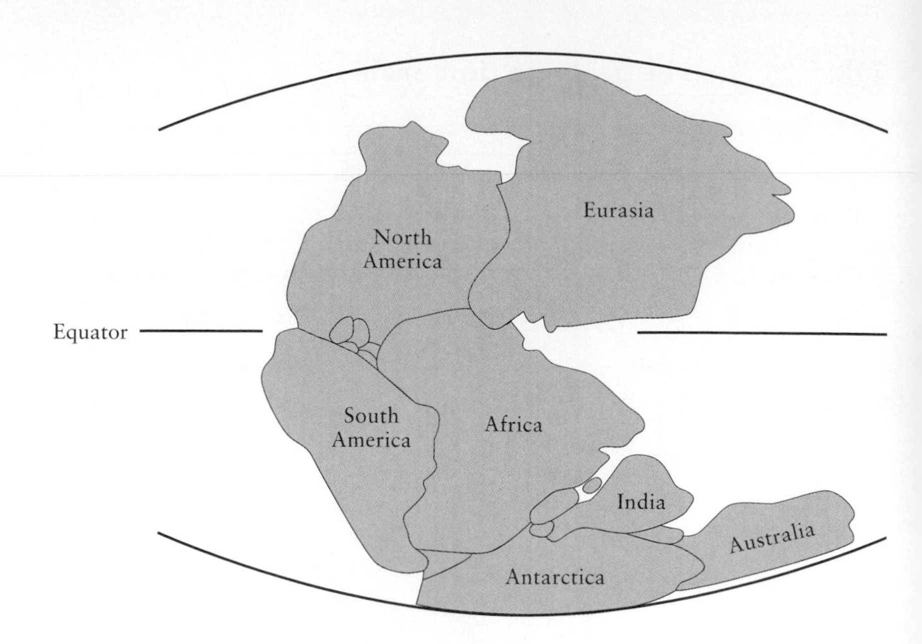

Figure 25.1 Approximate positions of continents at the beginning of the Mesozoic era, when all land had coalesced into a single landmass, known as Pangaea. After E. C. Pielou, *Biogeography,* Wiley, New York (1979).

called **continental drift.** The process has two extremely important consequences for ecological systems: First, the positions of continents and major ocean basins profoundly influence weather patterns, as we shall see below. Second, continental drift creates and breaks down barriers to dispersal, alternately connecting and separating evolving biotas in different regions of the earth.

Because we are interested in the origin of modern ecological systems, we shall consider continental drift beginning with the early part of the Mesozoic era, about 200 million years ago, when all the continents were joined in a giant landmass known as **Pangaea** (Figure 25.1). By 135 million years ago, at the beginning of the Cretaceous period, the northern continents, which collectively made up **Laurasia,** had separated from the southern continents, **Gondwana.** In addition, Gondwana itself had begun to break up into three parts: West Gondwana, including Africa and South America; East Gondwana, including Antarctica and Australia; and India, which had separated from present-day Africa and was drifting toward Asia

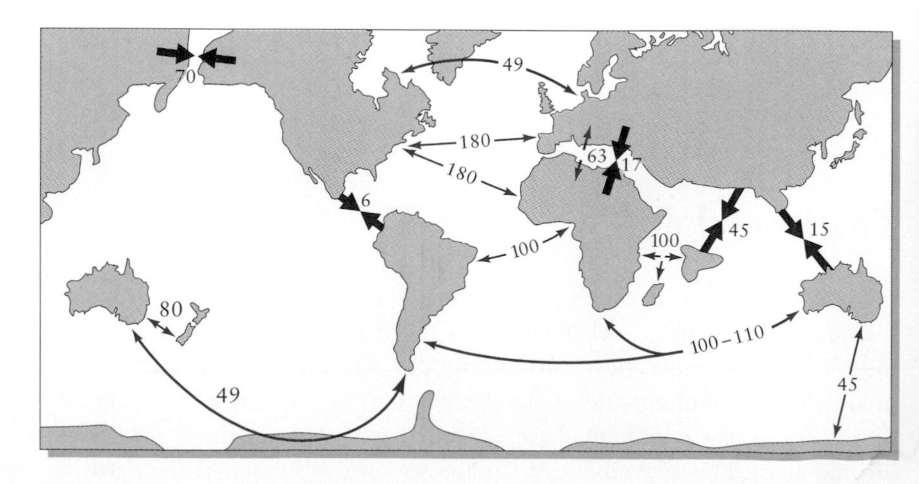

Figure 25.2 Estimates of times (in millions of years before present) at which direct dispersal routes between landmasses were created or broken. Black converging arrows show joins (the joining between Australia and Asia refers to the narrowing of a gap and the appearance of stepping-stone islands). Colored diverging arrows show separations. After E. C. Pielou, *Biogeography,* Wiley, New York (1979).

TABLE 25.2 **Estimated times of some of the major biogeographic events in earth history caused by continental drift**

PERIOD OR EPOCH	TIME (MILLION YEARS BEFORE PRESENT)	EVENT
Early Triassic	200	The continental crust formed a single continent, Pangaea
Late Triassic	180	West Laurasia (North America) ↔ Africa; West Gondwana (Africa + South America) ↔ India ↔ East Gondwana (Australia + Antarctica)
Early Cretaceous	135–125	South America ↔ Africa in far south because of rotational movement
Mid-Cretaceous	110–100	South America ↔ Africa at latitude of Brazil; Africa ↔ Madagascar ↔ India; Africa, India, and Australia all drifting northward
Late Cretaceous	80	North America ↔ (Europe + Greenland); (Antarctica + Australia) ↔ New Zealand + New Caledonia)
Very late Cretaceous	70	Contact made between northwestern North America and northeastern Siberia
Very early Paleocene	63	Africa ↔ Europe (temporarily)
Eocene	49	Dispersal route between North America and Eurasia, from being predominantly via North Atlantic, switches to Beringia because North Atlantic becomes wider and Beringia becomes warmer
Eocene	~49	Australia ↔ Antarctica
Eocene	45	India drifts into contact with Asia
Oligocene	~30	Turgai Strait (east of Ural Mountains) finally dries up
Miocene	17	Europe and Africa rejoined
Miocene	15	The narrowing gap between Australia and Southeast Asia, and the appearance of stepping-stone islands, permits plant dispersal
Pliocene	6	North America and South America joined by land bridge

Note: Double-headed arrows denote separations
Source: E. C. Pielou, *Biogeography,* Wiley, New York (1979).

(Table 25.2 and Figure 25.2). Seventy million years later, at the end of the Cretaceous period and the Mesozoic era, South America and Africa were widely separated. The connection between Australia and South America through a temperate Antarctica finally dissolved about 50 million years ago. At about the same time, in the Northern Hemisphere a widening Atlantic Ocean finally separated Europe and North America, but a land bridge had already formed on the other side of the world between North America and Asia.

Many details of continental drift have yet to be resolved, particularly in such complicated areas as the Caribbean Sea, Australasia, and the Mediterranean Sea–Persian Gulf region. Nevertheless, the past history of connections between the continents endures in the distributions of animals and plants. We have only to look at the distribution of the flightless ratite birds to see the connection between the southern continents: emus and cassowaries in Australia and New Guinea, rheas in South America, ostriches in Africa, and the extinct moas of New Zealand all descended from a common ancestor that inhabited Gondwana before its breakup.

Biogeographic regions

The distributions of animals suggested to the nineteenth-century naturalist Alfred Russel Wallace, codiscoverer with Darwin of the theory of evolution by natural selection, six major biogeographic regions (Figure 25.3). We now know that these correspond to landmasses isolated many millions of years ago by continental drift. Over the course of that isolation, animals and plants in each region developed distinctive characteristics independently of evolutionary changes in other regions. The **Nearctic** and **Palearctic** regions, corresponding roughly to North America and Eurasia, maintained connections across either what is now Greenland or the Bering Strait between Alaska and Siberia throughout most of the past 100 million years. As a result, these two areas share many groups of animals and plants. European forests seem familiar to travelers from North America; few species are the same, but both regions have representatives of many of the same genera and families.

The continents of the Southern Hemisphere, particularly Africa (**Ethiopian** region) and Australia (**Australian** region), experienced long histories of isolation from the rest of the terrestrial world, during which time many distinctive forms of life evolved in each. The **Oriental** region includes tropical areas of Southeast Asia, which were isolated from tropical areas of Africa and South America, in addition to contributions from the landmass of India, which drifted into contact with southern Asia about 45 million years ago. As one might expect, temperate and tropical Asia have closer affinities than temperate North America (Nearctic) and tropical South America (**Neotropical** region) because of the continuous land connection between them. Indeed, the temperate forests of Asia contain a

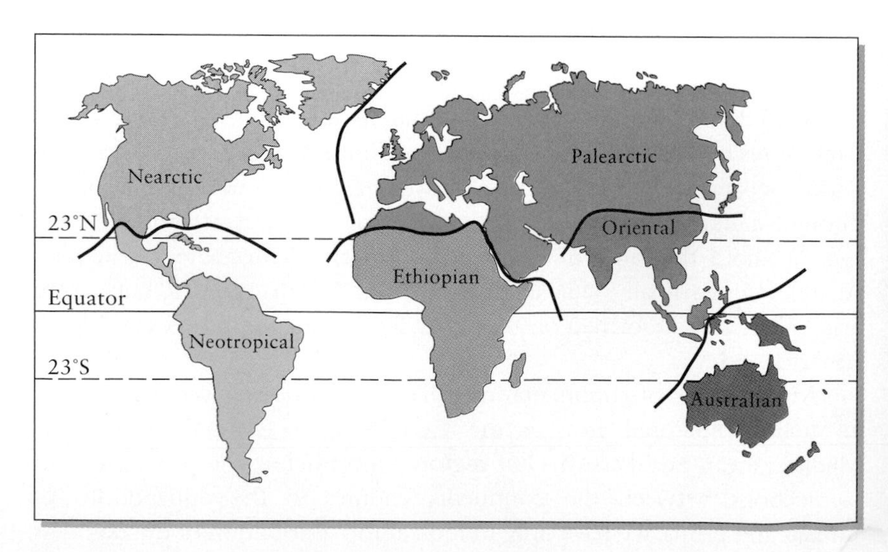

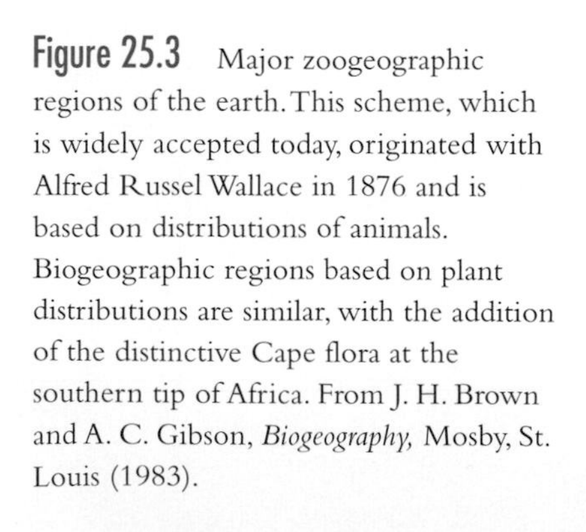
Figure 25.3 Major zoogeographic regions of the earth. This scheme, which is widely accepted today, originated with Alfred Russel Wallace in 1876 and is based on distributions of animals. Biogeographic regions based on plant distributions are similar, with the addition of the distinctive Cape flora at the southern tip of Africa. From J. H. Brown and A. C. Gibson, *Biogeography,* Mosby, St. Louis (1983).

high percentage of species derived primarily from tropical forests, whereas those of temperate North America lack such species.

Climate history

The climate patterns of the earth ultimately depend on energy received from the sun, which warms land and seas and evaporates water. Within this framework, the distribution of heat over the surface of the earth depends largely on the circulation of the oceans, which is driven by the rotation of the earth and constrained by the positions of the continents. In periods when polar regions, which receive relatively little solar energy, are covered by oceans with circulation to tropical areas, currents distribute heat rather evenly over the surface of the earth, and temperate climates extend quite close to the poles. When polar regions are occupied by landmasses or landlocked oceans, they can become very cold indeed, as they are at present. But this has not always been the case.

When the continents coalesced into one or a few large landmasses, ocean currents circulated quite freely to high latitudes, and the climate of the earth was much more equable than it is now. Fifty million years ago, large portions of North America and Europe were tropical, and the Antarctic land connection between South America and Australia supported luxuriant temperate vegetation and animal life, or so the fossils in Antarctic rocks tell us. However, as Antarctica drifted over the South Pole and as the northern polar ocean became trapped between North America and Eurasia, the earth's climate became more strongly differentiated into tropical (equatorial) and temperate (polar) zones. One consequence of the cooling (and drying) trend at high latitudes was the retreat to lower latitudes of plants and animals that could not tolerate freezing; this resulted in greater distinction between temperate and tropical biotas. During the early part of the Tertiary period, what is now temperate North America supported a mixture of tropical and temperate forms growing side by side. Today these plants and animals occupy different climate zones, greater stratification of climate being matched by greater stratification of the biota.

These gradual changes in climate had profound effects on geographic distributions of plants and animals. During the past 2 million years, however, the gradual cooling of the earth gave way to a series of violent oscillations in climate that had immense effects on habitats and organisms in most parts of the world. This was the Ice Age, or Pleistocene epoch—alternating periods of cooling and warming that led to the advance and retreat of ice sheets at high latitudes over much of the Northern Hemisphere and caused cycles of wet and dry or cool climates in the Tropics. Ice came as far south as Ohio and Pennsylvania in North America and covered much of northern Europe, driving vegetation zones southward, restricting tropical forest to isolated refuges of moist conditions, and generally disrupting biological communities all over the world.

One of the most striking and best-documented examples of this disruption concerns the migration of forest trees of eastern North America.

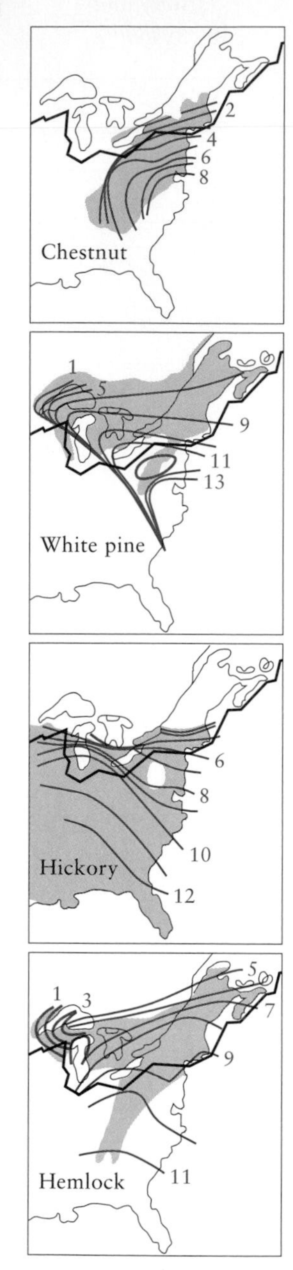

Figure 25.4 Migration of four species of trees in eastern North America from their Pleistocene refuges to their present distributions (shaded areas) following the retreat of the glaciers. Numerals associated with the distribution lines indicate thousands of years before the present. The bold line represents the maximum extent of glaciers about 18,000 years ago. From M. B. Davis, *Geoscience and Man* 13:13–26 (1976).

Pollen grains deposited in the lakes and bogs left by retreating glaciers record the coming and going of plant species in the northeastern United States. These records show plainly that the composition of plant associations changed as species migrated over different routes across the landscape. After the latest glaciers began to retreat about 12,000 years ago, the general pattern of reforestation began with spruce forest, which dominated the area until about 10,000 years ago, followed by associations of pine and birch, which were later replaced by more temperate species such as elm and oak.

The migrations of tree species from their southern refuges since the height of the last glaciation are mapped for some representative species in Figure 25.4. For species such as hemlock and hickory, postglacial migration involved northerly range extension from southern regions across most of the eastern United States. In contrast, white pine and chestnut appear to have emerged from refuges in the Carolinas and expanded their ranges to the west as much as to the north. Thus, the composition of forests during the past 10,000 years has included combinations of species that do not occur anywhere in eastern North America at the present time, and has lacked combinations of species that do occur at present. For some species, the environment changed too rapidly during the Ice Age, and they disappeared altogether.

Catastrophes in earth history

Although the Ice Age brought dramatic changes in climate and spelled the extinction of many forms of plants and animals, it pales beside the total disruption occasionally visited upon the earth by collisions with asteroids and other extraterrestrial bodies or resulting from major geologic upheavals in the earth's crust. Collisions have occurred many times in earth history, with consequences in direct proportion to the energy liberated by the impact. On occasion, such impacts have caused widespread destruction of ecosystems and extinction of many kinds of life on earth.

The most famous of these impacts occurred about 65 million years ago. Evidence now points to the Yucatán Peninsula of Mexico as the point of impact, and indications of the explosion and its aftermath appear as a layer of clay in geologic strata all over the world. Scientists have estimated that an asteroid 10 km in diameter traveling 25 km per second may have been responsible. Such a collision would have released enough energy to cause massive tidal waves around the world, start fires on an unprecedented scale, and throw enough dust into the air to block the sun and cool the earth's surface for years. As a result, much of the biomass of the earth would have been destroyed, either immediately by the direct effects of the impact or more slowly during the weeks and months afterward, and plant production in the oceans and on the land would have slowed to a standstill. All the killing left a thin band of carbon preserved in sedimentary rocks of the time, along with thick deposits resulting from massive erosion in some areas.

One of the results of the impact was the extinction of a large fraction of the species on earth and of many higher taxa as well. Such episodes are referred to as **mass extinctions.** Not all plants and animals felt the impact equally. All the dinosaurs disappeared, as did some large groups of marine organisms. Most higher taxa of plants survived, perhaps many of them as seeds in the soil, and mammals and birds survived to fill ecological vacancies left by the dinosaurs.

Catastrophes of such magnitude have happened infrequently—perhaps at intervals of tens or hundreds of millions of years—yet often enough to disrupt ecosystems and change the course of community development. Each major catastrophe, whether it originates within the earth through geologic upheavals or in outer space, brings about a period of extreme environmental stress. Geologists have found evidence in the geochemistry of sediments formed after such catastrophes that thousands of years may be required for environmental conditions to return to normal. The fossil record shows that some ecosystems—tropical reefs are a case in point—may disappear for millions of years, sometimes to be rebuilt by new kinds of reef-forming organisms.

From the perspective of community development, catastrophes have several important consequences. They may eliminate species and higher taxa and thus greatly reduce diversity in most systems. They may foster rapid evolutionary responses to new types of conditions, and these changes many remain long after conditions have returned to "normal." Finally, they may create opportunities for development of new types of biological associations. Although their effects cannot be easily identified or interpreted from present-day conditions and communities, such unique events in the past reach down through history to influence the present.

Organismal convergence

Just as long periods of isolation have led to unique life forms in many regions of the earth, similar environmental conditions in each of these regions have also led to the evolution of similar solutions to common problems. Plants inhabiting areas with Mediterranean climates in western North America and in southern Africa have different evolutionary origins reflecting their more than 100 million years of isolation, but they share similar growth forms and similar adaptations to winter rainfall–summer drought conditions. Thus, the different evolutionary histories and taxonomic affinities of plants and animals of the earth's regions are in part obliterated by convergence in form and function.

Convergence is the process whereby unrelated species living under similar ecological conditions come to resemble one another more than their ancestors resembled one another. There are many examples of convergent form and function. Where woodpeckers are absent from a fauna, as they are from many isolated islands, other species may adapt to fill their role and become convergent on the woodpecker lifestyle (Figure 25.5). Rain forests in Africa and South America are inhabited by plants and

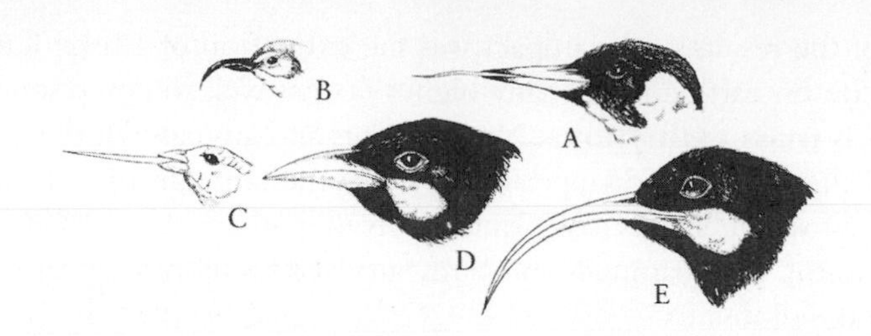

Figure 25.5 Unrelated birds that have become convergently adapted to extract insects from wood. (a) European green woodpeckers excavate with their beaks and probe with their long tongues. (b) Hawaiian honeycreepers *(Heterorhynchus)* tap with their short lower mandibles and probe with their long upper mandibles. (c) Galápagos woodpecker-finches trench with their beaks and probe with cactus spines. New Zealand huias (now extinct) divided foraging roles between the sexes. Males (d) excavated with their short beaks, and females (e) probed with their long beaks. After D. Lack, *Darwin's Finches,* Cambridge University Press, Cambridge (1947).

animals that have different evolutionary origins but are remarkably similar in appearance (Figure 25.6). Plants and animals of North and South American deserts resemble each other morphologically more than one would expect from their different phylogenetic origins. Similarities have also been noted in the behavior and ecology of Australian and North American lizards, despite the fact that they belong to different families and have evolved independently for perhaps 100 million years. Dolphins and penguins evolved from terrestrial ancestors, but have body shapes more closely resembling those of tuna.

Convergence exists, and it reinforces our belief that adaptations of organisms to their environments obey certain general rules governing structure and function. However, detailed studies often turn up remarkable differences between the plants and animals in superficially similar environments. Despite striking convergences among desert-dwelling organisms, for example, the ancient Monte Desert of South America lacks bipedal, seed-eating, water-independent rodents like the kangaroo rats of North America and the gerbils of Asia. Among frogs and toads, several South American forms have carried adaptation to desert environments a step further than their North American counterparts: they construct nests of foam to protect their eggs from drying out. Differences between the Australian agamid lizard *Amphibolurus inermis* and its North American iguanid analog, *Dipsosuarus dorsalis,* include diet, optimal temperature for activity, burrowing behavior, and annual cycle, even though at first glance the species are dead ringers for each other. Such differences are thought to reflect unique aspects of the evolutionary history of organisms in different regions, or perhaps subtle differences in the environments of the two regions.

An example drawn from the dispersal of seeds by ants illustrates how difficult it is to interpret similarities and differences in the attributes of organisms in different geographic regions. The seed-ant relationship—a type of mutualism—is encouraged by edible appendages, called elaiosomes, on the seeds. As we saw in Chapter 21, ants gather these elaiosomes, with the seeds attached, and carry them into their underground nests. In so doing, they effectively disperse and plant the seeds. This seed trait is uncommon in most of the world, being restricted primarily to a few species of trees in mesic environments. In Australia and the Cape region of South Africa, however, the trait is well represented among xerophytic shrubs, and it is associated with ecological and morphological features lacking in ant-dispersed plants elsewhere.

Ecologists have not resolved whether this difference between plants in Australia and South Africa and those elsewhere is a consequence of the unique evolutionary histories of Australian and Cape floras. Indeed, it has been suggested that the poor soils of Australia and South Africa make it costly for plants to produce the nutritionally expensive fleshy fruits that are dispersed by birds and mammals in most parts of the world. Thus, dispersal of seeds by ants in Australia and South Africa may represent an accident of local geology rather than reflect the unique historical origins of the flora and fauna. Similar reasoning suggests that many of the distinctive attributes of the reptile fauna of Australia have resulted not from its unique evolutionary history, but from the absence of avian predators, whose scarcity could be attributed ultimately to the poor nutrient status of the vegetation and the resulting paucity of insects. These issues have yet to be fully resolved. Certainly, ecologists must ensure that the comparisons they make involve habitats with closely matched physical characteristics. Otherwise, they cannot conclude unequivocally that differences in

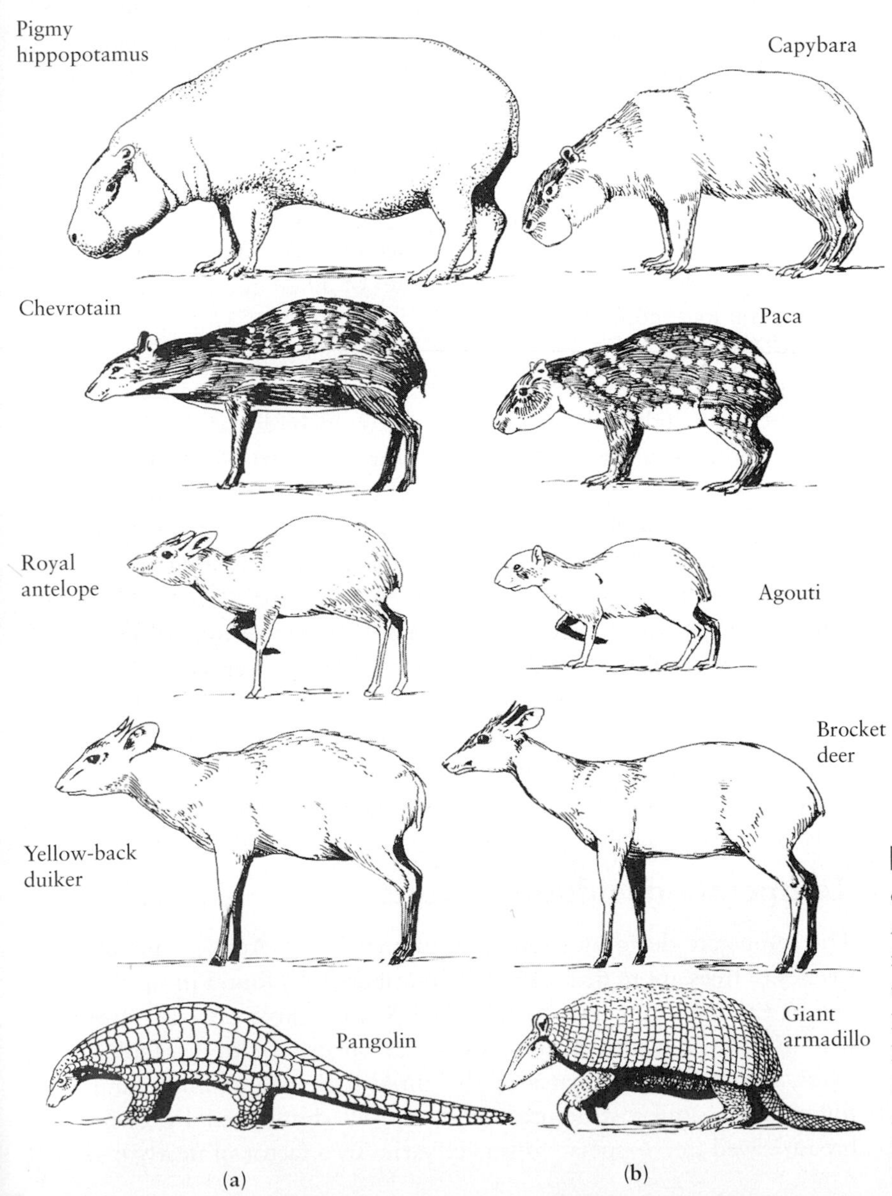

Figure 25.6 Morphological convergence between unrelated (a) African and (b) South American rain forest mammals. Each pair is drawn to the same scale. After F. Bourliere, in B. J. Meggars, E. S. Ayensu, and W. D. Duckworth (eds.), *Tropical Forest Ecosystems in Africa and South America: A Comparative Review,* Smithsonian Institution Press, Washington, D.C. (1973), pp. 279–292.

structure or function have resulted from different histories rather than different environments. On the whole, however, convergence of form and function under similar environmental conditions is a broadly applicable principle of ecology and evolution. Difficulties with the concept in particular situations are more likely to reflect a lack of adequate data about environment and evolutionary history than difficulties with the principle itself. Thus, the principle of convergence has also found application to tests of local determination of species diversity and other attributes of ecological communities.

Community convergence

According to the principle of convergence, we would expect independently derived communities that occupy similar habitats in different regions to have similar numbers of species. We can test this principle quite simply by comparing biodiversity in similar habitats in different biogeographic regions having different regional diversity. When local diversity is the same, in spite of differing regional diversity, it is likely that local factors have predominated in determining the local coexistence of species. However, when local diversity varies in parallel with regional diversity, in spite of similarity in local environments, then one must conclude that regional processes and the unique histories of different regions have left an imprint on local community diversity.

Where ecologists have tested for convergence of community properties by interregional comparison, they have had mixed results. In one of the first such studies, the relationship between bird species diversity and habitat complexity was found to be similar in eastern North America and Australia (Figure 25.7). This result agreed with the idea that species diversity depends on habitat type, that it is ultimately determined by interactions among species within the habitat, and that it is insensitive to the influences of historical differences between continents. One problem with this conclusion is that regional diversities of birds in Australia and North America are similar; so one would expect local diversity to be similar under either local or regional determination of local diversity. Other studies have revealed strong differences in both regional and local diversities between continents, suggesting that regional and historical factors also play a role in determining diversity. We shall look at just two examples here.

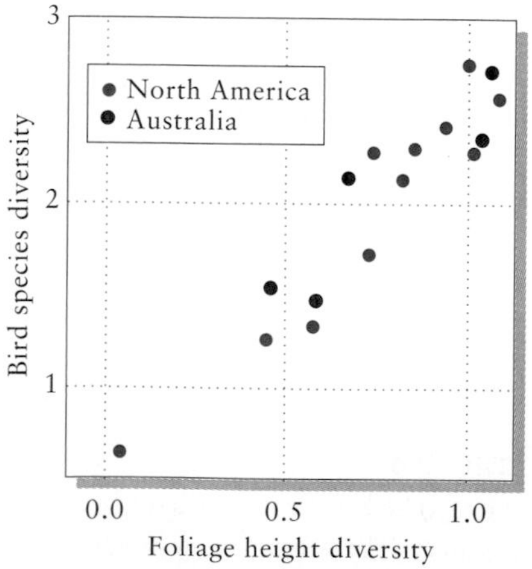

Figure 25.7 Relationship between bird species diversity and structural complexity of habitat, as indicated by foliage height diversity, in moist temperate habitats of Australia and North America. From H. F. Recher, *Am. Nat.* 103:75–80 (1969).

Temperate deciduous forests

The temperate deciduous forests of eastern North America include 253 species of trees, more than twice the number (124) found in similar habitats in Europe. Temperate eastern Asia, whose climate is also similar to that of eastern North America, has 729 species of trees (Table 25.3). Thus, although the climates of the three regions are similar and their forests have similar structures—they are all dominated by deciduous, broad-leaved trees—species diversity varies by a factor of nearly 6 among

TABLE 25.3 Taxonomic diversity of trees in the temperate deciduous broad-leaved forests of eastern North America, Europe, and Eastern Asia

	NUMBER OF TREE TAXA IN		
TAXA	Europe	Eastern North America	Eastern Asia
Orders	16	26	37
Families	21	46	67
Genera	43	90	177
Species	124	253	729
Percentage of genera predominantly tropical	5	14	32
Number of genera in fossil record	130	60	122

Source: R. E. Latham and R. E. Ricklefs, in R. E. Ricklefs and D. Schluter (eds.), *Species Diversity in Ecological Communities: Historical and Geographical Perspectives,* University of Chicago Press, Chicago (1993), pp. 294–314.

the different regions. These figures represent the total diversity of each region, but local diversity within small areas of uniform habitat exhibits parallel differences. Thus regional diversity and local diversity appear to be closely related.

Part of the greater diversity in Asia results from a greater proportion of species (32%) belonging to predominantly tropical genera. A continuous corridor of forest habitat from the Tropics of Southeast Asia to the north has facilitated invasion over evolutionary time scales of temperate ecosystems by tropical plants and animals. In the Americas, the wet Tropics of Central America are separated from the moist, temperate areas of North America by a broad subtropical band of dry vegetation. In Europe, the Mediterranean Sea and arid North Africa effectively separate temperate ecosystems from tropical Africa.

The fossil record suggests an ancient origin for the diversity anomaly among eastern North America, Europe, and eastern Asia. Almost twice as many genera of trees occur as fossils in eastern Asia and Europe as in North America (Table 25.3). However, although far more genera of trees occur in the fossil record of Europe than in that of North America, a large proportion of the European genera became extinct in association with the cooling of north temperate climates leading to the Ice Age, while few genera of North American trees disappeared. As Europe cooled, the Alps and the Mediterranean Sea posed effective barriers to southward movement, and many cold-intolerant plant taxa died out. In North America, southward migration to areas bordering the Gulf of Mexico was always possible during cold periods.

Tropical mangrove ecosystems

A similar anomaly exists in the species diversity of mangrove forests between the Caribbean region and the Indo-West Pacific region. Mangroves are tropical forests that occur within tidal zones along coastlines and river deltas (Figure 25.8). Mangrove trees tolerate high salt concentrations and anaerobic conditions in the water-saturated sediments in which they take root. Fifteen lineages of terrestrial trees have independently colonized mangrove habitat, and several of these have subsequently diversified there. We cannot explain the much greater diversity of Indo-West Pacific mangroves on the basis of habitat, because both regions have roughly equal areas of a similar variety of mangrove habitats (Figure 25.9). Within the Indo-Pacific region, the small areal extent of suitable mangrove habitat in East Africa and Madagascar (region D) may explain low mangrove diversity there.

The large diversity anomaly in mangroves appears to have resulted from plant taxa invading mangrove habitat more frequently in the Indo-West Pacific than in the Caribbean region, although the reasons for this are not clear. Possibly, the terrestrial habitats fringing much of the Caribbean were arid during the latter part of the Cenozoic era. As a result, wet terrestrial forest vegetation would have had little direct contact with mangrove habitat and thus there would have been few opportunities for terrestrial taxa to adapt gradually to mangrove conditions. This has not been a limiting factor in Southeast Asia, where wet conditions have prevailed in tropical habitats throughout most of the period of evolution of modern trees. In addition, much of Malaysia consists of islands of various size scattered on a shallow continental shelf, perhaps affording ideal conditions for isolation of populations in mangrove habitat and formation of new species of mangrove specialists.

Figure 25.8 Mangrove vegetation in an estuary on the Pacific coast of Costa Rica. Note the prop roots of *Rhizophora* trees at left and the buttressed trunks of *Pelliciera* at right. These trees are established on mud substrate within the tidal zone; hence the soil is flooded periodically with salt water.

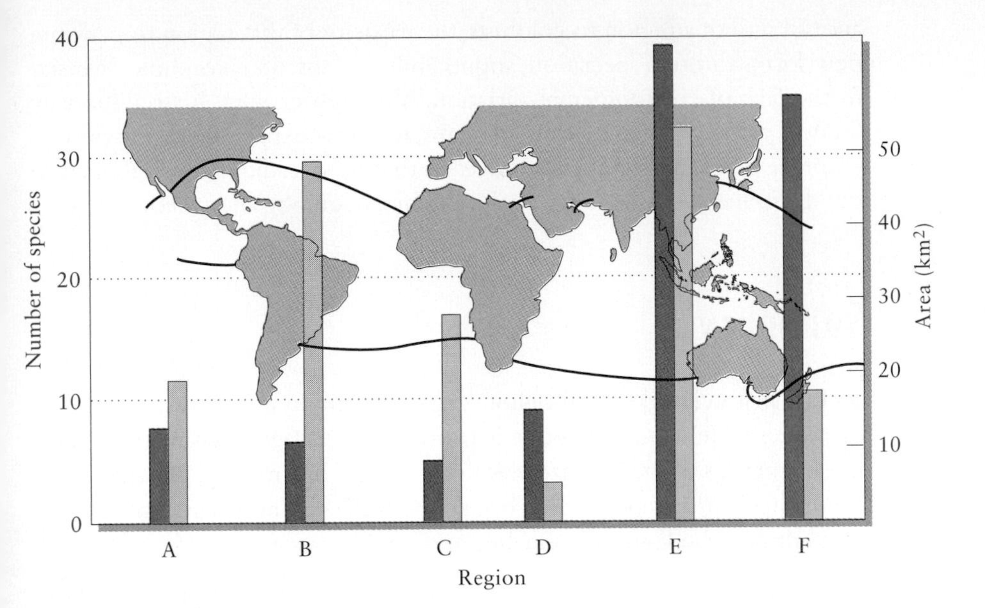

Figure 25.9 The limits of distribution of mangrove vegetation (black lines) along coasts of continents and islands of the world. Bars show the areal extent of mangrove habitat (gray) and numbers of species of mangrove trees and shrubs (colored) in each of six regions: (A) eastern Pacific Ocean, (B) Caribbean Sea and western Atlantic Ocean, (C) eastern Atlantic Ocean, (D) western Indian Ocean, (E) eastern Indian Ocean, and (F) western Pacific Ocean. After V. J. Chapman, *Tropical Ecol.* 11:1–19 (1970); P. Saenger, E. J. Hegerl, and J. D. S. Davie, *Environmentalist* 3 (suppl. 3):1–88 (1983).

The scales of processes regulating biodiversity

Perhaps the temperate forest and mangrove examples that we have just considered represent extremes, although similar disparities between the more diverse Indo-West Pacific region and the less diverse Caribbean region are found in marine communities, and numerous less striking examples have been reported. Nonetheless, these examples emphasize that the history and biogeographic position of a region may influence the diversity of both the entire region and its local habitats. Interactions of species within local habitats make up only half of the diversity equation.

Several processes are important to the regulation of biodiversity, each with a different characteristic scale of time and space. Scale in space varies from the activity ranges of individuals, through geographic and ecological dispersal of individuals within populations, to the expansion and contraction of geographic ranges of populations. Scale in time varies according to rates of individual and population movements, population interactions, and selective replacement of genotypes within populations (evolution). Both local, contemporary processes and regional, historical processes shape community attributes. The fate of a local population depends partly on the tendencies of physical conditions, interspecific competition, and predation to reduce local population size. Balancing these tendencies is the immigration of individuals from surrounding areas of population surplus. The persistence of a local population depends on the balance between these factors.

Local diversity of species depends on local rates of extinction—resulting from predators, disease, competitive exclusion, and changes in the physical environment—and regional rates of species production and immigration. Every point on earth has limited accessibility, via dispersal, to sources of colonizing species. Local diversity depends not only on the capacities of environments to support a variety of species, but also on the accessibility of a region to colonists, the capacity of that region to generate new forms through speciation, and its ability to sustain taxonomic diversity in the face of environmental variation. Although ecology has traditionally focused on local, contemporary systems, it is now expanding its purview to embrace global and historical processes that have traditionally belonged to the disciplines of systematics, evolution, biogeography, and paleontology.

Summary

1. Life first evolved several billion years ago, but an abundant fossil record of modern life appeared about 570 million years ago, a point that marks the beginning of the Paleozoic era of geologic history. The Mesozoic era, dominated on land by reptiles, began about 220 million years ago; the age of mammals, the Cenozoic era, began 65 million years ago.

2. The positions of the continents have changed continuously throughout the evolution of life, opening and closing different pathways of dispersal between continental landmasses and ocean basins and greatly altering climates on earth.

3. Because animals and plants have evolved independently on different continents during prolonged periods of geographic isolation, we can distinguish six major biogeographic regions: the Neotropical, Ethiopian, and Australian regions, derived from the former landmass of Gondwana, and the Oriental, Palearctic, and Nearctic regions, derived for the most part from the Northern Hemisphere landmass of Laurasia.

4. The climate of the earth cooled considerably during the Cenozoic era, causing tropical environments to contract to a narrower equatorial band and causing temperate and arctic environments, which are dominated by freezing temperatures for a part of the year, to expand in extent.

5. The Cenozoic cooling trend culminated in the Ice Age (alternating periods of glacial advance and retreat in the Northern Hemisphere), which caused extinctions of many species of plants and animals.

6. Infrequent global catastrophes have punctuated the development of life when large extraterrestrial bodies have hit the earth. One of the best known of these events occurred 65 million years ago, causing the extinction of the dinosaurs and other higher taxa of animals and bringing the Mesozoic era to a close. Such catastrophic changes in environments and their inhabitants have opened up new opportunities for evolution and have resulted in drastic reorganizations of biological communities.

7. The principle of convergence of form and function states that, despite their different histories of independent evolution, inhabitants of similar environments on different continents often resemble one another because they adapt to similar ecological factors.

8. If community diversity were regulated only by local interactions among species, whose outcome is determined primarily by environmental condi-

tions, then biodiversity would also exhibit convergence between regions. Several examples of nonconvergence in the diversity of temperate forests and mangrove forests demonstrate that the unique histories and biogeographic settings of each continent also influence local species diversity.

9. Biodiversity reflects a broad array of local, regional, and historical processes and events operating on a hierarchy of temporal and spatial scales. Thus, understanding patterns of species diversity requires consideration of the history of a region and integration of ecological study with the related disciplines of systematics, evolution, biogeography, and paleontology.

Suggested readings

Ben-Avraham, Z. 1981. The movement of continents. *American Scientist* 69: 291–299.

Brooks, D. R., and D. A. McLennan. 1991. *Phylogeny, Ecology and Behavior: A Research Program in Comparative Biology.* University of Chicago Press, Chicago.

Brown, J. H. 1995. *Macroecology.* University of Chicago Press, Chicago.

Brown, J. H., and A. C. Gibson. 1983. *Biogeography.* C.V. Mosby, St. Louis.

Carlquist, S. 1981. Chance dispersal. *American Scientist* 69:509–516.

Farrell, B. D., C. Mitter, and D. J. Futuyma. 1992. Diversification at the insect-plant interface. *BioScience* 42:34–42.

Flessa, K. W. 1986. Causes and consequences of extinction. In D. M. Raup and D. Jablonski (eds.), *Patterns and Processes in the History of Life,* pp. 234–257. Springer-Verlag, Heidelberg and New York.

Marshall, L. G. 1988. Land mammals and the Great American Interchange. *American Scientist* 76:380–388.

Orians, G. H., and R. T. Paine. 1983. Convergent evolution at the community level. In D. J. Futuyma and M. Slatkin (eds.), *Coevolution,* pp. 431–458. Sinauer Associates, Sunderland, Mass.

Otte, D., and J. Endler (eds.). 1989. *Speciation and Its Consequences.* Sinauer Associates, Sunderland, Mass.

Pielou, E. C. 1991. *After the Ice Age.* University of Chicago Press, Chicago.

Ricklefs, R. E., and G. W. Cox. 1972. Taxon cycles in the West Indian avifauna. *American Naturalist* 106:195–219.

Ricklefs, R. E., and R. E. Latham. 1992. Intercontinental correlation of geographical ranges suggests stasis in ecological traits of relict genera of temperate perennial herbs. *American Naturalist* 139:1305–1321.

Ricklefs, R. E., and D. Schluter (eds.). 1993. *Species Diversity in Ecological Communities: Historical and Geographical Perspectives.* University of Chicago Press, Chicago and London.

Stucky, R. K. 1990. Evolution of land mammal diversity in North America during the Cenozoic. *Current Mammalogy* 2:375–432.

Vermeij, G. J. 1991. When biotas meet: Understanding biotic interchange. *Science* 253:1099–1104.

PART VII

ECOLOGICAL APPLICATIONS

CHAPTER 26

EXTINCTION AND CONSERVATION

Humans have an immense impact on the earth. There are so many of us (the 1996 population of 5.8 billion is increasing at a rate of almost 2% a year), and each individual uses so much energy and so many resources, that our activities influence virtually everything in nature. Most of the land surface of the earth and, increasingly, the oceans have come under the direct control of humankind. Virtually all areas within temperate latitudes that are suitable for agriculture have been brought under the plow or fenced. Worldwide, fully 35% of the land area is used for crops or permanent pastures; countless additional hectares are grazed by livestock. Tropical forests are being felled at the alarming rate of 17 million hectares (almost 2% of the remaining primary stands) each year. Semiarid subtropical regions, particularly in subsaharan Africa, have been turned into deserts by overgrazing and collection of firewood. Rivers and lakes overflow with the wastes of a consuming society. Our atmosphere reeks of gases produced by chemical industries and burning of fossil fuels.

We are fouling our nest, and we are still rushing to exploit much of what remains to be taken. Inevitably, this deterioration of the environment will lead to a declining quality of life for all human inhabitants of the earth, as it already has for many. The animals and plants with which we share this planet, and on which we depend for all kinds of sustenance, feel the impact of human life all the more. They have been pushed aside as we have taken over land and water for our own living space and for the production of food. Their environments have been poisoned by our wastes. Entire species have succumbed to habitat destruction, hunting, and other forms of persecution.

This deterioration need not continue. Humans can live in a clean and sustaining world, but only by placing support for our own population into balance with preservation of other species and of the ecological processes that nurture us. Legislation in many countries has already led to cleaner air and water, more efficient use of energy and material resources, and the rescuing of endangered species from further decline. The science of ecology has much to say about rational development and management of the natural world as a sustainable, self-replenishing system. What we have learned about the adaptations of organisms, the dynamics of populations, and the processes that occur in ecosystems suggests simple but urgent guidelines for living in reasonable harmony with the natural world.

First, environmental problems can never be brought under control as long as the human population continues to increase. Certainly the earth could support many more individuals than it does at present, but quality of life would be drastically reduced in the short term, and there would be little prospect for sustainability in the long term. Even the present-day human population cannot maintain itself on a sustainable basis. Reforestation cannot keep pace with growing demands for lumber, paper, and fuelwood, and so vast amounts of previously uncut forest are being harvested each year. Most of the important fisheries of the Northern Hemisphere have collapsed and yield only a fraction of their previous production. Large areas of deteriorated land are lost to agriculture every year. As the human population increases, such demands on the environment will only increase.

Studies of natural populations show that their control depends on factors that act in a density-dependent fashion; these factors (which include food shortage, disease, predation, and social strife) reduce fecundity, or increase mortality, or both, as populations grow. If the human population were to come under such external control, the toll in human suffering—disease, famine, warfare—would be enormous. Thus, maintaining individual quality of life at a high level will require first of all that humans exhibit a reproductive restraint that defies the entire history of evolution, during which "fitness" has been measured in terms of breeding success rather than quality of life. Only an increasing appreciation of the negative economic and environmental consequences of overpopulation will cause humanity to value individual human experience over numbers of progeny as the two become increasingly incompatible.

Second, individual consumption of energy, resources, and food produced at higher trophic levels must be reduced. It is inconceivable that the earth could sustain the resource and energy depletion that would result if everyone consumed at the level now exhibited by affluent citizens of developed countries. Efficiency can be increased and superfluous consumption reduced without impairing comfort or enjoyment of life. Insistence on a high-energy lifestyle magnifies the strain inflicted by overpopulation on the world's resources and on the quality of the environment. Each individual human can reduce her or his impact by eating lower on the food chain (reducing meat consumption, for example), investing in energy- and resource-efficient technologies, and living closer to equilibrium with the physical world (lowering the thermostat setting in winter and raising it in summer are simple but effective).

Third, although it is inevitable that most of the world will come under human management, systems should be maintained in as close to their natural state as possible to keep natural ecosystem processes intact. As a general rule, the less we alter nature, the easier it will be to sustain the environment in a healthy condition. For example, many areas covered by tropical forests are unsuitable for grazing or agriculture because these activities upset natural processes of ecosystem maintenance and cause the land to deteriorate. Such areas should be left in forest as reserves, recreation areas, or sites for sustained exploitation of forest products. Deserts can be watered, and they often become tremendously productive for certain types of agriculture. But the costs of maintaining such managed systems can be extremely high as soils accumulate salts from irrigation water and aquifers become depleted. Living with nature is always preferable to, and less costly than, going against its grain.

This chapter and the next elaborate on these themes, each addressing different aspects of applied ecology. In this chapter, we shall consider the problem of conserving species—that is, preventing their populations from dwindling to extinction. In the following chapter, we shall discuss the rational development of the natural world, which includes maintaining natural populations and ecosystem processes so that we and future generations will benefit from them. The problems that will arise in both efforts can be understood by applying basic principles of ecology. We must remember, however, that although solutions can be advanced from an ecological point of view, implementing them will require concerted social, political, and economic action.

Biological diversity

More than 1,400,000 species of plants and animals have been described and given Latin names (Figure 26.1). Insects account for about half of these. Many more species, particularly in poorly explored regions of the Tropics, await scientific discovery. Some experts have estimated that the final species count could reach between 10 and 30 million. Such estimates may be inflated, but it is incontestable that we share this planet with several million other kinds of plants, animals, and microbes.

Making lists of species names is one way of tabulating diversity, but such lists represent only part of the concept of biological diversity, or **biodiversity,** which includes the many unique attributes of all living things.

Although each species differs from every other in the name that science has assigned it, it also differs in the way its adaptations define its place in the ecosystem. For example, different species of plants have dissimilar tolerances for soil conditions and water stress and have disparate defenses against herbivores; they also differ in growth form and in strategies for pollination and seed dispersal. Animals too vary in their own obvious ways. These variations constitute **ecological diversity.**

These important differences between species result from genetic changes, or evolution. And evolution requires genetic variation within populations; otherwise no change can occur. Because genetic variability is crucial to the continued evolutionary response of populations to changes in the environment, **genetic diversity,** which occurs both between and within species, is another important component of biodiversity.

Finally, biodiversity has a geographic component. Different regions have different numbers of species, and if diversity were a contest, tropical rain forests and coral reefs would be the clear winners. Equally important, however, is the fact that some regions boast unique species found nowhere else. Species whose distributions are limited to small areas are called **endemics,** and regions with large numbers of endemic species are said to possess a high level of **endemism.** Clearly, conservation of global biodiversity is best served by directing efforts toward areas of high endemism as well as high diversity. Oceanic islands are well known for harboring unique forms; virtually all the birds, plants, and insects of such isolated islands as the Hawaiian and Galápagos archipelagoes occur nowhere else (Figure 26.2). As a result, when habitat destruction, hunting, or the introduction of alien species results in a loss of local populations in such places, this is more likely to signify a global loss of species in these areas of high endemism

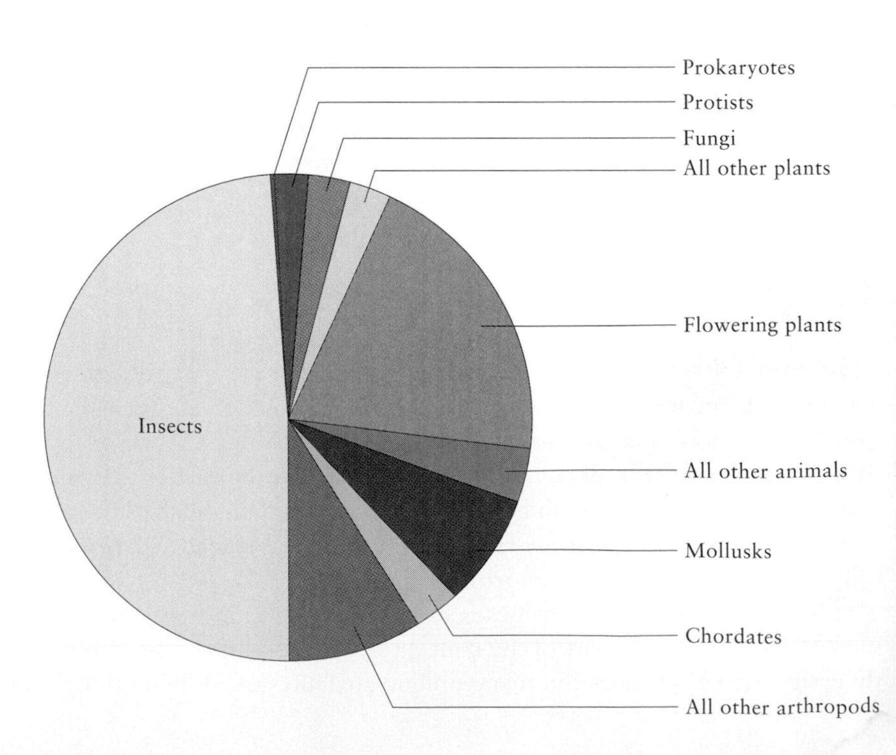

Figure 26.1 Proportions of the 1,445,850 described species that belong to the major taxonomic groups of the five kingdoms of organisms. Data from V. Grant, *The Evolutionary Process,* Columbia University Press, New York (1985).

Figure 26.2 The Hawaiian silversword, found only at high elevation on Haleakala Volcano on the island of Maui, Hawaii. Courtesy of the U.S. National Park Service.

than elsewhere. Fossil-bearing deposits have shown that more than half of the avifauna of Hawaii has disappeared since human colonization of the islands. These birds occurred nowhere else; now they are gone forever. So are the giant moas (ostrich relatives) of New Zealand. Steller's sea cow (a giant relative of dugongs and manatees), which was endemic to the Bering Sea, became extinct by 1768, less than 30 years after it was discovered and first hunted by Europeans.

Human activity now affects all regions of the earth (no refuge, not even in the deepest ocean abyss, exists in a pristine state), so more and more species are now vulnerable to anthropogenic extinction. Many will disappear before they become known to science. Some will carry to their graves valuable and irreplaceable genetic resources. Others will be missed because their presence on earth enriches our own lives. A great effort will be required to slow the loss of biodiversity.

The value of diversity

Why do we care? What concern is it of ours if a species of beetle disappears from South America? Many species already are gone. Do we really miss them? In fact, extinction occurs normally in natural systems. Why should we try to stop it?

Of course, our concern does not center on natural extinction. The rate of disappearance of certain kinds of species, particularly those most vulnerable to hunting, pollution, and destruction of habitat, is probably now at an all-time high in the history of the earth. Some estimates suggest the disappearance of more than one species each day, most of them tropical rain forest insects. This accelerated loss of species is directly linked to the growth and technological capacities of the human population.

The rationale for conserving biodiversity depends on the value we place on individual species. This value arises from many considerations related to our own personal interest and involvement. For many people extinction raises a moral issue. Some take the position that because humankind affects all of nature, it is our moral responsibility to protect nature. If morality derives from a natural law—that is, if morality is intrinsic to life itself—then we may presume that the rights of nonhuman individuals and species are as legitimate as the rights of individuals within human society. Of course, no species is guaranteed a right to perpetual existence, just as no human is guaranteed immortality. But extinction by unrestrained hunting, pollution, habitat destruction, and irresponsible spread of disease may be analogous to murder, manslaughter, genocide, and other infringements of individual human rights.

Throughout history, humans have shown even less sense of responsibility toward nature than toward each other. Whether or not species have natural rights, such rights have not been recognized in the past, nor are they likely to be generally accepted in the future. For many people, the practical problems of personal survival make it difficult to see nature in any other way than as a source of food and fuel; for a few, morality is dictated more by personal greed than by concern for others—whether human or nonhuman.

In the absence of moral protection, the value of individual species can be argued only from the standpoint of their economic and recreational benefits to humankind. This case rests on a number of factors. Individual species have obvious economic importance as food resources, game species, and sources of forest and other natural products, drugs, and many organic chemicals, particularly oils and fragrances (Figure 26.3). For example, more than a hundred important medicinal drugs (including codeine, colchicine, digitalin, L-dopa, morphine, quinine, strychnine, and vinblastine), which account for about one-quarter of all prescriptions filled in the United States, are extracted directly from flowering plants. Some species of economic importance have been cultivated or domesticated and then selectively bred to enhance their desirable qualities. These species are not in danger of extinction, but making room for their cultivation on a large scale has often endangered other species that are perceived as having lesser value. An example is the classic conflict between sheep ranchers and wolves, which occasionally kill sheep and other livestock. Wolves were driven out of most of North America, often with handsome bounties on their heads, and often with the result that herds of deer and other herbivores became so large as to damage the environment, including, ironically, its value for grazing sheep. The point is that assigning economic value to species favors some over others and often does not address the issue of conserving biodiversity in a general sense.

Figure 26.3 Any public market in a tropical country, such as this one in Nairobi, Kenya, offers hundreds of varieties of local plant products—fruits, fibers, medicinals—many harvested from natural ecosystems, others from species cultivated locally or throughout the world.

At times, we may argue for conservation of a particular habitat by comparing the economic values of native species that occur there with value accrued from altering or otherwise managing the habitat. Under many circumstances, however, the short-term gains of converting forest to agriculture, for example, or of excessively exploiting a marine resource are assumed to outweigh any long-term value of conserving the natural system for sustained income. The value of conserved species and habitats usually becomes apparent only when the long-term costs of overexploitation or habitat conversion are properly accounted for, a practice that is encouraged neither by the pragmatism of desperation nor by the notorious shortsightedness of politics.

Species diversity in ecological systems may have intrinsic value for stabilizing ecosystem function. An increasing number of studies are showing that diverse systems are better able to maintain high productivity in the face of environmental variations. For example, on experimental plots of Minnesota prairie containing differing numbers of species, David Tilman and J. A. Downing at the University of Minnesota demonstrated that biomass production was less affected by severe drought on high-diversity plots than on low-diversity plots. Such results can be explained by positing that higher-diversity systems are more likely to include some species that can withstand particular stresses. As the environment changes, different species can take over the roles of predominant producers in a system. Such switching among species is less likely to occur in less diverse systems.

High value may be placed on some individual species because they attract tourists to an area. The practice of visiting an area to see its unspoiled habitats and the animals and plants that live in them is referred to as **ecotourism.** Many tropical countries have capitalized on this attraction by establishing parks and support services for tourists (Figure 26.4). In Latin America, spectacular quetzals, macaws, and monkeys draw tourists to many areas where these species are protected. Diversity itself is often the attraction in tropical rain forests and coral reefs, with their hundreds of different species of trees, birds, or fish. In East Africa, lions, elephants, and rhinoceroses have great value because of the tourist dollars, pounds, francs, and yen they bring into countries that are badly in need of foreign currencies (Figure 26.5). Unfortunately, a few self-serving individuals prize elephants more for the value of their ivory and rhinoceroses for the value of their horn, which is made into dagger handles in some Arab countries and is believed by some Asians to have aphrodisiac properties. These considerations set up conflicts between the economic interests of the many and those of the selfish few, and the lawless who pander to them. Such conflicts escalate the costs of conservation to the point that blood has been shed in confrontations between poachers and government wardens.

An interesting illustration of the intensity of poaching comes from a study in a national park in Zambia, where the fraction of tuskless female elephants increased from 10% in 1969 to 38% in 1989 as a direct result of selective illegal ivory hunting. Tusklessness in females is a genetic trait, and because poachers kill only individuals with tusks, poaching strongly favors tusklessness in a population. A change in the frequency of a trait from 10% to 40% within one generation is strong selection indeed.

Ecotourism has been responsible for the development and maintenance, including better protection from poaching, of an increasing number of parks and reserves in many parts of the world, and its impact will expand as more people become aware of the gratification that comes from experiencing nature directly, even from the comparative luxury of ecotourist hotels and camps. The capacity of ecotourism to confer enough

(a)

(b)

Figure 26.4 Ecotourists on "safari" in Kenya, East Africa, bring in millions of dollars in internationally traded "hard" currencies and provide employment for many of the local people as guides and wardens, as well as in hotels and restaurants and businesses that support them.

(a)

(b)

(c)

Figure 26.5 The inspiring diversity of wildlife in Africa is as much a part of its attraction as any particular species. Nowhere else can one see elephants, cheetahs, zebras, and dozens of other large mammals in their natural environments.

value on species to guarantee their protection is, however, finite. People have limited money to spend, and merely increasing reserve systems will not necessarily generate more tourism. Furthermore, some areas of immense biological importance, with high diversity and endemism, either are not attractive to or are inaccessible to most tourists. Deserts, semiarid regions, many islands, and most marine ecosystems fall into this category. Of course, on a simple numerical basis, most species simply are not very interesting or even perceptible to the general public. Their preservation will depend on their living in association with more highly valued species or habitats.

Individual species may have considerable value as indicators of broad and far-reaching environmental change. During the 1950s and 1960s, populations of many predatory and fish-eating birds in the United States (particularly the peregrine falcon, bald eagle, osprey, and brown pelican) declined drastically to the point that several of these species had disappeared from large areas, the peregrine from the entire United States. The causes of these population declines were traced to pollution of aquatic habitats by breakdown products (residues) of DDT, a pesticide that was widely used after World War II. These pesticide residues resisted degradation and entered aquatic food chains, where they accumulated in fatty tissues of animals and were concentrated with each step in the food chain. The high doses consumed by top predators interfered with their physiology and reproduction, causing overly thin eggshells and deaths of embryos. Breeding success plummeted, and populations followed. The viability of the peregrine population is a sensitive indicator of the general health of the environment. Its demise sounded the alarm to environmentalists; Rachel Carson warned of a "silent spring" when no birds would be left to sing.

The United States government responded by banning DDT and related pesticides, and chemical companies have since devised alternatives that have less drastic environmental effects. Bald eagles and ospreys are becoming familiar sights once again; and, thanks to the helping hands of dedicated biologists who reared birds obtained from other parts of the geographic range and released them in the eastern United States, peregrine falcons have staged a spectacular comeback. This was a major victory, not only for the peregrine and the cause of species conservation, but also for the general quality of our own environment. Unfortunately, this silver lining draws attention to a dark, ugly cloud: manufacture and export of DDT to foreign countries is still legal in the United States and remains a source of great profit to a few. DDT will bring a bleaker future to the many who live in countries that have yet to ban the toxin. It may even have played a role in the decline of songbirds that breed in North America and Eurasia but migrate to more tropical latitudes during winter.

This general background illustrates the difficulties the conservation movement faces—difficulties grounded in ignorance and greed as well as in legitimate compromises between conflicting values. Be this as it may, the ecological considerations bearing on the conservation of species usually are relatively straightforward. To understand them, it will help to review what we know about extinction itself.

Types of extinction

It is useful to distinguish three types of extinction. **Background extinction** reflects the fact that as ecosystems change, some species disappear and others take their places. This turnover of species, at a relatively low rate, appears to be a normal characteristic of natural systems. **Mass extinction** refers to the dying off of large numbers of species as a result of natural catastrophes. Volcanoes, hurricanes, and meteor impacts happen occasionally. Some occur locally, others affect the entire globe, and species that happen to be in the way disappear. **Anthropogenic extinction**—extinction caused by humans—is similar to mass extinction in the number of taxa affected and in its global dimensions and catastrophic nature. Anthropogenic extinction differs from mass extinction, however, in that its causes theoretically are under our control.

Most information on background and mass extinction comes from the fossil record, which reveals appearances and disappearances of species through geologic time. Disappearances may occur in two ways. First, species may evolve sufficiently that individuals are no longer recognized as belonging to the same taxon as their ancestors and are given a different scientific name. True extinction has not taken place, and such instances are therefore referred to as **pseudoextinctions.** Second, a population may cease to exist, in which case its disappearance from the fossil record is a case of true extinction. The finer the resolution of the fossil record, the greater the probability of distinguishing between the two.

Where true extinction can be demonstrated, the life spans of species in the fossil record vary according to taxon, but they generally fall within the range of 1 to 10 million years. Thus, on average, the probability that a particular species will go extinct in a single year is in the range of 1 in a million to 1 in 10 million. This is the background rate of species extinction: 10^{-6} to 10^{-7} per year. If, as conservative estimates have it, on the order of 1 to 10 million species inhabit the earth, this would amount to about 1 species extinction per year at the background rate.

Mass extinctions occupy the other end of the spectrum. Natural catastrophes may cause the disappearance of a substantial proportion of species locally or globally, depending on the severity and geographic extent of the catastrophe. Such catastrophes may include prolonged drought, hurricanes of great force, and volcanic eruptions. When Krakatau, a volcanic island in the East Indies, exploded on 26 August 1883, not an organism was left alive; any that survived the initial explosion were buried under a thick layer of volcanic debris and ash. Whether any unique species disappeared in this catastrophe cannot be known because the island had not been well surveyed for its biodiversity prior to the explosion.

Some mass extinctions detectable in the fossil record are thought to have been caused by the impacts of large comets or asteroids (collectively referred to as bolides), associated tidal waves and fires, and prolonged darkness resulting from dust and smoke in the atmosphere. As we have seen in the previous chapter, spectacular examples of mass extinctions occurred at the end of the Paleozoic era (Permian period) and the

Mesozoic era (Cretaceous period). The first involved the disappearance of perhaps 95% of species and numerous higher taxa. The second is most famous for the extinction of the dinosaurs, but some other major groups, notably ammonites (predaceous, nautilus-like mollusks) disappeared as well. Whatever the exact cause, these extinctions were associated with discrete, calamitous events.

And anthropogenic extinction? Are we to be regarded as a "human bolide" in terms of our impact on the environment? Well, not yet. Many extinctions have undoubtedly gone unrecorded, and rates of extinction in many groups (particularly among large animals hunted for food and among island forms) are far above background levels. Nevertheless, if humankind turns out to be a disaster for global biodiversity, the full force of the impact will come in the future. Most important, it is preventable. Examining the causes of extinction will enable us to see why this is so.

Causes of extinction

Species disappear when deaths exceed births over a prolonged period. This much is obvious, but the statement also emphasizes that extinction may result from a variety of mechanisms that influence birth and death processes within a population. It has also been said that extinction represents failure to adapt to changing conditions, whether because the changes occur too fast or because a population is evolutionarily unresponsive. We shall discuss four general types of factors that can cause population decline: (1) climate change, (2) reduction of habitat area, (3) declining habitat quality, and (4) overexploitation.

Climate change

Climate determines physical conditions and habitat structure, which are critical to the well-being of every population. Over the long history of the earth, changes in global climate have been brought about by the drifting of continents and associated changes in oceanic circulation. Where physical barriers to dispersal prevent distributions of species from following shifts in climate belts, local populations may become extinct, being replaced by others that are better suited for survival in the new climate and habitat type. Drastic changes in climate during the Ice Age, combined with barriers to dispersal in southern Europe, were responsible for today's impoverished European flora and fauna. Local changes in climate and habitat are brought about by changing landforms, which may create rain shadows and redirect the drainages of rivers. For endemic species, these changes may bring about extinction.

At present, the burning of wood and fossil fuels is increasing the carbon dioxide concentration in the atmosphere, and thereby increasing the average temperature of the earth by enhancing the so-called greenhouse effect, which we shall discuss in the next chapter. This anthropogenic change in climate, which may amount to between 2° and 6°C, could equal the warming of the earth's climate since the last glaciation, only 50 times faster. It is likely to cause the extinction of many species, particularly plants, with narrow temperature tolerances.

Habitat area and population size

Large areas of habitat support large populations, which are less susceptible than small populations to extinction due to small-scale catastrophic events or random variations in population size. Just by chance, every population experiences variations in births and deaths during any particular period. These cause what is known as stochastic, or random, variation in population size. The magnitude of this variation varies inversely with the number of individuals in a population. Very small populations, such as those isolated in restricted fragments of habitat, may become extinct just by chance if they suffer a series of very unlucky years. This phenomenon is referred to as **stochastic extinction,** and although it is relatively unlikely except in the smallest populations, its probability increases with fragmentation of suitable habitat, and it is particularly troublesome for species, such as large predators, that have low population densities.

Small population size may further increase the probability of extinction by reducing genetic variation in a population. Fewer individuals contain a smaller proportion of the total genetic variation of a larger population. Furthermore, **inbreeding** (mating among close relatives) tends to reduce genetic variation. When a population goes through a period of small population size and, as a result, exhibits reduced genetic diversity, it is said to have passed through a **bottleneck.** Such populations may not have the capacity to respond to rapid change in the environment, which may favor some genotypes in some years and other genotypes in other years. Small glades of xeric habitat on rocky outcrops in the Ozark Mountains of Missouri, for example, support restricted populations, generally 20–50 individuals, of the collared lizard, a resident of Southwestern deserts that colonized the Ozarks during a period of hot, dry climate 4,000 to 8,000 years ago. Genetic surveys have shown that these lizards are genetically uniform within populations but differ between populations. This is exactly the pattern expected to result from random loss of genetic diversity within small populations.

It is difficult to generalize about problems resulting from population bottlenecks because there are several cases of species that have been reduced to near extinction and have lost much of their genetic variability, but have recovered with spectacular growth when protected. The northern elephant seal is a case in point. By 1890, hunting had reduced its once numerous population to about 20 individuals. Since then the population has increased explosively, passing 30,000 in 1970 and extending throughout much of the species' former range in California and Mexico. Several years ago, investigators could not detect any genetic differences between individuals within the species, though they used tests that reveal ample genetic variation in other species of mammals. Similarly, one of Africa's large cats, the cheetah, has no detectable genetic variation within its population. This genetic uniformity suggests that cheetahs may have gone through a population bottleneck sometime in their recent past.

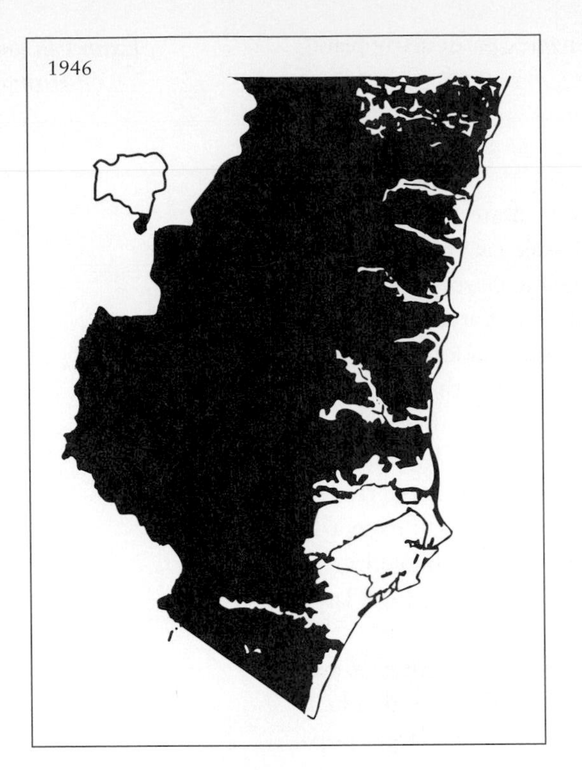

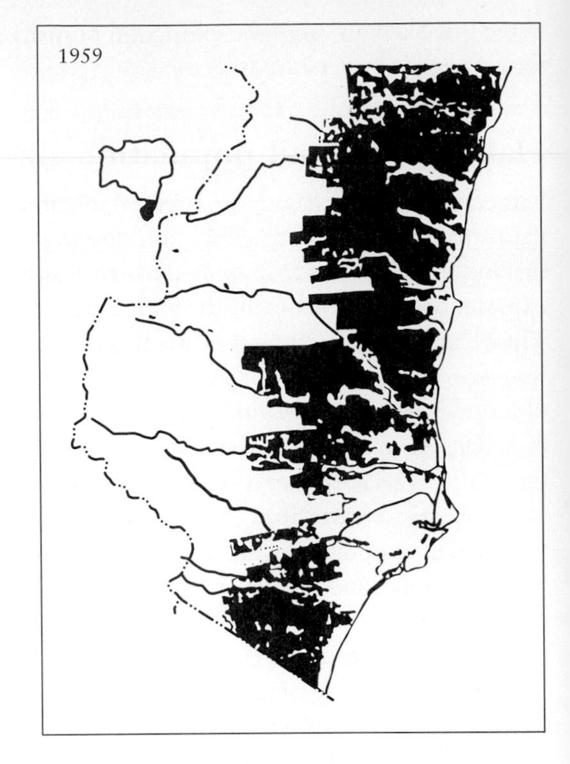

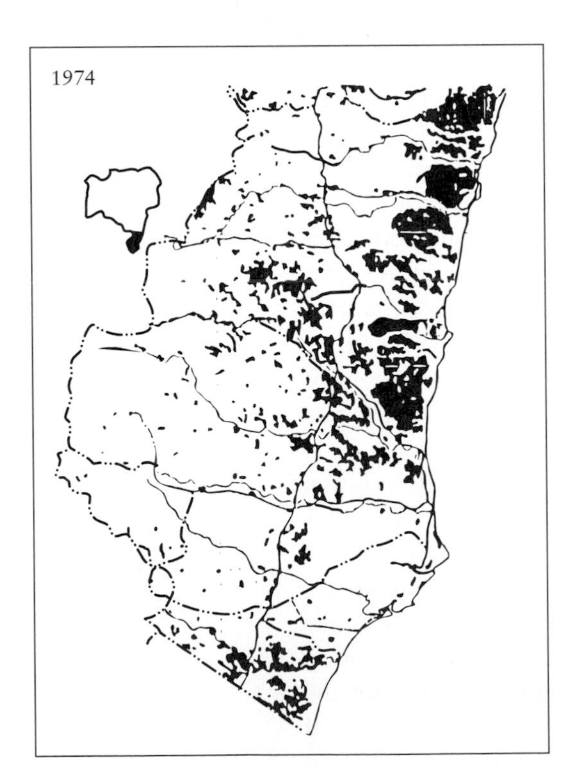

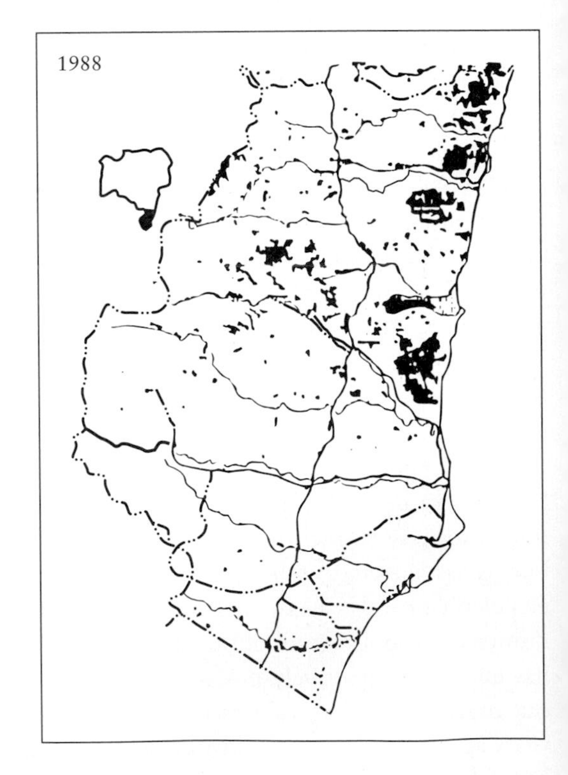

Figure 26.6 Maps of forest habitat documenting the decimation of Atlantic coastal forests in Bahia, southern Brazil, during the past 50 years. Several unique, endemic species have disappeared from this area, and others are gravely threatened. Maps by J. R. Mendonça, Projeto Mata Atlântica Mordeste, Convênio CEPLAC/Jardim Botânico de Nova Lorque.

Nonetheless, although its reproductive success appears to be somewhat impaired in captivity, the cheetah population appears to be healthy and self-sustaining where it is not persecuted by humans.

Reduction of habitat and, especially, fragmentation of habitat into small remnants poses a tremendous threat to wildlife. For example, the Atlantic coastal forests of Brazil have been reduced to only a small percentage of their former extent (Figure 26.6), with the extinction of many endemic birds and mammals. The remaining fauna is now coming under intense conservation efforts, but too late for many of the region's inhabitants. Even in North America, habitat fragmentation is causing population declines. Our largest national parks have lost many of their mammal species in the past 50 years, suggesting that these reserves are insufficient in size to maintain viable populations. The loss of many native songbird species from fragmented temperate forest habitats is partly a consequence of small population size and stochastic local extinction, but fragmentation has also increased access to forest habitat by some predators and nest parasites that are more typical of fields and agricultural lands, with drastic consequences for songbird survival and reproductive success in some areas. Thus, habitat fragmentation also causes a deterioration of habitat quality.

Habitat quality

Ultimately, of course, habitat loss may cause extinction by wiping out suitable places to live. Animals of the forest will disappear when all the forest has been cut. Even when suitable habitat remains, however, conditions within the habitat may change and cause a population to begin a decline toward extinction. Frequently, a decrease in habitat quality can be traced to the introduction of predators, competitors, or disease organisms—that is to say, biological agents of change. In many cases, the habitats affected are those of isolated islands to which organisms have been introduced from more diverse continental biotas. The brown tree snake, introduced from Asia, has literally eaten most of the native land birds of Guam to extinction. Most of these birds were endemics found nowhere else. The Hawaiian Islands have also suffered greatly from introductions of aliens, which have resulted in the extinction of a large proportion of native birds and other groups, including land snails. The major agents of avian mortality have been malaria and pox virus—which would not have been a problem if the mosquito that transmits these diseases had not also been introduced to the islands. Native forests of Hawaii have also suffered from invasion by aggressive, weedy species (Figure 26.7). Finally, although Hawaii is not severely affected, habitat quality in many parts of the world is worsened by various forms of pollution originating at a distance, including smog and acid rain.

Overexploitation

Weapons and other harvesting tools, such as kilometer-long drift nets, have made humans such efficient hunters that many species have literally been hunted to extinction. Within recent history, North American fatalities have

(a)

(b)

Figure 26.7 (a) Exclusion of cattle and goats from fenced areas on the island of Hawaii allows reestablishment of native vegetation. The extreme disturbance caused by grazers allows introduced weedy species to become established. (b) Feral boars, originally introduced to Hawaii as a game species, have also caused the disappearance of native species by routing seedlings out of the soil. Courtesy of the U.S. Department of Agriculture, Soil Conservation Service.

included the Steller's sea cow, great auk, passenger pigeon, and Labrador duck—all formerly abundant species, all prized for food, all vulnerable, and all slaughtered mercilessly until the last were gone. Extinction by overhunting and overfishing is not, however, a recent phenomenon. Wherever humans have colonized new regions, some elements of the fauna have suffered. For example, shortly after aboriginal people colonized Australia some 50,000 years ago, several large marsupial mammals, flightless birds, and a tortoise disappeared from the island continent. The advent of humans in the Americas about 12,000 years ago was accompanied by the rapid extinction of 56 species in 27 genera of large mammals, including horses, a giant ground sloth, camels, elephants, the saber-toothed tiger, a lion, and others.

Madagascar, a large island off the southeastern coast of Africa, received its first human inhabitants only 1,500 years ago, yet this event brought the demise of 14 of 24 species of lemurs (mostly large species suitable for food) and between 6 and 12 species of elephant birds, flightless giants found only on Madagascar. Similar extinctions occurred widely in the Polynesian islands and Hawaii as humans spread throughout the region. In each of these cases, a technologically superior species encountered populations unaccustomed to hunting pressures. Their lack of defenses and failure even to recognize danger spelled disaster; lack of restraint on the part of their hunters turned disaster into extinction.

Vulnerability to extinction

Why do some species seem more vulnerable to anthropogenic extinction than others? This question has been difficult to answer. Clearly, species that attract the attention of human exploiters are brought under great pressure. In addition, species that have evolved in the absence of hunting (particularly those on remote islands lacking most types of predators) and in the absence of disease organisms seem to fare poorly after the arrival of humans. Vulnerability is also associated with limited geographic range, restricted habitat distribution, and small local population size.

But what makes one species rare and locally distributed when a close relative that exhibits superficially similar adaptations is abundant throughout a wide geographic distribution? The difference between success and failure in natural systems may hinge on a very few percentage points of breeding success or longevity—perhaps too few for us to detect in studies of natural populations. Most species persist for a million or more years, so their populations must be fully self-sustaining and capable of recovering from setbacks inflicted by a variable world. Whatever causes a population to embark on a slow decline to extinction may be very subtle indeed. So far, ecologists have been able to say little on this point. What they can address, however, is the problem of reversing the declining population trends of species that, in the absence of anthropogenic pressures, would be self-sustaining.

Strategies for conservation of species

The simple way to maintain a population of a particular species is to guarantee the existence of a sufficient area of suitable habitat that can be kept free of alien competitors, predators, and diseases. In practice, the design of such preserves must take into account the ecological requirements of the species and the minimum size of a population that can sustain itself in the face of environmental variation. This size is called the **minimum viable population,** or MVP. The MVP must be large enough to remain out of danger of stochastic extinction brought about by chance events. The population also must be distributed widely enough so that local calamitous events such as hurricanes and fires cannot affect the entire species. At the same time, some degree of population subdivision may prevent the spread of disease from one part of a population to another.

Guaranteeing suitable habitat becomes more complex when a population has different habitat requirements during different seasons or when it undertakes large-scale seasonal migrations. In the Serengeti ecosystem of eastern Africa, fantastically large populations of grazers, such as wildebeests, zebras, and gazelles, undertake long-distance seasonal movements in search of suitable grazing as the pattern of rainfall distribution varies seasonally within the region. It would not be possible to isolate a part of this area as a preserve, because populations need the entire area of the Serengeti ecosystem at different times of the year. Thundering herds of buffalo can never be restored to American prairies, because their migration routes are now blocked by miles of fencing and habitat converted to agriculture. Buffalo survive in a few small reserves in the American West—most notably the Greater Yellowstone ecosystem—but the natural environment of the buffalo has been irrecoverably lost.

Long-distance migration poses similar problems for the conservation of many types of birds. Wading birds, such as sandpipers, breed on arctic tundra, but maintenance of their populations also depends on conservation of the beaches and estuaries that they use during spring and fall migrations and as wintering grounds (Figure 26.8). Many American songbirds, whose

Figure 26.8 Sandpipers and gulls feed on eggs laid by horseshoe crabs during May along the shores of Delaware Bay. These eggs are a major food source for migrating shorebirds. Many of the horseshoe crabs in this photograph are stranded by the ebbing tide and will die of exposure.

populations have been declining during recent decades, spend their winters, wisely it would seem, in forests of Central and South America. Their populations have been placed in double jeopardy by forest fragmentation throughout much of their breeding range in North America and by extensive clearing of forests and spraying of pesticides such as DDT in Latin America. Migration systems between Europe and Africa and between Siberia and Southeast Asia face the same problems, exacerbated by the toxic pesticides used throughout these regions.

When threats of extinction come from the dwindling of suitable habitat for a particular species, conservation strategy is relatively straightforward, but it may be expensive and politically difficult to achieve. It is also impractical to develop a conservation strategy for every species, and the well-being of the majority will necessarily depend on conservation efforts directed toward a few of the most critically endangered or conspicuous species. As habitat becomes more and more the focus of conservation efforts, however, it becomes especially important to identify habitats that are most critical to maintaining species diversity as a whole and to determine the area of habitat required to maintain minimum viable populations of most species. Each decision about a species or a habitat will depend on value judgments. What determines which species should be saved? How is their "value" measured?

What makes an area critical for conservation? The most valuable areas are those that provide havens for the largest number of species not represented elsewhere; thus value reflects a combination of local diversity and endemism. As a rule, endemism is highest on oceanic islands, in the Tropics, and in mountainous regions. Thus, such localities as Madagascar and the Hawaiian, Galápagos, and Canary Islands are extremely critical ones for conservation. Extensive surveys of biodiversity resources in continental regions are beginning to identify critical areas of concern for conservation, but these efforts are continually hampered by lack of detailed information and by conflicting values attached to different components of biodiversity.

On a continent, preserves, whose number and area are necessarily limited by economic considerations, must target habitats and areas of special biological interest. From the standpoint of conservation of biodiversity, more is to be gained by setting aside several small reserves spread out over a variety of habitats and areas of high endemism than by preserving an equal area within a single habitat type. Values other than biodiversity may dictate the setting aside of large preserves, however, and we shall discuss some of these factors in the next chapter. One thing is certain: the cost of setting aside larger and larger amounts of habitat increases out of proportion to the area itself. This is so simply because land that is least expensive in terms of economic, social, and political values is set aside first. As more land is added to a preserve system, the cost of acquiring it, in terms of purchase price and potential resources forfeited, invariably increases. It is no accident that most parks and reserves are located in remote, underpopulated areas and that establishing a conservation area becomes more difficult when it conflicts with economic interests.

One example of this conflict involved the setting aside of a large tract of old-growth redwood forest in northern California as Redwoods National Park, which was vigorously opposed by the timber industry. In this case, the uniqueness of the redwood habitat and its rapid conversion into managed tree farms greatly increased the value to society of setting aside a large area of this habitat for posterity. A similar controversy surrounds the old-growth Douglas fir forests of Washington and Oregon, where the fates of such unique inhabitants as spotted owls and marbled murrelets are pitted against the local lumber economy.

Many tropical countries, particularly in Central and South America, are in the enviable position of having large tracts of uncut forest and relatively undisturbed tropical habitats of other kinds. These have been protected fortuitously in the past by their geographic remoteness and by the small size of local human populations. It is still possible to set aside large parks and reserves in such countries as Brazil, Ecuador, Peru, and Bolivia, and several governments have moved rapidly during the past decade to preserve tracts of what remains. The problem is complicated, however, by the rapid growth of the human population, by increased exploitation of forest products, and by the conversion of forests to agriculture. Such exploitation is justified by a legitimate need to feed people and generate export income for economic development. Thus, the price of conservation is rising very rapidly in much of the world, and many developing countries are unable to foot the bill. Even when lands are set aside "on paper," many countries cannot afford to protect them from squatters, poachers, and self-serving politicians who may grant mining and lumbering concessions within protected lands in order to extract short-term profits. For this reason, conservation must be an international effort, and the wealth of the developed countries must be shared globally to protect the earth's biodiversity. The "haves" must simply have less so that the "have nots" can have enough without destroying the varied habitats and creatures that are our common heritage and our common trust.

Design of nature preserves

In many cases, the boundaries of biological preserves are dictated by available land area and economic considerations; basically, all that can be done is to set aside whatever remains in a relatively pristine state. In other situations, those who design preserves may have more latitude in deciding just how to draw the boundaries of a park or conservation area. Here, ecological principles derived from the **theory of island biogeography** can help planners to arrive at the best solution to the boundary problem. There are two guiding principles: the **species-area relationship** and the **edge effect.** Large areas support more species than small areas because large population sizes of individual species reduce the chances of stochastic extinction, promote genetic diversity within populations, and buffer populations against disturbances. Edges should be minimized because the effects of habitat alteration extend for some distance beyond the areas directly altered. For example, predators and parasites that inhabit agricultural areas, such as cowbirds (which are nest parasites of other birds), rats, and feral cats, may venture into the edges of forest habitats. As we have seen, the nesting success of many songbirds in the United States has decreased dramatically as forest fragmentation has increased. Physical conditions in forests adjacent to cleared land are further altered by increased wind and sunlight, and these changes may also affect the productivity of forest species, especially understory plants.

According to these considerations, when a preserve is to be carved from an area of uniform habitat, such as a broad expanse of tropical rain forest, (1) larger is better than smaller, (2) one large area is better than several smaller areas that add up to the same total size, (3) corridors connecting isolated areas are desirable, and (4) circular areas are better than elongate ones with much edge. However, faced with choosing between a single large area of uniform habitat and several smaller areas, each in a different habitat, planners should remember that the smaller areas will often contain a greater total number of species among them because endemic species may be found in one habitat but not the others.

As always, nature preserves must be designed in accordance with the habits of their inhabitants, and requirements for special features (such as nesting sites, water holes, and salt licks) must be taken into account. In mountainous areas, many species undertake altitudinal migrations over the seasonal cycle, and so preserves set aside at different elevations must be connected by suitable corridors for travel. Roads and pipelines set in the way of migratory movements or dispersal must be bridged in some manner to allow passage.

Rescues from the brink of extinction

There have been many times when a particular species has come so close to extinction that its preservation has required exceptional human intervention. Such efforts, which may cost millions of dollars, usually are directed toward species that appeal to the public. Some may question the wisdom of spending several million dollars (as happened recently) to free three gray whales trapped in Arctic ice. But the incident dramatized the empathy that many humans feel for the plight of some other creatures. And many people, though perhaps less enthusiastic about spending so much to rescue individuals, are willing to devote considerable resources to the preservation of species.

In recent decades, zoological parks have become increasingly involved in maintaining viable, genetically diverse populations of species that are endangered or even extinct in the wild. Eventually, with the development of suitable preserves, many of these populations could be reintroduced into natural settings. As the population of California condors in southern California dwindled below thirty and then below twenty individuals in the wild during the 1970s and early 1980s, management personnel made the difficult decision to bring the entire population into captivity. In specially constructed breeding facilities located at the Los Angeles and San Diego Zoos, the birds are protected from several mortality factors that were destroying the wild population: indiscriminate shooting, lead poisoning from slugs left in deer carcasses upon which condors fed, and poisons and traps set out for coyote and rodent control, to which condors were attracted by baits or poisoned carcasses. In addition, condors in captivity can be induced to lay up to three eggs a year, instead of the usual one, and most of the chicks are reared successfully. The objective of such a captive rearing program is to produce young that can be reintroduced into their native habitat. Such a program is costly, and its success ultimately depends on controlling those mortality factors that threatened the population in the first place, which often requires legislation, land purchases, and public education. In the particular case of the California condor, the program's success can be properly judged only after 30 or 40 years and the expenditure of tens of millions of dollars.

Like many other endangered species, the California condor can be saved from extinction. So can the Hawaiian crow and the black-footed ferret. The experience gained through these captive breeding and release programs will be useful to similar efforts in the future. The California condor program, like other such programs, has heightened local residents' awareness of conservation issues and has resulted in the preservation of large tracts of habitat in mountainous regions of southern California and elsewhere. People have also come to understand that as long as care is taken, viable condor populations are compatible with other land uses, such as recreation (as long as human access to nesting sites is restricted), hunting (as long as steel rather than lead bullets are used), and ranching (as long as coyote and rodent control programs, if they are to persist at all, are condor-safe). Concessions to condors are neither difficult nor expensive. Making them simply depends on instilling values that acknowledge natural systems as an integral part of the environment of humankind.

Summary

1. Humankind has an immense impact on the earth, managing or otherwise affecting most of its land surface and waters. Human activities have

caused deterioration in ecological systems and extinctions of many species. The repercussions are accelerating as the human population grows toward 6 billion individuals and the per capita consumption of energy and resources increases apace.

2. The environmental crisis cannot be fully resolved until human population growth is stopped, consumption of energy and resources declines, and economic development takes ecological values into consideration.

3. Of immediate concern is the preservation of biodiversity, which encompasses the variety of living beings—plants, animals, and microbes—on earth. The concept of biodiversity recognizes genetic diversity within and between populations and acknowledges the special value of areas of endemism that are inhabited by species with restricted geographic ranges.

4. The value of individual species is rooted in generalized moral considerations, in aesthetics, in the economic and recreational benefits we derive from them, and in their role as indicators of environmental deterioration. Diversity itself may also help to stabilize ecosystem function in the face of environmental variation.

5. Background extinction consists of natural extinctions resulting from environmental change and from the evolutionary turnover of species within communities. Mass extinctions, which appear episodically in the fossil record, reflect calamitous events in earth history, particularly impacts of meteors or other extraterrestrial bodies. Anthropogenic extinction is the disappearance of species as a result of human activities: habitat destruction, overexploitation, introduction of predators and disease organisms, and pollution of various kinds.

6. Reduction of habitat area may hasten a population's decline toward extinction by making it more vulnerable to stochastic, or random, changes in population size or by causing reduced genetic variability and thereby impairing the capacity of the population to survive environmental change.

7. Optimally designed nature preserves should include a high proportion of endemic species. For a given area of uniform habitat, preserves should be amalgamated (rather than dispersed in several small areas) to reduce the chances of stochastic extinction due to small population size, and they should be close to circular in shape to reduce edge effects.

8. In extreme cases, individual species can be rescued from the brink of extinction by massive recovery efforts that may include habitat restoration and captive breeding. Such costly programs, although they are focused on individual species, often highlight more general conservation problems and result in the conservation of habitat whose ecological value greatly exceeds that of the individual species it was preserved to save.

Suggested readings

Angermeier, P. L., and J. R. Karr. 1994. Biological integrity versus biological diversity as policy directives. *BioScience* 44:690–697.

Burney, D. A. 1993. Recent animal extinctions: Recipes for disaster. *American Scientist* 81:240–251.

Caro, T. M., and M. K. Laurenson. 1994. Ecological and genetic factors in conservation: A cautionary tale. *Science* 263:485–486.

Ceballos, G., and J. H. Brown. 1995. Global patterns of mammalian diversity, endemism, and endangerment. *Conservation Biology* 9:559–568.

Diamond, J., and T. J. Case. 1986. Overview: Introductions, extinctions, exterminations, and invasions. In J. Diamond and T. J. Case (eds.), *Community Ecology,* pp. 65–79. Harper & Row, New York.

Eisner, T., J. Lubchenco, E. O. Wilson, D. S. Wilcove, and M. J. Bean. 1995. Building a scientifically sound policy for protecting endangered species. *Science* 268: 1231–1232.

Glen, W. 1990. What killed the dinosaurs? *American Scientist* 78:354–370.

Hansen, A. J., T. A. Spies, F. J. Swanson, and J. L. Ohmann. 1991. Conserving biodiversity in managed forests. *BioScience* 41:382–392.

Mills, L. S., M. E. Soulé, and D. F. Doak. 1993. The keystone-species concept in ecology and conservation. *BioScience* 43:219–224.

Myers, J. P., et al. 1987. Conservation strategy for migratory species. *American Scientist* 75:18–26.

Pimm, S. L. 1991. *The Balance of Nature? Ecological Issues in the Conservation of Species and Communities.* University of Chicago Press, Chicago.

Primack, R. L. 1993. *Essentials of Conservation Biology.* Sinauer Associates, Sunderland, Mass.

Redford, K. H. 1992. The empty forest. *BioScience* 42:412–422.

Robinson, S. K., et al. 1995. Regional forest fragmentation and the nesting success of migratory birds. *Science* 267:1987–1990.

Rolston, H., III. 1985. Duties to endangered species. *BioScience* 35:718–726.

Simons, T., S. K. Sherrod, M. W. Collopy, and M. A. Jenkins. 1988. Restoring the bald eagle. *American Scientist* 76:252–260.

Soulé, M. E. 1985. What is conservation biology? *BioScience* 35:727–734.

Soulé, M. E. (ed.). 1986. *Conservation Biology: The Science of Scarcity and Diversity.* Sinauer Associates, Sunderland, Mass.

Terborgh, J. 1974. *Preservation of natural diversity: The problem of species extinction.* BioScience 24:715–722.

Terborgh, J. 1989. *Where Have All the Birds Gone?* Princeton University Press, Princeton, N.J.

Terborgh, J. 1992. *Diversity and the Tropical Rain Forest.* Scientific American Library, W. H. Freeman, New York.

Tilman, D., and J. A. Downing. 1994. Biodiversity and stability in grasslands. *Nature* 367:363–365.

Western, D., and M. C. Pearl (eds.). 1989. *Conservation for the Twenty-First Century.* Oxford University Press, Oxford.

Westman, W. E. 1990. Managing for biodiversity. *BioScience* 40:26–33.

Wilson, E. O. (ed.). 1988. *Biodiversity.* National Academy Press, Washington, D.C.

CHAPTER 27

DEVELOPMENT AND GLOBAL ECOLOGY

The way to preserve biodiversity is to set aside large areas of the variety of natural habitats found on earth and maintain their capacity for supporting species. Basically, this means minimizing human impacts of all kinds over representative areas of the earth's surface. As the human population grows, however, this goal recedes farther into the distance and becomes, in the eyes of most people, less pressing than the problem of maintaining basic life-supporting systems for humans. Resolution of the conflict between natural and human values will ultimately depend on erasing the distinctions between them and somehow making them more compatible.

For many species, and perhaps for most of the biodiversity of the Tropics, we must set aside either pristine reserves or areas carefully managed to sustain certain populations. In the end, these will be natural plant and animal parks where remnant viable populations of hundreds of thousands—perhaps millions—of species will cling to their last footholds on this planet. Setting aside these reserves will be justified by moral and aesthetic considerations and by their considerable economic value as destinations for tourists, as protected watersheds, and as regions that can assimilate a part of the excess carbon dioxide we produce by burning fossil fuels (more will be said about this later).

What of the 90% or so of the rest of the earth that has been, or soon will be, converted to supporting the human population—devoted to living space, food production, forestry, mineral production, hunting, and so on? What about our own ecology? Can the earth sustain an expanding human population indefinitely at a high quality of life? To what degree are human values compatible with natural values? That is, can natural and managed ecosystems intergrade, or are reserves the only alternative to completely altered environments dominated by humans and their domesticated species?

Clearly, a sustainable biosphere will never be achieved as long as the human population continues to grow. The earth offers no new regions to colonize. Except for portions of the wet Tropics, much of which cannot support dense human populations, most of the habitable areas of the earth have been filled. Further population increase will lead to further crowding, tearing at not only the fabric of human society but also the life-supporting systems of the environment.

It is easy to be pessimistic about the future, but there is also plenty of room for optimism. Many programs for cleaning up the environment and protecting endangered species have been undeniable successes, and these have not been limited to the developed countries. Environmental concerns are shared by people all over the globe. Moreover, relatively straightforward ecological and engineering solutions exist for most environmental problems. To put these solutions into effect, however, we must develop the will to value long-term sustainable use of the earth's resources above the quick profits of rapid and unplanned development. We must also recognize certain undeniable facts that govern the rational development of natural resources:

1. We must accept the fact that the human population of the earth will continue to increase, at least for the near future, and that most of the surface of the earth and the oceans will be devoted to supporting that population.

2. Given these premises, we must manage the planet so as to maintain natural processes in a healthy state. By paying attention to the basic principles of ecology, it is possible to implement management practices that minimize interference with the ability of ecosystems to maintain themselves and respond to perturbation while maximizing their production for human use.

3. We must realize that different ecosystems have different optimal uses and that certain exploitation and management practices are environment-friendly whereas others are not.

4. The most productive regions of the earth do not necessarily correspond to the regions of greatest human population density. These imbalances can be overcome by transport of foods, materials, and energy from one region to another, which will require a high level of international communication, cooperation, and sharing of wealth.

5. The goal of maintaining a sustainable biosphere can be met only if the costs, both short-term and long-term, of population growth and ecological mismanagement can be fully evaluated and assigned to the goods and services that produce them.

Ecological processes

Throughout this book, we have discussed various processes involved in biological production and in the regulation of communities and ecosystems. These processes occur in managed as well as natural ecosystems. Two key aspects of ecosystem function are the harnessing of energy and the continual recycling of materials. In natural systems, the primary source of energy is sunlight; recycling is accomplished by a variety of regenerative processes, some of them physical or chemical, some of them biological. In any of these processes, an imbalance that leads to the accumulation or depletion of some component of an ecosystem normally sets in motion restorative mechanisms that push the system back to a self-maintaining steady state. For example, when dead organic matter accumulates within a system, decomposing organisms increase in number and consume the excess detritus. When herbivores increase to high levels and begin to deplete their food resources, declining birth rates and increasing mortality check population growth and restore a sustainable relationship between consumer and resource.

Restorative processes may be physical, but more often they involve biological transformations. From the composition of the atmosphere to the most basic character of many habitats, plants, animals, and microbes have greatly modified the condition of the earth's land surfaces and waters and are responsible for maintaining their qualities. When natural processes are disrupted, environments may undergo drastic change. Worse, they may lose their capacity to respond to perturbation and become permanently degraded. Thus, maintaining a sustainable biosphere requires that we conserve the ecological processes responsible for its productivity.

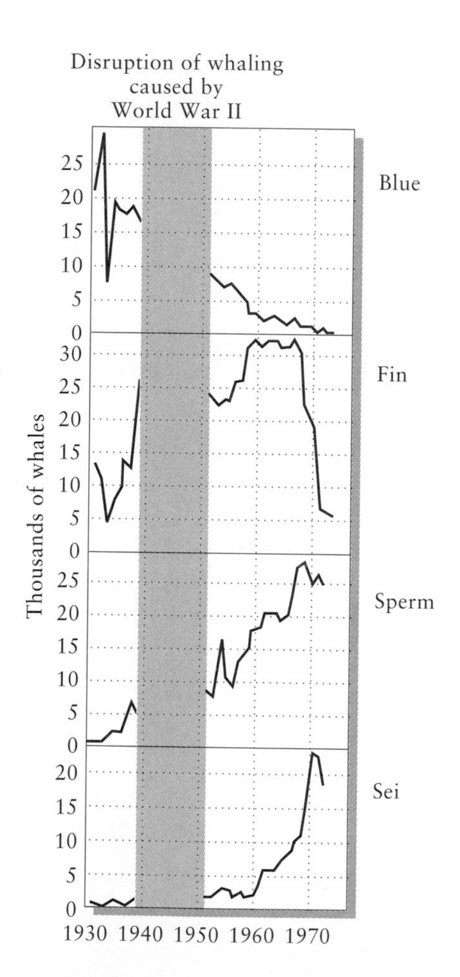

Figure 27.1 Commercial catches of four types of whales, illustrating shifts to new, less profitable species as populations of more profitable species are hunted to low levels. Based on R. Payne, in W. Jackson (ed.), *Man and the Environment,* 2d ed., W. C. Brown, Dubuque, Iowa (1973), p. 143.

Threats to ecological processes

All human activities have consequences for the environment. Fishing is a good example. The goal is to harvest a food resource for human consumption. However, when we simply maximize short-term returns from a fishery—getting it while we can—fish stocks are reduced or even collapse, the fishery goes bankrupt, and our attention turns to other exploitable populations. We have seen this pattern in the commercial whale fishery, in which one species after another was hunted to near extinction. Humpback, right, bowhead, and gray whale populations were decimated during the nineteenth century. During the twentieth century, the whaling industry turned its attention to profitable blue whales, following which the commercial catch declined to nil during the middle of the century; moved on to fin whales, which declined precipitously between 1965 and 1975; and progressed to less and less profitable species, such as sperm and, finally, sei whales (Figure 27.1). Under the strict protection now afforded, some populations of whales are increasing. Other fisheries, such as the once immensely profitable sardine fishery of western North America and the anchovy fishery of Peru, have not recovered. It is possible that these fish stocks were pushed below the level from which they could recover and that the entire ecosystem has shifted so that it no longer includes the sardine as a prominent link in the food chain.

Often the consequences of human activities are less direct and more unexpected; these consequences may be difficult to detect or may be far removed in time or space. To provide a relatively straightforward example, the clearing of land for agriculture or lumber frequently leads to erosion and deposition of silt downstream from a watershed over long periods. Thus riverine habitats may be altered and reservoirs behind dams filled in. Erosion of logged land in Queensland, Australia, is causing damage to streams and to the Great Barrier Reef off the coast. We'll look at some more examples as we consider different kinds of threats to ecological processes.

Overexploitation

Fishing, hunting, grazing, fuelwood gathering, lumbering, and the like are classic consumer-resource interactions. In most natural systems, such interactions achieve steady states because as a resource becomes scarce, efficiency of exploitation plummets; consumer populations then begin to decline or seek alternative resources until consumers and their first resource are brought back into balance. Efficiency of exploitation and ability of resources to resist exploitation are characteristics of consumers and resources that have evolved over long periods of interaction.

In economic systems, consumer-resource interactions may also come into balance because as a resource becomes scarce and its price increases, demand for that resource drops; people either do without or find cheaper alternatives. However, because the human population's ability to exploit natural systems has been escalated out of all proportion by its ability to use tools, renewable resources may not become scarce until they are very nearly depleted and are unable to sustain even reduced exploitation. Technological skills have advanced too rapidly for nature to keep pace; humans have gained the upper hand with their weapons, plows, and chain saws. As a result, many ecosystems that historically supported the growth of the human population, such as the vast forests and prairies of North America, have been converted to other uses. Where the fertility of the land itself has been exhausted (as in drier parts of Africa and increasingly elsewhere), this leaves the prospect of a large human population without a resource base to support itself. Where the population can no longer move to other areas or shift to new food sources, the prospect of population control by starvation and associated diseases and social strife portends a grim future.

That future is already upon us in some parts of the world. In sub-Saharan Africa, overgrazing and fuelwood gathering have left little vegetation to support human or any other life. In other parts of the world, vast areas of formerly productive land have been laid waste by ecologically ill-conceived practices (Figure 27.2). Although much of the Tropics can sustain intensive agriculture, particularly in mountainous regions with

(a)

(b)

Figure 27.2 (a) Severely overgrazed rangeland in Sandoval County, New Mexico, where the vegetation has been reduced to unproductive, largely inedible species. (b) Maintenance of large goat populations by providing water in Greenlee County, Arizona, results in overgrazing in the immediate vicinity. Courtesy of the U.S. Department of Agriculture, Soil Conservation Service.

volcanic soils, the high natural productivity of large portions of the Tropics depends on the presence of native vegetation. Old lowland soils are deeply weathered and deficient in the clay particles that retain many soil nutrients. Bedrock lies so far beneath the surface that new mineral nutrients do not enter upper layers of soil; nutrients that do exist there are susceptible to leaching. The natural fertility of many tropical ecosystems is maintained by constant recycling of nutrients between detritus and living plants. Break the cycle—by clear-cutting the forest, for example—and the nutrients are lost. Over much of the Amazon Basin, forested land cleared for grazing becomes so infertile that it must be abandoned after 3 years of ranching. After vegetation has been cleared and burned, harvesting cattle takes away the last remaining nutrients from the system. To be sure, the forests will regenerate, but many decades or even centuries must pass before the natural fertility of the ecosystem is restored.

Many human populations in the Tropics maintain themselves by the practice of "shifting agriculture," in which small patches of forest are cut and burned to release nutrients into the soil, planted for 2 or 3 years, and then abandoned in favor of a new patch. A particular patch typically recovers sufficiently to repeat this process in 50–100 years. Accordingly, as long as only 1–2% of the forest is cut each year, and thus perhaps 2–6% is under cultivation at any one time, the land can sustain this practice. This type of agriculture requires little input of labor, materials, or energy and takes advantage of the natural successional processes of tropical forests, but it supports only sparse human populations. When the land is cultivated more intensively by tilling, fertilization, watering, and weeding, productivity and long-term sustainability may increase greatly, but so do the inputs of labor and materials.

The situation in the Republic of Panama is typical of that in many tropical countries. Prior to the arrival of Europeans in Central America shortly after 1500, shifting agriculture and locally more intense cultivation of certain crops supported a human population that probably equaled the present-day population of about 2 million. The pre-European population existed in a long-term equilibrium with the environment; the present-day population most certainly does not. Why the difference?

In pre-European times most of Central America was intensively used, and the human population was distributed very widely. The indigenous people of Panama were greatly reduced in number following contact with Europeans, through disease, armed conflict, and disruption of society—a pattern typical of much of the earth. Following the depopulation of Panama and much of the rest of Central America, agricultural land was abandoned and forests regrew. Thus, the vast forests, virtually devoid of human impact, that existed 50–100 years ago were a relatively recent development. The present human population of Panama has grown tremendously from a relatively small size, and it has not experienced the long-term equilibrium enjoyed by former indigenous inhabitants. The redevelopment of this land that has accompanied the recent population increase has taken a different, more highly mechanized and energy-intensive course.

Patterns of food consumption have changed dramatically. Individual men and women are larger and require more food, and meat makes up a greater part of the diet. As a result, vast areas of forest have been converted to relatively unproductive rangeland to feed cattle; this practice was originally restricted to drier parts of the country but now occurs virtually everywhere. In addition, as in many tropical countries, land has been cleared to grow export products shipped to other parts of the world: coffee, sugar, bananas, and beef, to name a few. Many tropical regions lack abundant mineral resources, and their people must rely on agricultural exports to pay for imports of the manufactured goods that have become essential to a high standard of living. The result for much of the Tropics is a decline in the natural productivity of the environment, which precludes alternative sustainable land uses.

The solution? Reduce intake, eat lower on the food chain, determine maximum sustainable yields for resource populations, consider alternative sustainable uses of land, increase agricultural intensity on land that will bear it, and improve distribution of food between areas of production and areas of need. Most of these solutions carry a price tag. Planning for sustained use cuts short-term returns, and these losses must be made up elsewhere to maintain the wealth of a country. Increasing agricultural intensity requires disproportionately greater inputs of energy, labor, and chemical fertilizers, each contributing its own problems. Relying on crops selected for high food production often makes agriculture more vulnerable to outbreaks of pests and disease. Segments of the human population that are forced by impoverished land to import food also must earn money to buy it; otherwise, they will be reduced to welfare status at a cost to other segments of the population. As long as local growth of the human population is not tied to the ability of local resources to support population growth, imbalances between human consumers and their resources will continue to proliferate.

Introductions of alien species

Both intentionally and unintentionally, humans have taken other species everywhere they have traveled. Aborigines brought dingoes (semidomesticated dogs) to Australia; Polynesians brought rats to Hawaii. Of course, the global movements of species by human agency have increased immensely since Europeans began colonizing most of the world some 500 years ago. Introductions have included edible and horticultural varieties of plants, and their pests; commercially valuable trees; domesticated animals for work or meat; familiar backyard animals, especially birds; disease organisms; and such commensals of human habitation and transportation as the ubiquitous cockroach and dandelion. The result has been a globally distributed flora and fauna of alien species that have displaced or otherwise wreaked havoc with local biotas.

To take an extreme example, most of the area of the island of New Zealand is occupied by introduced plants and animals (Figure 27.3). The native forest was cut long ago and replaced by pines from North America and eucalyptus from Australia; moas (large, flightless birds) were killed off by Maori natives before Europeans arrived, and sheep now take their place; most birds of the countryside are those that were transplanted from England to stave off the homesickness of early colonists. Only at the southern tip of New Zealand do native forests of southern beech persist in wet and remote fiordlands. Of the total New Zealand flora of 2,500 species, fully 500 are naturalized introductions, and they account for most of the present vegetation. These aliens prospered for a variety of reasons. Most of the natural habitat in New Zealand had been greatly disturbed by lumbering, farming, and ranching, which made invasion by weedy European species, accustomed to disturbance and intensely cultivated landscapes, relatively easy. Also, because of their comparatively low diversity and simple community structure, island ecosystems tend to be easier to invade than

(a)

(b)

Figure 27.3 Native forests over most of New Zealand (a) have been replaced by agricultural landscapes featuring mostly plants and animals introduced from Europe (b).

continental ecosystems—there are simply fewer native species to provide effective competition.

Aliens can have drastic effects on native habitat, as we have seen in previous chapters. In Australia, introduced prickly pear cactus invaded millions of hectares of native grassland, turning it into impenetrable thickets. In Hawaii, feral pigs rooting through forest litter have interfered with the regeneration of native forests; populations of birds in the same forests have been decimated by introduced bird pox and malaria transmitted by introduced mosquitoes.

In many cases, alien species can bring benefits. After all, our crops and domesticated animals are alien to most of their present distributions. In other cases, even though aliens might displace native plants and animals, they do not necessarily disrupt ecosystem function. However, the effects of introduced species are often difficult to predict. In aquatic systems in particular, introduced consumers at high levels in the food chain have seriously disrupted ecosystem function and have caused basic changes in community structure. Efficient predators, such as the Nile perch introduced to Lake Victoria in East Africa and the peacock bass introduced to Gatun Lake in Panama, can virtually eliminate entire trophic levels of smaller planktivorous fish. As a result, densities of zooplankton increase dramatically and algae are cleared from the water, reducing the overall productivity of the environment. In this way, a single predator—referred to as a **keystone species** in this case because of its critical place in ecosystem function—can shift the character of a habitat from one state to a qualitatively different state.

Habitat conversion

Altering the basic nature of a habitat often upsets natural processes of regeneration and control and brings about disastrous consequences.

(a)

(b)

Figure 27.4 Examples of soil erosion and gully formation on plowed farmland (a, Macon County, North Carolina) and heavily grazed pasture (b, near Bethany, Missouri). Courtesy of the U.S. Department of Agriculture, Soil Conservation Service.

Cutting a tropical forest on impoverished soil breaks the tight cycling of nutrients that maintains forest productivity and greatly alters the physical structure of the soil by exposing it to increased leaching and sunlight. As a result, the productivity of the land decreases precipitously, and soil erosion may increase tenfold or more. By some estimates, up to 1% of the earth's topsoil is lost to erosion every year (Figure 27.4). In the Amazon Basin, for example, erosion rates increased from 6–10 metric tons (T) ha^{-1} yr^{-1} in

Figure 27.5 Farmstead in Oklahoma abandoned during the height of the Dust Bowl period in 1937. Courtesy of the U.S. Department of Agriculture.

1960 to 18–190 T ha^{-1} yr^{-1} by 1985, largely as a result of deforestation and overgrazing.

Problems associated with habitat conversion are not restricted to tropical forests. On the American prairies, plowing destroyed the dense root mats of perennial herbs that formerly held the soil together. A prolonged drought in the central United States during the 1920s and 1930s turned former prairies converted to agriculture into a devastated "dust bowl" of blowing soil (Figure 27.5). Mangrove forests provide natural protection for coastlines in many parts of the Tropics. Where these have been cleared for fuelwood and land reclamation, coasts have been laid bare to rampaging hurricane-driven floodwaters. Damming rivers brings the benefits of flood control, irrigation water, and power generation, but also increases silt transport, blocks fish migrations, alters downstream water conditions, and may even change the local weather.

Irrigation

Water makes the desert bloom. Humankind has employed various irrigation schemes to increase the productivity of land since the beginning of agriculture (Figure 27.6). Only recently, however, has the ancient practice of irrigation been applied on immense scales to land that would otherwise be totally unsuitable for agriculture. The benefits are tremendous, but so are the costs, many of which surface only after years of profitable irrigation. The primary costs are the environmental effects of developing the dams, wells, canals, and dike work required to support irrigation; lowered water tables where wells are the source of irrigation water; reduction of groundwater quality through the introduction of pesticides and fertilizers or the concentration of naturally occurring toxic elements; the

(a)

(b)

Figure 27.6 Irrigation can turn desert into productive farmland (a, furrow irrigation of cotton in the Imperial Valley of southern California). However, the salting that accompanies irrigation requires periodic flooding to flush salts from the soil (b, Maricopa County, Arizona). Courtesy of the U.S. Department of Agriculture.

accumulation of salt in irrigated soils in arid zones; and transmission of diseases by aquatic organisms. In most cases, the costs of delivering water to crops, including the burden of future environmental problems, are underwritten by the population at large through taxes and other subsidies: rarely does irrigation pay its own way. Why, then, is there so much irrigated land? According to M. P. Reisner, in his book *Cadillac Desert,* many of the largest water projects in the western half of the United States have been the result of overzealous federal agencies pandering to large agribusiness interests.

Fertilization and eutrophication

Any substance that enhances the productivity of a habitat may be considered a fertilizer. We apply fertilizers to agricultural lands to increase crop production, but a portion of these chemicals make their way into groundwater and from there into rivers, lakes, and eventually the ocean. Nitrates, phosphates, and other inorganic fertilizers have the same effect on rivers and lakes as they do on agricultural lands: they increase biological production. A consequence of this artificial fertilization, often called **eutrophication,** is change in the chemical and biological conditions of a body of water. Although increased production is not necessarily bad, it may cause a change in the species composition of a river or lake. Input of inorganic fertilizers may also turn clear, oligotrophic waters into turbid environments that are less attractive for recreation. Often, nutrient inputs upset seasonal cycles of nutrient use and regeneration in natural bodies of water, leading to accumulation of organic material, high rates of bacterial decomposition, and deoxygenation of the water. Under such conditions, fish may suffocate and contribute further to the load of organic material in the water.

Direct input of organic wastes, such as sewage and runoff from feedlots, poses a greater problem for water quality. Suspended or dissolved organic materials in water create what is known as **biological oxygen demand,** meaning that the decomposition of these materials by bacteria uses oxygen present in the water. When organic materials are added from outside a system, they may completely alter the natural balance of oxygen production by photosynthesis and oxygen consumption by respiration, because these organic inputs are unrelated to the natural productivity of the system. Under these conditions, a stream or lake may become anoxic for long periods and unsuitable for many forms of life.

Before water pollution came under strict controls in North America and Europe, large sections of major rivers became completely anoxic, killing off local fish populations and preventing the migration of other species, such as shad and salmon, between the ocean and their headwater spawning grounds (Figure 27.7). The costs of such moribund rivers to fisheries and recreation, not to mention to aesthetic sensibility, were enormous. In general, natural conditions can be restored by cutting off sources of organic nutrients, either by diverting the inputs to larger bodies of water that can absorb them or by improving treatment of sewage. The costs of these solutions have been more than repaid in the long run by the benefits of enhanced water quality for fisheries, public health, and recreation.

Accumulation of toxins in the environment

Toxins are poisons that kill animals and plants by interfering with their normal physiological functions. Toxic substances can be divided into several classes: acids, heavy metals, organics, and radiation are the most notable. Toxins are by no means novel. Plants have evolved chemical defenses that kill or sicken herbivores, and many animals use toxins to kill

Figure 27.7 Fish die-off in a stream near Colorado Springs, Colorado, caused by oxygen deficiency related to organic pollution. Courtesy of the U.S. Department of Agriculture, Soil Conservation Service.

their prey. Human technology, however, has produced a stupefying array of nonnatural chemical substances with adverse effects on life. Many are produced specifically because of their toxic effects, among them a variety of insecticides, rodenticides, and herbicides. Some of these, such as the insecticide DDT (the use of which is now banned in the United States), have had damaging effects far beyond their intended victims.

Acids. Acids are very reactive substances that produce hydrogen ions (H+) and may be extremely toxic at high concentration (low pH). Acids affect organisms directly by interfering with physiological functions and indirectly through their influence on nutrient availability and regeneration. In particular, high acidity reduces the solubility of phosphorus in soils and waters, which tends to reduce productivity.

All environments contain natural acids, which are produced when carbon dioxide dissolves in water to form carbonic acid and when bacterial metabolism forms organic acids. Anthropogenic sources of acid are primarily of two kinds. The first occurs in coal mining areas where reduced sulfur compounds associated with coal are exposed to atmospheric oxygen. Sulfur bacteria oxidize pure sulfur and thiol (reduced) forms of sulfur to sulfates, which may then be converted to sulfuric acid in streams that drain mining areas—hence the term **acid mine drainage.** In some places, the water becomes so acid as to sterilize the aquatic environment (Figure 27.8).

A more widespread problem is **acid rain.** Coal and oil are not pure hydrocarbons; they contain sulfur and nitrogen compounds as well. After all, these fossil fuels are the remains of plants and animals that, when living, contained nitrogen and sulfur in proteins and other organic molecules. Burning coal and oil, in addition to producing carbon dioxide and water vapor, spews nitrous oxides and sulfur dioxide into the atmosphere. When

Figure 27.8 Streams draining from the refuse of coal mining may be extremely acid. Courtesy of the U.S. Department of Agriculture.

these gases dissolve in raindrops, they are converted to acid and cause acid rain. The pH of natural rain is about 6—slightly on the acid side of neutral (pH 7)—because of naturally occurring carbonic acid. In highly industrialized areas, the pH of rain may drop to between 3 and 4, which is 100 to 1,000 times the acidity of natural rain.

The consequences of acid rain have been severe in some regions, such as the northeastern United States, Canada, and Scandinavia, where rivers and lakes tend to be oligotrophic and thus do not contain dissolved bases to buffer acid inputs. As a result, pH may drop to as low as 4, acid enough to stunt growth or even cause mortality of fish and other organisms. Acid rain may also lower the pH of soil, which increases the rate of leaching of soil nutrients and precipitates phosphorus compounds, making them unavailable for uptake by plant roots. Acid directly affects the foliage of plants as well, making leaves more susceptible to disease and frost damage. The ecological scope of the acid rain problem includes its adverse effects on productivity and wildlife in both aquatic and terrestrial habitats. The solutions are primarily technological and economic: scrubbing offending gases from the effluents of power plants and automobiles, finding alternatives to burning fossil fuels for energy, and reducing total demand for energy.

Heavy metals. Even in low concentrations, mercury, arsenic, lead, copper, nickel, zinc, and other heavy metals are toxic to most forms of life. They are introduced into the environment in a variety of ways, principally as refuse from mining and mineral smelting, as waste products of manufacturing processes, as fungicides (such as lead arsenate), and through the burning of leaded fuel. Although banned in the United States, leaded gasoline is still used widely throughout the world. The effects of heavy metals are varied but include interference with neurological function in vertebrates. Ecological studies have revealed the movement and concentration of heavy metals through food chains and the persistence and transformation of these elements in ecosystems.

Many toxic metals, including copper and nickel particulates released into the atmosphere by smelters, eventually accumulate in soil. The concentration of copper averages about 30 parts per million (ppm) in unpolluted temperate zone soils. Concentrations in excess of 100 ppm adversely affect mosses, lichens, and large fungi, earthworm abundance drops off dramatically above 1,000 ppm, and most species of vascular plants cannot tolerate concentrations above 5,000 ppm (0.5%). As fungi die out, decomposition of organic matter and nitrification of organic nitrogen in the soil decrease. In one study in Sweden, at copper levels of 2,000 ppm, fungal populations had fallen to only 20–30% of their natural levels. Concentrations exceeding 1,000 ppm may extend for 10–20 km from sites of metal smelting, with predictable effects on the diversity and productivity of local communities. These effects can be mitigated to some degree by taller smokestacks, which distribute wastes over larger areas at lower concentrations, but ultimately, solving the problem will require a change in the technology of metal production to reduce toxic by-products.

High concentrations of metals in some soils have given biologists an unusual opportunity to observe evolutionary responses of plants. Refuse material from mining operations often contains concentrations of copper, lead, zinc, or arsenic in excess of 1,000 ppm (and, in some cases, in excess of 10,000 ppm), which are toxic to most plants. These mine tailings, however, are often colonized by grasses and other plants that, when analyzed, reveal unusually high tolerances for toxic metals. Specific tolerances are the result of genetic factors that occur as rare mutants in populations living on normal soils. On mine soils, these mutant individuals have higher evolutionary fitness than intolerant forms, so their progeny proliferate and spread to form a genetically distinct subpopulation restricted to the area of contamination. Selection has also favored reduced outcrossing in these populations, which prevents fertilization of metal-tolerant genotypes by pollen from intolerant plants growing on nearby uncontaminated soils and lacking the genes for metal tolerance! This example illustrates one facet of the immense capacity of natural systems to respond to perturbations of all sorts. Nature is extremely forgiving of human excesses—perhaps too much so for our own good.

Toxic organic compounds. Toxic organic compounds are widespread in nature as chemical defenses against herbivores and as metabolic by-products of various microorganisms, such as the bacterium that causes botulism and the dinoflagellate that causes toxic red tides. Although agricultural pesticides include some natural compounds, such as nicotine and pyrethrins, most are far more deadly concoctions produced in the laboratory, to which pests have had no previous exposure and no opportunity to evolve resistance. The latter include organomercurials (such as methylmercury), chlorinated hydrocarbons (DDT, lindane, chlordane, dieldrin), organophosphorus compounds (parathion, malathion), carbamate insecticides, and triazine herbicides. Although these compounds do their job in agriculture and pest management, many accumulate in other parts of the ecosystem where they adversely affect plant production and wildlife populations. These undesirable side effects occur because of the difficulty of delivering pesticides to particular targets without other species getting in the way, because of the indiscriminate action of most pesticides, which are designed to attack basic physiological functions, and because many of these substances persist in the environment and may be concentrated as organisms feed on others that contain them.

The design of modern pesticides and delivery systems takes these problems into account, and many of the worst problems have lessened. Unfortunately, this progress is offset by increasingly widespread and more intense use of chemicals of all kinds in agriculture. Overuse and misuse of pesticides can be addressed in part by applying them properly and in the smallest effective amounts. But because insects and other pests may evolve tolerances to pesticides, just as many plants have evolved tolerances to toxic elements in soil, applications of these chemicals often produce only short-term benefits, and their amounts must be increased to achieve continued results. Through ecological research, we can assess the vulnerability of natural systems to these pollutants, prescribe safe applications, and—perhaps most important—find suitable alternatives to humankind's chemical warfare with the environment. Some microorganisms—some of them genetically engineered for special biochemical properties—can be used to metabolize pesticides and other toxic compounds to innocuous by-products. This type of approach, using biological agents to clean up the environment and help restore habitats, is referred to as **bioremediation.**

Biological control of pests, which relies on the principles of predator-prey interactions, often provides effective regulation of insect populations without the adverse effects of chemical pesticides. We have seen how parasitoid wasps were used to control scale insects in California citrus groves, where cyanide fumigation was becoming ineffective because of evolved cyanide tolerance. Similarly, cactus moths effectively reduced the prickly pear cactus where it had been introduced in Australia. Some types of herbivore damage can be alleviated by inducing natural resistance in crop plants. A relatively innocuous, herbivorous mite introduced to cotton crops can induce chemical resistance that depresses subsequent infestations by more damaging species of mites. This **cross-resistance** to pest organisms is reminiscent of the immunity to dangerous diseases that sometimes develops following infection by related organisms that produce only mild symptoms.

Another type of toxic environmental pollution caused by organic compounds is that resulting from **oil spills.** Crude oil is a complex mixture of hydrocarbons, with concentrations of nitrogen up to 1% and of sulfur up to 5%. Oil pollution occurs at the source in areas of oil production, rarely as a result of breaks in the nearly 100,000 km of oil pipeline in the world, and most frequently in the ocean from offshore drilling and the wreckage of oil tankers (Figure 27.9).

Hydrocarbons are produced naturally in the marine system by algae at a rate of 20–30 million ($20–30 \times 10^6$) metric tons (T) per year. These nonpetroleum hydrocarbons are very widely dispersed and do not constitute pollution, in the sense that they are natural products in low concentration. Contamination of oceans by petroleum hydrocarbons from natural seeps amounts to about $0.2–0.6 \times 10^6$ T annually; various types of anthropogenic oil spills dump $3–6 \times 10^6$ T into the environment each year—that is, 0.1–0.2% of global oil production. This total is small, but its local effects can be devastating. Petroleum kills by coating the surfaces of organisms and, because hydrocarbons are organic solvents, by disrupting biological membranes. Over time, oil slicks disperse by evaporation of lighter fractions, emulsion of other fractions in water, and weathering and microbial breakdown of the rest. But certain types of sensitive ecosystems, such as coral reefs, may take decades to recover fully.

Radiation. Radiation comes in a broad spectrum of energy intensities, ranging from generally harmless long-wavelength radio and infrared radiation, through damaging ultraviolet radiation of shorter wavelength, to the extremely high energy of cosmic rays and subatomic particles released by the disintegration of atomic nuclei (radioactive decay). Natural sources

(a)

(b)

Figure 27.9 (a) Oil slick from a damaged tanker approaching the Caribbean coast of Panama. (b) Fringe of dead mangroves killed by the oil spill. Photographs by C. H. Hansen. Courtesy of the Smithsonian Tropical Research Institute.

create an unavoidable background level of radiation. Under some circumstances, radioactive substances, such as radon gas present in soils in regions having granitic bedrock, can become concentrated and pose public health hazards. Such dangers are minor, however, compared with the extreme radiation hazards that could result from accidents at nuclear power plants, such as those that occurred at Three Mile Island, in Pennsylvania, and Chernobyl, in Ukraine, from the waste products of nuclear power generation, and from nuclear war. The effects of intense radiation on life forms can be seen clearly in the results of experiments in which habitats are exposed to radiation sources.

The possibility of nuclear war is diminishing as the superpowers dismantle their nuclear arsenals and nations turn to economic, social, and environmental problems more pressing than national defense. Still, radioactive wastes produced by peaceful uses of the atom pose tremendous disposal problems. Depending on the waste product, radiation does not decline to harmless levels for thousands or even millions of years, far beyond the life span of waste containers, not to mention that of the institutions to which their care is entrusted. The waste disposal problem may ultimately limit the use of nuclear power, at least until new methods can be devised to deal with radioactive wastes.

Ultraviolet light is another form of radiation that is becoming a serious threat to the environment as the protective mantle of ozone in the atmosphere shows signs of depletion. We shall consider this problem in detail as we examine the atmospheric factors that threaten the entire globe.

Global threats

Because of the circulation of the atmosphere and oceans, certain types of pollution have global consequences: their effects extend far beyond the sources of the pollution itself. By far the most worrisome of these changes in the environment are the destruction of the ozone layer in the upper atmosphere and the increase in carbon dioxide and other "greenhouse" gases.

The ozone layer and ultraviolet radiation

Ozone (O_3) is a molecular form of oxygen that is highly reactive as an oxidant, capable of chemically oxidizing organic molecules and destroying their proper functioning. As a consequence, ozone is toxic to animal and plant life even in small concentrations. Near the earth's surface, ozone is produced by photochemical oxidation of molecular oxygen (O_2) in the presence of nitrous oxide (NO_2). Because NO_2 is a product of gasoline combustion, ozone can reach high levels in the exhaust fumes that pollute cities, particularly where there is strong sunlight. In Los Angeles, for example, which is well known for its smog, ozone concentrations in the atmosphere at ground level can reach 0.5 ppm, which is perhaps 20–50 times the normal level and is damaging to health, crops, and natural vegetation.

Photochemical oxidation in the upper atmosphere also produces ozone, where it has the extremely beneficial effect of shielding the surface of the earth from ultraviolet radiation by absorbing solar radiation of short wavelength (especially in the range of 200–300 nm). The concentration of ozone is maintained at an approximate equilibrium by photodissociation of O_3 to O_2 and O. Unfortunately, this dissociation is accelerated by certain substances, among them chlorine atoms and simple chlorine compounds. Levels of chlorine in the upper atmosphere have been increasing because of the release into the atmosphere of chlorofluorocarbons (CFCs), which are used as propellants in spray cans and as coolants in air conditioning and refrigeration systems. Decreases in stratospheric ozone of 50% or more (so-called **ozone holes**) have been observed at high latitudes in both hemispheres.

The resulting elevation of ultraviolet radiation at the earth's surface will increase the incidence of skin cancer, because DNA also absorbs ultraviolet radiation between 280 and 320 nm in wavelength, and the energy of the absorbed radiation causes damage to DNA molecules (mutation). Of greater concern is the fact that ultraviolet radiation also damages the photosynthetic apparatus of plants and may cause a reduction of primary production—the base of the food chain for the entire ecosystem. This has already been observed in the oceans surrounding Antarctica. The threat is so great that the international community, through the Vienna Convention for the Protection of the Ozone Layer (1985) and the Montreal Protocol (1987), has agreed to phase out the use of all CFCs by the end of the twentieth century, and by 1997 in the European Community. Even though it is too early to see the result, we can anticipate that this action will reverse the damage that has already been done and allow atmospheric ozone to return to its natural equilibrium level, perhaps within a century.

Carbon dioxide and the greenhouse effect

Carbon dioxide (CO_2) occurs naturally in the atmosphere. Without it, the earth would be a very cold place, because most of the sunlight absorbed by the earth's surface would be reradiated into the cold depths of space. As it is now, as short-wavelength radiation from the sun warms the surface of the earth, the surface radiates thermal energy back into the atmosphere. Much of this energy, which occurs as long-wavelength infrared radiation peaking at 10,000 nm, is absorbed by carbon dioxide in the atmosphere. As the CO_2 heats up, its thermal energy is reradiated at even longer wavelengths—about half of it back toward the surface of the earth. Thus CO_2 forms an insulative blanket over the earth's surface that lets short-wavelength ultraviolet and visible light from the sun pass through, but retards loss of heat as longer-wavelength infrared radiation (Figure 27.10). Glass in greenhouses works on the same principle, and so the function of CO_2 in the atmosphere is known as the **greenhouse effect.**

At times in the distant past, the concentration of CO_2 in the atmosphere was far greater than during recent human experience, and the average temperature of the earth was correspondingly much warmer. During the early part of the Paleozoic period, roughly 500–350 million years ago, CO_2 levels were 10–15 times greater than at present. After some variation, CO_2 levels declined steadily over the past 100 million years from about 5 times the present level to the present level itself. During this time, the earth experienced a gradual cooling resulting in an expansion of temperate and boreal climate zones and culminating in the Ice Ages of the past million years. The problem we face in global warming is not that the earth has never been so warm, but that the climate will change so quickly that ecological systems, not to mention the human population, will not be able to keep up.

Before 1850, the concentration of CO_2 in the atmosphere was on the order of 280 ppm (0.028%). During the last 150 years, which have witnessed tremendous increases in the burning of wood, coal, oil, and gas for

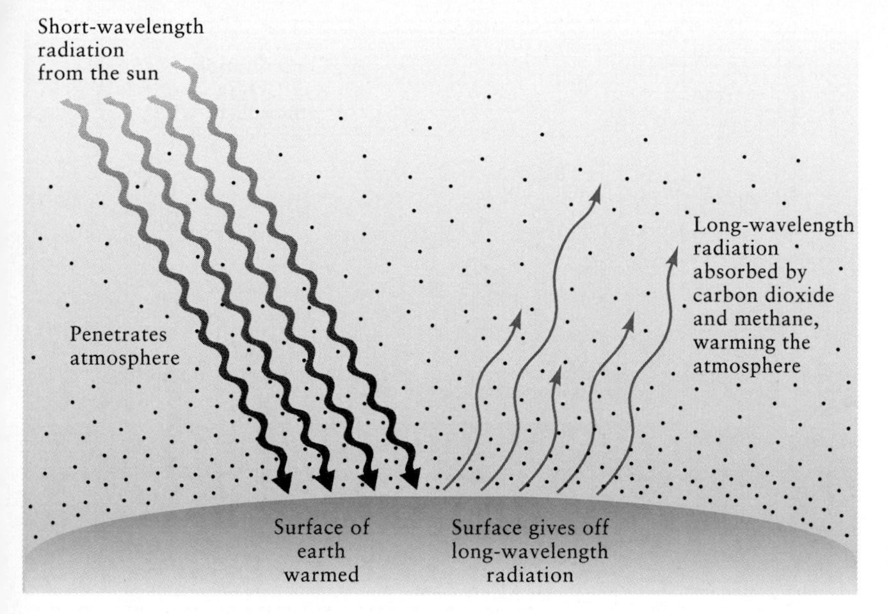

Figure 27.10 Schematic depiction of the greenhouse effect, illustrating how greenhouse gases, such as carbon dioxide and methane, allow short-wavelength light to pass through the atmosphere but absorb longer-wavelength (infrared) radiation given off by the earth.

energy production, it has increased to 345 ppm (Figure 27.11). Half of this increase has occurred during the last 30 years, and the rate of increase appears to be rising. This change in the chemistry of the atmosphere has created fears of a major warming of the earth's climate, with disastrous consequences resulting from the melting of ice caps, a rise in sea level owing to the expansion of warmed ocean water, the drying of agricultural lands, and the shifting of ecological zones faster than plants can extend their geographic ranges.

Carbon dioxide is not the only greenhouse gas. Methane, produced in abundance by the ruminant fermentation of cattle, absorbs infrared radiation 20 times more effectively than does CO_2; the chlorofluorocarbon $CFCl_3$ is 10,000 times more effective than CO_2 and, even in its low concentration of 0.0002 ppm, is responsible for almost one-tenth of the enhanced greenhouse gas effect. Although carbon dioxide continues to be the most important greenhouse gas, and is responsible for most of the increase in the greenhouse properties of the atmosphere, the relative contributions of other gases, particularly CFCs and methane, have increased greatly during the twentieth century (Figure 27.12).

If present levels of carbon dioxide warm the earth, more CO_2 will warm it more. At least that is the view of many climatologists. The question is, how much? And with what consequences? Current estimates of global warming in the next century suggest a rise in average global temperature of anywhere between 2° and 6°C. These predictions are based on models of the behavior of carbon dioxide and heat in the atmosphere, but scientists do not fully agree on the details of these models. Hence the differences in opinion that we hear in scientific discussions and the popular press.

The level of carbon dioxide in the atmosphere represents a balance between processes that add CO_2 and those that remove it. Before the Industrial Revolution, addition of CO_2 to the atmosphere by the

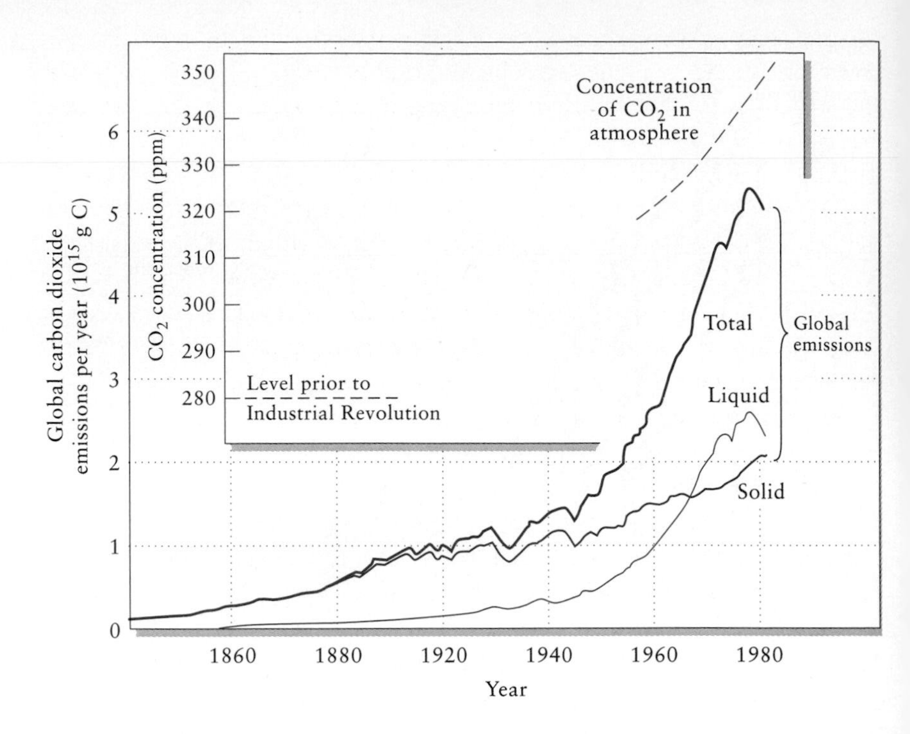

Figure 27.11 Global emissions of carbon dioxide from combustion of solid (coal) and liquid (oil) fossil fuels since 1860, and the resulting increase in the concentration of carbon dioxide in the atmosphere. Accurate measurements of carbon dioxide concentration have been taken at Mauna Loa Observatory, Hawaii, which is far from regions of concentrated fossil fuel consumption, only since the late 1950s. Data from A. M. Solomon, J. R. Trabolka, D. E. Reichle, and L. D. Voorhees, R. M. Rotty, and C. M. Masters, in *Atmospheric Carbon Dioxide and the Global Carbon Dioxide Cycle,* U.S. Department of Energy, Washington, D.C. (1985), pp. 1–13 and 63–80.

respiration of terrestrial organisms (approximately 120×10^9 T [10^{15} g] of carbon per year) was balanced by the gross primary production of terrestrial vegetation, and the total amount in the atmosphere was maintained in equilibrium. During the past 150 years, increasing amounts of CO_2 have been added to the atmosphere by the burning of fossil fuels and the clearing and burning of forests. At present, forest clearing accounts for the addition of about 2×10^9 T of carbon to the atmosphere annually; burning of fossil fuels accounts for the addition of about 5×10^9 T. Thus human activities have increased carbon flux into the atmosphere by about 6%—perhaps more, perhaps less, depending on whose figures we use.

The atmosphere exchanges carbon dioxide with the oceans, where excess carbon is precipitated as calcium carbonate sediments. It has been estimated that at present the oceans absorb more carbon than they release to the atmosphere by about 2.4×10^9 T annually. In spite of the fact that the oceans are a net carbon dioxide sink, their capacity to absorb carbon is *less than half* of the anthropogenic input to the atmosphere, and part of their capacity may be used to offset excess production of carbon dioxide by terrestrial systems. No matter how the arithmetic is done, it shows carbon dioxide concentrations in the atmosphere to be rising very rapidly.

And what of the consequences of increased carbon dioxide and global warming? Increased CO_2 could increase plant production, particularly in productive regions where neither water nor soil nutrients limit photosynthesis. In more arid regions, a greater concentration of CO_2 could decrease the water requirements of plants, because it would allow carbon dioxide to enter leaves more rapidly and stomates could then be closed to reduce water loss.

Warmer temperatures caused by the greenhouse effect will have mixed effects, some of them potentially disastrous. On the positive side, warmer temperatures lengthen the growing season and speed metabolism, and thereby tend to enhance production in moist environments. Balancing this benefit is the likelihood of increasing drought stress in arid environments, which may reduce production and accelerate the conversion of overused grazing lands and croplands to useless wasteland (desertification).

Many ecologists are concerned that global warming will cause climate belts to shift geographically so fast that plant populations will not be able to keep up. Distributions of plants expand by colonization, whose rate is limited by the dispersal distances of seeds. As glaciers retreated from North America and Europe at the end of the Pleistocene epoch, beginning about 18,000 years ago, distributions of plants shifted great distances. Boundaries of tree distributions advanced, following the receding edges of the glaciers, by as much as 100 m per year. The concern of ecologists is that the present global warming may shift climate belts more rapidly than this, greatly outpacing the ability of plants to keep up. The result will be widespread disruption of natural ecological communities and, in a worst-case scenario, extinction of a part of the flora and its associated fauna. It is impossible to gauge the accuracy of these predictions, but the potential for ecosystem deterioration, a probable rise in sea levels, and changes in agriculture resulting from global warming, combined with the current rate of depletion of fossil fuel resources and the problem of disposal of toxic materials arising from fossil fuel combustion, should prompt immediate action to reduce the burning of forests and of fossil fuels and to accelerate the search for alternative sources of energy.

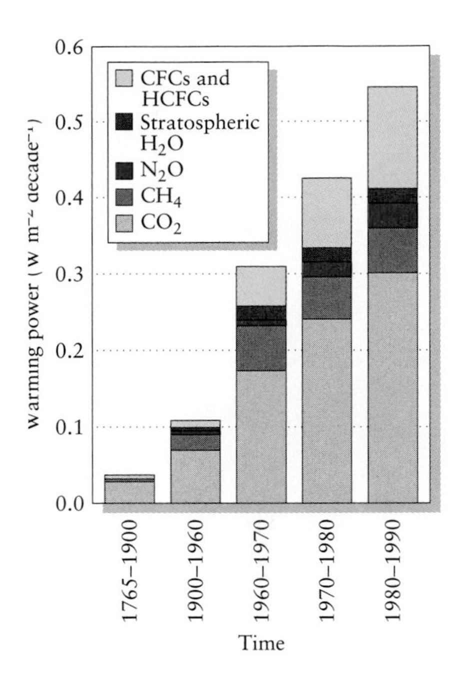

Figure 27.12 Increases in radiative forcing—the warming power of the atmosphere—and changes in the proportions of the greenhouse gases responsible since the middle of the eighteenth century. From T. E. Graedel and P. J. Crutzen, *Atmosphere, Climate, and Change,* Scientific American Library, W. H. Freeman, New York (1995).

Human ecology

The human population is increasing worldwide at a rate of almost 2% per year. Even if population growth were to stop today, staggering problems would remain. The present human population is consuming resources faster than new resources are being regenerated by the biosphere, all the while pouring forth so much waste that the quality of the environment in most regions of the earth is deteriorating at an alarming rate. If we are to leave a habitable world for future generations, our top priority must be to achieve a sustainable, equilibrial relationship with the rest of the biosphere. This will require putting an end to population growth, developing sustainable energy sources, providing for regeneration of nutrients and other materials, and restoring deteriorated habitats.

In many respects, humankind has stepped beyond the bounds of the usual ecological mechanisms of restraint and regeneration. Our ability to tap nonrenewable sources of energy in the form of coal, oil, and gas deposits has temporarily removed conventional food-energy limitations on our population growth. No longer is most of the human population supported by the land it occupies. Our technological ability to reach out and continually usurp new land and resources has pushed density-dependent population feedbacks well into the future.

Our present course, however, leads in a predictable direction. It is not an inviting one: increasing energy, material, and food shortages; most of the population living in abject poverty, filth, and disease; the environment so badly polluted as to leave the earth gasping; social and political strife escalating as the rich and powerful try to defend their high standard of living. These are the inevitable mechanisms of population control that will eventually come into play, as they do for every species.

The future need not be like this. Where we have escaped natural restraint, we must substitute our own restraint. Where we produce waste products that cannot be regenerated by ecological systems, we must find ways to recycle them ourselves. Energy consumption must be scaled back, and production must be based increasingly on renewable energy sources, such as the sun and wind. Achieving these goals will require a consensus among human social, economic, and political institutions. The hope for such a consensus lies in making people aware of the global deterioration of the quality of human life and educating them in the basic ecological principles that must form the foundation of a self-sustaining system.

Above all, humankind has the choice of adopting a new attitude toward its relationship with nature. We are a part of nature, not apart from nature. To the extent that our intelligence, culture, and technology have given us the power to dominate nature, we must also use these abilities to impose self-regulation and self-restraint. This is the greatest challenge facing us. We have succeeded famously in becoming the technological species. Our survival now depends on our becoming the ecological species and taking our proper place in the economy of nature.

Summary

1. The key to survival of the human population is development of sustainable interactions with the biosphere. This will require control of human population growth, an increasing reliance on renewable energy sources, and total recycling of material wastes.

2. Maintaining a sustainable biosphere means that we must conserve the ecological processes responsible for its productivity.

3. The principal local threats to natural processes are overexploitation of resources, introductions of alien species, habitat conversion, enrichment with organic wastes (eutrophication), and production of toxic materials.

4. On a global scale, various airborne pollutants, especially chlorofluorocarbons, have seriously reduced the concentrations of ozone (O_3) in the upper atmosphere, allowing damaging ultraviolet radiation to reach the surface of the earth at increased intensities.

5. Increasing levels of carbon dioxide in the atmosphere, produced by the burning of fossil fuels, threaten to increase the average temperature of the earth by 2°–6°C, with potentially disastrous consequences for natural ecosystems and agriculture. In addition, coastal settlements may be inundated by rising sea levels fed by melting polar ice caps and expansion of warmed ocean water.

6. Solutions to the environmental crisis will require new attitudes promoting sustainability and self-restraint.

Suggested readings

Arrow, K., et al. 1995. Economic growth, carrying capacity, and the environment. *Science* 268:520–521.

Ausubel, J. H. 1991. A second look at the impacts of climate change. *American Scientist* 79:210–221.

Bongaarts, J. 1994. Can the growing human population feed itself? *Scientific American* 270:36–42.

Bongaarts, J. 1994. Population policy options in the developing world. *Science* 263:771–776.

Browder, J. O. 1992. The limits of extractivism. *BioScience* 42:174–182.

Clark, W. C. 1989. Managing planet earth. *Scientific American* 261:46–64. (See also other articles in the same issue.)

Daily, G. C., and P. R. Ehrlich. 1992. Population, sustainability, and Earth's carrying capacity. *BioScience* 42:761–771.

Davis, G. R. 1990. Energy for planet earth. *Scientific American* 263:54–62.

Goldemberg, J. 1995. Energy needs in developing countries and sustainability. *Science* 269:1058–1059.

Graedel, T. E., and P. J. Crutzen. 1995. *Atmosphere, Climate, and Change.* Scientific American Library, W. H. Freeman, New York.

Graham, R. L., M. G. Turner, and V. H. Dale. 1990. How increasing CO_2 and climate change affect forests. *BioScience* 40:575–587.

Hecht, S. B. 1993. The logic of livestock and deforestation in Amazonia. *BioScience* 43:687–695.

Houghton, R. A., and G. M. Woodwell. 1989. Global climate change. *Scientific American* 260:36–44.

Hughes, T. P. 1994. Catastrophes, phase shifts, and large-scale degradation of a Caribbean coral reef. *Science* 265:1547–1551.

Katzman, M. T., and W. G. Cale, Jr. 1990. Tropical forest preservation using economic incentives. *BioScience* 40:827–832.

Pimentel, D., et al. 1987. World agriculture and soil erosion. *BioScience* 37:277–283.

Pimentel, D., et al. 1991. Environmental and economic effects of reducing pesticide use. *BioScience* 41:402–409.

Reisner, M. P. 1986. *Cadillac Desert.* Viking, New York.

Repetto, R. 1990. Deforestation in the Tropics. *Scientific American* 262:36–42.

Rowland, F. S. 1989. Chlorofluorocarbons and the depletion of stratospheric ozone. *American Scientist* 77:36–45.

Schneider, S. H. 1989. The greenhouse effect: Science and policy. *Science* 243: 771–782.

Vitousek, P. M., P. R. Ehrlich, A. H. Ehrlich, and P. A. Matson. 1986. Human appropriation of the products of photosynthesis. *BioScience* 36:368–373.

White, R. M. 1990. The great climate debate. *Scientific American* 263:36–43.

APPENDIX A
INTERNATIONAL SYSTEM OF UNITS

The International System of Units (abbreviated SI after the French *Système international*) is described in detail in such publications as the National Bureau of Standards Special Publications 330, 1977 edition (U.S. Dept. of Commerce, Washington, D.C.), and M. H. Green, *Metric Conversion Handbook* (Chemical Publ. Co., New York, 1978).

The seven basic SI units are as follows.

CATEGORY OF MEASUREMENT	NAME OF SI UNIT	ABBREVIATION
Length	meter	m
Mass	kilogram	kg
Time	second	s
Electric current	ampere	A
Temperature	kelvin	K
Amount of substance	mole	mol
Luminous intensity	candela	cd

The ampere, mole, and candela are not often used in ecology and will not be discussed here. One familiar unit of time is the second. The SI units of length and mass are metric, with simple conversions to and from the English system; for example, 1 meter equals 3.28 feet, and 1 kilogram equals 2.205 pounds. The kelvin is identical to the degree Celsius (°C) except that zero on the kelvin scale is absolute zero, the coldest temperature possible (–273°C). Most ecologists use °C instead of K, because of the convenient designations for the freezing and boiling points of water (0°C and 100°C).

In the SI system, all other measurements are derived from the basic seven. For example, the SI unit of flow is the cubic meter per second, which is derived from the base units meter and second. Some of these derived units frequently employed in ecology are as follows.

CATEGORY OF MEASUREMENT	NAME OF SI UNIT	ABBREVIATION	EXPRESSION IN TERMS OF	
			Other units	*Base units*
Area	square meter	m^2		m^2
Volume	cubic meter	m^3		m^3
Velocity	meter per second	$m\ s^{-1}$		$m\ s^{-1}$
Flow	cubic meter per second	$m^3\ s^{-1}$		$m^3\ s^{-1}$
Density	kilogram per cubic meter	$kg\ m^{-3}$		$kg\ m^{-3}$
Force	newton	N		$kg\ m\ s^{-2}$
Pressure	pascal	Pa	$N\ m^{-2}$	$kg\ m^{-1}\ s^{-2}$
Energy, heat	joule	J	N m	$kg\ m^2\ s^{-2}$
Power	watt	W	$J\ s^{-1}$	$kg\ m^2\ s^{-1}$

There are also a number of other units that are not part of SI but nonetheless are widely used: the liter (L), a unit of volume equal to 1/1000 cubic meter; the are (a), a unit equal to 100 square meters, and the more often used hectare (ha), equal to 100 a and 10,000 m^2; and the minute, hour, and day, familiar units of time. The calorie (cal), a familiar unit of heat energy, is not derived from SI base units and should be abandoned in favor of the joule (=0.239 cal), even though ecologists continue to use the calorie and kilocalorie.

Multiples of 10 of the SI units are given special names by adding the following prefixes to the unit name.

FACTOR	PREFIX	ABBREVIATION	NUMBER	NAME
10^9	giga	G	1,000,000,000	Billion*
10^6	mega	M	1,000,000	Million
10^3	kilo	k	1000	Thousand
10^2	hecto	h	100	Hundred
10^1	deka	da	10	Ten
10^{-1}	deci	d	0.1	Tenth
10^{-2}	centi	c	0.01	Hundreth
10^{-3}	milli	m	0.001	Thousandth
10^{-6}	micro	μ	0.000 001	Millionth
10^{-9}	nano	n	0.000 000 0001	Billionth

*Milliard in the United Kingdom; the British billion is a million million (10^{12}), equivalent to the U.S. trillion, commonly used only in astronomy and the federal budget.

Hence 1000 meters becomes a kilometer (km) and $^1/_{1000}$ meter is a millimeter (mm); 1000 joules is a kilojoule (kJ) and 1 million watts is a megawatt (MW).

APPENDIX B
CONVERSION FACTORS

Length
1 meter (m) = 39.4 inches (in.)
1 meter = 3.28 feet (ft)
1 kilometer (km) = 3281 feet
1 kilometer = 0.621 mile (mi)
1 micron (μ) = 10^{-6} meter
1 inch = 2.54 centimeters (cm)
1 foot = 30.5 centimeters
1 mile = 1609 meters
1 angstrom (Å) = 10^{-10} meter
1 millimicron (mμ) = 10^{-9} meter

Area
1 square centimeter (cm^2) = 0.155 square inches ($in.^2$)
1 square meter (m^2) = 10.76 square feet (ft^2)
1 hectare (ha) = 2.47 acres (a)
1 hectare = 10,000 square meters
1 hectare = 0.01 square kilometer (km^2)
1 square kilometer = 0.386 square mile
1 square mile = 2.59 square kilometers
1 square inch = 6.45 square centimeters
1 square foot = 929 square centimeters
1 square yard (yd^2) = 0.836 square meter
1 acre = 0.407 hectare

Mass
1 gram (g) = 15.43 grains (gr)
1 kilogram (kg) = 35.3 ounces (oz.)
1 kilogram = 2.205 pounds (lb)
1 metric ton (T) = 2204.6 pounds
1 ounce = 28.35 grams
1 pound = 453.6 grams
1 short ton = 907 kilograms

Time
1 year (yr) = 8760 hours (hr)
1 day (d) = 86,400 seconds (s)

Volume
1 cubic centimeter (cc or cm^3) = 0.061 cubic inch ($in.^3$)
1 cubic inch = 16.4 cubic centimeters
1 liter (L) = 1000 cubic centimeters
1 liter = 33.8 U.S. fluid ounces (oz.)
1 liter = 1.057 U.S. quarts (qt)
1 liter = 0.264 U.S. gallon (gal)
1 U.S. gallon = 3.79 liters

1 Brit. gallon = 4.55 liters
1 cubic foot (ft^3) = 28.3 liters
1 milliliter (ml) = 1 cubic centimeter
1 U.S. fluid ounce = 29.57 milliliters
1 Brit. fluid ounce = 28.4 milliliters
1 quart = 0.946 liter

Velocity
1 meter per second ($m\ s^{-1}$) = 2.24 miles per hour (mph)
1 foot per second ($ft\ s^{-1}$) = 1.097 kilometers per hour
1 kilometer per hour = 0.278 meter per second
1 mile per hour = 0.447 meter per second
1 mile per hour = 1.467 feet per second

Energy
1 joule (J) = 0.239 calorie (cal)
1 calorie = 4.184 joules
1 kilowatt-hour (kWh) = 860 kilocalories (kcal)
1 kilowatt-hour = 3600 kilojoules
1 British thermal unit (Btu) = 252.0 calories
1 British thermal unit = 1054 joules
1 kilocalorie = 1000 calories

Power
1 kilowatt (kW) = 0.239 kilocalorie per second
1 kilowatt = 860 kilocalories per hour
1 horsepower (hp) = 746 watts
1 horsepower = 15,397 kilocalories per day
1 horsepower = 641.5 kilocalories per hour

Energy per unit area
1 calorie per square centimeter = 3.69 British thermal units per square foot
1 British thermal unit per square foot = 0.271 calorie per square centimeter
1 calorie per square cenitmeter = 10 kilocalories per square meter

Power per unit area
1 kilocalorie per square meter per minute = 52.56 kilocalories per hectare per year
1 footcandle (fc) = 1.30 calories per square foot per hour at 555 nm wavelength
1 footcandle = 10.76 lux (lx)
1 lux = 1.30 calories per square meter per hour at 55 nm wavelength

Metabolic energy equivalents
1 gram of carbohydrate = 4.2 kilocalories
1 gram of protein = 4.2 kilocalories
1 gram of fat = 9.5 kilocalories

Miscellaneous
1 gram per square meter = 0.1 kilogram per hectare
1 gram per square meter = 8.97 pounds per acre
1 kilogram per square meter = 4.485 short tons per acre
1 metric ton per hectare = 0.446 short ton per acre

GLOSSARY

Acclimation. A reversible change in the morphology or physiology of an organism in response to environmental change; also called acclimatization.
Acid mine drainage. Water runoff from surface mining, usually coal strip mining, containing sulfuric acid, which forms when organic sulfur is oxidized on contact with the atmosphere.
Acid rain. Precipitation with high acidity (pH < 4) caused by the solution of certain atmospheric gases (sulfur dioxide and nitrous oxide) produced by combustion of fossil fuels.
Activity space. The range of environmental conditions suitable for the activity of an organism.
Adaptation. A genetically determined characteristic that enhances the ability of an individual to cope with its environment; an evolutionary process by which organisms become better suited to their environments.
Adaptive landscape. The relationship of fitness to the phenotypes of organisms, portrayed as peaks and valleys of fitness values.
Adaptive radiation. The evolution of a variety of forms from a single ancestral stock; often occurs after organisms colonize an island group or enter a new adaptive zone.
Adaptive zone. A set of environmental conditions or portion of the ecological niche that requires particular adaptations for living.
Additive genetic variance (V_A). Variation in a phenotypic value within a population due to the difference in expression of alleles in the homozygous state.
Age class. The individuals in a population of a particular age.
Age-specific. Pertaining to attributes, such as survival or fecundity, that vary as a function of age.
Age structure. The distribution of individuals among age classes within a population.
Alkaloids. Nitrogen-containing compounds, such as morphine and nicotine, that are produced by plants and are toxic to many herbivores.
Allele. One of several alternative forms of a gene.
Allelopathy. Direct inhibition of one species by another using noxious or toxic chemicals.
Allochthonous. Referring to materials transported into a system; particularly minerals and organic matter transported into streams and lakes. *Compare with* Autochthonous.
Allometric constant. Slope of the relationship between the logarithm of one measurement of an organism and the logarithm of another, usually its overall size.
Allometry. A relative increase in a part of an organism or a measure of its physiology or behavior in relation to some other measure, usually its overall size.

Allopatric. Occurring in different places; usually referring to geographic separation of populations.
Alluvial. Referring to sediment deposited by running water.
Alpha diversity. The variety of organisms occurring in a particular place or habitat; often called local diversity.
Altruism. In an evolutionary sense, enhancing the fitness of another individual by acts that reduce the evolutionary fitness of the altruistic individual.
Ambient. Referring to conditions of the environment surrounding the organism.
Amino acid. Any one of about thirty organic acids containing the amino group NH_2, which are the building blocks of proteins.
Ammonification. Metabolic breakdown of proteins and amino acids with ammonia as an excreted by-product.
Anaerobic. Without oxygen.
Anion. A part of a dissociated molecule carrying a negative electric charge.
Anisogamous. Having gametes unequal in size and behavior; usually a large, sedentary (female) and a small, motile (male) gamete. *Compare with* Isogamous.
Annual. Referring to an organism that completes its life cycle, from birth or germination to death, within a year.
Anoxic. Lacking oxygen; anaerobic.
Anthropogenic extinction. Extinction caused by the activities of humans, either through direct exploitation of a population or destruction of its habitat.
Antibody. A protein produced by the body's immune system that binds to foreign proteins (antigens), especially those on the surfaces of pathogens.
Antigen. A foreign protein or other organic compound that stimulates the body's immune response.
Aphotic zone. In lakes and oceans, the water layer below the depth to which light penetrates.
Aposematism. *See* Warning coloration.
Asexual reproduction. Reproduction without the benefit of the sexual union of gametes (fertilization).
Aspect diversity. The variety of outward appearances of species that live in the same habitat and are eaten by visually hunting predators.
Assimilation. Incorporation of any material into the tissues, cells, and fluids of an organism.
Assimilation efficiency. A percentage expressing the proportion of ingested energy that is absorbed into the bloodstream.
Assimilatory. Referring to a biochemical transformation that results in the reduction of an element to an organic form, and hence its gain by the biological compartment of the ecosystem.
Association. A group of species living in the same place.
Assortative mating. Preferential mating between individuals having either similar appearance or genotypes (positive assortative mating) or dissimilar appearance or genotypes (negative assortative mating).
Asymmetrical competition. An interaction between two species in which one exploits a particular resource more efficiently than the other; the second may persist by better avoiding predation or by subsisting on a different resource.
Atoll. A coral reef surrounding a lagoon, often built upon a sunken volcano.
Autecology. The study of organisms in relation to their physical environment. *Compare with* Synecology.
Autochthonous. Referring to materials produced within a system; particularly organic matter produced, and minerals cycled, within streams and lakes. *Compare with* Allochthonous.
Autotroph. An organism that assimilates energy from either sunlight (green plants) or inorganic compounds (sulfur bacteria). *Compare with* Heterotroph.
Background extinction. Extinction of species or higher taxa during periods without rapid environmental change.
Barren. An area with sparse vegetation owing to some physical or chemical property of the soil.
Basal metabolic rate (BMR). The energy expenditure of an organism that is at rest, fasting, and in a thermally neutral environment.
Batesian mimicry. Resemblance of an unpalatable species (model) by an edible species (mimic) to deceive predators.
Benthic. On or within the bottom of a river, lake, or ocean.
Benthos. The environment on or within the bottom sediments of rivers, lakes, and oceans; also, the organisms that live there.
Beta diversity. The variety of organisms within a region arising from turnover of species between habitats.
Bet hedging. In a life history, reducing the risk of mortality or reproductive failure in a variable environment by adopting an intermediate strategy or several alternative strategies simultaneously, or by spreading one's risk over time and space, (e.g., by perennial rather than annual reproduction).
Biennial. Requiring 2 years to complete the life cycle.
Biodiversity. A measure of the variety of organisms within a local area or region, often including genetic variation, taxonomic uniqueness, and endemism. *See* Diversity.
Biological community. *See* Community.

Biological control. Use of natural enemies, particularly parasitoid insects, bacteria, and viruses, to control pest organisms.

Biological oxygen demand (BOD). The amount of oxygen required to oxidize the organic material in a water sample; high values in aquatic habitats often indicate pollution by sewage and other sources of organic wastes, or the overproduction of plant material resulting from overenrichment by mineral nutrients.

Biomass. Weight of living material, usually expressed as a dry weight, in all or part of an organism, population, or community. Commonly presented as weight per unit of area, a biomass density.

Biomass accumulation ratio. The ratio of weight to annual production.

Biome. A major type of ecological community (e.g., the grassland biome).

Bioremediation. Restoration of natural habitats or ecological conditions by use of biological agents (e.g., bacterial degradation of spilled oil or other pollutants).

Biota. Fauna and flora together.

Birth rate (*b*). The average number of offspring produced per individual per unit of time, often expressed as a function of age (x).

Boreal. Northern; often refers to the coniferous forest regions that stretch across Canada, northern Europe, and Asia.

Bottleneck. *See* Population bottleneck.

Boundary layer. A layer of still or slow-moving water or air close to the surface of an object.

Breeding system. Degree of polygyny, outcrossing, and selective mating within a population; the adaptations by which organisms adjust these attributes.

Brood parasite. An organism that lays its eggs in the nest of another species or that of another individual of the same species.

C_3 photosynthesis. Photosynthetic pathway in which carbon dioxide is initially assimilated into a three-carbon compound, phosphoglyceraldehyde (PGA), in the Calvin cycle.

C_4 photosynthesis. Photosynthetic pathway in which carbon dioxide is initially assimilated into a four-carbon compound, such as oxaloacetic acid (OAA) or malate.

Calcification. Deposition of calcium and other soluble salts in soils where evaporation greatly exceeds precipitation.

Calvin cycle. The basic assimilatory sequence of photosynthesis during which an atom of carbon is added to the five-carbon ribulose bisphosphate (RuBP) molecule to produce phosphoglyceraldehyde (PGA) and then glucose.

CAM photosynthesis. Photosynthetic pathway in which the initial assimilation of carbon dioxide into a four-carbon compound occurs at night; found in some succulent plants in arid habitats.

Carbonate ion. An anion (CO_3^{2-}) formed by the dissociation of carbonic acid or one of its salts.

Carbonic acid. A weak acid (H_2CO_3) formed when carbon dioxide dissolves in water.

Carnivore. An organism that consumes mostly flesh.

Carrying capacity (*K*). The number of individuals in a population that the resources of a habitat can support; the asymptote, or plateau, of the logistic and other sigmoid equations for population growth.

Caste. Individuals within a social group sharing a specialized form or behavior.

Cation. A part of a dissociated molecule carrying a positive electrical charge.

Cation exchange capacity. The ability of soil particles to absorb positively charged ions, such as hydrogen (H^+) and calcium (Ca^{2+}).

Cellulose. A long-chain molecule made up of glucose subunits, found in the cell walls and fibrous structures of plants.

Chaos. Erratic change in the size of populations governed by difference equations and having high intrinsic rates of growth.

Character displacement. Divergence in the characteristics of two otherwise similar species where their ranges overlap, caused by the selective effects of competition between the species in the area of overlap.

Chemoautotroph. An organism that oxidizes inorganic compounds (often hydrogen sulfide) to obtain energy for synthesis of organic compounds (e.g., sulfur bacteria).

Clade. A group of organisms that includes all the descendants of a single common ancestor; a monophyletic group.

Cladistics. A system for describing relationships among species or other taxa based on shared, derived characteristics (synapomorphies) indicating a common and unique ancestry.

Cladogram. A representation of the relationships among a monophyletic group of species or other taxa based on a cladistic analysis; each branch in a cladogram is defined by uniquely shared, derived characteristics (synapomorphies).

Clay. A fine-grained component of soil, formed by the weathering of granitic rock and composed primarily of hydrous aluminum silicates.

Cleaning symbiosis. A mutualistic arrangement between two species in which one grooms the other to remove parasites.

Climax. The end point of a successional sequence, or sere; a community that has reached a steady state under a particular set of environmental conditions.

Cline. Gradual change in population characteristics or adaptations over a geographic area.

Closed community concept. The idea, popularized by F. C. Clements, that communities are distinctive associations of highly interdependent species.

Clumped distribution. A distribution of individuals in space indicating a tendency for association.

Clutch size. Number of eggs per set; usually with reference to the nests of birds.

Coadaptation. Evolution of characteristics of each of two or more species in response to changes in the other(s), often to mutual advantage. *See* Coevolution.

Coarse-grained. Referring to qualities of the environment that occur in large patches with respect to the activity patterns of an organism and, therefore, among which the organism can select. *Compare with* Fine-grained.

Codon. A sequence of three nucleotides in DNA or RNA that specifies which amino acid will be placed at a particular position in a protein.

Coefficient of relationship (*r*). The probability that one individual shares with another a genetic factor inherited from a common ancestor.

Coevolution. The occurrence of genetically determined traits (adaptations) in two or more species selected by the mutual interactions controlled by these traits.

Coexistence. Occurrence of two or more species in the same habitat; usually applied to potentially competing species.

Cohort. A group of individuals of the same age recruited into a population at the same time.

Cohort life table. *See* Dynamic life table.

Community. An association of interacting populations, usually defined by the nature of their interaction or the place in which they live.

Compartmentalization. Subdivision of a food web into groups of strongly interacting species somewhat isolated from other such groups.

Compartment model. A representation of a system in which the various parts are portrayed as units (compartments) that receive inputs from, and provide outputs to, other such units.

Compensation point. The depth of water or level of light at which respiration and photosynthesis balance each other; the lower limit of the euphotic zone.

Competition. Use or defense of a resource by one individual that reduces the availability of that resource to other individuals, whether of the same species (intraspecific competition) or other species (interspecific competition).

Competition coefficient (*a*). A measure of the degree to which one consumer uses the resources of another, expressed in terms of the population consequences of the interaction.

Competitive exclusion principle. The hypothesis that two or more species cannot coexist on a single resource that is scarce relative to demand for it.

Condition. A physical or chemical attribute of the environment that, while not being consumed, influences biological processes and population growth (e.g., temperature, salinity, acidity). *Compare with* Resource.

Conductance. The capacity of heat, electricity, or a substance to pass through a particular material.

Conduction. The ability of heat to pass through a substance.

Conformer. An organism that allows its internal environment to vary with external conditions.

Congeneric. Belonging to the same genus.

Connectance (*C*). The proportion of interspecific interactions in a community matrix not equal to zero.

Connectedness food web. A depiction of the feeding relationships among species within a community.

Conspecific. Belonging to the same species.

Constraint. In a life history, a functional or genetic relationship between two or more characters that have opposing effects on evolutionary fitness and which therefore restrict evolutionary response.

Consumer. An individual or population that uses a particular resource.

Consumer chain. *See* Food chain.

Consumer-resource interaction. Any ecological or evolutionary interaction between species in which one preys upon or otherwise consumes the other.

Contagion. The transmission of a disease by direct or indirect contact.

Continental climate. A climate lacking the tempering effect of the ocean, usually exhibiting extremes of temperature. *Compare with* Maritime climate.

Continental drift. Movement of continents on the surface of the earth over geologic time; rates of drift are on the order of centimeters per year.

Continuum. A gradient of environmental characteristics or of change in the composition of communities.

Continuum index. A scale of an environmental gradient based on changes in physical characteristics or community composition along that gradient.

Convection. Transfer of heat by the movement of a fluid (for example, air or water).

Convergence. Resemblance among organisms belonging to different taxonomic groups resulting from adaptation to similar environments.

Cooperation. Association or social interaction among individuals of the same or different species for mutual benefit.

Correlated response. A response of the phenotypic value of one trait to selection upon another trait.
Cost of meiosis. The fact that gamete production and fertilization results in a female parent's contributing only one-half of her genotype to each of her offspring.
Countercurrent circulation. Movement of fluids in opposite directions on either side of a separating barrier through which heat or dissolved substances can pass.
Crassulacean acid metabolism. *See* CAM photosynthesis.
Cross-resistance. Resistance or immunity to one disease organism resulting from infection by another, usually closely related, organism.
Crypsis. An aspect of the appearance of an organism whereby it avoids detection by others.
Cybernetic. Pertaining to feedback controls and communication within systems.
Cycle. Recurrent variation in a system periodically returning to its starting point.
Cyclic climax. A steady-state, cyclic sequence of communities, none of which by itself is stable.
Cyclic succession. Continual community change through a repeated sequence of stages.
Damped oscillation. Cycling with progressively smaller amplitude, as in some populations approaching their equilibria.
Death rate (d_x). The percentage of newborn individuals dying during a specified interval. *Compare with* Mortality.
Defensive mutualism. A relationship between two species in which one defends the other against some enemy and usually receives some type of nourishment or living space in return.
Demographic. Pertaining to populations, particularly their growth rate and age structure.
Demography. Study of the age structure and growth rate of populations.
Denitrification. Biochemical reduction, primarily by microorganisms, of nitrogen from nitrate (NO_3^-) eventually to molecular nitrogen (N_2).
Density. Referring to a population, the number of individuals per unit of area or volume; referring to a substance, the weight per unit of volume.
Density compensation. Increase in population size in response to a reduction in the number of competing populations; often observed on islands.
Density dependent. Having an influence on the individuals in a population that varies with the degree of crowding within the population.
Density independent. Having an influence on the individuals in a population that does not vary with the degree of crowding within the population.
Deoxyribonucleic acid (DNA). A long macromolecule whose sequence of subunits (nucleotides) encodes genetic information.
Deterministic. Referring to the outcome of a process that is not subject to stochastic (random) variation.
Detoxification. Biochemical conversion of a toxic substance to harmless by-products.
Detritivore. An organism that feeds on freshly dead or partially decomposed organic matter.
Detritus. Freshly dead or partially decomposed organic matter.
Developmental response. Acquisition of one of several alternative forms by an organism depending on the environmental conditions under which it grows.
Diapause. Temporary interruption in the development of insect eggs or larvae, usually associated with a dormant period.
Diffusion. Movement of particles of gas or liquid from regions of high concentration to regions of low concentration by means of their own spontaneous motion.
Dimorphism. The occurrence of two forms of individuals within a population.
Dioecy. In plants, the occurrence of reproductive organs of the male and female sex on different individuals. *Compare with* Monoecy.
Diploid. Having two sets of chromosomes. *See* Haploid; Meiosis.
Directional selection. Differential survival or reproduction within a population favoring an extreme phenotype, resulting in an evolutionary shift in the population mean toward that phenotype.
Disease. An unhealthy condition of an individual organism that impairs a vital function.
Disease organism. An organism, usually endoparasitic, that causes a disease condition in another organism.
Dispersal. Movement of organisms away from their place of birth or away from centers of population density.
Dispersion. The spatial pattern of distribution of individuals within populations.
Disruptive selection. Differential survival or reproduction within a population favoring two or more extreme phenotypes, tending to promote genetic polymorphism.
Dissimilatory. Referring to a biochemical transformation that results in the oxidation of the organic form of an element, and hence its loss from the biological compartment of the ecosystem.
Dissolution. The entry of a substance into solution with water.
Distribution. The geographic extent of a population or other ecological unit.

Diversity. The number of taxa in a local area (alpha diversity) or region (gamma diversity). Also, a measure of the variety of taxa in a community that takes into account the relative abundance of each one.
Diversity index. A measure of the variety of taxa in a community that takes into account the relative abundance of each one.
DNA. *See* Deoxyribonucleic acid.
Dominance (species). The numerical superiority of a species over others within a community or association.
Dominance hierarchy. The orderly ranking of individuals in a group based on the outcome of aggressive encounters.
Dominance variance (V_D). Variation in a phenotypic value within a population due to the unequal expression of alleles in the heterozygous state.
Dominant. Pertaining to an allele that masks the expression of another (recessive) allele of the same gene.
Dominant (species). Pertaining to species that are abundant or exert great ecological influence within an ecological system.
Donor. The individual that is the active party in a particular behavioral interaction.
Dormancy. An inactive state, such as hibernation, diapause, or seed dormancy, usually assumed during an inhospitable period.
Downstream drift. The tendency of riverine plants and animals, or their propagules, to be carried downstream by currents.
Dynamic life table. The age-specific survival and fecundity of a cohort of individuals in a population followed from birth to the death of the last individual; cohort life table.
Dynamic steady state. Condition in which fluxes of energy or materials into and out of a system are balanced.
Ecocline. A geographic gradient of vegetation structure associated with one or more environmental variables.
Ecological diversity. A measure of diversity taking into account the varied ecological roles of different species.
Ecological efficiency. The percentage of energy in the biomass produced by one trophic level that is incorporated into the biomass produced by the next higher trophic level.
Ecological range. The distribution of an individual or species with respect to one or a variety of ecological conditions or resource characteristics.
Ecological release. Expansion of habitat and resource use by populations in regions of low species diversity resulting from low interspecific competition.
Ecological system. A regularly interacting or interdependent group of biological items forming a unified whole that functions in an ecological context.
Ecology. The study of the natural environment and of the relations of organisms to each other and to their surroundings.
Ecomorphology. The study of the relationship between the ecological relations of an individual and its morphology.
Ecosystem. All the interacting parts of the physical and biological worlds.
Ecotone. A habitat created by the juxtaposition of distinctly different habitats; an edge habitat; a zone of transition between habitat types.
Ecotourism. Travel for the recreational purpose of observing unusual species or ecological habitats and landscapes.
Ecotype. A genetically differentiated subpopulation that is restricted to a specific habitat.
Ectomycorrhizae. Mutualistic associations of fungi with the roots of plants in which the fungus forms a sheath around the outside of the root.
Ectoparasite. A parasite that lives on or attached to the host's surface (e.g., a tick).
Ectothermy. The capacity to maintain body temperature by gaining heat from the environment, either by conduction or by absorbing radiation.
Edaphic. Pertaining to or influenced by the soil.
Edge effect. A change in conditions or species composition within an otherwise uniform habitat as one approaches a boundary with a different habitat.
Egestion. Elimination of undigested food material.
Electric potential (Eh). The relative capacity, measured in volts, of one substance to oxidize another.
Electron acceptor. A substance that readily accepts electrons and thus is capable of oxidizing another substance.
El Niño. A warm current from the Tropics that intrudes each winter along the west coast of northern South America.
Eluviation. The downward movement of dissolved soil materials, carried by percolating water, from the topmost (A) horizon.
Emergent. Referring to any plant that rises above a surface, as of water or the canopy of a forest.
Emigration. Movement of individuals out of a population. *Compare with* Immigration.
Endemic. Confined to a certain region; with respect to disease, present at a low level within a local population.
Endemism. The quality of belonging to a particular region.
Endomycorrhizae. Mutualistic associations of fungi with the roots of plants in which part of the fungus resides within the root tissues.
Endoparasite. A parasite that lives within the tissues or bloodstream of its host.

Endothermy. The capacity to maintain body temperature by the metabolic generation of heat.
Energetic efficiency. The ratio of useful work or energy storage to energy intake.
Energy. Capacity for doing work.
Energy flow food web. A depiction of the feeding relationships among species in a community based on the quantity of energy transferred as food.
ENSO. El Niño-Southern Oscillation: an occasional shift in winds and ocean currents, centered in the South Pacific region, that has worldwide consequences for climate and biological systems.
Environment. The surroundings of an organism, including the plants, animals, and microbes with which it interacts.
Environmental grain. The spatial or temporal heterogeneity of the environment relative to the activities of an organism.
Environmental variance (V_E). Variation in a phenotypic value within a population due to the influence of environmental factors.
Enzyme. An organic compound in a living cell or secreted by it that accelerates a specific biochemical transformation without itself being affected.
Epidemiology. The study of factors influencing the spread of disease through a population.
Epifaunal. Referring to animals living on the surface of a substrate.
Epilimnion. The warm, oxygen-rich surface layers of a lake or other body of water. *Compare with* Hypolimnion.
Epiphyte. A plant that grows on another plant and derives its moisture and nutrients from the air and rain.
Equilibrium. A state of balance between opposing forces.
Equilibrium isocline. A line on a population graph designating combinations of competing populations, or predator and prey populations, for which the growth rate of one of the populations is zero.
Equilibrium theory of island biogeography. The idea that the number of species on an island exists as a balance between colonization by new immigrant species and extinction of resident species.
Equitability. Uniformity of abundance in an assemblage of species. Equitability is greatest when species are equally abundant.
Escape space. Refuge from predators and parasites; often reflected in the adaptations of prey organisms enabling them to fight, flee, or escape detection.
Estuary. A semi-enclosed coastal water body, often at the mouth of a river, having a high input of fresh water and great fluctuation in salinity.
Euphotic zone. The surface layer of water to the depth of light penetration at which photosynthesis balances respiration. *See* Compensation point.
Eusociality. The complex social organization of termites, ants, and many wasps and bees, dominated by an egg-laying queen that is tended by nonreproductive offspring.
Eutrophic. Rich in the mineral nutrients required by green plants; pertaining to an aquatic habitat or soil with high productivity.
Eutrophication. Enrichment of water by nutrients required for plant growth; often refers to overenrichment caused by sewage and runoff from fertilized agricultural lands and resulting in excessive bacterial growth and oxygen depletion.
Evapotranspiration. The sum of transpiration by plants and evaporation from the soil. *See also* Potential evapotranspiration.
Evolution. Change in the heritable traits of organisms through the replacement of genotypes within a population.
Evolutionarily stable strategy (ESS). A strategy such that, if all members of a population adopt it, no alternative strategy can invade.
Evolutionary ecology. The integrated study of evolution, genetics, adaptation, and ecology; interpretation of the structure and function of organisms, communities, and ecosystems in the context of evolutionary theory.
Excretion. Elimination from the body, by way of the kidneys, gills, and dermal glands, of excess salts, nitrogenous waste products, and other substances.
Expectation of further life (e_x). The average remaining lifetime of an individual of age x.
Experiment. A controlled manipulation of a system to determine the effect of a change in one or more factors.
Exploitation. Removal of individuals or biomass from a population by consumers.
Exploitative competition. Competition between individuals by way of their reduction of shared resources.
Exponential growth. Continuous increase or decrease in a population in which the rate of change is proportional to the number of individuals at any given time. *See* Geometric growth.
Exponential rate of increase (r). The rate at which a population is growing at a particular time, expressed as a proportional increase per unit of time. *See* Geometric rate of increase.
External factor. In systems modeling, a material input from outside the system or a condition of the environment of the system that influences its structure and function.
External loading. Input of nutrients to a lake or stream from outside the system, especially sewage and runoff from agricultural lands. *Compare with* Internal loading.

Extinction. Disappearance of a species or other taxon from a region or biota.

Extra-pair copulation (EPC). Mating received by a female from a male that is not her mate.

Facilitation. Enhancement of a population of one species by the activities of another, particularly during early succession.

Facultative. Referring to the ability to adjust to a variety of conditions or circumstances; optional for the organism. *Compare with* Obligate.

Fall bloom. The rapid growth of algae in temperate lakes following the autumnal breakdown of thermal stratification and mixing of water layers.

Fall overturn. The vertical mixing of water layers in temperate lakes in autumn following breakdown of thermal stratification.

Fecundity. The rate at which an individual produces offspring.

Feedback control. *See* Internal control.

Field capacity. The amount of water that soil can hold against the pull of gravity.

Filter feeder. An organism that strains tiny food particles from its aqueous environment by means of sievelike structures (e.g., clams and baleen whales).

Fine-grained. Referring to qualities of the environment that occur in small patches with respect to the activity patterns of an organism, and among which the organism cannot usefully distinguish. *Compare with* Coarse-grained.

Fitness. The genetic contribution by an individual's descendants to future generations of a population.

Fixation. An increase in the proportion of an allele to unity, resulting in the elimination of all alternative alleles.

Floristic. Referring to the species composition of plant communities.

Fluvial. Pertaining to a stream or river.

Flux. Movement of energy or materials into or out of a system.

Food chain. A representation of the passage of energy from a primary producer through a series of consumers at progressively higher trophic (feeding) levels.

Food chain efficiency. *See* Ecological efficiency.

Food web. A representation of the various paths of energy flow through populations in the community, taking into account the fact that each population shares resources and consumers with other populations.

Forbs. Herbaceous, broad-leaved vegetation (i.e., other than grasses) consumed by grazers.

Founder event. Colonization of an island or patch by a small number of individuals that possess less genetic variation than the parent population. *Compare with* Population bottleneck.

Frequency dependence. The condition in which the expression of a process varies with the relative proportions of phenotypes in a population.

Frequency-dependent selection. The condition in which the fitness of a genetic trait or phenotype depends on its frequency in the population.

Front. A meeting of two air or water masses having different characteristics.

Functional food web. A depiction of the relationships among species within a community based on the influence of consumers on the dynamics of resource populations.

Functional response. A change in the rate of exploitation of prey by an individual predator as a result of a change in prey density. Type I: Exploitation is directly proportional to prey density. Type II: Exploitation levels off at high prey density (predator satiation). Type III: As in Type II, but exploitation is additionally low at low prey density owing to lack of a search image or to efficient prey escape mechanisms. *See also* Numerical response.

Gaia hypothesis. The idea that the biosphere has evolved to optimize the conditions for life.

Gamete. A haploid cell that fuses with another haploid cell of the opposite sex during fertilization to form a zygote. In animals, the male gamete is called the sperm and the female gamete is called the egg or ovum.

Gamma diversity. The inclusive diversity of all the habitat types within an area; regional diversity.

Gene. Generally, a unit of genetic inheritance. In biochemistry, refers to a part of the DNA molecule that encodes a single enzyme or structural protein.

Gene flow. The exchange of genetic traits between populations by movement of individuals, gametes, or spores.

Gene frequency. The proportion of a particular allele of a gene in the gene pool of a population.

Gene pool. The whole body of genes in an interbreeding population.

Generalist. A species with broad food or habitat preferences.

Generation time. The average age at which a female gives birth to her offspring, or the average time for a population to increase by a factor equal to the net reproductive rate.

Genetic death. The death of an individual owing to a deficiency in its genetic makeup.

Genetic diversity. A measure of the variety of genetic factors in the gene pool of a population.

Genetic drift. Change in allele frequencies due to random variations in fecundity and mortality in a population.

Genetic load. Deaths sustained by a population resulting from genetic factors conveying low fitness to individuals.
Genetic variance (V_G). Variation in a phenotypic value within a population due to the expression of genetic factors.
Genotype. All the genetic characteristics that determine the structure and functioning of an organism; often applied to a single gene locus to distinguish one allele, or combination of alleles, from another.
Genotype-environment interaction. Variation in the relative expression of alternative genetic factors (alleles) depending on the environment.
Geometric growth. Periodic increase or decrease in a population in which the increment is proportional to the number of individuals at the beginning of the period, often the breeding season. *See* Exponential growth.
Geometric rate of increase (λ). The factor by which the size of a population changes over a specified period. *See* Exponential rate of increase.
Giving-up time (GUT). *See* Optimum giving-up time.
Gondwana. A giant landmass in the Southern Hemisphere during the early Mesozoic era made up of present-day South America, Africa, India, Australia, and Antarctica.
Gonochoristic. In animals, having separate male and female individuals; in plants, called dioecious.
Gradient analysis. The portrayal and interpretation of the abundances of species along gradients of physical conditions.
Grain. The scale of heterogeneity of habitats in relation to the activities of organisms.
Greenhouse effect. The warming of the earth's climate because of the increased concentration of carbon dioxide and certain other pollutants in the atmosphere.
Gross production. The total energy or nutrients assimilated by an organism, a population, or an entire community. *See* also Net production.
Gross production efficiency. The percentage of ingested food used for growth and reproduction by an organism.
Group selection. Elimination of groups of individuals with a detrimental genetic trait, caused by competition with other groups lacking the trait; often called intergroup selection, and associated with the evolution of altruism.
Growing season. The period of the year during which conditions are suitable for plant growth; in temperate regions, generally between the first and last frosts.
Guild. A group of species occupying similar ecological positions within the same habitat.
Habitat. The place where an animal or plant normally lives, often characterized by a dominant plant form or physical characteristic (that is, a stream habitat, a forest habitat).
Habitat compression. Restriction of habitat distribution in response to an increase in the number of competing species.
Habitat expansion. An increase in the average breadth of habitat distribution of species in depauperate biotas, especially on islands, as compared with species in more diverse biotas.
Habitat selection. Choice of, or preference for, certain habitats.
Handicap principle. The idea that elaborate, sexually selected displays and adornments act as handicaps that demonstrate the generally high fitness of the bearer.
Haplodiploidy. A sex-determining mechanism by which females develop from fertilized eggs and males from unfertilized eggs.
Haploid. Having one set of chromosomes.
Hardy–Weinberg equilibrium. The mathematical proposition that the frequencies of genes and genotypes within a population remain unchanged in the absence of selection, mutation, genetic drift, and assortative mating.
Heat. A measure of the kinetic energy of the atoms or molecules in a substance.
Heat of melting. The amount of heat energy that must be added to a substance to make it melt.
Heat of vaporization. The amount of energy that must be added to a substance to make it vaporize.
Herbivore. An organism that consumes living plants or their parts.
Heritability (h^2). The proportion of variance in a phenotypic trait that is due to the effects of additive genetic factors.
Hermaphrodite. An organism that has the reproductive organs of both sexes.
Heterogeneity. The variety of qualities found in an environment (habitat patchiness) or a population (genotypic variation).
Heterotroph. An organism that uses organic materials as a source of energy and nutrients. *Compare with* Autotroph.
Heterozygous. Containing two forms (alleles) of a gene, one derived from each parent.
Hibernation. A state of winter dormancy involving lowered body temperature and metabolism.
Homeostasis. The maintenance of constant internal conditions in the face of a varying external environment.
Homeothermy. The ability to maintain a constant body temperature in the face of a fluctuating environmental temperature; warm-bloodedness. *Compare with* Poikilothermy.
Homologous. Having similar evolutionary origins.
Homoplasy. Independent evolutionary acquisition of similar derived structures. *See* Convergence.
Homozygous. Containing two identical alleles at a gene locus.

Horizon. A layer of soil distinguished by its physical and chemical properties.
Humus. Fine particles of organic detritus in soil.
Hydrological cycle. The movement of water throughout the ecosystem.
Hydrolysis. A biochemical process by which a molecule is split into parts by the addition of the parts of water molecules.
Hygroscopic water. Water held tightly by surface adhesion to particles in the soil, generally unavailable to plants.
Hyperdispersion. A pattern of distribution in which distances between individuals are more even than expected from random placement; overdispersion.
Hyperosmotic. Having an osmotic potential (generally, salt concentration) greater than that of the surrounding medium.
Hypertonic. Having a salt concentration greater than that of the surrounding medium.
Hyphae. The threadlike filaments that make up the mycelium, or major part of the body, of a fungus.
Hypolimnion. The cold, oxygen-depleted part of a lake or other body of water that lies below the zone of rapid change in water temperature (thermocline). *Compare with* Epilimnion.
Hypo-osmotic. Having an osmotic potential (generally, salt concentration) less than that of the surrounding medium.
Hypothesis. A conjecture about or explanation for a pattern or relationship embracing a mechanism for its occurrence.
Hypotonic. Having a salt concentration less than that of the surrounding medium.
Ideal free distribution. The distribution of individuals across resource patches of different intrinsic quality that equalizes the net rate of gain of each individual when competition is taken into account.
Identity by descent. The probability that homologous genes in different individuals are direct copies of the same gene in a common ancestor.
Illuviation. The accumulation of dissolved substances within a soil layer, usually the middle (B) horizon.
Immigration. Movement of individuals into a population. *Compare with* Emigration.
Inbreeding. Mating between closely related individuals.
Inclusive fitness. The fitness of an individual plus the fitnesses of its relatives, the latter weighted according to degree of relationship; usually applied to the consequences of social interaction between relatives.
Individual distance. The distance within which one individual does not tolerate the presence of another.
Individualistic concept. The idea espoused by H. A. Gleason that the distributions of species reflect their tolerances of physical factors and not interactions between species.
Inducible response. Any change in the state of an organism caused by an external factor; usually reserved for the response of organisms to parasitism and herbivory.
Industrial melanism. The evolution of dark coloration by cryptic organisms in response to industrial pollution, especially by soot, in their environments.
Information center. A group of organisms (i.e., a roost or breeding colony) within which information about distant feeding sites or other ecological conditions is conveyed.
Infrared (IR) radiation. Electromagnetic radiation having a wavelength longer than about 700 nm.
Inhibition. The suppression of a colonizing population by another that is already established, especially during successional sequences.
Innate capacity for increase (r_0). The intrinsic growth rate of a population under ideal conditions without the restraining effects of competition.
Interaction variance (V_G). Variation in a phenotypic value within a population due to the influence of alleles at other genetic loci on the expression of a gene.
Intergroup selection. *See* Group selection.
Intermediate disturbance hypothesis. The idea that species diversity is greatest in habitats with moderate amounts of physical disturbance, owing to the coexistence of early and late successional species.
Intermediate host. A host that harbors an asexual stage of the life cycle of a parasite or disease organism.
Internal control. In systems modeling, the influence of one component of a system on other components within the system.
Internal loading. Regeneration of nutrients within a system, usually referring to the sediments of a lake or river. *Compare with* External loading.
Interspecific competition. Competition between individuals of different species.
Intrasexual competition. Competition between members of the same sex, as in the case of combat between males.
Intraspecific competition. Competition between individuals of the same species.
Intrinsic rate of increase (r_m). Exponential growth rate of a population with a stable age distribution; that is, under constant conditions.
Ion. One of the dissociated parts of a molecule, each of which carries an electric charge, either positive (cation) or negative (anion).
Irruption. A sudden upsurge in the numbers of a population, especially when natural ecological controls are disturbed.

Isogamous. Having gametes similar in size and behavior; not differentiated into unequal (male and female) gamete types. *Compare with* Anisogamous.
Iteroparity. The condition of reproducing repeatedly during the lifetime. *Compare with* Semelparity.
Joint equilibrium. The combination of population sizes at which two or more populations are in equilibrium with respect to each other.
Key factor. An environmental factor that is particularly responsible for change in the size of a population.
Key factor analysis. A statistical treatment of population data designed to identify factors most responsible for change in population size.
Keystone species. A species, often a predator, having a dominating influence on the composition of a community, which may be revealed when the keystone species is removed.
Kinetic energy. Energy associated with motion.
Kin selection. Differential reproduction among lineages of closely related individuals based on genetic variation in social behavior.
Kranz anatomy. An arrangement of tissues in the leaves of C_4 plants in which photosynthetic cells containing chloroplasts are grouped in sheaths around vascular bundles.
Laterite. A hard substance rich in oxides of iron and aluminum, frequently formed when tropical soils weather under alkaline conditions.
Laterization. Leaching of silica from soil, usually in warm, moist regions with an alkaline soil reaction.
Laurasia. A large landmass in the Northern Hemisphere during the Mesozoic era consisting of what is presently North America, Europe, and most of Asia.
Leaching. Removal of soluble compounds from leaf litter or soil by water.
Lek. A communal courtship area on which several males hold courtship territories to attract and mate with females; sometimes called an arena.
Liana. A climbing plant of tropical rain forests, usually woody, that roots in the ground.
Liebig's law of the minimum. The idea that the growth of an individual or population is limited by the essential nutrient present in the lowest amount relative to requirement.
Life history. The adaptations of an organism that more or less directly influence life table values of age-specific survival and fecundity (e.g., reproductive rate, age at maturity, reproductive risk).
Life table. A summary by age of the survivorship and fecundity of individuals in a population.
Life zone. A more or less distinct belt of vegetation occurring within, and characteristic of, a particular latitude or range of elevation.
Lignin. A long-chain, nitrogen-containing molecule made up of phenolic subunits that occurs in woody structures of plants and is highly resistant to digestion by herbivores.
Limestone. A rock formed chiefly by the sedimentation of shells and the precipitation and sedimentation of calcium carbonate ($CaCO_3$) in marine systems.
Limit cycle. An oscillation of predator and prey populations that occurs when stabilizing and destabilizing tendencies of their interaction balance.
Limiting resource. A resource that is scarce relative to demand for it.
Limnetic. Of or inhabiting the open water of a lake.
Limnology. The study of freshwater habitats and communities, particularly lakes, ponds, and other standing waters.
Littoral. Pertaining to the shore of the sea, especially the intertidal zone, and often including waters to the depth limit of emergent vegetation.
Loam. Soil that is a mixture of coarse sand particles, fine silt, clay particles, and organic matter.
Local mate competition. Direct interactions between males that compete to mate with females, particularly at or near their place of birth, hence potentially involving mating with close relatives.
Locus. The segment of a chromosome on which a gene resides.
Logistic equation. The mathematical expression for a particular sigmoid growth curve in which the percentage rate of increase decreases in linear fashion as population size increases.
Lognormal distribution. A characterization of the number of species in logarithmically scaled abundance classes, according to which most species have moderate abundance and fewer have either extremely high or low abundance.
Lottery hypothesis. The idea that habitat patches are colonized at random from the pool of species in an area, thus maintaining the diversity of the assemblage; generally applied to coral reef habitats.
Lower critical temperature (T_c). The ambient temperature below which warm-blooded animals must generate heat to maintain their body temperature.
Luxury consumption. Uptake of a nutrient in excess of need when the nutrient is abundant.
Maritime climate. A climate subject to the tempering influence of the ocean, usually exhibiting a narrow range of temperature. *Compare with* Continental climate.
Mark-recapture method. A method of estimating the size of a population by the recapture of marked individuals.
Mass extinction. Abrupt disappearance of a large fraction of a biota, thought to be caused by such environmental catastrophes as meteor impacts; significant mass extinctions occurred at the end of the Permian and Cretaceous periods.

Mate choice. Selection of a mate based on the characteristics of its phenotype or territory.

Mate guarding. Close association of males with their female mates to prevent mating by other males.

Mating system. The pattern of matings between individuals in a population, including number of simultaneous mates, permanence of pair bond, and degree of inbreeding.

Maximum sustainable yield (MSY). The highest rate at which individuals may be harvested from a population without reducing the size of the population—that is, at which recruitment equals or exceeds harvesting.

Mediterranean climate. A pattern of climate found in middle latitudes, characterized by cool, wet winters and warm, dry summers.

Meiofauna. Animals that live within a substrate, such as soil or aquatic sediment.

Meiosis. A series of two divisions by cells destined to produce gametes, involving pairing and segregation of homologous chromosomes and a reduction of chromosome number from diploid to haploid.

Melanism. Occurrence of black pigment, usually melanin.

Mesic. Referring to habitats with plentiful rainfall and well-drained soils.

Mesophyte. A plant that requires moderate amounts of moisture.

Metabolism. Biochemical transformations responsible for the building up and breaking down of tissues and the release of energy by the organism.

Metamorphosis. An abrupt change in form during development that fundamentally alters the function of the organism.

Metapopulation. A population that is divided into subpopulations between which individuals migrate from time to time. Habitat fragmentation is causing many species to assume a metapopulation structure.

Micelle. A complex soil particle resulting from the association of humus and clay particles, with negative electric charges at its surface.

Microcosm. A small, simplified system, often maintained in a laboratory, that contains the essential features of a larger natural system.

Microenvironment. The conditions within a microhabitat, that is, experienced by an individual at a particular time.

Microhabitat. The particular part of the habitat that an individual encounters at any particular time in the course of its activities.

Migration. The movement of individuals between one place and another, or between subpopulations in a metapopulation.

Mimic. An organism adapted to resemble another organism or an object.

Mimicry. Resemblance of an organism to some other organism or an object in the environment, evolved to deceive predators or prey into confusing the organism with that which it mimics.

Mineralization. Transformation of elements from organic to inorganic forms, often by dissimilatory oxidations.

Minimum viable population (MVP). The minimum number of individuals necessary to prevent a population from losing genetic variation or suffering stochastic extinction over an acceptably long period.

Mitosis. The division of a cell into two identical daughter cells, involving replication and equal partitioning of the cell's DNA.

Mixed evolutionarily stable strategy. An evolutionarily stable strategy involving more than one phenotype within a population; generally the outcome of frequency-dependent fitnesses of the phenotypes.

Mixed reproductive strategy. The presence in a population of individuals of the same gender with different reproductive behaviors, often larger territorial and smaller nonterritorial males.

Model. An organism, usually unpalatable or otherwise noxious to predators, upon which a mimic is patterned.

Model (mathematical). A quantitative representation of the relationships among the entities in a system (e.g., among the species in a community).

Monoclimax theory. The idea that all successional seres within a climate zone lead to the same single climax community.

Monoecy. In plants, the occurrence of male and female reproductive organs in different flowers on the same individual. *Compare with* Dioecy.

Monogamy. A mating system in which each individual mates with only one individual of the opposite sex, generally involving a strong and lasting pair bond. *Compare with* Polygamy.

Monophyletic group. All of the species descended from a single common ancestor.

Morph. A specific form, shape, or structure.

Mortality (m_x). Ratio of the number of deaths to the number of individuals at risk, often described as a function of age (x). *Compare with* Death rate.

Müllerian mimicry. Mutual resemblance of two or more conspicuously marked, unpalatable species to enhance predator avoidance.

Mutation. Any change in the genotype of an organism occurring at the gene, chromosome, or genome level; usually applied to changes in genes to new allelic forms.

Mutualism. A relationship between two species that benefits both.
Mycelium. The rootlike network of filaments (hyphae) making up the nonreproductive part of the body of a fungus.
Mycorrhizae. Close associations of fungi and tree roots in the soil that facilitate the uptake of minerals by trees.
Myxomatosis. A viral disease of some mammals, transmitted by mosquitoes, that causes a fibrous cancer of the skin in susceptible individuals.
Natural selection. Change in the frequency of genetic traits in a population through differential survival and reproduction of individuals bearing those traits.
Nectar. A sugary solution secreted by the flowers of plants to attract potential pollinators.
Negative assortative mating. Preferential mating between individuals having differing appearances or genotypes.
Negative feedback. The tendency of a system to counteract externally imposed change and return to a stable state.
Neighborhood size. The number of individuals in a population included within the dispersal distance of a single individual.
Neritic zone. The region of shallow water adjoining a seacoast.
Net aboveground productivity (NAP). Accumulation of biomass in aboveground parts of plants (trunks, branches, leaves, flowers, and fruits) over a specified period; usually expressed on an annual basis (NAAP).
Net production. The total energy or nutrients accumulated as biomass by an organism, a population, or an entire community by growth and reproduction; gross production minus respiration.
Net production efficiency. The percentage of assimilated food used for growth and reproduction by an organism.
Net reproductive rate (R_0). The expected number of offspring of a female during her lifetime.
Neutral equilibrium. The particular state of a system that has no forces acting upon it.
Newton's law of cooling. The principle that the rate at which a body loses heat by conduction varies in direct proportion to the difference between its temperature and that of its surroundings.
Niche. The ecological role of a species in the community; the ranges of many conditions and resource qualities within which the organism or species persists, often conceived as a multidimensional space.
Niche breadth. The variety of resources used and range of conditions tolerated by an individual, population, or species.
Niche overlap. The sharing of niche space by two or more species; similarity of resource requirements and tolerance of ecological conditions.
Niche preemption. A model in which species successively procure a proportion of the available resources, leaving less for the next species.
Nitrification. Breakdown of nitrogen-containing organic compounds by microorganisms, yielding nitrates and nitrites.
Nitrogen fixation. Biological assimilation of atmospheric nitrogen to form organic nitrogen-containing compounds.
Nonrenewable resource. A resource present in fixed quantity that can be used completely by consumers (e.g., space).
Normal distribution. A bell-shaped statistical distribution in which the probability density varies in proportion to $\exp(-x^2/2)$, where x is a distance (the standard deviation) from the mean.
Nucleotide. Any of several chemical compounds forming the structural units of RNA and DNA and consisting of a purine or pyrimidine base, ribose (RNA) or deoxyribose (DNA) sugar, and phosphoric acid.
Null model. A set of rules for generating community patterns, presupposing no interaction between species, against which observed community patterns can be compared statistically.
Numerical response. A change in the population size of a predatory species as a result of a change in the density of its prey. *See also* Functional response.
Nutrient. Any substance required by organisms for normal growth and maintenance.
Nutrient cycle. The path of an element as it moves through the ecosystem, including its assimilation by organisms and its regeneration in a reusable inorganic form.
Nutrient use efficiency (NUE). Ratio of production to uptake of a required nutrient.
Obligate. Referring to a way of life or response to particular conditions without alternatives. *Compare with* Facultative.
Oceanic zone. Region of the ocean beyond the continental shelves.
Oligotrophic. Poor in the mineral nutrients required by green plants; pertaining to an aquatic habitat with low productivity.
Omnivore. An organism whose diet is broad, including both plant and animal foods; specifically, an organism that feeds on more than one trophic level.
Omnivory. In the sense of food-web analysis, feeding on more than one trophic level.
Open community. A local association of species having independent and only partially overlapping ecological distributions.
Open community concept. The idea, advocated by H. A. Gleason and R. H. Whittaker, that communities are the

local expression of the independent geographic distributions of species.

Optimal foraging. A set of rules, including breadth of diet, by which organisms maximize food intake per unit of time or minimize the time needed to meet their food requirements; risk of predation may also enter the equation for optimal foraging.

Optimal outcrossing distance. The distance from which pollen is received by a plant that maximizes seed set.

Optimum giving-up time (GUT). The time that an organism should spend within a patch of resources before moving on to the next in order to maximize its rate of food intake.

Order of magnitude. A factor of 10; for example, three orders of magnitude is a factor of 1,000.

Ordination. A set of mathematical methods by which communities are ordered along physical gradients or along derived axes over which distance is related to dissimilarity in species composition.

Organismal viewpoint. The idea that a community is a discrete, highly integrated association of species within which the function of each species is subservient to the whole.

Oscillation. Regular fluctuation through a fixed cycle above and below some mean value.

Osmoregulation. Regulation of the salt concentration in cells and body fluids.

Osmosis. Diffusion of substances in aqueous solution across the membrane of a cell.

Osmotic potential. The attraction of water to an aqueous solution owing to its concentration of ions and other small molecules; usually expressed as a pressure.

Outcrossing. Mating with unrelated individuals within a population.

Overdispersion. *See* Hyperdispersion.

Overlapping generations. The co-occurrence of parents and offspring in the same population as reproducing adults.

Oxic. Having oxygen.

Oxidation. Removal of one or more electrons from an atom, ion, or molecule. *Compare with* Reduction.

Oxygen dissociation curve. The relationship between the fraction of the maximum potential binding of oxygen to hemoglobin and the partial pressure of oxygen in blood; an indication of the affinity of oxygen for hemoglobin.

Oxygen tension. The partial pressure of oxygen dissolved in an aqueous solution, such as blood.

Ozone. A molecule consisting of three atoms of oxygen (O_3), which, in the upper atmosphere, blocks the penetration of ultraviolet light to the earth's surface.

Ozone hole. A region of severe ozone depletion in the upper atmosphere, usually at high latitude.

Pangaea. A supercontinent existing at the end of the Paleozoic era that included practically all of the earth's landmasses, including the future Laurasia and Gondwana.

Parasite. An organism that consumes part of the blood or tissues of its host, usually without killing the host.

Parasitoid. Any of a number of insects whose larvae live within and consume their host, usually another insect.

Parental investment. An act of parental care that enhances the survival of individual offspring or increases their number.

Parent material. Unweathered rock from which soil is, in part, derived.

Parent-offspring conflict. The situation arising when the optimum level of parental investment in a particular offspring differs from the viewpoints of the parent and that offspring. This conflict derives from the fact that offspring are genetically equivalent from the point of view of their parents, but siblings carry identical copies of only half of one another's genes.

Parity. The age-specific pattern of giving birth.

Parthenogenesis. Reproduction without fertilization by male gametes, usually involving the formation of diploid eggs whose development is initiated spontaneously.

Partial pressure. The proportional contribution of a particular gas to the total pressure of a mixture.

Pattern-climax theory. The idea that the species composition of the climax community continuously varies geographically in response to changing physical conditions of the environment.

Pelagic. Pertaining to the open sea.

Per capita. Expressed on a per individual basis.

Perennial. Referring to an organism that lives for more than 1 year; lasting throughout the year.

Perfect flower. A flower having both male and female sexual organs (anthers and pistils).

Permafrost. A layer of permanently frozen ground in very cold regions, especially the Arctic and Antarctic.

Permeability. The capacity of a material to pass through something, such as a biological membrane.

Pest pressure hypothesis. The idea that individuals are vulnerable to pests and pathogens when crowded in the vicinity of their parents, permitting the coexistence of many different types of species that are attacked by different pests.

pH. A scale of acidity or alkalinity; the logarithm of the concentration of hydrogen ions.

Phenolic. Pertaining to compounds, such as lignin, based on the phenol chemical structure, a hydroxylated 6-carbon ring (C_6H_5OH).

Phenolics. Aromatic hydrocarbons produced by plants, many of which exhibit antimicrobial properties.

Phenotype. The physical expression in an organism of the interaction between its genotype and its environment; the outward appearance and behavior of the organism.
Phenotypic value. The measure of a particular phenotypic trait in a particular organism.
Phenotypic variance (V_P). A statistical measure of the variation in a structure or function (phenotypic value) among individuals in a population.
Pheromones. Chemical substances used for communication between individuals.
Photic. Pertaining to surface waters to the depth of light penetration.
Photoautotroph. An organism that uses sunlight as its primary energy source for the synthesis of organic compounds.
Photoperiod. The length of the daylight period each day.
Photorespiration. Oxidation of carbohydrates to carbon dioxide and water by the enzyme responsible for CO_2 assimilation, in the presence of bright light.
Photosynthates. The organic products of photosynthesis; that is, simple carbohydrates.
Photosynthesis. Use of the energy of light to combine carbon dioxide and water into simple sugars.
Photosynthetic efficiency. Percentage of light energy assimilated by plants, based either on net production (net photosynthetic efficiency) or on gross production (gross photosynthetic efficiency).
Phylogenetic effect. Resemblance of the morphology or ecology of species resulting from their common ancestry.
Phylogenetic reconstruction. Inferring the evolutionary relationships among taxa from an analysis of derived character traits, phenotypic similarities, or genetic distances.
Phylogeny. The evolutionary relationships among species or other taxa.
Phytoplankton. Microscopic floating aquatic plants.
Plankton. Microscopic floating aquatic plants (phytoplankton) and animals (zooplankton).
Pleiotropy. The influence of one gene on the expression of more than one trait in the phenotype.
Podsolization. Breakdown and removal of clay particles from the acidic soils of cold, moist regions.
Poikilothermy. Inability to regulate body temperature; cold-bloodedness. *Compare with* Homeothermy.
Poisson distribution. A statistical description of the random distribution of items among categories, often applied to the distribution of individuals among sampling plots.
Polyandry. A mating pattern in which a female mates with more than one male at the same time or in quick succession.
Polygamy. A mating system in which a male pairs with more than one female (polygyny) or a female pairs with more than one male (polyandry) at the same time. *Compare with* Monogamy.
Polygyny. A mating pattern in which a male mates with more than one female at the same time or in quick succession.
Polygyny threshold. The difference between the intrinsic values of territories or males such that the realized values of an unmated male on a poorer territory and a mated male on a better territory are equal in the eyes of an unmated female.
Polymorphism. The occurrence of more than one distinct form of individual or genotype in a population.
Pool. The quantity of a particular material within a compartment of the ecosystem.
Population. The group of organisms of a particular species inhabiting a particular area.
Population bottleneck. A period of extremely small population size during which a population is vulnerable to the loss of genetic diversity through genetic drift. *Compare with* Founder event.
Population genetics. The study of changes in the frequencies of genes and genotypes within a population.
Positive assortative mating. Preferential mating between individuals having similar appearance or genotypes.
Potential evapotranspiration (PE). The amount of transpiration by plants and evaporation from the soil that would occur, given the local temperature and humidity, if water were not limited.
Prairie. An extensive area of level or rolling, almost treeless grassland in central North America.
Precipitation. Rainfall or snowfall. Also, the change of a compound from a dissolved form to a solid form.
Predator. An animal (rarely, a plant) that kills and eats animals.
Prediction. A logical consequence of a hypothesis or outcome of a model describing some aspect of a system.
Primary consumer. An herbivore, the lowermost consumer on the food chain.
Primary host. A host that harbors the sexual stage of the life cycle of a parasite or disease organism.
Primary producer. A green plant or other autotroph that assimilates the energy of light to synthesize organic compounds.
Primary production. Assimilation (gross primary production) or accumulation (net primary production) of energy and nutrients by green plants and other autotrophs.
Primary succession. The sequence of communities developing in a newly exposed habitat devoid of life.
Production. Accumulation of energy or biomass.

Promiscuity. Mating with many individuals within a population, generally without the formation of strong or lasting pair bonds.

Protandry. A course of development of an individual during which its sex changes from male to female.

Protogyny. A course of development of an individual during which its sex changes from female to male.

Proximate factor. An aspect of the environment that the organism uses as a cue for behavior (e.g., day length). Proximate factors often are not directly important to the organism's well-being. *Compare with* Ultimate factor.

Pseudoextinction. Disappearance of a species from the fossil record by evolutionary transformation into a form described as a different species.

Pyramid of energy. The concept that the energy flux through any given link in the food chain decreases with progressively higher trophic levels.

Pyramid of numbers. Charles Elton's concept that the sizes of populations decrease with progressively higher trophic levels; that is, as one progresses along the food chain.

Quantitative genetics. Study of the inheritance and response to selection of quantitative traits.

Quantitative trait. A trait having continuous variation within a population and revealing the expression of many gene loci.

Queen. In a bee, wasp, ant, or termite colony, a fertile, fully developed female whose function is to lay eggs and who is the mother of all the members of the colony.

Radiation. Energy emitted in the form of electromagnetic waves.

Rain shadow. A dry area on the leeward side of a mountain range.

Random dispersion. The condition in which each individual occurs without regard to the position of other individuals.

Rarefaction. A method of determining the relationship between species diversity and sample size by randomly deleting individuals from a sample.

Reaction norm. The relationship between the appearance of a phenotype and variations in the environment produced by a particular genotype.

Recessive. Pertaining to an allele whose expression is masked by an alternative (dominant) allele in a diploid, heterozygous state.

Reciprocal altruism. The exchange of altruistic acts between individuals.

Reciprocal transplant experiment. An exchange of individuals of the same species between two different habitats or regions to determine the relative contributions of genotype and environment to the phenotype.

Recombination. The formation of offspring with combinations of genes that did not occur in either parent, by crossing over between chromosomes and independent assortment and mixing of maternal and paternal chromosomes.

Recruitment. Addition of new individuals to a population by reproduction; often restricted to the addition of breeding individuals.

Redox potential. The relative capacity of an atom or compound to donate or accept electrons, expressed in volts (electric potential). Higher values indicate more powerful oxidizers.

Reduction. Addition of one or more electrons to an atom, ion, or molecule. *Compare with* Oxidation.

Refugium. A place where a species or community can persist in the face of environmental change over the remainder of its distribution.

Region. A geographic area without significant internal barriers to dispersal, but generally much larger than the dispersal distance of an individual.

Regional diversity. The number of species or other taxa occurring within a large geographic area encompassing many biomes or habitats; gamma diversity.

Regulator. An organism that maintains an internal environment different from external conditions.

Regulatory response. A rapid, reversible physiological or behavioral response by an organism to change in its environment.

Relative abundance. Proportional representation of a species in a sample or a community.

Renewable resource. A resource that is continually supplied to the system such that it cannot be fully depleted by consumers.

Reproductive effort. The allocation of time or resources, or the assumption of risk, in order to increase fecundity.

Rescue effect. Prevention of the extinction of a local population by immigration of individuals from elsewhere, often from a more productive habitat.

Residence time. The ratio of the size of a compartment to the flux through it, expressed in units of time; thus, the average time spent by energy or a substance in the compartment.

Resource. A substance or object required by an organism for normal maintenance, growth, and reproduction. If the resource is scarce relative to demand, it is referred to as a limiting resource. Nonrenewable resources (such as space) occur in fixed amounts and can be fully used; renewable resources (such as food) are produced at a rate that may be partially determined by their use.

Respiration. Use of oxygen to break down organic compounds metabolically to release chemical energy.
Resting metabolic rate (RMR). *See* Basal metabolic rate (BMR).
Rhizome. An underground, usually horizontal stem of a plant that produces both roots and aboveground shoots, and which may be modified to store carbohydrate nutrient reserves.
Riffle. A shallow stretch of fast-moving, often rough water between quieter pools in a stream.
Riparian. Along the bank of a river or lake.
Ritualized behavior. Behavior, especially antagonistic interactions, adapted to a conventionalized form within a species.
River continuum concept. The idea that a river system encompasses a continuum of conditions from the headwaters to the mouth, characterized by increasing streambed size and water flow, and interconnected by the movement of nutrients and organisms with downstream currents.
RuBP. Ribulose bisphosphate, a five-carbon carbohydrate to which a carbon atom is attached during the assimilatory step of the Calvin cycle in photosynthesis.
RuBP carboxylase. An enzyme in the Calvin cycle of photosynthesis responsible for the reaction of ribulose bisphosphate and carbon dioxide to form two molecules of phosphoglyceraldehyde.
Ruderal. Pertaining to or inhabiting highly disturbed sites. *See* Weed.
Rumen. An elaboration of the forepart of the stomach of certain ungulate (hoofed) mammals within which cellulose is broken down by symbiotic bacteria.
Runaway sexual selection. The situation in which females persistently choose the most extreme male phenotypes in a population, leading to continuous elaboration of secondary sexual characteristics.
Saturation point. With respect to primary production, the amount of light that causes photosynthesis to attain its maximum rate.
Sclerophyllous. Referring to the tough, hard, often small leaves of drought-adapted vegetation.
Search image. A behavioral selection mechanism that enables predators to increase searching efficiency for prey that are abundant and worth capturing.
Secondary plant compounds. Chemical products of plant metabolism that are produced specifically for the purpose of defense against herbivores and disease organisms.
Secondary sexual characteristics. Traits, other than the sexual organs themselves, that distinguish the sexes, including, for example, the beard of human males.
Secondary succession. Progression of communities in habitats where the climax community has been disturbed or removed.
Sedimentation. The settling of particulate material to the bottom of an ocean, lake, or other body of water.
Selection. Differential survival or reproduction of individuals in a population owing to phenotypic differences among them.
Selection differential (*S*). The difference between the mean phenotypic value of a group of selected individuals and that of the population from which they were drawn.
Selection response (*R*). The difference between the mean phenotypic value of the offspring of selected individuals and that of the population from which the parents were drawn.
Selective death. Death or forgone fecundity resulting from the lower fitness of an individual having a deficient genotype.
Self-incompatible. Unable to mate with oneself because of structural, biochemical or timing factors.
Selfing. Mating with oneself; applicable, of course, only to individuals (usually plants) having both male and female sexual organs (hermaphrodites).
Selfish behavior. Behavior that conveys an advantage to the individual without regard to others.
Self-regulation. The idea that population size is regulated with respect to its resources by individual restraint from overexploitation.
Self-thinning curve. In populations of plants limited by space or other resources, the characteristic relationship between average plant weight and density.
Semelparity. The condition of having only one reproductive episode during the lifetime. *Compare with* Iteroparity.
Senescence. Gradual deterioration of function in an organism with age, leading to increased probability of death; aging.
Sensory exploitation hypothesis. The idea that female mate choice is based upon preexisting preferences resulting from sensory sensitivities to particular signal forms.
Sequential hermaphroditism. The condition in which an individual is first one sex and then another. *See* Protandry and Protogyny.
Sere. A series of stages of community change in a particular area leading toward a stable state.
Serial polygamy. A mating pattern in which an individual mates with several individuals of the opposite sex in succession; the offspring of each mating are generally cared for, in part, simultaneously.
Serpentine. An igneous rock rich in magnesium that forms soils toxic to many plants.

Set point. In a homeostatic mechanism, the reference condition to which the actual body condition is compared.
Sex ratio. The ratio of the number of individuals of one sex to that of the other sex in a population.
Sexual dimorphism. The condition in which the males and females of a species differ in appearance.
Sexual reproduction. Reproduction by means of the union of two gametes (fertilization) to form a zygote.
Sexual selection. Selection by one sex for specific characteristics in individuals of the opposite sex, usually exercised through courtship behavior.
Shannon-Weaver index (*H*). A logarithmic measure of the diversity of species weighted by the relative abundance of each.
Simpson's index (*D*). A measure of the diversity of species weighted by the relative abundance of each.
Simultaneous hermaphroditism. The condition in which an individual has both male and female sexual organs at the same time.
Sink population. A population that would continually decrease in size owing to high mortality, low reproduction, or both if it were not maintained by immigration from other populations.
Skewed sex ratio. Any departure of the proportion of males or females in a population from 50%, or a 1:1 sex ratio.
Social behavior. Any direct interaction among distantly related individuals of the same species; usually does not include courtship, mating, parent-offspring, and sibling interactions.
Social dominance. Physical domination of one individual over another, initiated and sustained by aggression within a population.
Sociobiology. The study of the biological basis of social behavior.
Soil. The solid substrate of terrestrial communities, resulting from the interaction of weather and biological activities with the underlying geologic formation.
Soil horizon. A layer of soil formed at a characteristic depth and distinguished by its physical and chemical properties.
Soil profile. A characterization of the structure of soil vertically through its various horizons, or layers.
Soil skeleton. The physical structure of mineral soil, referring principally to sand grains and silt particles.
Solar equator. The parallel of latitude that lies directly under the sun at any given season.
Source population. A population that produces an excess of individuals over the number needed for self-maintenance and from which there is net emigration.
Spaced distribution. A distribution in which each individual maintains a minimum distance between itself and its neighbors.
Specialist. A species that uses a restricted range of habitats or resources.
Specialization. An adaptation of form or function that suits an individual particularly well to a restricted range of habitats, resources, or environmental conditions; the evolutionary process of such restriction.
Speciation. The division of a single species into two or more reproductively incompatible daughter species.
Species. A group of actually or potentially interbreeding populations that are reproductively isolated from all other kinds of organisms.
Species–area relationship. The pattern of increase in the number of species with respect to the area of the habitat, island, or region being considered.
Species diversity. *See* Diversity.
Species pool. The entire group of species within a source area from which colonists of an island or habitat are drawn.
Species richness. A simple count of the number of species.
Specific heat. The amount of energy that must be added or removed to change the temperature of a substance by a specific amount. By definition, 1 calorie of energy is required to raise the temperature of 1 gram of water by 1 degree Celsius.
Spitefulness. A social interaction in which the donor of a behavior incurs a cost in order to reduce the fitness of a recipient.
Spring bloom. An increase in phytoplankton growth during early spring in temperate lakes associated with vertical mixing of the water column.
Spring overturn. The vertical mixing of water layers in temperate lakes in spring as surface ice disappears.
Stabilizing selection. Differential survival or reproduction of phenotypes closest to the mean value of a population; selection against extreme phenotypes tending to maintain the population mean.
Stable age distribution. The proportion of individuals in various age classes in a population that has been growing at a constant rate.
Stable equilibrium. The particular state to which a system returns if displaced by an outside force.
Standard deviation (s or σ). A measure of the variability among items in a sample, such as individuals in a population; the square root of the variance, and hence the square root of the average squared deviation from the mean.

Static life table. The age-specific survival and fecundity of individuals of different age classes within a population at a given time; a time-specific life table.
Steady state. The condition of a system in which opposing forces or fluxes are balanced.
Steppe. Usually treeless plains, especially in southeastern Europe and Asia in regions of extreme temperature range and sandy soil.
Stochastic. Referring to patterns resulting from random effects.
Stochastic extinction. Decrease of a population's size to zero resulting from random fluctuations in births and deaths.
Stratification. The establishment of distinct layers of temperature or salinity in bodies of water based on the different densities of warm and cold water or saline and fresh water.
Stress tolerance. One of Grime's three types of plant life-history strategies, in which adaption emphasizes physiological mechanisms to cope with stressful environmental conditions.
Succession. Replacement of populations in a habitat through a regular progression to a stable state.
Superorganism. An association of individuals in which the function of each promotes the well-being of the entire system.
Superparasitism. Occurrence of more than one individual of a particular parasitoid species per host.
Survival (l_x). Proportion of newborn individuals alive at age x; also called survivorship.
Switching. A change in diet to favor items of increasing suitability or abundance.
Symbiosis. An intimate, and often obligatory, association of two species, usually involving coevolution. Symbiotic relationships can be parasitic or mutualistic.
Sympatric. Occurring in the same place; usually referring to areas of overlap in species distributions.
Synapomorphy. A shared, derived trait; that is, a trait uniquely inherited by a monophyletic group of species from their common ancestor.
Synecology. The study of the relationships of organisms and populations to biotic factors in the environment. *Compare with* Autecology.
Synergism. The interaction of two causes such that their total effect is greater than the sum of the effects of the two acting independently.
Systematics. The classification of organisms into a hierarchical set of categories (taxa) emphasizing their evolutionary interrelationships.
Systems ecology. The study of an ecological structure as a set of components linked by fluxes of energy and nutrients or by population interactions; frequently applied to ecosystems.
Systems modeling. Portrayal of a system's function by mathematical functions describing the interactions among its components.
Taiga. A moist coniferous forest bordering the arctic zone, dominated by spruce and fir trees.
Tannins. Polyphenolic compounds, produced by most plants, that bind proteins, thereby impairing digestion by herbivores and inhibiting microbes.
Taxonomy. The description, naming, and classification of organisms.
Temperature profile. The relationship of temperature to depth below the surface of the water or soil, or height above the ground.
Territoriality. The situation in which individuals defend exclusive spaces, or territories.
Territory. Any area defended by one or more individuals against intrusion by others of the same or different species.
Theory of island biogeography. *See* Equilibrium theory of island biogeography.
Thermal conductance. The rate at which heat passes through a substance.
Thermal stratification. Sharp delineation of layers of water by temperature, the warmer layer generally lying over the top of the colder layer.
Thermocline. The zone of water depth within which temperature changes rapidly between the upper warm water layer (epilimnion) and lower cold water layer (hypolimnion).
Thermodynamic. Relating to heat and motion.
Thermophilic. Preferring warm or hot environments.
Three-halves power law. A generalization proposing that the relationship between the logarithms of the biomass and density of a population of plants has a slope of $-3/2$.
Time lag. A delay in the response of a population or other system to change in the environment; also called time delay.
Time-specific life table. *See* Static life table.
Tit-for-tat strategy. A game theory strategy in which each player is cooperative or selfish toward each other player depending on its past experience with that player.
Tolerance. In reference to succession, the indifference of establishment of one species to the presence of others.
Torpor. Loss of the power of motion and feeling, usually accompanied by a greatly reduced rate of respiration.
Trade-off. A balancing of factors, all of which are not attainable at the same time, especially traits that tend to increase the evolutionary fitness or performance of the individual.
Transit time. The average time that a substance or energy remains in the biological realm or any compartment of a system; ratio of biomass to productivity.

Transpiration. Evaporation of water from leaves and other parts of plants.
Transpiration efficiency. The ratio of net primary production to transpiration of water by a plant, usually expressed as grams per kilogram of water; water use efficiency.
Trophic. Pertaining to food or nutrition.
Trophic level. Position in the food chain, determined by the number of energy-transfer steps to that level.
Trophic mutualism. A symbiotic relationship between two species based on complementary methods of obtaining energy and nutrients.
Trophic structure. The organization of a community based on the feeding relationships of populations.
Type I, II, and III functional responses. *See* Functional response.
Ultimate factor. An aspect of the environment that is directly important to the well-being of an organism (e.g., food). *Compare with* Proximate factor.
Ultraviolet (UV) radiation. Electromagnetic radiation having a wavelength shorter than about 400 nm.
Upwelling. Vertical movement of water, usually near coasts and driven by offshore winds, that brings nutrients from the depths of the ocean to surface layers.
Vapor pressure. The pressure exerted by a vapor that is in equilibrium with its liquid form.
Variance (*V*). A statistical measure of the dispersion of a set of values about its mean.
Veil line. The point on a lognormal distribution of species abundances below which one expects fewer than one individual per species; hence the veil line separates observed from potentially observed species.
Vertical mixing. Exchange of water between deep and surface layers.
Vesicular-arbuscular mycorrhizae. A type of endomycorrhizal association of fungi with the roots of plants distinguished by the branching pattern of growth of the fungus within the root tissues.
Viscosity. The quality of a fluid that resists internal flow.
Wahlund effect. An excess of homozygotes compared with heterozygotes in a population, relative to Hardy–Weinberg frequencies, caused by the mixing of two populations with different equilibrium allele frequencies.
Warning coloration. Conspicuous patterns or colors adopted by noxious organisms to advertise their noxiousness or dangerousness to potential predators; aposematism.
Water potential. The force by which water is held in the soil by capillary and hygroscopic attraction.
Watershed. The drainage area of a stream or river.
Water use efficiency. *See* Transpiration efficiency.
Weathering. Physical and chemical breakdown of rock and its component minerals at the base of the soil.
Weed. A plant or animal, generally having high powers of dispersal, capable of living in highly disturbed habitats.
Wilting coefficient. The minimum water content of the soil at which plants can obtain water.
Xeric. Referring to habitats in which plant production is limited by the availability of water.
Xerophyte. A plant that tolerates dry (xeric) conditions.
Zonation. The distribution of organisms in bands or regions corresponding to changes in ecological conditions along a continuum, for example, intertidal zonation and elevational zonation.
Zooplankton. Tiny floating aquatic animals.
Zygote. A diploid cell formed by the union of male and female gametes during fertilization.

INDEX

Please note that *italicized* page numbers indicate references to boldfaced words or phrases in the text. **Bold-faced** page numbers indicate pages with illustrations.